# The Eastern Mediterranean as a Laboratory Basin for the Assessment of Contrasting Ecosystems

# NATO Science Series

*A Series presenting the results of activities sponsored by the NATO Science Committee. The Series is published by IOS Press and Kluwer Academic Publishers, in conjunction with the NATO Scientific Affairs Division.*

*General Sub-Series*

| | |
|---|---|
| **A. Life Sciences** | IOS Press |
| **B. Physics** | Kluwer Academic Publishers |
| **C. Mathematical and Physical Sciences** | Kluwer Academic Publishers |
| **D. Behavioural and Social Sciences** | Kluwer Academic Publishers |
| **E. Applied Sciences** | Kluwer Academic Publishers |
| **F. Computer and Systems Sciences** | IOS Press |

*Partnership Sub-Series*

| | |
|---|---|
| **1. Disarmament Technologies** | Kluwer Academic Publishers |
| **2. Environmental Security** | Kluwer Academic Publishers |
| **3. High Technology** | Kluwer Academic Publishers |
| **4. Science and Technology Policy** | IOS Press |
| **5. Computer Networking** | IOS Press |

*The Partnership Sub-Series incorporates activities undertaken in collaboration with NATO's Partners in the Euro-Atlantic Partnership Council – countries of the CIS and Central and Eastern Europe – in Priority Areas of concern to those countries.*

**NATO-PCO-DATA BASE**

The NATO Science Series continues the series of books published formerly in the NATO ASI Series. An electronic index to the NATO ASI Series provides full bibliographical references (with keywords and/or abstracts) to more than 50000 contributions from international scientists published in all sections of the NATO ASI Series.
Access to the NATO-PCO-DATA BASE is possible via CD-ROM "NATO-PCO-DATA BASE" with user-friendly retrieval software in English, French and German (© WTV GmbH and DATAWARE Technologies Inc. 1989).

The CD-ROM of the NATO ASI Series can be ordered from: PCO, Overijse, Belgium.

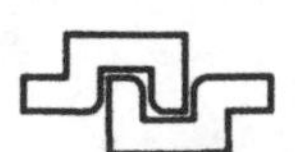

**Series 2: Environmental Security – Vol. 51**

**Kluwer Academic Publishers**

Dordrecht / Boston / London

Published in cooperation with NATO Scientific Affairs Division

e NATO Advanced Research Wor
diterranean as a Laboratory Basin fo

e
27, 1998

. Catalogue record for this book is available from the

ISBN 0-7923-5585-7 (HB)
ISBN 0-7923-5586-5 (PB)

Published by Kluwer Academic Publishers,
P.O. Box 17, 3300 AA Dordrecht, The Netherlands.

Sold and distributed in North, Central and South America
by Kluwer Academic Publishers,
101 Philip Drive, Norwell, MA 02061, U.S.A.

In all other countries, sold and distributed
by Kluwer Academic Publishers,
P.O. Box 322, 3300 AH Dordrecht, The Netherlands.

*Printed on acid-free paper*



Printed in the Netherlands

## TABLE OF CONTENTS

# The Eastern Mediterranean as a Laboratory Basin for the Assessment of Contrasting Ecosystems

edited by

Paola Malanotte-Rizzoli
Massachusetts Institute of Technology,
Cambridge, MA, U.S.A.

and

Valery N. Eremeev
Marine Hydrophysical Institute,
Sevastopol, Ukraine

**Kluwer Academic Publishers**
Dordrecht / Boston / London

Published in cooperation with NATO Scientific Affairs Division

Proceedings of the NATO Advanced Research Workshop on
The Eastern Mediterranean as a Laboratory Basin for the Assessment of Contrasting Ecosystems
Kiev, Ukraine
March 23–27, 1998

A C.I.P. Catalogue record for this book is available from the Library of Congress

ISBN 0-7923-5585-7 (HB)
ISBN 0-7923-5586-5 (PB)

Published by Kluwer Academic Publishers,
P.O. Box 17, 3300 AA Dordrecht, The Netherlands.

Sold and distributed in North, Central and South America
by Kluwer Academic Publishers,
101 Philip Drive, Norwell, MA 02061, U.S.A.

In all other countries, sold and distributed
by Kluwer Academic Publishers,
P.O. Box 322, 3300 AH Dordrecht, The Netherlands.

*Printed on acid-free paper*



Printed in the Netherlands

## TABLE OF CONTENTS

## PREFACE

This book is the outcome of a NATO Advanced Research Workshop on "The Eastern Mediterranean as a laboratory basin for the assessment of contrasting ecosystems" that was held in Kiev, Ukraine, March 23-27, 1998. The scientific rationale of the workshop can be summarized as follows.

The Eastern Mediterranean is the most nutrient impoverished and oligotrophic large water body known. There is a well-defined eastward trend in nutrient ratios over the entire Mediterranean that starts at the Gibraltar Straits and, through the western basin, proceeds to the Ionian and Levantine Seas. Supply of nutrients to the entire Mediterranean is limited by inputs from the North Atlantic and various river systems along the sea. The unique feature of the Mediterranean is the presence of an eastward longitudinal trend in available nitrate/phosphate ratios. This apparently induces a west-to-east variation in the structure of the pelagic food web and trophic interactions. In this context the Mediterranean, and in particular its Eastern basin, provides probably a unique platform to explore the hypotheses related to the suggested phosphate-limitation on production and to the shift between "microbial" and "classical" modes of operation of the photic food web.

The major exception of the overall oligotrophic nature of the Eastern Mediterranean is the highly eutrophic system of the Northern Adriatic Sea. Here, during the last two decades the discharges of the northern rivers (especially of the Po), together with municipal sewage, have led to a very marked increase of nutrients and subsequent imponent eutrophication events. The northern Adriatic eutrophic ecosystem occupies the entire continental shelf, ~600 km long, and is very similar to the Black Sea northwestern shelf ecosystem, even though not as extreme. The crucial assessment of the degree of eutrophism of the Northern Adriatic shelf has been the objective of a major programme supported by the European Union, with field work that extended from February 1993 to June 1995, the ELNA project (Eutrophic Limits of the Northern Adriatic, T.S. Hopkins, 1996). It must be pointed out that eutrophication and pollutant transports are a major concern in all of the coastal areas of the Eastern Mediterranean in particular all along the northern Turkey coastline of the Levantine basin. It is important to identify the differences and similarities of the nature of eutrophism and trophic-dynamic characteristics of these regions within the framework of the large-scale oceanography of the entire Eastern basin.

The above ongoing research activities comprising extensive field work and modeling efforts clearly demonstrate the unique feature of the Eastern Mediterranean as a laboratory basin of contrasting ecosystems, from extreme eutrophic (Northern Adriatic) to extreme oligotrophic (Levantine basin) and from ecosystems driven by river inputs (Northern Adriatic) to ecosystems driven by wind forcing and deep water formation processes (Levantine basin).

The specific objectives of the ARW were:

1) To assess the state-of-the-art knowledge of the contrasting ecosystems of the Eastern Mediterranean on the basis of ongoing

national/international programmes and from the phenomenological as well as the modeling perspective.

2) To identify the major environmental issues, both on the regional and basin scale, in relation to eutrophication, transport of pollutants and health of the system.

The papers collected in this volume address these objectives in detail and constitute the most up-to-date synthesis of the state-of-the-art research in the considered basins.

Primary support for the Workshop was provided by NATO under the auspices of the Scientific Affairs Division, NATO Science Programme and Cooperation Partners, and we thank Dr. L.V. da Cunha, Director of the Division. It is with particular pleasure that we thank Prof. Evgeny Nikiforovich, of the Institute of Hydromechanics of the National Academy of Sciences in Ukraine, Chairman of the local organizing Committee, for his invaluable assistance in all the organizational phases before, during and after the Workshop and for providing a friendly environment to foster interactions and future collaborations between scientists from NATO and from Partner Countries. We also thank Ms. Lisa McFarren, secretary to Prof. Malanotte-Rizzoli at MIT for her assistance in the production of the publicity material and documents for NATO. Finally, special thanks are due to all the contributors to this volume who provided an exciting forum for discussion and the excellent series of lectures that make this book a stimulating record of events.

Paola Malanotte-Rizzoli
Valery N. Eremeev

October 1998

## LIST OF CONTRIBUTORS

I. Aboucora
Higher Institute of Applied Sciences
Damascus, SYRIA

F. Abousamra
Higher Institute of Applied Sciences
Damascus, SYRIA

E.G. Arashkevich
P.P. Shirov Institute of Oceanology
Russian Academy of Sciences
Moscow, RUSSIA

M. Baker
Higher Institute of Applied Sciences
Damascus, SYRIA

E.P. Bitukov
Institute of Biology
of the Southern Seas
Ukrainian Academy of Sciences
2, Nahimov Avenue
335000 Sevastopol
Crimea, UKRAINE

K. Bouras
Higher Institute of Applied Sciences
Damascus, SYRIA

S. Brenner
Israel Oceanographic Limnological
Research
Haifa 31080
ISRAEL

V.I. Burenkov
P.P. Shirshov Institute of Oceanology
Russian Academy of Sciences
Nakhimovski prospect 36
Moscow 117851, RUSSIA

E.D. Christou
National Centre for Marine Research
Kosmas, Helliniko
16604 Athens, GREECE

G. Civitarese
CNR-Instituto Sperimentale Talassografico
V. le R. Gessi 2
I-34123, Trieste
ITALY

Alessandro Crise
Dipartimento Oceanologia e Geofisica
Ambientale
Borgo Grotta Gigante
P.O. Box 2011 (Opicina)
34016 Trieste, ITALY

G. Crispi
Osservatorio Geofisico Sperimentale
P.O. Box 2011
34016 Trieste, ITALY

Maurizio D'Alcala
Stazione Zoologica
"A. Dohrn"
villa Comunale
Naples 80121, ITALY

D. Ediger
Middle East Technical University
Institute of Marine Sciences
P.O. Box 28
Erdemli 33731
Icel, TURKEY

Vladimir V. Efimov
Marine Hydrophysical Institute
Ukrainian Academy of Sciences
2, Kapitanskaya St.
335000 Sevastopol
Crimea, UKRAINE

Valery N. Eremeev
Marine Hydrophysical Institute
Ukrainian National Academy of Sciences
2 Kapitanskaya St.
Sevastopol, CRIMEA 335000

S.V. Ershova
P.P. Shirshov Institute of Oceanology
Russian Academy of Sciences
Nakhimovski prospect 36
Moscow 117851, RUSSIA

M.A. Evdoshenko
P.P. Shirshov Institute of Oceanology
Russian Academy of Sciences
Nakhimovski prospect 36
Moscow 117851, RUSSIA

I.K. Evstigneeva
Institute of Biology of the Southern Seas (IBSS)
Sevastopol 335011
Crimea, UKRAINE

Saker Fayes
Higher Institute of Applied Sciences & Technology
Damascus, SYRIA

Zosim Z. Finenko
Institute of Biology of Southern Seas
National Academy of Sciences of Ukraine
Sevastopol, UKRAINE

Mikhail V. Flint
P.P. Shirshov Institute of Oceanology
Russian Academy of Sciences
Nakhimovski prospect 36
Moscow 117851, RUSSIA

Miroslav Gacic
Osservatorio Geofisico Sperimentale
Dipartimento Oceanologia e Geofisica Ambientale, Borgo Grotta Gigante
P.O. Box 2011 (Opicina)
34016 Trieste, ITALY

Alessandra Giorgetti
Osservatorio Geofisico Sperimentale
P.O. Box 2011
34016 Trieste, ITALY

O. Gotsis-Skretas
National Centre for Marine Research
Kosmas, Helliniko
16604 Athens, GREECE

Y.P. Ilyn
Dept. of Shelf Hydrophysical Institute
Marine Hydrophysical Institute
Ukrainian Academy of Sciences
2, Kapitanskaya St.
335000 Sevastopol
Crimea, UKRAINE

V.A. Ivanov
Marine Hydrophysical Institute, NASU
2, Kapitanskay St.
Sevastopol, 335000
UKRAINE

A.E. Kideys
Institute of Marine Sciences
P.O. Box 28
Erdemli 33731, TURKEY

Harilaos Kontoyannis
National Center for Marine Research
Aghios Kosmas
Hellinikon 16604
Athens, GREECE

O.V. Kopelevich
P.P. Shirshov Institute of Oceanology
Russian Academy of Sciences
Nakhimovski prospect 36
Moscow 117851, RUSSIA

R.D.Kos'yan
Southern Branch of P.P. Shirshov
Institute of Oceanology
Gelendzhik-7
353 470, RUSSIA

Alexander V. Kovalev
Institute of Biology
of the Southern Seas
Ukrainian Academy of Sciences
2, Nahimov Avenue
335000 Sevastopol
Crimea, UKRAINE

E. Krasakopoulou
National Centre for Marine Research
Aghios Kosmas
Ellinikon 16604, GREECE

Vladimir G.Krivosheya
Southern Branch of P.P. Shirshov
Institute of Oceanology
Russian Academy of Sciences
353470 Gelendzhik-7, RUSSIA

Alexander I. Kubryakov
Marine Hydrophysical Institute
Ukrainian Academy of Sciences
2, Kapitanskaya St.
335000 Sevastopol
Crimea, UKRAINE

Nikita V. Kucheruk
P.P. Shirshov Institute of Oceanology
Russian Academy of Sciences
Nakhimovski prospect 36
Moscow 117851, RUSSIA

A.P. Kuznetsov
P.P. Shirshov Institute of Oceanology
Russian Academy of Sciences
Nakhimovski prospect 36
Moscow 117851, RUSSIA

Alexander Lascaratos
University of Athens
Dept. of Applied Physics
University Campus Building PHYS-V
15784, Athens
GREECE

E.M. Lemeshko
Marine Hydrophysical Institute
of Ukrainian Academy of Sciences
2, Kapitanska St.
335000 Sevastopol, UKRAINE

Olga V. Maksimova
P.P. Shirshov Institute of Oceanology
Russian Academy of Sciences
Nakhimovski Prospect 36
117851 Moscow, RUSSIA

Paola Malanotte-Rizzoli
Massachusetts Institute of Technology
54-1416 Department of Earth, Atmospheric
and Planetary Sciences
Cambridge, MA 02139 USA

Beniamino Manca
Osservatorio Geofisico Sperimentale
Dipartimento Oceanologia e Geofisica
Ambientale
Borgo Grotta Gigante, P.O. Box 2011
Opicina, 34016
Trieste, ITALY

V.I. Mankovsky
Marine Hydrophysical Institute
National Academy of Science of Ukraine
2, Kapitanskaya St.
335000 Sevastopol
Crimea, UKRAINE

Ivona Marasovic
Institute of Oceanography & Fisheries
P.O. Box 500
Split 21001, CROATIA

Maria Grazia Mazzocchi
Laboratorio di Oceanografia Biologica
Stazione Zoologica "A. Dohrn", Villa Comunale
80121 Napoli, ITALY

Natalya Milchakova
The A.O. Kovalevsky Institute of Biology of the Southern Seas
Urkainian Academy of Sciences
2, Nahimov Avenue
335000 Sevastopol
Crimea, UKRAINE

N.A. Milchakova
Institute of Biology of the Southern Seas (IBSS)
Sevastopol 335011
Crimea, UKRAINE

A.V. Mishonov
Marine Hydrophysical Institute
National Academy of Science of Ukraine
2, Kapitanskaya St.
335000 Sevastopol
Crimea, UKRAINE

Alex Yu Mitropolsky
Institute of Geological Sciences
Ukrainian Academy of Sciences
335000 Sevastopol
Crimea, UKRAINE

L.V. Moskalenko
Southern Branch of P.P. Shirshov Institute of Oceanology
Gelendzhik-7
353 470, RUSSIA

Ernesto Napolitano
Institute of Meteorology & Oceanography
Istituto Universitario Navale
Corss Umberto I 179
8013P Naples, ITALY

N.P. Nezlin
P.P. Shirshov Institute of Oceanology
Russian Academy of Sciences
Nakhimovski prospect 36
Moscow 117851, RUSSIA

Z. Nincevic
Institute of Oceanography & Fisheries
P.O. Box 500
Split 21001, CROATIA

Kostas Nittis
National Center for Marine Research
Aghios Kosmas
Hellinikon 16604
Athens, GREECE

Temel Oguz
Middle East Technical University
Institute of Marine Sciences
P.O. Box 28
Erdemli 33731, Icel
TURKEY

S.P. Olshtynsky
Institute of Geological Sciences NAS
Kiev, UKRAINE

A. Ostrovskaya
Institute of Biology of Southern Seas
National Academy of Sciences of Ukraine
Sevastopol, UKRAINE

I.M. Ovchinnikov
Southern Branch of P.P. Shirshov Institute of Oceanology
Gelendzhik-7
353 470, RUSSIA

Emin Ozsoy
Middle East Technical University
Institute of Marine Sciences
P.O. Box 28
Erdemli 33731, Icel
TURKEY

K. Pagou
National Centre for Marine Research
Kosmas, Helliniko
16604 Athens, GREECE

E.V. Pavlova
Institute of Biology of Southern Seas
National Academy of Sciences of Ukraine
Sevastopol, UKRAINE

S.A. Piontkovski
Institute of Biology of the Southern Seas
335011 Sevatsopol, UKRAINE

A.Y. Plis
Yuri M. Plis
Ukrainian Scientific & Research
Institute of Ecological Problems
6 Bakulina Str
Kharkov 310166
UKRAINE

C.S. Polat
Istanbul University
Institute of Marine Sciences &
Management
Muskule Sok., No. 10
Vefa-Istanbul, TURKEY

A.V. Prusov
Marine Hydrophysical Institute
Of Ukrainian Academy of Sciences
Sevastopol, Kapitanskaya St., 2
335000 Crimea, UKRAINE

P.V. Rybnikov
P.P. Shirshov Institute of Oceanology
Russian Academy of Sciences
Nakhimovski prospect 36
Moscow 117851, RUSSIA

I. Salihoglu
Middle East Technical University
Institute of Marine Sciences
P.O. Box 28
33731 Erdemli
Icel, TURKEY

Emilio Sansone
Istituto Universitario Navale
Via Acton 38
80133 Napoli, ITALY

Naum B. Shapiro
Marine Hydrophysical Institute
Ukrainian National Academy of Sciences
Sevastopol, Crimea
UKRAINE

S.V. Sheberstov
P.P. Shirshov Institute of Oceanology
Russian Academy of Sciences
Nakhimovski prospect 36
Moscow 117851, RUSSIA

M.V. Shokurov
Marine Hydrophysical Institute
of Ukrainian Academy of Sciences
Sevastopol, Kapitanskaya St., 2
335000, Crimea
UKRAINE

Georgy E. Shulman
The A.O. Kovalevsky Institute
of Biology of the Southern Seas
Ukrainian Academy of Sciences
2, Nahimov Ave.
335000 Sevastopol
Crimea, UKRAINE

Ioanna Siokou-Frangou
National Center for Marine Research
Aghios Kosmas
Hellinikon 166 04
Athens, GREECE

A.A. Shmeleva
Institute of Biology of Southern Seas
National Academy of Sciences of Ukraine
Sevastopol, UKRAINE

G.E. Shulman
The A.O. Kovalevsky Institute of Biology
Of the Southern Seas
National Academy of Sciences of Ukraine
2 Nakhimov Ave.
Sevastopol 335011
Crimea, UKRAINE

V.A. Skryabin
Institute of Biology of Southern Seas
National Academy of Sciences of Ukraine
Sevastopol, UKRAINE

B.G. Sokolov
Institute of Biology of the Southern Seas
335011 Sevatsopol, UKRAINE

C. Solidoro
Osservatorio Geofisico Sperimentale
P.O. Box 2011
34016 Trieste, ITALY

D.M. Soloviev
M.V. Solov'ev
Marine Hydrophysical Institute
of Ukrainian Academy of Sciences
2, Kapitanska St.
335000 Sevastopol, UKRAINE

Eraterini Souvermezoglou
National Center for Marine Research
Aghios Kosmas
Hellinikon 166 04
Athens, GREECE

S.V. Stanichny
Marine Hydrophysical Institute
of Ukrainian Academy of Sciences
2, Kapitanska St.
335000 Sevastopol, UKRAINE

N.M. Stashchuk
Marine Hydrophysical Institute, NASU
2, Kapitanskay St.
Sevastopol, 335000
UKRAINE

Irina N. Sukhanova
P.P. Shirshov Institute of Oceanology
Russian Academy of Sciences
Nakhimovski prospect 36
Moscow 117851, RUSSIA

A.M. Suvorov
Marine Hydrophysical Institute of
Ukrainian
National Academy of Sciences
2 Kapitanskaya St.
Sevastopol, Crimea 335000
UKRAINE

H. Taljo
Higher Institute of Applied Sciences
Damascus, SYRIA

I.N. Tankovskaya
Institute of Biology of the Southern Seas
(IBSS)
Sevastopol 335011
Crimea, UKRAINE

Alexander Theocharis
National Center for Marine Research
Aghios Kosmas
Hellinikon 16604
Athens, GREECE

V.B. Titov
Southern Branch of P.P. Shirshov
Institute of Oceanology
Gelendzhik-7
353 470, RUSSIA

Yyriy Tokarev
Institute of Biology
of the Southern Seas
Ukrainian Academy of Sciences
2, Nahimov Avenue
335000 Sevastopol
Crimea, UKRAINE

S. Tugrul
Middle East Technical University
Institute of Marine Sciences
P.O. Box 28
33731 Erdemli
Icel, TURKEY

L. Ursella
Osservatorio Geofisico Sperimentale
P.O. Box 2011
I-34016, Trieste
ITALY

O. Uslu
Institute of Marine Sciences & Technology
1884/8, Sokak N010
35340 Inciralti-Izmir, TURKEY

Z. Uysal
Institute of Marine Sciences
P.O. Box 28
Erdemli 33731, TURKEY

V.L. Vladimirov
Marine Hydrophysical Institute
National Academy of Science of Ukraine
Kapitanskaya St. 2
Sevatsopol 335000
Crimea, UKRAINE

V.I. Vlasenko
Marine Hydrophysical Institute, NASU
2, Kapitanskay St.
Sevastopol, 335000
UKRAINE

V.I. Vasilenko
Institute of Biology of the Southern Seas
335011 Sevastopol, UKRAINE

D. Vilicic
Faculty of Science
University of Zagreb
10000 Zagreb
Rosseveltov trg 6
CROATIA

R. Williams
Plymouth Marine Laboratory
Prospect Place
Plymouth PL1 3DH, UKRAINE

V.G. Yakubenko
Southern Branch of P.P. Shirov
Institute of Oceanology
Gelendzhik-7
353 470, RUSSIA

A. Yilmaz
Middle East Technical University
Institute of Marine Sciences
P.O. Box 28
33731 Erdemli
Icel, TURKEY

Oleg Yunev
Institute of Biology
of the Southern Seas
2, Nahimov Avenue
335000 Sevastopol
Crimea, UKRAINE

# THE EASTERN MEDITERRANEAN IN THE 80'S AND IN THE 90'S: THE BIG TRANSITION EMERGED FROM THE POEM-BC OBSERVATIONAL EVIDENCE

Paola Malanotte-Rizzoli[a], Beniamino Manca[b],
Maurizio Ribera d'Alcala[c], Alexander Theocharis[d]

[a]Department of Earth, Atmospheric and Planetary Sciences, Massachusetts Institute of Technology, Cambridge, MA 02139 U.S.A.
[b]Osservatorio Geofisico Sperimentale, P.O. Box 2011, 34016 Opicina-Trieste, Italy
[c]Stazione Zoologica "A. Dohrn", Villa Comunale I, 80121 Napoli, Italy
[d]National Centre for Marine Research, Aghios Kosmas, Hellinikon 16604, Athens, Greece

## 1. Abstract

The thermohaline circulation of the Eastern Mediterranean underwent a dramatic change between 1987 and 1995. In 1987 the "engine" of the Eastern Mediterranean "conveyor belt" was the convective cell of the Southern Adriatic, while in 1995 the active convection region moved to the Aegean Sea. This change actually started as early as 1991. The phenomenological evidence of the POEM programme shows that in 1987 the source of Levantine Intermediate Water (LIW) mass was the Levantine basin and the bottom water mass was formed in the Southern Adriatic. In 1991 all the intermediate/deep water masses on the horizons $\sigma_\theta = 29.00$ to 29.18 kg/m$^3$ were formed inside the Aegean sea, from which they spread out into the entire Eastern Mediterranean through the Cretan Arc Straits.

In the last decade the Eastern Mediterranean has been the object of the multinational collaborative programme P.O.E.M. (Physical Oceanography of the Eastern Mediterranean) sponsored by UNESCO, IOC and CIESM. Under this programme a series of general hydrographic surveys was carried out by the R/V of Greece-Israel-Italy and Turkey in the period 1985-1987, culminating in POEM-AS87 in which the German R/V Meteor covered the entire Eastern Mediterranean with a basin-wide station network. In 1987 the regular CTD surveys were implemented by a transient-tracer survey (1). The observational dataset collected in these surveys was intercalibrated, pooled and distributed to all the participating scientists in a series of UNESCO sponsored workshops. The joint analyses and interpretation led first to a group paper summarizing the new findings that included extended modeling results (2); second a special issue of Deep-Sea Research was devoted to this POEM-Phase 1 research (3). Recently, the entire POEM-Phase 1 dataset has been revisited for the Ionian sea with an in-depth complete reanalysis



*P. Malanotte-Rizzoli and V.N. Eremeev (eds.),*
*The Eastern Mediterranean as a Laboratory Basin for the Assessment of Contrasting Ecosystems*, 1–6.

that has led to important new findings (4). These include the first detailed definition of the upper thermocline circulation in the Ionian sea, with the discovery of the strong Mid-Ionian Jet (MIJ) crossing the basin interior in north/south direction and then becoming the Mid-Mediterranean Jet (MMJ); and the first definition of the pathways of the intermediate LIW and of the Eastern Mediterranean Deep Water (EMDW).

In 1990 POEM evolved into POEM-BC (Biology and Chemistry) a fully interdisciplinary programme, with the major overall objective of establishing the phenomenology of the 90's for the chemical and biological parameters together with a reassessment of the phenomenology of the physical properties, contrasted to that of the 80's, (POEM-Phase 1). The first interdisciplinary multi-ship general survey of the entire basin was carried out in October 1991, POEM-BC-O91, followed by a more restricted survey in April 1992 (the Ionian basin only) and a final basin-wide survey by the R/V Meteor in January 1995, with a second transient tracer network of stations. This was part of the LIWEX experiment aimed to investigate the successive phases of the LIW formation and concentrated in the Northern Levantine region of the Rhodes gyre during the successive months, February through April 1995. The analysis of the Meteor cruise, including the transient tracer observations revealed a very important, dramatic change in the deep thermocline circulation, the Eastern Mediterranean "conveyor belt". Specifically, in 1987 the driving engine of the deep, closed thermohaline cell was the Southern Adriatic, where deep convection leads to the formation of the Adriatic Deep Water (ADW) that exists from the Otranto Straits, becomes EMDW and spreads throughout the eastern Levantine in the bottom layer. General upwelling to the intermediate transitional layer (below 1,000 m) provides the return pathways to the Southern Adriatic closing the cell (1). In winter 1995 the situation was completely different: the engine of the deep thermohaline circulation was now the Aegean sea, with deep, denser water masses exiting from the Cretan Arc Straits, spreading throughout the entire basin and pushing to the west, while simultaneously lifting, the less dense EMDW of southern Adriatic origin (5).

We present here the first observational evidence that this dramatic change in the Eastern Mediterranean circulation actually started in 1991 and involved not only the deep water mass pathways but the intermediate ones as well, specifically the LIW origin and pathways. This evidence is based on the first joint analysis of the POEM-BC-O91 general survey. This analysis revealed first that the upper thermocline circulation (upper ~250 dbar) was actually extremely similar in the 80's and 90's. Most important the MIJ, emanating from the Atlantic Ionian Stream (AIS) entering the Sicily Straits, is quite strong, crossing the Ionian interior from North to South and surrounding a general anticyclonic region in the Southwestern Ionian both in the 80's and 90's. Recent results based on drifter observations confirm the persistence of the MIJ from the 80's throughout 1995-96 (6).

On the other hand, a dramatic change is observed from 1987 to 1991 in all the intermediate and deep water mass pathways. In 1987 the LIW was formed in the proper Levantine basin, entered the Cretan passage south of Crete, with a major branch proceeding directly to the Sicily Straits and successive branches "peeling off" and being

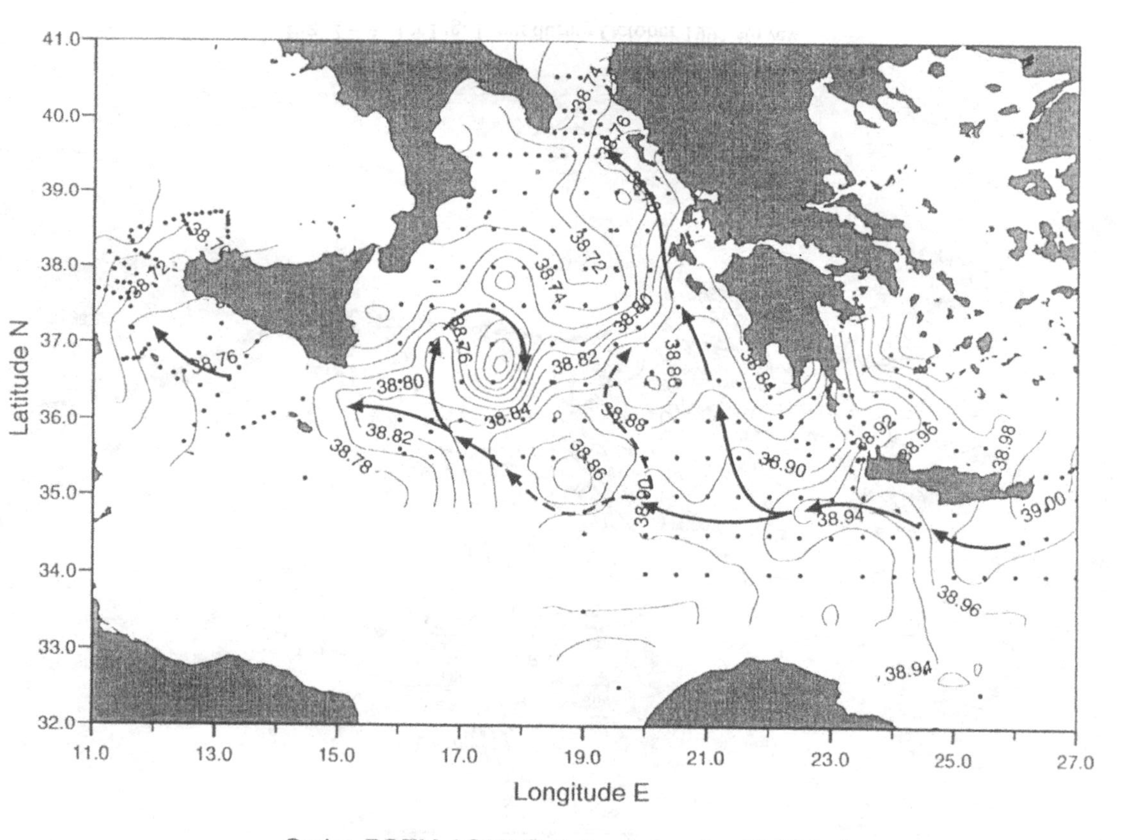


Fig. 1 – Distribution of salinity at the 29.05 kg•m$^{-3}$ isopycnal horizon during August-September 1987 survey.

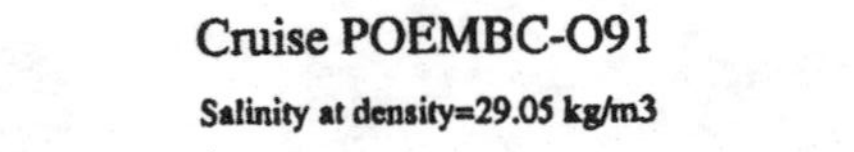


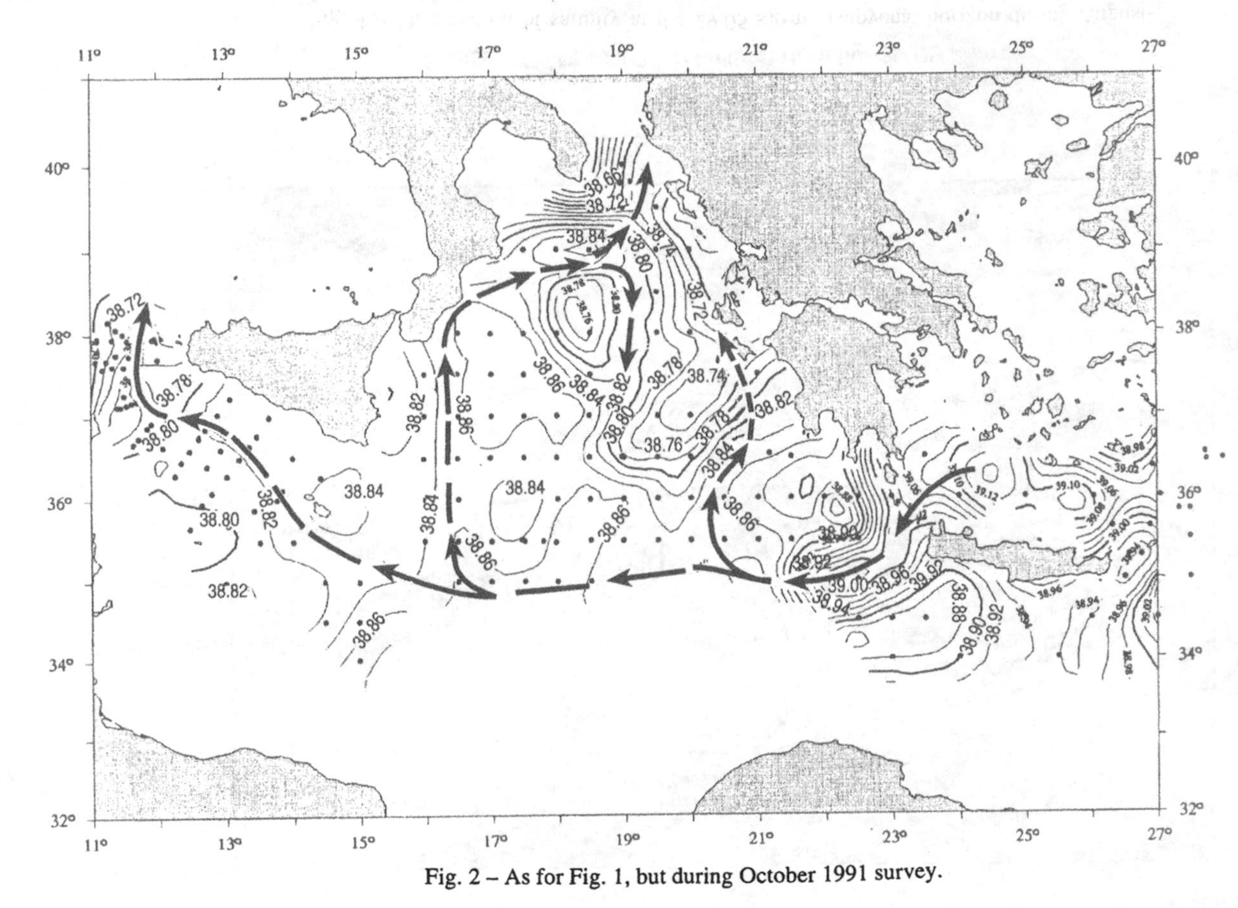


Fig. 2 – As for Fig. 1, but during October 1991 survey.

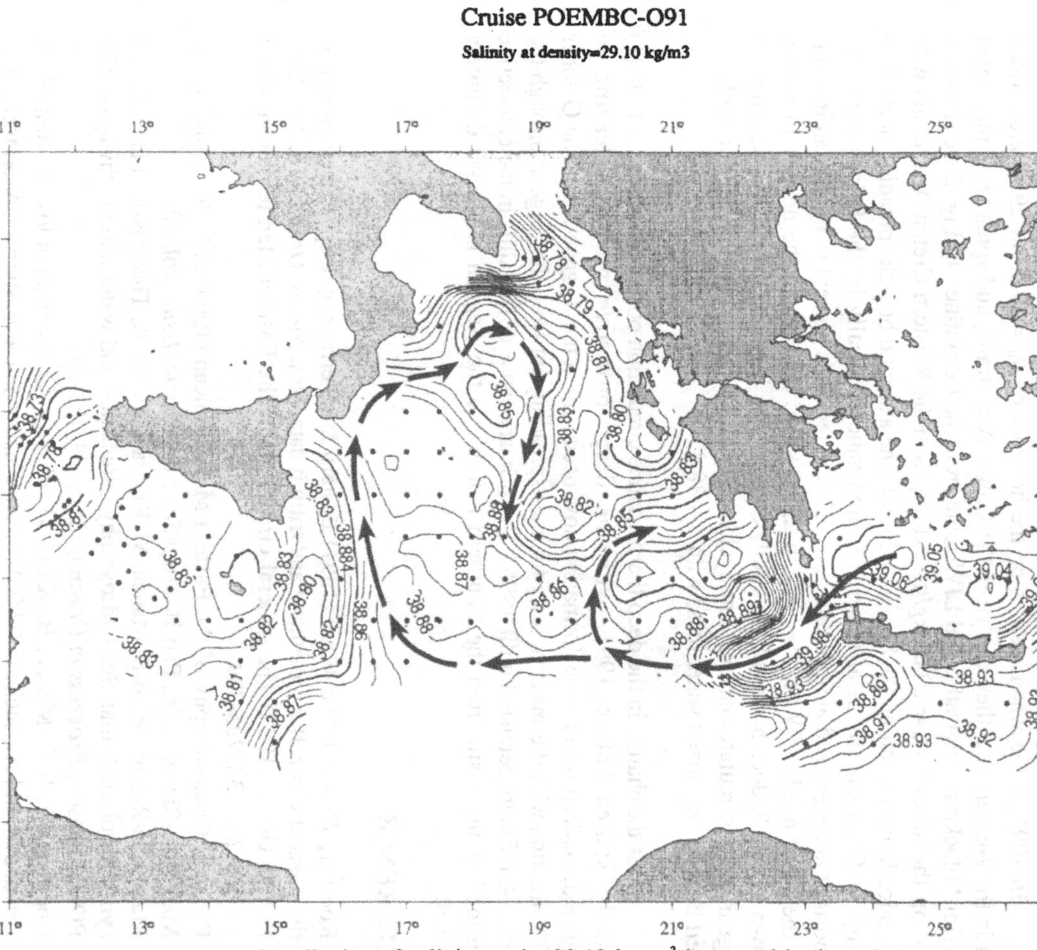

Fig. 3 – Distribution of salinity at the 29.10 kg•m$^{-3}$ isopycnal horizon during October 1991 survey.

veered northward by the anticyclonic gyres of the Ionian interior. The LIW pathway on its typical horizon, the isopycnal surface $\sigma_\theta$ = 29.05 kg/m$^3$, is shown in Fig. 1. This behavior is observed consistently on all the isopycnal surfaces from 29.00 to 29.15 kg/m$^3$.

In 1991 all the water masses on the horizons of $\sigma_\theta$ = 29.00 to 29.18 are formed inside the Aegean Sea, they exit from the Cretan Arc Straits, and spread into the Ionian interior "blocking" the traditional LIW route from the Levantine. In Fig. 2 we show the LIW on the horizon $\sigma_\theta$ = 29.05 kg/m$^3$ exiting from the Western Cretan Arc Straits with the major branch reaching the Straits of Sicily, and a second branch spreading northward along the Greek coastlines after being veered anticyclonically by the Pelops gyre. An important homogenized anticyclonic recirculation region is present in the Ionian interior. On the other hand, LIW on the horizon 29.10 kg/m$^3$ (and on all the deeper horizons), shown in Fig. 3, does not reach the Sicily Straits. The intrusive tongue exiting from the Aegean Sea recirculates entirely anticyclonically all around the Ionian interior, which is filled by a homogenized water mass with S ~ 38.87.

Thus the change in the deep thermohaline circulation observed in Winter 1995 (1) actually started before 1991, with the deep convective processes leading to intermediate/deep water mass formation now occurring in the Aegean Sea. The October 1997 situation with the massive salty tongues spreading out from the Aegean through the Cretan Arc Straits persisted until 1995. The primary cause of this shift in the convective "engine" of the basin from the Southern Adriatic to the Aegean is not yet definitely studied.

**REFERENCE**

1. Roether, W. and R. Schlitzer (1991) Eastern Mediterranean deep water renewal on the basis of chlorofluoromethane and tritium data, *Dyn. Atmos. Oceans*, **15**, 333-354.
2. POEM Group (1992) The general circulation of the Eastern Mediterranean, *Earth Science Rev.*, **32**, 285-309.
3. Physical Oceanography of the Eastern Mediterranean (1993) A.R. Robinson and P. Malanotte-Rizzoli, eds., Part II, *Deep-Sea Res., Special Issue*, **vol. 40**.
4. Malanotte-Rizzoli, P., B.B. Manca, M. Ribera d'Alcala, A. Theocharis et al. (1997) A synthesis of the Ionian Sea hydrography, circulation and water mass pathways during POEM Phase I, *Progress in Oceanography*, **39**, 153-204.
5. Roether, W., B.B. Manca, B. Klein, D. Bregant, D. Georgopoulos, V. Beitzel, V. Kovacevic and A. Luchetta (1996) Recent changes in the Eastern Mediterranean deep water, *Science*, **271**, 333-335.
6. Poulain, P.-M. (1998) Lagrangian measurements of surface circulation in the Adriatic and Ionian seas between November 1995 and March 1997, *Rapp. Comm. Int. Mer. Medit.*, **35**, 190-195.

# OCEANOGRAPHIC DATA HOLDINGS IN THE UKRAINIAN MARINE CENTERS FOR THE MEDITERRANEAN SEA

V.N. EREMEEV, A.M.SUVOROV
*Marine Hydrophysical Institute of Ukrainian*
*National Academy of Sciences*
*2 Kapitanskaya St., Sevastopol, Crimea 335000, Ukraine*
*e-mail: eremeev@mhi2.sebastopol.ua; suvorov@alpha.mhi.iuf.net*

**Abstract.** This paper provides a survey of the oceanographic data holdings in the Ukrainian institutions for the Mediterranean Sea. So, the Ukrainian marine centers hold hydrological, hydrophisical, hydrochemical, hydrobiological, hydrometeorological, and other Mediterranean data collected for more then 40 years from about 500 cruises and 17.000 oceanographic station.

## 1. Introduction

In Ukraine, activities aimed at the establishment of a national system for compiling, transfer, storage, analysis and dissemination of oceanologic data and information are being conducted in the framework of the project "National Bank of Oceanologic Data". This project is one of several specific research projects of the National program for the study and use of resources of the Azov Sea - Black Sea basin and other World Ocean regions. Financial support is provided by the Ministry of Science and Technologies of the Ukraine (fig. 1).

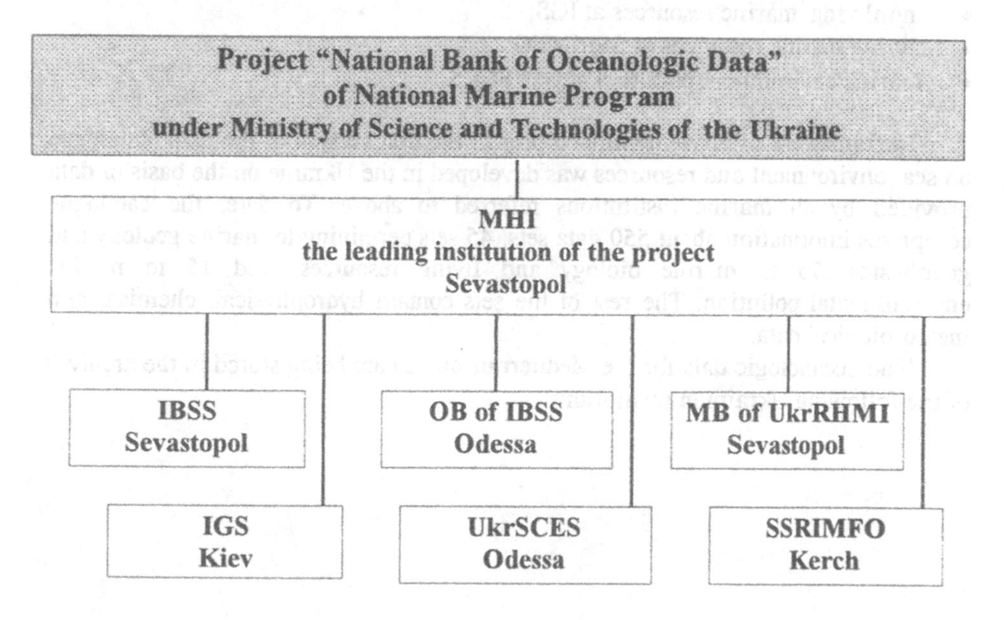


*Figure 1.* Participants of the project "National Bank of Oceanologic Data"



*P. Malanotte-Rizzoli and V.N. Eremeev (eds.),*
*The Eastern Mediterranean as a Laboratory Basin for the Assessment of Contrasting Ecosystems*, 7–18.

The leading institution of the project is

- Marine Hydrophysical Institute (MHI) of the Ukrainian National Academy of Sciences (Sevastopol).

Other participants of the project are:

- Institute of Biology of Southern Seas (IBSS) of the Ukrainian National Academy of Sciences (Sevastopol);
- Institute of Geological Sciences (IGS) of the Ukrainian National Academy of Sciences (Kiev);
- Ukrainian Scientific Center of the Ecology of Sea (UkrSCES) of the Ministry of Nuclear Safety and Environment of the Ukraine (Odessa);
- Southern Scientific Research Institute of Marine Fisheries and Oceanography (SSRIMFO) of the State Committee of Fisheries of the Ukraine (Kerch);
- Marine Branch of Ukrainian Research Hydrometeorological Institute (MB of UkrRHMI) of the State Committee of Hydrometeorology of the Ukraine (Sevastopol);
- Odessa Branch of the Institute of Biology of Southern Seas (OB of IBSS) of the Ukrainian National Academy of Sciences (Odessa).

The conception and principles of constructing a national distributed oceanologic information system have been developed, which included the foundation of four oceanologic data centers:

- oceanographic and satellite data at MHI;
- nonliving marine resources at IGS;
- living marine resources at SSRIMFO;
- marine environment pollution at Ukr SCES.

In framework of abovementioned project the first version of the data catalogue on sea environment and resources was developed in the Ukraine on the basis of data provided by all marine institutions referred to above. To date, the catalogue comprises information about 550 data sets: 45 sets pertaining to marine geology and geophysics, 75 to marine biology and living resources, and 15 to marine environmental pollution. The rest of the sets contain hydrophysical, chemical and meteorological data.

The oceanologic data for the Mediterranean Sea are being stored in the archives of the following Ukrainian institutions.

## 2. Marine Hydrophysical Institute (MHI)

MHI research vessels accomplished measurements in the Mediterranean Sea during 48 cruises (in total, there were 176 MHI cruises in all the regions of the World Ocean). Most of the stations were located in the Eastern part of the sea (the Aegean and Levant Seas), in the Alboran Sea and in the Gulf of Lions. It was made about 1600 stations. The measurements were conducted from 1968 till 1994 during all seasons (fig. 2). The following parameters were measured:

- temperature and salinity (Nansen bottles, MBT, CTD);
- currents (moorings);
- meteorological data:
- actinometric data;
- chemical parameters (dissolved oxygen, pH, alkalinity, phosphate, nitrate, nitrite, ammonium, silicate);
- hydrooptical parameters (Secchi disk depth, vertical profiles of spectral transparency, color index, scattering function, radiance index spectra, bioluminescence);
- radioactivity (Sr-90, Cs-134, Cs-137, Ce-144, Rn-222);
- biological data (chlorophyll, primary production, phyto-, zoo-, and ichthyoplankton, squid, etc.).

Biological data were obtained by scientists of the Institute of the Biology of Southern Seas and were kept there. Additional measurements were made in some cruises, for example, continuous measurements of the fine structure of temperature in the surface layer.

MHI has a special reference database including all information about R/V cruises (time, scientific staff, types of measurements, used instruments, coordinates of stations).

Brief information about cruises and measurements in the Mediterranean Sea is shown in Table 1. Cruises of R/V "Mikhail Lomonosov" (ML), R/V "Academic Vernadsky" (AV) and R/V "Professor Kolesnikov" (PK) are listed in the table. During these cruises the measurements were done in the basins of the Black Sea (BS), the Mediterranean Sea (MS) and the Atlantic Ocean (AO). The main types of the performed researches are indicated in the table: M - Meteorology, A - Actinometry, C - Chemistry, B - Biology, O - Optics, CTD - CTD-casts, MBT - Mechanical Bathytermographs, OSD - Nansen's bottles. The total number of stations in the Mediterranean Sea is shown in the last column.

Regretfully, only a part of the data collected by the research vessels after 1983 is stored in a computer readable form. The data from the previous years are being kept in the form of tables, punched cards, reports, etc. For the present moment, only about 20% of the data have been transferred onto modern data carriers.

(a)

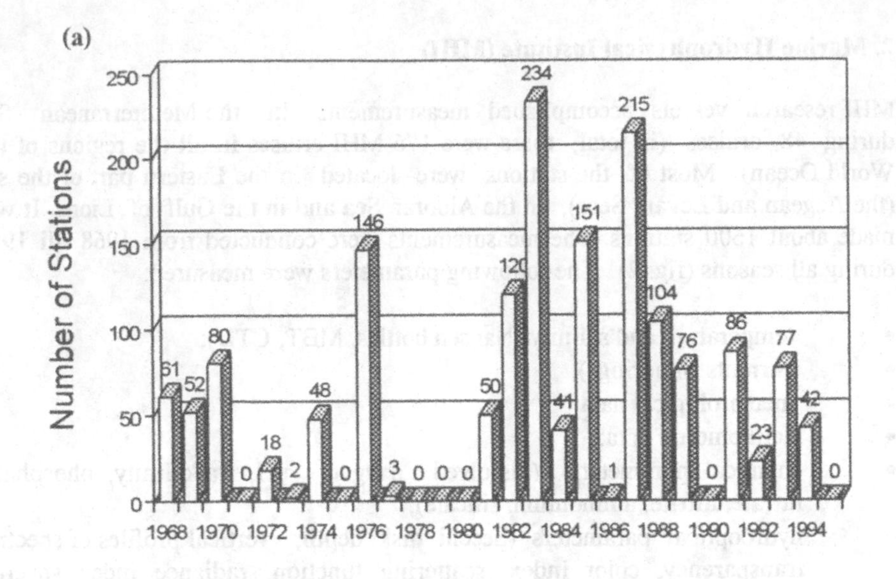


(b)

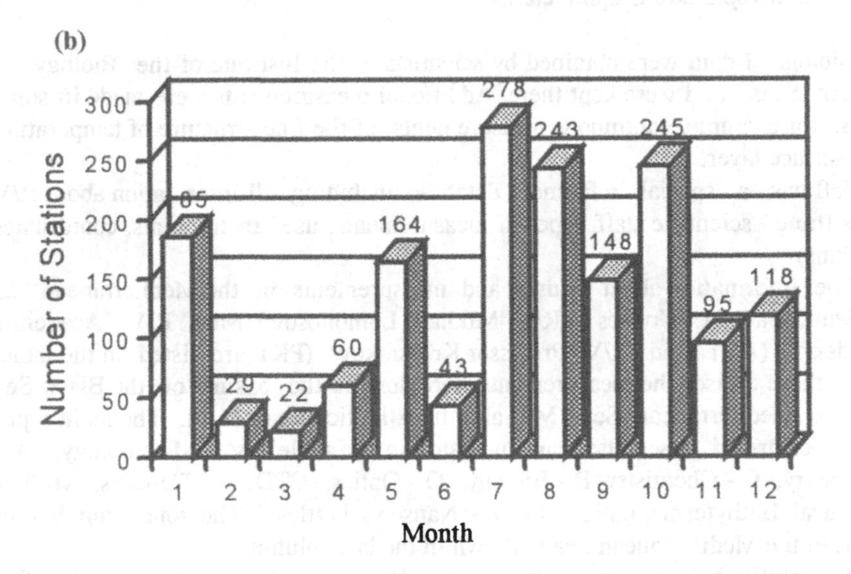


*Figure 2* . **Distribution of the MHI's oceanographic stations on years (a) and months (b)**

TABLE 1. Cruises of the MHI's Research Vessels in the Mediterranean Sea

| RV/ Cruise | Cruise Time Start-Finish | Regions | | | Parameters | | | | | T,S Measure. | | | Stat. Number |
|---|---|---|---|---|---|---|---|---|---|---|---|---|---|
| | | BS | MS | AO | M | A | B | C | O | CTD | MBT | OSD | |
| ML/21 | 28.04.68-24.08.68 | | * | * | * | | * | * | * | 0 | 0 | 192 | 58 |
| ML/22 | 25.12.68-24.04.69 | * | * | * | * | * | | * | * | 160 | 0 | 73 | 14 |
| AV/1-1 | 11.02.69-15.04.69 | | * | * | | | * | * | * | 0 | 11 | 11 | 6 |
| AV/1-2 | 08.06.69-04.07.69 | | * | * | | | | | | 0 | 10 | 10 | 1 |
| ML/23 | 24.06.69-28.08.69 | | * | | * | * | | * | * | 0 | 0 | 27 | 29 |
| ML/24 | 11.12.70-10.04.71 | | * | * | | * | * | * | * | 0 | 37 | 41 | 14 |
| AV/2-1 | 16.02.70-27.04.70 | | * | * | | | * | | * | 0 | 29 | 40 | 11 |
| AV/2-2 | 04.06.70-03.09.70 | | * | * | * | * | * | * | * | 0 | 85 | 67 | 2 |
| ML/25 | 29.07.70-27.10.70 | | * | * | * | * | | * | * | 0 | 0 | 40 | 39 |
| AV/3-2 | 01.03.71-21.03.71 | | * | * | | * | * | * | | 0 | 0 | 0 | 0 |
| ML/26-1 | 09.03.72-24.06.72 | | * | * | | * | * | * | * | 0 | 91 | 56 | 1 |
| ML/27 | 14.12.72-24.04.73 | * | * | * | | * | * | * | * | 0 | 158 | 66 | 9 |
| ML/28 | 22.01.74-05.05.74 | * | * | * | | * | * | * | * | 190 | 45 | 45 | 8 |
| AV/9 | 09.10.74-30.11.74 | | * | * | * | | * | * | * | 458 | 54 | 53 | 37 |
| ML/30 | 17.04.76-28.08.76 | * | * | * | * | | * | * | * | 86 | 168 | 57 | 7 |
| AV/13 | 18.06.76-12.08.76 | | * | | * | * | | * | * | 330 | 64 | 70 | 138 |
| ML/32 | 04.06.77-30.09.77 | * | * | * | * | | * | * | * | 216 | 54 | 119 | 3 |
| ML/42 | 04.07.81-16.09.81 | | * | * | * | * | * | | * | 125 | 0 | 24 | 50 |
| PK/3-1 | 21.03.82-09.06.82 | * | * | | | | | * | * | 174 | 0 | 92 | 74 |
| PK/4 | 02.07.82-10.08.82 | * | * | | | | * | | * | 114 | 0 | 0 | 20 |
| PK/5-1 | 01.09.82-12.11.82 | * | * | | * | | | | * | 124 | 0 | 0 | 26 |
| PK/5-2 | 29.12.82-04.03.83 | * | * | | | * | | | | 108 | 0 | 0 | 30 |
| PK/6-1 | 30.03.83-31.05.83 | * | * | * | * | | | | * | 147 | 0 | 0 | 11 |
| PK/6-2 | 24.06.83-25.08.83 | * | * | | | | | | | 146 | 0 | 0 | 69 |
| PK/7 | 20.09.83-14.11.83 | * | * | | * | * | * | | * | 180 | 0 | 0 | 124 |
| AV28 | 30.12.83-13.05.84 | * | * | * | * | * | | | | 235 | 0 | 43 | 3 |
| PK/9 | 20.06.84-23.09.84 | * | * | | * | * | | * | * | 650 | 0 | 0 | 38 |
| PK/10 | 06.11.84-09.02.85 | * | * | * | | | | | | 283 | 0 | 0 | 114 |
| PK/11 | 06.03.85-09.06.85 | * | * | * | * | | * | * | * | 269 | 0 | 0 | 1 |
| PK/12-1 | 26.07.85-28.08.85 | * | * | | | * | * | | * | 131 | 0 | 0 | 33 |
| ML/45 | 14.11.85-25.01.86 | | * | * | * | | * | * | | 137 | 0 | 0 | 3 |
| AV/33 | 18.02.86-03.06.86 | | * | * | * | * | * | * | | 230 | 0 | 0 | 1 |
| PK/14-1 | 07.08.86-06.11.86 | * | * | * | | | | | | 385 | 0 | 0 | 103 |
| PK/16 | 17.06.87-10.09.87 | * | * | | | * | | | * | 52 | 0 | 0 | 51 |
| PK/17-1 | 01.10.87-06.11.87 | * | * | | | | | | | 165 | 0 | 0 | 8 |
| ML/49 | 16.10.87-19.02.88 | | * | * | * | | | * | * | 193 | 0 | 0 | 14 |
| PK/17-2 | 12.11.87.-20.12.87 | * | * | | | | * | | * | 160 | 0 | 0 | 39 |
| PK/18 | 10.04.88-30.05.88 | * | * | | | | * | * | | 220 | 0 | 0 | 26 |
| PK/19 | 02.08.88-06.10.88 | * | * | | * | * | | | | 195 | 0 | 0 | 27 |
| AV/38 | 28.10.88-23.02.89 | | * | * | | | * | * | | 180 | 0 | 0 | 51 |
| PK/20 | 16.11.88.-06.03.89 | * | * | * | | | | * | | 480 | 0 | 0 | 10 |
| PK/21 | 31.03.89-15.05.89 | * | * | | * | * | | * | * | 340 | 0 | 0 | 24 |
| PK/23 | 04.08.89-03.10.89 | * | * | | | * | * | | * | 184 | 0 | 0 | 42 |
| ML/52-a | 04.05.90-30.05.90 | * | * | | * | * | | | * | 0 | 0 | 0 | 0 |
| PK/27 | 20.07.90-25/08.91 | * | * | | | | | * | | 117 | 0 | 0 | 86 |
| ML-55 | 30.09.92-02.11.92 | * | * | | | | | * | | 21 | 0 | 0 | 23 |
| PK/31 | 16.11.93-14.01.94 | * | * | | * | | * | * | * | 108 | 0 | 0 | 119 |

## 3. Institute of Biology of Southern Seas (IBSS)

Research vessels of IBSS carried out measurements in the Mediterranean Sea during 63 cruises. The measurements were made by research vessels " Akademik Kovalevsky" (1958-1990) and "Professor Vodyanitsky" (1977-1989) at approximately 1000 stations in the Aegean, Ionic, Adriatic, Tyrrhenian and Ligurian seas. The following parameters were measured:

- temperature and salinity (Nansen bottles, MBT, CTD);
- currents;
- chemical parameters;
- hydroacoustics;
- bioluminescence;
- primary production;
- phyto-, zoo-. bacterio- and ichtiyoplankton;
- ichthyology;
- phyto- ,and zoobenthos;
- sanitary hydrobiology;
- radioactive and chemical hydrobiology.

Brief information about cruises and measurements in the Mediterranean Sea is shown in Table 2. Cruises of R/V " Akademic Kovalevsky " (AK), and R/V "Professor Vodyanitsky " (PV) are listed in the table. The main types of the performed researches are indicated in the table: H-Hydrology, C- Currents, Hc - Hydrochemistry, Ha - Hydroacoustics, Bl - Bioluminescence, P - Primary production, Ph - Phytoplankton, Bp - bacterioplankton, Zp - zooplankton , Ip - ichtiyoplankton, I - ichthyology, Pb - phytobenthos, Zb - zoobenthos, S - sanitary hydrobiology, R- radioactive and chemical hydrobiology.

TABLE 2 . Cruises of the IBSS's Research Vessels in the Mediterranean Sea

| RV/ Cruise | Cruise Time Start-Finish | Parameters | | | | | | | | | | | | | | |
|---|---|---|---|---|---|---|---|---|---|---|---|---|---|---|---|---|
| | | H | C | Hc | Ha | Bl | P | Ph | Bp | Zp | Ip | I | Pb | Zb | S | R |
| AK/1 | 25.08.58 - 24.10.58 | * | | | | | | | | | * | * | | | | |
| AK/2 | 12.07.59 - 24.10.59 | * | * | | | | | | | | * | * | | | | |
| AK/3 | 28.05.60 - 13.08.60 | * | * | | | | | * | | * | * | * | | | | |
| AK/4 | 11.10.60 - 23.02.61 | * | * | | | | * | * | | * | * | * | * | | | |
| AK/5 | 03.12.61 - 03.03.62 | * | * | | | | | * | | * | | * | * | | | |
| AK/6 | 29.09.63. - 12.12.63 | * | * | | | | * | * | | * | * | * | * | | | |
| AK/8 | 08.66 - 10.66 | | | | | | | | | | * | | | | | |
| AK61 | 27.09.67 - 11.12.67 | * | * | | | | | | | | | | | | | * |
| AK/62 | 30.04.68 - 02.07.68 | * | * | | | | * | * | * | * | * | * | | | | * |
| AK/63 | 14.09.68 - 05.11.68 | | | | | | | | | | | | | | | * |
| AK/64 | 15.08.69 - 29.10.69 | | | | | | | | | | * | * | | | | |
| AK/65 | 28.12.69 - 13.03.70 | * | * | | | | * | * | | * | | | | | | |
| AK/66 | 19.05.70 - 29.07.70 | | | | | | | | | | * | | * | | | |
| AK/67 | 25.09.70 - 28.10.70 | * | | * | | | * | * | | * | | | | | | |
| AK/68 | 23.06.71 - 07.09.71 | | | | | | | | | | * | * | | | | |

TABLE 2 (cont.)

| RV/ Cruise | Cruise Time Start-Finish | Parameters | | | | | | | | | | | | | | |
|---|---|---|---|---|---|---|---|---|---|---|---|---|---|---|---|---|
| | | H | C | Hc | Ha | Bl | P | Ph | Bp | Zp | Ip | I | Pb | Zb | S | R |
| AK/69 | 02.11.71 - 25.01.72 | * | | * | | | | | * | | | | | | | |
| AK/70 | 07.05.72 - 16.07.72 | | | | | | | * | | * | | * | | | | * |
| AK/71 | 28.08.72 - 29.09. 72 | * | | * | | | * | * | | * | | | | | | |
| AK/72 | 17.07.73 - 05.10.73 | | | | | | | | | | * | * | | | | |
| AK/74 | 21.04.74 - 29.06.74 | | | | | | | * | * | | * | * | | * | | |
| AK/75 | 23.08.74 - 28.10.74 | * | | | * | * | | * | | * | | | | | | |
| AK/76 | 07.06.75 - 04.08.75 | * | | | | | | * | | * | | | | | | |
| AK/78 | 07.09.75 - 20.11.75 | | | | | | | | * | | | | | | | |
| AK/82 | 06.09.77 - 05.11.77 | | | | | | | | | | * | * | | | | |
| AK/83 | 18.03.78 - 17.05.78 | | | | | | | | | | * | * | | | | |
| AK/87 | 18.08.79 - 17.10.79 | | | | | | | | | * | * | * | | * | | * |
| AK/89 | 20.03.80 - 03.06.80 | | | | | | | | | * | * | * | * | * | | |
| AK/90 | 12.08.80 - 26.10.80 | | | | | | | | | * | | | * | * | | |
| AK/95 | 08.09.83 - 28.10.83 | * | | * | | | | | | * | | * | * | | | |
| AK/96 | 10.11.83 - 30.12.83 | * | | | * | * | | * | * | * | | | * | * | * | |
| AK/97 | 11.05.84 - 23.07.84 | * | | | * | * | | * | | * | | | | | | |
| AK/98 | 24.08.84 - 13.10.84 | | | | | | | * | | | | | * | | | |
| AK/100 | 25.06.85 - 24.08.85 | * | | | | | | | | * | | * | * | | | |
| AK/101 | 11.09.85 - 10.11.85 | * | | * | | | | | | * | | | | | | |
| AK/102 | 12.04.86. - 27.05.86 | * | | * | | | | | | | | | * | | | |
| AK/103 | 25.07.86 - 08.09.86 | * | | * | | | | | | | | | * | | | |
| AK/104 | 11.10.86 - 28.11.86 | * | | * | | | | | | | | | * | | * | |
| AK/105 | 03.06.87 - 18.07.87 | | | | | | | | | | * | * | | | | |
| AK/107 | 29.08.87 - 13.10.87 | | | * | | | | * | | * | | | | | | * |
| AK/109 | 03.11.87 - 18.12.87 | * | | | * | * | | * | | | | | | | | |
| AK/111 | 22.07.88 - 05.09.88 | | | | | | | | | | * | * | | | | * |
| AK/112 | 22.09.88 - 05.11.88 | | | * | | | | | | | | | | | | * |
| AK/114 | 28.04.89 - 12.06.89 | * | | | | | | | | | * | * | | | * | |
| AK/115 | 29.06.89 - 18.08.89 | * | | * | | | | | | | | | * | | | |
| AK/116 | 08.09.89 - 28.10.89 | * | | | * | * | | * | | * | | | | | | |
| AK/118 | 25.06.90 - 09.08.90 | * | | | | | | | | | * | * | * | | | |
| AK/119 | 31.08.90 - 20.10.90 | * | | * | | | | | | | | | * | * | | |
| PV/1 | 18.12.76 - 04.03.77 | | | | | | | | | | * | * | | | * | |
| PV/3 | 14.10.77 - 28.11.77 | | | | | | | | | | | * | | | * | |
| PV/6 | 27.07.79 - 30.10.79 | * | * | * | | | | | | | | * | | | * | |
| PV/7 | 23.11.79 - 06.02.80 | * | | | | | * | * | | * | | | | | | |
| PV/9 | 16.08.80 - 09.10.80 | * | * | * | | | | * | | * | | | | | * | |
| PV/11 | 15.11.81 - 15.02.82 | * | | | * | * | | | * | * | | * | | | | |
| PV/12 | 26.06.82 - 01.06.82 | * | | * | | | | | * | * | | | | | * | |
| PV/15 | 31.05.83 - 14.08.83 | * | | * | | | * | | | * | | | | | | |
| PV/7 | 05.06.84 - 13.09.89 | * | * | | | | * | | * | | | | | | * | |
| PV/19 | 15.05.85 - 29.07.85 | * | * | | | | | * | | * | | | | | | |
| PV/22 | 07.12.86 - 27.12.86 | * | | * | | | | | | | | | | | | * |
| PV/27 | 15.07.88 - 13.09.88 | * | | * | | | * | | | * | * | * | | * | | |
| PV/28-1 | 27.04.89 - 21.06.89 | * | | * | | | * | | | * | | | | * | * | * |
| PV/28-2 | 08.07.89 -31.08.89 | * | | * | | | | * | | * | | * | | | * | |

Only 10% of data are kept in the institute archives, other data are kept by the principal investigators. An inventory of data is currently being prepared. All data are still stored in form of tables and reports. For loading in a database the data have to be checked and transferred onto modern computer carriers .

**4. Institute of Geological Sciences (IGS)**

IGS (Department of Modern Marine Sediments Genesis ) has the following materials on Geology and Geochemistry of the Eastern Mediterranean Sea:

1) Experimental data from
   cruise r/v "Faras el Bahr" (1969);
   cruise 19 r/v "Akademic Vernadsky" (1978-1979);
   cruise r/v "Gidrolog" (1981);
   cruise 24 r/v "Professor Kolesnikov" (1989-1990)
   cruise 31 r/v "Professor Kolesnikov" (1993-1994),

represented with columns of:

- bottom sediments;
- bottom surface samples;
- Aeolian and water suspension;
- porous waters.

2) Analytical materials:
represented with the results of granulemetrical analysis, mineralogical analysis, chemical analysis for determination of $CaCO_3$, $SiO_2$, $C_{org}$ and Fe, Mn, Ti, Na, K, Cr, Cu, Zn, Ni, Cd, Zr, Co, V, Mo mikroelements, as well as emission spectral, atomic absorption, Roentgen fluorescent, neutron activation, chromatographic and other kinds of physical analysis.

The analytical base includes about 500 analyses.

3) IGS possesses materials on bottom sediments from the Nile Estuary, the Abu-Kyr Bay, the Levant Sea, the Aegean Sea and Izmir Bay.

4) On the base of the data obtained, a monograph and about 10 articles were published, lithological (on sediments types), geochemical (on elements), geoecological (landscape-geochemical) maps, as well as sections and profiles were constructed.

A part of the obtained materials is not brought into concrete data bases, nor systematized, nor digitized. Including into the National data bank on Marine Geology and Geophysics in a state ready for application by users needs additional financing and serious development.

**5. Ukrainian Scientific Center of the Ecology of Sea (UkrSCES)**

Research vessels of UkrSCES carried out measurements in the Mediterranean Sea during 334 cruises:

◊ 310 cruises of R/V type "Passat";
◊ 24 cruises of R/V "J. Gakkel" and "V. Parshin".

For R/V type "Passat" the measurements were performed from 1968 up to 1992 during all seasons. 8000 stations had been made. The following parameters were measured:

- temperature and salinity (Nansen battles, MBT, CTD);
- currents;
- meteorological data;
- aerological data;
- chemical parameters (dissolved oxygen, pH, alkalinity, phosphates, nitrates, ammonium, silicates);
- pollution data;
- MARPOLMON pollution data (oil slicks, tar bulls, dissolved oil);
- microlayer data (chemistry, pollution);
- hydrooptical parameters (Secchi disk, depth);
- radioactivity (Sr-90, Cs-134, Cs-137, Ce-144, Rn-222) in the atmosphere.

For R/V "J. Gakkel" and "V. Parshin" the measurements were performed from 1978 till 1992 in 24 cruises. More than 3400 station had been made.
The following parameters were measured:

- temperature and salinity;
- currents;
- meteorological data;
- chemical parameters;
- pollution data;
- MARPOLMON pollution data;
- hydrooptical parameters;
- radioactivity in the atmosphere.

Additional special measurements were made in cruises, for example, continuous measurements of the fine structure of temperature in the surface layer. UkrSCES has a database including information of R/V cruises (time, types of measurements, used instruments). Only part of the data collected by research vessels after 1976 is stored in a computer readable form. The data are being kept in the form of tables, cards, reports, etc. Only 15% of the data up today have been transferred onto modern carriers.

**6. Southern Scientific Research Institute of Marine Fisheries and Oceanography (SSRIMFO)**

Research vessels of SSRIMFO carried out measurements in the Mediterranean Sea from 1959 till 1984 during 8 cruises. The following parameters were measured:

- temperature - 339 stations;
- salinity - 325 stations;
- dissolved oxygen - 286 stations;
- phosphates - 228 stations;
- silicates - 230 stations.

The Mediterranean Sea was not of the main scientific interest for SSRIMFO and all the data are still stored in the form of tables. For loading into a database these data have to be checked and transferred onto modern computer data carriers.

## 7. Marine Branch of Ukrainian Research Hydrometeorological Institute (MB of UkrRHMI)

In 1973-1990 research vessels of State Oceanographic Institute (SOI) conducted investigations in the Aegean Sea and Eastern part of Mediterranean (table 3).

The main goals were to explore exchange of pollutants between the Black Sea and Mediterranean Sea and to specify balance of contamination substances. Staff from Marine Branch of Ukrainian Research Hydrometeorological Institute (former Sevastopol Branch SOI) took part in 16 cruises mentioned above. Nowdays these data of amount about 1200 stations are stored in an archive of Marine Branch. List of variables includes:

- temperature;
- salinity;
- currents;
- $O_2$, pH, $PO_4$, P, $NO_2$, $NO_3$. Si, As, Cu, Hg, Cd, Ag, Hg, Pb, Cr, Co, Zn, Se, Au, Sb, Cs, Ba;
- oil hydrocarbons and detergents.

TABLE 3. Research cruises carried out in the Aegean Sea and Eastern Part of Mediterranean

| R/V | Cruise No | Period |
|---|---|---|
| Mgla | 1 Med | May-August 1973 |
| Mgla | 2 Med | August-October 1974 |
| Mgla | 3 Med | March-April 1975 |
| Mgla | 4 Med | July-August 1975 |
| Yakov Gakkel | 2 | August-October 1976 |
| Yakov Gakkel | 3 | February-April 1977 |
| Yakov Gakkel | 5 | May-July 1978 |
| Yakov Gakkel | 9 | August-September 1979 |
| Yakov Gakkel | 10 | October-December 1980 |
| Yakov Gakkel | 11 | February-April 1981 |
| Yakov Gakkel | 28 | March-May 1987 |
| Yakov Gakkel | 29 | August-October 1987 |
| Yakov Gakkel | 31 | February-April 1988 |
| Yakov Gakkel | 33 | February-April 1989 |
| V.Parshin | 1 | June-July 1989 |
| V.Parshin | 4 | March-April 1990 |

## 8. Odessa Branch of the Institute of Biology of Southern Seas (OB of IBSS)

Research vessels of the OB of IBSS carried out measurements in the Mediterranean Sea during 65 cruises. Most measurements were made in the Aegean Sea from 1972 up to 1990. The following parameters were measured:

- temperature and salinity:
- currents:
- waves;
- Secchi disk and Forel Ule color scale;
- chemical parameters (dissolved oxygen, pH, alkalinity, phosphates, nitrates, organic nitrogen, ammonium, silicates, organic carbon);
- pollutants in particulate and dissolved forms;
- primary production, phytoplankton pigments;
- benthic organisms;
- phyto-, zoo-, and ichtyoplankton;
- microbiological parameters;
- biochemical parameters.

Only 10% of the data have been transferred onto modern carriers, the other data are being kept in the form of tables and reports.

## 9. Conclusions and recommendations

The Ukrainian marine centers (MHI, IBSS, IGS, UkrSCES, SSRIMFO, MB of UkrRHMI, OB of IBSS) hold hydrological, hydrophysical, hydrochemical, hydrobiological, hydrometeorological, and other Mediterranean data collected for more than 40 years from about 530 cruises and 17 040 oceanographic stations (table 4).

TABLE 4 . Number of oceanographic cruises and stations of the Ukrainian marine centers

| Institute | Number of cruises | Number of stations |
|---|---|---|
| MHI | 48 | 1600 |
| IBSS | 65 | 1000 |
| SSRIMFO | 8 | 340 |
| UkrSCES | 334 | 11400 |
| MB of UkrRHMI | 16 | 1200 |
| OB of IBSS | 60 | 1000 |
| IGS | 5 | 500 |
| | Total: | 17 040 |

Unfortunately, only a part of these data (about 15-20%) is now presented in a computer readable form, the other data are kept in a form of tables, reports, punched cards, etc. The great part of these data is at risk of being lost to future because of media degradation. The sole copies of manuscript data could be easily lost affected by the environment conditions or accidents, for instance, by fires. Additionally, manuscript data are hardly used by researchers who require data in digital form with all pertinent meta-data in order to perform the most comprehensive studies possible. To meet the requests, it is necessary to make the Mediterranean data available both for the national and international scientific communities in a common exchange format, on modern computer data carriers, as well as to prevent any obstacles for further using of the data.

Creation of the more precise and full Mediterranean data catalogue is planned in Ukraine for the current year in the frames of "National bank of Oceanologic data" project mentioned above. But to convert analog and tabular Mediterranean data into modern digital forms, to carry out the quality control of these data, to merge them into national and international databases, the Ukrainian marine centers need financial and technical support.

## 10. Acknowledgments

We would like to thank all our colleagues from Marine Hydrophysical Institute of the Ukrainian National Academy of Sciences (Sevastopol), Institute of Biology of Southern Seas of the Ukrainian National Academy of Sciences (Sevastopol), Institute of Geological Sciences of the Ukrainian National Academy of Sciences (Kiev), Ukrainian Scientific Center of the Ecology of Sea of the Ministry of Nuclear Safety and Environment of the Ukraine (Odessa), Southern Scientific Research Institute of Marine Fisheries and Oceanography of the State Committee of Fisheries of the Ukraine (Kerch), Marine Branch of Ukrainian Research Hydrometeorological Institute of the State Committee of Hydrometeorology of the Ukraine (Sevastopol), Odessa Branch of the Institute of Biology of Southern Seas of the Ukrainian National Academy of Sciences (Odessa) for materials provided.

# The study of a persistent warm core eddy in winter 1993 and the water mass formation in the Eastern Mediterranean Sea.

ALEXANDER I. KUBRYAKOV, NAUM B. SHAPIRO
*Marine Hydrophysical Institute, Ukrainian National Academy of Sciences, Sevastopol, Crimea*

**Abstract.** Survey data from the second leg of the 31 cruise of R/V "Professor Kolesnikov" covering December 6-19 1993 are analyzing to study a persistent warm core eddy located to the southeast of the Cyprus. Although the survey has been carried out in winter, vertical distribution of temperature and salinity has the most likely typical summer-fall character. The core of the eddy consists of warmer and saltier water than the environment and is characterized by an isothermal, isohaline lens of water wedged between the seasonal and permanent thermoclines. The eddy has no apparent signal at the surface, the maximum signal occurs near the base of the thermostad at the depth of 400 m approximately. The signal is shifted a little to north at the depth of 100 m compared with deeper levels. Possible mechanisms for the generation of the eddy are suggested. Some speculations about mechanism formation of intermediate water masses are discussed, being based on multi-layered three-dimensional model of ocean circulation with isopycnal coordinates. Finally, the most favorable conditions for formation of the Levantine Intermediate Water masse are indicated.

## 1. Introduction

The 31 cruise of R/V "Professor Kolesnikov" went through November 1993 - January 1994. The program of the cruise had included investigation of regions of the Black Sea and the Izmir Bay, which are traditional for us at the last time, and very interesting from oceanographic point of view the region of Eastern Mediterranean Sea - Levantine Basin.

The multinational program Physical Oceanography of the Eastern Mediterranean (POEM), which has been ongoing since 1985, has provided a totally description of the basin-wide circulation [7,8,11-14] and its variability, which are formed by processes of three predominant and interacting scales: basin scale, subbasin scale and mesoscale. In particular, the circulation in the Levantine Basin consists of a complex system of subbasin scale gyres interconnected by being possessed of various intensity and meandering currents. And the Levantine Basin is the source of the saline and warm water mass which spreads at intermediate depths over entire Mediterranean Sea and is the major contributor to the salty and warm Mediterranean outflow into the Atlantic



*P. Malanotte-Rizzoli and V.N. Eremeev (eds.),*
*The Eastern Mediterranean as a Laboratory Basin for the Assessment of Contrasting Ecosystems,* 19–31.

Ocean. The data collected during POEM reveal one feature which is a persistent or recurrent warm core eddy located to the southeast of the island of Cyprus. This eddy, called the Cyprus eddy [1-3], is one of several warm core eddies which comprise the subbasin scale Shikmona gyre [12].

In the following discussion we present results obtained from observation data of the Cyprus eddy during the cruise, then we discuss some our speculations about formation of intermediate water masses.

## 2. Observation of the Cyprus eddy

The main goal of the second leg of the 31 cruise of R/V "Professor Kolesnikov" was the study of the Shikmona gyre. The data for this study consist of CTD profiles from the survey conducted during December 6-19 1993. Measurements were made with a CTD-zond "SHIK-03" designed and made in Marine Hydrophysical Institute. The accuracy of this probe is 0.02 $^0$C and 0.02 practical salinity units (psu) through temperature and salinity respectively, and resolving ability is 0.005 $^0$C and 0.001 psu. The casts were made to a depth of 700 m or 1000 m. The survey includes 55 stations and consists of 3 zonal and 1 meridional sections. Fig. 1 shows the locations of these stations. Typical station spacing ranged from 8 to 10 miles. The survey was enough wide, with good resolution and has carried out in winter period during strong storms and navigational troubles.

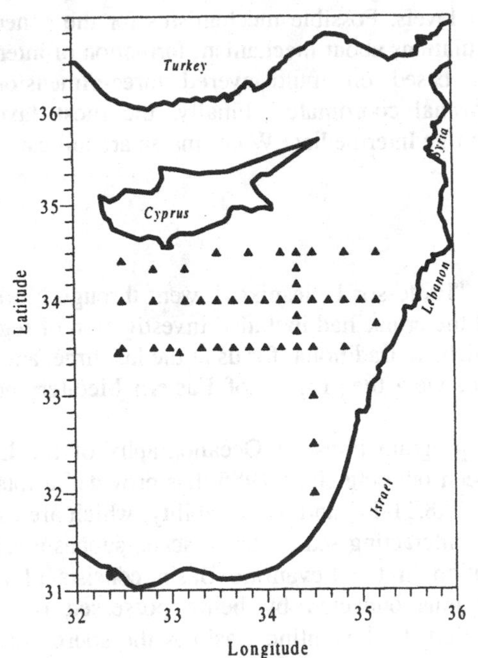


Fig. 1. Locations of the stations sampled during December 1993

In Fig. 2 we present the mean temperature and salinity profiles on the polygon. Although the survey has been carried out in winter, these profiles have the most likely typical summer-fall character and consist of four water masses: high salinity and warm Levantine Surface Water (LSW) in the the upper mixed layer of 50 m depth approximately, formed probably by excessive evaporation and wind mixing in the summer; low salinity Atlantic Subsurface Water (ASW) up to 150 m associated with entering water at surface through the Gibraltar Strait (Lacombe and Tchernia, 1972); high salinity and warm Levantine Intermediate Water (LIW) up to 400 m associated with the subduction processes; and low salinity and cold Deep Water (DW) below 600 m formed in the Adriatic [8,16].

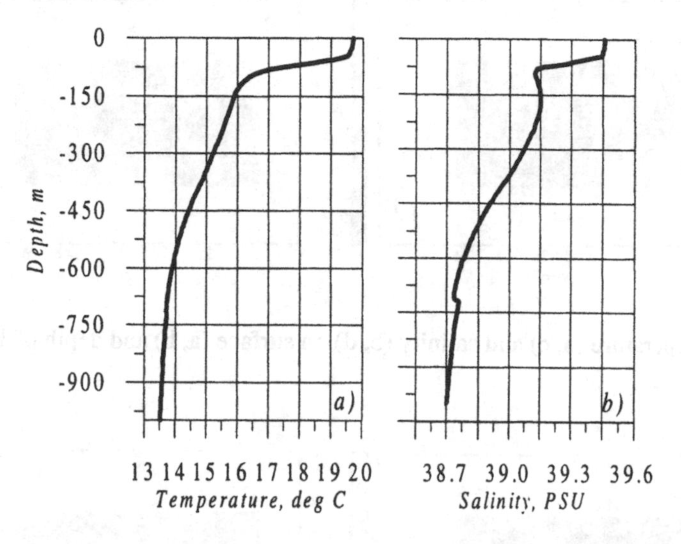


Fig. 2. Mean profiles of temperature (a) and salinity (b) on the polygon.

Fig. 3 shows the horizontal distribution of temperature and salinity at the surface and the 100 m depth. At the surface the temperature varies from 18.0 $^0$C to 20.8 $^0$C ($\Delta$ T=2.8 $^0$C) and salinity varies from 39.37 to 39.54 psu ($\Delta$ S=0.17 psu ). Note, we do not see any surface signal of eddy here. The range magnitude of temperature variability at the 100 m depth is the same (15.8 - 18.4 $^0$C, $\Delta$ T=2.6 $^0$C), but salinity one is more larger (38.98 - 39.40 psu, $\Delta$ S = 0.42 psu). And here we can see the signal of the eddy in the region of 34° 25'E. In Fig. 4 the horizontal distribution of temperature and salinity at the 400 m and 1000 m depths is presented. We see distinctly the signal caused by eddy in state of warmer and saltier water not only at the 400 m depth but at the 1000 m also. The coordinates of the center of the signal are 34° N and 34° 25'E approximately. Note that this signal is shifted a little to south compared to the depth of 100 m.

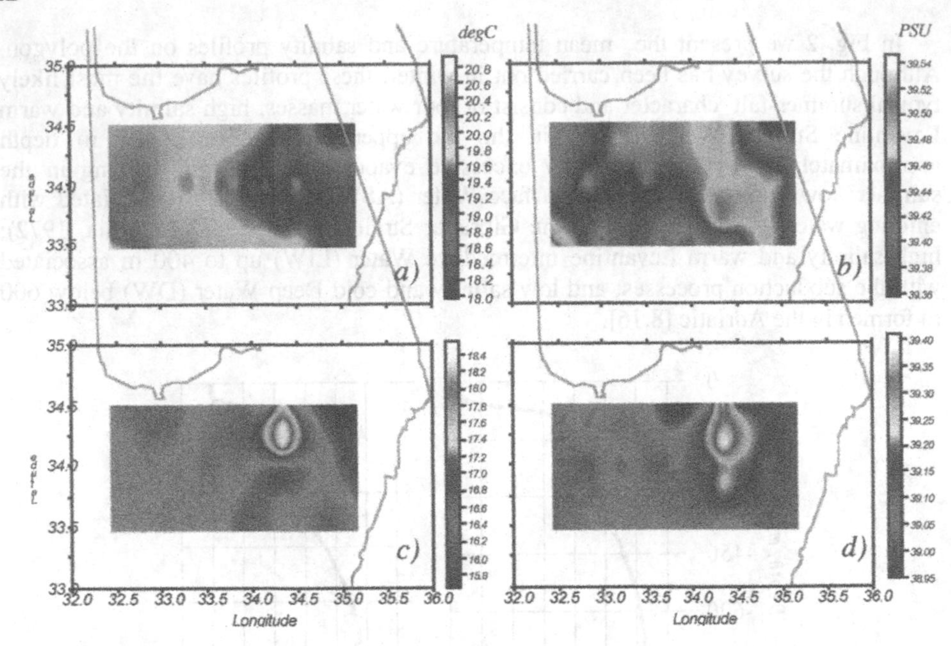


**Fig.3. Temperature (a, c) and salinity (b, d) on surface (a, b) and depth of 100 m (c, d)**

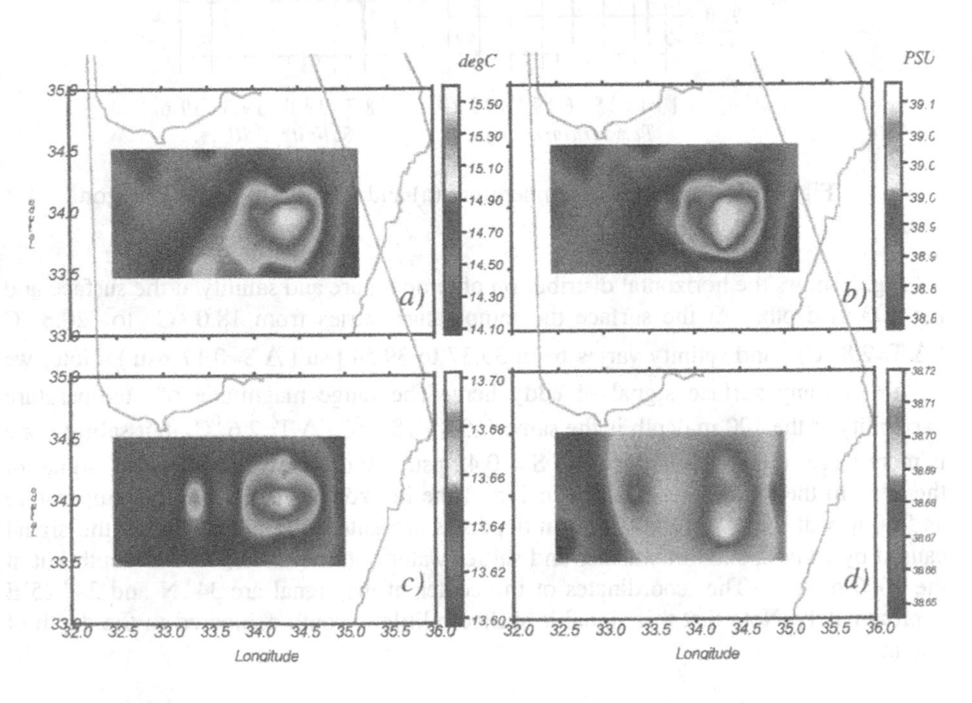


**Fig.4. Temperature (a, c) and salinity (b, d) on the depth (a, b) and of 100 m (c, d)**

In Fig. 5 the temperature and salinity profiles from the stations at the center and at the edge of the eddy are showed. We see the presence of warmer and saltier water in the center of eddy from 100 m to 700 m at least. The distinctive feature of the center profiles is the nearly isothermal and isohaline layer extending from 150 m to nearly 350 m, so called thermostad. The maximum horizontal temperature and salinity contrasts of 2-2.2°C and 0.15-0.20 psu respectively occur in the layer 300-500 m, that is deeper than the earlier data showed [3].

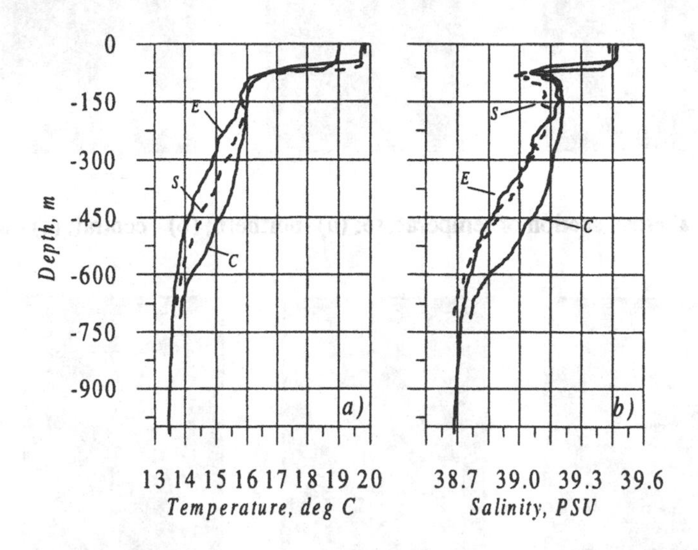


Fig. 5. The profiles of temperature (a) and salinity (b) of the stations at the center (*C*), eastern (*E*) and southern (*S*) edge of the eddy.

The characteristics of the thermostad can be seen good in the zonal sections of temperature, salinity and density (Figs. 6, 7 and 8, respectively). The core of the eddy consists of a lens of water trapped between the 15.5°C and 16.5°C isotherms (Fig. 6) and the 39.1 psu and 39.2 psu (Fig. 7). This lens appears as a broadening layer which separates the seasonal thermocline above and the permanent thermocline below. The core is isothermal but the surrounding water column is in the smooth transition zone between the seasonal and permanent thermocline. This water mass is just Levantine Intermediate Water.

Let us note, at that place when the lower boundary of eddy is most concave, the lower boundary of seasonal thermocline is convex just above. And it seems to be that the axis of eddy is a bit of sloping. We see the same picture still better at the meridional section (Fig. 9). The isotherms are most concave in the vicinity of 34° N from 350 m and deeper, and they are convex just above. From 140 m and upper, the isotherms are concave to the north of 34° 12'N and therefore the signal of eddy appears to more north at the depth of 100 m (Fig. 3).

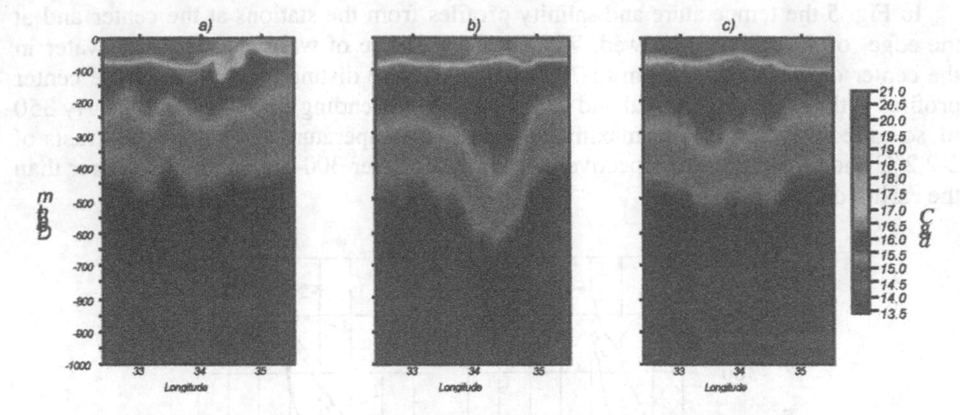


**Fig. 6. Zonal cross-section of temperature: (a) -northern; (b) - central; (c) - southern.**

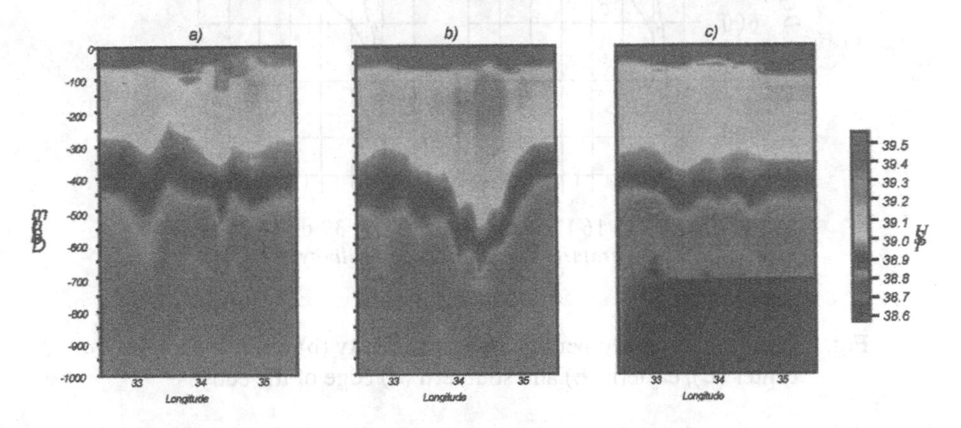


**Fig. 7. Zonal cross-section of salinity: (a) -northern; (b) - central; (c) - southern.**

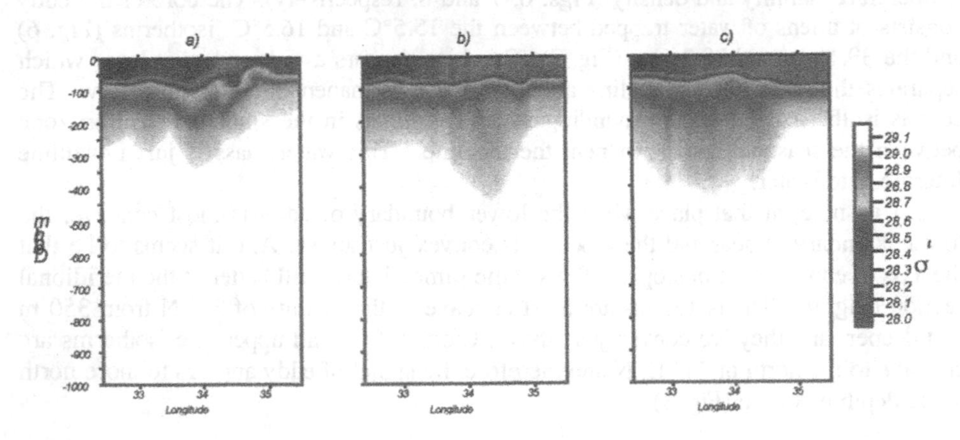


**Fig. 8. Zonal cross-section of density: (a) -northern; (b) - central; (c) - southern.**

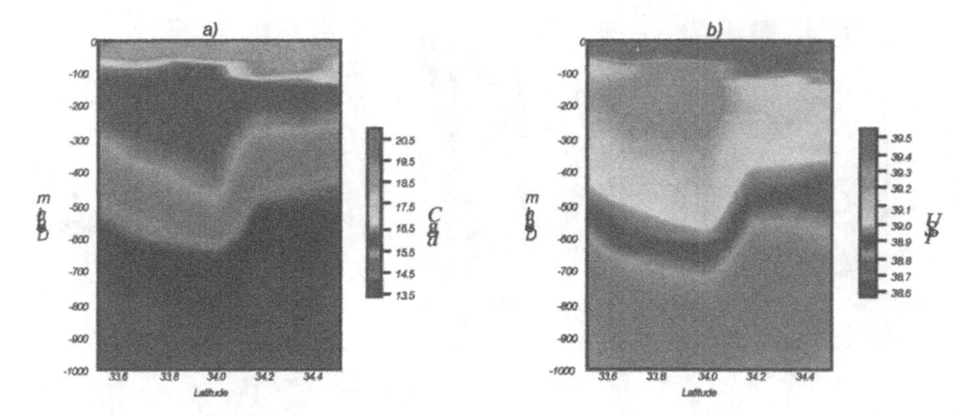


Fig. 9. Meridional cross-section of temperature (a) and salinity (b).

We can good look around the configuration of the eddy at the pictures of bedding of the isotherm surface of the 15.0°C and the isohaline surface of the 38.9 (Fig. 10). We see here the nearly regular inverted cone.

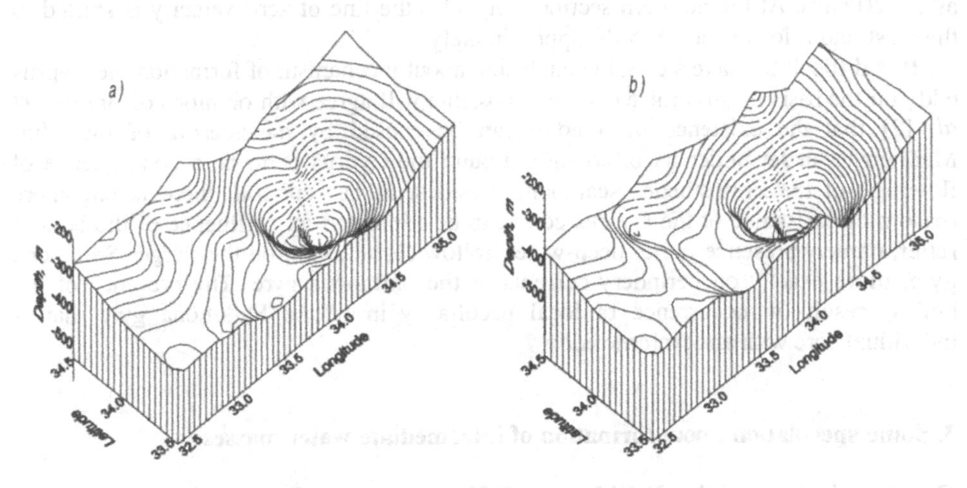


Fig. 10. The isosurfaces of the 15.0 °C temperature (a) and the 38.9 psu salinity (b).

To estimate the dynamics of the eddy the geostrophic velocities have calculated which present a balance between the pressure gradient and the Coriolis force. These velocities have been calculated by dynamic method relative to the reference level (or level of no motion) of 700 m. We have adopted this level because the most of casts have been made to a depth of 700 m. The change of this reference level has not practically affected on the results of calculations. In Fig. 11 the distribution of meridional component of geostrophic velocity is shown along the three zonal cross-sections. As might be expected, the position of main flows corresponds to the eddy position. At the central and southern sections (Fig. 11 b,c) there is the southward flow

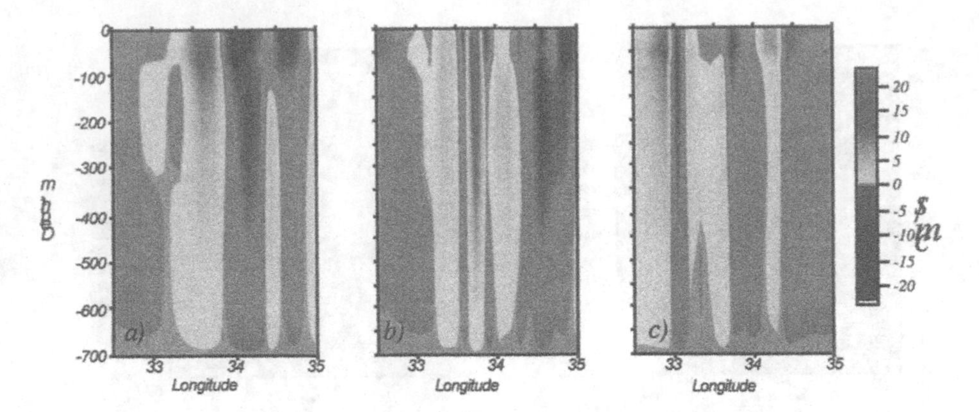


Fig. 11. Zonal cross-sections of geostrophic velocity: (a) -northern; (b) - central; (c) - southern. Positive values indicate northward flow and negative values are southward. Reference depth is 700 m.

to the right from 34°45' E and the northward flow to the left. The maximum velocities attain 20 cm/s. At the northern section (Fig. 11a) the line of zero velocity is shifted to the west and is located at 33° 50'E approximately.

It is difficult to make a certain conclusion about mechanism of formation the Cyprus eddy on the basis of this cruise data. It is possible will agree with opinions of Brenner *et al.* [2] that the existence of a eddy can be caused by meandering of the Mid-Mediterranean Jet under its bifurcation around the Cyprus or to be a consequence of flowing of the Eratosthenes seamount. Besides, in our opinion, there is one more reason: the existence of the Cyprus eddy can be stipulated for influence of the bottom relief, namely presence rather deep-water hollow (H > 2 km), on the larger Shikmona gyre, more exactly on periphery currents of the Shikmona gyre. The presence of the hollow results in occurrence of local peculiarity in of the Shikmona gyre that is individual core with anticyclonic eddy.

## 3. Some speculation about formation of intermediate water masses

The investigations of the POEM group [13] are a clear indication that the Levantine Intermediate Water is formed in the centers of anticyclonic eddies. To describe the mechanism of the formation thermostad and the surface signal of such eddies Brenner *et al.* [2] have used one-dimensional model. Also Lascaratos *et al.* [6] have used the one-dimensional model to examine how the formation of the LIW is controlled by the preconditioning of the hydrography and the surface fluxes. Such models simulate the subsurface low salinity water only with a parameterized horizontal advection because this process is substantially three-dimensional. And besides, the one-dimensional model can not take into account other effects connected with the three-dimensional circulation, in particular with the presence of eddy.

We shall describe briefly mechanism of formation of intermediate water masses, being based on original multi-layered three-dimensional model of ocean circulation with isopycnal coordinates [9,15]. The model was used to study the subduction processes in the Atlantic Ocean and the Black Sea.

According to this model the ocean water column consists of the upper mixed layer (UML) and internal layers possessing appointed density values. Internal layers do not exchange water with each other. The water can penetrate only to the UML from the close lower layer or from the UML to one. Evolution of the UML occurs through two regimes: entrainment and subduction. When the entrainment takes place the UML captures water from underlying layer. The subduction is the process when water leaves the UML and becomes a part of underlying layer. So the water of underlying layer acquires the properties - temperature, salinity, potential vorticity - which formed earlier in the UML. It is important that the water from the UML comes only to certain internal layers, so the density of incoming water correspond to appointed density of the internal layer. Thus different water masses can be formed which have the same density but rather any (up to the state equation) temperature and salinity. As further this water spreads in the internal layer, it actually represents intermediate water mass. Note one more , we consider the water mass has been formed after it had left the UML and actually entered the main thermocline. Some of this subducted water may be reentrained into the UML in the following winter.

The regime of evolution of UML depends on alignment of fluxes of buoyancy through sea surface, mechanical energy expended on mixing and advection. We'll display the conditions which promote the formation of intermediate water masses. For our goal it is quite enough to consider only two equations of the model: the equation of turbulent kinetic energy balance and the continuity equation. The equation of turbulent kinetic energy balance in the UML (traditionally, in local and stationary approximation, see [10]) is

$$h[B - W_e \cdot \Delta b] = 2mu_*^3 \qquad (1)$$

where $h$ is the thickness of the UML;

$B = a{\cdot}B_o + \varepsilon$ is the effective buoyancy flux;

$B_o$ is the buoyancy flux through sea surface;

$$a = \begin{cases} 1, & if\, B_o > 0; \\ a_1 < 1, & if\, B_o < 0; \end{cases}$$

$\varepsilon$ is the background dissipation of turbulent energy in the UML;

$\Delta b = b_o - b_1$;

$b_o$, $b_1$ are buoyancy in the UML and underlying layer respectively;

$m = 1.25$ is empirical constant;

$u_*$ is dynamic friction velocity proportional to wind speed;

$W_e = \min(W, 0) \le 0$ is the entrainment velocity, when $W < 0$;

$W_s = \max(W, 0) \ge 0$ is the subduction velocity, when $W > 0$;

$W$ is the normal velocity through the base of the UML, so that $W = W_e + W_s$.

The continuity equation is integrated over the UML

$$\frac{\partial h}{\partial t}+\frac{\partial U}{\partial x}+\frac{\partial V}{\partial y}+W-W_o=0 \qquad (2)$$

where $U$, $V$ are components of total stream in the UML;
$W_o = (Precipitation - Evaporation)$ is normal water velocity through sea surface.

The buoyancy flux $B_o$ is connected with heat flux $\tilde{A}_o$ and salt flux $R_o$ by relation

$$B_o = \alpha\tilde{A}_o + \beta R_o$$

where $\alpha = \frac{\partial b}{\partial T} > 0$; $\beta = \frac{\partial b}{\partial S} < 0$ - are coefficients of the thermal expansion and the haline contraction respectively. The heat flux can be presented as [4]

$$\tilde{A}_o = \lambda\,(T_a - T_o)$$

and salt flux equals

$$R_o = -W_o S_o$$

where $\lambda > 0$; $T_a$ is the effective air temperature; $T_o$ and $S_o$ are the sea surface temperature and salinity respectively.

As it follows from the equation (1) of turbulent kinetic energy balance (neglecting of the background dissipation $\varepsilon$), if $B_o < 0$ then the entrainment will take place only. For example, it will because of strong cooling of sea surface ($\tilde{A}_o < 0$, and $\alpha|\tilde{A}_o| > \beta R_o$) or because of strong evaporation ($R_o > 0$; $|\beta R_o| > \lambda\tilde{A}_o$). If $B_o > 0$ then the entrainment can occur as well as the subduction.

Thus, the condition $B_o > 0$ is the necessary condition of formation of intermediate water masses.

We accept the finite-difference approximation (numerical analogue) of equations (1), (2) in form

$$h^{n+1}\left[B^n-(W_e)^{n+1}\cdot\Delta b^n\right]=2mu_*^3\ , \qquad (3)$$

$$\frac{h^{n+1}-h^n}{\Delta t}+div(\vec{V}_{UML})^n+W^{n+1}-W_o=0 \qquad (4)$$

Here, $div(\vec{V}_{UML})=\frac{\partial U}{\partial x}+\frac{\partial V}{\partial y}$, $\Delta t$ is time step, index $n$ specifies a time moment.

We believe that functions at the $n$ moment of time are known. When the subduction takes place, the thickness of the UML can be found from equation (3) directly (hereafter the background dissipation $\varepsilon$ is neglected)

$$h^{n+1}=\frac{2mu_*^3}{B^n} \qquad (5)$$

So, the necessary and sufficient condition of occurrence of subduction (at $B_o > 0$, of course) can be easy got from equations (4) using (5)

$$\frac{2mu_*^3}{B^n} < h^n - \Delta t \cdot div(\bar{V}_{UML})^n + W_o \cdot \Delta t \quad (6)$$

We can see from (6) that increase of $B$, weaker intensity of winds and convergence of currents in the UML promote subduction. If the cooling takes place ($\tilde{A}_o < 0$), then $B_o$ can be more zero ($B_o > 0$) only because of intensive precipitation ($R_o > 0$). In this case the fresh and cold water will be formed at the sea surface which will fall into appropriate underlying layers later. Such situation is possible in ocean at northern latitudes, but hardly it takes place in the Mediterranean Sea. A situation is more pertinent for Mediterranean Sea, when intensive heating ($\tilde{A}_o > 0$) and evaporation occurs. Though the evaporation does not promote that buoyancy flux would be positively, but it results to increase of surface water density, so getting warmer, but becoming heavier (due to salting) water will gradually fill surface layers to form the intermediate water masses. Thus, the surface water with high salinity (due to evaporation) can penetrate inside to the sea only owing to intensive heating ($T_a >> T_o$) of the sea surface, i.e. in the spring. If the heating is insufficiently intensive and the entrainment takes place, the following picture will: the water at the surface layer is more salty, but the underlying layer with original small salinity is deeper (but with the greater density, of course) and again more salty water is still deeper.

Because of intensive heating the water will come to subsurface layers from the UML, which have been formed in the UML earlier, when the entrainment took place. The entrainment can occur, for example, in winter, when the strong cooling is, the bottom of the UML is deepening, the water is involved to the UML from the thermocline and is mixed by convection. The layer of local minimum of salinity is not present in winter; temperature and salinity are constant through depth up to the bottom of the thermocline.

Thus, the mixing of waters occurs at any time of a year, but renewal (formation) of intermediate water masses, i.e. the incoming of the appropriate waters to thermocline, occurs only owing to subduction process, which takes place in certain season and evidently at certain locations.

The relation between divergence and vorticity can be illustrated if, for example, we adopt the quasi-stationary motion equations without nonlinear inertial members and with non-geostrophic approximation dynamics made conditionally by Rayleigh friction. In this case the relation looks like

$$f \cdot div(\mathbf{v}) = -r \cdot rot(\mathbf{v}) \quad (7)$$

Here, $f$ is the Coriolis parameter, $\mathbf{v}$ is the horizontal velocity, $r$ is the friction coefficient. As since $rot(\mathbf{v}) < 0$ in the anticyclonic eddy, then $div(\mathbf{v}) > 0$. Let the ocean consists of the three layers: the first is the UML, the second is the main thermocline, where the eddy is located, and the third is a motionless layer. (Note, the nonlinearity in the pressure gradient can be kept in this case). The equation

$$div(\vec{V}) = 0 \qquad (8)$$

follows (in the "rigid lid" approximation) from the continuity equation of the whole of the water column, here $\vec{V} = \vec{V}_{UML} + \vec{V}_{MT}$ are the sum of total streams in the UML and the main thermocline respectively. The relation

$$div(\vec{V}_{UML}) = -div(\vec{V}_{MT}) \qquad (9)$$

results from (8). If the eddy is axis-symmetrical, then in the center of the eddy the sign of the total stream divergence $div(\vec{V}_{MT})$ coincides with the sign of the velocity divergence $div(\vec{v})$. Therefore, in the center of the anticyclonic eddy $div(\vec{V}_{MT}) > 0$ and just above, into the UML, $div(\vec{V}_{UML}) < 0$, i.e. the convergence takes place, which promotes the subduction and, as though, engages the water from the UML. So our reasoning explain the fact, received from observation data, of preferable formation of intermediate water masses into anticyclonic eddies.

The quantitative description of thermohaline circulation and processes of formation of intermediate water masses requires realization of special numerical experiments.

**Acknowledgments**

The authors would like to express their gratitude to Paola Malanotte-Rizzoli for her friendship and for given an opportunity to take part in such a very fruitful workshop.

**References**

1. Brenner, S. (1989) Structure and evolution of warm core eddies in the eastern Mediterranean Levantine Basin. *Journal of Geophysical Research,* **94**, 12593-12602.
2. Brenner, S., Rozentraub, Z. Bichop, J. and Krom, M. (1991) The mixed-layer/thermocline cycle of a persistent warm core eddy in the eastern Mediterranean. *Dynamics of Atmospheres and Oceans,* **15**, 457-476.
3. Brenner, S. (1993) Long-term evolution and dynamics of a persistent warm core eddy in the Eastern Mediterranean Sea. *Deep-Sea Research II,* **40**, 1193-1206.
4. Haney, R. L. (1971) Surface thermal condition for ocean circulation models. *Journal of Physical Oceanography,* 1, 241-248
5. Lacombe, H., Tchernia, P. (1972) Caracters hydrologiques et circulation des eaux en Mediterranee, in D. J. Stanley (ed.), *The Mediterranean Sea: A Natural Sedimentation Laboratory*, Dowden, Hutshinson and Ross, Stroudsbourg, Pa., pp. 25-36.
6. Lascaratos, A., Williams, R., G., Tragou, E., (1993) A Mixed Layer Study of the Formation of Levantine Intermediate Water. *Journal of Geophysical Research,* **98**, 14739-14749.

7. Malanotte-Rizzoli, P. and Robinson, A. R., (1988) POEM: Physical Oceanography of the Eastern Mediterranean. *EOS,* The Oceanography Report, **69**(15).
8. Malanotte-Rizzoli, P. and Hecht, A., (1988) Large scale properties of the Eastern Mediterranean: a review. *Oceanologica Acta*, **11**, 323-335.
9. Michailova, E. N., Shapiro, N. B. (1992) Hydrodynamical multilayered ocean model. (in russ.). *MHI Ukrainian NAS*, pp. 40
10. Niiler P. P., and Kraus, E. B., (1977) One-dimensional models of the upper ocean, in E. B. Kraus (ed.), *Modelling and Predictions of the Upper Layers of the Ocean*, Pergamon, Elmsford, N. J., pp. 325.
11. Ozsoy E., Hecht, A., Unluata, U., Brenner, S., Oguz, T., Bishop, J., Latif, M. A. and Z. Rozentraub, Z., (1991) A review of the Levantine Basin Circulation and its variability during 1985-1988. *Dynamics of Atmospheres and Oceans*, **15**, 421-456.
12. Ozsoy E., Hecht, A., Unluata, U., Brenner, S., Sur, H.I., Bishop, J., Latif, M. A., Rozentraub, Z. and Oguz, T., (1993) A synthesis of the Levantine Basin circulation and hydrography 1985-1990. *Deep-Sea Research II,* **40**, 1075-1119.
13. POEM Group (1992) General circulation of the Eastern Mediterranean. *Earth-Sclence Reviews*, **32**, 285-309.
14. Robinson A. R., Golnaraghi, M., Leslie, W. G., Artegiani, A., Hecht, A., Lazzoni, E., Michelato, A., Sansone, E. and Unluata, U., (1991) The eastern Mediterranean general circulation: features, structure and variability. *Dynamics of Atmospheres and Oceans*, **15**, 215-240.
15. Ryabtsev, Yu. N., Shapiro, N. B. (1996) Modelling of the generation and evolution of the cold intermediate layer in the Black Sea. *Physical Oceanography,* 7, 49-63.
16. Zore-Armanda, M. (1974) Formation of Eastern Mediterranean deep waters in the Adriatic, in *Colloques Internationaux du C. N. R. S. No 215. Processus de Formation des Eaux Oceaniques Profondes,* pp. 128-233.

# INTRASEASONAL LARGE-SCALE ANOMALIES OF THE MEDITERRANEAN SEA SURFACE TEMPERATURE

V.V.EFIMOV, A.V.PRUSOV, M.V.SHOKUROV
*Marine Hydrophysical Institute of Ukrainian Academy of Sciences, Sevastopol, Kapitanskaya St., 2, 335000, Crimea, Ukraine.*

ABSTRACT.
**For studying the intraseasonal sea surface temperature anomalies in the Mediterranean Sea COADS data were used, which are analyzed in terms of characteristics specific to separate seasons of the year. A method of local Lyapunov coefficients calculated from the changes of the SST anomalies during one-month time intervals was applied. It was found that the coefficients change during a year in such a manner, that maximal values occur in the spring and autumn. That means that SST anomalies are growing in these periods of a year and are decaying in the intermediate intervals. The spatial distributions of the growing and decaying anomalies with relation to the annual cycle of SST in the Mediterranean Sea are discussed.**

## 1. Introduction.

In the World Ocean intraseasonal variations of the sea surface temperature, defined as oscillations occurring in separate seasons or months of the year with spatial scales exceeding a thousand kilometers and temporal scales being of the order of a month, represent an important aspect of the atmosphere-ocean circulation. As an example, they may be related to the longer-period interannual anomalies, such as El-Niño–Southern Oscillation [1]. Physical mechanisms responsible for the generation of intraseasonal SST anomalies include both the local processes of the atmosphere-ocean interaction due to variability of thermal fluxes, and the vertical/horizontal advection connected with the large-scale inhomogeneities of temperature and current velocity fields. The study of such anomalies calls for the use of state-of-the-art numerical models for the air-sea interaction.

For the inland seas, such as the Mediterranean, the study of intraseasonal SST anomalies seems to be a less complicated problem. SST anomalies there having spatial scales of the order the dimensions of the sea, are coupled only with the processes of air-sea interaction and the variability of sea circulation, but not with the horizontal advection from adjacent regions. There are vast literature about Mediterranean weather and climate components, that can effect on SST, such as pressure [2], atmosphere circulation in general [3,4], rainfall [5]. Some results on decadal time scales were obtained for SST itself in [6,7,8].

This study presents the results of processing of SST data from the widely-known COADS dataset [9]. The method applied to study nonlinear systems by means of the finite-time Lyapunov coefficients was used to interpret observational data, rather than to scrutinize the theoretical hydrodynamical model. As a result, some characteristic parameters describing the temporal evolution of the SST anomalies and their spatial distribution were computed, similar to the local Lyapunov coefficients and eigenvectors. The behavior of these characteristic parameters during the annual cycle is discussed.



*P. Malanotte-Rizzoli and V.N. Eremeev (eds.),*
*The Eastern Mediterranean as a Laboratory Basin for the Assessment of Contrasting Ecosystems,* 33–47.

## 2. Methods of Analysis and Data.

To study the dynamics of a nonlinear system, such as the ocean-atmosphere system, known method for evaluating Lyapunov characteristic exponents is used. Assume, that we are analyzing an evolution of the system, described by the autonomous differential equation [10] in $n$-dimensional space

$$d\mathbf{x}/dt = \mathbf{F}(\mathbf{x}),\ \mathbf{x} = (x_1,\ldots,x_n) \qquad (1)$$

We take the arbitrary particular solution of this equation, trajectory $\mathbf{x}_0(t)$, and determine the behaviour of trajectories $\mathbf{x}(t)$, close to this trajectory. This could be done by linearization (1) in relation to small deviations $\xi = \mathbf{x} - \mathbf{x}_0$:

$$d\xi/dt = \mathbf{A}'\xi \qquad (2)$$

where $A'_{ij} = \partial F_i/\partial x_j$ at $\mathbf{x} = \mathbf{x}_0$ is the Jacobian matrix of the map $\mathbf{F}(\mathbf{x})$. Integration of the equation (2) from $t$ to $t+\tau$ gives solution

$$\xi(t+\tau) = \mathbf{A}(t,\tau)\xi(t) \qquad (3)$$

The eigenvalues of matrix **A** define the compression and stretching factors in the directions of the respective eigenvectors for real eigenvalues, and the angular rotation velocity in the respective planes for complex-conjugate pairs of eigenvalues. Generally matrices **A*A** and **AA***, where * indicates transpose, rather then matrix **A** proper are considered to define Lyapunov exponents. It can be shown [11] that these matrices defines a coupled eugenvalue problem such that

$$\mathbf{A}^*\mathbf{A}\,\zeta_s = \lambda_s^2\,\zeta_s,\ \mathbf{A}\mathbf{A}^*\,\eta_s = \lambda_s^2\,\eta_s \qquad (4)$$

where $\lambda_s^2$ are eugenvalues of those matrices. They determine the local, or finite-time Lyapunov numbers $\lambda_s$ which characterize the exponential loss of correlation between the two close trajectories. Alongside $\lambda_s$ local Lyapunov exponents $\Lambda_s=1/\tau\ln\lambda_s$ are considered.

Orthogonal eugenvectors $\zeta_s$, termed the "local Lypunov vectors", evolves to orthogonal vectors $\lambda_s\eta_s$ during time interval $(t,\ t+\tau)$ (sometimes they are named "optimal disturbances" [12]). A small sphere of radius $\varepsilon$ centered at point $\mathbf{x}_0(t)$ elongates into an ellipsoid centered at point $\mathbf{x}_0(t+\tau)$ with the principal axes $\varepsilon\lambda_s$, directed along $\eta_s$. All above mentioned values are defined only locally in state space and in time and depends on $t$ and $\tau$ [13]. At infinite time limit local Lyapunov exponents become usual (global) ones. It is well known that even one positive global Lyapunov exponent means the developing of instability and leads to the deterministic chaos because of the exponential scatterring of close trajectories in the respective direction.

Let us consider the description of the interannual large-scale variability of SST, using the conception of Lyapunov exponents. It is well recognized that SST has a pronounced annual cycle forced by external periodic excitation, so the equation (1), describing the SST dynamics, is not autonomous. Hence, the linear equation (2), which describes the local dynamics of SST anomalies in relation to the annual cycle, is not autonomous either, and matrix **A'** is a periodic function of time. During months with positive Lyapunov exponents, instability is developing, and the state of the system is changing. During steady months, when all $\Lambda_s$ are negative, the anomalies are exponentially decreasing. The value and the sign of the anomaly, which developes during unstable months, depends on the previous state of system, which depends on the decreasing during the months of stability. Thus, a chaos on the interannual scales may be generated.

The existence of positive local Lyapunov exponents during certain seasons is very important in terms of dynamics. However, it is impossible to evaluate these exponents theoretically using primitive equations for the ocean-atmosphere system, because of the complexity of this problem. Thus, the detection of unstable directions via analysis of a dataset is of interest beyond all doubt, because it permits to explain qualitatively the generation of interannual variability in the ocean-atmosphere system.

The data used in this study have been taken from the known COADS dataset and consist of the mean monthly SST over the Mediterranean Sea during 45 years (1945-1989) [9]. The spatial resolution is 1°×1°, and the real dimension of state space is high, number of squares for the Mediterranean Sea is $N$=252 (in [14] data from COADS Release 1 were used). To reduce the dimension of the problem, prior to the analysis, the data were subjected to a truncated empirical orthogonal functions (EOF) expansion.

The distribution of variance in state space varies during a year, and the optimal basis has to be choosen separately for every season. We have used the EOF-basis, constructed for each pair of the consecutive months $t$ and $t$+1 ($t$=1,...,12). The basis, constructed for the whole year, in contrast to the bimonthly one, has not yielded any positive results, because of the strong seasonal variations of SST variance over the Mediterranean.

In general, from dynamical point of view the EOF-basis has physical meaning only for systems described by certain linear stochastic equatuions, but for nonlinear deterministic systems it loses the sense. However in small part of phase space of high-dimensional nonlinear system the behavior of trajectories can be considered as linear in corresponding tangent space. At low enough level of excitation trajectories effectively employ and fill some low-dimensional subspace of entire tangent space. Due to this filling local EOF-basis, constructed using this small part of phase space, is representative and complete one in this local tangent subspace. Another part of phase space can have another local basis due to curvature of cloud of trajectories. In terms of SST clymatic system this mean that during different seasons different forms of anomalies are developing.

For every $t$=1,...,12 construct the matrix of SST anomalies, including months $t$ and $t$+1, with dimension $N\times 2m$, where $N$ is the number of points in a physical space, $m$=45 is the number of years. After the singular value decomposition of this matrix we obtained EOFs $a_{ki}$, principal components $b_{pi}$ and eigenvalues $\alpha_i$. This decomposition converges fast enough, and some first leading eigenvectors describe a sufficient part of the initial data variance and yield a good reproduction of the large-scale patterns of SST anomalies. Therefore, we will consider hereinafter reduced data

$$\Delta T_{kp}=\sum_{i=1}^{n} \alpha_i a_{ki} b_{pi} \tag{5}$$

**where $n$ is the dimension of EOFs basis in state space, which is small enough ($n<<N$). In this new basis, we have $m$ points for the current month $t$ $\xi_{ip}{}^{t} = b_{pi}$ and $m$ points for the following month $t+1$ $\xi_{ip}{}^{t+1} = b_{p+m,i}$ ($p=1,\ldots,m$, $i=1,\ldots,n$). For convenience above mentioned entities are shown in Fig.1.**

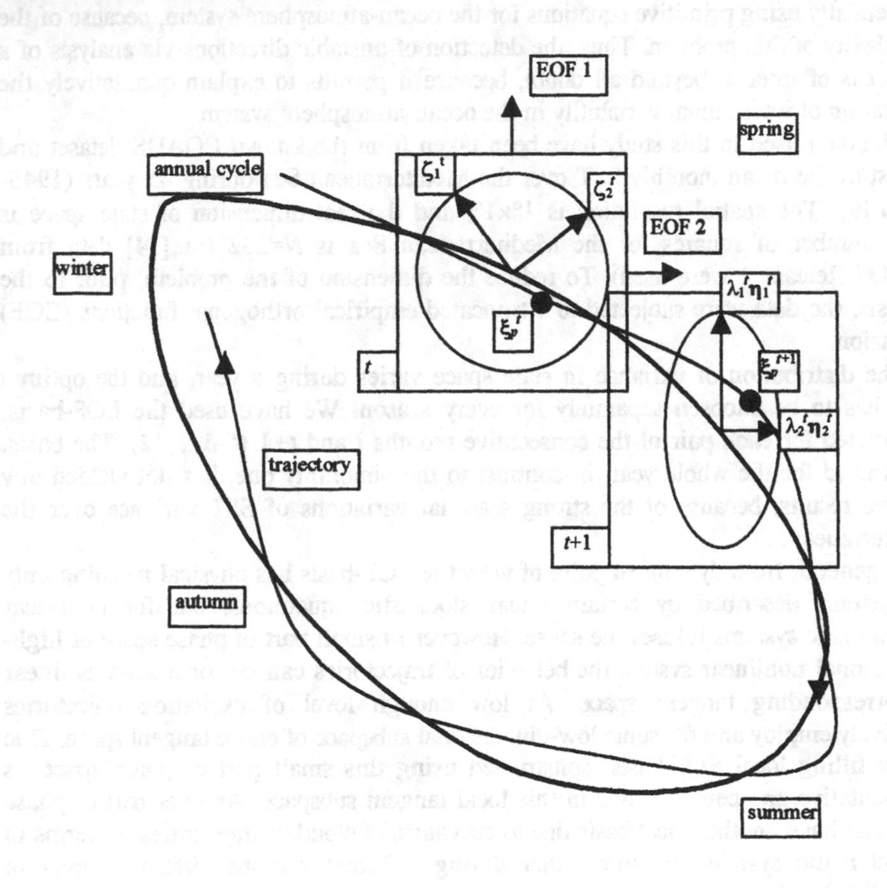


*Figure 1.* **Schematic diagram representing the trajectory of a dynamical system in which there is an near-periodic annual cycle. The heavy-line trajectory is the normal annual cycle. Cross sections of the attractor by planes at the two consecutive months $t$ and $t$+1, spanned by two leading EOFs, are shown. Points $\xi^t_p$ and $\xi^{t+1}_p$ indicate intersections of the trajectory with these planes. Lyapunov vectors $\zeta^t_1$, $\zeta^t_2$ and the vectors $\lambda^t_1\eta^t_1$, $\lambda^t_2\eta^t_2$ are on these planes.**

Thus, further in instability analysis $\tau$ equaled 1 month is considered and dependence on $t$ is denoted by superscript. To construct a linear map $\xi^{t+1}=\mathbf{A}^t\xi^t$, the least-squares fit method is used. For every $i=1,\ldots,n$, the equation

$$\xi_i^{t+1}=\sum_{j=1}^{n} A_{ij}{}^t\xi_j^t \tag{6}$$

defines the $n$-dimensional hyperplane in $(n+1)$-dimensional vector space $\mathbf{y}^i=(\xi_i^{t+1},\xi_1^t,\ldots,\xi_n^t)$. Usually in linear regression analysis the noise is added to the right-hand side of (6) and its variance is minimized. It implies that the SST data for the precede and following month, that is $\xi^t$ and $\xi^{t+1}$, are known exactly, but the regression connection between them is only approximate due to the many real physical factors not taken into account and described as the noise. We assume, that regression equation is valid exactly, but there are data errors (measrements errors, gaps and so on), and $\mathbf{y}^i$ are inexact values. This difference is rather essential. The first method is used in the parameter estimating in linear stochastic models, where noise has dynamical meaning. The second method is used for Lyapunov exponents estimating in numerical simulation of differential equations, where the noise reflects errors of numerical scheme [10]. So we have to find a minimum sum of squares of deviations in direction normal to hyperplane (6)

$$S_i(A_{ij}{}^t)=\sum_{p=1}^{m}(\xi_{ip}{}^{t+1}-\sum_{j=1}^{n}A_{ij}{}^t\xi_{jp}{}^t)^2\,[1+\sum_{j=1}^{n}(A_{ij}{}^t)^2]^{-1} \tag{7}$$

The minimization of $S_i$, relative to $A_{ij}{}^t$, is equivalent to the diagonalization of the covariance matrix of vector $\mathbf{y}^i$

$$R_{js}{}^i=\sum_{p=1}^{m} y_{jp}{}^i\,y_{sp}{}^i \tag{8}$$

which has $n+1$ eigenvalues $r_q{}^i$, ranged in the decreasing order, and $n+1$ orthogonal eigenvectors $\mathbf{z}^i{}_q$ $(q=1,\ldots,n+1)$. The hyperplane, spanned over the first $n$ eigenvectors, i.e. orthogonal to the last eigenvector $\mathbf{z}^i{}_{n+1}$, is defined by the equation

$$(\mathbf{z}^i{}_{n+1},\mathbf{y}^i)=z^i{}_{n+1,1}\xi_i^{t+1}+\sum_{j=1}^{n} z^i{}_{n+1,j+1}\xi_j^t \tag{9}$$

Comparing it with (6) one can obtain the value of $A_{ij}{}^t$

$$A_{ij}{}^t=-z^i{}_{n+1,j+1}/z^i{}_{n+1,1} \tag{10}$$

To estimate confidence intervals, we took the typical least-squares fit sample value of variance $A_{ij}{}^t$

$$\sigma^2(A_{ij}{}^t)=\frac{n}{m}\frac{r^i_{n+1}}{(tr\mathbf{R}^i-r^i_{n+1})} \tag{11}$$

where $r^i{}_{n+1}$ is orthogonal to the hyperplane residual variance, that is equal to the minimum of $S_i$ in (7); $tr\mathbf{R}^i$ is the trace of matrix $\mathbf{R}^i$ in (8) or the total variance of $\mathbf{y}^i$. The explained variance, $(tr\mathbf{R}^i-r^i{}_{n+1})$ is shared equally between all $n$ components $\xi_j^t$, because their initial variances are nearly the same.

Assuming a normal probability distribution for $\xi^t$, we obtain for the large m normal distribution for $A_{ij}^t$, with 95%-confidence interval:

$$A_{ij}^t \pm 2\sigma(A_{ij}^t) \quad (12)$$

The knowledge of $\sigma(A_{ij}^t)$ permit to find out whether significant growing components of vector $\xi^t$ exist and whether there is a significant relationship between them. The local Lyapunov vectors $\zeta_s^t$ provide the spatial distribution of evolving modes, reconstructed from EOF space to the physical domain

$$\Delta T_{sk}^{\zeta} = \sum_{i=1}^{n} a_{ki} \zeta_{si}^{t} \quad (13)$$

The spatial distribution of disturbances in physical space, which initially corresponded to the Lyapunov vectors, after the one-month interval of evolution is

$$\Delta T_{sk}^{\eta} = \sum_{i=1}^{n} a_{ki} \eta_{si}^{t} \quad (14)$$

Thus, the Lyapunov analysis has proven to be an easy-to-use method to extract unstable patterns from observational data. It should be emphasized however, that this method for the study of SST anomalies instability requires that such a basis be choosen, which would describe satisfactorily the spatial variability of anomalies. For example, to analyse the whole World Ocean, it is necessary to retain a large number of EOFs, since the areas of instability can be localized in physical space. Otherwise, the SST decomposition in small areas can prove unsuitable due to the low percentages of leading EOFs, because of the presence of various meso-scale processes in the field of SST anomalies, having noisy character. Therefore, it is necessary to find a compromise between the number of EOFs for the description of geographical features of SST and the growth of errors of the $A_{ij}^t$ estimating.

## 3. Spatial and Temporal Distribution of Intraseasonal Anomalies.

Table 1 lists the total variances of SST anomalies for a two-month interval and contributions of the leading EOFs (in percent) to them. For comparison, it also shows the total variance and EOF contributions for a yearly time interval. As it is seen, the changes of $\alpha_i^2$ during a year are rather significant. For instance, for the first EOF, $\alpha_1^2$ varies from 36 to 62%, and for the second one from 18 to 38%. Thus for analysis of SST intraseasonal anomalies in the Mediterranean Sea, it was sufficient to be restricted by the first three leading EOFs determined for the successive two-month time intervals.

TABLE 1. Variances for every pair of consecutive months $\sigma^2_{t,t+1}$ (°C)$^2$, percentages of the first three EOFs $\alpha_i^2$ and their sum $\alpha_1^2+\alpha_2^2+\alpha_3^2$. In the last row are the same on a yearly basis.

| $t,t+1$ | $\sigma^2_{t,t+1}$ (°C)$^2$ | $\alpha_1^2$(%) | $\alpha_2^2$(%) | $\alpha_3^2$(%) | $\alpha_1^2+\alpha_2^2+\alpha_3^2$(%) |
|---|---|---|---|---|---|
| 1-2 | 0.175 | 41.3 | 28.1 | 9.1 | 78.5 |
| 2-3 | 0.149 | 46.0 | 23.5 | 7.5 | 77.0 |
| 3-4 | 0.222 | 49.0 | 26.9 | 5.5 | 81.4 |
| 4-5 | 0.361 | 54.6 | 23.4 | 7.1 | 85.1 |
| 5-6 | 0.485 | 58.3 | 20.1 | 6.3 | 84.7 |
| 6-7 | 0.518 | 61.6 | 18.3 | 6.1 | 86.0 |
| 7-8 | 0.494 | 60.7 | 20.4 | 6.4 | 87.5 |
| 8-9 | 0.463 | 53.4 | 26.3 | 6.0 | 85.7 |
| 9-10 | 0.485 | 47.3 | 31.6 | 6.0 | 84.9 |
| 10-11 | 0.519 | 46.1 | 31.6 | 6.8 | 84.5 |
| 11-12 | 0.421 | 41.6 | 33.5 | 8.5 | 83.6 |
| 12-1 | 0.272 | 35.6 | 38.1 | 8.3 | 82.0 |
| Year | 0.381 | 49.7 | 26.6 | 7.0 | 83.3 |

Fig.2 shows the spatial distributions of SST anomaly variance and of the leading EOFs for the two-month basis, as well as for the yearly basis. The first EOF has the same sign all over the sea for both the two-month and the one-year intervals; the second EOF changes its sign between the western and the eastern part of the sea. The third mode and the higher modes have, naturally, a finer structure, however, we will not consider them here, due to their small contribution to the total variance. Further, our results will cover the decomposition into two leading EOFs for every two-month interval, and only for comparison the third EOF will be taken into account in some cases.

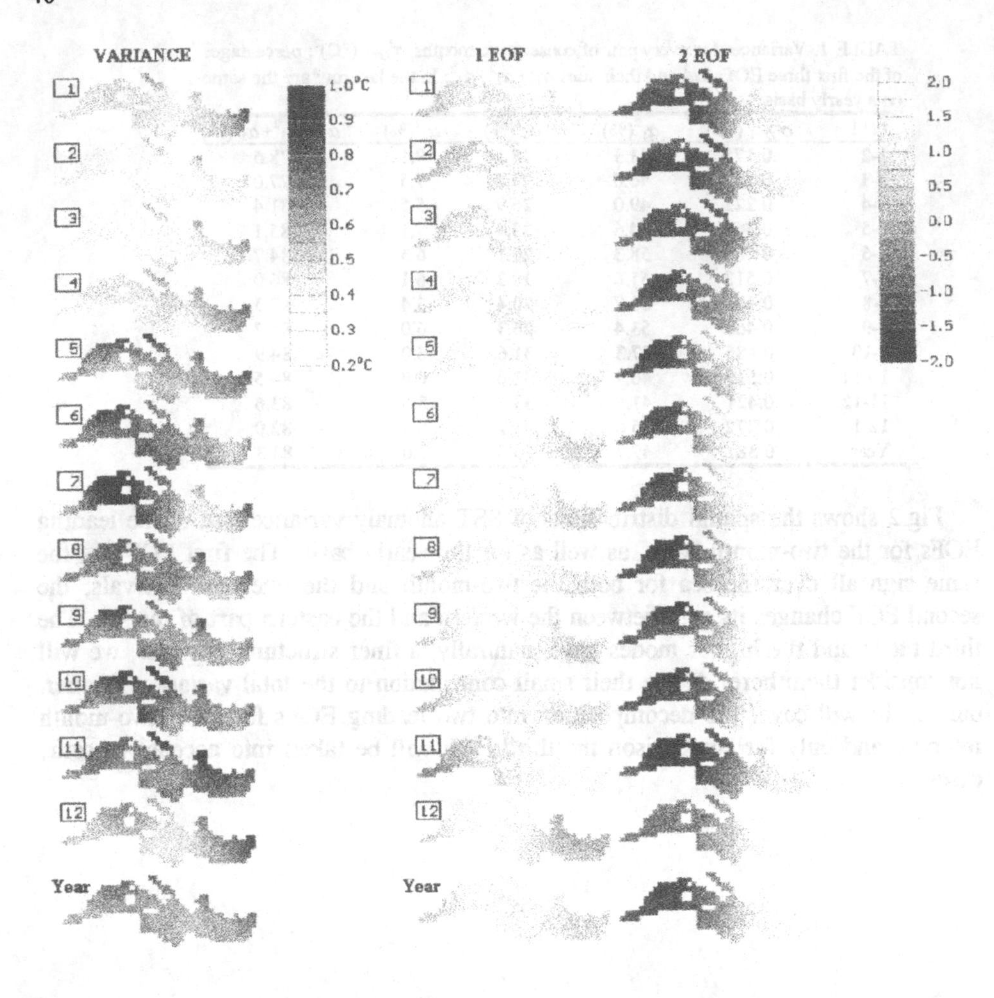


*Figure 2.* **Left fragment: distribution of SST anomalies variance (square root of it) for every month $t$ = 1,...,12 and (bottom) for the whole year. Right fragment: leading two EOFs for every pair of consecutive months $t$, $t$+1 and (bottom) for the whole year.**

The matrix $\mathbf{A}^t$ and the results of its transformations are displayed in Tab.2. We include 95% confidence intervals for $A^t_{11}$ and $A^t_{22}$, the two largest Lyapunov coefficients, with their 95% confidence interval, and the leading Lyapunov vectors $\zeta_s^t$ and the vectors $\eta_s^t$. The diagonal elements of matrix $A^t_{ij}$ and coefficients $\lambda^t_1$, taken from Tab.2, are also shown in Fig. 3. Note, that diagonal elements of $A^t_{ij}$ indicate the change of the given EOF, and the non-diagonal ones ($i \neq j$) indicate an interaction between the EOF, having number $i$, and the EOF, having number $j$, that is, an interaction between different EOFs.

TABLE 2. Elements of the matrix $A^t$ and Lyapunov coefficients $\lambda^t_i$ with the 95% confidence interval in brackets. Components of the first Lyapunov vectors $\zeta^t_1$ and corresponding vectors $\eta^t_1$ are in the last two columns.

| t,t+1 | $A^t_{11}$ (95%) | $A^t_{22}$ (95%) | $A^t_{12}$ | $A^t_{21}$ | $\lambda^t_1$ $\lambda^t_2$ (95%) | $\zeta^t_{11}$ $\zeta^t_{12}$ | $\eta^t_{11}$ $\eta^t_{12}$ |
|---|---|---|---|---|---|---|---|
| 1-2 | 0.57 (0.14) | 0.56 (0.13) | 0.04 | 0.06 | 0.61 0.51 (0.14) | -0.74 0.67 | 0.73 0.68 |
| 2-3 | 1.59 (0.15) | 1.10 (0.16) | 0.22 | -0.65 | 1.86 0.86 (0.16) | 0.90 0.43 | 0.82 0.57 |
| 3-4 | 1.34 (0.14) | 1.73 (0.21) | 0.40 | -0.32 | 1.77 1.38 (0.17) | 0.01 1.00 | -0.22 0.98 |
| 4-5 | 1.79 (0.16) | 2.10 (0.21) | -0.24 | -1.83 | 3.15 1.05 (0.19) | 0.79 -0.62 | 0.49 -0.87 |
| 5-6 | 1.17 (0.15) | 1.10 (0.21) | -0.39 | -0.47 | 1.57 0.71 (0.18) | 0.75 -0.67 | 0.72 -0.69 |
| 6-7 | 1.08 (0.16) | 1.00 (0.22) | -0.48 | -0.61 | 1.59 0.49 (0.19) | 0.76 -0.66 | 0.71 -0.70 |
| 7-8 | 0.84 (0.15) | 1.29 (0.17) | -0.08 | 0.49 | 1.41 0.80 (0.16) | 0.49 0.87 | 0.24 0.97 |
| 8-9 | 0.83 (0.12) | 1.43 (0.21) | 0.20 | 0.45 | 1.57 0.70 (0.17) | 0.45 0.89 | 0.35 0.94 |
| 9-10 | 1.23 (0.21) | 1.12 (0.19) | -0.14 | -0.35 | 1.43 0.93 (0.20) | 0.81 -0.59 | 0.75 -0.66 |
| 10-11 | 1.01 (0.15) | 0.99 (0.18) | 0.12 | 0.37 | 1.25 0.76 (0.17) | 0.76 0.65 | 0.68 0.73 |
| 11-12 | 0.57 (0.12) | 0.89 (0.15) | 0.03 | 0.15 | 0.89 0.55 (0.14) | 0.30 0.95 | 0.22 0.97 |
| 12-1 | 0.81 (0.17) | 0.64 (0.14) | 0.26 | -0.03 | 0.84 0.59 (0.16) | 0.36 0.93 | 0.54 0.84 |

Fig.3 shows, that the maximum growth of SST anomalies in spring occurs in April-May and during autumn in August-September, whereas in the interim, in the winter and summer, the anomalies decay. It is of interest, that the growth and the decay of anomalies is accompanied by a change in the relationship between diagonal and non-diagonal elements of the matrix $A^t_{ij}$, which implies change in the relationship between different EOFs during the evolution of anomalies.

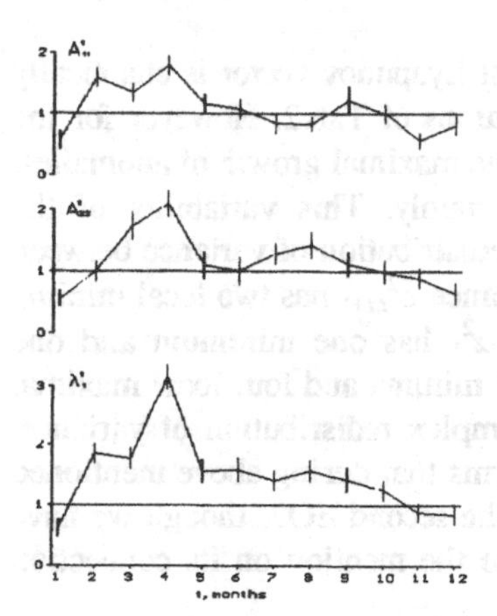


*Figure 3.* Diagonal elements of the matrix $A^t$: $A^t_{11}$ and $A^t_{22}$ and maximum Lyapunov coefficient $\lambda^t_1$ for every $t$. Bars indicate the 95% confidence interval.

In Fig.4 a spatial distribution of the leading Lyapunov vectors, determined according to the equation (13), is shown for the two most distinctive months of a year. Fig.4 also shows the spatial distribution of vectors $\eta^t_1$, determined according to the equation (14). It is seen from Tab.2, that the growing SST anomalies in April are formed mainly by the second EOF (the ratio between the first EOF and the second one is 0.49 to 0.87). This implies, that the unstable growing SST anomalies in the western and eastern parts of the Mediterranean Sea have the opposite signs (Fig.4). The decaying SST anomalies in December, conversely, are basically generated by the first EOF. Thus, these anomalies all over the Mediterranean Sea have the same sign.

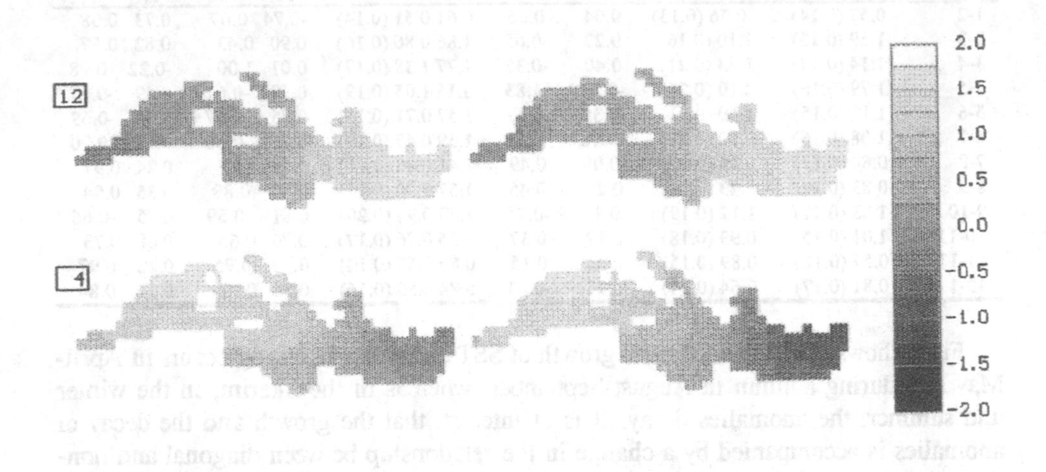


*Figure 4.* Spatial distribution of Lyapunov vectors $\zeta^t{}_1$ (left) and the vectors $\eta^t{}_1$ (right) for two months: December ($t$ = 12) and April ($t$ = 4).

It is worth to emphasize, that behaviour of the first Lyapunov vector is chaotically enough during a year according to the last two coloumns of Tab.2. However for the three lines in Tab.2 (3-4, 4-5, 7-8), corresponding to the maximal growth of anomalies, the Lyapunov vector consists of the second EOF mainly. This variability of the Lyapunov vector is in accordance with the complex redistribution of variance between EOFs during a year. According to Tab.1, the total variance $\sigma^2_{t,t+1}$ has two local minima and two local maxima, the share of the first EOF $\alpha^2_1$ has one minimum and one maximum, the share of the second EOF has four local minima and four local maxima. Lyapunov vectors describe the linear part of this complex redisribution of variance, that reflects on their shape. Our main conclusion affirms that during above mentioned months linear evolution consists of strong growth of the second EOF, though we have no reasonable explaination for this fact in addition to the mention on its connection with ocean-atmosphere interaction.

There is one important question, connected with the low frequency trend in SST. It is known from literature [7], that the Mediterranean SST has significant trend for the COADS period, moreover the trend is different for different seasons and for different parts of the sea. We also estimated the trend for every point separately for each month of a year and found that results are in accordance with known ones. Trend is stronger in summer, reaching 20% of total root-mean-square amplitude. Next important question appears on this stage. The behaviour of the montly anomalies (matrix $\mathbf{A}^t$ in our case) may depend on the climatic (averaged over several years) trend of the SST. Then we must choose time intervals having constant climatic characteristics and estimate the matrix $\mathbf{A}^t$ sepatately for each such interval. Otherwise one can remove the climatic trend as an additive term and consider the residual as a stationary process. Apriori, there are no objective reasons to prefer any variant. Practically, it is rather difficult to select the time intervals without the climatic trend over the whole sea, that is why the second variant was choosen. All results were recalculated for such detrended data, starting with the EOF expansion and finishing to the Lyapunov vectors. Results left unchanged within confidence intervals. Naturally, the variance decreases on 10-20% depending on season. One can conclude, that the behaviour of the SST anomalies for intraseasonal time scales doesn't depend essentially on low-frequecy climatic trends.

## 4. Relationships between SST and Local Heat-Flux Anomalies.

Both changes in the heat fluxes through the surface and the horizontal and vertical advection determine SST anomalies. For the open ocean conditions over large spatial scales (>1000km meridionally, and >3000km zonally), SST anomalies are dominated by the local changes in the heat fluxes, in comparison with advective influences [15]. In the extratropics, over monthly and intraseasonal temporal scales, the contribution of net solar and infrared fluxes to the total anomalous heat budget is small [16]. Thus, in the extratropical regions of oceans, latent and sensible heat fluxes are the major components in producing monthly thermal anomalies. Note that latent and sensible monthly flux anomalies are correlated (cooler air is usually drier), so that they usually reinforce [16].

It is of interest to consider changes in the Mediterranean SST anomalies on intraseasonal time scales in relation to local heat-flux anomalies. We use a highly simplified slab-model of the well-mixed upper ocean:

$$\frac{d\Delta T}{dt} = -\frac{\Delta Q}{\rho_w C_{pw} h} \tag{15}$$

and the bulk formulas for latent and sensible sea-air heat fluxes:

$$Q=\rho_a u[LC_E(q_w^{sat}-q_a)+C_{pa}C_H(T_w-T_a)] \tag{16}$$

where $u$ is scalar wind speed; $\rho_a$, $q_a$, $T_a$ and $C_{pa}$ are density, specific humidity, temperature, and specific heat at constant pressure of air at the observation level, $\rho_w$ and $C_{pw}$ are the same for the water; $q_w^{sat}$ and $T_w$ denote the saturation specific humidity and sea surface temperature respectively; $L$ is the latent heat of evaporation of water; $C_E$ and $C_H$ are the transfer coefficients, respectively, $h$ is the depth of the slab.

In this one-dimensional description of the mixed layer temperature, horizontal temperature advection and horizontal and vertical mixing have been omitted, so that only the effect of the anomalous sensible and latent heating $\Delta Q$ on the SST anomaly is considered.

Separating all the quantities in (16) into climatic quantities, denoted by an overbar, and anomalous monthly quantities, denoted by $\Delta$,

$$\Delta Q = \rho_a u[LC_E(\Delta q_w^{sat} - \Delta q_a) + C_{pa}C_H(\Delta T_w - \Delta T_a)] \tag{17}$$

**dropping small terms and supposing that anomalous temperature and humidity differences are correlated, especially in extratropics [16],**

$$\Delta q_w^{sat} = \left(\frac{\partial q^{sat}}{\partial T}\right)_{T=\overline{T}_w} \Delta T_w, \quad \Delta q_a = \overline{r}\left(\frac{\partial q^{sat}}{\partial T}\right)_{T=\overline{T}_a} \Delta T_a \tag{18}$$

**where $\overline{r} \approx \overline{q}_a / \overline{q}_a^{sat}$ is climatic mean relative humidity, one can use only two fundamental surface variables $\Delta T_w$ and $\Delta T_a$ to obtain the anomalous flux $\Delta Q$.**

$$\frac{d\Delta T_w}{dt} + a\Delta T_w = b\Delta T_a, \tag{19}$$

**where**

$$a = \frac{\rho_a \overline{u}}{\rho_w C_{pw} \overline{h}}\left[LC_E\left(\frac{\partial q^{sat}}{\partial T}\right)_{T=\overline{T}_w} + C_{pa}C_H\right],$$

$$b = \frac{\rho_a \overline{u}}{\rho_w C_{pw} \overline{h}}\left[LC_E \overline{r}\left(\frac{\partial q^{sat}}{\partial T}\right)_{T=\overline{T}_a} + C_{pa}C_H\right] \tag{20}$$

**Next, we test the equation (19), using the dataset for $T_w$ and $T_a$. A finite-difference approximation of the equation (19) can be obtained by means of integration over the time interval $\Delta t$ for constant $a$ and $b$:**

$$\Delta T_w^{t+\Delta t} = \alpha \Delta T_w^t + \beta \Delta T_a^{t+\Delta t} + \gamma \Delta T_a^t, \tag{21}$$

**where**

$$\alpha = \exp(-a\Delta t); \quad \beta = \frac{b}{a}\left[1 - \frac{1-\alpha}{a\Delta t}\right]; \ldots\ldots \gamma = \frac{b}{a}\left[-\alpha + \frac{1-\alpha}{a\Delta t}\right]. \tag{23}$$

**The monthly averaged values over the time interval $\Delta t$=1month are in equation (21). The equation (21) was tested using, as above, the EOF-basis, constructed for each pair of the consecutive months $t$ and $t$+1. Results of the least-square fitting of the regression coefficients $\alpha$ and $\beta$ for the first EOF are listed in Tab.3. The contribution of the two parts of the equation (21), that is $\alpha\Delta T_w^t$ and $\beta\Delta T_a^{t+\Delta t}$, to the total variance of EOF for the month given (in percent) is also given in Tab.3. The third term $\gamma\Delta T_a^t$ on the right-hand side of the equation (21) gives negligible contribution to the variance explained and is not significant. As one can see, the regression coefficients $\alpha$ and $\beta$ for the first EOF have a pronounced annual cycle, having extreme values in June-July: $\alpha$=0.21, $\beta$=0.85 (92% of the explained variance) and in December-January: $\alpha$=0.48, $\beta$=0.26 (56% of the explained variance). Parameters for the second EOF have a similar annual behavior.**

TABLE 3. Variances of sea surface and air temperature anomalies, coefficients $\alpha$ and $\beta$ for the first EOF, and variance explained for every pair of months $t$, $t$+1.

| $t,t$+1 | $\sigma^2_{t+1}(T_w)$ | $\sigma^2_{t+1}(T_a)$ | $\alpha$ | $\beta$ | Var. expl. |
|---|---|---|---|---|---|
| 1-2 | 0.127 | 0.894 | 0.497 | 0.212 | 66.9 |
| 2-3 | 0.170 | 0.596 | 0.403 | 0.362 | 73.6 |
| 3-4 | 0.274 | 0.546 | 0.435 | 0.550 | 84.2 |
| 4-5 | 0.448 | 0.736 | 0.364 | 0.699 | 84.2 |
| 5-6 | 0.521 | 0.712 | 0.235 | 0.831 | 89.8 |
| 6-7 | 0.514 | 0.595 | 0.207 | 0.853 | 92.3 |
| 7-8 | 0.473 | 0.569 | 0.198 | 0.773 | 91.0 |
| 8-9 | 0.452 | 0.697 | 0.167 | 0.645 | 85.8 |
| 9-10 | 0.518 | 0.819 | 0.185 | 0.689 | 82.5 |
| 10-11 | 0.521 | 0.961 | 0.426 | 0.545 | 83.2 |
| 11-12 | 0.321 | 0.822 | 0.463 | 0.234 | 62.4 |
| 12-1 | 0.223 | 0.757 | 0.480 | 0.259 | 55.7 |

The parameter $\tau = 1/a = -\Delta t/\ln(\alpha)$ has a simple physical meaning: it determines the time of adjustment of the upper sea to the external forcing. For instance, for the first EOF, $\tau$ = 1.36 month in January and $\tau$ = 0.61 month in June. In winter, the upper mixed layer is deeper, and specific humidity is less. As a result, despite the stronger mean wind, the adjustment time is larger than in summer. Choosing the characteristic monthly values $\overline{T}_w$ = 10°C, $\overline{u}$ = 10m/s, for winter, and $\overline{T}_w$ = 25°C, $\overline{u}$ = 5m/s, for summer, from (20), we can obtain reasonable estimates of the mixed layer depth: $\overline{h}$ = 43m for winter and $\overline{h}$ = 15m for summer.

The annual variations of the adjustment time, referred to above, give a reasonable explanation of the variations of monthly SST anomalies. In summer, $\Delta T_w$ responds to the changes of $\Delta T_a$ sufficiently quickly, and according to the equation (19), $\Delta T_w$ has enough time to approach $\Delta T_a$. In reality , the variances of $\Delta T_w$ and $\Delta T_a$ are almost equal in summer (Tab.3). In winter $\Delta T_w$ has insufficient time to adjust to the anomalies $\Delta T_a$ and, as a result, the variance of $\Delta T_w$ is significantly lesser than that of $\Delta T_a$. It is of interest, that in winter, the percentage of the variance of $\Delta T_w$, explained by the equation (21), decreases up to 60-70%. That means, that a significant contribution to the SST anomalies is given by the other heat balance components, neglected in equation (19). Also, it means that the highly simplified slab-model ceases to be valid.

The values of $\beta$ given in Tab.3 vary in the annual cycle in accordance with the speculations given above. Because of $\overline{T}_a < \overline{T}_w$, following the equation (20), in winter, $b<a$, and $b>a$ in summer. For the second EOF, the yearly variations of the parameters obtained, by and large are the same, and the rather small unexplained variance of $\Delta T_w$ in summer also corresponds to equation (21).

In this section we gave a simple interpretation of the SST anomaly variance increase during a spring, obtained in the previous Section. There SST anomalies were considered as an autonomous system, having the behavior locked to the annual cycle. Some features of this behaviour, namely the spring variance increase, is explained by the change of mixed layer depth. Note, that we did not have all the data sets nesessary for deriving the total heat balance of the upper sea layer. That is why model (19) is rather simplified, but, nonetheless, it satisfactorily governs monthly SST anomalies, especially, in summertime. The principal purpose of this analysis is to give evidence of the local origin of intraseasonal SST anomalies revealed during springtime. Being local in space and time, the nature of forcing of these anomalies, consistent with the simple equation (19), indicates atmospheric processes as the source of these oscillations.

## 5. Summary.

We have presented the results of analysis of monthly SST data with a purpose of identifying intraseasonal anomalies in relation to the annual cycle. The main results can be summarized as follows:

There are periods in a year, when large-scale anomalies in the Mediterranean Sea are growing. The most unstable anomalies occur in April, and the less unstable ones in August-September. In the intermediate months in December and June, the anomalies are decaying.

Spatial distributions of the growing and decaying anomalies are different. The growing disturbances have the form similar to the second EOF, and the decaying ones are reminiscent of the first EOF.

The relationship between monthly anomalies of sea surface temperature and atmospheric temperature corresponds to a simple equation of regression with the coefficients varying in a yearly cycle.

## References.

1. Efimov, V.V., Prusov, A.V., and Shokurov, M.V. (1995) Seasonal instability of Pacific sea surface temperature anomalies, *Q. Journal Roy. Met. Soc.* **121**, 1651-1679.
2. Bartzokas, A. (1989) Annual variation of pressure over the Mediterranean Sea, *Theor. Appl. Climatol.* **40**, 135-146.
3. Alpert, P., Neeman, B.U., and Shay-El, Y. (1990) Intermonthly variability of cyclone tracks in the Mediterranean, *J. Climate* **3**, 1474-1478.
4. Flocas A.A. (1984) The annual and seasonal distribution of fronts over central-southern Europe and the Mediterranean, *J. Climatol.*, **4**, 255-267.
5. Goosens, C. (1985) Principal component analysis of Mediterranean rainfall. *J. Climatol.* **5**, 379-388.
6. Sahsamanoglou, H.S. and Makrogiannis, T.J. (1992) Temperature trends over the Mediterranean region 1950-88, *Theor. Appl. Climatol.* **45**, 183-192.
7. Metaxas, D.A., Bartzokas A., and Vibras A. (1991) Temperature fluctuations in the Mediterranean area during the last 120 years, *Int. J. Climatol.* **11**, 897-908.
8. Bartzokas, A., Metaxas, D.A., and Ganas, I.S. (1994) Spatial and temporal sea-surface temperature covariances in the Mediterranean, *International Journal of Climatology* **14**, 201-213.

9. da Silva, A.M., Young C.C., and Levitus S. (1994) *Atlas of surface marine data 1994, volume 1: Algorithms and procedures.* NOAA Atlas NESDIS 6, Washington, D.C.
10. Eckmann J. -P. and Ruelle D. (1985) Ergodic theory of chaos and strange attractors. *Reviews of Modern Physics,* **57**, 617-656.
11. Lanczos, C. (1961) *Differential operators*, Van Nestned, Reinhold, New York.
12. Farrell B.F. (1990) Small error dynamics and the predictability of atmospheric flows. *J. Atm. Sci.* **47**, 2409-2416.
13. Yoden, S. and Nomura, M. (1993) Finite-time Lyapunov stability analysis and its application to atmospheric predictability, *J. Atm. Sci.* **50**, 1531-1543.
14. Efimov, V.V., Prusov, A.V., and Shokurov, M.V. (1997) On large-scale intraseasonal anomalies of the Mediterranean sea surface temperature, *The Global Atmosphere and Ocean System* **5**, 229-246.
15. Gill, A.E. and Niiler, P.P. (1973) The theory of the seasonal variability in the ocean. *Deep-Sea Res.* **20**, 141-177.
16. Cayan, D.R. (1992) Latent and sensible heat flux anomalies over the Northern oceans: driving the sea surface temperature, *J. Phys. Oceanogr.* **22**, 859-881.

# ECOLOGICAL PHYSIOGNOMY OF THE EASTERN MEDITERRANEAN

MAURIZIO RIBERA D'ALCALÀ
*Laboratorio di Oceanografia Biologica*
*Stazione Zoologica "A. Dohrn", Villa Comunale*
*80121 Napoli, Italy*
*E-mail: maurizio@alpha.szn.it*

MARIA GRAZIA MAZZOCCHI
*Laboratorio di Oceanografia Biologica*
*Stazione Zoologica "A. Dohrn", Villa Comunale*
*80121 Napoli, Italy*
*E-mail: grazia@alpha.szn.it*

## 1. Premise

The Eastern Mediterranean (EMED) is connected with the open ocean through three different sills and a wide buffering basin such as the Western Mediterranean. It is in fact almost completely enclosed by land, with the exceptions of the Sicily Channel and the very shallow connections of Bosphorus with the Marmara Sea and the more recent one with the Red Sea, which dates 1869. Though, it is a large basin with a maximum depth of more than 5000 m. Because of these characteristics, many relevant processes took place in the past and others are occurring today. To quote a few, the frequent changes in the pelagic system ([1] and references therein), the introduction of exotic species through the Suez Canal [2], the recent changes in thermohaline circulation [3] and their implications. Many topics have been thoroughly analyzed in this workshop and elsewhere. Instead, we will comment on some basic features of the pelagic realm, and the main forcings that make the area as it is now and eventually differentiate one region from another. Our contribution will mostly focus on inputs and internal transfers, which indeed cover only partially the functioning of an ecosystem. However, we think that these basic processes deserve attention and our aim is to promote a discussion on them. Moreover we will restrict our comments to the EMED proper, i.e. Adriatic and Aegean Seas will be excluded, whenever possible, from our analysis.



*P. Malanotte-Rizzoli and V.N. Eremeev (eds.),*
*The Eastern Mediterranean as a Laboratory Basin for the Assessment of Contrasting Ecosystems*, 49–64.

## 2. Background

The EMED is generally perceived as an extremely impoverished area for what biological production concerns. In a schematic report, Azov [4] called it a "marine desert" and ranked the EMED among the most oligotrophic regions of the world ocean. This statement holds true in terms of both average standing stocks (e.g. chlorophyll *a*, mesozooplankton numbers) and fluxes (e.g. primary production, fish catch). Though, this status is part of the recent (on geological time scale) history of the EMED.

Because of its character of semi-enclosed sea, EMED as well as the whole Mediterranean, underwent dramatic changes in the past, to such an extent that the general features we are going to sketch in this contribution can only refer to the present EMED.

Numerous geological records document oscillations in the functioning of the EMED ecosystem and the present oligotrophic condition seems to be an exceptional status in the history of the basin. This is considered by Rohling [1] who analysed the Pleistocene-Holocene period of EMED history, mostly on the basis of deep-sea cores sampled in the basin. It is well know that the strongest signal found in the EMED sediments is the recurrence of layers of anoxidized carbon, to be related with deep-water anoxia. These layers are termed 'sapropels' and were firstly discovered during the Swedish Deep Sea Expedition ([5], in [1]). Sapropel formation, and associated deep anoxia, can be due either to a decrease in ventilation of deep layers or to a significant increase of primary production in the surface layer, or both. The ongoing discussion has not yet completely solved the problem of which atmospheric and ocean dynamics can better explain the observed anoxic crises. Nevertheless, the evidence of frequent eutrophic periods of the EMED is indisputable.

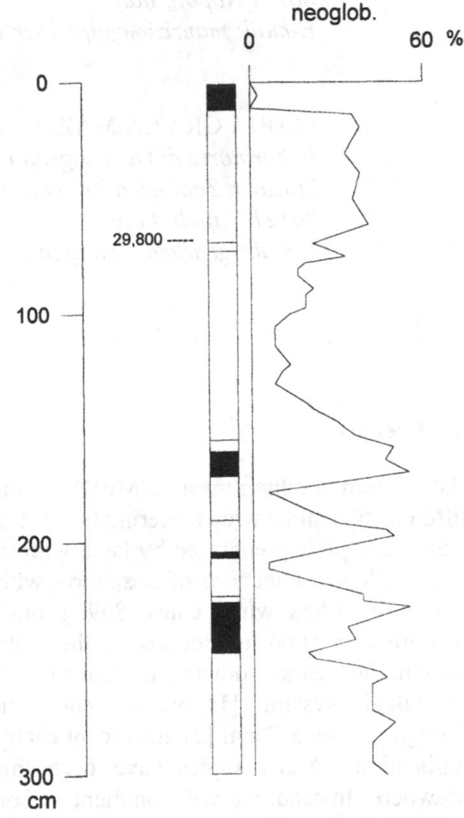


*Figure 1* - Downcore neogloboquanidrinid percentages in one core from North Levantine basin compared with sapropel occurrence (dark and shaded stacked bars). Time reference reported on the left. (Redrawn from [1]).

The layering of strata of different characteristics is shown in Fig.1, where the recurrent periods of higher exported production are indicated by highest peaks of foraminifer abundance. The last sapropel crisis (the top dark bar in Fig. 1) probably ended 6000 years B.P. It is quite fascinating to think that 6000 years ago (4000 B.C.) the EMED was much more turbid in the surface layer than today, and this was the EMED that the most advanced Neolithic civilizations, spread along Mediterranean coasts and islands, experienced. A few centuries later, Kurgan invasions from the Caspian regions would have given origin to the Indo-European phase of western culture, unfortunately destroying significant part of the heritage of the pre-existing cultures [6, 7]. A different perception of the Sea could probably be traced in old traditions of Mediterranean populations.

## 3. Present status

The present EMED claims for a Secchi disk world record (53 m) [8], which reinforces the widespread view of a basin with an extremely low production.
It is well known that the high variability of autotrophic processes precludes any simple estimate of yearly primary production at basin scale, even in quasi-steady state environments, such as subtropical systems. With the advent of CZCS satellite observations, a better synopticity was obtained. However, the primary productivity models based on chlorophyll fields and incident light are still inadequate. Several attempts have been conducted to obtain estimates of annual primary production for the EMED. Different approaches have also been utilized and in the following table are reported the results of these studies.

TABLE I. Estimates of annual primary production in the EMED pelagic system

| Total production ($g\ C \cdot m^{-2} \cdot y^{-1}$) | New Production ($g\ C \cdot m^{-2} \cdot y^{-1}$) | Method | Source |
|---|---|---|---|
| (18.2) | 6.0 | Phosphorus Budget | [9] |
| 20.3 | 11.0 | *In situ* data and *f* ratio | [10] |
| (36.4) | 12.0 | Oxygen utilization | [11] |
| 95.4 | (31.5) | CZCS data series | [12] |
| <11.5 | <4.7 | Coupled 3D model | [13] |

The table deserves a few explanatory comments.
The values in parentheses have been calculated assuming an *f* ratio of 0.33, which is probably high for the EMED [12] but has been an outcome of some of the papers listed. Bethoux data [9], [11] include Adriatic and Aegean Seas, because of the method used.
The estimates are generally in good agreement toward the low values, the only exception being the analysis based on the CZCS data. For such analysis, it is very critical, besides the algorithm used also the depth of the active photic zone. This might be the case in the Antoine and co-authors' estimate [12], but such a value cannot be ruled out unconditionally.
The first value by Bethoux [9] is based on the hypothesis that all the surface inputs of phosphorus are transferred to the intermediate layers by biota, i.e. he computed the export production. Apart from ignoring the upward flux of nutrient, what he was clearly aware of, input data used are quite uncertain, especially for the runoff, and no atmospheric contribution is considered. Bethoux revisited his estimate using the oxygen consumption at a basin scale, from which he also determined the upward flux of phosphate in the surface layer [11]. The approach relies very much on good water flux data and on the assumption that oxygen is utilized in a Redfieldian ratio. Our unpublished data collected in the EMED in the framework of the POEM-BC

international program show that the latter assumption is not strictly applicable. In addition, the deep water volume formed each year (0.75 Sv) is high at least by a factor of two if Adriatic Sea is the only source considered [14], [15]. One would then take that value as an upper bound.
The estimate by Dugdale & Wilkerson [10] also supports this conclusion, but they base their results on very few *in situ* measurements taken in their regional context.
Is it possible to reconcile the two extremes in the estimates?
A tentative way is to assume that the *f* ratio in the basin is very low, say 0.1 as also considered by Antoine and co-authors [12]. This in turn would confirm that export production is very small in the basin, and that most of the carbon fixed is recycled in the surface layer. Such being the case, even a new production of 10 g C $\cdot$ $m^{-2}$ $\cdot$ $y^{-1}$ would correspond to a total production of 100 g C $\cdot$ $m^{-2}$ $\cdot$ $y^{-1}$, which is in fact the value given by Antoine and co-authors. This picture would not contradict the transparency of the waters because pigmented protists would always be in small numbers. In addition, this is the way most of the ocean works [16].
The only result against this preliminary conclusion is the independent estimate, unfortunately limited to the Ionian Sea, derived by using a General Circulation Model coupled with an ecological, nitrogen based, model by Civitarese et al. [13]. Their total production estimate is 11.5 g C $\cdot$ $m^{-2}$ $\cdot$ $y^{-1}$ and for the new production 4.7 g C $\cdot$ $m^{-2}$ $\cdot$ $y^{-1}$.
Crise et al. (this volume) reported that a further implementation of the model also shows a general decrease of biomass, and presumably production, from the Ionian to the Levantine basin. Their estimate for the Ionian should then be lowered to have the average for the whole EMED. The reason for this discrepancy can reside in a strong underestimation of the consumption term, which is not explicitly commented in the paper. But it would be very interesting to try another estimate introducing a relevant consumption term in the upper layer.
Having this picture in mind, low export production but relatively high total production, the question arises on what are the dominant factors for this.

## 4. How the system works

In the News section of an early 1998 issue of Science, it is reported: “Nutrients are scarce in the Mediterranean, compared with the rest of the world ocean, because the sea’s main input comes from the surface waters in the Atlantic, which flows in through the Straits of Gibraltar” [17].
This sentence synthesizes the basic paradigm of the Mediterranean oligotrophy, i.e. the dominance of an inverse estuarine circulation. Is that statement absolutely tenable?
We do not think so or, at least, we think that the above statement can be very misleading.
Because the EMED is an inverse estuary in respect to WMED, the stronger oligotrophy of the former is used as a support to the general argument.
We made a quantitative comparison of different sources of nutrients (namely nitrogen) to sustain new production in the EMED proper.
Numbers are reported in Table II.

TABLE II. Sources of Dissolved Inorganic Nitrogen to the upper layer of the EMED proper

| Origin | Input value ($Mmol \cdot y^{-1}$) | Source | Notes |
|---|---|---|---|
| Sicily Channel | 55000 | This study | 1.4 Sv mean surface transport |
| Otranto Strait | -19000 | [18] | Inflow = 39000; Outflow=20000 |
| Atmosphere | 12600 | [19] | DIN = 0.5·TN |
| Runoff | 10000 | [9] | |
| Eddy diffusion | 8000÷160000 | This study | $K_v$=1.5÷10 $m^2 \cdot d^{-1}$; $\partial[N]/\partial z$=1.5÷10 $mmol \cdot m^{-4}$ |
| Upward flux | 180000 | [20] | $\Delta[NO_3]$=4.0 $\mu mol \cdot dm^{-3}$ |

For the fluxes at the Strait of Otranto, the water transports have been determined by interpolating current velocities at 5 mooring locations. This method does not take into account recirculation at scale smaller then the sampling grid.
The atmospheric input includes only the inorganic part, but the organic could also be utilizable by organisms.
Runoff data exclude the Adriatic rivers, because their real contribution to the EMED is much better determined at the Strait of Otranto.
For the eddy diffusion, we took the most reasonable range of both $K_v$ and the nutrient vertical gradient.
Our value for surface flux at the Sicily Channel is slightly different from the one used by Bethoux (7%) and we did not take into account this difference.
There are no data for vertical advection (e.g. upwelling). The indirect estimate of upward-downward flux given by Bethoux [20] has to be considered the sum of both processes. Interestingly, our values for internal vertical flux are quite in agreement on the upper side.
To summarize the outcome of the table, it is evident that the other sources contribute much more than the Atlantic Water input at the Sicily Channel.
This water mass has for sure a low nutrient load. But this happens for most of the tropical and subtropical oceans, where it is not the advective term that drives the system but all the others. If surface Atlantic Water were richer in nutrients, then Mediterranean sea surface waters would be correspondingly richer. But this cannot be considered as the cause for Mediterranean oligotrophy; it is just the cause of Mediterranean not being eutrophied by lateral advection. In our opinion, the EMED oligotrophy is due to modest terrestrial contribution to the open sea and a quite low upward flux of nutrients, because of restricted areas of upwelling and weak vertical nutrient gradients. This last aspect is quite relevant, because is one of the terms that determine the eddy diffusivity. In the EMED the gradient is small because the main nutricline is generally deep (300-400 m) [21] [22].
Apart from mesoscale activity which has a shorter time and space scale, most of the studies recently conducted in the basin have confirmed that in the areas where sub-basin structures are weaker or absent, the distribution of autotrophic organisms resembles very much a Typical Tropical Structure. Where gyres are permanent or semi-permanent, the vertical dynamics enhances vertical fluxes in different ways, thus increasing the potential biomass load of the system.
This has been observed in the cyclonic Rhodes Gyre as well as in the anticyclonic Cyprus Gyre, and in many other structures of the Levantine or the Ionian [22].

The only difference with the global ocean we can see is in the different scale of mesoscale dynamics [23] and in the fact that variance in the velocity field is higher than the average term. This should cause a higher variability in time and space of primary production processes. And this might be the reason why when comparing bacterial consumption and primary production the latter is apparently not sufficient to support the former [24]. Single episodes of short duration could explain the gap, as in most of the Tropical Ocean.

## 5. Trophic interplay in the upper layer

The trophic structure in oligotrophic oceans is typically represented by an inverted biomass pyramid [25, 26], with heterotrophic organisms having a dominant role in the pelagic system, also by controlling autotrophic populations [27]. In such environments, the trophic pathway is dominated by the microbial loop, which mainly comprises cyanobacteria, heterotrophic bacteria and protozoa. The available information on pelagic compartments in the open EMED seems to support this scenario. The autotrophic populations are dominated by picoplankton [28-30], which can account for more than 60% of total chlorophyll *a* in the 0-50 m layer [31]. In the offshore Ionian Sea, the phytoplankton community is mainly composed by nanoflagellates, naked dinoflagellates and coccolithophorids [29]. However, the complete size structure of pelagic communities and the nature of trophic interactions between organisms, as well as the role played by protozoans and metazoans in the pelagic food webs have still to be assessed for the EMED. A better knowledge of the taxonomy, biology and behaviour of organisms is necessary for a better depiction and understanding of their role in the pelagic food webs. Only very recently, basic information on the biology and feeding behaviour has been acquired for a small oceanic copepod [32] that dominates the mesozooplankton assemblages in the EMED epipelagic waters in autumn [33].
Some features recently recorded in the plankton distribution in the EMED suggest that some functional aspects in the pelagic domain differ at sub-basin scale. The presence of meso-scale oceanographic features clearly affects plankton distribution and dynamics when the physical forcing is strong and persistent in time [34]. The Rhodes Gyre highly contributes to increase the biological activity in the area southeast of the Rhodes island [21], [33], [34], [35]. The Cyprus Eddy, a quasi-stationary, persistent warm-core eddy southeast of Cyprus, constitutes a site of deep phytoplankton blooms and enhanced productivity [36-38].
During the POEM-BC synoptic cruises conducted in autumn 1991 in different EMED regions, copepods accounted for the highest mesozooplankton abundances in the first 300 m of the water column everywhere in the basin [34]. However, the relative contribution of other taxonomic groups to total numbers seemed to indicate regional differences in the structure of the pelagic food webs. High abundances of appendicularians co-occurred with low numbers of chaetognaths. Due to the different ecological features of these animals, their presence suggests alternative paths in the energy flow in the pelagic communities. Appendicularians occurred second after copepods in a rank order of abundance only in the westernmost part of the Sicily Channel and in the Rhodes Gyre, the two more productive areas in the open EMED [34]. Appendicularians trap small organisms such as pico- and nanoplankton in their

pharyngeal filters. They are also important consumers of POC and may remove colloidal DOC in the water at high clearance rates [39]. They are frequent in the diet of some larval and adult fish and contribute to the direct transformation of short-lived microbial carbon into long-lived harvestable resources [40]. Appendicularians' discarded houses are important sources of particulate organic aggregates, which become enriched microcosms [41] and play important role in the midwater carbon flux [42]. The trophic scenario in the open waters of the Sicily Channel and in the Rhodes area was likely based on short food chains at high turnover, since appendicularians have been suggested to mediate a substantial energy flow from DOC to fish [39]. By contrast, the structure of mesozooplankton communities in the Ionian Sea and at the neighbouring stations of the Cretan Passage and the Cretan Sea showed a higher contribution of chaetognaths to total zooplankton, a feature that differentiated the area from other regions in the basin. Chaetognaths feed mainly on copepods [43] and represent the most important predators in zooplankton assemblages collected with nets. Their relevance in the Ionian Sea was not easily explainable on the basis of the available information regarding that region. However, it did strongly suggest that the energy flow was there mainly channeled through a more complex food web, implying longer carbon recycling. In the pelagic environment, both the organism distribution and their trophic interactions result from complex biological and hydrographic dynamics at various spatial and temporal scales. For this reason, we are still far from achieving a coherent picture of the fluxes through the food webs in a patchy environment such as the EMED. Moreover, for this basin we still lack data on distribution, life cycles and rate processes of zooplankters such as heterotrophic protists, ostracods, gelatinous organisms (among others). For example, the vertical distribution of micrometazoans has been only very recently studied in the EMED (Levantine Sea), by sampling with fine nets (0.05 mm mesh size), which have not been traditionally used in oceanic deep waters [44].

## 6. The Deep Conversion Cycle

The problem of how deep pelagic communities receive food supplies has always puzzled biological oceanographers. This is particularly true for oligotrophic environments where the combined effect of high rate of recycling and generally low primary production in the upper layer prevents high flux of usable carbon to reach the bathypelagos. Likewise, the vertical distribution of organisms in mesopelagic and bathypelagic realms has frequently risen questions, because of discrepancies often observed among fluxes, metabolic needs and biomass. The EMED could be a case of study in this respect. Scotto di Carlo and co-authors [45] compared zooplankton abundances in the Tyrrhenian Sea (the Mediterranean sub-basin for which vertical distribution data where available) with other tropical and subtropical regions and concluded that Mediterranean was very similar to those as regards the vertical trend. In addition, deep-sea fauna abundance appeared to be strongly related to surface primary production, at least according to the scanty data available at time. They also proposed a quantitative description of abundance versus depth, fitting the data with a negative exponential function of the latter, as reported below:

$$\log_{10} N = a \cdot z + b$$

where $N$ is the number of individuals per cubic meter, $z$ the depth in meters and $a$ and $b$ are constants with typical values in the order of -0.0006 ÷-0.0017 and 1÷2, respectively. For the Tyrrhenian Sea, the coefficients had the following values: $a$ = -0.0010; $b$ = 1.5803. Using the same function, Weikert and Tinkhaus [46] derived the coefficients for the Levantine Sea ($a$ = -0.00107; $b$ = 1.52624). Thus they concluded that Levantine basin was definitely poorer than Tyrrhenian Sea which, in turn, Scotto di Carlo et al. [45] considered the poorest in the Western Mediterranean, at least for deep zooplankton abundances.

Both papers assumed that the slopes of the regressions could be considered as a measure of the rate of vertical flux of organic material reaching the depths. This in turn would imply a similar relation, i.e. an exponential decrease, for vertical flux of organic material that would maintain a steady supply to the bathypelagic fauna and, eventually, to the benthos.

A few years later, Weikert and Koppelmann [46] revisited the former analyses, using better resolved data from other areas and data from the Levantine Sea. They found that a power function fitted definitely better the vertical trend. In the logarithmic form the expression is:

$$\log_{10} N = a \cdot \log_{10} z + b$$

Typical values for $a$ and $b$ are in the order of –2 ÷ -6 and 10 ÷ 20, respectively. For the Levantine Sea, the values are $a$ = -3.101; $b$ = 12.5 (note that N units are ind.· 1000 $m^{-3}$). The relevant difference in the trend is the fact that the number of organisms decreases more slowly with depth, i.e. tends toward a more constant value, thus calling for an enhanced supply of usable carbon for their feeding as compared to the previous findings. The two curves are reported in Fig. 2.

The authors speculated about the possible explanation of this trend and put forward three hypotheses: 1) resuspension of organic matter from the bottom, 2) very specialized and more efficient bathypelagic fauna (“new fauna” in the paper), 3) weak pressure by predators.

All the three hypotheses would have ecological implications on the functioning of the system. It is evident that the basin cannot support high values of deep biomass. It is worth noting that, in their comparison, EMED resulted poorer than NE Atlantic but richer than SE Pacific, which appeared to be definitely the most deprived basin in terms of deep mesozooplankton. Even if not explicitly declared, the authors implied that the vertical flux of particulate carbon might be not sufficient to support the survival of the bathial organisms, particularly if an exponential decay in particle flux is assumed. This is not the case. According to Betzer [47], also the vertical flux of POC can be fitted with a power function similar the one reported above. An additional variable is included in the Betzer’s relationship, i.e. the surface production. The expression is the following:

$$\log_{10} F = \log_{10} A + a \cdot \log_{10} z + \log_{10} P$$

where $F$ is the vertical flux of POC, $P$ is the surface primary production and $A$, $a$ and $b$ are constants. Because of the uncertainty in $P$, we use the only available value of POC flux for the EMED at 880 m depth [48] and the same Betzer constant $a$ (indeed measured values for $F$ at 200 and 1000 m in the WMED [49] fit very well the Betzer equation). The value reported is ~ 1 g · m$^{-2}$ ·y$^{-1}$ which, by the way, does not have to be taken as a lower bound for the flux.

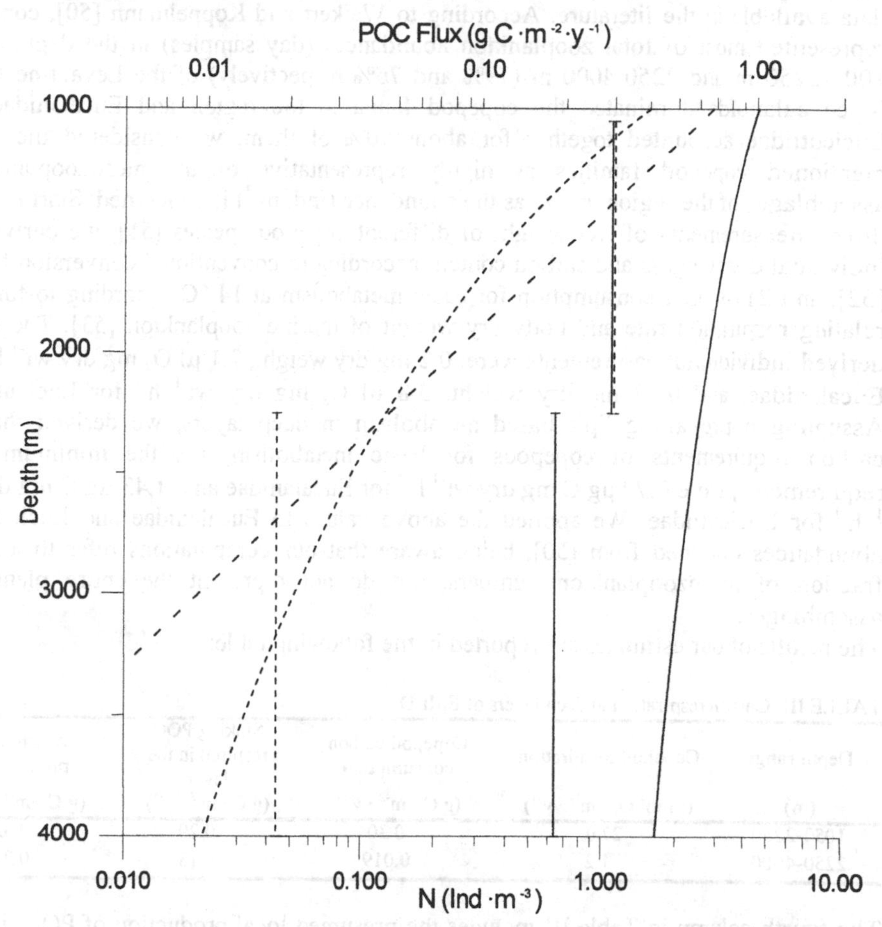


*Figure 2* - Estimated vertical trend of zooplankton abundance and POC flux in the 1000-3000 m layer for the Eastern Mediterrnean. Large dashed line: exponential fit of zooplankton numbers; fine dashed line: power fit of zooplankton numbers; continuous line: power fit of POC flux. The vertical bars represent zooplankton carbon needs (fine dashed) and POC consumed (large dashed) in the depth layers 1050-2250 m and 2250-4000 m.

By a simple transformation of the above equation we have:

$$F_{z_2} = F_{z_1} \cdot (z_2 / z_1)^a$$

The resulting vertical trend is reported in Fig. 3 ($a$ = -0.628; $F_{1000}$ = 1.01 g · m.$^2$ ·y$^{-1}$).

It is worth noting that POC flux and vertical distribution of biomass follow an analogous trend, which could suggest a strict relationship among the two.
To test the hypothesis that sinking POC would not be sufficient to sustain deep mesozooplankters [46], we took into account copepod distribution and physiological data available in the literature. According to Weikert and Koppelmann [50], copepods represented most of total zooplankton abundances (day samples) in the depth layers 1000-2250 m and 2250-4000 m (79% and 76% respectively) of the Levantine Basin. Since calanoids dominated the copepod fauna in the region and Eucalanidae and Lucicutiidae accounted together for about 90% of them, we considered the above mentioned copepod families as highly representative of the mesozooplanktonic assemblages of the region, as far as the abundance (ind. $m^{-3}$) is concerned. Starting from direct measurements of wet weight of different copepod species [51], we derived: 1) individual dry weights and carbon content according to conventional conversion factors [52], and 2) oxygen consumption for basic metabolism at 14 °C according to function relating respiration rate and body dry weight of marine zooplankton [53]. The above derived individual measurements were: 0.2 mg dry weight, 2.1 μl $O_2$ mg dry $wt^{-1}$ $h^{-1}$ for Eucalanidae, and 0.02 mg dry weight, 3.8 μl $O_2$ mg dry $wt^{-1}$ $h^{-1}$ for Lucicutiidae. Assuming a prevailing lipid-based metabolism in deep layers, we derived that the carbon requirements of copepods for basic metabolism, i.e. the minimum food requirements, were 0.79 μg C mg dry $wt^{-1}$ $h^{-1}$ for Eucalanidae and 1.43 μg C mg dry $wt^{-1}$ $h^{-1}$ for Lucicutiidae. We applied the above values to Eucalanidae and Lucicutiidae abundances obtained from [50], being aware that our computations refer to a major fraction of mesozooplankton numbers, but do not represent the entire planktonic assemblages.
The results of our estimates are reported in the following table:

TABLE III. Carbon respiration in deep layers of EMED

| Depth range (m) | Copepod Respiation ($\mu mol\ O_2 \cdot m^{-3} \cdot y^{-1}$) | Copepod carbon consumption ($g\ C \cdot m^{-2} \cdot y^{-1}$) | Sinking POC respired in the layer ($g\ C \cdot m^{-2} \cdot y^{-1}$) | Bacterial production ($g\ C \cdot m^{-2} \cdot y^{-1}$) |
|---|---|---|---|---|
| 1050-2250 | 27.0 | 0.30 | 0.29 | 0.32 |
| 2250-4000 | 1.2 | 0.019 | 0.18 | 0.23 |

The fourth column in Table III includes the presumed local production of POC, i.e. the "non sinking" POC [54]. Numbers about bacterial production and turnover times are quite controversial. To the best of our knowledge, no data are available for deep layers in the EMED. We used the cell counts for the 1000 m depth reported in table 2 of Robarts and co-authors [24] and applied the log-log relationship between abundance and production of Dufour and Torréton [55] for the layers below 1000 m.
In more detail, we assumed the abundance of $0.2 \cdot 10^8$ cells·$dm^{-3}$ to be representative of the 1050-2250 layer and half of this value for the lower layer considered, i.e. a vertical trend similar to that observed by Dufour and Torréton [55] for oligotrophic areas. The production was then derived by the formula (Table 1 in the paper):

$$\log_{10} P = 2.5 \cdot (\log_{10} B - 1.40)$$

Units are µg C · $dm^{-3}$· $h^{-1}$ and µg C · $dm^{-3}$ for *P* and *B* respectively, whereas *B* is computed as N (cells · $dm^{-3}$) times 20 ·$10^{-9}$ µg·$cell^{-1}$. It is worth noting that for the upper bathypelagic layer (1200-2250 m), copepod consumption and sinking POC flux are very similar, which would not leave too much carbon for bacteria utilization. However, our estimates are definitely crude. For the deeper layer, which in the basin is relatively more restricted in space, copepod carbon utilization is definitely less than POC flux. For the sake of simplicity we do not differentiate between attached and free living bacteria, but the table suggest that bacteria production could enter somehow in the carbon utilization of copepods.

We consider even more relevant the following considerations.

POC respiration accounts for ~ 30 µmol $O_2$ · $m^{-3}$ · $y^{-1}$ as an upper limit, very similar to copepod respiration, which decreases much faster with depth. Assuming a growth efficiency of 10% for deep bacteria as compared to the 40% generally in use [56], bacteria respiration accounts for ~ 100 µmol·$m^{-3}$·$y^{-1}$. It is well established in the literature [57], [49] that the difference between potential respiration due to POC is significantly less than total respiration and current opinion is that the additional carbon to be respired is DOC, at least its labile part.

Lefèvre et al. [49] estimated for the NW-MED organic carbon oxidation rate for the depth interval 1000-3500 m to be 9.2 g C · $m^{-2}$ · $y^{-1}$. By contrast, the oxidizable carbon deriving from POC flux as deducible from sediment traps was in the order of 1 g C · $m^{-2}$ · $y^{-1}$. The above authors did not consider bacterial respiration but ETS derived oxygen consumption.

Even considering a possible overestimation of respiration from ETS measurements, it appears that most of metabolic $CO_2$ comes from DOC oxidation, whose chemical composition and reactivity is far from being determined but whose concentration is always higher than 60 µmol · $dm^{-3}$.

To confirm such a discrepancy, we derived from the literature reasonable values of oxygen utilization rate, obtained with different approaches.

From a parallel analysis of freon and oxygen data Schlitzer et al. [15] reported a value of 500 µmol·$m^{-3}$·$y^{-1}$. A similar value, 305 µmol·$m^{-3}$·$y^{-1}$, can be obtained assuming a turnover time of EMDW of ~ 200 years [14] and an oxygen utilization of 61000 µmol·$m^3$ [11] in deep water from the source through the end of the path in the Levantine basin. Finally ETS measurements allow an estimated value of 1000 µmol·$m^{-3}$·$y^{-1}$ for the Levantine (M. Azzaro and B. LaFerla, pers. comm.).

It is quite evident that neither sinking POC flux neither the derived biotic respiration (copepods and sampled bacteria) do account for the oxygen utilization rate independently measured. They account apparently for no more then 20%. Within this scenario (POC falling from sub-photic zone, more or less enriched with bacteria scavenged along the path, contributing very little to deep consumption), additional pathways have to be hypothesized.

Taking a 10% efficiency of the respired oxygen which we conservatively assume to be in the order of 400 µmol·$m^{-3}$·$y^{-1}$, ~ 250 µmol·$m^{-3}$·$y^{-1}$ respired by other organisms should produce more than 0.2 mg·$m^{-3}$·$y^{-1}$ of new biomass. This in turn equals to a virtual flux of more than 0.2 g·$m^{-2}$·$y^{-1}$ on a layer of 1000 m, a similar amount of what derives from the surface. This undetected carbon could be steadily consumed by other components of the deep community which sustain a deep conversion cycle at the expenses of DOC.

This cycle is sketched in Fig. 3.
The outlined scheme emphasizes the relevance of DOC to be converted in "new" POC, which includes in this cartoon also biomass of higher organisms. We hypothesize that bacteria still play a major role in this, but possible additional steps, as in the upper layers, have to be taken into account. For example, mesopelagic and bathypelagic gelatinous zooplankters (e.g. salps, appendicularians, doliolids) might contribute to DOC transformation by rapidly removing and repacking colloidal DOC (> 0.2 μm particle size) [39], and by exporting particles ranging from colloid to picoplankton size with fecal pellets and discarded houses [58]. The contribution to the marine carbon flux of zooplankters which are not adequately collected with the traditional sampling devices has been acknowledged long ago. However, their role in mesopelagic fluxes has been only recently assessed, by using a submersible ROV (remotely operated vehicle) [59]. In addition, because of the EMED smaller volume as compared with the open ocean, and the correspondent faster turnover rate of deep water masses, terrestrial DOC, imported at the time of deep water formation, can enhance deep consumption, reinforcing the relevance of the boundary in the basin.

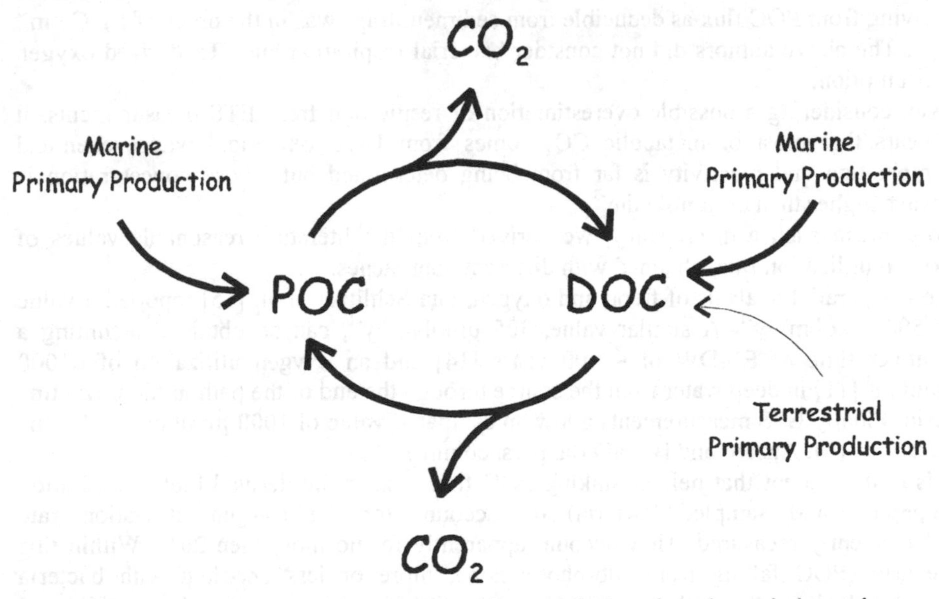


*Figure 3* – Schematics of the deep conversion cycle. Different kind of transformations to be imputed to different organisms are not shown for clarity purposes

Interestingly the DOC to POC conversion is implicitly assumed in numerous papers on carbon cycling in the ocean (see [54], among the others) but has never been explicitly considered as a significant process for sustaining the whole deep fauna.

## 7. Conclusions

In this contribution, we addressed only a few among the many problems pertaining to the EMED ecological physiognomy. In particular, we focused on its renowned oligotrophy, because this has always been the starting point of any debate on the EMED.

Oligotrophy has not always been the status of the basin. Geological records (and maybe also records from human history) demonstrate that past climate fluctuations caused much higher levels of export production than today values. The present EMED is oligotrophic, but because of the short (in geological terms) time scales of changes taking place in the basin it cannot even be assumed that the next future of the basin will be like the present one.

On the other side, it is apparent that its oligotrophy could be relieved by an increase of nutrient input from Gibraltar, but this is much less likely to happen because it should involve a change in the general circulation at global scale. So to blame the Atlantic Ocean for not being enough generous with the Mediterranean misses the point. EMED oligotrophy could indeed be relieved if terrestrial and atmospheric inputs would change and/or a general redistribution of nutrient would be forced by modification of the thermohaline circulation. By the way this process might be occurring in these years because of the recent transient.

In this scenario, the sub-basin structures play a relevant role but they do not reverse the trend. Though, much more has to be investigated about pulses in biological activity and also on its internal phase-lags.

Finally, the EMED gave us the opportunity to address the question of how deep sea works. Because of its high boundary to volume ratio, the hypothesized deep conversion cycle could have a major role in this basin.

## 8. References


1. Rohling, E.J. (1994) Review and new aspects concerning the formation of Eastern Mediterranean sapropels, *Marine Geology* **122**, 1-28.
2. Por, F.D. (1978) Lesseptian Migration. *Ecological Studies* **23**, pp.1-215.
3. Roether, W., Manca, B., Klein, B., Bregant, D., Geogopulos, D., Beitzel, V., Kovacevic, V., and Luchetta, A. (1996) Recent changes in Eastern Mediterranean Deep Waters, *Science* **271**, 333-335.
4. Azov, Y. (1991) Eastern Mediterranean - a Marine Desert?, *EMECS '90* **23**, 225-232.
5. Kullenberg, B. (1952) On the salinity of the water contained in marine sediments, *Medd. Oceanogr. Inst. Goteborg* **21**, 1-38.
6. Gimbutas, M. (1973) Old Europe c. 7000-3500 B.C.: the earliest European civilization before the infiltration of the Indo-European peoples, *Journal of Indo-European Studies* **1**, 1-20.
7. Gimbutas, M. (1973) The beginning of the Bronze Age in Europe and the Indo-Europeans, *Journal of Indo-European Studies* **1**, 163-215.
8. Berman, T., Walline, P.W., Schneller, A., Rothenburg, J., and Towsend, D.W. (1985) Secchi disk record: a claim from the Eastern Mediterranean, *Limnol. Oceanogr.* **30**, 449-450.
9. Bethoux, J.P. (1981) Le phosphore et l'azote en Mer Méditerranée, bilans et fertilité potentielle, *Mar. Chem.* **10**, 141-158.
10. Dugdale, R.C. and Wilkerson, F.P. (1988) Nutrient sources and primary production in the Eastern Mediterranean, *Oceanol. Acta* **9**, 179-184.
11. Bethoux, J.P. (1989) Oxygen consumption, new production, vertical advection and environmental evolution in the Mediterranean Sea, *Deep-Sea Res.* **36**, 769-781.

12. Antoine, D., Morel, A., and André, J.M. (1995) Algal pigment distribution and primary production in the Eastern Mediterranean as derived from coastal zone color scanner observations, *J. Geophys. Research* **100**, 16193-16209.
13. Civitarese, G., Crise, A., Crispi, G., and Mosetti, R. (1996) Circulation Effects on nitrogen dynamics in the Ionian Sea, *Oceanol. Acta* **19**, 609-622.
14. Hopkins, T.S. (1978) Physical Processes in the Mediterranean Basins, in B. Kjerfve (ed.), *Estuarine Transport Processes*, Univ. of South Carolina Press, pp. 269-310.
15. Schlitzer, R., Roether, W., Oster, H., Junghans, H.-G., Hausmann, M., Johannsen, H., and Michelato, A. (1991) Chlorofluoromethane and oxygen in the Eastern Mediterranean, *Deep-Sea Res.* **38**, 1531-1551.
16. Harrison, W.G. (1992) Regeneration of nutrients, in *Primary productivity and biogeochemical cycles in the sea*, P.G. Falkowski and A.D. Woodhead (eds.), Plenum Press, New York, pp. 385-407.
17. Williams, N. (1998) The Mediterranean beckons to Europe's oceanographers, *Science* **279**, 483-484.
18. Civitarese, G., Gacic, M., Vetrano, A., Boldrin, A., Bregant, D., Rabitti, S., and Souvermezoglou, E. (1998) Biogeochemical fluxes through the Strait of Otranto (Eastern Mediterranean), *Cont. Shelf Res.*, in press.
19. Guerzoni, S., Chester, R., Dulac, F., Barak, H., Loye-Pilot, M.-D., Measures, E., Moulin, C., Rossini, P., Saydam, C., Soudine, A., and Ziveri, P. (1998) The role of atmospheric deposition in the biogeochemistry of the Mediterranean Sea, *Prog. Oceanogr.*, submitted.
20. Bethoux, J.P., Morin, P., Madec, C., and Gentili, B. (1992) Phosphorus and nitrogen behaviour in the Mediterranean Sea, *Deep-Sea Res.* **39**, 1641-1654.
21. Salihoglu, I., Saydam, C., Bastürk, Ö., Yilmaz, K., Göcmen, D., Hatipoglu, E., and Yilmaz, A. (1990) Transport and distribution of nutrients and chlorophyll-*a* by mesoscale eddies in the Northeastern Mediterranean, *Mar. Chem.* **29**, 175-390.
22. Yilmaz, A. and Tugrul, S. (1998) The effect of cold- and warm-core eddies on the distribution and stoichiometry of dissolved nutrients in the Northeastern Mediterranean, *J. Mar. Syst.*, submitted.
23. Robinson, A.R., Hecht, A., Pinardi, N., Bishop, J., Leslie, W.G., Rosenthroub, Z., Mariano, A.J., and Brenner, S. (1987) Small synoptic/mesoscale eddies and energetic variability of the eastern Levantine Basin, *Nature* **327**, 131-134.
24. Robarts, R.D., Zohary, T., Waiser, M.J., and Yacobi, Y.Z. (1996) Bacterial abundance, biomass, and production in relation to phytoplankton biomass in the Levantine Basin of the southeastern Mediterranean Sea, *Mar. Ecol. Prog. Ser.* **137**, 237-281.
25. Dortch, Q. and Packard, T.T. (1989) Differences in biomass structure between oligotrophic and eutrophic marine ecosystems, *Deep-Sea Res.* **36**, 223-240.
26. Gasol, J.M., del Giogio, P.A., and Duarte, C.M. (1997) Biomass distribution in marine planktonic communities, *Limnol. Oceanogr.* **42**, 1353-1363.
27. Banse, K. (1995) Zooplankton: pivotal role in the control of ocean production, *ICES J. mar. Sci.* **52**, 265-277.
28. Li, W.K.W., Zohary, T., Yacobi, Y.Z., and Wood, A.M. (1993) Ultraphytoplankton in the Eastern Mediterranean Sea: towards deriving phytoplankton biomass from flow cytometric measurements of abundance, fluorescence and light scatter, *Mar. Ecol. Prog. Ser.* **102**, 79-97.
29. Rabitti, S., Bianchi, F., Boldrin, A., Daros, L., Socal, G., and Totti, C. (1994) Particulate matter and phytoplankton in the Ionian Sea, *Oceanol. Acta* **17**, 297-307.
30. Yacobi, Y.Z., Zohary, T., Kress, N., Hecht, A., Robarts, R.D., Waiser, M., Wood, A.M., and Li, W.K.W. (1995) Chlorophyll distribution throughout the southeastern Mediterranean in relation to the physical structure of the water mass, *J. Mar. Syst.* **6**, 179-190.
31. Berland, B.R., Burlakova, Z.P., Georgieva, L.V., Izmestieva, M., Kholodov, V.I., Krupatkina, D.K., Maestrini, S.Y., and Zaika, V.E. (1987) Phytoplancton estival de la Mer du Levant, biomasse et facteurs limitants, in *Production et Relations Trophiques dans les Ecosystèmes marins. 2e Coll. Franco-Soviétique*, IFREMER, Yalta, Act. Coll. n. 5, pp. 61-83.
32. Mazzocchi, M.G. and Paffenhöfer, G.-A. (1998) First observations on the biology of *Clausocalanus furcatus* (Copepoda, Calanoida), *J. Plankton Res.* **20**, 331-342.
33. Siokou-Frangou, I., Christou, E.D., Fragopoulu, N., and Mazzocchi, M.G. (1997) Mesozooplankton distribution from Sicily to Cyprus (Eastern Mediterranean): II. Copepod assemblages, *Oceanol. Acta* **20**, 537-548.

34. Mazzocchi, M.G., Christou, E.D., Fragopoulu, N., and Siokou-Frangou, I. (1997) Mesozooplankton distribution from Sicily to Cyprus (Eastern Mediterranean): I. General aspects, *Oceanol. Acta* **20**, 521-535.
35. Christou, E.D., Siokou-Frangou, I., Mazzocchi, M.G., and Aguzzi, L. (1998) Mesozooplankton abundance in the Eastern Mediterranean during spring 1992, *Rapp. Comm. int. Mer Médit.* **35**, 410-411.
36. Krom, M.D., Brenner, S., Kress, N., Neori, A., and Gordon, L.I. (1992) Nutrient dynamics and new production in a warm-core eddy from the Eastern Mediterranean Sea, *Deep-Sea Res.* **39**, 467-480.
37. Krom, M.D., Brenner, S., Kress, N., Neori, A., and Gordon, L.I. (1993) Nutrient distributios during an annual cycle across a warm-core eddy from the Eastern Mediterranean Sea, *Deep-Sea Res.* **40**, 805-825.
38. Zohary, T., Brenner, S., Krom, M.D., Angel, D.L., Kress, N., Li, W.K.W., Neori, A., and Yacobi, Y.Z. (1998) Buildup of microbial biomass during deep winter mixing in a Mediterranean warm-core eddy, *Mar. Ecol. Prog. Ser.* **167**, 47-57.
39. Flood, P.R., Deibel, D., and Morris, C.C. (1992) Filtration of colloidal melanin from sea water by planktonic tunicates, *Nature* **355**, 630-632.
40. Fortier, L., Le Fèvre, J., and Legendre, L. (1994) Export of biogenic carbon to fish and to deep ocean: the role of large planktonic microphages, *J. Plankton Res.* **16**, 809-839.
41. Alldredge, A.L. (1976) Discarded appendicularian houses as source of food, surface habitats, and particulate organic matter in planktonic environments, *Limnol. Oceanogr.* **21**, 14-23.
42. Gorsky, G., Lins da Silva, N., Dallot, S., Laval, P., Braconnot, J.C., and Prieur, L. (1991) Midwater tunicates: are they related to the permanent front of the Ligurian Sea (NW Mediterranean)?, *Mar. Ecol. Prog. Ser.* **74**, 195-204.
43. Feigenbaum, D.L. and Maris, R.C. (1984) Feeding in the Chaetognatha, *Oceanogr. Mar. Biol. Ann. Rev.* **22**, 343-392.
44. Böttger-Schnack, R. (1997) Vertical structure of small metazoan plankton, especially non-calanoid copepods. II. Deep Eastern Mediterranean (Levantine Sea), *Oceanol. Acta* **20**, 399-419.
45. Scotto di Carlo, B., Ianora, A., Fresi, E., and Hure, J. (1984) Vertical zonation patterns for Mediterranean copepods from surface to 3000 m at a fixed station in the Tyrrhenian Sea, *J. Plankton Res.* **6**, 1031-1056.
46. Weikert, H. and Koppelmann, R. (1996) Mid-water zooplankton profiles from the temperate ocean and partially landlocked seas. A re-evaluation of interoceanic differences, *Oceanol. Acta* **19**, 657-664.
47. Betzer, P.R., Showers, W.J., Laws, E.A., Winn, C.D., DiTullio, G.R., and Kroopnick, P.M. (1984) Primary productivity and particle fluxes on a transect of the equator at 153° W in the Pacific Ocean, *Deep-Sea Res.* **31**, 1-11.
48. Heussner, S. and Price, B.N. (1997) Synthesis of the results from the CINCS, EMPS, EUROMARGE-AS, EUROMARGE-NB, MERMAIDS, OTRANTO & PELAGOS Projects. Biogeochemical budgets, in E. Lipiatou (ed.), *Interdisciplinary Research in the Mediterranean Sea*, European Community, Luxembourg, pp. 298-314.
49. Lefèvre, D., Denis, M., Lambert, C.E., and Miquel, J.-C. (1996) Is DOC the main source of organic matter remineralization in the ocean water column?, *J. Mar. Syst.* **7**, 281-291.
50. Weikert, H. and Koppelmann, R. (1993) Vertical structural patterns of deep-living zooplankton in the NE Atlantic, the Levantine Sea and the Red Sea: a comparison, *Oceanol. Acta* **16**, 163-177.
51. Shmeleva, A.A. (1965) Weight characteristics of the zooplankton of the Adriatic Sea, *Bull. Inst. oceanogr. Monaco* **65**, 1-24.
52. Parsons, T. and Takahashi, M. (1973) *Oceanographic processes*, Pergamon Press, Oxford.
53. Omori, M. and Ikeda, T. (1984) *Methods in Marine Zooplankton Ecology*, Iohn Wiley & Sons, New York.
54. Druffel, E.R.M., Williams, P.M., Bauer, J.E., and Ertel, J.R. (1992) Cycling of dissolved and particulate organic matter in the open ocean, *J. Geophys. Res.* **97**, 15639-15659.
55. Dufour, P. and Torréton, J.-P (1996) Bottom-up and top-down control of bacterioplankton from eutrophic to oligotrophic sites in the tropical northeastern Atlantic Ocean, *Deep-Sea Res.* **43**, 1305-1320.
56. Bjørnsen, P.K. and Kuparinen, J. (1991) Determination of bacterioplankton biomass, net production and growth efficiency in the Southern Ocean, *Mar. Ecol. Prog. Ser.* **71**, 185-194.

57. Christensen, J.P., Packard, T.T., Dortch, Q.F., Minas, H.J., Gascard, J.C., Richez, C., and Garfield, P.C. (1989) Carbon oxidation in the Deep Mediterranean Sea: evidence for dissolved organic carbon source, *Global Biogechem. Cycles* **3**, 315-335.
58. Bedo, A.W., Acuna, J.L., Robins, D., and Harris, R.P. (1993) Grazing in the micron and sub-micron particle size range: the case of *Oikopleura dioica* (Appendicularia), *Bull. Mar. Sci.* **53**, 2-14.
59. Steinberg, D.K., Silver, M.W., and Pilskaln, C.H. (1997) Role of mesopelagic zooplankton in the community metabolism of giant larvacean house detritus in Monterey Bay, California, USA, *Mar. Ecol. Prog. Ser.* **147**, 167-179.

# TYPE DEFINITION OF VERTICAL CHLOROPHYLL DISTRIBUTION IN CONTRASTING ZONES OF THE MEDITERRANEAN SEA

OLEG A. YUNEV, NADEZHDA A. OSTROVSKAYA,
ZOSIM Z. FINENKO

*Institute of Biology of Southern Seas, National Academy of Sciences of Ukraine, Sevastopol/UKRAINE*

**Abstract** The analysis of vertical distribution of chlorophyll concentration (CHL) in deep-water regions of the Mediterranean Sea was carried out on the basis of historical data received during 1979 - 1987 in western and eastern parts of the sea. It was shown that for a deep-water part of the Mediterranean sea presence of two types of a vertical structure: one-modal, satisfactorily described by the Gaussian curve, and quasi-homogeneaus uniform distribution from a surface up to certain depth and subsequent downturn of concentration up to a minimum was characteristic for various regions and seasons. Parameters describing both types of vertical CHL distribution in the Mediterranean Sea in researched period were computed. Type of a vertical CHL profile and the parameters, describing its form, in each particular case in general corresponds to character of vertical distribution of pigment in moderate latitudes with expressed seasonal variability.

## 1. Introduction

From oceanographic studies carried out in the past twenty years it is clear that the mechanisms controlling biochemical cycle are not uniform throughout the World Ocean. Phytoplankton, the major component responsible for primary production in the sea, has a principal influence on biochemical cycling from local to global scales. A major objective in oceanography today is estimating the mean and the variance of primary production on a global basis [1]. Biological variability is poorly sampled by classical shipboard operations and it does not allow more accurate estimations of phytoplankton biomass over large scale.

Satellite observations of ocean colour mode of the CZCS and SeaWiFS provide a synoptic view of phytoplankton pigment concentrations of a global scale at high spatial resolution with good time coverage. But these approaches are limited to provide information only one extinction depth. In the pelagic ecosystems of the ocean most part of phytoplankton occurs below this layer. This limitation have to be removed.



*P. Malanotte-Rizzoli and V.N. Eremeev (eds.),*
*The Eastern Mediterranean as a Laboratory Basin for the Assessment of Contrasting Ecosystems*, 65–80.

The first steps in this direction have been done by Platt and co-workers [2]. They proposed a generalised chlorophyll profile to be used in models of predicting primary production. Morrel and Berton [3] have suggested that the shape of the pigment profile be related to the surface biomass and the vertical profile can thus be recovered from satellite data.

The aim of this investigation is to provide mathematical representation of the vertical profile of chlorophyll in the Mediterranean Sea. In the present study historical data obtained during last years on the vertical chlorophyll distribution in the Mediterranean Sea were analysed and parameterised. Such representation can then be implemented to existent models to determine of primary production.

## 2. Material and methods

Type definition of vertical chlorophyll distribution in the Mediterranean Sea (depth > 200 m) (Fig. 1) was carried out on the basis of the analysis of 10 data sets, distinguished by number of stations, localisation and time of survey realisation (Table 1). The data were obtained during cruises of the R/V of Institute of Biology of the Southern Seas and Marine Hydrophysical Institute (Sevastopol) from 1979 to 1987 and being kept in a databank of a Department of Ecological Physiology of Phytoplankton of IBSS (Sevastopol).

Chlorophyll-a concentration (CHL) was determined either spectrophotometrically (SP) according to SCOR-UNESCO [4], or fluorometrically (FLU) according to JGOFS-Protocols [5]. In the second case total concentration of chlorophyll-a and phaeopigments-a was analysed, because the use of standard spectrophotometrical method did not provide separate determination of pigments and phaeopigments.

The samples of water from discrete depths in a layer from a surface up to 150-200 m were collected with plastic flow batometers. For the description of vertical CHL distribution profiles having no less than 5 depths were used, that has made 95 % from all profiles obtained in cruises.

For parameterisation of one-modal vertical CHL distribution the Gauss's formula was used in the following form [6]:

$$C(Z) = B_0 + L * \exp [ - (Z-Z_m)^2 / 2 * \sigma^2 ] \quad (1)$$

where the second term on the right-hand side of the equation (1) is a Gaussian curve (representing the CHL maximum) superimposed on constant for each separate profile background $B_0$ (Fig. 2, A), L - amplitude of a maximum and $\sigma$ - reflection of width of the maximum background, equal $4\sigma$.

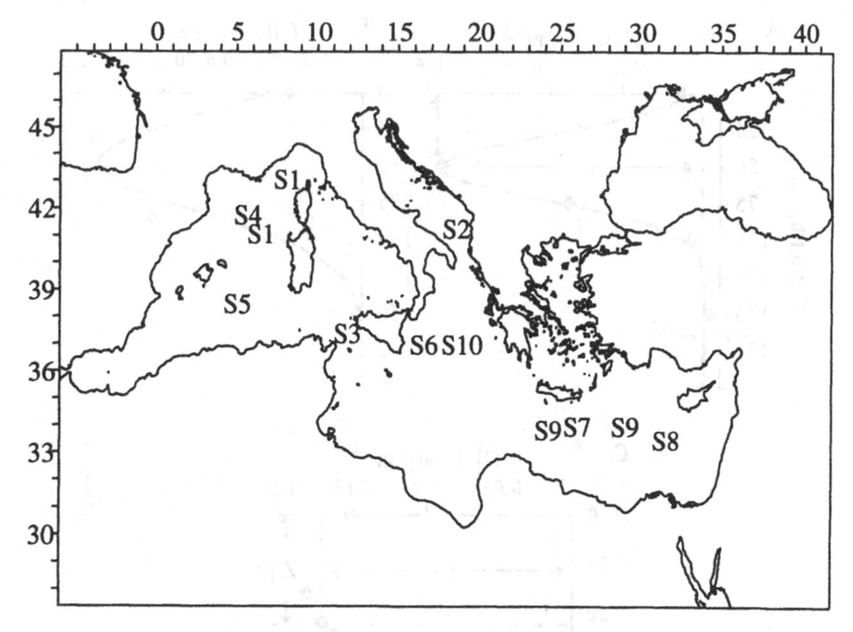


*Figure 1.* Location of data sets (S1 - S10) for type definition of vertical chlorophyll distribution in different regions of the Mediterranean Sea.

TABLE 1. Areas of study for type definition of vertical chlorophyll distribution and relevant information concerning the cruises.

| NN | Area of study | Cruise | Month year | Number of profiles | Method of CHL determination |
|---|---|---|---|---|---|
| S1 | Ligurian and Balearic Seas | 7 PV*) | Dec 1979 | 20 | SP |
| S2 | Southern Adriatic Sea | 12 PV | Mar 1982 | 5 | SP |
| S3 | Sicilian Strait | 12 PV | Mar 1982 | 6 | SP |
| S4 | Balearic Sea | 12 PV | Apr 1982 | 14 | SP |
| S5 | Alboran Sea | 15 PV | June 1983 | 3 | FLU |
| S6 | Western Ionian Sea | 15 PV | June 1983 | 3 | FLU |
| S7 | Crete-African Strait | 15 PV | July 1983 | 2 | FLU |
| S8 | Levantine Sea | 15 PV | July 1983 | 1 | FLU |
| S9 | Levantine Sea | 7 PK | Oct 1983 | 22 | FLU |
| S10 | Ionian Sea | 17 PK | Dec 1987 | 15 | FLU |

*) Abbreviation of R/V name: PV - Professor Vodynitsky, PK - Professor Kolesnikov.

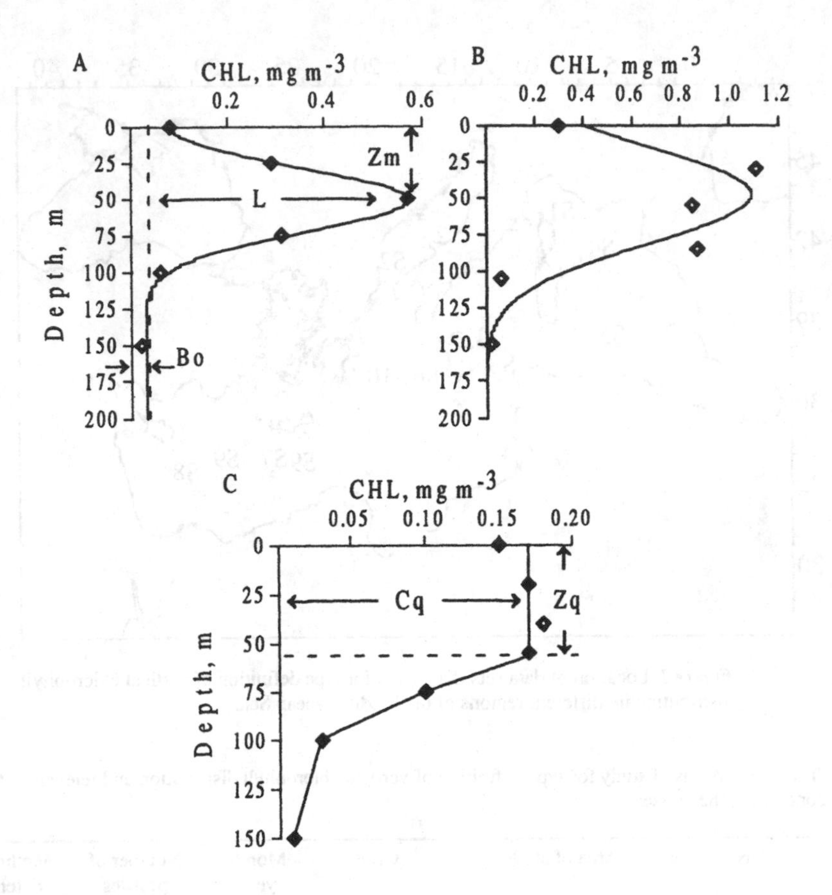


*Figure. 2.* **Examples of types of vertical CHL distribution in the Ligurian and Balearic Seas in December 1979 (set S1).**
**A-1-st type (station 8), described with maximal fit ($r^2$=0.99, F=230, SEF=0.02);**
**B-1-st type (station 7), described with minimal fit ($r^2$=0.82, F=3, SEF=0.30);**
**C-2-nd type (station 11).**

The calculation of parameters of the equation (1) was carried out with use of a package of the applied programs SigmaPlot for Windows. The statistical estimation of correspondence of received profiles and computed curves was carried out by the program ANOVA. In work we adduce the following parameters of ANOVA: coefficient of determination ($r^2$), F-relation (F) and standard error of fit (SEF).

As the integration characteristic of vertical one-modal CHL distribution the integrating of CHL in a layer of 0-150 m was applied by use of profile computed according to the equation (1).

**3. Results**

The analysis of vertical CHL distribution in various deep-water regions of the Mediterranean Sea has shown existence of two types of a profile: one-modal, for the description of which it was possible to use a Gaussian curve (1-st type), and quasi-homogeneaus uniform distribution of CHL ($C_q$) from a surface up to some depth $Z_q$ with the subsequent downturn of concentration up to a minimum (2-nd type) (Fig. 2).

Presence of maximum on CHL profile (Fig. 2, A and B) was defined by fulfilment of conditions (1):
$C_m / C_a > 1.3$ and $Z_m \neq 0$, where $C_m$ - maximal and $C_a$- average in a layer 0 - $Z_m$ concentrations, $Z_m$ - depth of a maximum of concentration. The value of 1.3 was taken according to results of Bio-optical Expert Group of the TU Black Sea Data Base, created in Erdemli (Turkey), which has established an error of CHL determination by SP or FLU equal 30 % from average concentration on the data of repeated determinations [7].

Fulfilment of a condition (2): $0.7 < C(Z) / C_a < 1.3$, where C (Z) - any and $C_a$ the average concentrations in a layer 0-$Z_q$ was the basis for classifying a profile as 2-nd type of vertical CHL distribution (Fig. 2, C).

The statistical estimation of conformity of parameters of a Gaussian curve to the 1-st type of vertical CHL distribution by use of ANOVA has shown, that the statistical characteristics can vary in very wide limits. So for example, in the first data set (S1) F-relation varied from 3 up to 230 and SEF - from 0.02 up to 0.30 mg $m^{-3}$ (Fig. 2, A and B), but $r^2$ was not lower 0.8 in all investigated data sets.

The second type of vertical CHL distribution was characterised by parameters $C_q$ and $Z_q$ and CHL integrated in a layer 0-150 m ($C_i$). Conformity of $C_q$ and $Z_q$ to the second type of vertical CHL profile was estimated only by fulfilment of the condition (2). The integrating of CHL was carried out in this case on really measured concentrations.

Distribution of both types of vertical CHL profiles in separate regions of the Mediterranean Sea in limits of one month (except for data sets S5 - S8, where the researches were carried out on one station during one week) is shown on Fig. 3.

S1. LIGURIAN AND BALEARIC SEAS, DECEMBER 1979

In the Ligurian and Balearic Seas in December 1979 approximately identical number of vertical CHL profiles of both types was received . The existence of a maximum on a vertical CHL profiles was marked mainly in the south-western part of region, where

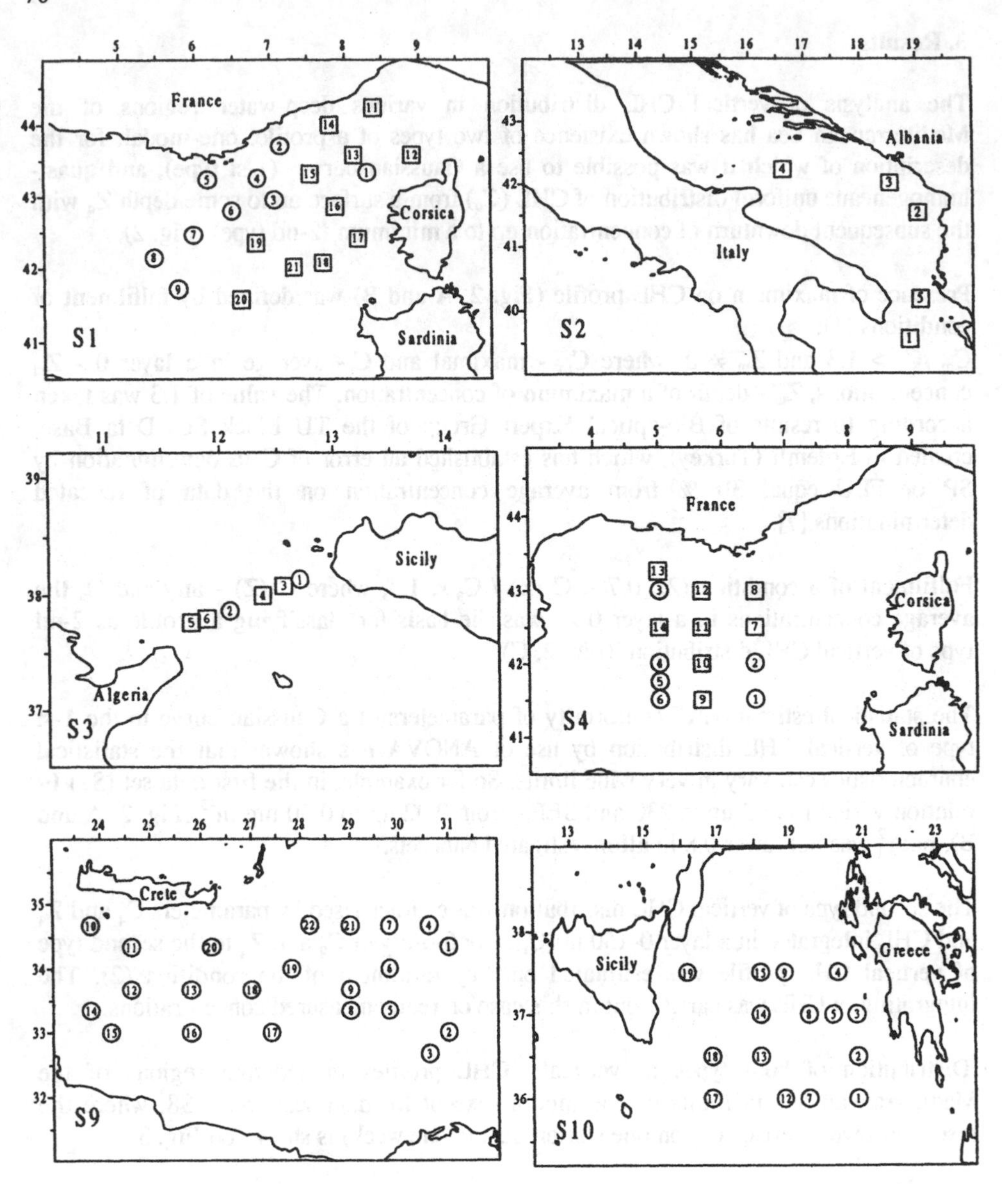


*Figure 3.* Location of stations with two types (circles-1-st and squares- 2-nd type) of vertical CHL distribution in different regions of the Mediterranean Sea. Figures in circles and squares mean number of station.

significant variability of all parameters of a Gaussian curve, surface CHL ($C_o$) and $C_i$ was observed (Table 2). $C_o$ and $C_m$ changed approximately in 7-8 times. Width of maximum peak changed from 36 up to 136 m and depth of a maximum - from 32 up to 53 m. $C_i$ changed from 14.3 up to 85.0 mg $m^{-2}$

Table 2. The characteristic of vertical CHL distribution in the Ligurian and Balearic Seas in December 1979. $C_o$ (mg $m^{-3}$) and $C_i$ (mg $m^{-2}$) - surface and integrated CHL accordingly; $B_0$, L, $Z_m$ and σ- parameters of the 1-st type profile; $r^2$, F and SEF - statistical parameters of ANOVA; $C_q$ and $Z_q$ - parameters of the 2-nd type profile of CHL.

| 1-st type | | | | | | | | | | 2-nd type | | | |
|---|---|---|---|---|---|---|---|---|---|---|---|---|---|
| N st. | $C_0$ | $B_0$ | L | $Z_m$ | σ | $C_i$ | $r^2$ | F | SEF | N st. | $C_q$ | $Z_q$ | $C_i$ |
| 1 | 0.21 | 0 | 0.32 | 33 | 30 | 20.7 | 0.93 | 9 | 0.05 | 10 | 0.21 | 30 | 13.0 |
| 2 | 0.12 | 0.05 | 0.25 | 53 | 33 | 26.9 | 0.94 | 11 | 0.04 | 11 | 0.17 | 55 | 14.8 |
| 3 | 0.1 | 0.06 | 1.69 | 48 | 10 | 51.4 | 0.98 | 26 | 0.05 | 12 | 0.49 | 25 | 27.1 |
| 4 | 0.1 | 0.05 | 0.32 | 32 | 9 | 14.7 | 0.95 | 6 | 0.06 | 13 | 0.17 | 20 | 8.2 |
| 5 | 0.12 | 0.07 | 0.25 | 52 | 15 | 19.5 | 0.91 | 6 | 0.05 | 14 | 0.16 | 60 | 16.7 |
| 6 | 0.19 | 0.01 | 0.68 | 48 | 14 | 25.3 | 0.95 | 7 | 0.13 | 15 | 0.16 | 50 | 13.9 |
| 7 | 0.3 | 0 | 1.09 | 48 | 34 | 85.3 | 0.82 | 3 | 0.30 | 16 | 0.12 | 70 | 11.1 |
| 8 | 0.08 | 0.03 | 0.54 | 51 | 22 | 33.8 | 0.99 | 231 | 0.02 | 17 | 0.1 | 50 | 7.3 |
| 9 | 0.04 | 0.01 | 0.27 | 46 | 19 | 14.3 | 0.99 | 76 | 0.02 | 18 | 0.1 | 75 | 9.4 |
| | | | | | | | | | | 19 | 0.09 | 65 | 6.4 |
| | | | | | | | | | | 20 | 0.17 | 50 | 10.8 |

For the second type of a profile there was a smaller variability of CHL - $C_q$ changed from 0.09 till 0.49 mg $m^{-3}$ and $C_i$ - from 6.4 up to 27.1 mg $m^{-2}$, but at the same time width of quasi-homogeneous uniform layer of CHL changed from 20 up to 75 m.

S2. SOUTHERN ADRIATIC SEA, MARCH 1982

On five stations in a researched part of the Adriatic Sea in March 1982 only uniform vertical distribution of CHL was received. The concentration of pigment of 0.08-0.18 mg m$^{-3}$ was observed from a surface up to 38-100 m (Table 3). $C_i$ changed from 6.9 up to 24.4 mg m$^{-2}$.

Table 3. The characteristic of vertical CHL distribution in the southern Adriatic Sea in March 1982. Abbreviations are the same as in Table 2.

| N st. | $C_q$ | $Z_q$ | $C_i$ |
|---|---|---|---|
| 1 | 0.18 | 94 | 24.4 |
| 2 | 0.08 | 62 | 6.9 |
| 3 | 0.08 | 100 | 9.4 |
| 4 | 0.1 | 100 | 11.7 |
| 5 | 0.15 | 38 | 11.1 |

## S3. SICILIAN STRAIT, MARCH 1982

In the Sicilian Strait in March 1982 mainly uniform CHL distribution with significant variability of $Z_q$ was received (Table 4). In two vertical profiles with one-modal CHL distribution small increasing of concentration in a maximum on depths 9 and 26 m in comparison with a surface was observed . On the average $C_i$ in both cases were close and changed from 11.0 up to 36.4 mg m$^{-2}$.

Table 4. The characteristic of vertical CHL distribution in the Sicilian Strait in March 1982. Abbreviations are the same as in Table 2.

| 1-st type | | | | | | | | | | 2-nd type | | | |
|---|---|---|---|---|---|---|---|---|---|---|---|---|---|
| N st. | $C_0$ | $B_0$ | L | $Z_m$ | $\sigma$ | $C_i$ | $r^2$ | F | SEF | N st. | $C_q$ | $Z_q$ | $C_i$ |
| 1 | 0.15 | 0.05 | 0.25 | 9 | 6 | 11.0 | 0.94 | 17 | 0.03 | 3 | 0.22 | 85 | 22.0 |
| 2 | 0.24 | 0.05 | 0.35 | 26 | 18 | 22.9 | 0.82 | 5 | 0.07 | 4 | 0.31 | 46 | 23.3 |
| | | | | | | | | | | 5 | 0.34 | 46 | 36.4 |
| | | | | | | | | | | 6 | 0.16 | 23 | 11.0 |

## S4. BALEARIC SEA, APRIL 1982

The mosaic picture of vertical CHL distribution was observed in the northern Balearic Sea in April 1982 like that was there in December 1979. On polygon with the sizes $2^0$ x $2^0$ approximately identical number of vertical CHL profiles with the maximum on depth 23-50 m and with uniform distribution of high concentration up to 18-36 m was received. Significant differences of CHL in a surface layer on stations with a different type of vertical profiles were also observed (Table 5). $C_0$ of 1-st type profiles varied from 0.11 till 0.54 mg $m^{-3}$, whereas $C_q$- from 0.68 up to 1.06 mg $m^3$. An the same time $C_i$ of both types of profile differed insignificantly and changed from 15 up to 72 mg $m^{-2}$.

Table 5. The characteristic of vertical CHL distribution in the Balearic Sea in April 1982. Abbreviations are the same as in Table 2.

| 1-st type | | | | | | | | | | 2-nd type | | | |
|---|---|---|---|---|---|---|---|---|---|---|---|---|---|
| N st. | $C_0$ | $B_0$ | L | $Z_m$ | $\sigma$ | $C_i$ | $r^2$ | F | SEF | N st. | $C_q$ | $Z_q$ | $C_i$ |
| 1 | 0.11 | 0.08 | 0.18 | 46 | 9 | 15.7 | 0.82 | 3 | 0.05 | 7 | 0.89 | 18 | 26.5 |
| 2 | 0.24 | 0.03 | 0.35 | 23 | 28 | 23.9 | 0.82 | 3 | 0.08 | 8 | 1.03 | 31 | 49.2 |
| 3 | 0.54 | 0.05 | 0.83 | 31 | 40 | 72.0 | 0.83 | 7 | 0.03 | 9 | 0.75 | 36 | 59.2 |
| 4 | 0.19 | 0.18 | 0.7 | 33 | 11 | 46.2 | 0.91 | 9 | 0.08 | 10 | 0.68 | 18 | 31.4 |
| 5 | 0.25 | 0 | 0.96 | 50 | 28 | 64.7 | 0.96 | 9 | 0.08 | 11 | 0.79 | 36 | 70.2 |
| 6 | 0.15 | 0.02 | 0.22 | 50 | 40 | 22.4 | 0.86 | 6 | 0.03 | 12 | 0.87 | 36 | 63.2 |
| | | | | | | | | | | 13 | 0.98 | 21 | 55.4 |
| | | | | | | | | | | 14 | 1.06 | 21 | 40.6 |

## S5-S8. ALBORAN SEA, WESTERN IONIAN SEA, CRETE-AFRICAN STRAIT AND LEVANTINE SEA, JUNE-JULY 1983

In summer months on all repeated stations in investigated regions of the Mediterranean Sea vertical CHL profiles with a well expressed maximum on depth of 70-108 m were received. The concentration in a maximum differed from surface on all stations on the average in 6 times. On the contrary, insignificant distinctions of parameters of repeated profiles and high values of the statistical characteristics of ANOVA $r^2$ and F were observed (Table 6).

Table 6. The characteristic of vertical CHL distribution in the Alboran Sea, western Ionian Sea, Crete-African Strait and Levantine Sea in June-July 1983. Abbreviations are the same as in Table 2.

| Set | Date | $C_0$ | $B_0$ | L | $Z_m$ | $\sigma$ | $C_i$ | $r^2$ | F | SEF |
|---|---|---|---|---|---|---|---|---|---|---|
| S5 | 8.06 | 0.01 | 0.01 | 0.23 | 92 | 23 | 14.5 | 0.96 | 50 | 0.02 |
| | 10.06 | 0.07 | 0.07 | 0.27 | 102 | 28 | 28.3 | 0.91 | 16 | 0.05 |
| | 13.06 | 0.08 | 0.08 | 0.27 | 108 | 19 | 24.5 | 0.99 | 624 | 0.01 |
| S6 | 20.06 | 0.07 | 0.08 | 0.3 | 113 | 27 | 30.2 | 0.95 | 24 | 0.04 |
| | 24.06 | 0.09 | 0.1 | 0.23 | 108 | 25 | 28.8 | 0.97 | 46 | 0.02 |
| | 27.06 | 0.12 | 0.1 | 0.28 | 104 | 21 | 29.5 | 0.98 | 91 | 0.02 |
| S7 | 19.07 | 0.1 | 0.15 | 0.6 | 70 | 5 | 30.0 | 0.95 | 31 | 0.04 |
| | 26.07 | 0.12 | 0.16 | 0.8 | 77 | 5 | 34.0 | 0.92 | 20 | 0.09 |
| S8 | 30.07 | 0.08 | 0.07 | 0.35 | 85 | 23 | 30.0 | 0.98 | 119 | 0.02 |

## S9. LEVANTINE SEA AND CRETE-AFRICAN STRAIT, OCTOBER 1983

On all stations of investigated region of the sea and strait vertical CHL profiles with one-modal distribution of pigment were received. The peak of CHL with the maximal concentration of 0.2 - 0.7 mg $m^{-3}$ was on depths from 78 up to 116 m (Table 7). Width of peak changed from 50 to 120 m. In most cases high values $r^2$ and F were received. The integrated concentration changed in a range of 12.0-32.0 mg $m^{-2}$.

## S10. IONIAN SEA, DECEMBER 1987

In the Ionian Sea in December 1987 also one-modal vertical CHL distribution was received on all stations. Unlike the previous region the peak of a CHL maximum in the Ionian Sea was expressed much more poorly (Table 8). The relation of $C_m/C_0$ on the average for region has made 2.8, whereas in the Levantine Sea in October 1983 it was 5.1 In the Ionian Sea the widest range of variability of CHL maximum depth was received: from 28 up to 101 m. $C_i$ changed from 13.9 up to 25.4 mg $m^{-2}$ and insignificantly differed from concentration in the Levantine Sea in October 1983.

The monthly average changes of parameters of the 1-st and 2-nd types of vertical profiles and $C_i$ in various regions of the Mediterranean Sea for all period of research are shown in Tables 9- 11. In various seasons in the Mediterranean Sea very high variability of all parameters of both types of vertical CHL profiles, even in limits of separate region for one month, was received. So for example, in the Ionian Sea in December 1987 $Z_m$ varied from 28 up to 101 m (Table 9). The minimum value of $Z_m$ equal 9 m was received in Sicilian Strait in March 1982, which differed from the maximal one, received in the Levantine Sea in December 1982 more than in 10 times.

$C_i$ could change in a range of 14.3 - 85.3 mg m$^{-2}$ (data set S1) and 13.9 - 25.4 mg m$^{-2}$ (S10) in case of one-modal vertical CHL distribution.

Table 7. The characteristic of vertical CHL distribution in the Levantine Sea and Crete-African Strait in October 1983. Abbreviations are the same as in Table 2.

| N st. | $C_0$ | $B_0$ | L | $Z_m$ | $C_i$ | $\sigma$ | $r^2$ | F | SEF |
|---|---|---|---|---|---|---|---|---|---|
| 1 | 0.07 | 0.08 | 0.63 | 85 | 35.4 | 15 | 0.99 | 388 | 0.02 |
| 2 | 0.08 | 0.07 | 0.44 | 94 | 25.7 | 14 | 0.99 | 169 | 0.02 |
| 3 | 0.07 | 0.07 | 0.46 | 93 | 33.2 | 20 | 0.98 | 97 | 0.05 |
| 4 | 0.04 | 0.08 | 0.48 | 94 | 33.2 | 18 | 0.98 | 100 | 0.03 |
| 5 | 0.05 | 0.07 | 0.71 | 101 | 35.2 | 14 | 0.99 | 90 | 0.04 |
| 6 | 0.06 | 0.04 | 0.18 | 86 | 17.3 | 26 | 0.85 | 10 | 0.04 |
| 7 | 0.06 | 0.08 | 0.45 | 96 | 30.8 | 17 | 0.97 | 46 | 0.04 |
| 8 | 0.07 | 0.05 | 0.43 | 86 | 26.7 | 18 | 0.92 | 16 | 0.07 |
| 9 | 0.06 | 0.08 | 0.41 | 116 | 39.8 | 32 | 0.84 | 9 | 0.09 |
| 10 | 0.08 | 0.08 | 0.38 | 91 | 24.0 | 13 | 0.98 | 44 | 0.03 |
| 11 | 0.07 | 0.05 | 0.27 | 85 | 26.6 | 29 | 0.98 | 44 | 0.02 |
| 12 | 0.07 | 0.08 | 0.28 | 95 | 28.4 | 24 | 0.95 | 12 | 0.04 |
| 13 | 0.07 | 0.05 | 0.2 | 92 | 22.2 | 31 | 0.89 | 8 | 0.04 |
| 14 | 0.06 | 0.06 | 0.29 | 93 | 27.3 | 26 | 0.97 | 29 | 0.03 |
| 15 | 0.08 | 0.07 | 0.26 | 98 | 26.9 | 26 | 0.91 | 11 | 0.05 |
| 16 | 0.07 | 0.04 | 0.38 | 90 | 32.7 | 29 | 0.97 | 32 | 0.03 |
| 17 | 0.07 | 0.08 | 0.36 | 113 | 31.5 | 23 | 0.88 | 5 | 0.13 |
| 18 | 0.21 | 0.08 | 0.27 | 78 | 25.7 | 21 | 0.97 | 19 | 0.03 |
| 19 | 0.08 | 0.08 | 0.37 | 93 | 30.2 | 20 | 0.97 | 22 | 0.03 |
| 20 | 0.13 | 0.14 | 0.5 | 112 | 36.0 | 12 | 0.85 | 6 | 0.07 |
| 21 | 0.07 | 0.07 | 0.44 | 99 | 30.0 | 18 | 0.99 | 67 | 0.02 |
| 22 | 0.1 | 0.01 | 0.53 | 90 | 24.1 | 17 | 0.98 | 62 | 0.04 |

Besides in the Ligurian and Balearic Seas and the Sicilian Strait simultaneously with a 1-st tipe profiles, having maximum on 32-53 m, there were the stations with uniform CHL distribution from a surface up to 20-75 m and having $C_i$ of 6.4-27.1 mg $m^{-2}$.

Table 8. The characteristic of vertical CHL distribution in the Ionian Sea in December 1987. Abbreviations are the same as in Table 2.

| N st. | $C_0$ | $B_0$ | L | $Z_m$ | $\sigma$ | $C_i$ | $r^2$ | F | SEF |
|---|---|---|---|---|---|---|---|---|---|
| 1 | 0.03 | 0 | 0.19 | 52 | 32 | 14.2 | 0.91 | 7 | 0.03 |
| 2 | 0.03 | 0.05 | 0.27 | 83 | 15 | 17.5 | 0.95 | 13 | 0.03 |
| 3 | 0.13 | 0.07 | 0.29 | 28 | 16 | 21.5 | 0.95 | 14 | 0.04 |
| 4 | 0.14 | 0.08 | 0.53 | 41 | 7 | 21.1 | 0.86 | 4 | 0.06 |
| 5 | 0.14 | 0.08 | 0.26 | 53 | 15 | 21.5 | 0.88 | 5 | 0.06 |
| 6 | 0.12 | 0.07 | 0.31 | 42 | 7 | 15.9 | 0.83 | 3 | 0.05 |
| 7 | 0.12 | 0 | 0.17 | 41 | 39 | 13.9 | 0.94 | 10 | 0.03 |
| 8 | 0.15 | 0.09 | 0.61 | 65 | 7 | 24.0 | 0.82 | 3 | 0.08 |
| 9 | 0.12 | 0.06 | 0.32 | 64 | 21 | 25.4 | 0.92 | 7 | 0.06 |
| 10 | 0.08 | 0.06 | 0.21 | 101 | 20 | 19.2 | 0.89 | 8 | 0.04 |
| 11 | 0.14 | 0 | 0.19 | 61 | 59 | 21.8 | 0.88 | 7 | 0.04 |
| 12 | 0.03 | 0.04 | 0.32 | 71 | 12 | 15.4 | 0.97 | 26 | 0.03 |
| 13 | 0.07 | 0.04 | 0.21 | 83 | 30 | 21.2 | 0.96 | 26 | 0.02 |
| 14 | 0.12 | 0.05 | 0.44 | 68 | 10 | 18.3 | 0.85 | 6 | 0.07 |
| 15 | 0.10 | 0 | 0.19 | 41 | 42 | 16.5 | 0.83 | 4 | 0.04 |

## 4. Discussion .

The existence of two types of CHL profile in the Mediterranean Sea is defined in the basic by hydrophysical and hydrochemical conditions in each specific case. The development of a deep CHL maximum in open deep regions of the sea is a consequence of a number of the reasons. Usually such maximum will be formed in stratified oligotrophic waters, where nutrient depletion in the upper layers in result of consumption them by phytoplankton occurs [8]. Stratification of water mass is a barrier to penetration of additional nutrients from the bottom layers into the well-lit surface layer. The feeding of phytoplankton comes in basic at the expense of regeneration of organic substances [9]. The optimum conditions for formation of a deep CHL maximum in such situation exist at the bottom of a euphotic layer where

Table 9. Statistical characteristics of parameters of the 1-st type of CHL vertical distribution in various regions of the Mediterranean Sea. Avg- mean , Min- minimal and Max- maximal value; Var- coefficient of variation , N - number of profiles.

| Set | N | Co ($mg\ m^{-3}$) | | | | Bo ($mg\ m^{-3}$) | | | | L ($mg\ m^{-3}$) | | | | Zm (m) | | | | σ (m) | | | |
|---|---|---|---|---|---|---|---|---|---|---|---|---|---|---|---|---|---|---|---|---|---|
| | | Avg | Min | Max | Var | Avg | Min | Max | Var | Avg | Min | Max | Var | Avg | Min | Max | Var | Avg | Min | Max | Var |
| 1 | 10 | 0.14 | 0.04 | 0.30 | 57 | 0.03 | 0 | 0.07 | 87 | 0.60 | 0.25 | 1.69 | 82 | 46 | 32 | 53 | 17 | 21 | 9 | 34 | 47 |
| 3 | 2 | 0.20 | 0.15 | 0.24 | - | 0.05 | 0.05 | 0.05 | - | 0.30 | 0.25 | 0.35 | - | 18 | 9 | 26 | - | 12 | 6 | 18 | - |
| 4 | 6 | 0.25 | 0.11 | 0.54 | 62 | 0.06 | 0 | 0.18 | 108 | 0.50 | 0.18 | 0.96 | 62 | 39 | 23 | 50 | 29 | 9 | 3 | 20 | 72 |
| 5 | 3 | 0.05 | 0.01 | 0.08 | 71 | 0.05 | 0.01 | 0.08 | 71 | 0.26 | 0.23 | 0.27 | 9 | 101 | 92 | 108 | 19 | 23 | 19 | 28 | 19 |
| 6 | 3 | 0.09 | 0.07 | 0.12 | 27 | 0.09 | 0.08 | 0.10 | 12 | 0.27 | 0.23 | 0.30 | 13 | 108 | 104 | 113 | 4 | 24 | 21 | 27 | 13 |
| 7 | 2 | 0.11 | 0.10 | 0.12 | - | 0.16 | 0.15 | 0.16 | - | 0.70 | 0.60 | 0.80 | - | 74 | 79 | 77 | - | 5 | 5 | 5 | - |
| 8 | 1 | 0.08 | - | - | - | 0.07 | - | - | - | 0.35 | - | - | - | 85 | - | - | - | 23 | - | - | - |
| 9 | 22 | 0.08 | 0.04 | 0.21 | 44 | 0.07 | 0.01 | 0.14 | 35 | 0.40 | 0.18 | 0.71 | 33 | 95 | 78 | 116 | 10 | 21 | 12 | 32 | 29 |
| 10 | 15 | 0.10 | 0.03 | 0.15 | 42 | 0.05 | 0 | 0.09 | 70 | 0.30 | 0.17 | 0.61 | 44 | 60 | 28 | 101 | 33 | 22 | 7 | 59 | 69 |

Table 10. Statistical characteristics of parameters of the 2-nd type of CHL vertical distribution in various regions of the Mediterranean Sea. Abbreviations are the same as in Table 9.

| Set | N | Cq ($mg\ m^{-3}$) | | | | Zq (m) | | | |
|---|---|---|---|---|---|---|---|---|---|
| | | Avg | Min | Max | Var | Avg | Min | Max | Var |
| 1 | 11 | 0.18 | 0.09 | 0.49 | 63 | 50 | 20 | 75 | 36 |
| 2 | 5 | 0.12 | 0.08 | 0.18 | 38 | 79 | 38 | 100 | 35 |
| 3 | 4 | 0.26 | 0.16 | 0.34 | 32 | 50 | 23 | 85 | 51 |
| 4 | 8 | 0.88 | 0.68 | 1.06 | 15 | 27 | 18 | 36 | 8 |

there is light enough for photosynthesis and increased contents of nutrients. Usually the formation of a CHL maximum in moderate latitudes is characteristic of summer months [8].

Table 11. Statistical characteristics of $C_i$ (mg m$^{-2}$) in various regions of the Mediterranean Sea. Abbreviations are the same as in Table 9.

| Type | | S1 | S2 | S3 | S4 | S5 | S6 | S7 | S8 | S9 | S10 |
|---|---|---|---|---|---|---|---|---|---|---|---|
| 1 | Avg | 32.4 | - | 16.9 | 40.8 | 22.4 | 29.5 | 32.0 | 30 | 21.1 | 19.2 |
| | Min | 14.3 | - | 10.9 | 15.7 | 14.5 | 14.5 | 30.0 | - | 12.0 | 13.9 |
| | Max | 85.3 | - | 22.9 | 72.0 | 28.3 | 30.2 | 34.0 | - | 32.0 | 25.4 |
| | Var | 71 | - | - | 58 | 32 | 2 | - | - | 29 | 19 |
| | N | 9 | - | 2 | 6 | 3 | 3 | 2 | 1 | 22 | 15 |
| 2 | Avg | 12.6 | 12.7 | 23.2 | 49.5 | - | - | - | - | - | - |
| | Min | 6.4 | 6.9 | 11.0 | 26.5 | - | - | - | - | - | - |
| | Max | 27.1 | 24.4 | 36.4 | 70.2 | - | - | - | - | - | - |
| | Var | 46 | 54 | 45 | 31 | - | - | - | - | - | - |
| | N | 11 | 5 | 4 | 8 | - | - | - | - | - | - |

In those cases when there is the vertical mixing or active upward transporting nutrients from the bottom layers into euphotic layer, the uniform development of phytoplankton in all well-lit upper layer takes place. Such conditions, as a rule, are encountered in coastal upwelling areas or in an open part of the sea during winter convectional mixing and winter-spring blooming of phytoplankton at moderate and high latitudes [3].In the given situations a 2-nd type of CHL profile will be mainly formed.

Parameterisation of CHL profiles allows to carry out quantitative comparison of peculiarities of vertical distribution of pigment in different regions of the sea. For the description of one-modal vertical CHL distribution in the Mediterranean Sea a Gaussian curve [6] was used. From the literature it is known, that by use of three parameters of a Gaussian curve it is possible to describe a large part of vertical CHL profiles in the sea [2, 3, 8].

The degree of conformity of measured and computed profiles depends in basic on two circumstances: number and discreteness of sampling and conformity of a vertical CHL profile to normal distribution. The minimum errors at such accounts are possible to expect at the description of continuous chlorophyll fluorescence *in situ* profiles and localisation of a CHL maximum on large depths, when the CHL profile is the most close to normal distribution, just as in case of tropical regions of ocean, for which this formula was offered.

High variability of parameters and the small discreteness of sampling, making in most cases not more of 5-6 depths, have found reflection in significant variability of statistical estimations of the description of vertical CHL distribution. Presence in one region of two types of vertical CHL distribution increases variability of all parameters of a Gaussian curve, which was marked in the Ligurian and Balearic Seas and the Sicilian Strait (Table 9).

Despite received significant variability of practically all characteristics of vertical CHL profiles in the Mediterranean Sea it is possible to note some general regularities of vertical distribution of pigment in various regions of the sea. The high value of $C_0$ (on the average 0.14 -0.88 mg m$^{-3}$) were received in western part of the sea (in the Ligurian and Balearic Seas and the Sicilian Strait) in winter-spring months and coincided with phytoplankton blooming at this time of a year in moderate latitudes. In the same regions a significant part of vertical profiles with quasi-homogeneous vertical uniform CHL distribution in the upper layer, also characteristic for this season was received. In the Balearic Sea in April CHL of 0.68-1.06 mg m$^{-3}$ was distributed up to 18-36 m. For western part of the Mediterranean Sea in winter-spring months also the high value of $C_m$ of 1-st type of vertical CHL profile (on the average 0.2-1.7 mg m$^{-3}$), which did not locate deeper 50 m, were characteristic.

The low CHL (0.08-0.18 mg m$^{-3}$) and its uniform distribution up to 38-100 m were characteristic for a southern part of the Adriatic Sea in March 1983, that testified to intensive vertical mixing of the top 100 m layer, but absence of phytoplankton blooming in this part of the sea in investigated season.

In the summer 1983 in the Alboran Sea $C_0$ did not exceed 0.08 mg m$^{-3}$, maximum of concentration (on the average 0.26 mg m$^{-3}$) lowered up to 100 m. The similar situation, is characteristic for summer months, was observed also in other regions of the Mediterranean Sea in June - July 1983.: in western part of the Ionian Sea, the Crete-African Strait and the Levantine Sea. The maximal CHL in a maximum (0.6-0.8 mg m$^{-3}$) was observed in the Crete-African Strait and located on minimum (77-79 m) for obtained summer depth (Table 9).

Presence of a maximum of CHL on 85-95 m and the absence of vertical CHL profiles with uniform distribution of pigment in the upper layers was observed in the Levantine Sea in October 1983. The low $C_0$ and $C_m$ were very close to that was received in July (Table 9), that testified for proceeding of stratification in October and deficiency of nutrients for a phytoplankton feeding.

A situation different from above discussed was observed in the Ionian Sea in December 1987. The presence only of 1-st type of CHL vertical profiles, not high values of $C_0$ and $C_m$ testified to stratification of water mass, characteristic for summer months.

However, the wide range of $Z_m$ (from 28 up to 101 m) and significant its variability (Var = 33 % was maximal for this parameter among other regions) and insignificant increase of CHL in a maximum in comparison with a surface (in 2.8 times in comparison with 5.1, received in the Levantine Sea in October) could testify to, probably, warm winter and the absence of winter convectional mixing, delivering nutrients from bottom into upper layers. On the average $C_i$ equal 19.2 mgm$^{-2}$ in the Ionian Sea in December 1987 was lower than it was observed for summer and autumn in other regions of the Mediterranean Sea (Table 11).

Thus, type definition of vertical CHL profiles and determination of parameters, describing this distribution, allow to carry out a qualitative and quantitative estimation of vertical distribution of pigment in various regions of the Mediterranean Sea. Character of vertical distribution and the values of parameters, describing it, reflect in general hydrophysical, hydrochemical and meteorological conditions in these regions of the sea in different seasons.

## References


1. Platt, T. and Sathyendranath, S. (1982) Oceanic Primary production: estimation by remote sensing at local and regional scales, *Science* **241**, 1613-1620.
2. Platt, T., Caverhill, C. and Sathyendranath, S. (1991) Basin-scale estimates of oceanic primary production by remote sensing: the North Atlantic, *J. Geophys. Res.* **96**, 15,147-15,159.
3. Morel, A. And Berton J.-F. (1989) Surface pigments, algal biomass profiles, and potential production of the euphotic layer: Relationships reinvestigated in view of remote-sensing applications, *Limnol. Oceanogr.* **34**, 1545-1562.
4. SCOR-UNESCO (1966) Determination of photosynthetic pigments in sea water, Paris, 69 p.
5. JGOFS Protocols (1994) *Protocols for the Joint Global Ocean Flux Study JGOFS)* Core Measurements, Manual and Guides **29**, 97-100.
6. Lewis, M. R., Cullen, J. J. and Platt, T. (1983) Phytoplankton and thermal structure in the upper ocean: Consequences of nonuniformity in chlorophyll profile, *J. Geophys. Res.* **88**, 2565-2570.
7. Ivanov, L., Konovalov, S., Melnikov, V., Mikaelyan, A., Yunev, O., et al. (1998) Physical, Chemical and Biological Data Sets of the TU Black Sea Data Base: Description and Evaluation, *Diagnosis of the environmental condition in coastal and shelf waters of the Black Sea,* Mar. Hydrophys. Inst. NAN of Ukraine, Sevastopol, pp. 11-38.
8. Cullen, J. J. (1982) The deep chlorophyll maximum: Comparing vertical profiles of chlorophyll a, *Can. J. of Fish. and Aquatic Sci.* **39**, 791-803.
9. Goldman J.C. (1988) Spatial and temporal discontinuities of biological processes in pelagic surface waters, *Toward a Theory on Biological-Physical Interactions in the World Ocean.* Ed. B.J.Rothschild. Kluwer Academic Publishers, 273-296

## COMPOSITION AND ABUNDANCE OF ZOOPLANKTON OF THE EASTERN MEDITERRANEAN SEA

KOVALEV A. V.[1], A. E. KIDEYS [2], E. V. PAVLOVA[1], A. A. SHMELEVA [1], V. A. SKRYABIN[1], N. A. OSTROVSKAYA[1], AND Z. UYSAL[2]

*1) Institute of Biology of the Southern Seas, Nakhimov Ave. 2, Sevastopol, Ukraine*

*2) Institute of Marine Sciences, P.O. Box 28, Erdemli 33731 Turkey*

**Abstract**

Results of investigations on the composition and quantitative distribution of zooplankton of the eastern Mediterranean Sea were reviewed for the period 1950-1980. The studies reviewed comprise more than 250 stations (with about 2000 samples) mainly from the expeditions carried out by the Institute of Biology of the Southern Seas (IBSS), attached to the National Academy of Science of the Ukraine. Many of these stations are located in poorly studied deep-water regions. For comparison purposes data from the literature on the coastal regions were also used.

During these studies, >30 new species of zooplankton were identified from the eastern Mediterranean. The abundances of zooplankton from different locations in this region were evaluated. A trend was observed whereby the quantity of zooplankton was seen to decrease from the west to the east and from the north to the south. Such events correspond to an impoverishment of nutrients in the water.

The coastal regions, desalinized by river inflows, were poorer than the deep-water regions with respect to species number, but richer in abundance.

**Key words:** zooplankton, eastern Mediterranean, abundance, biomass, spatial changes.

### 1. Introduction

The Mediterranean Sea is conventionally divided into western and eastern regions. The border between them lies through the Appenin peninsula, the Island of Sicily and the relatively shallow strait of Tunisia [1, 2]. The eastern region includes the Ionic, Adriatic, Syrtis, Levantine and Aegean Seas. Some authors [3, 4] further divide the eastern



*P. Malanotte-Rizzoli and V.N. Eremeev (eds.),*
*The Eastern Mediterranean as a Laboratory Basin for the Assessment of Contrasting Ecosystems*, 81–95.

region according to the bottom topography and hydrological structure of the central (Syrtis, Ionic and Adriatic seas) and the eastern (Levantine and Aegean Seas) water bodies. However, the surface current system affects the whole eastern Mediterranean Sea [3], therefore, its planktonic fauna is sufficiently uniform [5], so in this study we have considered the eastern Mediterranean Sea as one unit. Of course, it is taken into consideration that in some regions, (e.g. river mouths), environmental factors and plankton differ greatly. But in the absence of geomorphological obstacles the waters of these regions are mixed with waters of the open Mediterranean.

The water exchange between the Mediterranean and Black seas occurs through the Marmara Sea. We consider the Marmara Sea as being of the same composition of the eastern Mediterranean Sea.

The eastern Mediterranean is distinguished from the western area by several factors namely; a reduced influence of the Atlantic Ocean whereas there occurs an increasing influence of the Indian Ocean, by the differences in river inflows (the Po, the Nile and rivers that flow into the Black and Azov seas), and by hydrological, hydrochemical and biological features [6, 3, 7, 8]. The reduced influence of the Atlantic ocean is expressed by the gradual decrease in the volume of Atlantic water in the North-African current. Due to evaporation the salinity of the water increases from 36.5‰ in the waters off Gibraltar to 39‰ in the eastern shores [3], where the temperature of the water near the surface in summer is raised from 20° to 27° C. The temperature and salinity of the water also greatly changes from the north to the south of the Mediterranean Sea. The concentration of the nutrients in the waters of the North-African current decreases five-fold from a westerly to an east direction [9, 10, 11]. Production values also decrease accordingly [7]. However the nutrient content and zooplankton abundance in the eastern Mediterranean are characterised by a rather significant unevenness of distribution. An increase in these values is noted in the Marmara Sea and the northern Aegean Sea, in the regions of river estuaries and some other areas.

Environmental factors pertaining to the Mediterranean, as mentioned here, play an important role in shaping the composition and abundance of plankton as a whole and zooplankton in particular.

Studies of zooplankton in the Mediterranean Sea, including its eastern sector, began in the second half of the 19th century. The first large expedition in the eastern Mediterranean Sea was carried out by an Austrian ship the R/V "Pola" between 1890-1894. Later a substantial contribution to the study of plankton was made by the Danish on board the R/V "Tor" during 1908-1910 and the R/V "Dana" in 1930. The quantitative study of plankton was initiated during these two expeditions. Jespersen [12] observed the relative poverty of macroplankton in the eastern Mediterranean Sea with respect to the west.

During 1911-1914, the faunistic investigations in the Adriatic Sea were conducted by the Austrian expedition on the R/V "Naiad" and the Italian expedition on the R/V "Vila Velebita". The Yugoslavian R/V "Hvar" worked in the Adriatic Sea in 1948-1949. In the Mediterranean Sea in 1958 multi-disciplinary oceanological investigations began on the R/Vs "Akademik A.Kovalevsky", "Crystal", and in 1977 on the R/V "Professor Vodyanitsky". Besides, biological studies were executed in a number of hydrophysical expeditions of Soviet vessels "Akademik S.Vavilov", "Mikhail Lomonosov" and others. Consequently, zooplankton studies in the central regions of the eastern Mediterranean basins over the last forty years were conducted most actively by Soviet scientists, mainly from the Institute of Biology of the Southern Seas (IBSS). Extensive studies of zooplankton were recently conducted by scientists of Mediterranean countries under the international program POEM-BC-091 [13, 14].

In some coastal regions of the Mediterranean Sea, near the biological stations, institutes or universities of Italy, Greece, Turkey, Syria, Lebanon, Israel and Egypt, regular seasonal and perennial studies of plankton are conducted.

A great number of zooplankton species had defined during early faunal studies. Here it is worth mentioning that scientists of the IBSS identified 50 species of copepods not previously noted in the Mediterranean Sea and another 32 species that are new to the science [5]. Faunistic data of zooplankton resulted in many articles and monographs from various biogeographical areas in the eastern Mediterranean [6, 7, 8, 15,]. The fauna of the coastal and central regions were compared in the book of Kovalev [8].

Regular studies performed in the eastern Mediterranean Sea had explained the seasonal changes in the taxonomic composition, quantity and biomass of plankton [7]. The results of such studies on quantitative indices of zooplankton of central regions of the sea were summarised in Greze et al. [16]. The results of the investigations carried out by scientists of the Mediterranean countries were presented in a review by Moraitou-Apostolopoulou [17].

In the present work, results of the IBSS investigations were re-evaluated for the open waters of the eastern Mediterranean Sea with a comparison among its subregions as well as changes in the zooplankton composition from the west to the east and from the north to the south.

## 2. Material and methods

From 1958 onwards, the IBSS periodically conducted zooplankton studies in different regions of the eastern Mediterranean Sea, usually in deep waters. Sampling information of these studies are presented in Table 1. Location of sampling stations was shown in Fig. 1.

Usually the upper 200 or 500 metre stratum of water from the standard layers of 0-10, 10-25, 25-50, 50-75, 75-100, 100-150, 150-200, 200-300 m was processed. For sampling a Juday net (mouth diameter 3 cm, mesh size 112 μ) was used.. During the 1970s - 1980s, additional samples were sometimes collected using a Bogorov-Rass net or oceanic model of the Juday net (mouth diameter 80 cm, mesh size >200 μ) and also a water sampler [18] was employed for specific purposes, for instance to study the size composition of the zooplankton [19, 20, 21].

After taxonomic identification, the abundance and biomass of organisms (given as per cubic meter) were calculated by means of weight-dimensional characteristics of plankton. For rare animals the whole sample was screened whilst for mass organisms subsampling was performed.

TABLE 1. Cruises of the IBSS for zooplankton in the eastern Mediterranean Sea.

| Year | Month | Vessel | Region | Layer (m) | #of stations | #ofsamples |
|---|---|---|---|---|---|---|
| 1958 | II, V, VII, XI | Krystall | Southern Adriatic | 0-200 | 84 (8 daily) | 920 |
| 1960 | VII | Akademik Kovalevsky | North Adriatic | 0-100 | | |
| 1982 | II - V | Professor Vodyanitsky | North Adriatic | 0-100 | 25 | 55 |
| 1967 | VII | Akademik Kovalevsky | North Adriatic | 0-200 | 16 | 48 |
| 1959 | VI - IX | Akademik Vavilov | Syrtis Sea | 0-200 | 3 | 21 |
| 1959 | VI - IX | Akademik Vavilov | Levantine Sea | 0-200 | 7 | 49 |
| 1958 | VIII - IX | Akademik Kovalevsky | Aegean Sea | 0-200, 0-500 | | |
| 1960 | VI - VII | Akademik Kovalevsky | Aegean Sea | 0-200, 0-500 | 20 (2 daily) | 151 |
| 1960-1961 | XII - I | Akademik Vavilov | Aegean Sea | 0-200, 0-500 | | |
| 1980 | VIII, XII | Akademik Kovalevsky | Aegean Sea | 0-200 | 4 | 16 |
| 1959 | VII - VIII | Akademik Kovalevsky | Ionian Sea | 0-200, 0-500 | | |
| 1960 | VII | Akademik Kovalevsky | Ionian Sea | 0-200, 0-500 | 14 | 156 |
| 1968 | V | Akademik Kovalevsky | Ionian Sea | 0-500 | 1 (daily) | 35 |
| 1972 | XII | Mikhail Lomonosov | Ionian Sea | 0-500 | 1 (daily) | |
| 1973 | IV | Mikhail Lomonosov | Ionian Sea | 0-500 | 1 (daily) | |
| 1976 | IV | Mikhail Lomonosov | Ionian Sea | 0-500 | 1 (daily) | 81 |
| 1976 | VIII | Mikhail Lomonosov | Ionian Sea | 0-500 | 1 (daily) | |
| 1987 | XI - XII | Professor Kolesnikov | Ionian Sea | 0-200 | 40 | 40 |
| 1970 | I - II | Akademik Kovalevsky | Tunisian strait | 0-200 | | |
| 1972 | VII - IX | Akademik Kovalevsky | Tunisian strait | 0-200 | 33 | 271 |
| 1980 | VIII - X | Akademik Kovalevsky | Tunisian strait | 0-200 | 5 | 20 |
| 1982 | II - V | Professor Vodyanitsky | Tunisian strait | 0-200 | 5 | 14 |

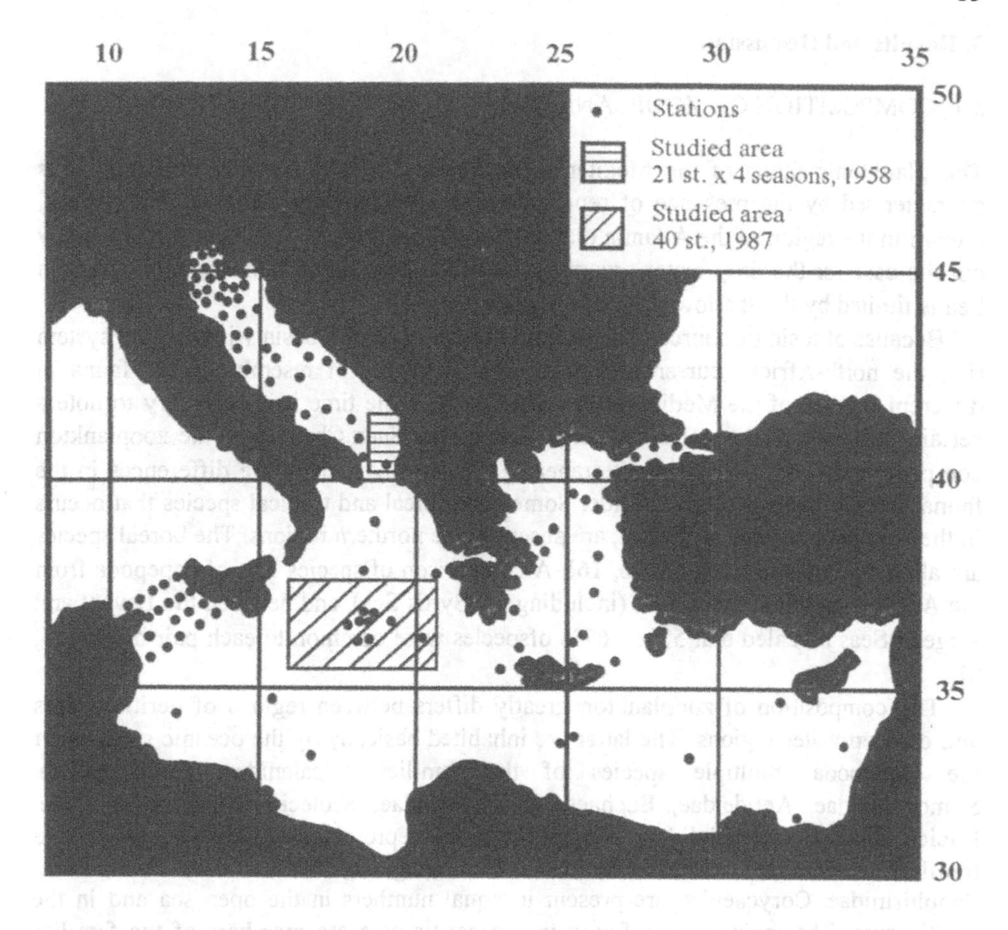


*Figure 1.* **Location of zooplankton stations of the IBSS in the eastern Mediterranean Sea during 1958-1987.**

## 3. Results and Discussion

### 3.1. COMPOSITION OF ZOOPLANKTON

The planktonic fauna of the Mediterranean Sea is basically Atlantic in origin. It is characterised by the presence of representatives of practically all taxonomic groups, known in the region of the Atlantic that adjoins Gibraltar [6, 8]. However this is usually not the case for the deep-water organisms of which penetration into the Mediterranean Sea is limited by the shallow strait of Gibraltar.

Because of a single source of fauna and the presence of a basin wide current system (i.e. the north-African current), there is a high degree of resemblance of fauna in different regions of the Mediterranean Sea. At the same time it is necessary to note a certain influence of the water exchange through the Suez Channel on the zooplankton composition of the eastern Mediterranean [15]. However, there are differences in the fauna that are dependent on latitude. Some subtropical and tropical species that occurs in the southern regions of the sea, are absent in the northern regions. The boreal species are absent in the south regions [6, 16]. A comparison of species lists of copepoda from the Adriatic and the Ionian Seas (including the Syrtis Sea), and between the Levant and Aegean Seas revealed that 57% - 67% ofspecies were common to each pair of seas [5, 8].

The composition of zooplankton greatly differs between regions of neritic waters and of deep-water regions. The latter are inhabited basically by the oceanic species. Of the Copepoda, multiple species of the families Eucalanidae, Calocalanidae, Spinocalanidae, Aetideidae, Euchaetidae, Phaennidae, Scolecithricidae, Metridiidae, Lucicutiidae, Heterorhabdidae, Augaptilidae are represented. Many species of the families Calanidae, Clausocalanidae, Centropagidae, Pontellidae, Oncaeidae, Sapphirinidae, Corycaeidae are present in equal numbers in the open sea and in the neritic area. The main species found in the neritic area are members of the families Paracalanidae (*Paracalanus parvus, P. denudatus, P. nanus*), Centropagidae (*Centropages ponticus*), Temoridae (*Temora stylifera, T. longicornis*), Pontellidae (*Labidocera brunescens, L. madurae*), Acartiidae (*Acartia discaudata, A. longiremis, A.latisetosa* and others), Oithonidae (*Oithona nana, O. hebes*) [8].

Deep-water regions of the sea like open regions of the oceanic tropics and subtropics are characterised by a high number of species, but a low abundance of organisms. In the Ionian Sea (grid position 36$^{0}$ 45' N; 18$^{0}$ 45' E) a series of observation at four stations were recorded in December, 1972 and April, 1973 (27th cruise of RV "Mikhail Lomonosov") and in April and August, 1976 (30th cruise RV "Mikhail Lomonosov") and findings were made of 188 copepod species. From these only 8 species were found in numbers exceeding 10 specimens/m3 [21b]. However, near the Egyptian coasts in depths of less than 110 m (120 samples) 132 species of copepoda were discovered [22]. At the same time from the entire Levantine Sea 171 species were known [5]. Nearer the coast, where the influence of Nile is strong, only 31 copepod species were occuring [23]. Among them were semi-oceanic or even oceanic species. Zooplankton in the coastal zone and in bays. is characterised by a low species number. Near Alexandria and in the Abu Qir bay, in depths of less then 15 m, 13 basically neritic species of Copepoda were noted before the Nile overflow [24].

Cladocera (*Evadne tergestina, E. nordmani, E. spinifera*) and other inhabitants of water with low salinity were quite numerous. During the Nile overflow when the salinity of the uppermost ten metre layer was reduced to 3-10%$_0$, the amount of planktonic species had decreased. The same changes in plankton composition between the open sea and the coastal zone and between bays and harbours are known from the literature for other regions of the eastern and western Mediterranean Sea [6, 25, 26, 27, 28, 29, 30]. A few representatives of the fresh-water fauna were noted only near estuaries of large rivers and in desalinised lagoons. It is in the plankton of the deep-water of the eastern Mediterranean that species of Copepoda are most numerous (Table 2).

TABLE 2. Number of copepod species in different seas of the eastern Mediterranean [8].

| Seas | Number of species |
|---|---|
| Adriatic | 234 |
| Ionian and Syrtis | 236 |
| Levantine | 171 |
| Aegean | 161 |
| Marmara | 53 |

Copepoda account for 70-90% of the total zooplankton abundance. Of the copepods, in the surface 50 m layer, *Paracalanus aculeatus, P. parvus, P. nanus, Calocalanus plumulosus, Temora stylifera, Centropages typicus, Acartia negligens, Oithona plumifera* dominate. In the layer 0-200 m *Calanus minor, C. tenuicornis, Calocalanus pavoninus, Clausocalanus furcatus, Centropages violaceus, Oithona setigera* and others were frequently observed. At the depths greater than 200 m, the species *Spinocalanus abyssalis, Aetideus armatus, Chiridius poppei, Scolicithricella abyssalis* were seen. In the surface layers Appendicularia, Ostracoda, Chaetognatha were rather abundant.

From the other groups a few specimens of Foraminifera, Tintinnoinea and Hydromedusae were found. Siphonophora, Polychaeta and Mollusca were also numerous. Of the Siphonophora, *Eudoxoides spiralis, Lensia subtilis, L.conoidea, Hippopodius hippopus, Abilopsis tetragona, Chelophies appendiculata* and others were found [6, 16, 31, 32, 33, 34].

On the shelf region adjacent to the eastern coasts of the Mediterranean Sea, unlike the region of Nile, zooplankton in different areas was distinguished with a higher number of species. For instance, in the region of Beirut during 25 years of studies 173 species of copepods were registered [35]. In Lattakia port (Syria) and in the estuary at three stations studied in the period from March to October, 1991, 54 species of copepods were registered [36].

From five stations sampled four times during 1983-1984 (August, November, February, May) around the northwest coast of Island of Rhodes (from a depth profile of 50-350 m), 85 copepod species were registered [37]. This high number of species in this region is probably related to the fact that the east branch of the North-African current flows past close to the coast causing upwelling. This current brings the Atlantic zooplankton there. In the zooplankton composition there are also species from the

Indian ocean, which are transported into the Mediterranean Sea through the Suez channel.

### 3.2. DISTRIBUTION OF ABUNDANCE AND BIOMASS OF ZOOPLANKTON

Analysis of the works on zooplankton of the deep eastern Mediterranean [7, 8, 13, 14, 16, 19, 20, 21b, 31, 32, 33, 34, 38, 39] shows that most of this region is characterised by a comparatively even distribution of abundance and biomass of zooplankton. Their values were not high. High values for zooplankton abundance were registered in the Adriatic and Aegean seas (Table 3).

TABLE 3. Average values of the number (ind./m$^3$) and biomass (mg/m$^3$) of zooplankton in the eastern Mediterranean Sea in the 0-200 m layer in summer [16].

| Seas | Number (ind./m$^3$) | Biomass (mg/m$^3$) |
|---|---|---|
| Mid and southern Adriatic | 1724 | 56 |
| Ionian | 1041 | 33 |
| Syrtis | 1007 | 15 |
| Levantine | 1438 | 18 |
| Aegean | 1032 | 23 |

As some authors consider [17], the eastern Mediterranean can be distinguish from the western region by zooplankton biomass values. However the results of our studies on mesozooplankton agree with the conclusion of Jespersen [12] that there is a significant reduction in biomass of macrozooplankton from the west to the east in the Mediterranean Sea. As it is well known, annual and seasonal variability in the number and biomass of zooplankton are great. So the most reliable material in the evaluation of spatial distribution are those sampled from different areas with close time intervals. At the time of such observations in 1972 from daily sampled stations, the average quantity of zooplankton in the Sardinian Sea was nearly twice of the Ionian Sea [21b]. More significant are the samples of the 90th cruise of the RV "Akademik A.Kovalevsky" (September-October, 1980), collected from the four different areas from Alboran to the Aegean Sea (Table 4).

The reduction in the number and biomass of zooplankton along the North-African current from the west to the east is also observed from results of investigations carried out in different years (Table 5).

A reduction in the number and biomass of mesozooplankton from the Tunisian Strait to the Ionian Sea and an increase in the Levantine Sea was observed during the research program POEM-BS-091 in 1991 [13, 14]. Thereby, the trend of changes in the amount of zooplankton from a western to an eastern directon seems conclusive. Such a reduction in the biomass of zooplankton complies with the reduction in nutrient concentrations and phytoplankton [7]. In the Levantine Sea the amounts of zooplankton were higher than in the Syrtis Sea [16, 34]. It may be connected with the complex dynamics of water in the Levantine Sea. It is characterised by the existence of three cyclonic and one anticyclonic large-scale cycles [3] and connected with them are areas of upwelling and downwelling [13]. In the southern Levantine Sea there occurs an influence of the Nile on the zooplankton. After the construction of the Aswan dam in 1965 this influence greatly decreased [40, 41]. In the Levantine Sea high concentrations of nutrients and nannoplankton were also noted [42].

TABLE 4. Changes in the number (ind./m$^3$) and biomass (mg/m$^3$) of net zooplankton from the west to the east of the Mediterranean Sea in September-October, 1980 in 0-100 m layer [8].

| Region | Number of observations | Number (ind./m$^3$) | Biomass (mg/m$^3$) |
|---|---|---|---|
| Alboran Sea | 19 | 1952 | 70.4 |
| Sardinian Sea | 6 | 2293 | 30.6 |
| Tunisian strait | 6 | 1050 | 27.0 |
| Aegean Sea (middle) | 6 | 867 | 22.0 |

TABLE 5. Average values of number (ind./m$^3$) and biomass (mg/m$^3$) of net zooplankton in the area of the North-African current in the Mediterranean Sea during summer [8].

| Regions, years | Number of observation | 0-100 m | | 100-200 m | |
|---|---|---|---|---|---|
| | | ind./m$^3$ | mg/m$^3$ | ind./m$^3$ | mg/m$^3$ |
| Alboran Sea (1970, 1974, 1980) | 34 | 1597 | 52 | 624 | 20 |
| Sardinian Sea (1976, 1980) | 18 | 1650 | 43 | 360 | 13 |
| Tunisian strait(1972, 1980) | 26 | 1005 | 28 | 282 | 11 |
| Syrtis Sea (1959) | 3 | 1265 | 17 | 735 | 9 |
| Levantine Sea(1959, 1960) | 7 | 1826 | 26 | 612 | 13 |

Other interesting changes affecting zooplankton biomass occur from north to south in two separate chain patterns of the seas: Adriatic-Ionian-Syrtis and Marmara-Aegean-Levantine Seas. Both chains in the north are characterised by the strong influence of river input, high dynamic activity and essential seasonal changes of the environment. This stimulates high productivity in the northern seas with a gradual reduction in productivity towards the south. As a result the number and biomass of zooplankton regularly and greatly decreases from the north to the south (Table 6, 7).

In the northernmost area of the Adriatic Sea, the number of zooplankton in the 0-25 m layer reaches 13000 specimen/m$^3$, biomass 300 mg/m$^3$ [16]. However, in a southward direction these values greatly decreased (Table 6). But along the coast of the Appenin peninsula they remain rather high.

The number and biomass of zooplankton in the Sea of Marmara were also very high (Table 7) [21]. Mass organisms seen were those typical for the Black Sea: Tintinnoidea, Hydromedusae, Appendicularia, larvae of Mollusca and Polychaeta, copepods *Paracalanus parvus, Acartia clausi,* and *Oithona similis,* etc.

TABLE 6. Changes in the average values in the number (ind./m$^3$) and biomass (mg/m$^3$) of zooplankton from the Adriatic to the Syrtis Sea during summer [8].

| Regions, years | Number of observations | 0-100 m | | 100-200 m | |
|---|---|---|---|---|---|
| | | ind./m$^3$ | mg/m$^3$ | ind./m$^3$ | mg/m$^3$ |
| Adriatic Sea, northern (1960, 1982)* | 25 | 3354 | 90 | - | - |
| Adriatic Sea, middle and southern (1958, 1963) | 47 | 1735 | 58 | 1464 | 54 |
| Ionian Sea (1959, 1976) | 18 | 1104 | 41 | 547 | 15 |
| Syrtis Sea (1959) | 3 | 1265 | 17 | 735 | 9 |

*[43]

TABLE 7. Changes in average values of the number (ind./m$^3$) and biomass (mg/m$^3$) of the net zooplankton from the Marmara to the Levantine Sea during summer and autumn [8].

| Regions, years | Number of observations | 0-100 m | |
|---|---|---|---|
| | | ind./m$^3$ | mg/m$^3$ |
| Sea of Marmara (winter 1969-1970) | 5 | 12000 | 90 |
| Aegean Sea, northern (1958) | 6 | 1473 | 47 |
| Aegean Sea (1958, 1960, 1980) | 17 | 1057 | 26.4 |
| Levantine Sea (1959, 1960) | 7 | 1826 | 26 |

In the Aegean Sea, the number and biomass of zooplankton decreased from the north to the south of the sea five fold during the 1959, 1960 and 1980 IBSS cruises. It is neccessary to note that in the Aegean Sea in the 0-100 m layer, despite low abundance, the biomass was as high as in the Levantine Sea (Table 7). This is explained by the fact that in the 0-100 m layer in the northern Aegean Sea the biomass increases due to large organisms, in particular gelatinous ones: Salpae, Medusae, and Siphonophora. It is known that in other regions of the World's oceans the biomass of the gelatinous organisms also increases with eutrophication of the water [44].

TABLE 8. Quantitative indices of zooplankton in some coastal and deep-water regions of the Mediterranean Sea in the 0-50 m layer [8].

| Regions | Number (ind./m$^3$) | Biomass (mg/m$^3$) | Reference |
|---|---|---|---|
| Coastal regions (average annual values) | | | |
| Triest harbour | 2560 | - | 45 |
| Region of Alexandria | | | |
| Easern harbour | 29700 | 191.0 | 46 |
| Coastal zone (<50m) | 7400 | - | 47 |
| Neritic zone (<20m) | 1850 | - | 47 |
| Neritic zone (>200m) | 1230 | - | 47 |
| Region of Beirut | 5000 | - | 48 |
| Near the sewage outputs | 175 | - | 48 |
| Saronicos harbour | 1287 | 21.0 - 65.0 | 49 |
| ibid | 9224 | - | 4 |
| ibid, 25km from sewage output | - | 135.7 | 50 |
| Elefsis harbour | - | 195 - 265 | 49 |
| Deep-water regions (in different seasons) | | | |
| Mid and southern Adriatic (III-IX. 1958, 1963, 1982) | - | 54.0 | 8, 16 |
| Tunisian strait (IX. 1980) | 794 | 26.6 | 8 |
| Ionian Sea (IV,VIII,XII. 1972,1973,1976) | 651 | 29. 5 | 8 |
| Syrtis Sea (summer, 1959) | 1108 | 19.7 | 34 |
| Levantine Sea (IX-XI. 1959, 1960) | 1419 | 19.6 | 34 |
| Aegean Sea (VIII. 1980) | 756 | 23.1 | 8 |
| Aegean Sea, south (VI-I. 1958, 1960, 1961) | 903 | 14.1 | 33 |
| Aegean Sea, north (summer, 1958) | 1473 | 47.0 | 33 |

Note: The plankton net with a mesh size of 200-250 μ which was used in the coastal areas, does not catch organisms less than 0.5 mm length. So the data obtained by the Juday net with a mesh size of 112 μ in the deep-water regions was reduced by 22% for number and by 6% for the biomass prior to tabulation. This corresponds to their average share in the samples of the net with mesh size 112 μ [8].

The number and biomass of zooplankton in coastal and deep-water regions of the sea are often greatly distinguished (Table 8). In coastal regions they are characterised by higher values. The main reason for this is the eutrophication of coastal waters by the continental shelf runoff.

## 4. Conclusions

The results of this study allow us to define the composition and large-scale distribution of zooplankton in the eastern Mediterranean. During the course of investigations reviewed here, the species list of zooplankton (mainly copepods) was greatly enriched.

The regular reduction in the quantity and biomass of zooplankton seen from the west to the east of the Mediterranean was shown. This is connected with an impoverishment of nutrients due to the Atlantic water. The same phenomenon was observed in the direction from the north towards the south. Waters of the northern Adriatic and the Sea of Marmara, which are rich in nutrients due to river inflow flow southward, mixing with nutrient-poor water and therefore lose their abundance of nutrients. This is accompanied by a reduced amount of planktonic organisms.

It is also shown that the deep-water regions are richer in terms of taxonomic composition but are poorer with respect to biomass and quantity of zooplankton in comparison with the coastal regions which are under the influence of continental runoff.

## 5. Acknowledgements

The present investigation was carried out with the support ofa NATO Linkage Grant. We thank to Mrs Alison M. Kideys for improving the English of the text.

## 6. References

1. Goncharov, V. P., and Mikhailov, V. V. (1963) New data on bottom relief of the Mediterranean Sea, *Oceanology*, **3**, 6, 1056-1060. (in Russian)
2. Kiortsis, V. (1985) Mediterranean marine ecosystems: establishment of zooplankton communities in transitional and partly isolated areas, in M. Moriatou-Apostoloupoulou and V. Kiortsis (eds.), *Mediterranean marine ecosystems*, *pp.* 377-385.
3. Ovchinnikov, I.M., Plakhin, E. A., and Moskalenko, L. V. (1976) *Hydrology of the Mediterranean Sea*, Hydrometeoizdat, Moskow, 375 p. (in Russian)
4. Moraitou-Apostolopoulou, M. (1981) The annual cycle of zooplankton in Elefsis Bay (Greece), *Rapp. et proc.-verb.reun.Commis. int. explor.sci.Mer Mediterr.*, Monaco, **27**, fasc.7, 105-106.
5. Kovalev, A. V. and Shmeleva, A. A. (1982) Fauna of Copepoda in the Mediterranean Sea, *Ecologiya morya*, **8**, 82-87. (in Russian)
6. Furnestin, M. -L. (1979) Aspects of the zoogeography of the Mediterranean plankton, in S. Van der Spoel and A.C. Pierrot-Bults (eds.), *Zoogeography and diversity in plankton,* pp. 191-253.

7. Greze, V. N. (1989) *Pelagial of the Mediterranean Sea as ecological system,* Naukova dumka, Kiev, 198 p. (in Russian)
8. Kovalev, A. V. (1991) *Structure of the zooplankton communities of the Atlantic Ocean and the Mediterranean basin,* Naukova dumka, Kiev, 141 p. (in Russian)
9. Thomsen, H. (1931) Nitrate and phosphate contens of Mediterranean water, *Rep. Dan. oceanogr. exp. 1908-1910 to Mediterr. and adjacent seas.* Copenhagen, **3**, **6**, 1-14.
10. McGill, D. A.. (1965) The relatives supplies of phosphate, nitrate and silicate in the Mediterranean Sea, *Rapp. et proc.-verb.reun.Commis. int. explor.sci.Mer Mediterr.*, Monaco, **18**, fasc.3, 737-744.
11. Egorova, V. A. (1966) Phosphates in waters of the Mediterranean Sea, *Chemical processes in the seas and oceans,* Moskow, Nauka, 102-110. (in Russian)
12. Jespersen, P. (1923) On the quantity of macroplankton in the Mediterranean and Atlantic, *Rep.Dan. Ocean.Exp. 1908-1910 to Mediterr. and adjacent seas,* Copenhagen, **3**, 1-17.
13. Mazzocchi, M. G., Christou, E. D., Fragopoulu, N., and Siokou-Frangou, I. (1997) Mesozooplankton distribution from Sicily to Cyprus (Eastern Mediterranean): I. General aspects, *Oceanologica acta,* **20**, 521-535.
14. Siokou-Frangou, I., Christou, E. D., Fragopoulu, N., and Mazzocchi, M. G. (1997) Mesozooplankton distribution from Sicily to Cyprus (Eastern Mediterranean): II Copepod assamblages, *Oceanologica acta*, **20**, 537-548.
15. Por, F. D. (1978) *Lessepsian migration,* Springer-Verlag, Berlin, 228 p.
16. Greze, V. N., Pavlova, E. V., Shmeleva, A. A., and Delalo, E. P. (1982) Zooplankton of the eastern Mediterranean and its quantitative distribution, *Ekologiya morya,* **8**, 37-46. (in Russian)
17. Moraitou-Apostolopoulou, M. (1985) The zooplankton communities of eastern Mediterranean (Levantine Bassin, Aegean Sea): influence of man-made factors, in M. Moraitou-Apostolopoulou and V. Kiortsis (eds.), *Mediterranean marine ecosystems,* 303-331.
18. Kovalev, A. V., Bileva, O. K., and Moryakova, V. K. (1977) Combined method of collecting and registration of marine zooplankton, *Biologiya morya,* **4**, 78-82. (in Russian)
19. Vodyanitsky, V. A. (1961) Some results of investigations of A.O. Kovalevsky Sevastopol station in the Mediterranean Sea in 1958-1960, *Oceanologiya,* **1**, **5**, 791-804. (in Russian)
20. Kovalev, A. V. (1971) Distribution of seston in the Tunisian strait, Ionic and Aegean seas, in V. A. Vodyanitsky (ed.), *Oceanographic investigations in the Tunisian strait,* Naukova dumka, Kiev, pp. 91-100. (in Russian)
21. Kovalev, A. V., Georgieva, L. V., and Baldina, E. P. (1976) Influence of water mass exchange from Bosphorus on the content and distribution of the plankton in the nearest seas, in V. I. Belyaev (ed.), *Investigation of water mass exchange from Tunis Chanel and Bosphorus,* Naukova dumka, Kiev, pp. 181-189. (in Russian)
21b. Bileva, O. K., Greze, V. N., Kovalev, A. V., Moryakova, V. K., and Skryabin, V. A. (1982) Comparative characterization and biological structure of zooplankton in the Ionian and Sardinian seas, *Ecologiya morya ,* **8**, 46-55. (in Russian)

22. Dowidar, N. M. and El-Maghraby, A. M. (1973) Notes on the occurence and distribution of some zooplankton species in the Mediterranean waters of UAR, *Rapp. et proc.-verb.reun.Commis. int. explor.sci.Mer Mediterr.*, Monaco, **21**, fasc.8, 521-525.
23. Salah, A. M. (1971) A preliminary check list of the plankton along the Egyptian Mediterranean coast, *Rapp. et proc.-verb.reun.Commis. int. explor.sci.Mer Mediterr. Monaco*, **20**, fasc.3, 317-322.
24. Dowidar, N. M. and El-Maghraby, A.M. (1971) Observation on the neritic zooplankton community in Abu Qir Bay during the flood season, *Rapp. et proc.-verb.reun.Commis.int.explor.sci.Mer Mediterr.*, Monaco. **20**, fasc.3, 385-389.
25. Lakkis, S. (1974) Considerations on the distribution of pelagic copepods in the eastern Mediterranean of the coast of Lebanon, *Acta Adriatica*, **18**, 39-52.
26. Kimor, B. and Wood, E. J. F. (1975) A plankton study in the eastern Mediterranean Sea, *Marine biology*, **29**, 321-333.
27. Pasteur, R., Berdugo, V., and Kimor, B. (1976) The abundance, composition and seasonal distribution of epizooplankton in coastal and offshore waters of the eastern Mediterranean, *Acta Adriatica*, **18(4)**, 55-80.
28. Regner, D. (1977) Investigations of copepods in coastal areas of Split and Sibenic, *Acta Adriatica*, **17(12)**, 3-19.
29. Specchi, M. and Fonda-Umani S. (1983) La communaute neritique de la region des emboushures du Po, *Rapp. et proc.-verb.reun.Commis. int. explor.sci.Mer Mediterr.*, Monaco, **28**, fasc.9, 197-199.
30. Siokou-Frangou, I. and Christou, E.D. (1995) Variabilite du cycle annuel du zooplancton dans la Baie d'Elefsis (Grece), *Rapp. du XXXIV congr. de la CIESM*, **34**, 217.
31. Greze, V. N. (1963) Zooplankton of the Ionian Sea, *Oceanologicheskiye Issledovaniya*, **9**, 42-59. (in Russian)
32. Shmeleva, A. A. (1963) State of food base of fish in South Adriatic in 1958, *Proceed. of Sevastopol Biological Station*, **16**, 138-152. (in Russian)
33. Pavlova, E. V. (1966) Composition and distribution of zooplankton in the Aegean Sea, *Investigations of plankton of the South Seas*, **7**, 38-61. (in Russian)
34. Delalo, E. P. (1966) Zooplankton in Eastern part of the Mediterranean Sea (of seas Levantine and Syrtis), *Investigation of Plankton in the Southern Seas* , Moscow, **7**, 62-81 (in Russian)
35. Lakkis, S. (1995) Cycle annuel du plancton cotier du Liban. Successions et variations saisonnieres des peuplements, *Rapp. du XXXIV congr. de la CIESM*, **34,** 212.
36. Baker, M. ( 1995) Contribution a l'etude de la biomasse zooplanctonique dans les eaux cotieres de la Lattaque (Syrie), R*app. du XXXIV congr. de la CIESM.* **34,** 203.
37. Siokou-Frangou, I. and Papathanassiou, E. (1989) Aspects du zooplancton cotier de l'ile de Rhodes (Mer Egee), *Vie Milieu*, **39(2)**, 77-85.
38. Kovalev, A. V. (1982) Zooplankton distribution in different water masses in the Tunisian strait, *Ecologiya morya*, **9**, 15-19. (in Russian)
39. Moryakova, V. K., Zaika, V. E., Bityukov, E. P., and Vasilenko, V.I. (1975) Distribution of planktonic animals and bioluminescence in Mediterranean Sea, in

V.E.Zaika (ed.), *Biological structure and productivity of planktonic communities of Mediterranean Sea*, Naukova dumka, Kiev, pp. 75-102. (in Russian)

40. Oren, O. H. (1969) Oceanographical and biological influenses of the Suez canal, the Nile and the Aswan dam on the Levant Basin, *Progr.Oceanogr.*, **5**, 161-167.
41. Wadie, W. F. (1984) The effect of regulation of the Nile river discharge on the oceanographic condition and productivity of the South-eastern part of the Mediterranean Sea, *Acta Adriatica*, **25**, **1/2**, 29-43.
42. Bernard, F. ( 1969) Distribution verticale des sels nutritifs et du phytoplancton en mer Mediterranee: essai sur l'epaisseur de la couche a photosynthese, *Rapp.et proc.-verb.reun.Commis.int.explor.sci.Mer Mediterr.*, Monaco, **15**, fasc.2, 283-292.
43. Greze, V. N., Bileva, O. K., Kovalev, A. V., and Shmeleva, A. A. (1985) Size and trophic structure of zooplankton in the Mediterranean Sea. *Biologiya morya*, **6**, 12-18. (in Russian)
44. Kovalev A. V, Niermann, U., Melnikov, V. V., Belokopytov, V., Uysal, Z., Kideys, A. E., Ünsal, M., and Altukhov, D. (1998) Long-term changes in the Black Sea zooplankton: the role of natural and anthropogenic factors, in L. Ivanov and T. Oguz (eds.), *NATO TU Black Sea Assessment Workshop: NATO TU-Black Sea Project: Symposium on Scientific Results*, Vol. 1, Kluwer Academic Publishers, Dordrecht, pp. 221-234.
45. Specchi, M., Fonda Umani, S., and Radini, G. (1981) Les fluctuations du zooplancton dans une station fixe du golfe de Trieste (Haute Adriatique), *Rapp. et proc.-verb.reun.Commis. int. explor.sci.Mer Mediterr.*, Monaco, **27**, fasc.7, 97-100.
46. Khalil, A. N., El-Maghraby, A. M., Dowidar N. M., and El-Zawawy, D. A. (1983) Seasonal variations of the Eastern Harbour of Alexandria, Egypt. *Rapp. et proc.-verb.reun.Commis. int. explor.sci.Mer Mediterr.*, Monaco, **28**, fasc.9, 217-218.
47. El-Maghraby, A. M. and Dowidar, N. M. (1973) Observations on the zooplankton community in the Egiptian Mediterranean waters, *Rapp. et proc.-verb.reun.Commis. int. explor.sci.Mer Mediterr.*, Monaco, **21**, fasc.8, 521-525.
48. Lakkis, S., and Kouyoumjian, H. (1974) Observations sur la composition et l'abondance du zooplancton aux embouchures d'effluents urbaince des eaux de Beyrouth, *Rapp. et proc.-verb.reun.Commis. int. explor.sci.Mer Mediterr.*, Monaco, **22**, fasc.9, 107-108.
49. Jannopoulos, C. (1976) The annual regeneration of the Elefsis bay zooplanktonic ecosystem, Saronicos gulf, *Rapp. et proc.-verb.reun.Commis. int. explor.sci.Mer Mediterr.*, Monaco, **23**, fasc.9, 109-111.
50. Verriopoulos, G., Moraitou-Apostolopoulou, M., and Hadzini-Kolaou, S. (1985) Quantitative and qualitative composition of zooplankton in the areas of Saronicos gulf, *Rapp. et proc.-verb.reun.Commis. int. explor.sci.Mer Mediterr.*, Monaco, **29**, fasc.9, 307-308.

# PHYSIOLOGICAL PARAMETERS AND PROBLEMS OF ENERGY BUDGET ESTIMATION IN MEDITERRANEAN AND BLACK SEA FISHES

G.E. SHULMAN
*The A.O. Kovalevsky Institute of Biology of the Southern Seas, National Academy of Sciences of Ukraine, 2.Nakhimov Ave., Sevastopol 335011,Crimea, Ukraine*

**Abstract**

During several expeditions in Mediterranean in 1970-1980s on the scientific vessel «Akademic Kovalevsky» (Institute of the Biology of the Southern Seas, Sevastopol) some parameters of energy metabolism of fishes were studied. It was shown that fat (triacylglicerol) accumulation, their utilization for energy metabolism, effect of food supply on these processes in Mediterranean fishes have features of similarity and difference with Black Sea ones. Principal scheme of energy balance for Black Sea fishes may be used for Mediterranean ones. At the exumple of horse mackerel Trachurus mediterranus energy consumption, utilization, expenditure for growth and metabolism was analysed. Intensity and efficiency of production formation was studied. Such investigation has perspectives for estination of energy and substance balance (budjet) not only for mass species of fishes but invertebrate animals of both water bodies too.



*P. Malanotte-Rizzoli and V.N. Eremeev (eds.),*
***The Eastern Mediterranean as a Laboratory Basin for the Assessment of Contrasting Ecosystems*, 97–112.**

## 1. Introduction

Department of Animal Physiology, organized in 1959 at Institute of Biology of the Southern Seas, has research interests focused primarily on energy and substance balance in mass species of marine organisms inhabiting the Southern seas [1-10]. Progress of the study implies the knowledge of physiological and biochemical aspects of species adaptations and life history, and the indication of organism and population condition . The main body of studies were conducted in the Black and Azov seas, while in the Mediterranean they were conducted sporadically during the cruises of the R/V Akademik Kovalevsky in 1969, 1971, 1973, 1976, 1977 and 1988. Obtained data have been partly published in a number of papers [4, 11-18]. The main investigation objects were fishes.

## 2. Results and discussion

Results of the compaj rison between physiological and biochemical features of fish from the Mediterranean, Black and Azov Seas are of considerable interest. For instance, examination of weight growth and energy accumulation in three races of anchovy, the most abundant pelagic fish (*Engraulis encrasicholus mediterraneus, E.e. ponticus and E.e. maeoticus*) has shown that these processes differ markedly between the races [12]. It should be noted that characteristics describing protein retention in the body, i.e. the protein growth, are the most reliable physiological parameters of weight increment, while the level of fatness is similarly reliable parameter of energy store. In Azov anchovy the accumulation of fat prevails over protein increase; in Black Sea anchovy the two processes keep balance; and in their Mediterranean counterpart it is protein increase that prevails. Our tentative hypothesis explained the specific metabolic responses of compared races by the difference in eating behaviour depending on availability of nutriment in the sea [12]. Indeed, mean biomass of zooplankton is 500 mg/m$^3$ in the Azov Sea, 100 mg/m$^3$ in the Black Sea and only 50 mg/m$^3$ in the Mediterranean [4]. We suggested, that owing to more abundant food Azov anchovy prey with less effort, and hence less energy expenditure, than Black Sea and Mediterranean anchovy do. Neutral lipids (triacylglycerols or fats) are the basic source of energy for pelagic fish, like anchovy. That is, Azov anchovy use up fat stored in the body most Mediterranean anchovy less efficiently. As a result, the latter cannot utilise as much food for storing up body fat as the former do. The above stated applies not to anchovy only, but also to many pelagic fishes of compared and other seas, e.g. the Caspian and Baltic seas [4]. Protein growth and increment are in opposition to fat accumulation: intensification of the latter by means of the endocrine system inhibits the former, and vice versa [19]. This results in smaller length of Azov anchovy (70-80 mm, mid-age groups) in comparison with Black Sea (90-100 mm) and Mediterranean (110-120) races.

Another point worthy of consideration is the relation between fat accumulation and the food supply of the fish. The high nutritive base of plankton-eating fishes is formed owing to nutrient-rich run-off of rivers which enter the sea and and provides the high primary and secondary production and hence the biomass of phyto- and zooplankton. This explains why anchovy from the eastern Azov Sea has the greatest fat content, as into this part of the sea the Don and the Kuban carry their water [4]. Similarly, the fatness of anchovy and sprat is highest in the northwestern Black Sea which receives waters of the Danube, the Dniper and the Dniester [10, 20, 21]. The effect of river run-off is distinctly manifested in the Mediterranean Sea too. Fig. 1 gives data on the fatness of fry of the red mullet *Mullus barbatus* estimated in different areas of the Adriatic Sea [11].

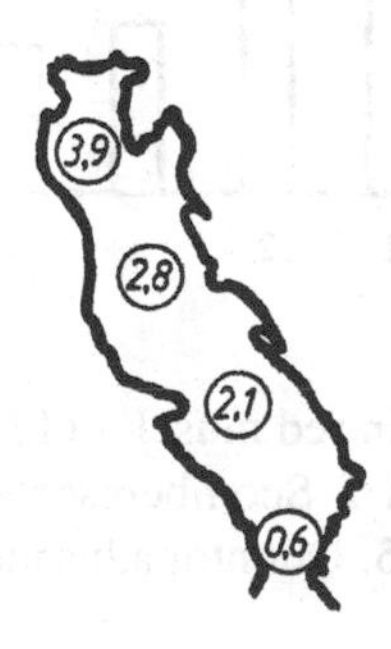


*Figure. 1.* Fat content of red mullet fry in Adriatic sea.

The figure clearly shows the gradual increase of fatness from the Otranto Strait towards the Po mouth. Similar trend was found in populations of the sprat *Sprattus sprattus* from the southern Adriatic Sea and the Bay of Venice [11]. These facts indicate the applicability of data about fat (triacylglycerol) content stored in the fish body for assessment of the condition of an organism or a population, and primarily, of the available food supply. These methods are widely used in studies of fish [22] and planktonic crustaceans of the Black and Azov seas [24,25]. Since recently, researches employ the cited parameters at evaluation of the food supply of larvae of the anchovy Engraulis mordax near the Californian coast and larvae of the spat Sprattus sprattus in the North Sea [26-28]. This approach may turn valid for mass species of the Mediterranean Sea. Moreover, establishing a continuous spatiotemporal monitoring of the state of populations of prevailing Mediterranean fishes and invertabrates like that organized on the Black Sea [23] is a timely and pertinent idea.

The importance of neutral lipids as an essential source of energy supporting vital activities of fish is well seen at examination of triacylglycerol content in red muscles of Mediterranean fishes with different motor activity (motile and sedentary) [16]. Fig.2 shows that the content of triacylglycerols is considerably higher in motile fishes in comparison with those of moderate and low swimming performance.

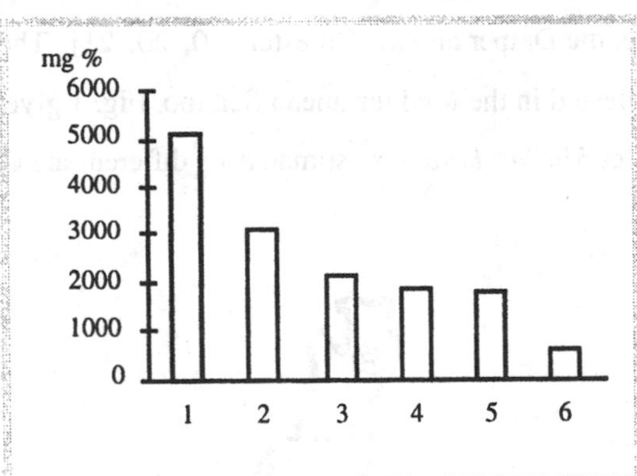


*Figure 2.* Triacylglycerol content in red muscles of Mediterranean fishes.
1. Scomber scombrus. 2 Scomberesox saurus. 3. Trachurus mediterraneus. 4. Diplpdus annularis. 5. Odontogadus merlangus. 6. Scorpaena porcus.

Our recent experiments on Mediterranean picarel show that this fish, classified as moderately motile, uses mostly protein, not fat (triacylglycerols) for satisfying its energy demands [[15]. Researches of our Department obtained similar results in studies conducted on Black Sea fishes [29].

In this paper we do not present all materials obtained about Mediterranean fishes. We intend to outline only those related to the problem of the energy balance (budget).Unexpectedly, the approach to settling this problem turned out a surprise even to us ourselves. We must again consider the difference in the nature of fat accumulation and protein growth in the three races of anchovy exhibit and the hypothesis explaining the phenomenon.

What surprised us most, was that from the sum of protein growth and fat accumulation caloric values it followed that all three races of anchovy have identical energy equivalents [8]. Then, the energy equivalent of food consumption is similar for these races. Therefore, it is only natural to assume that the explanation of the difference in fat accumulation roots not in the availability of food plankton but somewhere else. The next hypothesis we propose explains the difference by the duration of feeding and reproductive periods which, in their turn, depend on temperature conditions in the sea. In the Azov Sea water temperatures tolerated by warm-requiring fishes develop only during the short (May-November) warm season. For this half a year Azov anchovy must grow , maturate, spawn and feed in order to prepare for migration to Black Sea and wintering . That is why the fish has to accumulate such a considerable fat content. Contrary to Azov anchovy, the Mediterranean race does not store up much fat as the sea is warm nearly all year round that permits almost continuous feeding, maturating and spawning seasons. In this case, food consumed is used not for fat accumulation but rather for somatic and generative growth. Black Sea anchovy is intermediate between the Azov and Mediterranean races in examined characteristics. Then we drive to a conclusion that the principal distinction between the three seas compared using total energy equivalent and its protein and lipid components in the warm-requiring pelagic anchovy lies in the temperature of sea water rather than in the abundance of food zooplankton. Therefore, it is more pertinent to refer to the temperature capacity of the sea instead of its nutritive capacity. Then the abundance of food is a contributing, not a determining factor with regard to the specific lipid and protein metabolism. The above cited arguments do not imply greater significance of hydrological conditions for production characteristics of warm-requiring planktivorous fish in comparison with the trophic factor. Apparently the effect of the two factors on fish populations is not simple. Hydrological conditions of sea determine the term of feeding, maturation and spawning of fish and the «direction» of their metabolism. Trophic factor (the extent and density of nutritive base) accounts for specificity of protein and lipid metabolism and their correlation on the one hand, and plays essentially important role in formation of production, biomass and numbers of fish populations, on the other hand. The natural result is that energy potential of allied races and species of fish from the compared seas is similar on the organismic level, as we have shown, but different on the level of population.

Now we switch from the physiological and biochemical features of fish dwelling in the Mediterranean, Black and Azov seas to their energy balance. Problems encountered at estimation of energy balance were studied on mass Black Sea fishes, the anchovy *E.e.* ponticus, the sprat *Sprattus sprattus phalericus*, the horse-mackerel *Trachurus mediterraneus ponticus*, the red mullet *Mullus barbatus ponticus*, the picarel *Spicara flexuosa* and the whiting *Odontogadus merlangus euxinus* [8]. The research design was special at coupling studying of both energy and substance balances in order to trace the paths and varieties of energy accumulation, transformation and use, and to assess the energy cost of production processes – protein growth and fat accumulation in fish. This combinatory approach sets our study apart from the majority of researches in the energy balance which were made at other seas [30-37]. As we stated earlier, energy equivalents of allien varieties of fish almost coincide for the Black and Mediterranean Seas. This allows of employment of a number of parameters obtained on Black Sea fishes at studying the energy balance of Mediterranean fish.

An illustrative example is provided by the horse-mackerel *Trachurus mediterraneus* from the Mediterranean Sea and its Black Sea variety, *Tr.med. ponticus*. Energy potential is similar in these fishes , as our earlier studies showed [4-8]. An essential element of the balance is production, and from it we begin the analysis. Fig.3 presents data on the dynamics of the content of dry matter, protein, fat, glycogen, total inorganic substance and caloric content of the body of horse-mackerel over the annual cycle; the weighted means were calculated taking into account age variability. Fig.4 gives data on weight growth measured in all age groups of the fish.

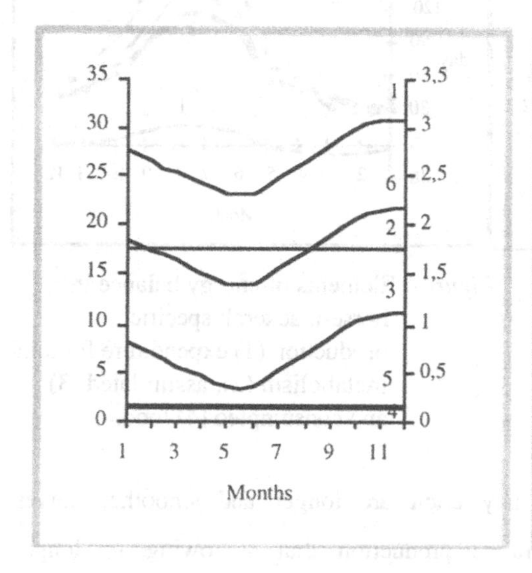


*Figure 3.* Proximate chemical composition of horse-mackerel: dry matter (1), protein (2), fat (3), sum of mineral substances (4), glycogen (5), energy equivalent (6).

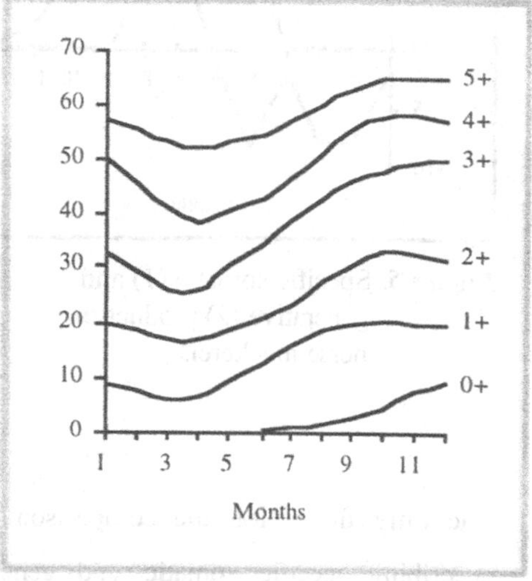


*Figure 4.* Total growth of wet mass of horse mackerel.

Knowing the scope of reproductive products formation during the intermittent spawning [8, 38, 39], it is easy to compute somatic and generative increase for each age group of the horse-mackerel over annual cycle [8]. Then, using the population age composition averages obtained for many years [40] one estimates average population specific somatic and generative production, i.t. estimates of the increase of investigated parameters per unit weight per unit of time. Fig. 5 and 6 shows specific production of horse-mackerel population represented as energy equivalent estimates [8]. Specific production cited in the diagram is in calories per gram of body weight per day. Evaluated per year, the total specific production ($P_{\Sigma}$) is 1.39 Kcal·g$^{-1}$, the somatic ($P_s$) and generative ($P_g$) are 0.48 and 0.91 Kcal·g$^{-1}$, correspondingly. Though these estimates were obtained for the Black Sea horse-mackerel subspecies, they may similarly apply to the Mediterranean horse-mackerel.

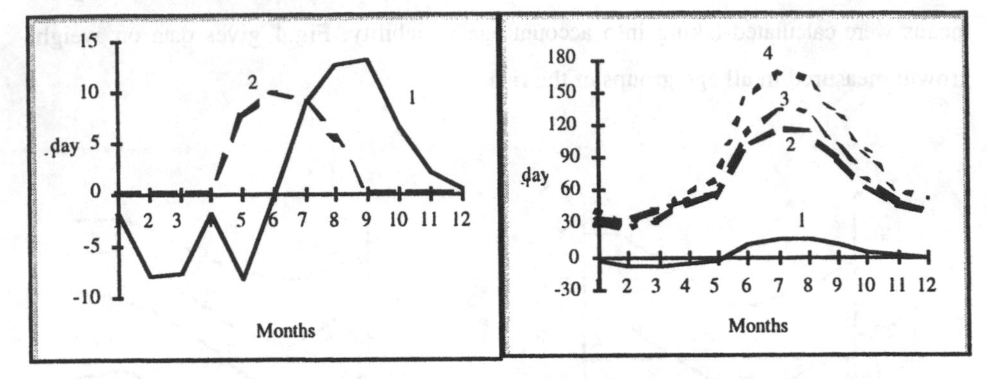


*Figure 5.* Specific somatic (1) and generative (2) production in horse-mackerel.

*Figure 6* Elements of energy balance in horse-mackerel: specific production (1) expenditure for total metabolism (2), assimilated (3) and consumpted (4) food.

The only distinction the comparison may elicit are longer and smoother curves describing specific somatic and generative production that is owing to longer maturation and feeding periods of annual cycle in the Mediterranean horse-mackerel.

Much more difficulty is encountered at comparing estimates of absolute production yielded by Mediterranean and Black Sea horse-mackerel, or other fishes examined. The number and biomass of prevailing pelagic fishes of the Black Sea were assessed regularly and reliably as early as in the 1970s.[41]. For instance, average annual biomass of Black Sea horse-mackerel makes up 36,000 t, with average annual absolute production of the fish being $4.95 \cdot 10^{10}$ Kcal (per population a year) and $P_{\Sigma}$ to B ratio 0.81 Kcal. However, we do not know corresponding data about Mediterranean horse-mackerel, as about regular and reliable estimation of biomass and numbers in mass fishes of Mediterranean Sea , and in particular in local populations of horse-mackerel. Studies in this field are necessary to provide determination of the energy balance of Mediterranean fishes.

Another essential trend at studying energy balance is evaluation of metabolicexpenditures. Researches performed at Department of Animal Physiology,

IBSS [9,42] measured the level of oxygen consumption in large number of Black Sea and Mediterranean fishes, including the two subspecies of the horse-mackerel *Tr. mediterraneus*. It was found that energy metabolism level is comparable in these ones. Carrying out our program, Yu.S. Belokopytin did a series of studies on extending data of the experiments to natural environment. The extrapolation procedure consisted of: 1) measuring oxygen consumption specific rate (intensity) in experiments (at the so-colled ''standard metabolism'') and computing the specific rate for an arbitrary fish with the body weight 1 g (coefficient 'a' in the popular equation $Q=aW^k$; Vinberg, 43); 2) specifying the dependence of this coefficient on temperature; 3) finding out the relationship between coefficient 'a' and temperature of water in which fish species inhabit; 4) determining relationship between the intensity of oxygen uptake and swimming velocity of the fish ('active metabolism'); 5) determining average daily velocity of swimming for fish in natural environment and in experiments; and
6) computing energy metabolism means in natural populations of fishes using the array of all resulting data. Employing results obtained by Yu.S. Belokopytin, we estimated expenditures on total metabolism over annual cycle for six prevailing fish species of the Black Sea, taking age and size structure of the population into [8]. We suppose, that estimates we obtained for Black Sea horse-mackerel are close to those in Mediterranean race.(Fig.6). Our calculations show that annual metabolic expenditures in population of horse-mackerel total 23.45 $Kcal.g^{-1}$.

With estimates of the specific production and specific metabolic expenditures available, specific food consumption can be easily computed.. Using equation C=P + Q + F (43), where C is consumption, P production, Q metabolic expenditures and F unassimilated food, one calculates P + Q which corresponds to assimilated food (A). It is known from literature that assimilated food makes up 80% on the average of food consumed [43, 44], that is, C = 1.25 A. It should be emphasized that the computation method for estimating food consumption in water organisms including fish usually yields results close to estimates obtained from experiments on direct studies of nutrition. Moreover, sometimes the computation method is more valid and reliable because experimental approach entails numerous methodical errors [8, 43, 45, 46]. Fig.6 shows food consumption intensity in horse-mackerel. Annual specific consumption of food in the horse-mackerel population was estimated 31.01 Kcal.g$^{-1}$ (converted to energy).

The performed study elicited important aspects related to food conversion and efficiency of its use for production processes. At first glance, study of the conversion of food items into elements of fish body is a very difficult task. But in fact the matter is not as grave as it seems. It is sufficient to know the numerical value of diet and the chemical composition of food and fish body. Naturally, this applies to terminal, not intermediate, metabolism (the latter is the subject of concern for biochemists). Fig.7 demonstrated how food is converted in horse-mackerel.

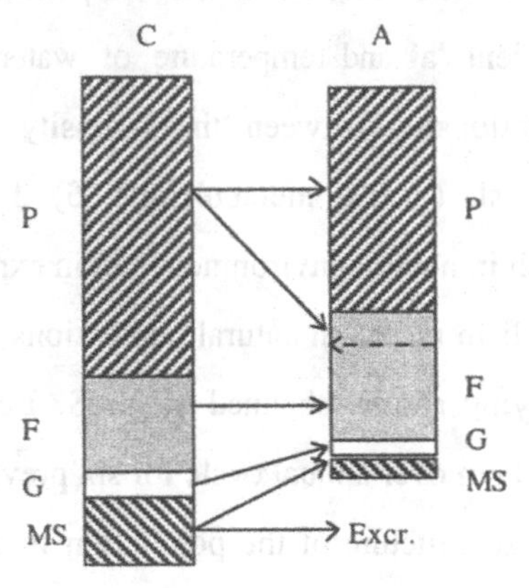


*Figure 7.* Scheme of food transformation in horse-mackerel: C-consumption, A- assimilation, P-protein, F-fat, G-glycogen, MS-mineral substances.

One can see from the diagram that in this fish food assimilated differs from consumed. Protein, glycogen and minerals, compared to the corresponding quantities in the food, considerably reduce when converted into the body tissues and organs, while fat increases drastically that is probably owing to protein conversion. This phenomenon takes place due to neutral lipids which provide active energy metabolism (expenditures for locomotion) and plastic metabolism(expenditures on the maturation of generative products) in the physiology of horse-mackerel.

The efficiency as the intensity of using nutriment for the constructive processes (growth and production) are equally important subjects. The first who addressed the problem of food utilization efficiency in water animals including fish was V.S.Ivlev, a famous Russian scholar [47].According to him the efficiency of food used for constructive processes may be estimated employing coefficients $K_1$ and $K_2$; the former is for food consumed and the latter is for assimilated food. Then, $K_1$=P/C=P/P + Q + F, and $K_2$ = P/A = P/P+Q .

In relevant Russian literature, it is coefficient $K_2$ that is especially widely applied. Table 1 gives estimates of the coefficient $K_2$ for six prevailing fishes of the Black Sea [8].

TABLE 1. Annual pattern of intensity and efficiency of food consumption in fish

| Species | Average annual rations | $K_2$ |
|---|---|---|
| Anchovy | 10.17 | 2.1 |
| Sprat | 6.16 | 4.6 |
| Horse mackerel | 5.43 | 5.6 |
| Red mullet | 1.55 | 25.1 |
| Pickerel | 1.74 | 13..9 |
| Whiting | 2.25 | 12.8 |

And their daily rations (as % of the body energy equivalent). All cited estimates are annual averages. Inspecting these data , one comes across a curious relationship. All the fishes examined form two distinct groups: the first embraces motile fishes, like anchovy, sprat and horde-mackerel, with high (10-20%) daily ration and low (2.5 - 5%) $K_2$. The another group are fishes of moderate and low motility (red mullet, pickerel and whiting) with low (1-4%) ration but high (12-25%) $K_2$. Thus, what we are facing are two opposing nutritional strategies. One of them is based on intensive consuming on food and the other on highly efficient consumption. . Here we give only o corresponding example obtained for a number of Mediterranean fishes by L.V.Tochilina and Yu.S. Belokopytin [48]. They reveald that fishes of the Mediterranean differ both in innate motility and the content of haemoglobin, the principal respiratory pigment of the blood. The difference is only natural as high rates of the energy metabolism in motile fishes require the efficient system of oxygen transport into tissues. What is not as natural as that, is the inverse proportion of white blood cells (leucocytes) to haemoglobin content: low motile fishes have highest content of white blood cells. The only sensible explanation to this strange phenomenon is that leukocytes provide more efficient assimilation of food consumed by the fish. Therefore, $K_2$ yields higher estimates in low motile fishes.

The totality of cited results and deductions allow to have an integrated insight into all components of the energy balance of the examined fishes. Figs 6 and 8 provide an illustrative example. The first shows the dynamics of these components over total annual cycle of horse-mackerel, and the second annual average of the energy for the fish population.

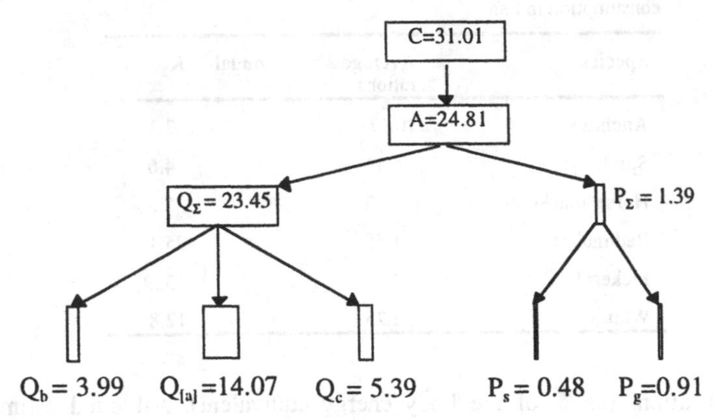


*Figure 8.* Annual flow of energy through the population of horse-mackerel.

In conclusion, it may be claimed that the knowledge we have accumulated about Black Sea fishes is quite applicable at studying the energy and substance balance in populations of Mediterranean fishes. Both earlier and recent data which have been gained at research institutes and laboratories of Mediterranean countries would promote the proposed comparative analysis. The only stumbling block on the way may be measuring and monitoring of the numbers and and biomass of mass species and mapping of local fish populations. Joint research programme and cruises of concerned Mediterranean and Black Sea countries would be to the progress of studies of the food balance and trophic dynamics in marine organisms of the Mediterranean. We believe, this approach will result in creation of models related to functioning of Mediterranean

**marine ecosystems based on both theoretical and empirical data, because only knowledge of facts gives the key to settling of complicated scientific problems.**

## 3. References


1. Ivlev, V.S. (1966) Elements of physiological hydrobiology (in Russ.), *in: Physiology of Marine Animals*, Nauka Publishing House, Moscow, pp. 3-45.
2. Sushchenya, L.M. (1972) *Intensity of Respiration in Crustacea* (in Russ.), Naukova dumka Publishing House, Kiev.
3. Sushchenya, L.M. (1975) *Quantitative Regularities of Crustacean Nutrition* (in Russ.), Nauka i Tekhnika, Minsk.
4. Shulman, G..E. (1972) *Physiologo-biochemical Features of AnnualCcycles ofFfish* (in Russ.), Pishchevaya promyshlennost, Moscow, (Translated into English and published as: Life cycles of Fish. Physiology and biochemistry.(1974) John Wiley and Sons, N.Y.-Toronto)
5. Khmeleva, N.N. (1973) *Biology and Energy Balance of MarineIizopods* (in Russ.) Naukova dumka Publishing House, Kiev.
6. Ivleva, I.V. (1981) *Temperature of Environment and Rate of Energy Metabolism in Water Animals* (in Russ.), Naukova dumka Publishing House, Kiev.
7. Belyaev, V.I., Nikolaev, V.M., Shulman, G.E., and Yuneva, T.V. (1983) *Tissue Metabolism in Fish* (in Russ.), Naukova dumka Publishing House, Kiev.
8. Shulman, G.E. and Urdenko, S.Yu. (1989) *Productivity of Black Sea Fishes* (in Russ.), Naukova dumka Publishing House, Kiev.
9. Belokopytin, Yu.S.(1993) *Energy Metabolism of Marine Fishes* (in Russ.).Kiev, Naukova dumka Publishing House.
10. Minyuk, G.S., Shulman, G.E., Shchepkin, V.Ya. and Yuneva, T.V. ((1997) *Black Sea Sprat (Relationship between Dynamics of Lipids and Biology and Fishery)* (in Russ.), Ecosea-Hydrophysica, Sevastopol.
11. Shulman, G.E.(1972a) On the level of fat stores in fishes at different regions of Central Mediterranean (in Russ.), *Voprosy Ichthyologii*, **12**, 33-40.
12. Shulman, G.E. (1978) Relationship of protein growth and fat accumulation with increase of mass and calority in fishes of genus Engraulis (in Russ.*), Biologiya Morya*, Vladivostok, **5**, 80-82.
13. Shchepkin, V.Ya. (1978) The study of succinate dehydrogenase activity of white skeletal, red muscles and liver of fishes with different ecophysiological features (in Russ.), *Biologiya Morya*, Sevastopol, **46,** 104-107.
14. Shchepkin, V.Ya., Shulman, G.E. and Sigaeva, T.G. (1976) The features of lipid composition of tissues in Mediterranean squids with different ecology (in Russ.), *Gydrobiologicheskiy Zhurnal*,**12**, 76-79.
15. Shchepkin, V.Ya., Belokopytin, Yu.S., Minyuk, G.S., and Shulman, G.E. (1994) The utilization of proteins, lipids and oxygen consumtion by pickarel at active swimming (in Russ.), *Gydrobiologicheskiy Zhurnal*, **30**, 58-61.

16. Shchepkin, V.Ya., and Shulman, G.E. (1978) The study of lipid composition of muscles and liver of mediterranean fish (in Russ.), *Zhurnal Evolyuzionnoy Biokhimii i Physiologii*, **14**, 230-235.

17. Shulman, G.E., and Yakovleva, K.K. (1978) New data about the level of fat stores in Mediterranean sprat (in Russ.), *Biologiya Morya*, Sevastopol, **46**, 100-104.

18. Muravskaya, Z.A., Pavlova, E.V. and Shulman, G.E. (1980) On the oxygen consumption and nitrogen excretion in *Calanus helgolandicus* and *Pontella mediterranea* (in Russ.), *.Ecologiya Morya*, **5**, 33-40

19. Sautin, Yu.Yu. (1989) Problem of regulation of adaptive changes of lipogenesis, lipolisis and transport of lipids in fish (in Russ.), *Uspekhi Sovremennoy Biologii*, **107**, 131-149.

20. Danilevsky, N.N. (1973) The fluctuation of Black sea anchovy'stock and methods of prediction of its possible catch (in Russ.), *Proceedings of All-Union Inst.of Marine Fishery and Oceanography*, **91**, 132-142.

21. Chashchin, A.K., and Akselev, O.I. (1990) Shoals' migrations and accessibility of Black sea anchovy for fishery at autumn-winter period (in Russ.), in: *Biological Resources of the Black Sea.*, All-Union Institute of Marine Fisheries and Oceanography, Moscow, pp. 80-93,

22. Lutz, G.I., Mikhman, A.S., Rogov, S.F., and Filchagin, N.K (1981) The nutrition of Azov sea pelagic fishes - tyulka and anchovy (in Russ.), *Gidrobiologichesky Zhurnal*, **17**, 26-31.

23. Shulman, G.E., Chashchin, A.K., Minyuk, G.S., Shchepkin, V.Ya., Nikolsky, V.N., Dobrovolov, I.S., Dobrovolova, S.G. and Zhigunenko, A.V. (1994) Long-term monitoring of Black sea sprat condition (in Russ.), *Doklady Rossiyskoy Akademii Nauk*, **335**, 124-126.

24. Yuneva, T.V., Svetlichny L.S., Yunev, O.A., Georgieva, L.V., and Senichkina, L.G. (1997) Spatial variability of *Calanus euxinus* lipid content in connection with chlorophyl concentration and phytoplankton biomass,*Okeanologiya*,.**37**, 745-752.

25. Shulman, G.E., Yuneva, T.V., Yunev, O.A., Anninsky, B.E. Svetlichny, L.S., Arashkevich, E.G., Romanova, Z.A., Uysal, Z., Bingel, F. and Kideys, A.E. (1997) Biochemical estimation of food provision for the copepod *Calanus euxinus* and the ctenophore *Pleurobranchia pileus* in the Black Sea, NATO TU-Black sea Project, Simposium on Scientific Results, Crimea, Ukraine, June 15-19, 1997, 120-122.

26. Hakanson, J.L. (1989) Analysis of lipid components for determining the condition of anchovy larvae, *Engraulis mordax*, *Marine Biology*, **102**, 143-152.

27. Hakanson, J.L.(1989a) Condition of larval anchovy (*Engraulis mordax*) in the Southern California Bight, as measured through lipid analysis, *Marine Biology*, **102**, 153-160.

28. Coombs, S.H., and Hakanson, J.L. (1991) Diel variation in lipid and elemental composition of sprat (*Sprattus sprattus*) larvae at mixed and stratified sites in the German Bight of the North Sea, *Int. Counc. Explor. Sea Comin. Meet*, L., **49**, pp. 1-11.

29. Muravskaya, Z.A., and Belokopytin,Yu.S. (1975) Effect of locomotion activity on the nitrogen excretion and oxygen consumption in pickarel (in Russ.), *Biologiya Morya*, **5**, 39-44.

30. Klovach, N.V. (1983) Effect of temperature and year's season on level of metabolism in Azov sea atherina (in Russ.), *Biologicheskiye Nauki*, **7**, 58-63.

31. Skazkina, E.P., and Kostyuchenko, V.A. (1968) Nutritive rations of Azov sea round goby (in Russ.), *Voprosy Ichthyologii*, **8**, 303-311.

32. Chekunova, V.N., and Naumov, A.G.( 1982) Energy metabolism and food requirements of Marmour Notothenia (in Russ.), *Voprosy Ichthyologii*,.**22** , 294-302.

33. Daan, N. (1975) Consumption and production in North sea coll an assessment of the ecological status of the stock, *Neth.J.Sea Res.*, **9**, 24-55.

34. Houde , E.D. and Schekter, R.C.(1983) Oxygen uptake and comparative energetics among eggs and larvae of three subtropical marine fishes, *Marine Biology*, **72**, 283-2.93.

35. MacKinnon, J.C.(1973) Analysis of energy flow and production in an unexpected marine flatfish population, *J. Fish. Res. Board* Canada, **30**, 1717-1728.

36. Ware D,.M. (1980) Bioenergetics of stock and recruitment, *Can. J. Fish. and Aquat. Sci.*, **37**, 1012-1024.

37. Wootton, R.J (1979) Energy costs of egg production and environmental determinants of fecundity in teleost fishes, in *Fish Physiology. Anabolic adaptiveness in teleost..*, Acad. Press, L, pp. 133-159.

38. Shulman, G.E., Oven, L.S. and Urdenko, S. Yu. (1983) Estimation of generative production in the Black Sea fishes (in Russ.), *Doklady Akademii Nauk SSSR*, **272**, 254-256.

39. Shulman G.E., and Urdenko, S.Yu. (1984) Relationship between generative and somatic production in Black sea fishes (in Russ.) *Doklady Akademii Nauk SSSR*, **275**, 1511-1513.

40. Kostyuchenko, V.A., Safyanova, T.E., and Revina, N.I (1979). Horse-mackerel (in Russ.), in *Raw resources of the Black sea.*, Pishchevaya Promyshlennost, Moscow, pp. 92-130.

41. Domashenko, G.P., Mikhailyuk, A.N., Chashchin, A.K., Shlyakhov V.A.,and Yuryev, G.S. (1985) Present state of industrial stocks of anchovy, horse-mackerel, sprat and whiting in Black sea, in *Oceanological and Fisheries Investigations in the Black Sea* (in Russ.), Agropromizdat, Moscow, pp. 87-100.

42. Belokopytin, Yu.S. (1978) The levels of energy metabolism in adult fishes (in Russ.), in G.E. Shulman (eds.), *Elements of Physiology and Biochemistry of total and active Metabolism in Fish*, Naukova dumka Publishing House, Kiev, pp. 46-63.

43. Vinberg, G.G .(1956) *Intensity of Metabolism and Nutritive Requirenments ofFfish* (in Russ.), BeloRuss. University Publishing House, Minsk.

44. Brett, J.R. and Groves, T.D.D. (1979) Physiological energetics, *in Fish Physiology. Bioenergetics and Growth*,. Acad.Press, N.Y.-L, **8**, 280-352.

45. Karzinkin, G.S. (1952). *The foundations of biological productivity of water bodies* (in Russ.), Pishchepromizdat, Moscow.

46. Manteifel, B.P., and Nikolsky, G.V. (1953) Problems of marine hydrobiology in the field of study the problem of utilization of fishery resources of open seas (in Russ.),*Voprosy Ichthiologii*, **1**, 18-23.

47. Ivlev, V.S. (1939) Energy balance in carp (in Russ.), *Zoologicheskiy Zhurnal*, **18**, 315-326.

48. Tochilina, L.V. and Belokopytin, Yu.S. (1992) The number of leicocytes in blood of Black sea and Mediterranean fishes (in Russ.), *Ekologiya Morya*, **42**, 41-45.

# THE FEATURES OF GEOSTROPHIC CIRCULATION AND CURRENTS IN THE EASTERN MEDITERRANEAN BETWEEN THE SYRIAN COAST AND CYPRUS ISLAND IN WINTER AND SUMMER

V.G.Krivosheya (1), I.M.Ovchinnikov (1), R.D.Kos'yan (1), V.B.Titov (1), L.V.Moskalenko (1), V.G.Yakubenko (1), F.Abousamra (2), I.Aboucora (2), K.Bouras (2)

[1]Southern Branch of P.P.Shirshov Institute of Oceanology. Gelendzhik-7, 353 470, Russia

[2]Higher Institute of Applied Sciences and Technology, Damascus, Syria

## Abstract

The geostrophic circulation and the structure of currents have been examined on the base of quasi-synchronous hydrophysical surveys and instrument current measurements at 7 mooring buoy stations carried out in this area for the first time in February 1992 and October 1993.

## 1. Introduction

In spite of more than secular history of oceanographic investigations of the Mediterranean Sea [7], the eastern part of the Levantine Basin between the Syrian coast and Cyprus was poorly studied until recently. Only not numerous separate data on currents over $1^0$ squares are available that were obtained by navigation method (according to ship drift) [12]. Besides that most general schemes of geostrophic circulation based on rather small amount of hydrological data, are available [1,6,7]. More recently in the Levantine Basin extensive studies were performed under the POEM Program [9,10,13], but even under this Program only a few hydrological stations were made in the Latakia Basin. Until 1992 there were no direct current measurements in this region.

Among the most significant factors that may influence upon water dynamics of the Latakia Basin, general circulation of the Eastern Mediterranean waters, wind regime and geomorphological features of the studied region are to be mentioned. The results of studies under the POEM Program [2,9,11] evidence that a part of Mid-Mediterranean Jet enters the Latakia Basin passing along the southern coast of Cyprus, and while approaching the southern boundary of the region under study



*P. Malanotte-Rizzoli and V.N. Eremeev (eds.),*
*The Eastern Mediterranean as a Laboratory Basin for the Assessment of Contrasting Ecosystems,* 113–126.

(30°N), it bifurcates. To the right of this jet the quasi-stationary gyre Shikmona is formed and to the left - the Latakian cyclonic eddy. Therefore in the studied region the northern currents have to prevail. The results of numeral modeling [14] evidence as well that the surface water transport between Cyprus and the Syrian coast is directed to the north.

The annual cycle of the wind field variability over the northern part of the Latakia Basin is divided into two periods: from April to September the western and south-western winds prevail, but from October to March the winds of northern and north-eastern directions are dominant. In the southern part of the region the winds of the south-western quarter prevail for the most part of the year [12]. So the character of the wind regime has to favour the surface water transport along the Syrian coast to the north.

The geomorphological features of the region involve the narrow shelf (from 2-4 to 13-15 km) and the steep continental slope (9-15 km off the shore the depth increases to 1200-1400 km) near the Syrian coast and two deep troughs: the Cyprian trough located in the north-west and the Latakian trough in the north with maximal depths equal to 1836 and 1504 m, respectively (Fig. 1). Between the troughs there exists an elevation of the bottom with minimal depth of 892 m, and in the western part of the area adjacent to Cyprus a plateau with 800-1000 m depth is located.

An essential contribution to the investigation of the Latakia Basin was made by two expeditions of the r/v "Vityaz" in winter 1992 and in autumn 1993. The expeditions were carried on jointly by the P.P.Shirshov Institute of Oceanology of the Russian Academy of Sciences and the Research Centre of the Syrian Arabian Republic. The investigations were complex: hydrophysics, hydrochemistry, geochemistry, hydrobiology [3,8]. For this purpose a "polygon" method was used, that is at first mooring buoy stations (MBS) with current measurements were set up, then two complex oceanographic surveys were performed over the dense network of stations (8-12 miles spacing). Between the surveys hydrobiological, hydrochemical and geochemical studies were carried on along one of the central sections.

In this work we intend to concentrate our attention upon analysing the data on water circulation and currents obtained in the course of expeditions. The volume of the data is presented in Tables 1 and 2 and the location of stations and sections is shown in Fig. 1.

**Table 1. Information about hydrophysical measurements. First expeditions**

| Hydrophysical surveys | |
|---|---|
| First survey<br>15.02-22.02.92; 66 STD stations | Second survey<br>05.03-08.03.92; 41 STD stations |

Current measurements at MBS

| №№ MBS | 1 | 2 | 3 | 4 | 5 | 6 | 7 |
|---|---|---|---|---|---|---|---|
| Date | 14.02-06.03.92 | 14.02-06.03.92 | 14.02-06.03.92 | 14.02-26.02.92 | 14.02-26.02.92 | 26.02-07.03.92 | 26.02-07.03.92 |
| Depth, m | 1400 | 525 | 62 | 72 | 1460 | 1430 | 70 |
| Levels, m | 15,50,100,250,500,1000 | 15,100,250,500 | 15,30 | 15,30 | 100,250,505 | 15,100,500,1000 | 15,30 |

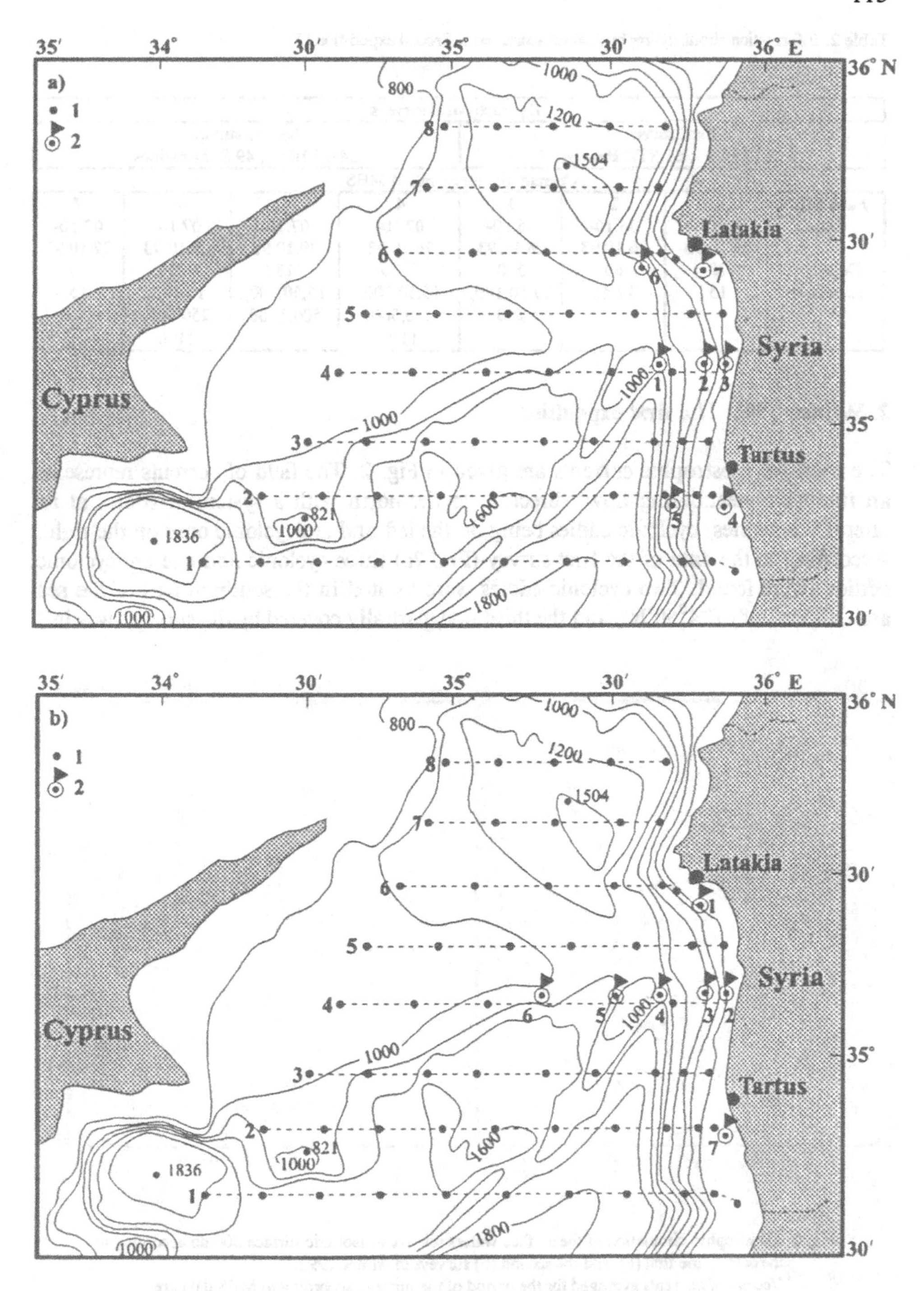


Fig. 1. Bottom relief and scheme of locations of hydrological stations (1) and mooring buoy stations - MBS (2). (a) February-March, 1992. (b) October, 1993.

Table 2. Information about hydrophysical measurements. Second expedition [5]

| Hydrophysical surveys | | | | | | | |
|---|---|---|---|---|---|---|---|
| First survey 8-12.10.93; 61 STD stations | | | | Second survey 24-28.10.93; 49 STD stations | | | |
| Current measurements at MBS | | | | | | | |
| №№ MBS | 1 | 2 | 3 | 4 | 5 | 6 | 7 |
| Date | 06.10-29.10.93 | 06.10-26.10.93 | 06.10-26.10.93 | 07.10-26.10.93 | 07.10-29.10.93 | 07.10-29.10.93 | 07.10-27.10.93 |
| Depth, m | 70 | 60 | 550 | 1230 | 1180 | 1235 | 73 |
| Levels, m | 15 | 17,45 | 15,50,100, 250 | 15,50,100, 250,500, 1150 | 15,50,100, 500,1100 | 15,100, 250,500, 1100 | 15 |

## 2. Winter 1992. The first expedition

The maps of geostrophic currents are given in Fig. 2. The field of currents represents an intensive meandering flow directed to the north with a system of eddies at its lateral boundaries, cyclonic eddies being on the left and anticyclonic ones on the right. According to the data of the first survey (Fig. 2a) three cyclonic and one anticyclonic eddies were found. Two cyclonic eddies were located in the southern part of the sea area under study (34°50'N), and the third one, partially covered by the survey, was in

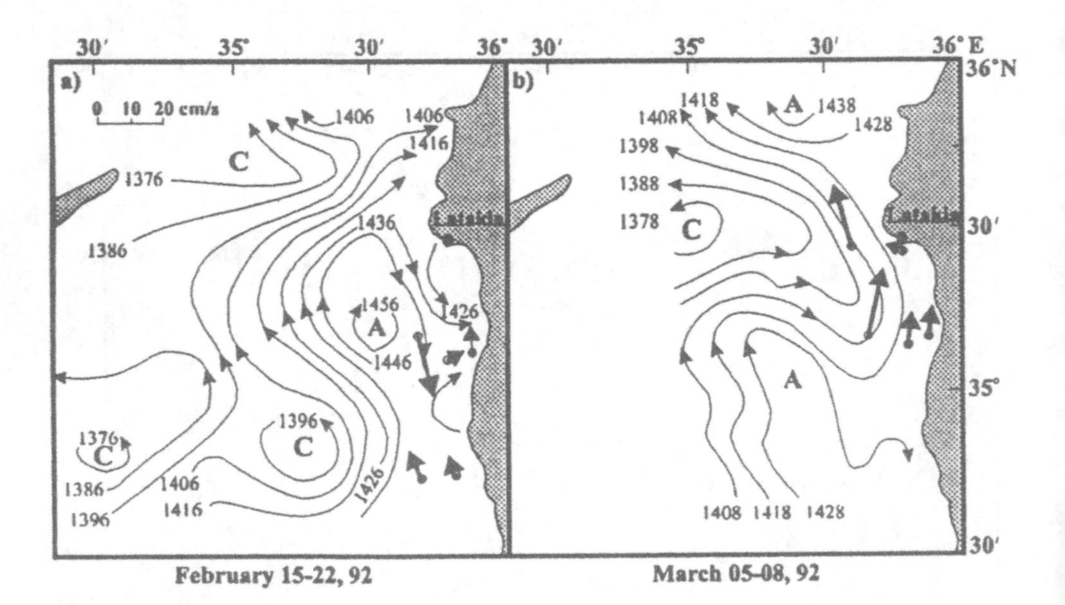


Fig. 2. Geostrophic circulation of the surface waters relative to isobaric surface 800 db according to the data of the first (a) and the second (b) surveys in winter 1992.
Vectors of currents averaged for the period of the surveys according to MBS data are shown by arrows. A - anticyclone, C - cyclone.

the north (35$^0$45'N). The anticyclonic eddy was located between the midstream of the flow and the Syrian shore in 35$^0$15'N. To the east of the anticyclone, just near the shore, a weakly defined secondary cyclonic eddy was outlined.

According to the data of the second survey which was limited by the meridian 35$^0$E (Fig. 2b), the dynamic picture was markedly changed. In the southern part, in place of the previous cyclonic eddy, there was an anticyclonic eddy elongated in meridional direction while the second anticyclonic eddy (its southern edge) was traced in the north. In the latitude of Latakia there was a cyclonic eddy.

This complex circulation pattern comprising meandering general flow with the system of eddies of different sign, reflects the influence of geomorphologic conditions on water dynamics. Clear regularity of eddy distribution relative to the midstream of the flow when cyclonic eddies are located to the left and anticyclonic to the right, indicates that the main reason of their formation lies in lateral shear (horizontal gradient) of velocity.

While comparing locations of peculiar points of the meander and the centres of eddies according to the results of two surveys, we may come to a conclusion that the whole dynamic system shifted to the north. Thus the southern cyclonic top of the meander together with the cyclonic eddy travelled to the north (as far as 35$^0$30'N), and the anticyclonic eddy moved into the most northern part of the polygon (35$^0$50'N). According to rough estimate the eddy structures shifted for a distance of about 40-50 miles in 17-18 days (between the middle dates of the surveys), that is their mean velocity was about 2.5-3 miles a day, the length of the meander wave being about 60 miles. But if a phase velocity and a sequence of performing polygon survey (from north to south, i.e. in direction of the meander movement) are considered, then during this survey (7 days) the meander traveled 15-20 miles northward, and therefore the real length of the meander wave considering this correction must be about 40-45 miles.

The meander shifting means that it is not stationary and is not "bound" to a local bottom relief. As it follows from the work [6], meanders and eddies typical of water circulation in the Eastern Mediterranean arrive there with the general water flow from the south and are transformed due to local geomorphological conditions.

Comparison of geostrophic circulation maps to current vectors at MBS in the surface layer which were averaged for the period of survey (Fig. 2a and 2b), reveals a good qualitative correlation of results (according to current direction). It was especially well-defined in the anticyclonic eddy area at MBS-1 where the stable currents were observed (Fig. 2a). But even in the area of the secondary weakly defined cyclonic eddy to the south of Latakia where the currents are weak and unstable, the instrumental measurements (MBS No 2 and 3) acknowledged the geostrophic current existence.

During the second survey when all the MBS were located in the area of a well-defined cyclonic meander, the instrumental measurements were in good agreement with the field of geostrophic currents (Fig. 2b). A good qualitative coincidence of measured and estimated currents testifies the realization of geostrophic relations there between a field of masses and a field of movement and allows to use widely a dynamic method for water circulation studies in this poorly investigated region.

Vertical structure of the current field is rather simple and is characterized by practically monodirectional movement through the whole water column. The changes in direction and velocity of the currents are synchronous at all the levels. These

peculiarities of the current field are typical of mesoscale eddies and meanders when the vertical water structure is quasi-homogenous (a result of density convection in winter), while the current changes are due to spatial migration of the eddies and slightly connected with a local wind.
Noted above is well confirmed by variability of the currents (Fig. 3), especially at

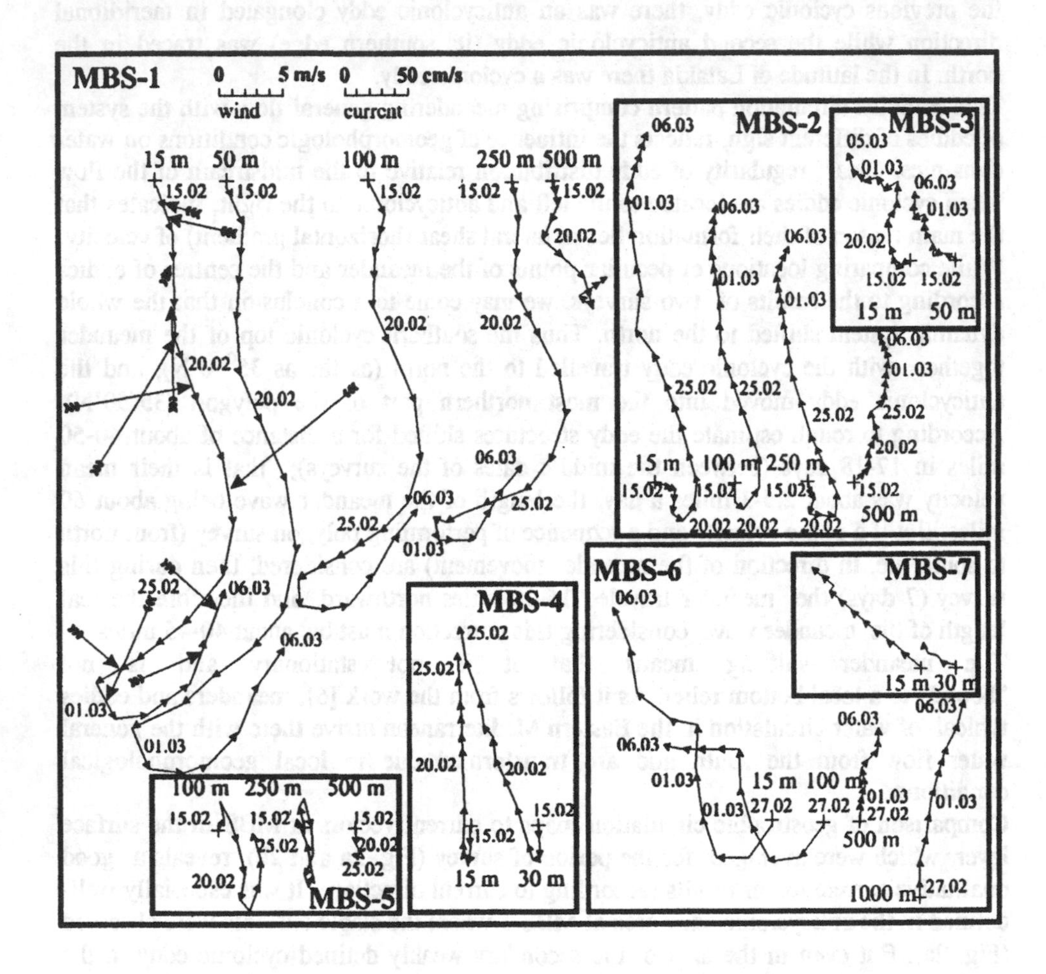


Fig. 3. The vertical structure and currents variability according to the data of instrumental measurements 15.02-06.03.92 (The Sequence arrows - mean diurnal vectors of the currents. The «plumage» arrows - mean diurnal vectors of the wind).

MBS-1 located in the south-eastern part of the anticyclonic eddy (Fig. 2a) where at the beginning (February 15-21) the currents were mainly of the southern direction. As the eddy was moving to the north, the current (after February 22) turned step by step south-westward. It became western (February 29) along the southern periphery of the eddy, but after the anticyclone leaving and under the influence of the cyclonic meander coming from the south (Fig. 2b) it turned north-eastward. As it is seen in Fig. 3, a

**connection between the current changing at the point and the local wind was practically absent. The analysis of successive changing of the currents (Fig. 3) allows to assure that the whole dynamic system (meander, eddies) was moving northward, in direction of the general water flow.**

**Quantitative characteristics of the current field are presented in Table 3. The most strong and stable currents were observed at the remoted from the shore points (MBS No 1, 2 and 6) located within well-defined dynamical structures. Mean current velocities in the upper layer were mostly 13-17 cm/s there, modal values were within the range of 31-40 cm/s, and maximal velocities reached 60-70 cm/s. The currents at those points were most stable, their coefficients of current stability being from 44-46% to 67-72%, and less changeable, their coefficients of variation being from 0.9-1.1 to 2.0-2.3. At the points near the shoreline and beyond the eddies the currents were weak and unstable. At point 5 the currents were most changeable and less stable, the coefficient of variation reached 15.1 there, and their stability fell to 8%.**

**It is to be noted that the obtained characteristics of currents (Table 3) were not stable in time for the mentioned points. As the whole dynamic system travelled in northern direction, characteristics of the currents at fixed points changed depending on that part of eddy or meander where they turned to be.**

**Table 3. Statistical characteristics of currents. First expedition**

| № MBS | Z,m | $\bar{\alpha}^0$ | $\bar{V}$ cm/s | σ(V) cm/s | $V_{max}$ cm/s | $\alpha^0$ max | $V_{mod}$ from-to | P(Vmod) % | $K_v$ | $K_s$ % |
|---|---|---|---|---|---|---|---|---|---|---|
| 1 | 15 | 174 | 15.2 | 31.9 | 70 | 160 | 31-35 | 16 | 2.1 | 46 |
| | 50 | 176 | 17.3 | 34.2 | 65 | 159 | 36-40 | 17 | 2.0 | 48 |
| | 100 | 174 | 12.1 | 27.6 | 60 | 150 | 31-35 | 19 | 2.3 | 44 |
| | 250 | 181 | 10.6 | 23.9 | 47 | 159 | 31-35 | 21 | 2.3 | 44 |
| | 500 | 156 | 4.2 | 13.3 | 40 | 123 | 1-5 | 27 | 3.2 | 39 |
| 2 | 15 | 1 | 13.4 | 18.9 | 59 | 9 | 16-20 | 16 | 1.5 | 66 |
| | 100 | 359 | 10.6 | 15.6 | 54 | 352 | 16-20 | 21 | 1.5 | 71 |
| | 250 | 258 | 8.9 | 16.1 | 49 | 344 | 6-10 | 19 | 1.8 | 60 |
| | 500 | 15 | 5.3 | 10.6 | 39 | 355 | 1-5 | 29 | 2.0 | 58 |
| 5 | 100 | 93 | 2.4 | 19.5 | 45 | 167 | 16-20 | 23 | 8.2 | 14 |
| | 250 | 337 | 1.2 | 18.8 | 42 | 174 | 16-20 | 22 | 15.1 | 8 |
| | 500 | 124 | 0.7 | 10.2 | 33 | 25 | 6-10 | 26 | 14.1 | 9 |
| 6 | 15 | 331 | 26.2 | 27.9 | 63 | 353 | 36-40 | 23 | 1.1 | 72 |
| | 100 | 301 | 15.9 | 21.3 | 55 | 350 | 26-30 | 17 | 1.3 | 67 |
| | 500 | 14 | 9.9 | 8.8 | 27 | 35 | 11-15 | 22 | 0.9 | 89 |
| | 1000 | 14 | 19.4 | 9.4 | 41 | 3 | 16-10 | 23 | 0.5 | 97 |

**Explanations for table 3.**

**$\bar{\alpha}^0$ and $\bar{V}$ - average values (for time of measurements) of current direction and velocity; σ(V) - standard deviation of current velocity; $V_{max}$ and $\alpha^0_{max}$ - maximal velocity of current and its direction; $V_{mod}$ - modal ranges of current velocity ; P($V_{mod}$)% - empirical probability of modal values; $K_v = \dfrac{\sigma(V)}{\bar{V}}$ - coefficient of current variation; $K_s$ - coefficient of current stability.**

**3. Late summer 1993. The second expedition**

As it is seen from the maps of geostrophic circulation (Fig.4 and 5), water dynamics in the studied region in late summer 1993 was characterized by the intensive space-time variability both on horizontal plane and vertically. In the period of the first survey (October 8-12, 1993), within the surface quasi-homogenous layer about 30-50 m in thickness, three large elements of circulation may be distinguished (Fig.4a): a weak northward flow, a strong anticyclonic eddy in the northern part of the region and a vast cyclonic eddy adjacent to the eastern shore of Cyprus. In latitude of 35°30'N the northern flow having met with the southern part of the anticyclonic eddy, declines under its forcing north-westward. In this case a part of the surface waters penetrates to the north along the western periphery of this eddy passing the northern extremity of Cyprus.

Somewhat deeper, at 100 m depth, the circulation differs essentially from that on the surface (Fig.4b). Only anticyclonic eddy in the north and cyclonic eddy in the west, in the vicinity of Cyprus, remained out of all the former elements of circulation. In the central part of the polygon a well-developed anticyclonic eddy of ellipsoid form was observed which missed on the sea surface (Fig.4a). The signs of its presence in that place occurred already at 50 m depth, i.e. at the lower boundary of the upper quasi-homogenous layer. Between the anticyclonic eddies (northern and central) a zone of cyclonic vorticity was formed. A weak southern flow was observed along the Syrian coast.

The circulation pattern observed at 100 m level remained at 250 m, but with some differences (Fig.4c). The northern anticyclonic eddy was shifted north-westward, beyond the studied region, and a cyclonic eddy was located in its place. A meandering flow to the south was traced between that eddy and the central anticyclonic one.

At 500 m depth the circulation was markedly changed (Fig.4d). It is represented by a meandering flow. Along its boundaries at the tops of meanders there was noted a tendency towards the formation of eddies: on the left (in flow direction) they were cyclonic and on the right anticyclonic. The western cyclonic eddy shrieked a great deal in its dimensions and became isolated in the south-western part of the polygon. The cyclonic eddy in latitude of 35°30' was clearly defined as before and well-developed.

According to the second survey data, 16 days later (October 24-28, 1993) all the former eddy structures remained in the upper quasi-homogenous layer, but their location was somewhat changed (Fig. 5a). The western cyclonic eddy moved to the south and became essentially weak. A cyclonic eddy was formed near Latakia. At 100 m depth the anticyclonic eddy in the centre of the polygon remained the only one of the large eddy structures (Fig.5b). The northern anticyclonic eddy shifted to the north. The cyclonic eddy neighbouring Cyprus was not traced. In the north near the Syrian coast a cyclonic vorticity was marked. A weak meandering flow southward was traced throughout the whole area. The same circulation pattern but with most clearly expressed midstream of meandering flow southward was observed at 250 m (Fig.5c). At 500 m depth (Fig.5d) the features of overlying layer circulation were retained.

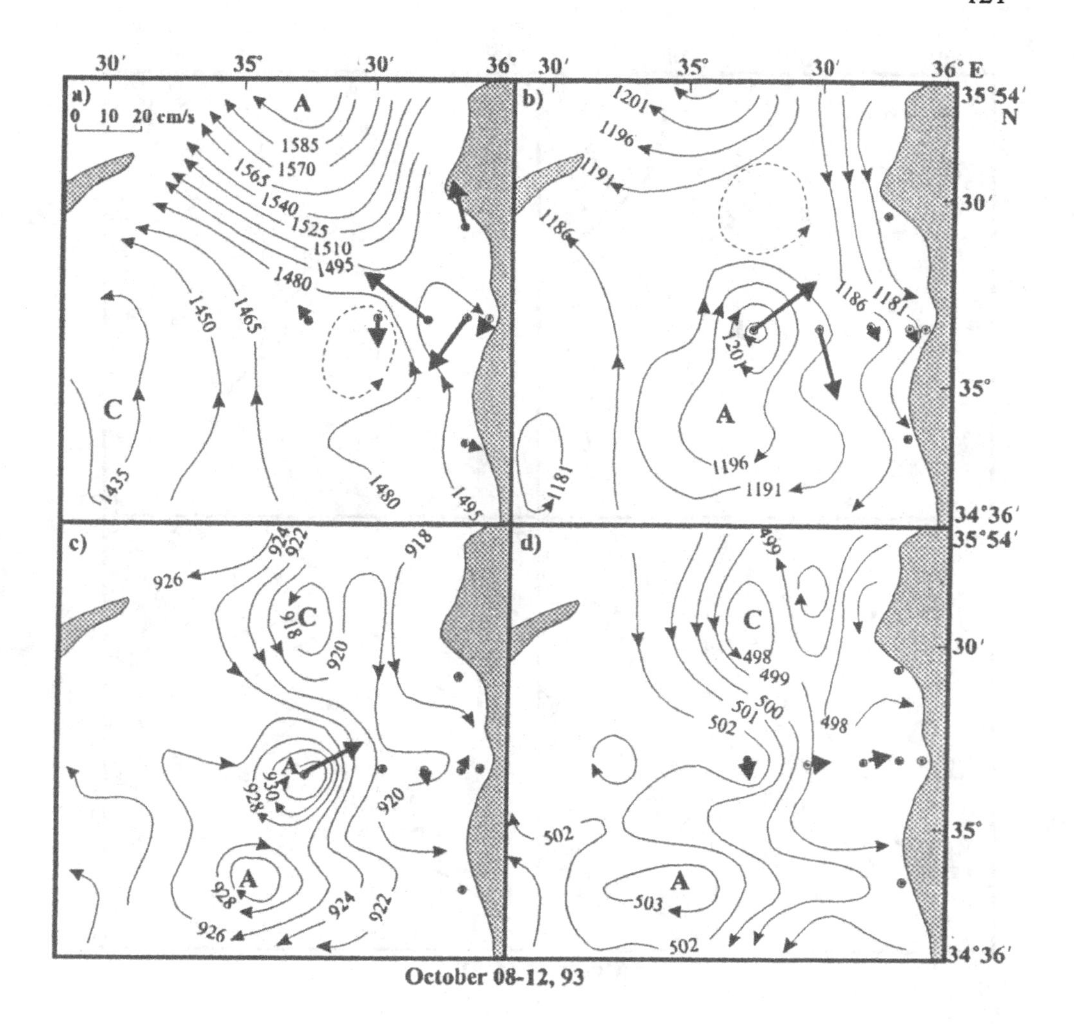


**Fig. 4. Geostrophic water circulation at the sea surface (a), 100 m level (b), 250 m (c) and 500 m (d) relative to isobaric surface 800 db according to the first survey data. Vectors of currents averaged for the period of the survey according to MBS data are shown by arrows.**
**A - anticyclone, C - cyclone.**

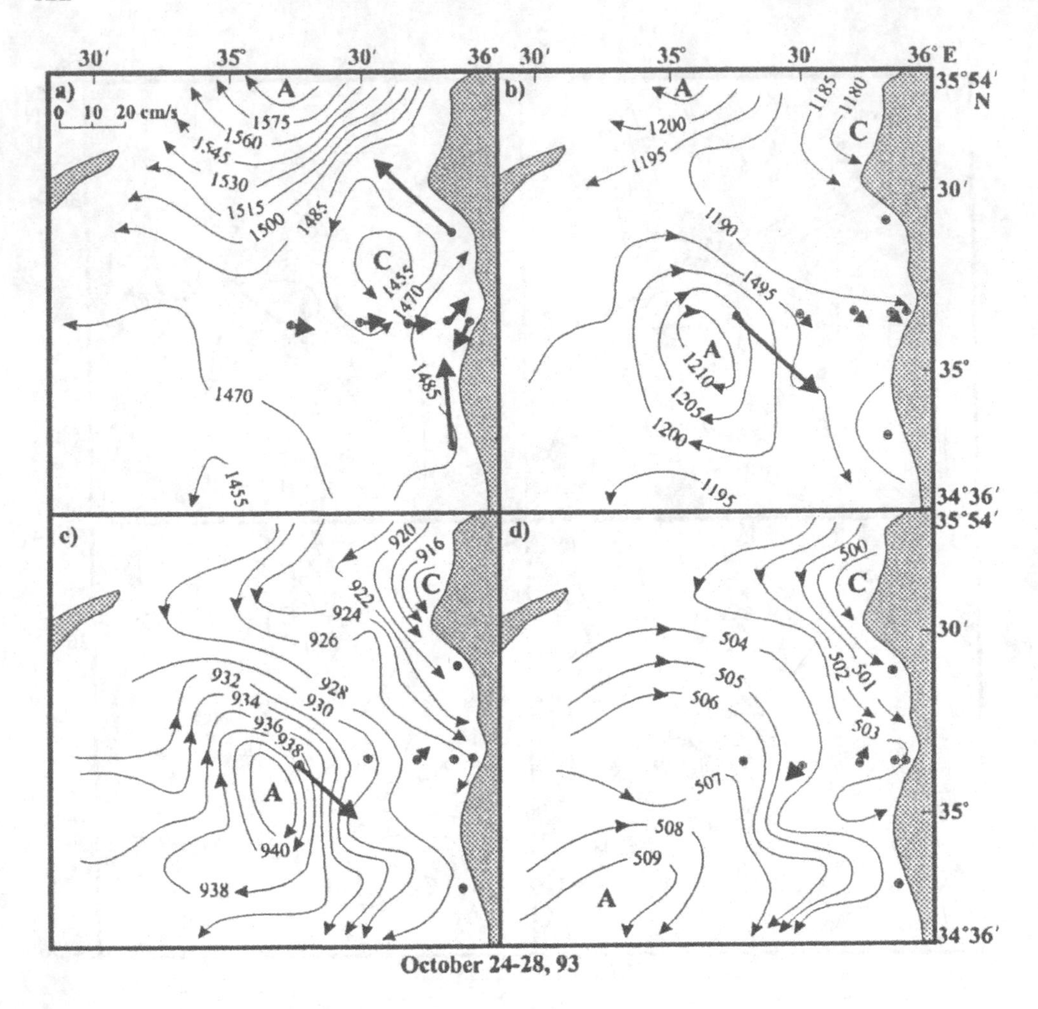


Fig. 5. Geostrophic water circulation at the sea surface (a), 100 m level (b), 250 m (c) and 500 m (d) relative to isobaric surface 800 db according to the second survey data. Vectors of currents averaged for the period of the survey according to MBS data are shown by arrows. A - anticyclone, C - cyclone.

In such a way the general water transport to the north which has to exist there the whole year [6], retained only in the most upper quasi-homogenous layer from 30 to 60 m in thickness. This transport was blocked to a great extent by a strong anticyclonic eddy at the northern boundary of the studied area. Within the deeper layer from 100 to 500 m (and even deeper) water transport was of opposite direction, i.e. from north to south.

Comparison of geostrophic current maps to instrumental measurements at MBS, in spite of intensive space-time variability of the current field, reveals their quite satisfactory correlation (Fig. 4 and 5), particularly where the current is well-defined (e.g. within the central anticyclonic eddy at MBS-6).

The statistical characteristics are given in Table 4. As it is seen from this table, velocities of the mean current vector which characterize water transport in general direction during the period of measurements, are not high and in most cases are less than 10 cm/s. Such low values of mean velocities are due to high variability of the current direction while eddies and meanders pass across the points of measurements. Only in the area of the central anticyclonic eddy (MBS-6, 100 and 200 m depths) where the currents were more stable, the mean velocities reach 19,5 and 12,1 cm/s, respectively.

Table 4. Statistical characteristics of currents. Second expedition

| № MBS | Z,m | $\bar{\alpha}^0$ | $\bar{V}$ cm/s | $\sigma(V)$ cm/s | $V_{max}$ cm/s | $\alpha^0$ max | $V_{mod}$ from-to | P(Vmod) % | $K_v$ | $K_s$ % |
|---|---|---|---|---|---|---|---|---|---|---|
| 3 | 15 | 359 | 3.0 | 16.1 | 59 | 359 | 11-15 | 21 | 5.4 | 22 |
| | 50 | 206 | 1.6 | 4.8 | 20 | 198 | 1-5 | 48 | 3.0 | 46 |
| | 100 | 190 | 2.2 | 6.1 | 17 | 289 | 1-5 | 40 | 2.8 | 44 |
| | 250 | 19 | 1.5 | 3.9 | 23 | 8 | 1-5 | 48 | 2.7 | 88 |
| 4 | 15 | 332 | 15.4 | 13.3 | 39 | 338 | 21-25 | 28 | 0.9 | 81 |
| | 50 | 96 | 0.9 | 6.4 | 20 | 309 | 1-5 | 40 | 7.4 | 18 |
| | 100 | 161 | 4.0 | 5.5 | 22 | 178 | 6-10 | 42 | 1.4 | 71 |
| | 250 | 189 | 1.6 | 4.8 | 20 | 210 | 1-5 | 51 | 3.0 | 45 |
| | 500 | 68 | 6.8 | 5.9 | 17 | 68 | 6-10 | 45 | 0.9 | 83 |
| | 1150 | 279 | 4.0 | 13.0 | 26 | 139 | 11-15 | 33 | 3.3 | 32 |
| 5 | 15 | 49 | 4.6 | 11.5 | 33 | 71 | 6-10 | 32 | 2.5 | 43 |
| | 50 | 137 | 3.0 | 6.3 | 21 | 203 | 1-5 | 45 | 2.1 | 58 |
| | 100 | 160 | 3.4 | 8.7 | 36 | 107 | 1-5 | 50 | 2.6 | 59 |
| | 500 | 14 | 2.1 | 6.6 | 16 | 16 | 6-10 | 53 | 3.1 | 34 |
| | 1000 | 215 | 7.4 | 10.7 | 27 | 215 | 11-15 | 32 | 1.4 | 65 |
| 6 | 15 | 119 | 3.4 | 7.8 | 26 | 118 | 1-5 | 35 | 2.3 | 53 |
| | 100 | 98 | 19.5 | 19.0 | 41 | 122 | 26-30 | 25 | 1.0 | 74 |
| | 250 | 98 | 12.1 | 15.1 | 33 | 130 | 21-25 | 25 | 1.2 | 70 |
| | 500 | 165 | 3.9 | 16.9 | 22 | 183 | 6-10 | 36 | 4.4 | 49 |
| | 1100 | 323 | 1.8 | 5.4 | 20 | 232 | 1-5 | 43 | 3.1 | 40 |

Explanations for table 4.

$\bar{\alpha}^0$ and $\bar{V}$ - average values (for time of measurements) of current direction and velocity; $\sigma(V)$ - standard deviation of current velocity; $V_{max}$ and $\alpha^0_{max}$ - maximal velocity of current and its direction; $V_{mod}$ - modal ranges of current velocity ; P(V)% and P($\alpha$)% - empirical probability of modal values; $K_v = \frac{\sigma(V)}{\bar{V}}$ - coefficient of current variation; $K_s$ - coefficient of current stability.

Maximal velocities in most of the points within the polygon were in the range from 16-20 cm/s to 36-41 cm/s. Only MBS-3 was an exception as the maximal velocity in the surface layer there reached 59 cm/s in the period of the first survey.
Among 20 levels of current measurements within the polygon the lowest modal range of velocity equal to 1-5 cm/s was observed at 10 levels, 6-10 cm/s range was at 7 levels, 21-25 cm/s range was at 2 levels. The highest modal range of velocity equal to 26-30 cm/s was fixed at one level (MBS-6, 100 m). As a rule, probability of velocity in the range 1-5 cm/s was equal to 40-50%, in the ranges 21-25 cm/s and 26-30 cm/s it was from 25 to 28%.
It is to be noted that at most points of measurements modal velocities of 1-5 cm/s range were found at 50 m, that is at the low boundary of seasonal thermocline (and pycnocline) with the maximal gradient of density.
The currents in the region under study were characterized by intensive time variability and low stability of direction. Variation coefficient of the currents at many levels was from 2 to 3 , and stability coefficient made up 40-60% (Table 3). The currents within well-defined eddies, in particular within the central anticyclone (MBS-6, 100 and 200 m), were most stable ($K_s$=70-80%) and less changeable ($K_v$=0,9-1,2). The highest variation coefficients (5,4 and 7,4) and the minimal stability of the currents (22 and 18%) were found in the nearshore zone at MBS-3 (15 m level) and MBS-4 (50 m level).
The analysis of current field variability according to the results of instrumental measurements showed that these variations are slightly connected with local winds. Thus in October the winds of the northern quarter of the horizon (from north-western to north-eastern) prevailed over the region under study. Nevertheless the water transport in the upper quasi-homogenous layer which is in direct contact with atmosphere, was directed to the north (Fig.4a, 5a), i.e. against the prevailing wind. Successive temporal variations of the currents at the points of measurements were not usually connected with the wind. At the same time these variations at all the levels correlated well with the location and shifting of the meandering and eddies which were just a main reason of the intensive space-time variability of the current field in the studied region.

## 4. Conclusion

In winter 1992 the circulation of waters in the area between the Syrian coast and Cyprus involved the meandering jet of the northern direction and the system of eddies at its boundaries, the cyclonic eddies being on the left and the anticyclonic on the right. The meanders and eddies were moving northward too with a phase velocity of 4-5 km/day. The peculiarity of the current field vertical structure was in one-directional movement throughout the whole water column. The changing of current directions at all the levels from the surface to the bottom was synchronous, that was favoured by a considerable homogeneity of waters over the depth. Maximal values of the vertical density gradient didn't exceed 0.0007 Sp.un/m.
In summer 1993 in connection with the well-developed seasonal pycnocline (maximal density gradients reached 0.0454 Sp.un/m) a two-layer structure of the water circulation was observed. Within the upper quasi-homogenous layer of 30-50

m in thickness the transport of waters from south to north was marked as a whole, but it was weaker than in winter 1992. This transport was blocked by a slightly migrating anticyclonic eddy at the northern boundary of the region under study. Under the pycnocline (deeper than 50-75 m) the southward movement was observed throughout the whole water thickness. It was most distinctly traced in the form of a meandering flow at the depths of 250 and 500 m. The main feature of the circulation in the intermediate layer (100-500 m) was a well-developed slightly migrating anticyclonic eddy located in the central part of the area. According to the data of direct measurements mean current velocities (for 20 days) within this eddy were 20 cm/s at 100 m depth, 12 cm/s at 250 m and 4 cm/s at 500 m.

The results of direct measurements evidence that the current velocities in winter 1992 were 2-5 times higher than in late summer 1993. At the same time in both seasons their considerable space and time variability was marked that was due to the meandering flows and the presence of eddy structures.

While comparing the general circulation of the Levantine Basin waters to the results of investigations in 1992 and 1993, it follows that the circulation of waters between the Syrian coast and Cyprus is formed and determined primarily under the influence of the water circulation in the region located to the south of the Lattakia Basin. In this place to the south of Cyprus a branch of the Mid-Mediterranean Jet passed [13]. The meandering flow traced near the Syrian coast in winter 1992 is most likely an extension of this jet. One of the reasons of generation of the slightly moving eddies in the Latakia Basin in summer lies probably in the general circulation of the Levantine Basin weakening in this season.

In order to understand better and to explain correctly the reasons of seasonal variability inherent to the general water circulation in the Latakia Basin and its interrelation with the general circulation of the Levantine Basin it is necessary to carry out synchronous studies in the Latakia Basin and the sea areas neighbouring it in the north and the south. In this way the spacing between the stations must not exceed 15 miles and 5-10 miles within the nearshore zone.

Acknowledgements

We appreciate very much the support of this work by the Department of Environment of the Ministry of Sciences and Technology of Russian Federation and the Scientific Studies Research Centre of Syria.

**References**

1. Atlas of oceanographic parameters of the Mediterranean Sea (1990) *GUNIO MO*, 450 p. (In Russian).
2. Brenner, P.S. (1989) Structure and Evolution of Warm Core Eddies in the Eastern Mediterranean Levantine Basin, *J.Geoph.Res. 94*, 12593-12602.
3. Kosyan, R.D., Krivosheya, V.G., Ovchinnikov, I.M., and Abousamra, F. (1995) Complex oceanographic investigations in the Mediterranean Sea and the Black Sea, *Oceanology 35(2)*, 302-305. (In Russian).
4. Krivosheya, V.G., Ovchinnikov, I.M., Titov, V.B., Udodov, A.I., Aboukora, I., Bouras, K., and Abousamra, F. (1996) Water Dynamics of the Eastern Mediterranean between Syrian Coast and the Island of Cyprus in Winter, *Oceanology 36(4)*, 512-519. (In Russian).

5. Krivoshya, V.G., Ovchinnikov, I.M., Titov, V.B., Udodov, A.I., Aboukora, I., Bouras, K., and Abousamra, F. (1997) Water Circulation and Currents variability in the Eastern Mediterranean between Coasts of Syria and Cyprus in Summer, *Oceanology 37(1)*, 27-34. (In Russian).
6. Moskalenko, L.V., and Ovchinnikov, I.M. (1991) Atlas of geostrophic circulation of the Mediterranean Sea, *Dep. in VINITI, No 3630-B91*, 33. (In Russian).
7. Ovchinnikov, I.M., Plakhin, E.A., Moskalenko, L.V., Neglyad, K.V., Osadchiy, A.S., Fedoseev, A.F., Krivosheya, V.G., and Voytova, K.V. (1976) The hydrology of the Mediterranean Sea, Hydrometeoizdat, 375 p. (In Russian).
8. Ovchinnikov, I.M., and Abousamra, F. (1994) The studies of winter regime of the Syrian waters in the Eastern Mediterranean, *Oceanology 34(3)*, 467-471. (In Russian).
9. Ozsoy, E., Hecht, A., and Unluata, U. (1989) Circulation and hydrography of the Levantine Basin: Results of POEM coordinated experiments 1985-1986, *Prog.Oceanogr. 22*, 125-170.
10. Ozsoy, E., Unluata, U., Oguz, T., Latif, M.A., Hecht, A., Brenner, S., Bishop, J., and Rozentroub, Z. (1991) A review of the Levantine Basin circulation and its variabilities during 1985-1988, *Dyn. Atmos.Oceans 15*, 421-456.
11. Robinson, A.R., Golnaraghi, M., Leslie, W.G., Artegiani, A., Hecht, A., Lazzoni, E., Michelato, A., Sansone, E.,Theocharis, A., and Unluata, U. (1991) Structure and variability of the Eastern Mediterranean general circulation, *Dyn. Atmos.Ocean. 15*, 215-240.
12. The Mediterranean oceanographic and meteorological data (1957) Staatsdrukkeriy-en Vitgeveriyt-s-Cravenhage. 91p.
13. The POEM Group (Robinson, A.R., Malanotte-Rissoli, P., Hecht, A., et al.) (1992) General circulation of the Eastern Mediterranean, *Earth-Science Reviews 32*, 285-309.
14. Tziperman, E., and Malanotte-Rizzoli, P. (1991) The climatological seasonal circulation of the Mediterranean Sea, *J.Mar.Res. 49*, 1-25.

# OPTICAL STUDIES BY THE MHI OF THE EASTERN MEDITERRANEAN: AVAILABLE DATA AND SOME RESULTS

V.I.MANKOVSKY, A.V.MISHONOV, V.L.VLADIMIROV, M.V.SOLOV'EV
*Marine Hydrophysical Institute, National Academy of Science of Ukraine*
*Kapitanskaya St. 2, Sevastopol, 335000, Crimea, Ukraine*

## 1. Abstract

Marine Hydrophysical Institute (MHI) was carried out the hydro-optical investigation in the east Mediterranean since 1972. These researches consist of the three main stages. During the first stage (1972-1980) the episodic measurements was provided in framework of the main Soviet program of marine research around all over the world, named "World Ocean". The second stage (1981-1989) of investigations was devoted to developing of the remote sensing methods of hydro-optical oceanographic investigations in frameworks of the national research program "Satellite oceanography". This stage includes the major part of data. The third stage (1990-1997) of research is characterised as episodic measurements provided in samples taken from sea surface and three special surveys of the Gulf of Izmir, which were done in co-operation with Institute of Marine Research (Izmir, Turkey). Comparative analyses of the hydro-optical parameters for Eastern Mediterranean seas allow to range these seas in order from hydro-optical point of view. This order is Levantine Sea, Ionian Sea, Aegean Sea, and Adriatic Sea.

## 2. Methods of measurements and data collected

During the entire period of investigation the next hydro-optical parameters were measured:

- Beam attenuation coefficient,
- Volume scattering function,
- Colour index,
- Radiance index of water mass,
- Secchi disk depth,
- Water colour by Forele-Ule scale.



*P. Malanotte-Rizzoli and V.N. Eremeev (eds.),*
*The Eastern Mediterranean as a Laboratory Basin for the Assessment of Contrasting Ecosystems*, 127–139.

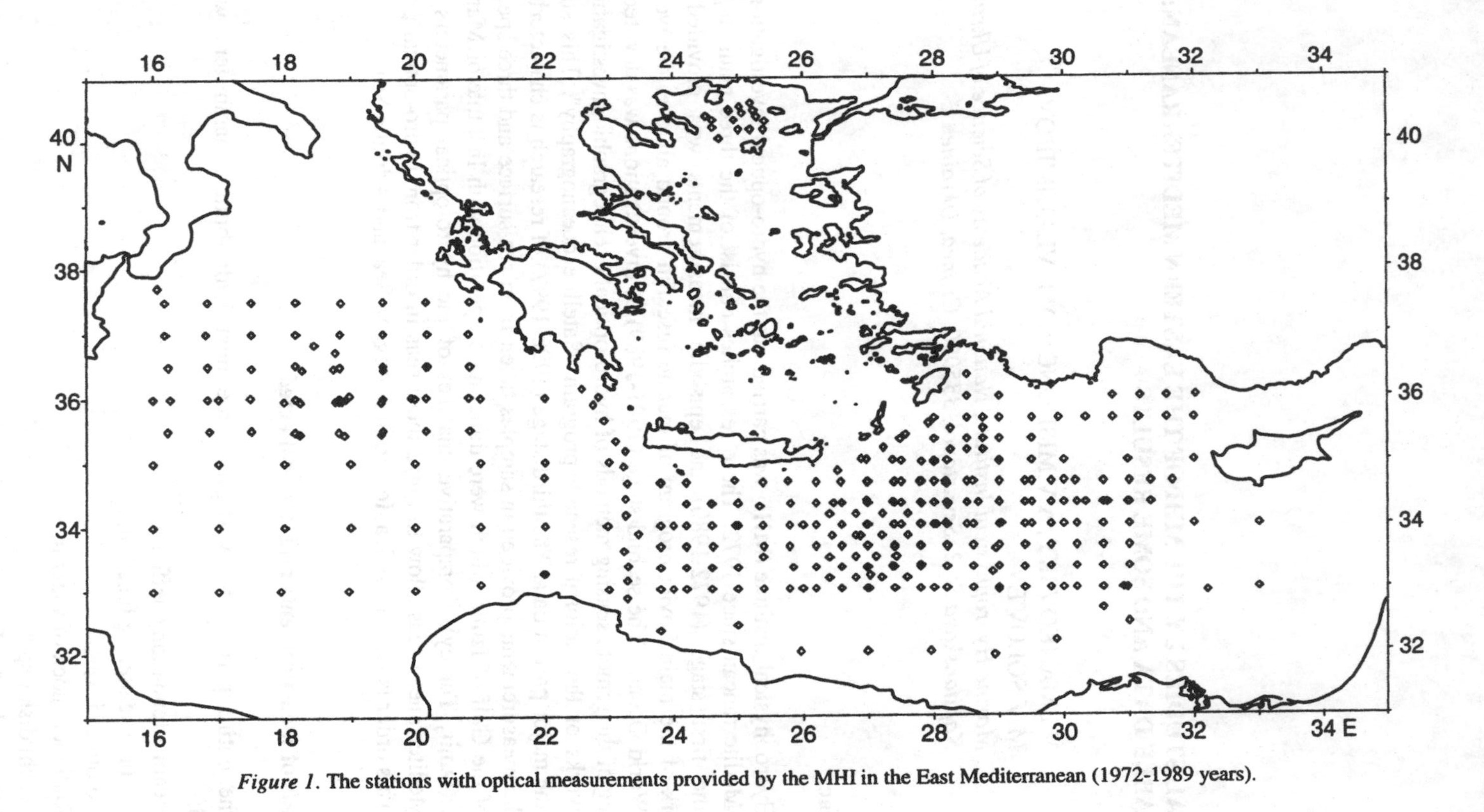


*Figure 1*. The stations with optical measurements provided by the MHI in the East Mediterranean (1972-1989 years).

TABLE 1. Optical measurements provided by the MHI in the East Mediterranean (1972-1997 years).

| Cruise/ RV [*)] | Terms of the cruise | | Region | Secchi Disk | Beam attenuation coefficient, $\varepsilon$ | | | | Colour Index | | # Radiance index | Volume scattering function |
|---|---|---|---|---|---|---|---|---|---|---|---|---|
| | | | | | Vertical profile | | | Measurements in samples | | | | |
| | Start | Finish | | Depth | Z (m) | $\lambda$ (nm) | # St. | $\lambda$ (nm) 416-700 | $\lambda$ (nm) | # St. | $\lambda$ (nm), 400-700 | 2°-162° |
| 27/*ML* | 18.12.1972 | 21.12.1972 | Ionian Sea | - | 200 | 480 | 5 | - | - | - | - | - |
| 9/*AV* | 14.10.1974 | 17.10.1974 | West from Crete | 3 | 350 | 525 | 14 | - | 554, 430 | 6 | - | - |
| 42/*ML* | 05.09.1981 | 08.09.1981 | Levantine Sea | 50 | 200 | 422, 493 | 24 | - | 564, 443 | 8 | 211 | - |
| 3_1/*Ayt* | 02.05.1982 | 04.06.1982 | Levantine Sea | 40 | 400 | 493 | 74 | - | 564, 443 | 30 | - | - |
| 4/*Ayt* | 28.07.1982 | 31.07.1982 | Levantine Sea | - | - | - | - | - | 564, 443 | 9 | - | - |
| 27/*AV* | 02.05.1983 | 05.05.1983 | Ionian & Aegean Seas | - | - | - | - | 5 | - | - | - | - |
| 7/*PK* | 01.10.1983 | 28.10.1983 | Levantine Sea | 58 | 80 | 422 | 124 | - | 564, 443 | 65 | 88 | - |
| 12/*PK* | 06.08.1985 | 21.08.1985 | Levantine Sea | 21 | 200 | 422 | 33 | - | - | - | - | - |
| 17/*PK* | 28.11.1987 | 06.12.1987 | Ionian Sea | 13 | 90 | 407 | 40 | - | - | - | - | - |
| 21/*PK* | 13.04.1989 | 16.04.1989 | Aegean Sea | 10 | 200 | 465, 540 | 24 | - | - | - | - | - |
| 23/*PK* | 20.08.1989 | 26.08.1989 | Levantine Sea | 25 | 200 | 457, 555 | 42 | - | - | - | 85 | - |
| 52/*ML* | 06.04.1990 | 08.04.1990 | Ionian & Aegean Seas | - | - | - | - | 5 | - | - | - | - |
| 55/*ML* | 07.10.1992 | 07.10.1992 | Ionian, Aegean & Adriatic Seas | 10 | - | - | - | 37 | - | - | - | 21 |
| - /*PR* | 23.09.1994 | 28.09.1994 | Gulf of Izmir | 33 | 75 | 418, 621 | 33 | - | - | - | - | - |
| - /*PR* | October 1996 | | Gulf of Izmir | 26 | 75 | 418, 659 | 38 | - | - | - | - | - |
| - /*PR* | April 1997 | | Gulf of Izmir | 13 | 75 | 418, 659 | 21 | - | - | - | - | - |
| | | | **Total:** | **302** | | | **472** | **47** | | **118** | **384** | **21** |
| | | | **In the Gulf of Izmir:** | **72** | | | **92** | | | | | |

[*)] *ML - Mikhail Lomonosov; AV - Akademik Vernadsky; Ayt - Aytodor; PK - Professor Kolesnikov; PR - Piri Reis.*

All these research (except last two) were done by means of original equipment designed in MHI [1-7]. Methods of investigation for these parameters are described below.

The vertical profiles of the beam attenuation coefficient were measured mostly on drift stations for one or two wavelength by means of jo-jo sounding transmittometers. Several stops on a different depths were provided for spectral scanning (six colour filters were applied) [1]. Measurements in water samples taken from the sea surface were done by means of laboratorial transmittometer using sixteen interference colour filters (spectral range from 416 to 700 nm) [2].

Volume scattering function values were measured in water samples by means of laboratorial nephelometer [3] on 520 nm wavelength.

The spectral reflection of the water mass may be characterised by means of colour index. This index can be calculated as:

$$I(\lambda_1, \lambda_2) = \frac{B_{\uparrow}(\lambda_1)}{B_{\uparrow}(\lambda_2)}, \tag{1}$$

were: $B_{\uparrow}(\lambda_1)$ and $B_{\uparrow}(\lambda_2)$ are values of the sun irradiation reflected by water mass measured under the water surface on two different wavelengths $\lambda_1$ and $\lambda_2$ respectively.

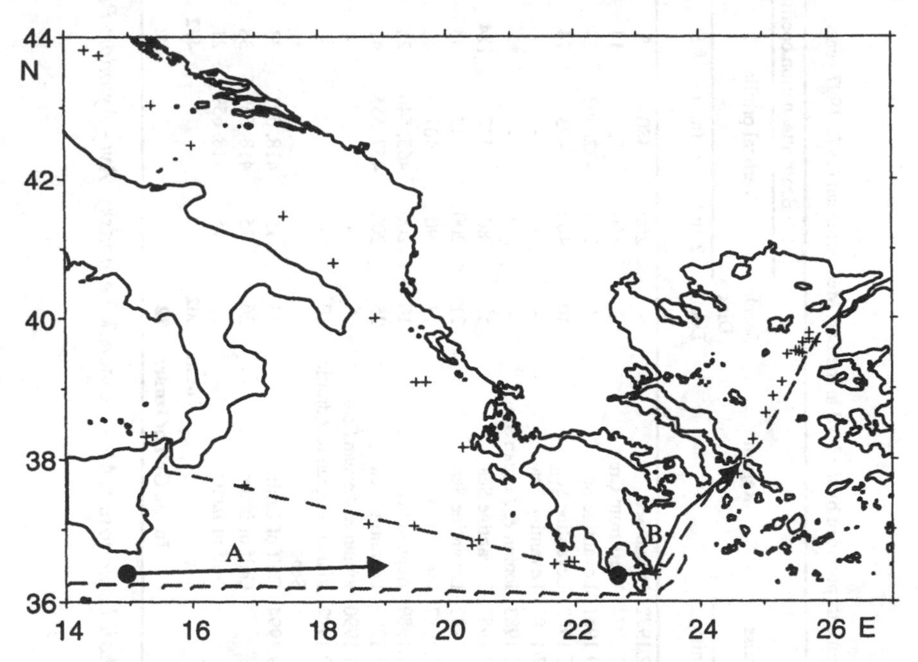


*Figure 2.* Surface samples (52, 55 cruises *Mikhail Lomonosov* & 27 cruise *Akademik Vernadsky* - rhombic symbols), and trace measurements of the colour index (8, 9, 10, 13 - 18, 20, 25 - 27, 30, 32, 41 cruises *Akademik Vernadsky* - dashed line). Tracks A and B are described below.

For these measurements the wavelength $\lambda_2$ usually selected in violet range of spectra where phytoplankton pigments absorption is maximal. The wavelength $\lambda_1$ is selected in green spectral range, so it is placing out of pigment's absorption lines. For such selection of wavelengths the colour index is rising up in correspondence with increasing of the water productivity. Correlation coefficient between colour index values and chlorophyll-a concentration in surface layer is about 0.86 with STD=±0.04 (n=175) [4]. That's why the colour index can be used for estimation of the surface chlorophyll-a concentration. The colour index values can be measured remotely and can be used for chlorophyll-a concentration assessment after atmosphere correction routine.

Colour index was measured *in situ* by means of the two techniques. In the *Mikhail Lomonosov*'s and *Professor Kolesnikov*'s cruises it was measured on 1-metre depth by means of deployed instrument during drift station. The research vessel *Akademik Vernadsky* is equipped with the special tube passing through the ship body. The colour index meter was placing into this tube on 6 metres depth and provides continuos measurement as well on stations as during ship moving [5, 6]. Measured values were recalculated from 6 meters depth to 1 metre depth by special empirical equation.

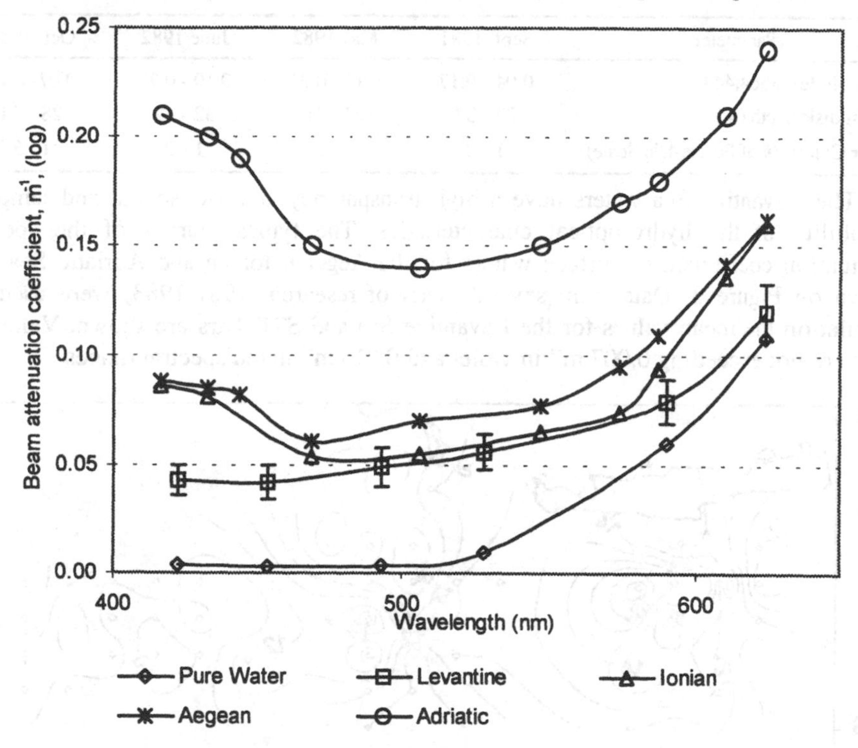


*Figure 3.* Spectral beam attenuation coefficient curves for surface waters in Eastern Mediterranean (data obtained in the Levantine Sea are averaged for 1981-1983 years, STD-bars are shown).

Radiance index of water mass was measured both on stations and during ship moving in spectral range from 400 to 700 nm [7]. The measuring instruments were placed on ship desk about 3-7 meters above the sea surface and measurements were taken during daytime.

Secchi disk depth and water colour were measured during daytime on drift stations.

Figure 1 shows the map with position of all stations taken. Figure 2 displays the sample points and colour index track (dashed line). All optical measurements provided in Eastern Mediterranean are shown in table 1.

## 3. Some results of the hydro-optical investigation in the Eastern Mediterranean

Main part on hydro-optical measurements was done in the Levantine Sea. This region was chosen as testing area for optical satellite instrument calibration [8]. Such decision was made due to some special features of the Levantine Sea, which were observed during the research and described below.

TABLE 2. Minimal and maximal values of the optical variables in the surface waters observed in the Levantine Sea.

| Parameter | Sept. 1981 | May 1982 | June 1982 | Oct. 1983 |
|---|---|---|---|---|
| Colour Index I(564,443) | 0.04 - 0.17 | 0.13 - 0.21 | 0.10 - 0.20 | 0.07- 0.17 |
| Secchi disk depth (m) | 33 - 37 | 27 - 41 | 32 - 42 | 28 - 44 |
| Water Colour (# of Forele-Ule Scale) | 1 - 2 | 1 - 2 | 1 - 2 | 1 - 3 |

The Levantine Sea waters have a high transparency and low spatial and temporal variability of the hydro-optical characteristics. The typical curves of the spectral attenuation coefficient of surface waters for the Aegean, Ionian and Adriatic Seas are shown on Figure 3. Data from several years of research (1981-1983) were used for calculation the mean values for the Levantine Sea and STD bars are shown. Values of STD are not exceeding 0.007 $m^{-1}$ in violet and 0.013 $m^{-1}$ in red spectral ranges.

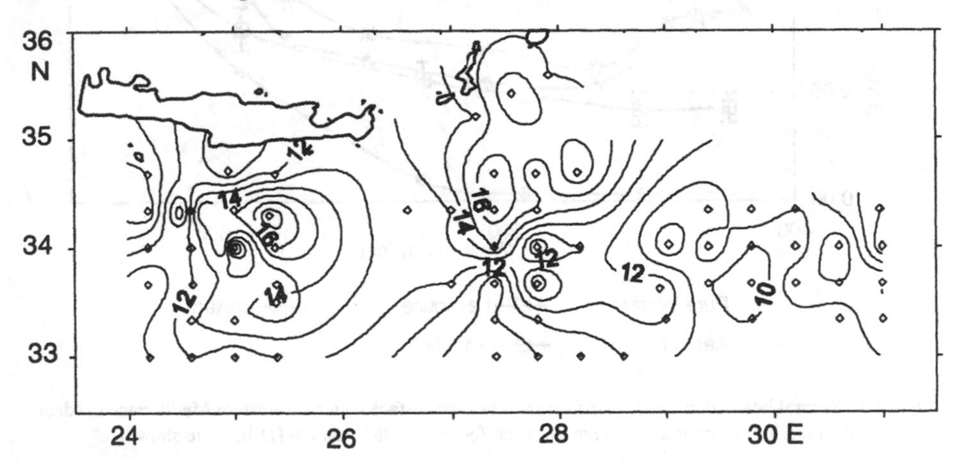


*Figure 4.* Colour Index I(564,443) $*10^2$ in the Levantine Sea, October 1983.

Comparison with other hydro-optical characteristics (table 2) also displays its low spatial-temporal variability in the Levantine Sea. Figure 4 and Figure 5 are shows the colour index values and Secchi disk depth fields for October 1983 survey.

The similar type of the vertical distribution of the beam attenuation coefficient was observed in the entire Levantine Sea (figure 6). It is characterised very homogeneous surface layer about 20m depth. Deeper in picnocline the layer with high values of the $\varepsilon$, was found.

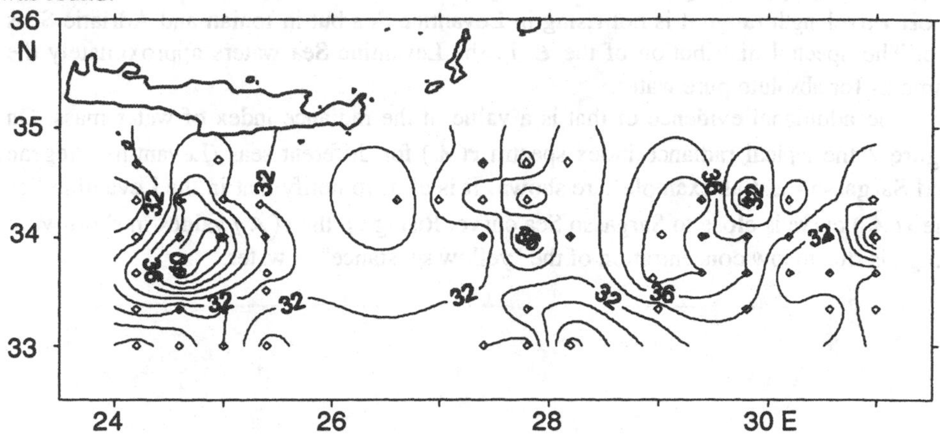


*Figure 5.* Secchi disk depth (m) in the Levantine Sea, October 1983.

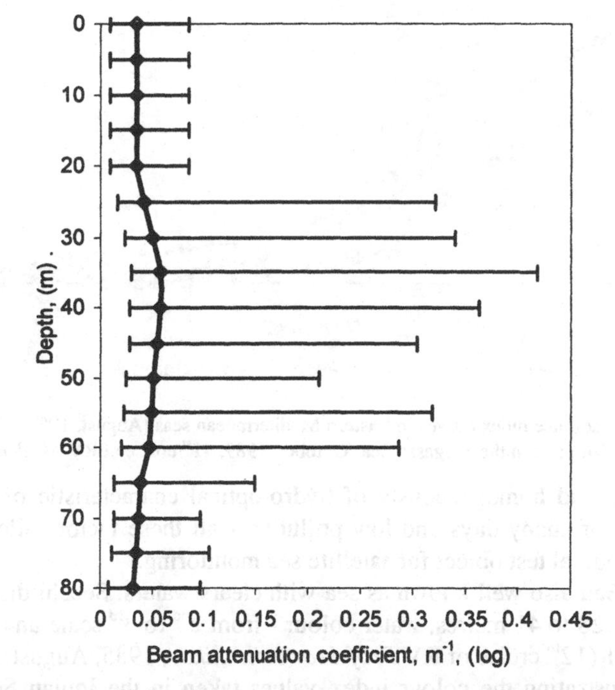


*Figure 6.* Beam attenuation coefficient (422 nm wavelength). Average profile, minimal and maximal values in the Levantine Sea (33°00' - 34°40'N, 23°48' - 31°00'E), October 1983.

The reason for low values of the beam attenuation coefficient in the surface waters of the Levantine Sea is low biological productivity of these waters. The total concentrations of chlorophyll-a and pheophityne pigments in the upper homogenous layer were from 0.05 to 0.08 mg·m$^{-3}$. These values are typical for pure oligothrophic water. It is well known that concentration of the "yellow substance" in these waters is low. It is displayed on figure 3 where the spectral distribution of $\varepsilon$ for surface waters in short wavelength range it is not rising in Levantine Sea but in Ionian and Adriatic Seas are. The spectral distribution of the $\varepsilon$ in the Levantine Sea waters approximately the same as for absolute pure water.

The additional evidence of that is a value of the radiance index of water mass. On figure 7 the typical radiance index spectra $\mathbf{r}(\lambda)$ for different seas (Levantine, Aegean and Sargasso - just as example) are shown. It is clear to notify that in the Levantine Sea the $\mathbf{r}(\lambda)$ curve is close to Sargasso Sea curve. Rising up the $\mathbf{r}(\lambda)$ values in short wave range is due to low concentration of the "yellow substance" in water.

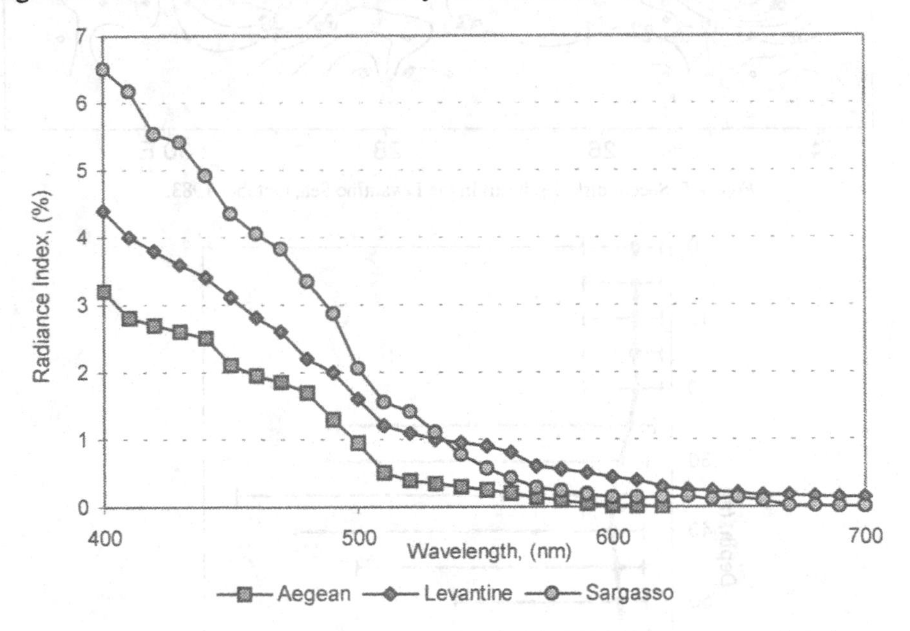


*Figure 7.* Typical radiance index curves in Eastern Mediterranean seas (August, 1985, 12th cruise of RV *Professor Kolesnikov*) and in the Sargasso Sea (October, 1989, 41st cruise of RV *Akademik Vernadsky*).

The stability and homogeneously of hydro-optical characteristic of the Levantine Sea, big amount of sunny days and low pollution - all these factors allows to use this area as the methodical test object for satellite sea monitoring.

The Ionian Sea also well known as sea with clears waters. Secchi disk depth values varied here from 28 to 41 metres, water colour - from 2nd to 3rd scale and colour index - from 0.10 to 0.11 (12th cruise of RV *Professor Kolesnikov*, 1985, August).

Figure 8 illustrating the colour index values taken in the Ionian Sea during 32nd cruise of RV *Akademik Vernadsky* (September 25, 1985). The calculated chlorophyll-a concentration values are shown here also. For this calculation the empirical equation obtained in [4] was used:

$$\log(C_{chl}) = 0.74 \cdot \log I(\lambda_1, \lambda_2) - 0.53\,, \qquad (2)$$

where: $\lambda_1$=550 nm; $\lambda_2$=440 nm. On the way about 150 nautical miles the colour index values were changed from 0.06 to 0.117 and calculated chlorophyll-a concentration - from 0.037 to 0.06 mg·m$^{-3}$. Next day the colour index values measured in Aegean Sea (figure 9) where its changed from 0.45 to 0.96 and calculated chlorophyll-a concentration values changed from 0.16 to 0.29 mg·m$^{-3}$.

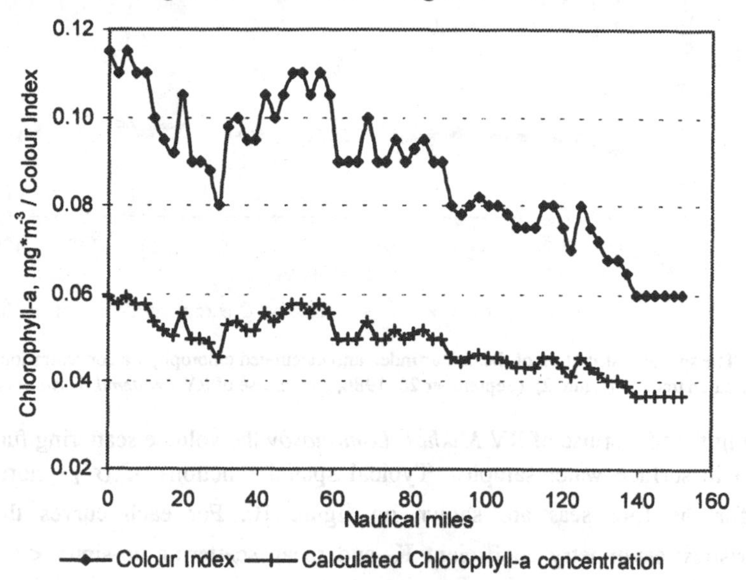


*Figure 8.* The spatial distribution of the colour index and calculated chlorophyll-a concentration in Ionian Sea (track A, figure 2), September 25, 1986, 32nd cruise of RV *Akademik Vernadsky*.

TABLE 3. Beam attenuation coefficient & colour index values obtained in surface water for the different Eastern Mediterranean seas

| Variables | Sea | | |
|---|---|---|---|
| | Aegean | Ionian | Adriatic |
| $\varepsilon$ ,m$^{-1}$ (log) $\lambda$ = 416 nm | 0.065-0.151 | 0.129-0.153 | 0.158-0.220 |
| I, (564/443) | 0.07-2.0 | 0.06-0.28 | - |

In table 3 the beam attenuation coefficient values obtained in surface water samples during 55th cruise of RV *Mikhail Lomonosov* (October 1992) for the different Eastern Mediterranean seas are shown. The highest values of $\varepsilon$ were observed in the Adriatic Sea, the lowest - in the Aegean Sea.

The colour index values obtained in 1974-1976 (for June-December) are also shown in table 3. These values were measured on the regular track of RV *Akademik Vernadsky* from Black Sea to Atlantic Ocean. In the shallow water in Aegean Sea high values of **I** were obtained (up to 2). In deep region these values are similar to Ionian Sea.

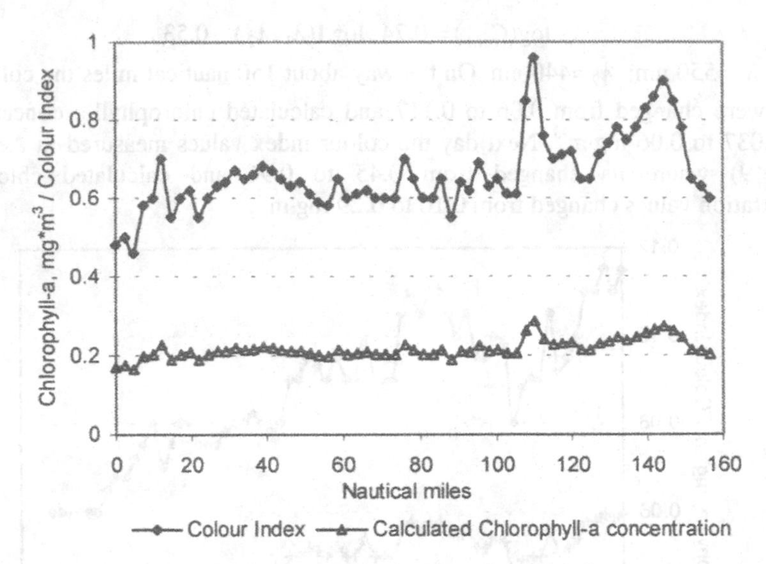


*Figure 9.* **The spatial distribution of the colour index and calculated chlorophyll-a concentration in Aegean Sea on track B (figure 2). (September 26, 1986, 32nd cruise of RV *Akademik Vernadsky*).**

**During the 55th cruise of RV *Mikhail Lomonosov* the volume scattering function was measured in surface water samples. Typical spatial functions of $\sigma(\gamma)$ normalised to $\sigma(90)$ for the four seas are shown on figure 10. For each curves the integral characteristics: asymmetry coefficient K, and mean scattering co-sinus $<\cos(\gamma)>$ are displayed. These values were calculated by equations:**

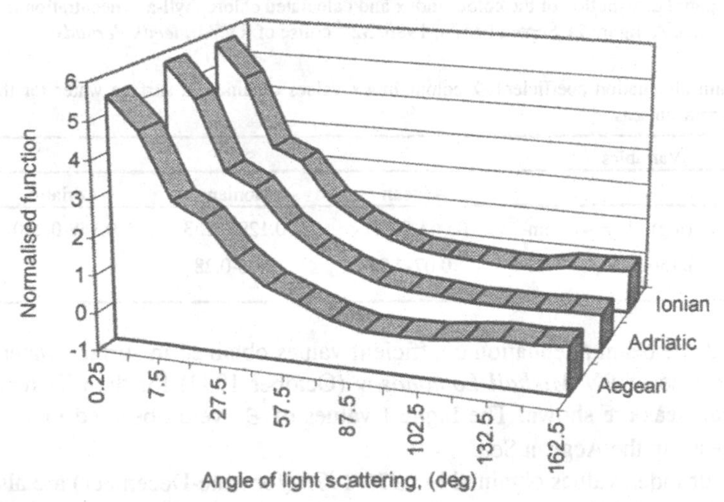


*Figure 10.* **Volume scattering function normalised to 90°: $\log[\sigma(\gamma)/\sigma(90)]$ of the Eastern Mediterranean surface waters (October, 1992, 55th cruise of RV *Mikhail Lomonosov*).**
**Adriatic Sea: K=164, $<\cos(\gamma)>$ =0.979; Ionian Sea (near coast of Greece): K=137, $<\cos(\gamma)>$ =0.977; Aegean Sea: K=92, $<\cos(\gamma)>$ =0.967.**

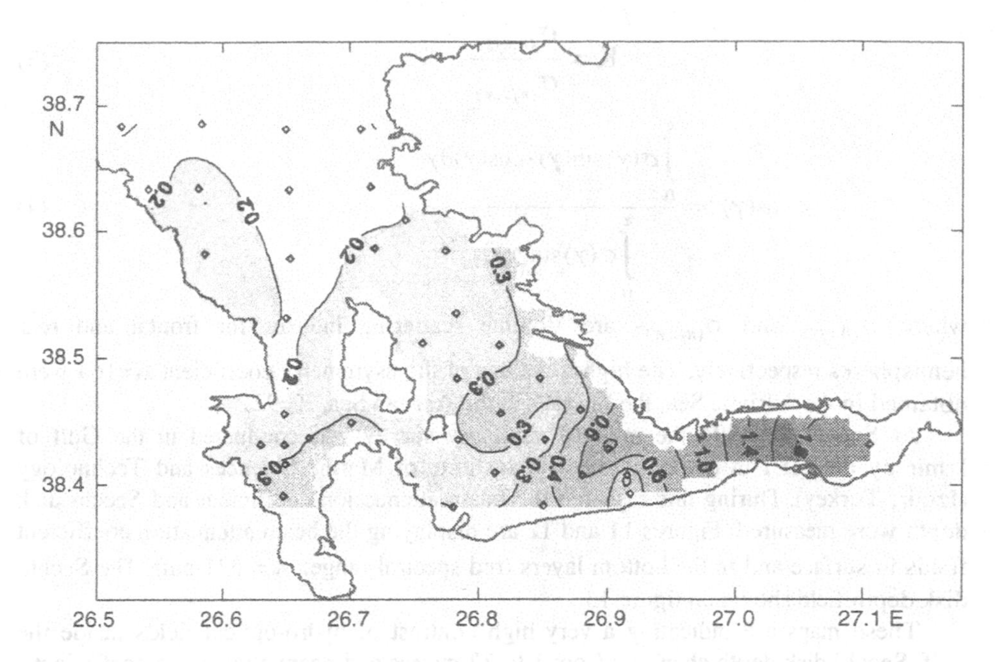


*Figure 11.* **Beam attenuation coefficient field (m$^{-1}$, log, $\lambda$ =621 nm). Gulf of Izmir, sea surface (September, 1994)**

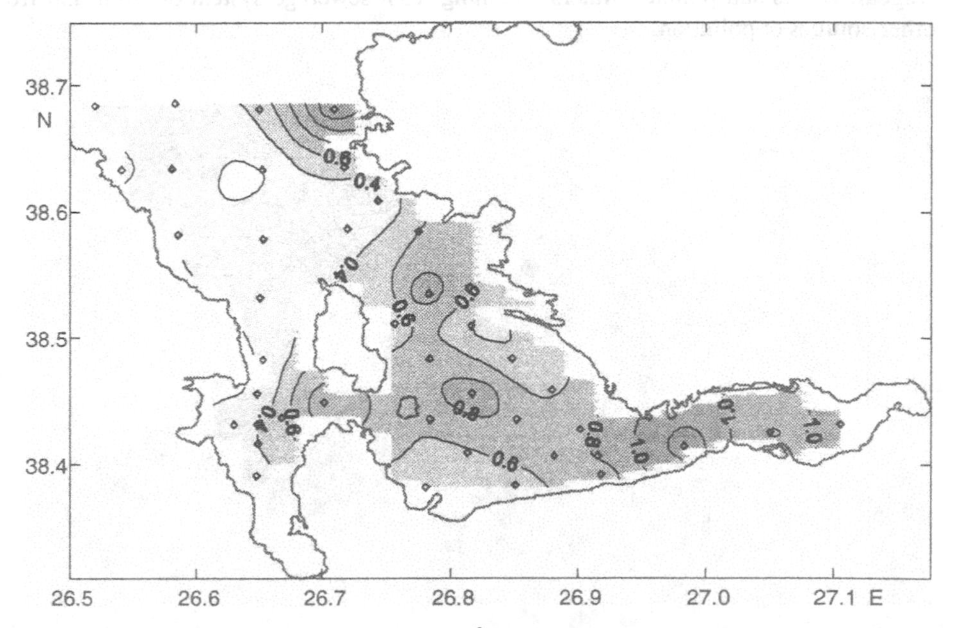


*Figure 12.* **Beam attenuation coefficient field (m$^{-1}$, log, $\lambda$ =621 nm). Gulf of Izmir, bottom level (September, 1994)**

$$K = \frac{\sigma_{(0,\pi/2)}}{\sigma_{(\pi/2,\pi)}}, \tag{3}$$

$$< \cos(\gamma) >= \frac{\int_0^{\pi} \sigma(\gamma)\sin(\gamma)\cdot\cos(\gamma)\partial\gamma}{\int_0^{\pi} \sigma(\gamma)\sin(\gamma)\partial\gamma}, \tag{4}$$

where $\sigma_{(0,\pi/2)}$ and $\sigma_{(\pi/2,\pi)}$ - are Volume scattering indexes for frontal and rear hemispheres respectively. The highest values of the asymmetry coefficient **K**=164 were obtained in the Adriatic Sea, the lowest - in the Aegean Sea, **K**=92.

At September 1994 the multy-disciplinary survey was conducted in the Gulf of Izmir by the MHI in co-operation with Institute of Marine Sciences and Technology (Izmir, Turkey). During this research the beam attenuation coefficient and Secchi disk depth were measured. Figures 11 and 12 are displaying the beam attenuation coefficient fields in surface and in the bottom layers (red spectral range: $\lambda = 621$ nm). The Sechhi disk depth field shown on figure 13.

These maps are indicating a very high contrast of hydro-optical fields inside the gulf. Secchi disk depth changing from 1 to 32 metres and beam attenuation coefficient - from 0.15 to 1.89 $m^{-1}$ (log-scale). Such differences are occurred due to mixture the pure Aegean waters and polluted waters incoming from sewerage system of Izmir and from other sources of pollution.

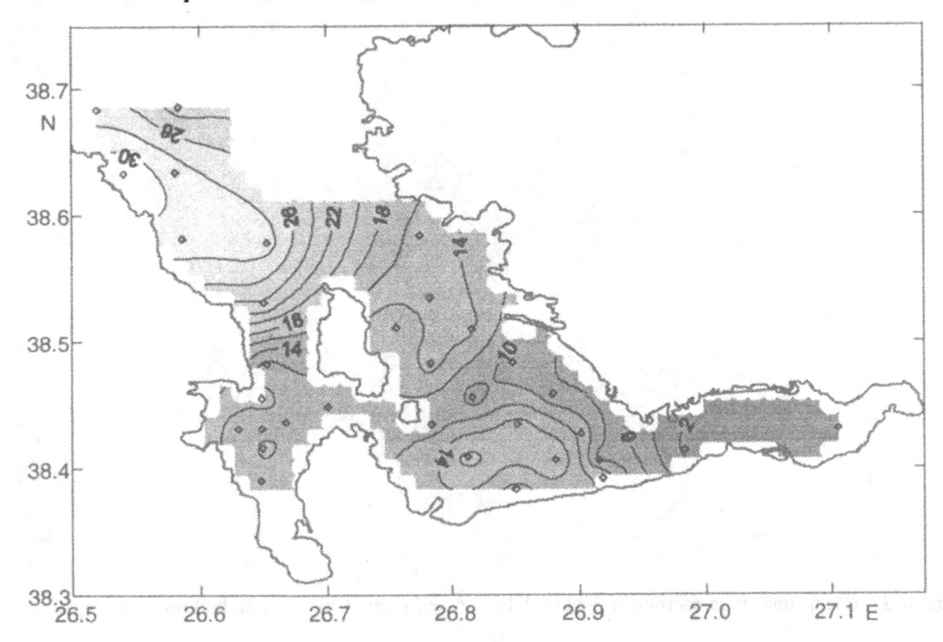


*Figure13*. Secchi disk depth (metres) in Gulf of Izmir (September, 1994)

Bottom beam attenuation field illustrated the flow of polluted water coming out from inner part of gulf to open sea along its north-east coast.

Comparative analyses of the hydro-optical parameters for Eastern Mediterranean seas allow range these seas in order from optical point of view. This order is Levantine Sea, Ionian Sea, Aegean Sea, and Adriatic Sea.

## 4. References


1. Lee, M.E. (1981) Photometer-transmittometer, in *Instruments for scientific investigations and automatic systems in Ukrainian Academy of Science*, Naukova Dumka, Kiev, pp. 89-90.
2. Man'kovsky, V.I. and Kaygorodov, M.N. (1980) Laboratorial auto-collimating transmittometer with variable base, in *Automatization of scientific research of the ocean and seas*, Marine Hydrophysical Institute, Sevastopol, pp. 91-92.
3. Man'kovsky, V.I. (1981) Marine impulse nephelometer, in *Instruments for scientific investigations and automatic systems in Ukrainian Academy of Science*, Naukova Dumka, Kiev, pp. 87-89.
4. Afonin, E.I., Berseneva, G.P., Krupatkina, D.K. et al. (1979) Assessment of the chlorophyll-a concentration in the sea surface layer by means of the colour index measurements, in *The light fields in the ocean*, Shirshov Institute of Oceanology Publishing, Moscow, pp. 191-196.
5. Neuymin, G.G. and Lee, M.E (1984) Optical methods for water biological productivity estimating, in *Limnology research automatization and light regime of the water reservoirs*, Nauka, Novosibirsk, pp. 10-21.
6. Afonin, E.I., Urdenko, V.A., Vladimirov, V.L., et al. (1987) About colour index meter using for under satellite investigations, in V.A. Urdenko & G. Zimmerman (eds.), *Remote sensing of the sea with atmosphere correction*, Institute of Space Research of the GDR, Moscow-Berlin-Sevastopol, v. 2, issue 2, pp. 85-121.
7. Afonin, E.I. and Kravtsov, G.L. (1984) Visible range Tele-photometer for sea up-welling radiation remote sensing investigations, in *The problems of ocean research from space*. MHI AS Ukrainian SSR, Sevastopol, pp. 74-83.
8. Vladimirov, V.L. and Urdenko, V.A. (1985) Possible test area for optical satellite instruments calibration, *Issledoniya Zemli iz Kosmosa* **2**, 85-89.

# PLANKTON COMMUNITIES IN THE EASTERN MEDITERRANEAN COASTAL WATERS

M. V. FLINT[1], I. N. SUKHANOVA[1], E. G. ARASHKEVICH[1], M. BAKER[2], H. TALJO[2]

[1] *P. P. Shirshov Institute of Oceanology, Russian Academy of Sciences, Moscow, Russia*

[2] *Higher Institute of Applied Sciences and Technology, Damascus, Syria*

**Plankton in the Eastern Mediterranean Sea has got a relatively poor study compared to that in the other regions of the basin. The available data mostly cover the central and the northern parts of the Levantine basin [5, 10, 13]. The data on the plankton in the Eastern Mediterranean coastal waters and closely adjacent areas are very few. Most of them were presented by Abi-Saab and Lakkis [1, 2, 6, 7, 8, 9]. The studies by Abi-Saab and Lakkis ere done in the Lebanese coastal waters and focused on taxonomic composition and temporal/spatial variability of the plankton community in the surface layer only.**

**The Eastern Mediterranean coastal waters lack any representative sampling of phyto- and zooplankton within the depth range sufficient for an adequate evaluation of their total numbers and biomass as well as peculiarities of their vertical distribution. In the Syrian coastal waters and adjacent areas no plankton studies had been done until the early 1990's.**



*P. Malanotte-Rizzoli and V.N. Eremeev (eds.),*
*The Eastern Mediterranean as a Laboratory Basin for the Assessment of Contrasting Ecosystems,* 141–158.

## 1. Materials and Methods

This paper bases on phyto- and zooplankton samplings done in the area between the Syrian coast and the Cyprus eastern coast , between 34°40′ N and 35°50′ N (see Krivosheya *et al.*, this volume). Samplings were done during two seasons: from 15.02.1992 through 08.03.1992 (24 cruise of R/V Vityaz) and from 30.09.1993 through 24.10.1993 (27 cruise of R/V Vityaz).

Mapping in 1992 included 5 latitudal sections (23 stations total), while in 1993 4 sections (21 stations) were done. Stations within the sections located 2-10 miles from each other over the shelf and 10-20 miles from each other in the deep-water area. Phytoplankton was sampled with 1.7 l Niskin bottles of Rosette multi-bottle system. In 1992 samples were taken from 0, 25, 50, 75, 100, 125, and 150 m levels and combined into a single integral sample for each station. Each integral sample was 5 l. This sampling approach stemmed from uniform distribution of temperature and salinity within the upper 350-400 m layerin the time of survey. In 1993 samples were taken from 8 to 10 levels within the upper 200 m layer. Levels for sampling were adjusted according to vertical distribution of water density and fluorescence (samples were taken from layers of the highest and the lowest density gradients and fluorescence intensity). As was done in 1992, samples were combined into a 5 l integral sample for each station.

Aside for the mappings, latitudinal sections along 35°20′ N were done both in 1992 and 1993, which surveyed vertical distribution of phytoplankton. The sections' area featured the most developed continental shelf in the whole region. Each section included 7 to 8 stations. In 1992 layers down to the bottom over the shelf and down to 350 m deep were sampled. Samples were taken from 0, 10, and 25 m levels and over each 25 m below. Samples were then combined into integral ones for every 25-50-meter layer. In 1993 layers down to the bottom over the shelf and down to 200 m deep beyond the shelf were sampled. Samples were taken from 16 to 18 levels and then combined into integral ones for every 5 to 7 levels. Levels for which sample were combined were selected as to cover the waters over the picnocline, the picnocline zone itself, the waters in the below and the layer with the strongest fluorescence signal.

Phytoplankton samples were concentrated with reverse filtration technique using 1 μm nuclear filters [12]. Samples were processed without preliminary fixation in a Nojeotte camera under a Jenalumar luminescent microscope (x400 magnification). Phytoplankton biomass was calculated according to the principle of cell shapes geometric similarity using a dedicated software by Institute of Oceanology RAS.

In 1992 and 1993 zooplankton was sampled at the same stations with phytoplankton. Judey type nets with mouth square 0.5 m$^2$ and filtering mesh size 180×180 μm were used for zooplankton sampling . The nets were equipped with Murena-2 electronic closing device by Institute of Oceanology RAS [3]. In 1992 samples were taken from 0-200 m (or 0-bottom over the shelf) and 200-500 m layers. In 1993 samples were taken from 0-50 m and 50-100 m (or 50-bottom in case of lower depths) and 100-350 m (or bottom) layers. At the section along 35°20′ N in 1992 samples were taken from 0-25, 25-50, 50-100, 100-150, 150-200, 200-300, 300-400, 400-500, 500-750, and 750-1000 m layers, or, respectively, from the layer down to the bottom in case of lower depths at a station. In 1993 the same layers were sampled, save that below 500 m deep only 500-1000 m layer was sampled.

Total zooplankton wet biomass was estimated by a displaced volume method.

Hydrophysical surveys, done at the same time with plankton samplings, suggested the data obtained in the second half of February and the early March 1992 attributed to hydrological winter conditions. The surface waters were mixed down to the main picnocline depth, 160-420 m, depending upon circulation pattern in a particular area, while no signs of seasonal water heating were observed. Water temperature down to the main picnocline was slightly above 16°C. In October 1993 the syrvey was done at the end of hydrological summer. The surface 30-50 m layer had temperature about 25°C. It resided upon a rigid seasonal thermocline, below which down to the main picnocline at about 350-500 m deep Levantine waters with the temperature of 16°C was found (for hydrological conditions see Krivosheya *et al.*, this volume).

**2. Phytoplankton**

Taxonomic composition of phytoplankton in winter and in the late summer showed high taxonomic diversity, typical for oligitrophic subtropical waters. In the Eastern Mediterranean the diversity stems from merging of the Mediterranean flora with a large group of introducents from the Indian Ocean and the Red Sea. The introducents account for up to 20% of species in these waters. All five species of 'the main tropical complex of Peridinea' were observed in both survey seasons: *Ceratum massiliense, C. carriense, C. trichoceros, Pyrocystis pseudonoctiluca,* and *P. fusiformis* [11]. Some species were observed in the Eastern Mediterranean for the first time. Among these was a group of Ceratum species, namely *C. hexacanthum, C. arietinum, C. symmetricum,* and *C. gibberum,* two large-size diatom species, namely *Gosleriella tropica* and *Rhizosolenia oceanica* and some other.

Dominating in the phytoplankton in winter were diatoms of neritic complex. The accounted for 40% to 80% from the total numbers and from 60% to 97% from the total biomass (Fig.1). The dominating diatoms were *Thalassiothrix frauenfeldianum, Asterionella japonica, Thalassionema nitzschioides, Chaetoceros curvisetus, C. socialis, C. affinis, Lauderia borealis,* and *Skeletonema costatum.* The second group by numbers and biomass were coccolithophorids, among which *Emeliania huxleyi* dominated. In winter taxonomic composition of phytoplankton was similar over the whole surveyed area.

Phytoplankton in the late summer featured domination of small dinoflagellates, the flagellates and coccolithophorids (Fig. 1). The most numerous dinoflagellates group everywhere were Gymnodinium and Oxytoxum genera species. The most numerous species among coccolithophorids were *Syracosphaera pulchra, Ophiaster hydroideus, Rhabdosphaera claviger, R. stylife* and others. Only the coastal stations over the depths less than 200 m featured significant role of neritic diatoms, the most numerous ones among which were various Chaetoceros species.

In winter 0-150 m layer's average phytoplankton numbers in different parts of the surveyed area varied from $23\times10^3$ cells/l to $127\times10^3$ cells/l with the average

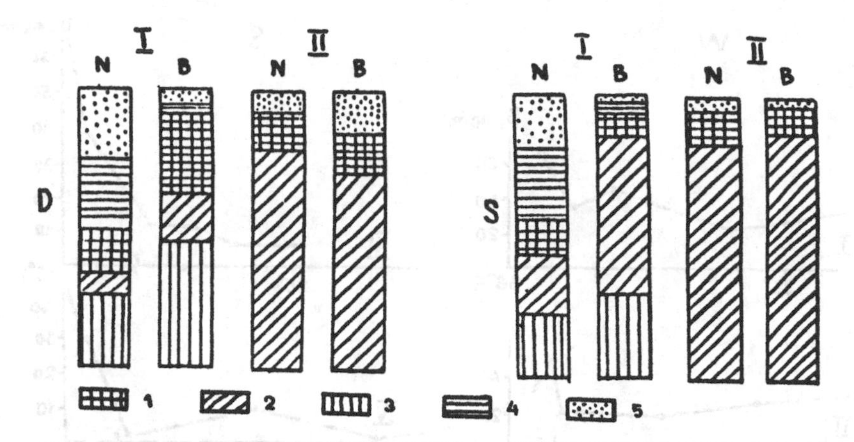


*Figure 1.* Percentage of the most abundant groups in total phytoplankton numbers (N) and biomass (B) in the deep water (D) and shelf (S) areas. I - late summer; II - winter. 1 - coccolithophorids; 2- diatoms; 3 - dinoflagellates; 4 - flagellates; 5 - the rest.

value of $50\times10^3$ cells/l. In the late summer the values were lower over the whole surveyed area. They varied from $5\times10^3$ cells/l to $50\times10^3$ cells/l with the average value of $19\times10^3$ cells/l. In both seasons the highest abundance was observed over the shelf. The number of cells here was 2-3 times as high as in the rest of the surveyed area. Both in winter and in the late summer the most northern part of the area exhibited high phytoplankton numbers. In the late summer this was due to diatoms *Rhizosolenia alata* and *R. calcar-avis* bloom.

Distribution of phytoplankton biomass in the surveyed area was similar to that of numbers. In winter the biomass in 0-150 m layer varied from 21 mg/m$^3$ to 164 mg/m$^3$ with the area's average value of 43 mg/m$^3$ or 8.6 g/m$^2$. The highest values were observed over the shelf (Fig. 2). A station done over the shelf at 35°10′ N on March 10 at the end of the winter survey and 10 days after the main mapping exhibited 3.5-fold growth of numbers and 5-fold growth of phytoplankton biomass. This was due to growing populations of two diatom species (*Lauderia borealis* and *Chaetoceros socialis*) and, most likely, marked the beginning of the early-spring diatom bloom.

In October phytoplankton biomass in the surveyed area within the 0-150 m (bottom) layer varied from 1.5 mg/m$^3$ to 43.5 mg/m$^3$ with the average value of 11.5 mg/m$^3$ or 1.7 g/m$^2$ (Fig. 2). The low biomass values were due to domination

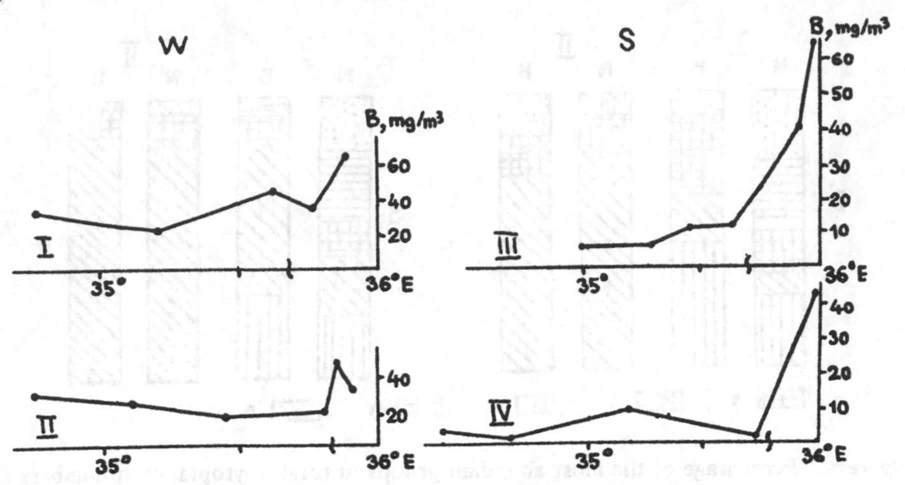


*Figure 2.* Distribution of phytoplankton biomass (B, mg/m3, mean values for 0-150[bottom] layer) along latitudinal sections normal to the Syrian coast.
W - winter; S - late summer. Vertical wavy line - shelf break location.
I and III - sections along 35°20' N; II and IV - sections along 35°00' N.

of small algae of nanoplankton size group with cell volume less than 500 μm³ in the deep-water parts of the region. In some parts of the area up to 40% from the numbers were made up by 10-30 μm³ cells. The highest biomass values were observed in the shelf areas where the most part of the phytoplankton was made up by diatoms. Similar to the numbers, the biomass values in the northern parts of the area were above the average level.

The obtained data suggested the surveyed area in February and early March as well as in October was oligotrophic.

Detailed studies of phytoplankton vertical distribution were done in both seasons at the section along 35°20′ N (Fig. 3, 4). In the late summer most phytoplankton stayed within the upper 50 m - i.e. within the mixed layer and the upper part of the seasonal picnocline. The numbers at the levels of maximum abundance varied from $30\times10^3$ cells/l to $270\times10^3$ cells/l. The biomass was within the range of 25 mg/m³ to 545 mg/m³. In the lower part of the picnocline and below the numbers dropped many times, while the biomass declined 1 or 2 orders of magnitude (Fig. 4, 5). This sharp decline of biomass along with depth was due to significant

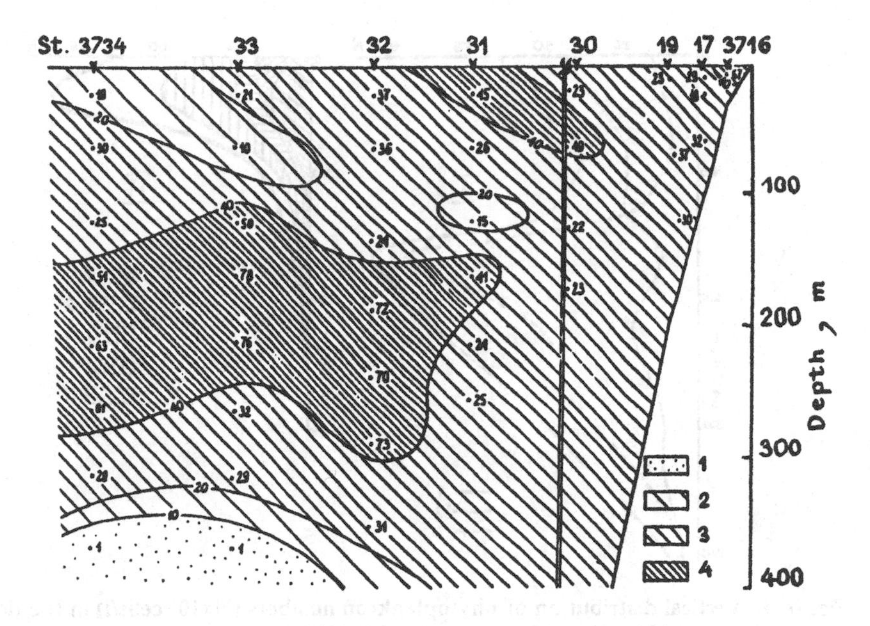


*Figure 3.* Vertical distribution of total phytoplankton biomass in the section along 35°20' N in winter of 1992. 1 - <10 mg/m³; 2 - 10-20 mg/m³; 3 - 20-40 mg/m³; 4 - >40 mg/m³.

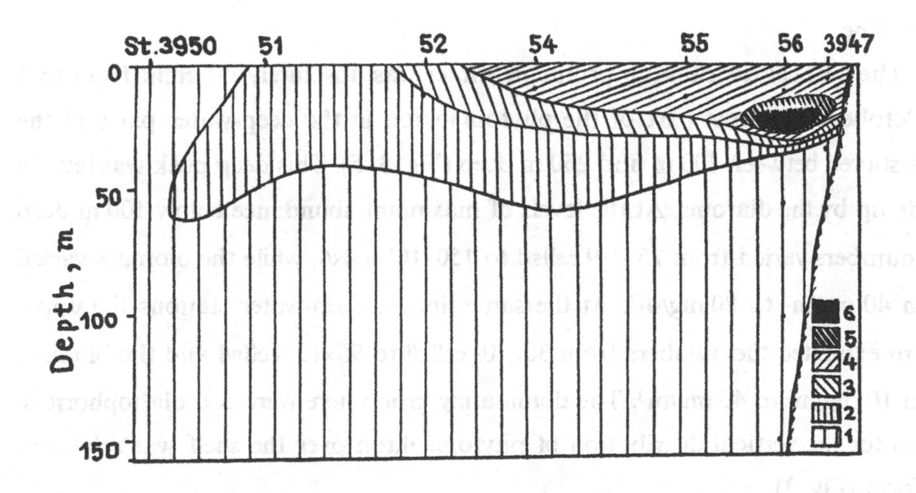


*Figure 4.* Vertical distribution of total phytoplankton biomass in the section along 35°20' N in late summer of 1992. 1 - <20 mg/m³; 2 - 20-50 mg/m³; 3 - 50-100 mg/m³; 4 - 100-200 mg/m³; 5 - 200-500 mg/m³; 6 - >500 mg/m³.

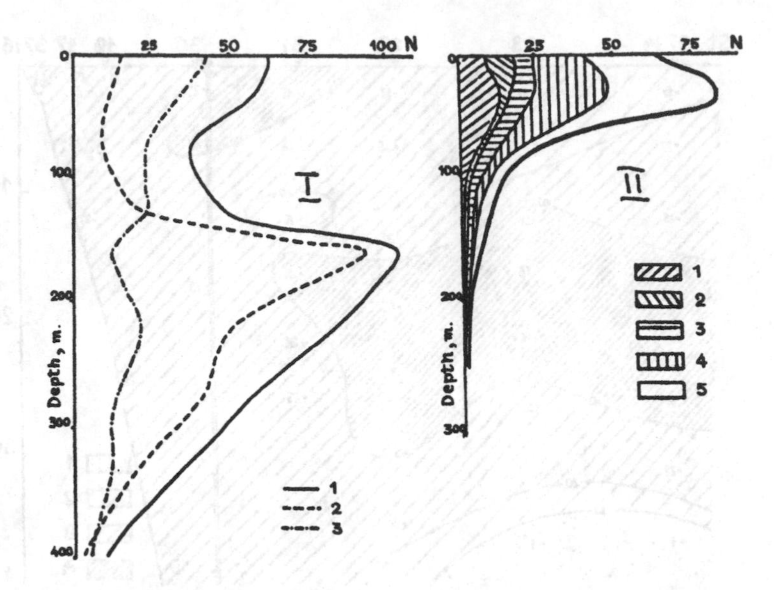


*Figure 5.* Vertical distribution of phytoplankton numbers (N×10³ cells/l) in the deep water area in winter of 1992 (I) and late summer of 1993 (II).
I - st. #3733; 1 - total phytoplankton, 2 - diatoms, 3 - coccolothophorids.
II - st. #3951; 1 - dinoflagellates, 2 - diatoms, 3 - coccolithophorids, 4 - flagellates, 5 - the rest.

increase of flagellates' and dinoflagellates' share in the phytoplankton below the picnocline.

The vertical distribution profile in winter was absolutely different from that in October. The most part of the phytoplankton in the deep-water parts of the area stayed between 100 m and 250 m deep (Fig. 3, 5). This deep peak was largely made up by the diatoms. At the levels of maximum abundance below 100 m deep the numbers varied from $75\times10^3$ cells/l to $150\times10^3$ cells/l, while the biomass varied from 40 mg/m³ to 80 mg/m³. At the same time in deep-water stations the upper 100 m exhibited the numbers from $30\times10^3$ cells/l to $95\times10^3$ cells/l and the biomass from 10 mg/m³ to 40 mg/m³. The dominating group here were coccolithophoridis. In winter the vertical distribution of phytoplankton over the shelf was relatively uniform (Fig. 3).

**3. Zoopankton**

In the second half of February and the early March in the winter zooplankton the dominating group by numbers and biomass were copepods. At stations over 300-400 m deep interzonal species *Pleuromamma abdominalis, P. gracilis, Eucalanus monachus, Neocalanus gracilis,* and *Euchatea marina* generally dominated by numbers. At some stations, especially in the central part of the area, the dominants included a large interzonal copepod *Euchirella sp.* In more shallow areas and in surface layers at deep water stations the most numerous were *Paracalanus sp.*, and some species of Clausiocalanus genus. Relatively numerous here were also *Calanus tenuicornis* and *Nannocalanus minor*. Copepod species with annual population cycle attributed to ontogenetic migrations and widespread in the northeastern and eastern parts of the Mediterranean Sea [4, 5]. were either few (*Rhincalanus cornutus, Eucalanus attenuatus*) or not found at all (*Calanus helgolandicus, Eucalanus elongatus*). Euphaseacea and Chaetognatha were found as few as under 10 ind/$m^2$. Jelly-body zooplankton was also very few. Only single individuals of *Thalia democratica* were observed more or less often. It should be noted that winter is the annual low-numbers period of salps and doliolids in the Eastern Mediterranean Sea [5]. Active growth of *T. democratica* population in the Syrian waters begins in April and the species soon becomes a dominant in zooplankton. Similar seasonal dynamics is observed for abundant species of Siphonophora.

At the end of the hydrological summer the taxonomic composition of copepods was the same as in winter. The most abundant copepod species were virtually the same: epipelagic forms *Paracalanus sp, Clausocalanus spp., Lucicutia sp., Nannocalanus minor* and interzonal *Pleuromamma abdominalis, P. gracilis, Euchaeta marina, Neocalanus gracilis,* and *Eucalanus monachus*. But generally copepods in October did not play a dominating role in zooplankton. Salps were numerous, dominating among which were *Thalia democratica*. The numbers of the species, represented by both blastozooids and oozids, in the upper 50 m layer reached 80-90 ind/m². Euphasiids and Chaetognatha, observed occasionally in winter, in October, reached 70-80 ind/m² and 100-110 ind/m², respectively. Siphonophora, observed in winter only occasionally, in late summer were found in almost all samples from the depths down to 1000 m.

Zooplankton biomass in the deep-water part of the surveyed area in winter averaged to 10±2.9 mg/m³ or to 5±1.5 g/m² in the 0-500 m layer. As studies of the vertical distribution indicated (see in the below), this layer contained nearly all zooplankton. Spatial distribution of the zooplankton beyond the shelf was quite uniform. The average biomass across the area varied from 7 mg/m³ to 16 mg/m³. The shelf area at all sections featured higher biomass values from 15 mg/m³ to 47 mg/m³, while the average value was 20±9.4 mg/m³ (Fig 6). The values obtained for the deep-water areas were close to those obtained by Grese [5] in other parts of the Mediterranean sea: for Algeria-Provance basin at 11.2 mg/m³, for the Tyrrhenian Sea at 10.8 mg/m³ and at 12.7 mg/m³ for the whole basin. Only for the northern part of the Adriatic Sea the study gives higher zooplankton biomass value at 17.9 mg/m³.

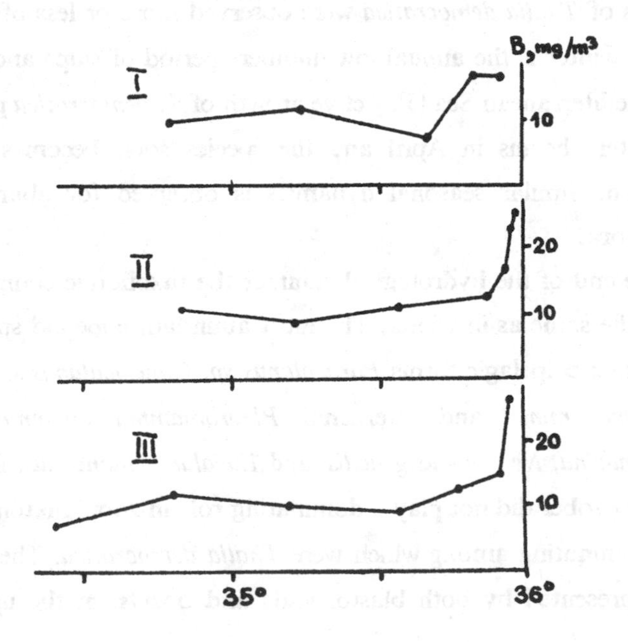


*Figure 6.* Distribution of total zooplankton biomass (B, mg/m3, mean values for 0-500[bottom] layer) along latitudinal sections normal to the Syrian coast in winter. I - section along 35°20' N; II - section along 35°00' N; III - section along 34°40' N.

Traditional approach to zooplankton biomass across the 0-200 m layer suggested the Syrian waters in winter were a little poorer than those in the other parts of the Mediterranean sea or close to them. We obtained average values at 17.7±6.3 mg/m³. In the western parts of the sea in winter Grese [5] obtained the values of 20 mg/m³, 33 mg/m³ in the eastern parts and the average value of 23.1 mg/m³ for the whole Mediterranean sea. Our studies said the waters over the shelf and the upper slope within the 500 m isobat were somewhat richer. The average biomass within the 0-200 m (bottom) layer here was 25.7±10.1 mg/m³.

The average zooplankton biomass in the late summer in the deep-water part of the surveyed area was estimated for the 0-350 m layer, which at that time held the majority of the plankton. It varied within the range of 7.4-15.4 mg/m³ (2.6-5.4 g/m²). The area's average value was 12.6±4.0 mg/m³ or 4.4±1.4 g/m², which practically matched the the values obtained at the end of winter. As was in winter, the water-column average zooplankton biomass over the shelf was higher

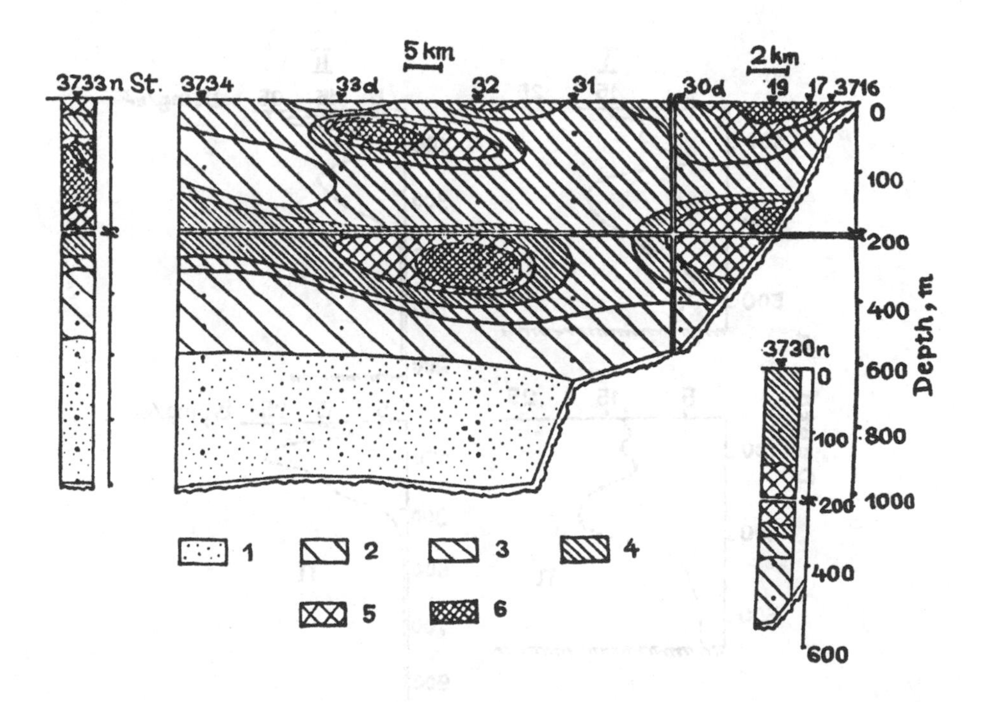


*Figure 7.* Distribution of total zooplankton biomass in the section along 32°20' N in winter. 1 - <1 mg/m³, 2 - 1-5 mg/m³, 3 - 5-10 mg/m³, 4 - 10-15 mg/m³, 5 - 15-20 mg/m³, 6 - >20 mg/m³. "d" and "n" after a station number mean "day" and "night", respectively

than that beyond. Over the depths of up to 100 m it varied from section to section within the range of 12 mg/m³ to 59 mg/m³ averaging at 33.3±10.7 mg/m³. The total zooplankton biomass in the shelf area varied from 0.5 g/m² to 4.7 g/m². In the late summer the zooplankton biomass over the shelf was a little higher than in winter.

A relatively uniform spatial distribution of zooplankton in the deep part of the surveyed area in winter and in the late summer suggests it is barely affected by peculiarities of mezoscale circulation, well pronounced in the region (see Krivosheya *et al.*, this volume). A relatively high zooplankton biomass over the Syrian shelf, observed in winter and summer, may hint the processes here are somewhat isolated. This assumption is also backed by clearly pronounced changes of N/P ratio, observed within the 300-500 m isobat area, as well as by earlier beginning of phytoplankton spring bloom over the shelf.

A peculiarity of the zooplankton vertical distribution in winter was a relatively high (compared to the late summer values) biomass found in the upper

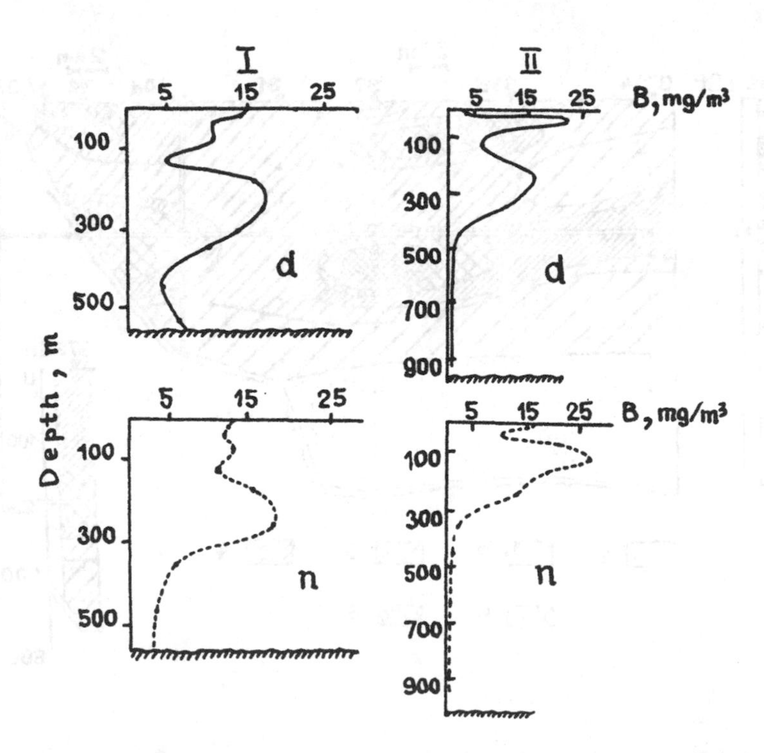


*Figure 8.* Day/night vertical distribution of total zooplankton biomass (B, mg/m³) at the stations of the section along 35°20' N in winter. I - shelf slope area (st. #3730); II - deep water area (st.#3733). "d" and "n" denote "daytime" and "nighttime" profiles, respectively.

mezopelagic zone between 200 m and 500 m (Fig. 7, 8). In the deep part of the area the average biomass across the layer was estimated as 7.5±3.2 mg/m³, while over the continental slope it was 13.3±5.8 mg/m³. The average biomass within the upper 200 m layer, as mentioned in the above, at deep stations was 17.7±6.3 mg/m³ , while at the shelf border it was 25±10.1 mg/m³.

The ratio between the zooplankton biomass in the 0-200 m layer and that in the 200-500 m layer at daytime in winter varied across the area from 1.1 to 8.3, while at night it varied from 1.3 to 11.5. The average values were 2.9±1.8 and 4.9±3.4, respectively. These significant variations of the above ratio stemmed from substantial differences between the biomass values in the 200-500 m layer at different stations: they differed 3.8 times in the daytime (from 4 mg/m³ to 15 mg/m³) and 5 times at night (from 2 mg/m³ to 10 mg/m³). This was because biomass estimations were affected by presence of large individuals, occasionally found in samples from 200-500 m depths, against the low background values of zooplankton abundance.

The trend for zooplankton biomass values growth in the upper 200 m layer, observed in daily variations in ratios between biomass values across the 0-200 m and across the 200-500 m layers, was not proved by analysis of daily zooplankton numbers variations across the 0-200 m layer in the surveyed area. In winter the daytime and the night values were nearly the same at 17.7±6.3 mg/m³ and 17.0±4.8 mg/m³, respectively. This fact along with a relatively high concentration of biomass within the 200-500m layer, compared to the summer values (see in the below), said the vertical migrations of interzonal species in winter were weak and they permanently resided beyond the photic layer. Daily migrations contributed very little into re-distribution of zooplankton biomass between the 0-200 m and 200-500 m layers in winter (Fig. 8). The contribution was barely visible against the background of weakly pronounced spatial non-uniformity of plankton distribution.

A detailed profiles of zooplankton vertical distribution in winter was obtained at the section along 35°20′ N (Fig. 7, 8). The peculiarity here was presence of two peaks - the surface one above 100 m deep and the deep-water one attributed to the 150-350 m layer. This was observed across the entire deep part of the section beginning from the shelf break zone (200 m). The exception was station #3731 only, located, according to hydrophysical mapping, near the core of the main current over the continental slope (about 670 m deep). Vertical distribution of zooplankton here was surprisingly uniform.

Mezoplankton biomass values at the surface and the deep peaks were very close to each other. For example, st. #3732 showed the values of 9 mg/m$^3$ and 12.5 mg/m$^3$, respectively (Fig. 7). The surface and the deep peaks were separated with a narrow layer of relatively low biomass, commonly attributed to the depth of 100-150(200) m. The zooplankton biomass here was about 5 mg/m$^3$ or less. At st. #3719 only over the shelf break it reached 9 mg/m$^3$ against the background of the surface and the deep peak biomass values of 27 mg/m$^3$ and 22.5 mg/m$^3$ respectively.

Another peculiarity of the zooplankton vertical distribution across the deep-water part of the surveyed area in winter was biomass drop below 500-600 m deep down to 0.5-0.9 mg/m$^3$ (Fig. 7, 8). The low values of biomass are generally common for Deep Mediterranean Water. But different position of the upper border of the water in different parts of the surveyed area did not result in respective displacement of the upper border of zooplankton-poor layer. For example, at st. #3733 the main picnocline and the upper border of the Deep Water were found about 100 m higher compared to those at the neighboring st. #3732 (see Krivosheya *et al.*, this volume). The border of the Deep Water was observed at 380 m and 500 m deep, respectively. But this made little effect on vertical distribution of zooplankton and location of the upper border of plankton-poor waters.

Detailed studies of daily changes in vertical distributions of zooplankton in winter based on daytime and nighttime sampling were done at st.st. #3730 and #3733 (Fig. 7, 8). The changes were small and are largely due to migrations of a certain part of zooplankton from the 200-300 m layer into the 50-150 m layer. Nighttime concentrations of zooplankton in the 150-300 m layer stayed relatively high, while sertain level of similarity of daytime and nighttime biomass vertical profiles persisted. A narrow minimum of biomass, which separates the surface and the deep-water layers of high concentrations in the daytime profiles, is lesser pronounced or not pronounced at all at night (Fig. 7, 8). Therefore, detailed studies of the daily dynamics of the zooplankton vertical distribution also say on absence of any significant changes in biomass profile and on permanent presence of relatively high zooplankton concentrations within the 150-300 m layer in winter.

Mezoplankton vertical distribution in the Syrian coastal waters in winter was quite uniform in space. This again hinted absence of any visible effect of the mezoscale circulation, well pronounced in the area, on the spatial structure of the zooplankton community. Stability of the situation is also proven by very similar biomass estimations, obtained in daytime and nighttime series at the same stations: st. #3730 showed 5523 $mg/m^2$ at daytime and 5539 $mg/m^2$ at nighttime, st. #3733 showed 5122 $mg/m^2$ and 5874 $mg/m^2$, respectively.

Vertical distribution of zooplankton biomass in the late summer dramatically differed from that in winter. Mapping clearly said that across the entire surveyed area the highest daytime concentrations of zooplankton were found within the 0-50 m layer. The biomass values here varied from 23 $mg/m^3$ to 88 $mg/m^3$ with the average value of $42.6 \pm 18.9$ $mg/m^3$. The average biomass within the 50-100 m layer was substantially lower at $17.2 \pm 7.8$ $mg/m^3$, while the ultimate values in most cases were 8 $mg/m^3$ and 32 $mg/m^3$. Below 100 m deep the biomass values dropped averaging at $5.2 \pm 1.9$ $mg/m^3$ and varying within the range from 3 $mg/m^3$ to 8 $mg/m^3$.

The same regularities in zooplankton vertical distributions were obtained in detailed study at the section along 35°20′ N (Fig. 9, 10). Most part of the plankton stayed within the upper 100 m layer, below which the biomass values were generally under 5 $mg/m^3$. Differences between the vertical distribution profiles in winter and the late summer can be seen from Fig.10. The differences stemmed from seasonal changes in taxonomic composition of the community. In winter the dominating role was played by large interzonal copepods, which mainly formed the deep-water biomass peak within the 150-350 m layer. In the late summer small epipelagic copepods dominated in the plankton. Salps that populated largely the surface 50 m layer also contributed significantly into the biomass.

Low biomass below 100 -150 m deep is typical for the Mediterranean waters in summer. Recalculation of the few data given by Grese [5] produced the

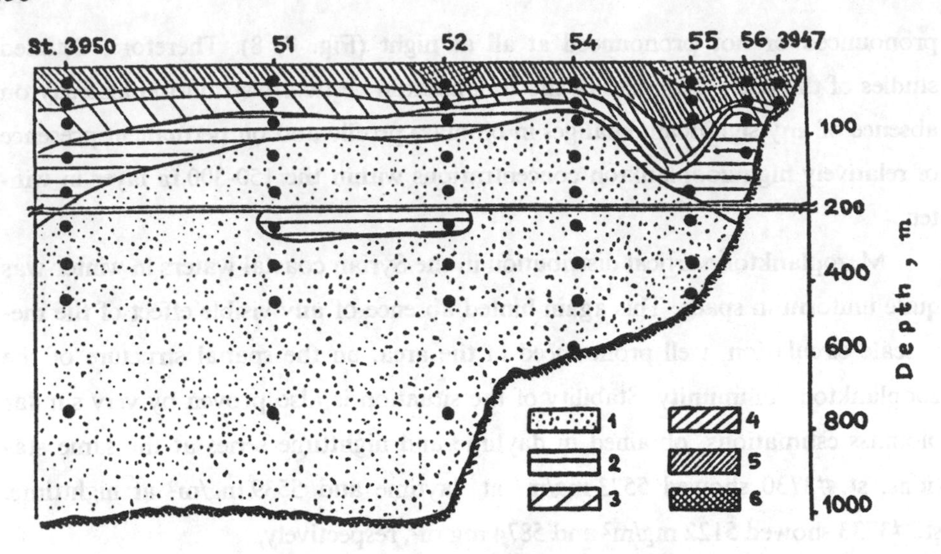


*Figure 9.* **Distribution of total zooplankton biomass in the section along 32°20' N in late summer. 1 - < 5 mg/m³, 2 - 5-10 mg/m³, 3 - 10-15 mg/m³, 4 - 15-20 mg/m³, 5 - 20-40 mg/m³, 6 - > 40 mg/m³.**

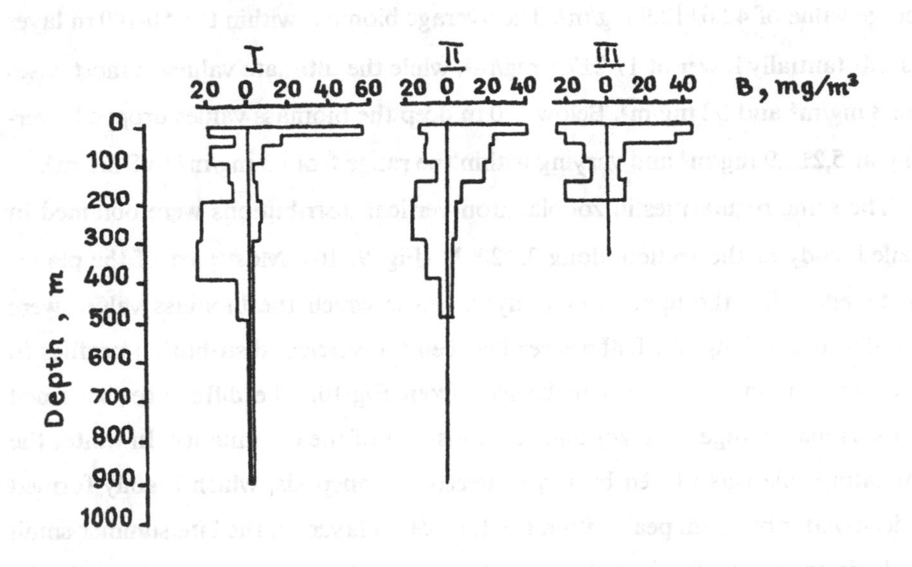


*Figure 10.* **Vertical distribution of total zooplankton biomass at the stations of the section along 35°20' N in winter (left profile) and late summer (right profile).**
**I - Deep water area (st. ## 3732 and 3952); II - area over lower part of the continental slope (st. ## 3730 and 3955); III - area over upper part of the continental slope ( st. ## 3716 and 3947).**

following values for summer: 3.5 mg/m$^3$ for the Algerian-Provance basin, 2.5 mg/m$^3$ for the Tyrrhenian sea, 4.1 mg/m$^3$ for the Adriatic Sea and 3.3 mg/m$^3$ in average for the Mediterranean Sea in general. The estimates are nearly the same as those we obtained for the Syrian waters.

Day and nighttime observations at the section showed very low variability of zooplankton biomass vertical distribution in the late summer. The average biomass in the upper 50 m layer at st. #3952 was 37 mg/m$^3$ at day and 47 mg/m$^3$ at night. Station #3955 showed 39 mg/m$^3$ and 31 mg/m$^3$, respectively. The durnal changes of biomass values at the stations within the 0-100 m layer were as follows: 23 mg/m$^3$ at day and 28 mg/m$^3$ at night at st. #3952; 24 mg/m$^3$ at day and 26 mg/m$^3$ at night at st. #3955.

## 4. References

1. Abi-Saab, M. (1985) Etude quattitative et qualitative du phytoplancton des eaux cotieres libanaises, *Lebenese Science Bulletin* ***1***, 197-222.
2. Abi-Saab, M. (1990) Variations quotidiennes des populations phytoplanctoniques durant une periode automnale en un point fuxe de la cote Libanaise, *Rapp. Pr.-Ver. Reun. CIESM.* **32**, 204.
3. Flint, M.V., Timonin, A.G. and Heptner, M.V. (1983) Electronic closer for plankton nets, in *Contemporary Methods of Quantitative Estimation of Distribution of Marine Plankton*, Nauka Publishing, Moscow, 73-81. (In Russian)
4. Grese, V.I. (1963) Peculiarities of bilogical structure of Ion basin pelagic zone, *Oceanology* **3**, 100-109. (In Russian)
5. Grese, V.I. (1989) *Pelagic Zone of the Mediterranean as an Ecological System,* Naukova Dumka Publishing, Kiev. (In Russian)
6. Lakkis, S. (1971) Contribution a l'etude du zooplancton des eaux libanaises, *Mar. Biol.,* **11**, 138-148.
7. Lakkis, S. (1973) Note preeliminaire sur la presence et la repartition des copepodes dans les eaux superticielles libanaises, *Rapp. Pr. - Ver. Reun.*

*CIESM.* **21**, 459-464.

8. Lakkis, S. and Novel-Lakkis, V. (1981) Composition, annual cycle and species diversity of the phytoplankton in Libanese coastal waters, *J. Plankt. Res.*, **3**, 123-136.
9. Lakkis, S. and Zeidane, R. (1983) Caractetistigues ecologiques et dynamiques du zooplancton des eaux cotieres libaaises, *Rapp. Pr.-Ver. Reun. CIESM,* **8**, 215-216.
10. Pancucci-Paradopoulou, M.-A., Siokou-Frangou, I. and Christou, E. (1990) On the vertical distribution and composition of deep-water Copepod population in the Eastern Mediterranean Sea, *Rapp. Pr. - Ver. Reun. CIESM*, **32**, 199.
11. Sukhanova, I.N. (1976) The quantitative composition and quantitative distribution of the phytoplankton in the northeastern Indian Ocean. Ecology and Biogeography of Plankton, *Proc. Inst. Ocean.*, **105**, 55-82. (In Russian)
12. Sukhanova, I. N. (1983) Concentrating of phytoplankton samples, in *Contemporary Methods of Quatitative Estimation of Marine Plankton Distribution*, Nauka Publishing, Moscow, 97-106. (In Russian)
13. Wiekert, H. (1990) Vertical distribution of zooplankton and micronekton in the deep Levantine Sea, *Rapp. Pr.-Ver. Reun. CIESM*, **32**, 199.

# COMPOSITION OF BOTTOM COMMUNITIES AND QUANTITATIVE DISTRIBUTION OF MACROZOOBENTHOS IN THE SYRIAN COASTAL ZONE

N.V.KUCHERUK [1], A.P.KUZNETSOV [1], P.V.RYBNIKOV [1],
SAKER FAYES [2].

[1] *P.P.Shirshov Institute of Oceanology Russian Academy of Science, Moscow, Russia.*

[2] *Higer Institute of Applied Sciences and Technology, Damascus, Syria.*

During February 1992 - October 1993 three research cruises were carried out in the eastern part of the Mediterranean. More than 200 quantitative samples of bottom fauna were collected by the grabb at the depths from 8 to 1200 m on the shelf of Syria (Figure 1), up to 6 grabbs at each station [7]. Basing on density and biomass, we calculated the values of a specific metabolic activity as a measure of species abundance [6].

The preliminary analysis of the data shows that the studied section of the Syrian shelf can be subdivided into two sharply distinct ecological areas. So called "**coralligenous bank**" [9] community was found at the southern part of the shelf at the depth of 25-100 m, deeper than the rocky bottom. Mass development of the calcigerous red macroalgae, underlaid by silty sediments is peculiar for this type of community. The crustose algae there form a firm crust, inhabited by diverse and abundant attached and motile fauna (Table 1-3). Deeper than 100 m community with dominance of sea-urchin *Cidaris cidaris* was found on the soft bottom.



*P. Malanotte-Rizzoli and V.N. Eremeev (eds.),*
*The Eastern Mediterranean as a Laboratory Basin for the Assessment of Contrasting Ecosystems*, 159–168.

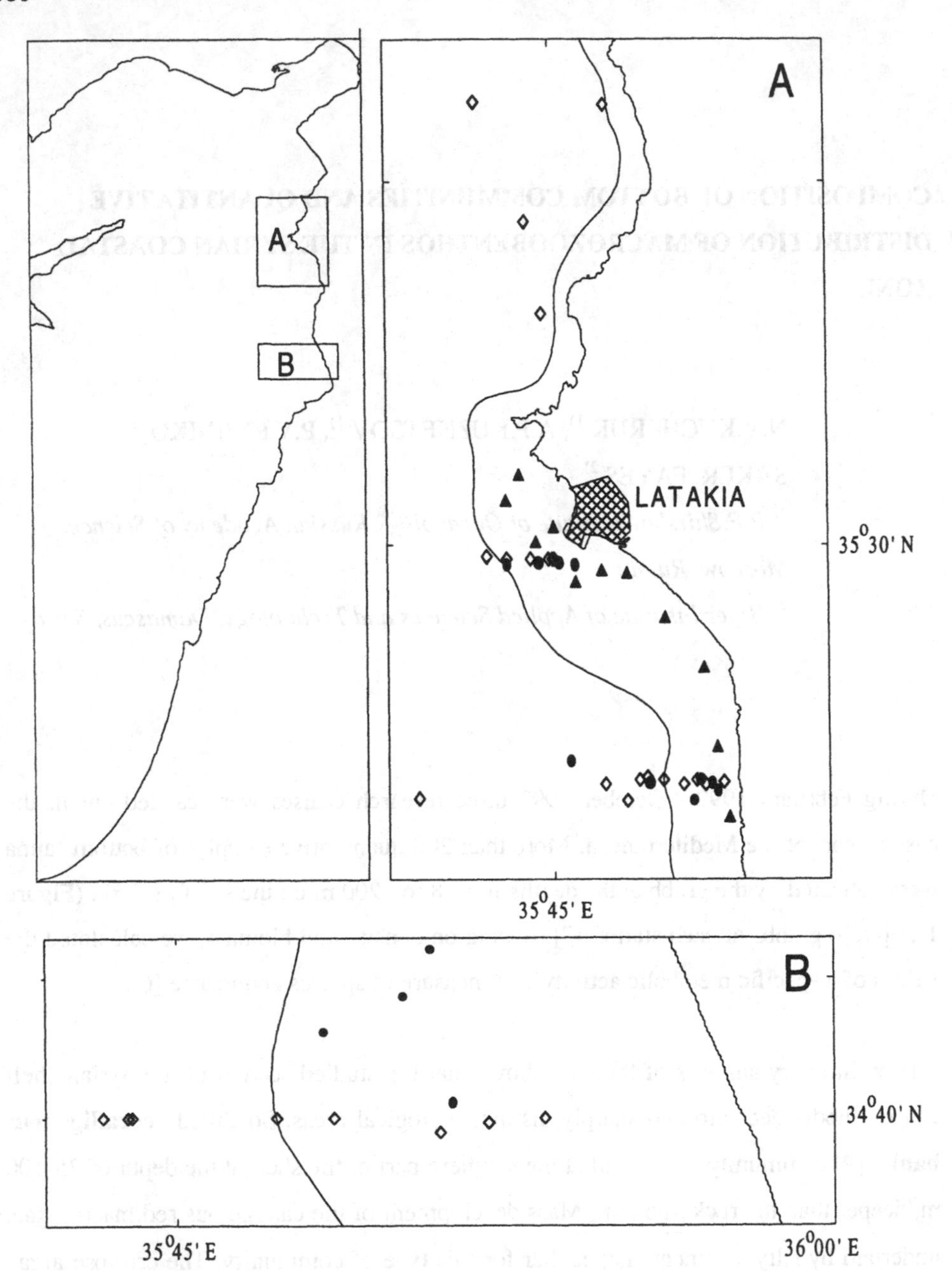


Figure 1. Maps of the benthic investigation regions along the Syrian coast.
A - north region, B - south region.
◇ - 24 cruise of r/v "Vityaz"
● - 27 cruize of r/v "Vityaz"
▲ - 11 cruize of r/v "Aqvanavt"

Table 1. Composition of macrozoobenthic community of "calciferous red algae" biocenosis at depth 25 m. Number of species - 54; biomass - 17,653 g/m$^2$.

| TAXONS | | Number of species | | Quantity, specmen/m$^2$ | | % of Bbiomass | | % of Respiration | |
|---|---|---|---|---|---|---|---|---|---|
| **SPONGIA** | | **1** | | **20** | | **0,0026** | | **0,004** | |
| **HYDROIDEA** | | **2** | | **20** | | **0,0070** | | **0,009** | |
| **POLYCHAETA** | | **25** | | **890** | | **0,3833** | | **0,463** | |
| | *Polynoidae* | | *2* | | *30* | | *0,0032* | | *0,007* |
| | *Nereidae* | | *2* | | *200* | | *0,0320* | | *0,055* |
| | *Amphinomidae* | | *2* | | *20* | | *0,0038* | | *0,006* |
| | *Chrysopetalidae* | | *2* | | *30* | | *0,0842* | | *0,073* |
| | *Eunicidae* | | *4* | | *140* | | *0,2073* | | *0,219* |
| | *Syllidae* | | *8* | | *360* | | *0,0340* | | *0,073* |
| | *Cirratulidae* | | *1* | | *20* | | *0,0019* | | *0,004* |
| | *Spionidae* | | *1* | | *20* | | *0,0006* | | *0,002* |
| | *Capitellidae* | | *1* | | *10* | | *0,0019* | | *0,003* |
| | *Chloramidae* | | *2* | | *60* | | *0,0147* | | *0,021* |
| **MOLLUSCA** | | **1** | | **10** | | **0,0510** | | **0,036** | |
| | *Loricata* | | *1* | | *10* | | *0,0510* | | *0,036* |
| **CRUSTACEA** | | **16** | | **600** | | **0,0595** | | **0,144** | |
| | *Isopoda* | | *3* | | *150* | | *0,0096* | | *0,026* |
| | *Amphipoda* | | *9* | | *380* | | *0,0365* | | *0,092* |
| | *Tanaidacea* | | *1* | | *20* | | *0,0038* | | *0,008* |
| | *Cumacea* | | *1* | | *10* | | *0,0006* | | *0,002* |
| | *Decapoda* | | *1* | | *20* | | *0,0038* | | *0,008* |
| **SIPUNCULOIDEA** | | **1** | | **20** | | **0,0051** | | **0,008** | |
| **ECHINODERMATA** | | **1** | | **90** | | **0,1202** | | **0,105** | |
| | *Ophiuroidea* | | *1* | | *90* | | *0,1202* | | *0,105* |
| **BRYOZOA** | | **4** | | **40** | | **0,4400** | | **0,250** | |
| **PANTOPODA** | | **2** | | **40** | | **0,0038** | | **0,008** | |

The northern part of the Syrian shelf was occupied by the communities of the soft sediments. The community with dominance of lessepsian migrant gastropod *Strombus persicus* and autochthonous bivalve *Acanthocardia tuberculata* was most common here on the sands. High biomasses (on the average 30 g/m$^2$) and rather high species richness (48 species per station, 13-16 species per sample) were typical for this community. These two dominants yielded their maximal biomasses at 20 m depth, and we have never found them below the summer thermocline. The most interesting peculiarity of

Table 2. Composition of macrozoobenthic community of "calciferous red algae" biocenosis at depth 50 m. Number of species - 28; biomass - 9,170 g/m$^2$.

| TAXONS | Number of species | Quantity, specmen/m$^2$ | % of Bbiomass | % of Respiration |
|---|---|---|---|---|
| **SPONGIA** | **1** | **40** | **4,330** | **0,333** |
| **NEMERTINI** | **1** | **20** | **0,030** | **0,003** |
| **POLYCHAETA** | **19** | **540** | **2,660** | **0,438** |
| *Polynoidae* | *1* | *10* | *0.010* | *0.003* |
| *Nereidae* | *1* | *10* | *0.060* | *0.011* |
| *Amphinomidae* | *2* | *60* | *0.400* | *0.072* |
| *Eunicidae* | *4* | *110* | *1,050* | *0,168* |
| *Syllidae* | *3* | *130* | *0,120* | *0,032* |
| *Pilargidae* | *1* | *10* | *0.020* | *0.005* |
| *Spionidae* | *1* | *20* | *0.100* | *0.019* |
| *Capitellidae* | *1* | *110* | *0.030* | *0.012* |
| *Maldanidae* | *2* | *30* | *0,280* | *0,039* |
| *Ampharetidae* | *1* | *20* | *0.120* | *0.022* |
| *Terebellidae* | *1* | *20* | *0.030* | *0.008* |
| *Hesionidae* | *1* | *10* | *0.440* | *0.049* |
| **CRUSTACEA** | **6** | **80** | **0,400** | **0,073** |
| *Amphipoda* | *4* | *50* | *0,040* | *0,013* |
| *Cumacea* | *1* | *10* | *0,010* | *0,003* |
| *Decapoda* | *1* | *20* | *0,350* | *0,057* |
| **SIPUNCULOIDEA** | **1** | **50** | **0,250** | **0,042** |

Table 3. Composition of macrozoobenthic community of "calciferous red algae" biocenosis at depth 90 m. Number of species - 17; biomass - 3,010 g/m$^2$.

| TAXONS | Number of species | Quantity, specmen/m$^2$ | % of Biomass | % of Respiration |
|---|---|---|---|---|
| **POLYCHAETA** | **14** | **480** | **1,890** | **0,589** |
| *Phyllodocidae* | *1* | *10* | *0,0033* | *0,006* |
| *Glyceridae* | *1* | *10* | *0,0266* | *0,027* |
| *Nereidae* | *1* | *30* | *0,0166* | *0,025* |
| *Eunicidae* | *3* | *30* | *0,2924* | *0,18* |
| *Syllidae* | *3* | *60* | *0,0100* | *0,02* |
| *Paraonidae* | *1* | *170* | *0,0199* | *0,044* |
| *Spionidae* | *1* | *10* | *0,0033* | *0,006* |
| *Capitellidae* | *1* | *10* | *0,0100* | *0,013* |
| *Chloramidae* | *1* | *40* | *0,0133* | *0,022* |
| *Maldanidae* | *2* | *110* | *0,2326* | *0,247* |
| **MOLLUSCA** | **1** | **10** | **0,100** | **0,0220** |
| *Bivalvia* | *1* | *10* | *0,1000* | *0,022* |
| **CRUSTACEA** | **1** | **30** | **0,020** | **0,0140** |
| *Amphipoda* | *1* | *30* | *0,0200* | *0,0140* |
| **SIPUNCULOIDEA** | **1** | **350** | **1,000** | **0,3750** |

the community was its trophic structure with prevailing of phytophagon (*S. persicus*). This phytophagous snail produced as high as 58% of the total community respiration (75% in total biomass, up to 93% at several stations), meanwhile mobile suspension-feeders (usually typical inhabitants of sands) provided altogether as low as 13% (1,2-30% per station) of community respiration (Table 4).

Table 4. Composition of macrozoobenthic community of "seagrass sand bottom" biocenosis at depth 20 m. Number of species - 64; biomass - 45,783 g/m$^2$.

| TAXONS | Number of species | Quantity, specmen/m$^2$ | % of Biomass | % of Respiration |
|---|---|---|---|---|
| **ACTINIARIA** | **2** | **24** | **0,0004** | **0,0038** |
| **NEMERTINI** | **1** | **2** | **0,00002** | **0,0002** |
| **POLYCHAETA** | **31** | **470** | **0,00754** | **0,0748** |
| *Phyllodocidae* | *2* | *8* | *0,00009* | *0,00091* |
| *Sigalionidae* | *1* | *2* | *0,00026* | *0,00176* |
| *Nereidae* | *2* | *8* | *0,00004* | *0,00061* |
| *Nephtiydae* | *3* | *28* | *0,00174* | *0,01087* |
| *Eunicidae* | *4* | *34* | *0,00090* | *0,00883* |
| *Poecilochaetidae* | *1* | *2* | *0,00001* | *0,00008* |
| *Ariciidae* | *1* | *2* | *0,00029* | *0,00194* |
| *Cirratulidae* | *3* | *136* | *0,00161* | *0,01897* |
| *Paraonidae* | *3* | *136* | *0,00161* | *0,01897* |
| *Spionidae* | *2* | *56* | *0,00030* | *0,00441* |
| *Capitellidae* | *2* | *10* | *0,00002* | *0,00035* |
| *Maldanidae* | *2* | *14* | *0,00013* | *0,00173* |
| *Terebellidae* | *1* | *12* | *0,00040* | *0,00379* |
| *Sabellidae* | *3* | *20* | *0,00013* | *0,00151* |
| *Hesionidae* | *1* | *2* | *0,00001* | *0,00008* |
| **MOLLUSCA** | **12** | **58** | **0,99081** | **0,9110** |
| *Gastropoda* | *3* | *6* | *0,74100* | *0,61315* |
| *Scaphopoda* | *1* | *2* | *0,00046* | *0,00296* |
| *Bivalvia* | *8* | *50* | *0,24935* | *0,29488* |
| **CRUSTACEA** | **15** | **160** | **0,00103** | **0,0162** |
| *Amphipoda* | *11* | *112* | *0,00081* | *0,01235* |
| *Tanaidacea* | *1* | *14* | *0,00004* | *0,00101* |
| *Cumacea* | *3* | *34* | *0,00018* | *0,00282* |
| **ECHINODERMATA** | **2** | **8** | **0,00145** | **0,0078** |
| *Ophiuroidea* | *2* | *8* | *0,00145* | *0,0078* |

Deeper, below the level of summer thermocline, water temperature drops by 15 degrees, and the rate of microbial decomposition of hydrocarbons decreases. The abundance and distribution of animals below this level likely depended on the oil impact. Let us compare communities situated on different distances from the oil refinery in the Baniyas town. They differ considerably. Ten miles from the town at the depth of 40 m total biomass there was as low as 5 $g/m^2$, with the dominance of capitellid polychaetes - well known indicator of the unfavorable environmental conditions due to anthropogenic pollution. The species diversity nevertheless remained high, fauna accounted as much as 37 species. Deeper (85 m) the biomass increased sharply up to 30 $g\ m^{-2}$, the species richness remained high - 33 species (Table 5). The sorting deposit-feeders (sea-urchins *Shizaster canaliferus*) and suspension-feeders (sea-pens *Pennatula rubra*) dominated. Going down to 120-175 m, both the species richness and the biomass drastically decreased (the biomass – 6 to 10 times, species number to 20-25 per station). Nevertheless, these values several times exceeded ones from northern, unpolluted part of the shelf. Near Latakia, at 20 miles from Banias refinery, influence of oil pollution was found to be negligible. So, we can see that both the biomass and species richness of benthos under the thermocline here were significantly lower and decreased with depth (Figure 2). Number of species here never exceeded 10. It looks like nowadays anthropogenic pollution appear to be stimulating rather than depressing factor for zoobenthos in Eastern Mediterranean.

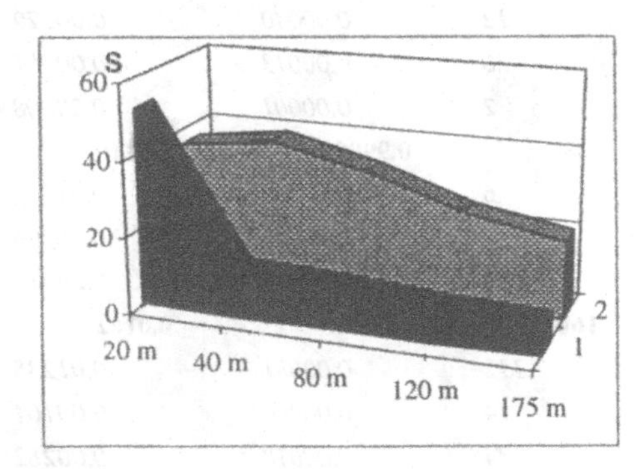


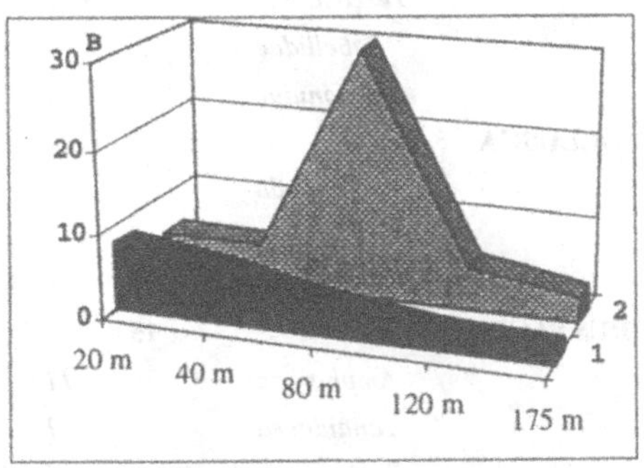


Figure 2. Macrobenthic species diversity and biomass depending on depth.

**S** - number of species; **B** - biomass, $g/m^2$.

1 - 20 miles from refinery, 2 - 10 miles from refinery.

Table 5. Composition of macrozoobenthic community of " silty bottom" biocenosis at depth 85 m. Number of species - 33; biomass - 30,388 g/m$^2$.

| TAXONS | Number of species | Quantity, specmen/m$^2$ | % of Biomass | % of Respiration |
|---|---|---|---|---|
| **COELENTERATA** | **3** | **8** | **0,15371** | **0,14558** |
| **NEMERTINI** | **1** | **4** | **0,00033** | **0,00163** |
| **POLYCHAETA** | **10** | **160** | **0,06475** | **0,14598** |
| *Pholoididae* | *1* | 2 | *0,00593* | *0,01197* |
| *Glyceridae* | *1* | *18* | *0,00527* | *0,01899* |
| *Nereidae* | *1* | 2 | *0,00007* | *0,00041* |
| *Nephtiidae* | *1* | *4* | *0,04281* | *0,06273* |
| *Eunicidae* | *1* | *12* | *0,00290* | *0,01096* |
| *Syllidae* | *1* | *4* | *0,00026* | *0,00138* |
| *Poecilochaetidae* | *1* | *10* | *0,00059* | *0,00318* |
| *Cirratulidae* | *1* | *6* | *0,00020* | *0,00123* |
| *Spionidae* | 2 | *102* | *0,00672* | *0,03513* |
| *Capitellidae* | *1* | *16* | *0,02035* | *0,05079* |
| *Chloramidae* | *1* | 2 | *0,00007* | *0,00041* |
| *Ampharetidae* | 2 | *6* | *0,00092* | *0,00383* |
| *Serpulidae* | *1* | 2 | *0,00013* | *0,00069* |
| *Hesionidae* | *1* | 2 | *0,00007* | *0,00041* |
| **MOLLUSCA** | **6** | **26** | **0,01799** | **0,0338** |
| *Bivalvia* | 6 | 14 | *0,00916* | *0,01737* |
| **CRUSTACEA** | **2** | **4** | **0,02825** | **0,05038** |
| *Cumacea* | *1* | 2 | *0,00494* | *0,01199* |
| *Decapoda* | *1* | 2 | *0,02331* | *0,03839* |
| **SIPUNCULOIDEA** | **2** | **32** | **0,0114** | **0,03871** |
| **ECHINODERMATA** | **2** | **12** | **0,71088** | **0,54421** |
| *Echinoidea* | *1* | *8* | *0,70337* | *0,52943* |
| *Ophiuroidea* | *1* | *4* | *0,00751* | *0,01478* |

Beyond the shelf edge, at the depths more then 200 m, the biomass of macrobenthos never exceeded 1 g m$^{-2}$, in some samples it was as low as 0.05 g m$^{-2}$. The small number of specimens (often only 2-3 per grabb) didn't allow to say anything about the structure of the slope communities. Nevertheless, the polychaeta *Spiochaetopteris tipicus* seemed to dominate at this depth [7]. Biodiversity is very low - 2 - 5 species per station. At depth more then 500 m only three taxons - Coelenterata, Polychaeta and Bivalvia were found (Table 6).

Table 6.Compozition of benthic communities in different depth zones (% of biomass)

| TAXONS | DEPHTH RANGE | | | | |
|---|---|---|---|---|---|
| | 5-40 • | 40-100 • | 100-200 • | 200-500 • | 500-1200 • |
| Spongia | - | 10,4 | <0,1 | 0,6 | - |
| Coelenterata | <0,1 | 17,0 | 12,0 | 1,0 | <1,0 |
| Nemertini | <0,1 | <1 | 33,2 | - | - |
| Polychaeta | 2,74 | 31,2 | 33,2 | 63,7 | 50,5 |
| Bryozoa | - | 17,0 | <1 | - | - |
| Sipunculida | <0,1 | 5,0 | 9,2 | 5,2 | - |
| Bivalvia | 33,9 | 4,0 | 5,4 | 22,5 | 34,5 |
| Gastropoda | 62,4 | - | - | - | - |
| Scaphopoda | 0,5 | - | - | - | - |
| Brachiopoda | - | - | - | <1 | - |
| Echinodermata | 0,2 | 4,0 | <0,1 | <0,1 | - |
| Crustacea: | 0,5 | 10,4 | 7,0 | 7,0 | - |
| Pogonophora | - | - | - | <0,1 | - |

Low benthic biomasses seem to be a typical feature of the shelf of Syria. And, besides, lessepsian migrant *Strombus persicus* produced up to 80% biomass of the richest near-shore community. This alien species could achieve such great ecological success just in case of presence of free ecological niche. *Strombus persicus* is the only mass species in the Eastern Mediterranean able to consume sea-grasses *Cymodocea nodosa* and *Halophila stipulacea.* This is remarkable that we could find just one alive specimen of carnivorous snail *Murex (Bolinus) brandaris*. In antient time the gastropods, living exactly by Eastern Mediterranean coast, were used for production of purple. Each snail yielded a tiny amount of paint, that's why thousands and thousands of mollusks had to e caught to paint only one toga of some rich Romanian. Millions of *Murex* were caught in the Eastern Mediterranean every year. However, several centuries ago this catching was stopped. Now only empty shells of *M. brandaris* are abundant at the Syrian beaches. The decline seems to happen recently. It is important to keep in mind that the main prey for *Murex,* autochthonous Mediterranean bivalve *Acanthocardia tuberculata* was numerous in eutrophicated localities only, for example, directly near the entrance to the port of Latakia.

We can suppose that the oligotrophy of the Eastern Mediterranean observed nowadays, results from decreasing of the Nile biogenic flux mostly due to the Asuan High Dam construction at 1965 [5,9]. The Nile waters were the only significant source of nutrients in the eastern part of the Mediterranean. Its cessation lead to impoverishment of coastal waters in biogenous elements [3,4,5,8]. Nowadays the high-productive area near delta of the Nile is even smaller than the zone of euthrophycation near health resorts of Tunis. As a results of drying up of the Nile flux the environmental conditions in the Levant Sea are more similar now to the olygothrophyc conditions in the Red Sea, in contrast to the western part of the Mediterranean. The autochthonous Mediterranean species appeared to be less adapted to the existence under high temperature and lack of food in comparison with preadapted to this environment lessepsian migrants. The most common and abundant shallow-water species along Syrian coast nowadays are lessepsian migrants bivalves *Pinctada radiata, Brachiodontes variabilis, Paphia textile,* crustaceans *Portunus pelagicus, Charybdis hellery* and several peneid shrimps [2,10]. The mollusk *Strombus persicus* was mentioned in the Mediterranean only in 1986 [1,11], its rapid distribution seems to be good example of aforesaid concept.

**References**

1. Barash, A., Danin, Z. (1986) Further additions to the knowledge of Indo-Pacific molluska in the Mediterranean sea (Lessepsian migrants), *Spixiana* **9, N 2**, 117-141.
2. Gilat, E. (1964) The macrobenthonic invertebrate communities of the Mediterranean continental shelf of Israil.), *Bull. Inst. oceonogr. Monaco* **62 (1290)**, 2-46.
3. Gorgy, S. (1966) Les pechcies et le mullieu marin dans la secteur medtterraneen de la Republique Arabie Unie, *Rev. Trav Inst. (sci. tech.) Peches Marit.* **30 (1)**, 25-80.
4. Halim, Y. (1960) Observations on the Nile bloom of phytoplankton in the Mediterranean, *J. Cons. Perm. Int. Explor. Mer* **26**, 57-67
5. Halim, Y., Guergues, S.K., Saleh, H.H. (1967) Hydrographic conditions and plankton in the south-east mediterretnean during the last normal Nile Flood (1964), *Intern. Revue Ges. Hydrobiol* **52 (3)**, 401-425.
6. Kucheruk, N.V. (1995) Upwelling and benthos: gulid structure of low species communities, *Report. RAS* **345 ( 2)**, 000-000 (in Russian, English translation).

7. Kuznetsov, A.P., Kucheruk, N.V., Fayes, S., Rybnikov, P.V. (1993) Benthic fauna of near-syrian region in the East Mediterranean, *Proc. RAS, ser. biol.* **4**, 600-612 (in Russian, English translation).
8. Oren, O.H. (1969) Oceanographic and biological influence of the Suez Canal, the Nile and the Aswan Dam on the Levant basin, *Rep. Progr. Oceanogr.* **5**, 161-16.
9. Pérès, J.M. (1967) The Mediterranean Benthos, *Oceanogr. Mar. Biol. Ann. Rev.* **5**, 449-533.
10. Por, F.D. (1978) *Lessepsian migration,* Springer Verlag, Berlin.
11. Por, F.D. (1990) Lessepsian migration. An appraisal and new data, *Bull. Inst. oceonogr. Monaco* **7**, 1-10.

# ABUNDANCE AND BIOMASS OF PHYTOPLANKTON AND CHLOROPHYLL CONCENTRATION IN EASTERN MEDITERRANEAN IN LATE SUMMER PERIOD

I. N. SUKHANOVA, O. V. MAKSIMOVA, and N. P. NEZLIN
*P. P. .Shirshov Institute of Oceanology, RAS,*
*36 Nakhimovskiy Avenue, Moscow, 117851, Russia*

**Abstract.** The data on taxonomic composition, abundance, and biomass of phytoplankton and chlorophyll *a* concentration in the eastern part of the Mediterranean Sea during late summer period are given in the paper. Different methods of measurements of the chlorophyll *a* concentration are compared. All the quantitative characteristics of phytoplankton manifest the oligotrophic level of water productivity during the period under study. The spatial distribution of the abundance and biomass of phytoplankton and chlorophyll *a* concentration are significantly influenced by the mesoscale eddies of typical size 10–50 miles. The quantitative characteristics of phytoplankton in the coastal zone were usually higher, as compared with the open sea regions. As for vertical distribution of micro- and nanophytoplankton, the maximum cell concentrations occurred within the upper homogenous layer and in the layer of pycnocline. The vertical profiles of phytoplankton and chlorophyll *a* were compared with the profiles of fluorescence; the maximum of fluorescence corresponded to the maximum of chlorophyll *a* concentration and seemed to result from high concentration of picophytoplankton. The maximum of fluorescence also manifested itself in the cases when the cell densities in the layers of maximum if micro- and nanophytoplankton exceeded $10^4$ cells per liter.

## 1. Introduction

This paper is based on the data obtained in October 1993 during Cruise 23 of R/V *Vityaz* in the eastern part of the Mediterranean Sea over the area between Cyprus and the coast of Syria (Figure 1). The studies were carried out within the scope of joint Russian-Syrian program.

The paper contains the data on abundance, biomass and chlorophyll *a* concentration; the comparative analysis of the results of direct and indirect measurements of phytoplankton biomass, using different methods.



*P. Malanotte-Rizzoli and V.N. Eremeev (eds.),*
*The Eastern Mediterranean as a Laboratory Basin for the Assessment of Contrasting Ecosystems,* 169–179.

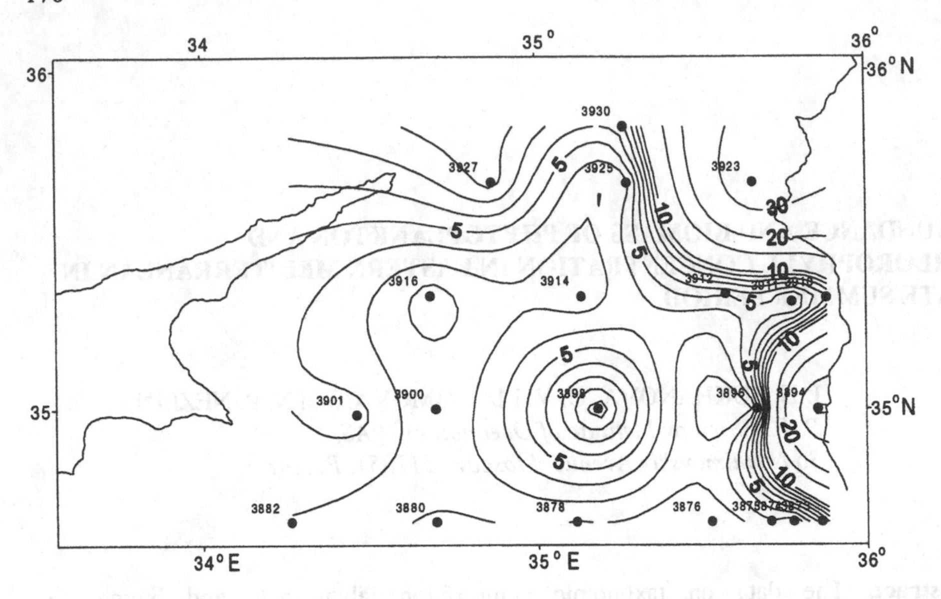


Figure 1. Spatial distribution of phytoplankton abundance ($10^3$ cells/l) over studied region.

## 2. Material and Methods

The following methods were used in this study:

1. Direct counts and measurements of cells and calculating of cell volume by the method of geometric similarity and transfer to a wet weight. Phytoplankton was counted and measured in a Nojeotte counting chamber (0,08 ml) and in a 1 ml counting chamber using a JENALUMAR luminescent microscope (Germany) at x375 and x90, accordingly. At the first step the samples of sea water (1,5-2.0 l) were concentrated by reverse filtration method using nuclear filters with pore size 1 μm [6].

2. The chlorophyll *a* concentration was estimated using fluorimethric method. The samples of 100-150 ml of sea water were filtered over glass-fiber filter GF/F (diameter 4.25 mm, "Whatman International Ltd", England). The samples were filtered immediately after collection using vacuum/pressure station of "Barnant Co" (USA) with vacuum of 15-20 mm of mercury column. After the end of filtration the filters were placed into freezing chamber (-15°C) and kept in exicator with silicogel by the end of the expedition. Then the filters were processed at the Biological Department of Moscow State University under the supervision of S. I. Pogosyan and N. M. Merzlyak. Pigments were extracted by 10 ml of the mixture of 90% acetone water solution and dimethylsulphoxide (3:2 volume ratio) at $4^0$ C during 4 to 6 hours in the dark. Fluorescence of the extract was measured using FP-550 fluorimeter (Japan, Jasco).

Water samples for measurements of phytoplankton biomass and chlorophyll *a* concentration were taken from the same plastic bottles of 1.7 l volume of the hydrophysical CTD-sound Neil-Braun.

3. The fluorescence (in arbitrary units) was measured using the fluoremeter "Sea Tech" connected with the CTD-sound Neil-Braun. The profiles of fluorescence were measured by the depth of 200–250 m at all the stations where phytoplankton samples were collected and chlorophyll *a* was measured.

The remote sensed data of surface concentration of plant pigments were also used. These data were obtained by CZCS radiometer. The CZCS radiometer at the American satellite “Nimbus-7” [1] collected information during 1978-1986; then the data were summarized in Goddard Space Flight Center (GSFC) and kindly granted us by our American colleagues within the framework of collaboration on the international SeaWiFS program. The data were obtained from the NASA Physical Oceanography Distributed Active Center at the Jet Propulsion Laboratory (California Institute of Technology).

The spatial resolution of CZCS data is about 20 km; the resolution of Sea WiFS data is about 10 km. The special software for selection of data from distinct regions was used; we already used it in our previous studies [3, 4].

The field studies consisted of two steps. At the first step (October 8–12, 1993) the samples were collected at the stations over the whole area under study. At this step the samples of phytoplankton and chlorophyll were collected from 10–12 levels and integrated (combined in distinct proportion) into one sample of 4–5 l volume; this sample characterized the water column from surface to 200 m or to bottom. The exception was the southern cross-section (stations 3873-3882). The samples there were collected from three layers: above the maximum of fluorescence, below it, and in the layer of maximum of fluorescence. At the second step the stations at the cross-section along the 35°20'N were repeated 10 days later, from October 19 to 24, 1993. The samples from 16–18 levels were integrated into 4–5 layers for the analysis of thin vertical structure of phytoplankton community. The following layers were studied: above the pycnocline, in the pycnocline, below it, and the layer of maximum of fluorescence.

## 3. Results

During the studies the temperature of the upper quasi-uniform layer (UQL) varied within the limits of 25 to 27°C, the salinity varied from 39.50 to 39.75 ppt, and the specific density from 26.1 to 26.9. The depth of UQL in the southern and central parts of the surveyed area varied from 20 to 50 m, in the northern part it was as deep as 75 m. The pronounced pycnocline occurred below the UQL. The vertical gradients at its upper boundary were 0.8°C/m, 0.05 ppt/m, and 0.07–0.12 $m^{-1}$. UQL and the layer of seasonal pycnocline belonged to the Atlantic water mass with the core at the depth of 50–90 m. The lower boundary of the Atlantic water mass coincided with the lower boundary of the seasonal pycnocline. Below the Levant water mass occurred with the upper boundary at 75–125 m, lower boundary at 180–400 m, and temperature of 15–17°C [9]. The stocks of nutrients during the period of studies were insignificant. The phosphate concentration by the depth of 100 m was 0.10–0.30 μg-at/l. The minimum concentrations occurred at the depth of 50–75 m. The nitrate concentration within the upper layer of 75 m was very low, often about analytic zero, except the coastal region between Latakia and Tartus (0.8–0.9 μg-at/l). At the depth of 150 m nitrate concentration increased by 5–6 μg-at/l. Nitrites occurred only in the layer of 75–150 m (by 0.4 μg-at/l). The concentrations of ammonia significantly varied from one station to another (0.25–2.0 μg-at/l). The distribution of this labile form of nitrogen is extremely unstable in all seas. The maximum concentrations of ammonia usually occur in the upper 100-m layer. The silica concentration in the upper layer varied from 1.3 to 2.25 μg-at/l. The silica concentration was evenly distributed down to 100 m. Below the silica concentration slowly increased and at 500 m it was 8–13 μg-at/l [2].

The composition of phytoplankton at the first step of studies indicated the late summer state of phytocenosis. At the station in open sea regions above 90% of total cell number consisted of dinoflagellates, coccolithophorids and flagellates. These groups also dominated in biomass. The dominating size fraction was small nanoplankton forms with cell volume <300–500 $\mu m^3$. By 40% of them consisted of smallest forms of 10–30 $\mu m^3$ volume. The contribution of diatoms into both abundance and biomass never exceeded 10%. Only at shallow shelf stations 3873, 3894, 3910, and 3923, and at one station in the open part of northern section (3927) the percentage of diatoms was from 16 to 28% of total abundance and from 15 to 89% of total biomass of phytoplankton. In the coastal zone diatoms were mainly represented by neretic species of genus *Chaetoceros* ( *C. curvisetus, C. socialis, C. compressus* and other), *Pseudo-nitzschia delicatissima*, and few benthic Pennatae. High biomass at the coastal and open sea stations resulted from large forms of diatoms, in particular *Rhizosolenia alata* and *R. calcar-avis*. These species are typical for spring and autumn phytoplankton of the Mediterranean and the Black Seas; they seem to indicate the beginning of autumn blooming. Ten days later, during the second step of studies, the abundance of these species increased at coastal stations, and they appeared in the central part of the area under study; it seems to confirm the beginning of the autumn blooming. At the same time the abundance of some species of *Chaetoceros* increased, and the autumn species *Cerataulina bergonii* appeared. The large species of diatoms occurred in phytoplankton during the second step of studies, it resulted in that diatoms dominated in total biomass, and the total abundance and biomass increased two–three-fold in the open sea regions and ten-fold over the shelf, as compared with the previous survey.

The abundance of phytoplankton during the late summer period varied within the limits of $5 \cdot 10^3$ to $50 \cdot 10^3$ cells/l in the layer 0–200 m or 0–bottom (Figure 1). At some levels the cell numbers were as high as $2–3 \cdot 10^5$ cells/l. The highest cell concentration occurred at coastal stations, in northern and central regions. The northern stations 3927 and 3930 were located within the zone of influence of stable anticyclonic eddy. A rarely observed phenomenon occurred in the center of the surveyed area. Two eddies of alternative direction occurred at two levels: the surface cyclonic eddy in the layer of 0–100 m was in contact with the subsurface anticyclonic eddy in the layer of 50–250, the latter was partly shifted under the cyclonic eddy. The water properties indicated the structure of lens type, with increased salinity and temperature [8, 9]. The maximum of phytoplankton abundance occurred in that lens. Slight increase of abundance occurred in the south-western region (stations 3882 and 3901), which was under the influence of quasi-stable cyclonic gyre.

The phytoplankton biomass varied within the limits of 1.2–43.2 μg/l (average for the layer of 0–200 m or 0–bottom). The high range of biomass variation resulted from the differences in composition and size structure of phytoplankton in open and coastal regions and the changes of these indices that occurred during the period between the first and the second steps of studies. In general, the biomass distribution was similar to the pattern of the distribution of phytoplankton abundance (Figure 1). The exception was the coastal station 3910, where the high density of cells did not result in the biomass increase due to the dominance of small flagellates, dinoflagellates and coccolithophorids, and the absence of typical coastal complex of diatoms. The average cell size at this station was 150 $\mu m^3$.

The spatial distribution of chlorophyll *a* concentration also had maxima near the shore, in the northern and south-eastern parts of the surveyed area (Figure 2); it corresponded to the distribution of phytoplankton abundance and biomass. In the central part of the surveyed area within the zone of influence of cyclonic eddy and high phytoplankton abundance and biomass the concentration of chlorophyll was the lowest (station 3898—0.085 μg/l). However, ten miles to north-east (station 3905), where only the layer of maximum of fluorescence was sampled, in the lens of increased salinity and temperature, the maximum chlorophyll concentration occurred (1.058 μg/l). The chlorophyll *a* concentration over the whole studied area was low and the averaged for the water column concentration varied from 0.054 to 0.251 μg/l. Similar values are typical to oligotrophic ocean regions.

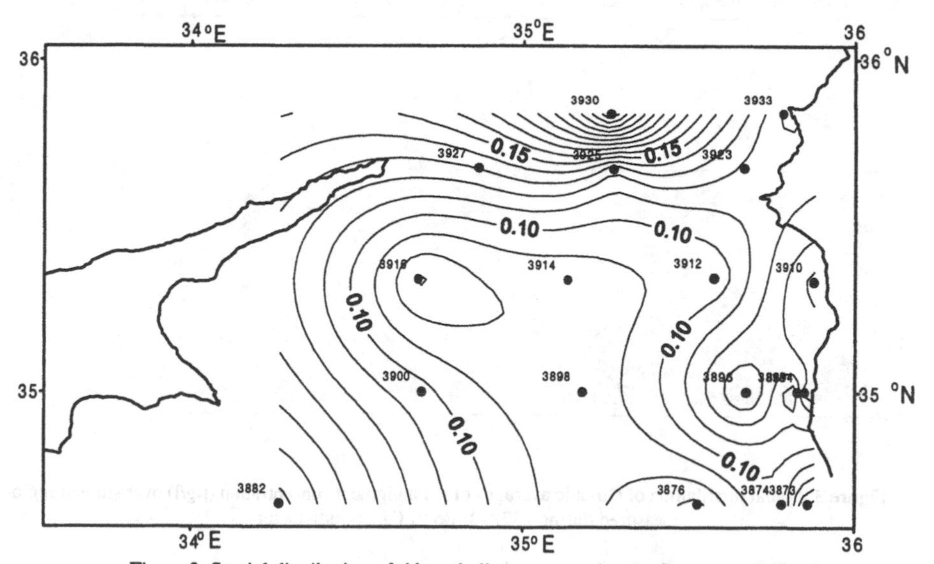


Figure 2. Spatial distribution of chlorophyll *a* concentration (μg/l) over studied region.

The remote data on chlorophyll *a* concentration illustrate the results comparable with the spectrophotometric measurements. The measured from the satellite surface chlorophyll *a* concentration in October varied in the eastern Mediterranean from 0.10 to 0.15 μg/l; in the coastal regions chlorophyll concentration increased by 0.20–0.30 μg/l (Figure 3).

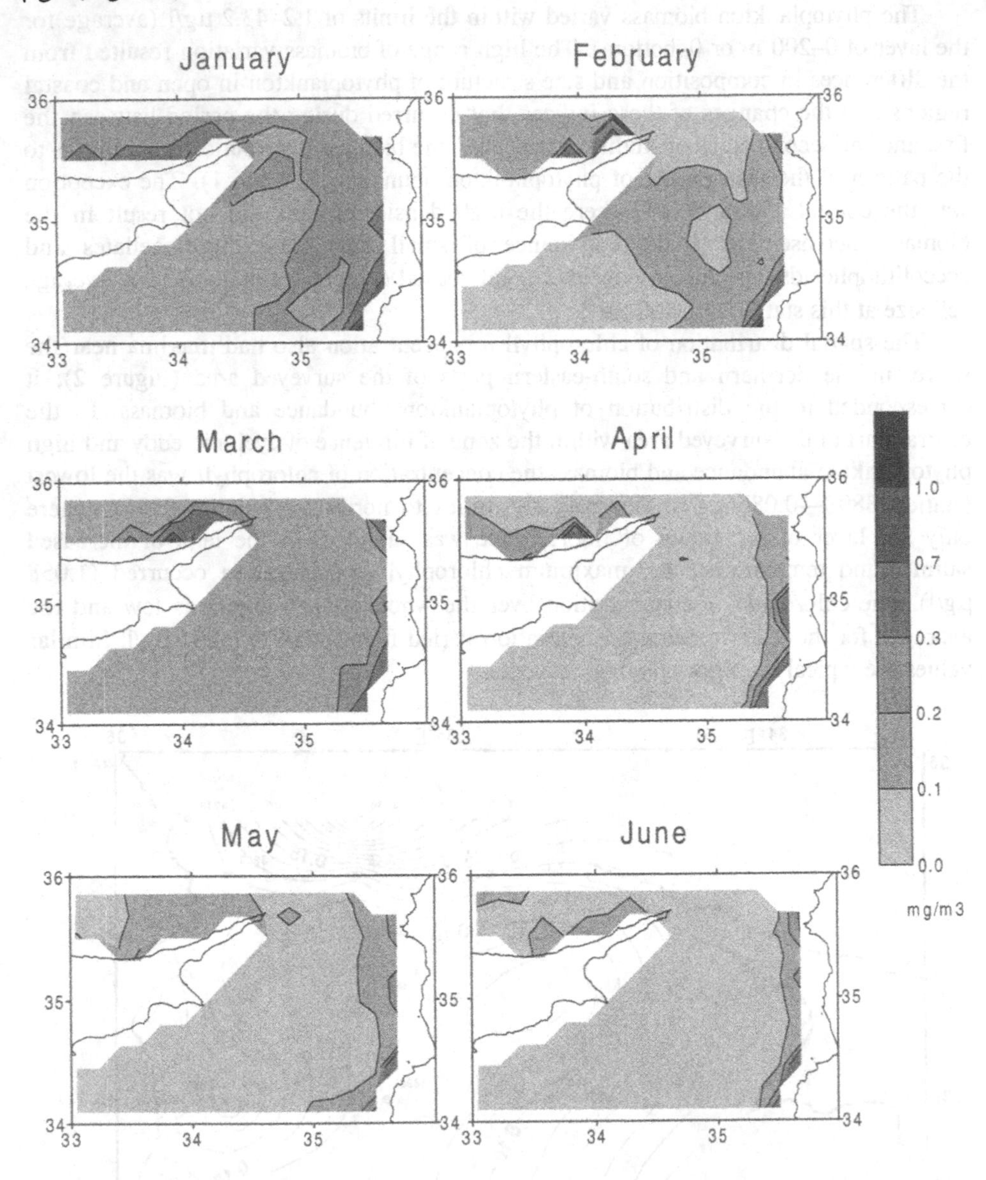


Figure 3. Spatial distribution of climatic averages of plant pigment concentration (μg/l) over studied region, measured during 1978–1986 by CZCS radiometer.

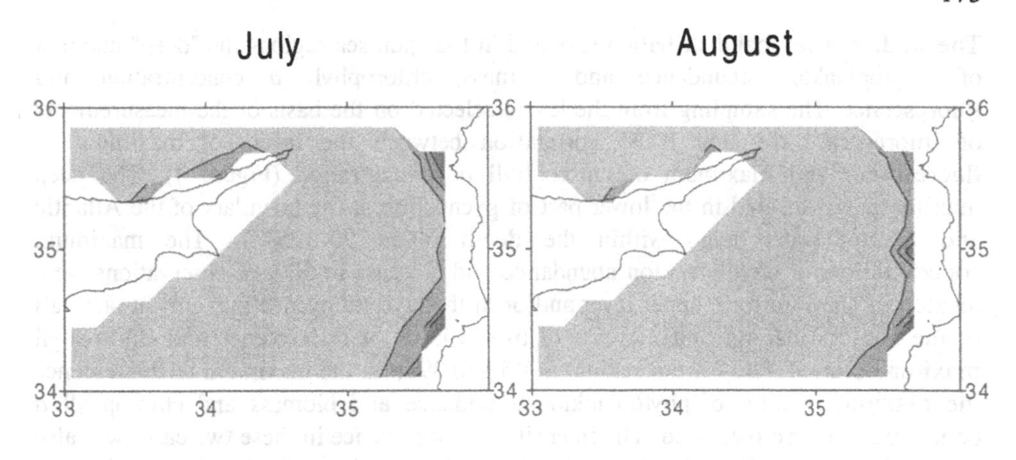


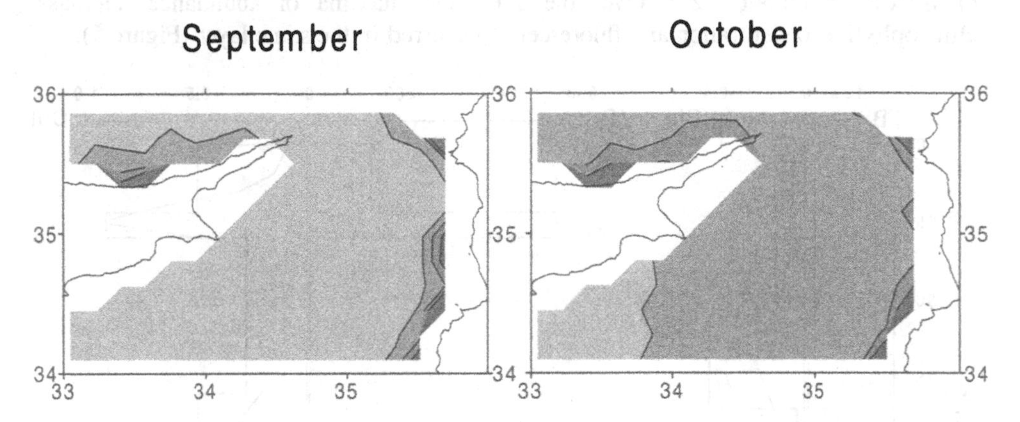


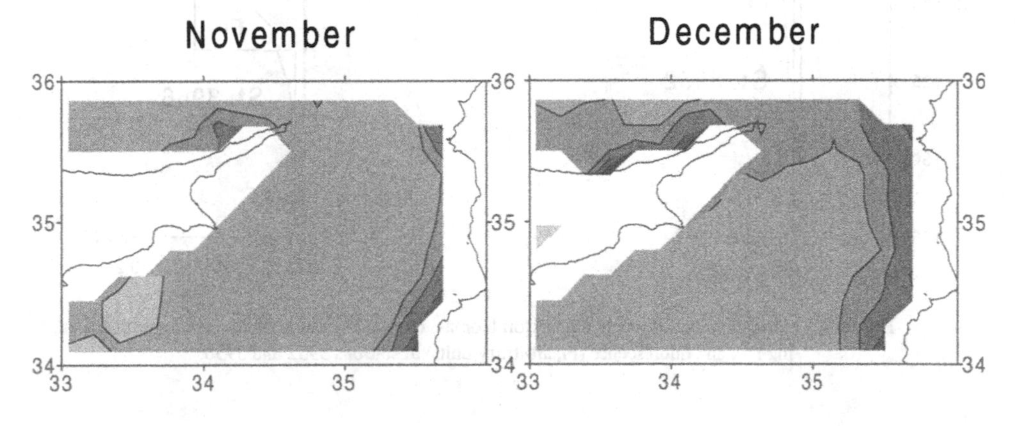


Figure 3 (continued).

The studies of vertical distribution revealed in the open sea regions the "deep" maxima of phytoplankton abundance and biomass, chlorophyll *a* concentration and fluorescence. The sampling from the levels selected on the basis of the measurements of fluorescence revealed 100% correlation between the layers of maximum of fluorescence and maximum of chlorophyll *a* concentration (Figure 4). The deep maximum was located in the lower part of pycnocline at the boundary of the Atlantic and Levant water mass within the depth range 90–125 m. The maximum concentrations of phytoplankton abundance and biomass in 50% of observations were located in quasi-uniform upper layer and/or in the layer of pycnocline; only at one-half of the stations insignificant increase of in the layer of fluorescence and chlorophyll maxima occurred. Only at the stations 3905 and 3956 in the maximum of fluorescence the maximum values of phytoplankton abundance and biomass and chlorophyll *a* concentration were measured. The intensity of fluorescence in these two cases was also of maximum (about 0.25). Over the shelf the maxima of abundance, biomass, chlorophyll *a* concentration and fluorescence occurred in the same layer (Figure 5).

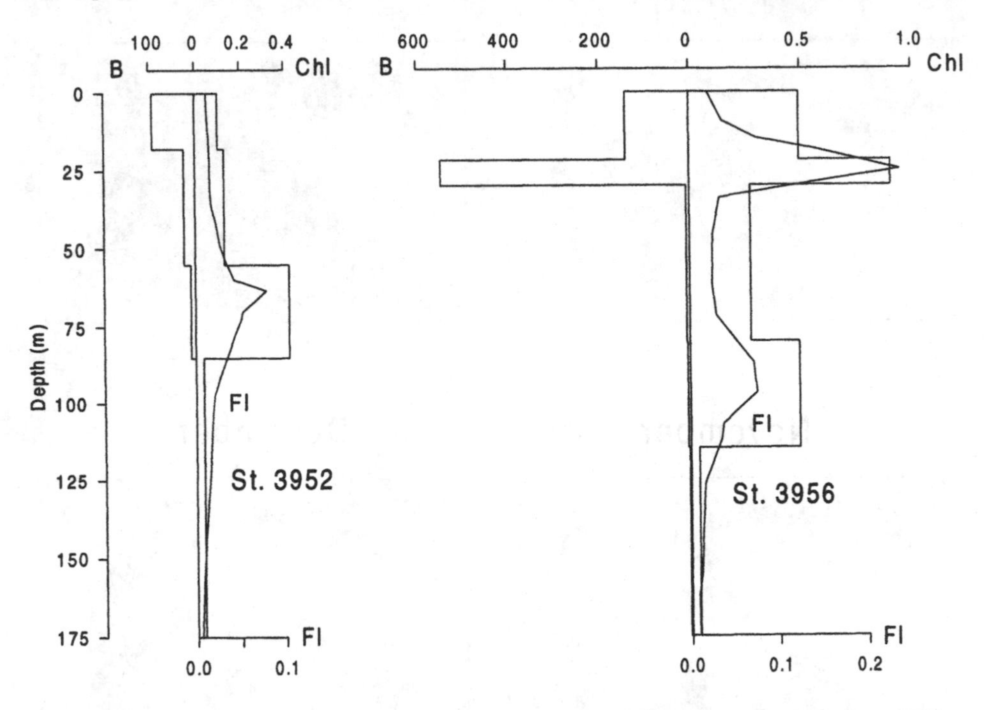


Figure 4. Vertical distribution of phytoplankton biomass (B, $mg \cdot l^{-1}$), chlorophyll *a* concentration (Chl, $mg \cdot l^{-1}$), and fluorescence (Fl, arbitrary units) at stations 3952 and 3956.

## 4. Discussion

The obtained results illustrate that in the open waters of eastern Mediterranean during late summer period the values of phytoplankton abundance and biomass and chlorophyll *a* concentration are similar to that observed in the oligotrophic ocean regions. Similar low values of phytoplankton abundance and biomass occurred in this region during winter season [7]. The mesoscale eddies are of great importance in formation of spatial heterogeneity of the quantitative indices of phytoplankton. Typical size of these eddies was 10–50 miles.

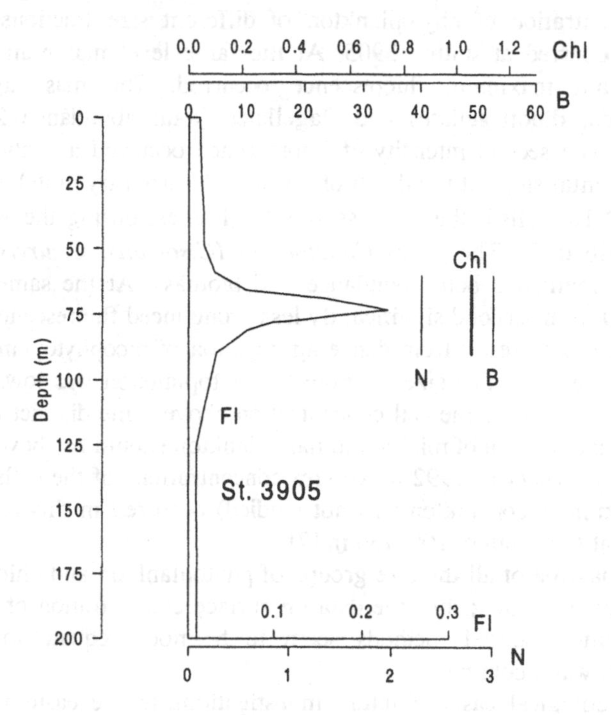


Figure 5. Vertical distribution of fluorescence (Fl, arbitrary units) at station 3905. The vertical lines indicate the levels of phytoplankton abundance (N, $10^5$ cells·$l^{-1}$) biomass (B, mg·$l^{-1}$), and chlorophyll *a* concentration (Chl, mg·$l^{-1}$) in the layer of fluorescence maximum.

The significant heterogeneity of vertical structure revealed narrow layers with biomass typical to mesotrophic waters; these values over the continental slope are similar to eutrophic waters. During the period of observations high biomass occurred in the upper mixed layer at the stations near the shelf (stations 3954, 3955) and in the layer of maximum density gradients (23–39 m) at the station 3956 over the slope. The layer-by-layer sampling revealed that in the open waters the deep maximum of chlorophyll *a* permanently occurred, which was well indicated by fluorimeter. The deep maximum seems to be formed by the smallest fraction of phytoplankton (picophytoplankton) and partly smallest cells of nanoplankton. This part of phytocenosis was not taken into account in our studies. The concentrations of picoplankton in the layer below the thermocline often occur in different regions of the World Ocean. In oligotrophic subtropical and tropical regions the transparency of water is high and the layers of maximum concentrations of not only picoplankton, but also large size fractions of phytoplankton occur at significant depth of 90–120 m.

During our studies the water transparency was very high: 1% of subsurface value of total quantum PAR irradiance was located at the layer of 65–83 m, and 0.1% was at 105–135 m. These light conditions are similar to the ones observed in the oligotrophic regions of the oceans. The optical properties of these waters corresponds to the waters of ocean types from 1A to 1B (classification of Jerlov [5]. According to our data, the maximum concentration of phytoplankton of different size fractions at significant depth (74 m) occurred at station 3905. At the same level maximum chlorophyll *a* concentration and maximum fluorescence occurred. The mass aggregations of coccolithophorids, dinoflagellates and flagellates (total abundance $2–3 \cdot 10^5$ cells/l) occurred there. The second intensity of fluorescence occurred at station 3956 in the region of continental slope at the depth of 26 m (0.24 arbitrary units). The abundance of cells was $2.4 \cdot 10^4$ cells/l, the biomass was the highest during the whole period of observations (546 µg/l). The genus *Chaetoceros (C.socialis, C.curvisetus, C.affinis)* and flagellates dominated both abundance and biomass. At the same station at the depth of 80–110 m the second significantly less pronounced fluorescence was observed (Figure 5); it seems to result from dence aggregation of picophytoplankton; however, the cell abundance of studied size fractions of phytoplankton was low. Evidently, the fluorescence appears under the cell concentration above some distinct level; this level for the cells of size fraction of micro- and nanoplankton should be above $2 \cdot 10^4$ cells/l.

During the February of 1992 maximum concentrations of the cells of micro- and nanophytoplankton (picoplankton was not studied) occurred in this region of eastern Mediterranean at the depth of 100–150 m [7].

The deep maxima of all the size groups of phytoplankton and chlorophyll *a* often occur in the studied region; thus the data on surface concentration of plant pigments obtained by remote sensed methods seem to be not adequate to chlorophyll *a* concentration in water column.

The methodological basis of future investigations in the eastern Mediterranean should include comprehensive studies of vertical structure of phytoplankton and chlorophyll *a* distribution combined with the measurements of primary production during different seasons.

## 5. Acknowledgments

The authors are grateful to S. I. Pogosyan and N. M. Merzlyak (Moscow State University) for methodological help in processing plant pigment data.

## 6. References

1. Feldman, G. C., Curing, N., Ng, C., Esaia, W., McClain, C. R., Elrod, J., Maynard, N., Edres, D., Evans, R., Brown, J., Walsh, S., Carle, M. and Podesta, G. (1989) Ocean colour: availability of the global data set, *Eos Trans. Amer. Geophys. Union.* **70**, 634-641.
2. Lukashev, Yu.F. (1993) The report of Hydrochemical Group, in *Report of Cruise 27 of R/V Vityaz. Part 1: Studies in Mediterranean Sea*, pp. 184–249.

3. Nezlin, N.P. (1997) Seasonal variation of surface pigment distribution in the Black Sea on CZCS data, in E. Ozsoy and A. Mikaelyan (eds.), *Sensitivity to Change: Black Sea, Baltic Sea and North Sea*, Kluwer Academic Publishers, Dordrecht, 131-138.
4. Nezlin, N.P., Musaeva, E.I. and Dyakonov, V.Yu. (1997) Estimation of Plankton Stocks in the Western Part of Bering and the Okhotsk Seas, *Oceanology. English Translation*, **37**, 370-375.
5. Nikolayev, V.P. (1993) The report of Hydrooptical Group, in *Report of Cruise 27 of R/V Vityaz. Part 1: Studies in Mediterranean Sea*, pp. 125–183.
6. Sukhanova, I.N. (1983) Concentration of phytoplankton in sample. in: *Sovremenniye metody kolichestvennoi otsenki raspredelenia morskogo planktona*, Nauka, Moscow, pp. 97-104. (In Russian).
7. Sukhanova, I.N. (1992) Report of Phytoplankton Group, in *Report of Cruise 24 of R/V Vityaz.* pp. 89–144.
8. Udodov, A.I. (1993) The report of Group of Currents Measurements, in *Report of Cruise 27 of R/V Vityaz. Part 1: Studies in Mediterranean Sea*, pp. 94–124.
9. Yakubenko, V.G. (1993) The report of Group of Hydrological Hydrocasts, in *Report of Cruise 27 of R/V Vityaz. Part 1: Studies in Mediterranean Sea*, pp. 31–87.

# MODELING PLANKTON PRODUCTION IN THE EASTERN MEDITERRANEAN: APPLICATION OF A 1-D VERTICALLY-RESOLVED PHYSICAL-BIOLOGICAL MODEL TO THE IONIAN AND RHODES BASINS

ERNESTO NAPOLITANO[1], TEMEL OGUZ[2],
PAOLA MALANOTTE-RIZZOLI[3], EMILIO SANSONE[1]

[1] *Institute of Meteorology and Oceanography, Istituto Universitario Navale, Naples, ITALY*
[2] *Middle East Technical University, Institute of Marine Sciences, Erdemli 33731, Icel, TURKEY*
[3] *Massachusettes Institute of Technology, Department of Earth, Atmospheric and Planetary Sciences, Cambridge, 02139, MA, USA .*

**Abstract** A one dimensional, coupled physical-biological model is used to study the biological production characteristics of the Rhodes and western Ionian basins of the Eastern Mediterranean. The biological model involves single aggregated compartments of phytoplankton, zooplankton, detritus as well as ammonium and nitrate forms of the inorganic nitrogen. The model simulations point to the importance of the contrasting dynamical characteristics of these two basins on affecting their yearly planktonic structures. The western Ionian basin is shown to possess only 10% of the Rhodes' productivity and therefore represent a most oligotrophic site in the Eastern Mediterranean. The Rhodes basin reveals a strong bloom in early spring, typically in March, a weaker bloom in early winter, typically in January, and a subsurface production below the seasonal thermocline during summer. This structure is slightly modified in the western Ionian basin, and the early winter and early spring blooms are merged to cover the entire winter. These results are supported favorably by the available observations both in their magnitudes and timing.



*P. Malanotte-Rizzoli and V.N. Eremeev (eds.),*
*The Eastern Mediterranean as a Laboratory Basin for the Assessment of Contrasting Ecosystems,* 181–204.

## 1. Introduction

In terms of its biochemical characteristics, the Eastern Mediterranean tends to be a poorly studied basin. Observations are confined generally to coastal areas as very little data are available for its offshore waters. The three dimensional, three compartment model by Crise et al (1998; hereafter referred to as CCM, 1998a) and Crispi et al. (1998; hereafter referred to as CCM, 1998b) is the only modeling work carried out insofar. This model investigates primarily the response of the circulation dynamics on the identification of different oligotrophic regimes of the Mediterranean sub-basins. When compared with the CZCS imagery, it seems to reproduce observed chlorophyll distributions according to the prevailing cyclonic and anticyclonic flow regimes.

The models with simplified biology using a limited number of state variables (usually four or five) are useful tools providing possible representation of pelagic ecosystem and its interactions with the upper layer dynamics. They have been applied successfully to different oceanic regimes (e.g. Fasham et al., 1991; McGillicuddy et al., 1995; McClain, 1995; Doney et al., 1996; Oguz et al., 1996; Kuhn and Radach, 1997). In this paper, we describe the characteristics of biological production of the Rhodes and the Ionian basins of the Eastern Mediterranean using such a simplified model. The Ionian and Rhodes basins are chosen particularly since they reflect two constrasting ecosystems; while the former represents a typical oligotrophic environment, the latter is known to be one of the most productive region of the Meditererranean Sea. These two basins also differ substantially by their dynamical regimes which govern ultimately their ecosystem characteristics.

The Rhodes cyclonic gyre is the most persistent feature of the Levantine basin circulation located in the northern Levantine basin to the west of Cyprus. Its dynamical and chemical characteristics are reasonably well-explored by a series of systematic surveys of the R.V. Bilim during the last decade (Salihoglu et al., 1990; Ozsoy et al., 1991, 1993; Sur et al., 1992; Yilmaz et al., 1994; Ediger and Yilmaz, 1996; Ediger et al., 1998; Yilmaz and Tugrul, 1998). On the other side, the Ionian Sea, identified as the region from the Sicily Strait to the Cretan passage, is a transition basin across which different water masses (e.g. Modified Atlantic Water, MAW; Levantine Intermediate water, LIW; Eastern Mediterranean Deep Water, EMDW) undergo transformations along their pathways between the Eastern and Western Mediterranean (Malanotte-

Rizzoli, et al., 1997). Here, we restrict our attention only to the western Ionian Sea characterized by the anticyclonic circulation and by the least dense upper layer water mass of the Eastern Mediterranean, MAW. It, therefore, offers completely opposite characteristics to those found in the Rhodes gyre. Practically, we perform here an intercomparision study to explore the biological characteristics of two regions known qualitatively as the most productive and the most oligotrophic sites of the Eastern Mediterranean and to relate them to the prevaling regional physical and dynamical characteristics. The paper is organized as follows. A brief description of the model formulation is first provided in Section 2. Its applications to the Rhodes and Ionian basins are then given in Sections 3 and 4, respectively. Conclusions are given in Section 5.

## 2. Model Description

The model is similar to the one described by Oguz et al. (1996) for studying the Black Sea annual plankton dynamics. It is a coupled physical-biological model including the level 2.5 Mellor and Yamada turbulence parameterization for vertical mixing. The biological state variables considered are phytoplankton biomass $P$, herbivorous zooplankton biomass $H$, and labile pelagic detritus $D$. Nitrate $N$ and ammonium $A$ constitute the other two state variables, since recent findings of relatively low N/P ratios (with respect to its Redfield value) in the euphotic zone of the northern Levantine imply nitrogen as the limiting nutrient for the primary production in the Rhodes gyre region (Ediger, 1995; Yilmaz and Tugrul, 1998).

The local changęs of the biological variables are expressed by a time and depth dependent advection-diffusion equation for transport, source and sinks in a one dimensional vertical water column. The general form of the equation is given by

$$\frac{\partial B}{\partial t} = \frac{\partial}{\partial z}\left[(K_h + \nu_h)\frac{\partial B}{\partial z}\right] + F_B \tag{1}$$

where $B$ represents any of the five biological variables, $t$ is time, $z$ is the vertical coordinate, $\partial$ denotes the partial differentiation. $F_B$ represents the biological interaction terms expressed for the phytoplankton, herbivore, detritus, ammonium, and nitrate equations, respectively, as

$$F_P = \Phi(I,N,A)\,P - G(P)\,H - m_p P \tag{2}$$

$$F_H = \gamma G(P)\,H - m_h H - \mu_h H \tag{3}$$

$$F_D = (1-\gamma)\,G(P)\,H + m_p P + m_h H - \epsilon D + w_s \frac{\partial D}{\partial z} \tag{4}$$

$$F_A = -\Phi_a(I,A)\,P + \mu_h H + \epsilon D - \Omega A \tag{5}$$

$$F_N = -\Phi_n(I,N)\,P + \Omega A \tag{6}$$

where the definition of parameters and their values used in the experiments are given in Table 1. The functions $\Phi(I,N,A)$ and $G(P)$ denote the light and nutrient limited phytoplankton growth and zooplankton grazing, respectively. The latter is represented by the maximum ingestion rate $\sigma_g$ multiplied by the Michaelis-Menten type limitation function

$$G(P) = \sigma_g \frac{P}{R_g + P} \tag{7}$$

where $R_g$ is the half saturation constant for the zooplankton grazing. Eq. (7) implies that zooplankton graze only phytoplankton. Their grazing on the detrital material is not taken into account since this usually constitutes only a small fraction of zooplankton diet (about 10%).

The phytoplankton production is parameterized as product of the maximum growth rate $\sigma_m$, minimum of the light and nutrient limitation functions and the phytoplankton standing crop. The temperature limitation of the phytoplankton growth is not included in the model, for simplicity. Then, the net growth rate $\Phi(I,N,A)$ is given by

$$\Phi(I,N,A) = \sigma_m min\left[\alpha(I), \beta_t(N,A)\right] \tag{8}$$

where $min$ refers to the minimum of either the light limitation function $\alpha(I)$ or the total nitrogen limitation function $\beta_t(N,A)$. The latter is expressed as the sum of ammonium and nitrate limitation functions, $\beta_a(A)$ and $\beta_n(N)$, respectively

$$\beta_t(N,A) = \beta_n(N) + \beta_a(A) \tag{9}$$

They are given by the Michaelis-Menten type uptake formulation as

$$\beta_a(A) = \frac{A}{R_a + A}, \qquad \beta_n(N) = \frac{N}{R_n + N}\exp(-\psi A) \tag{10}$$

where $R_n$ and $R_a$ are the half-saturation constants for nitrate and ammonium, respectively. The exponential term in eq. 10 represents the inhibiting effect of ammonium concentration on nitrate uptake, with $\psi$ signifying the inhibition parameter.

**Table 1. Model parameters used in the numerical experiments**

| Parameter | Definition | Value | Value | Unit |
|---|---|---|---|---|
| | | Rhodes | Ionian | |
| $a$ | Photosynthesis efficiency parameter | 0.01 | 0.02 | $m^2W^{-1}$ |
| $k_w$ | Light extinction coefficient for PAR | 0.05 | 0.04 | $m^{-1}$ |
| $k_c$ | Phytoplankton self-shading coefficient | 0.04 | 0.07 | $m^2$ (mmol N)$^{-1}$ |
| $\sigma_p$ | Maximum phytoplankton growth rate | 1.5 | 1.0 | day$^{-1}$ |
| $m_p$ | Phytoplankton mortality rate | 0.04 | 0.04 | day$^{-1}$ |
| $\sigma_g$ | Zooplankton maximum grazing rate | 0.6 | 0.8 | day$^{-1}$ |
| $m_h$ | Zooplankton mortality rate | 0.04 | 0.04 | day$^{-1}$ |
| $\mu_h$ | Zooplankton excretion rates | 0.07 | 0.04 | day$^{-1}$ |
| $\gamma_h$ | Food assimilation efficiency | 0.75 | 0.75 | Dimensionless |
| $R_n$ | Half saturation constant in nitrate uptake | 0.5 | 0.5 | mmol N m$^{-3}$ |
| $R_a$ | Half saturation constant in ammonium uptake | 0.2 | 0.2 | mmol N m$^{-3}$ |
| $R_g$ | Half saturation constant for zooplankton grazing | 0.5 | 0.5 | mmol-N m$^{-3}$ |
| $\psi$ | Ammonium inhibition parameter of nitrate uptake | 3 | 3 | (mmol N m$^{-3}$)$^{-1}$ |
| $\epsilon$ | Detritus decomposition rate | 0.1 | 0.1 | day$^{-1}$ |
| $w_s$ | Detrital sinking rate | 8.0 | 5.0 | m day$^{-1}$ |
| $\Omega_a$ | Nitrification rate | 0.05 | 0.05 | day$^{-1}$ |
| $\nu_b$ | Background kinematic diffusivity | 0.1 | 0.1 | cm$^2$s$^{-1}$ |

The individual contributions of the nitrate and ammonium uptakes to the phytoplankton production, given in eq's (5) and (6), are represented respectively by

$$\Phi_n(I,N) = \Phi(I,N,A)\left(\frac{\beta_n}{\beta_t}\right), \qquad \Phi_a(I,A) = \Phi(I,N,A)\left(\frac{\beta_a}{\beta_t}\right) \quad (11)$$

The light limitation is parameterized according to Jassby and Platt's (1976) hyperbolic tangent function and exponentially decaying irradiance with depth as

$$\alpha(I) = tanh\left[aI(z,t)\right] \tag{12}$$

$$I(z,t) = I_s \exp\left[-\left(k_w + k_c P\right) z\right] \tag{13}$$

where $a$ denotes photosynthesis efficiency parameter controlling the slope of $\alpha(I)$ versus the irradiance curve at low values of the Photosynthetically Active Radiance (PAR). $I_s$ denotes the surface intensity of PAR taken as the half of the climatological incoming solar radiation. $k_w$ is the light attenuation coefficient due to sea water, and $k_c$ is the phytoplankton self-shading coefficient. In the above formulation, $k_w$ and $k_c$ are taken to be constant with depth. The daily variation of the light irradiance, and hence the phytoplankton growth are neglected since the biological processes we consider have time scales much longer than a day.

## 2.1. Boundary Conditions

No-flux conditions described by

$$(K_h + \nu_h)\frac{\partial B}{\partial z} = 0 \tag{14}$$

are specified both at the surface and the bottom boundaries of the model for the variables $P, H, N$, and $A$. For detritus, it is modified to include the contribution of downward sinking flux

$$(K_h + \nu_h)\frac{\partial D}{\partial z} + w_s D = 0 \tag{15}$$

The bottom boundary of the model is taken at the 400 m depth for the Rhodes case and at the 300 m depth for the Ionian case, which are well below the euphotic zone comprising only the upper 100 m of the water column. Considering our choice of moderate detritus sinking rates (see Table 1), the advantage of locating the bottom boundary at considerable distance away from the euphotic layer is to allow complete remineralization of the detrital material until it reaches the lower boundary of the model. The complete remineralization was ensured by setting appropriate decomposition rate of detrital material in the model. This approach avoids prescription of the non-zero flux boundary condition in order to compensate the loss of detritus (if any) to the deep interior from the boundary. It is thus implicitely assumed here that the detrital pool in the model is formed by smaller size particles. Fast sinking large particles with a typical speed of the order of 100 m/day are therefore not modeled in the present work. We note that equations (2)-(6) together with the boundary conditions (14) and (15) provide a closed, fully conserved system. The state of the system at time $t$ is governed solely by internal dynamical processes without any contribution from the external sources.

The physical model is forced by daily climatological atmospheric fluxes. The wind stress forcing are taken from the ECMWF climatology. The heat flux data are provided by May (1982) whereas the data given by Antoine, Morel and Andre (1995) for the Ionian Sea is used for the specification of PAR for both simulations. The heat flux data is, however, adjusted slightly to provide the zero annual mean over the year. This adjustment is necessary to avoid the drift of the model from its perpetual state due to continuous warming/cooling of the water column during the time integration of the temperature equation. The resulting adjusted heat flux distributions over the year are shown in Fig. 1. The Rhodes basin receives greater cooling in winter (~200 W/m$^2$) as compared to the value of 130 W/m$^2$ for the Ionian Sea. In the summer, the warming in the Rhodes basin reaches 180 W/m$^2$ which is also higher from that of the Ionian basin by about 40 W/m$^2$. PAR attains its maximum intensity of 140 W/m$^2$ during the summer months, whereas its minimum value is set to 40 W/m$^2$ during December and January. Instead of prescribing the fresh water flux at the surface, the model is forced by the surface salinity whose annual variations are specified by the Mediterranean Oceanic Data Base (MODB) data set (Brasseur et al., 1996). The forcing by surface salinity instead of the fresh water flux using evaporation minus precipitation data is a matter of convinience here since it provides more realistic yearly salinity variations in the near-surface levels of the water column.

Although the atmospheric forcing functions used in the model are rather idealized, they are adequate for the purpose of the present work since the surface layer dynamics is introduced to the biological model only indirectly by specification of the vertical eddy diffusivity. There is no other feed back mechanism between the physical and biological models. The vertical eddy diffusion coefficient of the biological model is same with that of temperature or salinity computed from the turbulence closure parameterization (see Oguz et al., 1996 for details).

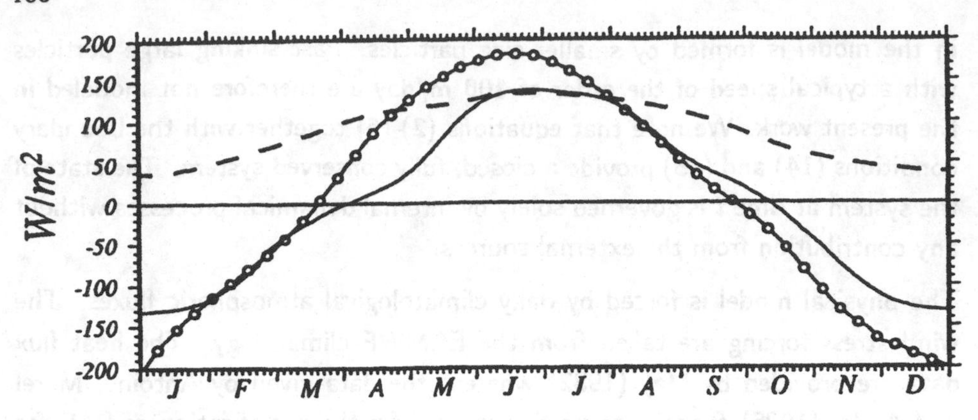


**Figure 1:** ***Climatological daily averaged heat fluxes; continuous line is for the Ionian basin, continuous line with open circles for the Rhodes basin, and photosynthetically available radiation (broken line) used in the simulation. Units are in*** $Wm^{-2}$

## 2.2. Initial Conditions

The model is initialized with the stably stratified upper ocean temperature and salinity profiles representative of the summer/autumn climatological conditions (Fig. 2a). The same initial temperature profile is used for both Rhodes and Ionian simulations since the subsurface structures of these two regions reflect characteristics of the LIW below 100-150 m depth. The salinity, on the other side, differs greatly between these two basins. The upper 150 m of the water column in the Rhodes gyre is occupied by much more saline water mass (with S>39.0 ppt) as compared to less saline modified Atlantic-based waters (minimum S~ 37.4 ppt) in the western Ionian basin (Fig. 2a).

In the biological model, the initial nitrate source drives the system as they are entrained and diffuse upward and utilized ultimately for the biological production. In the absence of any external source and sink as implied by the boundary conditions (see eq.'s 14 and 15), the model simply redistributes the initial nitrate source among the living and nonliving components of the sys-

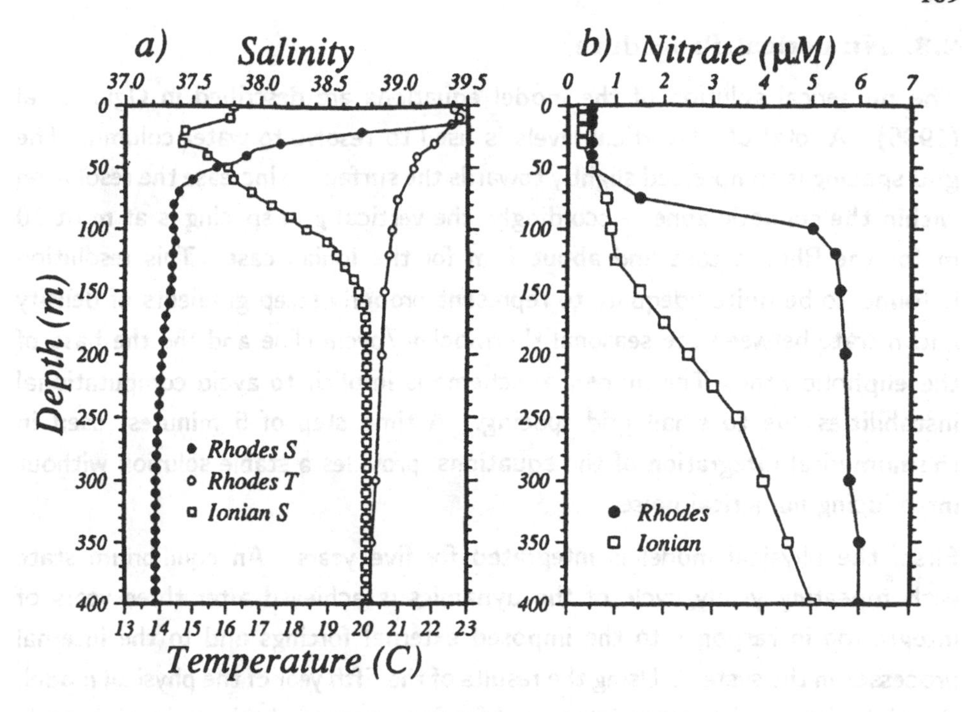


Figure 2: *The profiles of (a) temperature and salinity, and (b) nitrate used for the model initializations*

tem. Thus, specification of the initial phytoplankton, zooplankton, detritus and ammonium distributions is unimportant as they will be generated by the model during its transient evolution. The initial subsurface nitrate structure however will govern, to a large extent, the level of productivity in the euphotic zone. Accordingly, the state variable except nitrate are initialized by a vertically uniform small, non-zero values within the euphotic layer for both cases. The initial vertical nitrate structures are specified according to the data given by Berland et al. (1988) and Yilmaz and Tugrul (1998). These observed profiles, shown in Fig. 2b, demonstrate quite clearly how these two basins differ substantially in terms of their subsurface nitrate structure below the 100m depth. The Rhodes gyre reveals vertically uniform subsurface nitrate concentrations on the order of 5 $\mu M$. On the other side, the western Ionian subsurface nitrate concentrations increase more gradually with depth and reach the similar nitrate level of the Rhodes basin only below 500 m depth. As it will be presented in the following sections, these differences in the nitrate and salinity structures play crucial roles on the ultimate biological characteristics of these two regions.

### 2.3. Numerical Procedure

The numerical solution of the model equations are described in Oguz et al (1996). A total of 51 vertical levels is used to resolve to water column. The grid spacing is compressed slightly towards the surface to increase the resolution within the euphotic zone. Accordingly, the vertical grid spacing is at most 10 m for the Rhodes case and about 7 m for the Ionian case. This resolution is found to be quite adequate to represent properly steep gradients of density and nitrate between the seasonal thermocline/pycnocline and the the base of the euphotic zone. The numerical scheme is implicit to avoid computational instabilities due to small grid spacing. A time step of 5 minutes, used in the numerical integration of the equations, provides a stable solution without introducing numerical noise.

First, the physical model is integrated for five years. An equilibrium state with repeating yearly cycle of the dynamics is achieved after three years of integration in response to the imposed external forcings and to the internal processes in the system. Using the results of the fifth year of the physical model, the biological model is then integrated for four years to obtain repetative yearly cycles of the biological variables. The results of the biological model presented here are based on the fourth year of integration.

## 3. Simulation of Rhodes Basin ecosystem

### 3.1. Annual plankton structure

In agreement with its vertical mixing and stratification characteristics, the water column nitrate structure undergoes considerable variations during the year (Fig. 3a). The mixed layer waters of the entire summer and autumn seasons are extremely poor in nutrients, and characterized by only trace level nitrate concentrations of about 0.1 mmol/m$^3$. The nitrate depletion arises due to lack of supply from the subsurface levels because of the presence of strong seasonal thermocline/pycnocline. The zone of high stratification below the seasonal thermocline coincides with the strong nitrate variations (the so-called the nitracline). Approximately below 80-90 m depths, the nitrate attains its

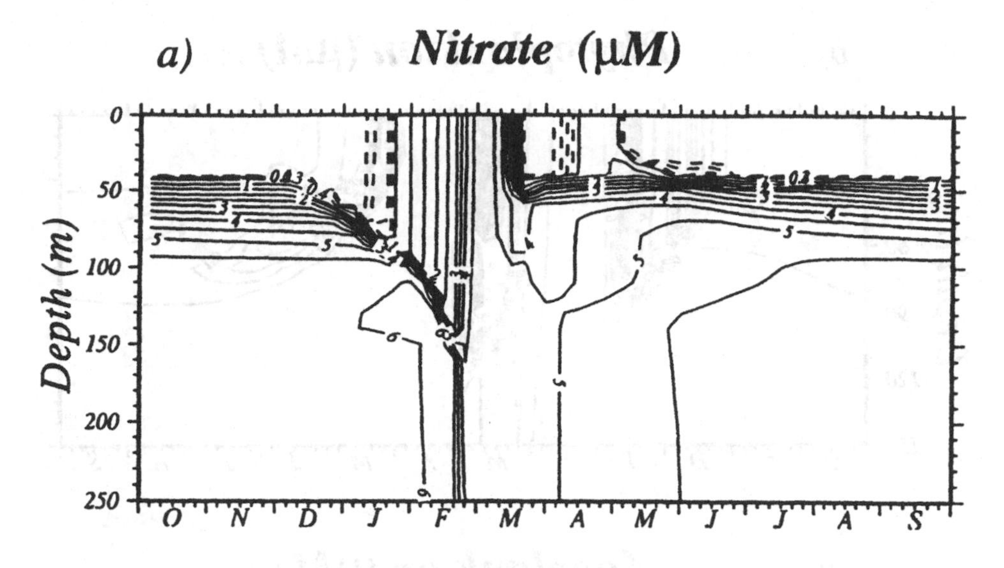


Figure 3: *Simulated annual distributions of (a) nitrate, (b) phytoplankton, (c) zooplankton, (d) detritus, (e) ammonium, within the upper 150 m depth for phyto- and zooplankton and the upper 250 m depth for others. Units are in μM. For nitrate, continuous lines are contours at intervals of 0.5 μM and broken lines at intervals of 0.1 μM. In all plots the time axis starts at October 1 and ends at September 30*

typical deep water values in excess of 5.0 mmol/m$^3$. This structure undergoes substantial modification during the winter months as the convective overturning mechanism brings the nitrate rich subsurface waters to near-surface levels. Under such conditions, nitrate concentrations attain their maximum values of 4.5 mmol/m$^3$ over the 400m deep homogeneous water column in February.

The phytoplankton structure exhibits a major algae production during the first half of March (Fig. 3b) immediately after the cessation of the strong mixing, shallowing of the mixed layer and higher rate of solar irradiance penetrating to deeper levels. Since the water column was already replenished by nitrate, all

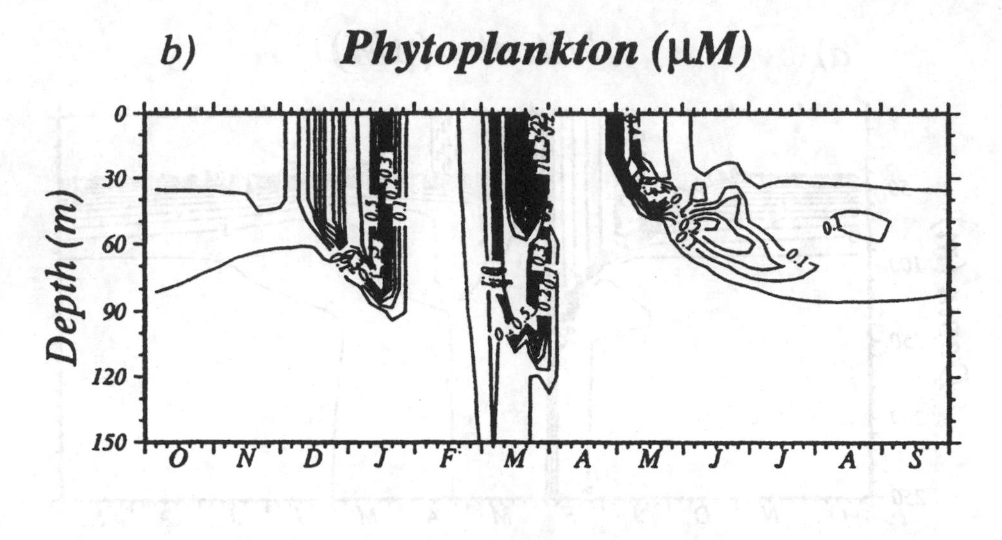
b) Phytoplankton (μM)
Depth (m)
0
30
60
90
120
150
O
N
D
J
F
M
A
M
J
J
A
S
0.1
0.2
0.5

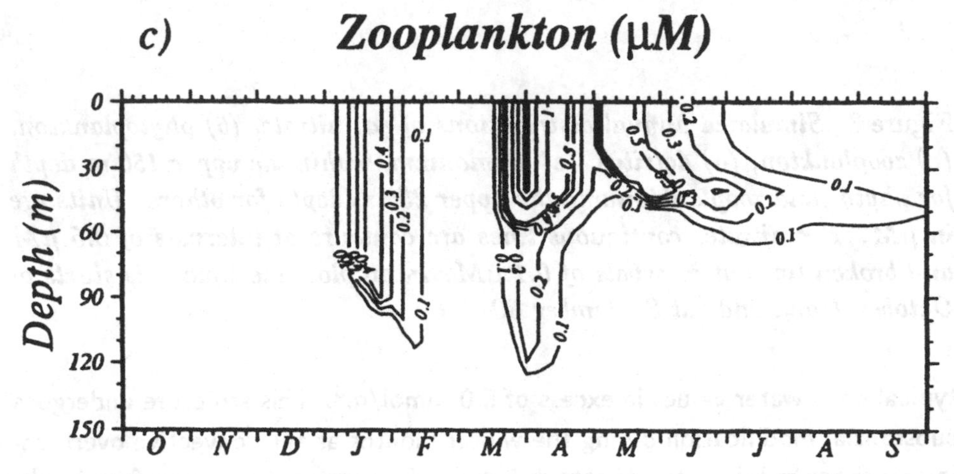
c) Zooplankton (μM)
Depth (m)
0
30
60
90
120
150
O
N
D
J
F
M
A
M
J
J
A
S
0.1
0.2
0.3
0.4
0.5

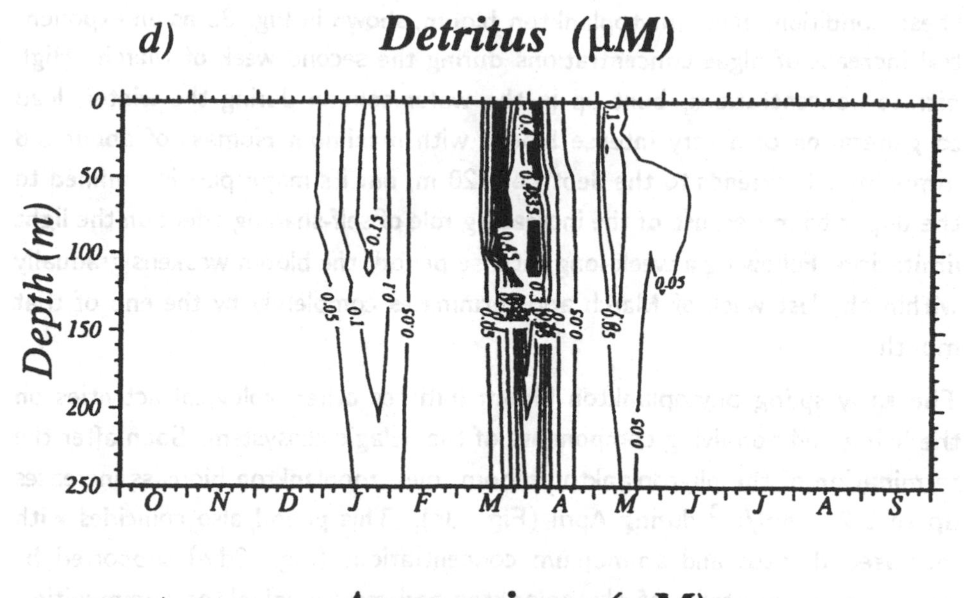
d) Detritus (μM)
Depth (m)
0
50
100
150
200
250
O N D J F M A M J J A S

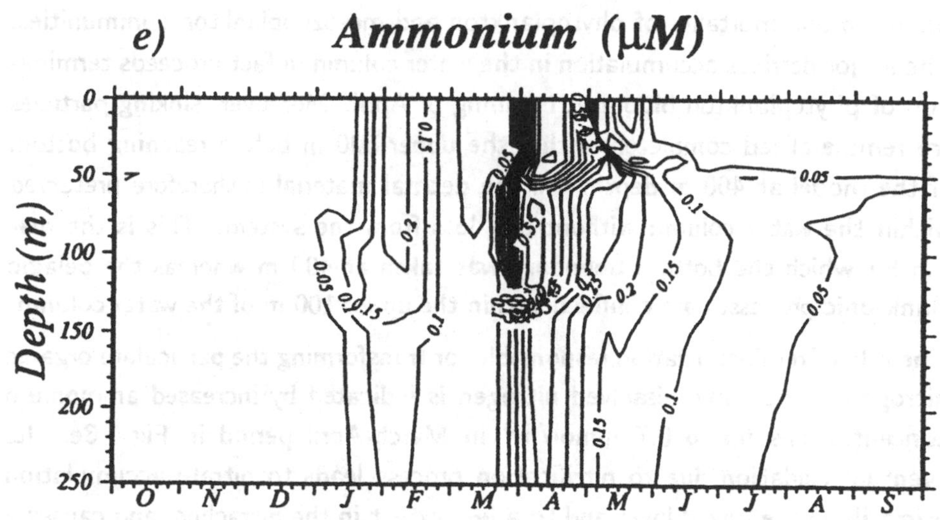
e) Ammonium (μM)
Depth (m)
0
50
100
150
200
250
O N D J F M A M J J A S

these conditions favor phytoplankton bloom, shown in Fig. 3b as an exponential increase of algae concentrations during the second week of March. High nitrate concentrations, built up in the water column during the winter, lead to generation of a very intense bloom with maximum biomass of about 3.8 mmol/m$^3$. It extends to the depth of 120 m, but its major part is confined to the upper 65 m because of the increasing role of self-shading effect on the light limitation. Following a week-long intense period, the bloom weakens gradually within the last week of March and terminates completely by the end of that month.

The early spring phytoplankton bloom initiates other biological activities on the living and non-living components of the pelagic ecosystem. Soon after the termination of the phytoplankton bloom, mesozooplankton biomass increases up to 2.2 mmol/m$^3$ during April (Fig. 3c). This period also coincides with increased detritus and ammonium concentrations (Fig. 3d,e) supported by excretion and mortality of phytoplankton and mesozooplankton communities. The major detritus accumulation in the water column in fact proceeds termination of phytoplankton bloom at begining of April. Moreover, sinking particles are remineralized completely within the upper 300 m before reaching bottom of the model at 400 m depth. All the detrital material is therefore preserved within the water column without any loss from the system. This is the reason for which the bottom boundary was taken at 400 m whereas the pelagic planktonic processes are confined within the upper 100 m of the water column.

The role of remineralization responsible for transforming the particulate organic nitrogen to inorganic dissolved nitrogen is indicated by increased ammonium concentrations up to 0.7 mmol/m$^3$ in March-April period in Fig. 3e. Its eventual oxidation due to nitrification process leads to nitrate accumulation primarily in the mixed layer and to a less extent in the nitracline, and causes a short-term increase in phytoplankton biomass up to about 0.5 mmol/m$^3$ within the mixed layer during the first half of May (Fig. 3b). As in the previous case, this secondary bloom is also followed by a small increase in mesozooplankton biomass, as well as in detritus and ammonium concentrations. The surface-intensified phytoplankton bloom event continues below the seasonal thermocline for another month by consuming available nitrate and ammonium within the nitracline zone. The subsurface biomass diminishes gradually towards the end of July as the contribution of losses from mesozooplankton grazing and

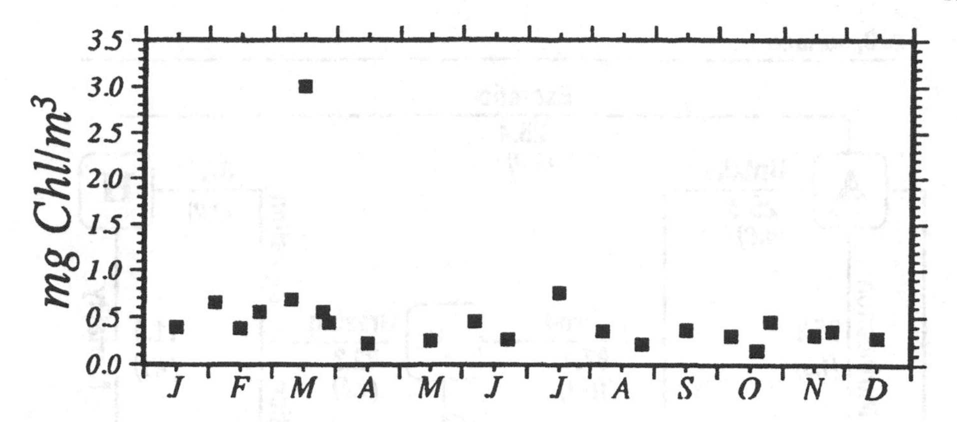


Figure 4: *A composite picture of the euphotic layer average chlorophyll concentrations (in mg Chl/m$^3$) within the year. The data compiled from the measurements of R.V. Bilim in the Rhodes gyre region during 1986-1995 period (A. Yilmaz, private com.)*

phytoplankton mortality exceeds production.

The annual phytoplankton structure exhibits another weak bloom from mid-December to mid-January. This is associated with the consumption of nitrate which are made readily available by the convective mixing initiated in the water column with the begining of cooling season. Once again, it is followed by increase in mesozooplankton stocks in January-February.

The presence of early spring bloom and other features of the model is supported fairly well by the observations. Fig. 4 shows the euphotic layer-averaged chlorophyll concentrations in the Rhodes gyre based on the data from 23 casts made by R.V. Bilim during the last 10 years. In this figure, the most noticeable feature is the chlorophyll value of 3.0 mg Chl/m$^3$ during March comparable with the values simulated from the model.

## 3.2. Annual Budget and Primary Production estimates

The annual mean intercompartmental transfer rates over the euphotic zone and the vertical fluxes across its base are shown in Fig. 5. They are expressed in terms of gC m$^{-2}$ yr$^{-1}$ and obtained by multiplying all the fluxes computed

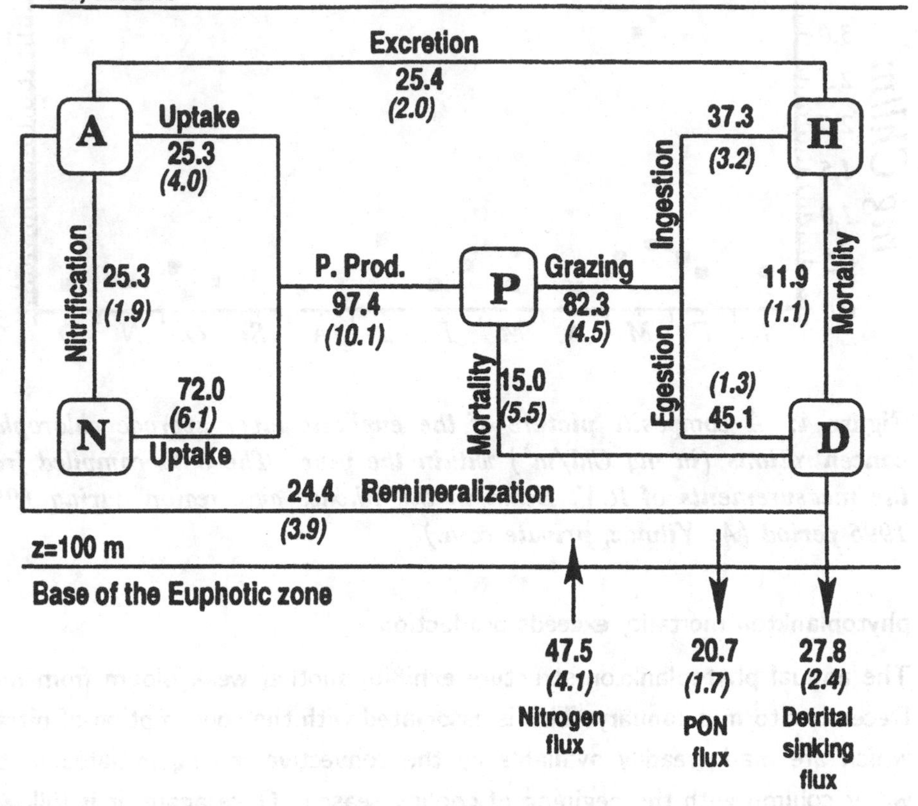


**Figure 5:** ***The annual nitrogen budget of the euphotic zone for both Rhodes and Ionian simulations. The latter is shown by the numbers in parantheses***

in *mmol* nitrogen units by the factor 0.12. At the base of the euphotic zone, the budget implies that the vertical diffusive nitrogen flux of 47.5 gC $m^{-2}$ $yr^{-1}$ from below is balanced by the sum of downward PON flux of 20.7 gC $m^{-2}$ $yr^{-1}$ and detrital sinking flux of 27.8 gC $m^{-2}$ $yr^{-1}$ leaving the euphotic zone. The total primary production (PP) is estimated as 97.4 gC $m^{-2}$ $yr^{-1}$ whose 25 % is met by ammonium uptake and the rest by the nitrate uptake. We, however, note that only two-third of the overall nitrate uptake of 72.0 gC $m^{-2}$ $yr^{-1}$ is supported from the subsurface levels and accounts for the new production, the rest is accounted by the recycling mechanism inside the euphotic zone. The budget suggests approximately 40% of the PP (37.3 gC $m^{-2}$ $yr^{-1}$) is utilized for the secondary production, whereas the rest goes directly into the detrital pool.

The PP estimation of 97 gC $m^{-2}$ $yr^{-1}$ agrees well with the value of 86.8 gC $m^{-2}$ $yr^{-1}$ computed by Antoine et al. (1995) on the basis of the CZCS analyses for the Northern Levantine. Using the data from 1986 and 1987 surveys of R.V. Bilim, Salihoglu et al. (1990) suggested a lower value of 60 gC $m^{-2}$ $yr^{-1}$. This estimate however does not cover the data from productive spring period, and therefore underestimates the annual rate. Moreover, our estimate is comparable with the value of 105 gC $m^{-2}$ $yr^{-1}$ obtained for the northwestern Mediterranean characterized by similar dynamical conditions (c.f. Levy et al., 1998).

## 4. Simulation of Ionian Basin ecosystem

Similar to their physical characteristics, a substantial difference in the water column nitrate structure of the western Ionian and Rhodes basins is indicated by Fig. 6a. The weak vertical mixing in the winter months imply lack of sufficient nitrate supply from the subsurface levels to support the biological production in the subsequent early spring season. Fig. 6a clearly shows no nitrate accumulation inside the mixed layer during the winter. Whatever nitrate is entrained into the mixed layer from the subsurface levels is consumed immediately in the phytoplankton production process. This is initiated during mid-January and gradually increased to its peak at the end of February (Fig. 6b). The phytoplankton biomass can attain in this period only 0.25 mmol $N/m^3$ which is an order of magnitude lower than the typical spring bloom values of the Rhodes simulation. The degradation of the phyoplankton bloom occurs during March. April is the period of intense nitrogen cycling followed by surface-intensified regenerated production during early May and its continuation at the subsurface levels below the thermocline in June with the maximum biomass of 0.16 $mmol/m^3$. The response of subsurface production can be traced in the nitrate field by the slight increase of isolines during early summer period (Fig. 6a). The annual distribution of zooplankton stock (Fig. 6c) reveals maximum biomass value of about 0.15 $mmol/m^3$ within the year. This is again one order of magnitude smaller than the values given in the Rhodes simulations.

The way in which the intensity of vertical mixing controls timing of the early spring phytoplankton bloom was described earlier in Oguz et al. (1996) in

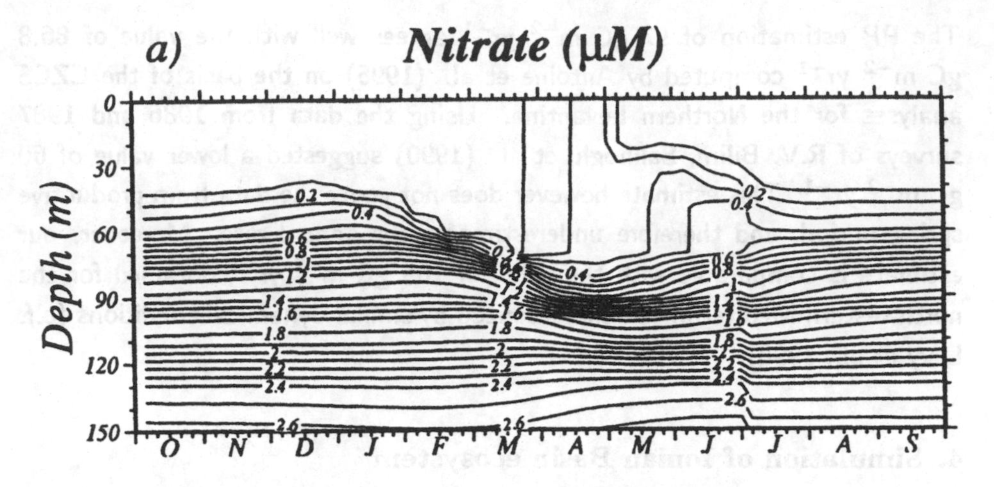


**Figure 6:** ***Simulated annual distributions of (a) nitrate, (b) phytoplankton and (c) zooplankton for the western Ionian basin within the upper 150 m depth of the water column. Units are in μM. For nitrate, the contours are at intervals of 0.1 μM, for phytoplankton at 0.02 μM and for zooplankton at 0.01 μM. In all plots the time axis starts at October 1 and ends at September 30***

the context of the Black Sea plankton production model. It was shown that even though the light and nutrient conditions may be favorable for initiation of the bloom in winter, it can be delayed depending on the intensity of the vertical mixing. In the absence of any zooplankton grazing pressure and other losses due to phytoplankton mortality and excretion during winter, the vertical mixing is the only sink term which can balance the production. The initiation of phytoplankton bloom therefore depends on the relative intensities of these two terms during winter-early-spring period. The biological production will therefore be initiated whenever the vertical mixing weakens and its magnitude is exceeded by that of the production term. In our simulations, the Rhodes case is a good example for the delay of the bloom until the weakening of the intense mixing event taking place during the winter. The other region having similar features is the northwestern Mediterranean deep water formation area where prevention of the bloom by strong vertical mixing was referred to as

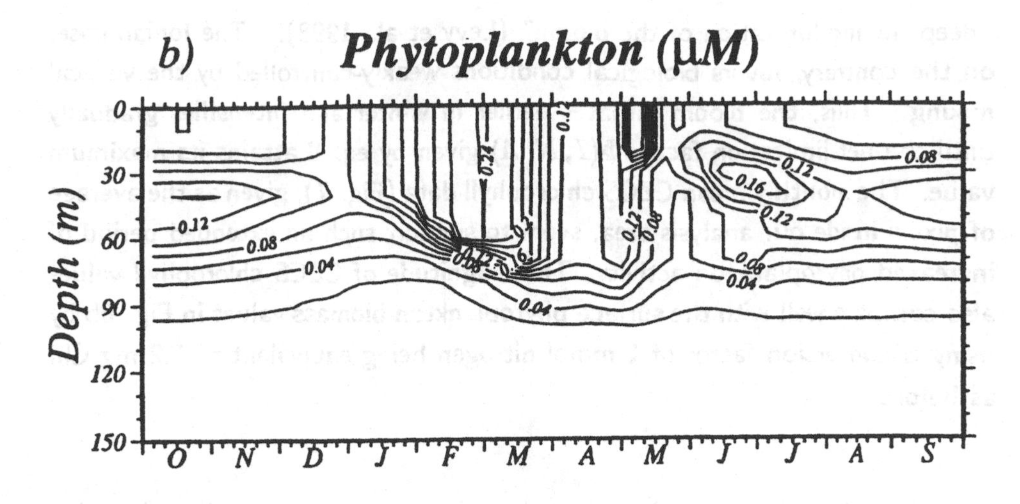
b) Phytoplankton (μM)
Depth (m)
0
30
60
90
120
150
O N D J F M A M J J A S
0.12
0.08
0.04
0.24
0.16


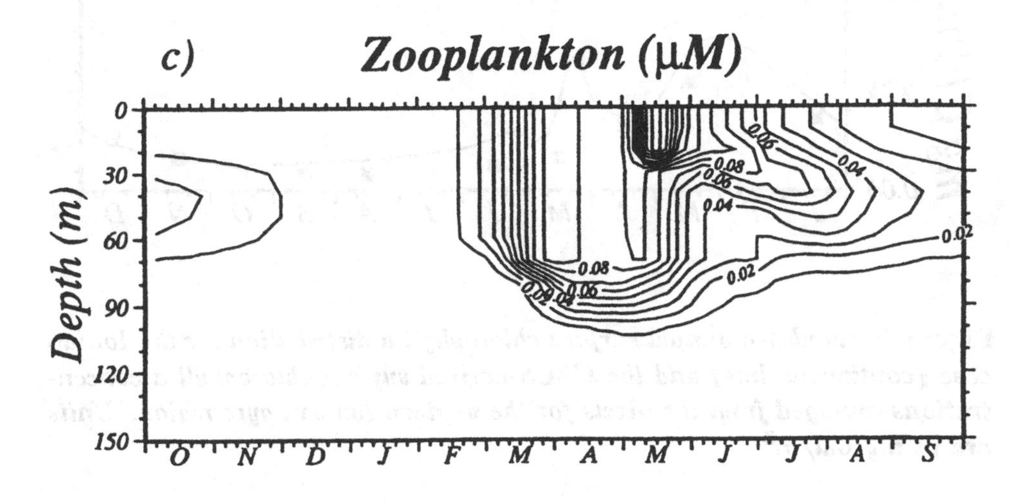
c) Zooplankton (μM)
Depth (m)
0
30
60
90
120
150
O N D J F M A M J J A S
0.08
0.06
0.04
0.02

"deep mixing limitation of the bloom" (Levy et al., 1998). The Ionian case, on the contrary, favors biological conditions weakly-controlled by the vertical mixing. Thus, the bloom initiates earlier in winter and intensifies gradually until the net limitation factor $\Phi(I, N, A)$ given by eq. 8 attains its maximum value. The monthly mean CZCS chlorophyll data (Fig. 7), given as the average of pixels inside our analysis area, seem to support such an extended period of increased phytoplankton activity. The magnitude of CZCS chlorophyll values also compare well with the surface phytoplankton biomass values in Fig. 6b by using a conversion factor of 1 mmol nitrogen being equivalent to 1.2 mg Chl as before.

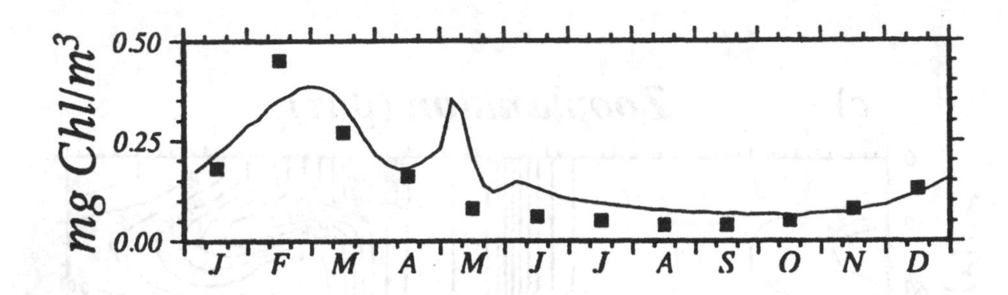


Figure 7: *Simulated annual surface chlorophyll-a distribution for the Ionian case (continuous line) and the CZCS-derived surface chlorophyll-a concentrations averaged from the pixels for the western Ionioan gyre region. Units are in mg Chl/m*$^3$

The annual budget and intercompartmental transfer rates for the Ionian simulation is also shown in Fig. 5 with the numbers in parantheses. The budget indicates that only 4.1 gC $m^{-2}$ $yr^{-1}$ nitrate flux enters into the euphotic zone from the lower layers. The contribution of recycled nitrogen through detrital breakdown and plankton excretion are 3.9 and 2.0 gC $m^{-2}$ $yr^{-1}$, respectively, which all together give rise to a total PP rate of 10.1 gC $m^{-2}$ $yr^{-1}$. This is an extremely low value, but is considered to be a typical for oligotrophic sites of the Eastern Mediterranean. Applying the CCM (1998a,b) model to the same region for the interpretation and analysis of the hydrographic and nutrient data collected during October 1991 and April 1992 surveys, Civitarese et al. (1996) obtained a very close mean PP value of 11.5 gC $m^{-2}$ $yr^{-1}$.

## 5. Conclusions

The biological productivity in two contrasting regions of the Eastern Mediterranean is studied using a one-dimensional vertically resolved pysical-biological model. These two regions are the Rhodes cyclonic gyre of the Northern Levantine Sea known to be a major, persistent sub-basin scale feature of the Eastern Mediterranean between the Rhodes and Cyprus Islands, and the anticyclonic gyre of the western Ionian Sea extending meridionally to the east of the Sicily Straits. In the Rhodes basin, the water column below the seasonal thermocline is typically characterized by relatively cold, saline and dense waters. On the contrary, the western anticyclonic Ionian gyre possesses very limited surface-intermediate water interactions because of the presence of less saline Modified Atlantic Waters (MAW) within the upper 100 m layer. The strong stratification introduced by the presence of MAW and LIW underneath prevents deep penetration of the vertical convection and, subsequently, the nutrient supply from deeper levels. Because of these two contrasting dynamical regimes, the two basins reveal quite different vertical nutrient structures, and subsequently quite different biological production characteristics. The annual primary production is estimated as 97.4 gC $m^{-2}$ $yr^{-1}$ in the Rhodes basin. On the other hand, the primary production in the western Ionian Sea amounts only 10% of the Rhodes' case. Therefore, these two basins represent biologically two end members of the Eastern Mediterranean. The annual production cycle also differs slightly in these two basins. The Rhodes gyre possesses a classical production cycle consisting of a strong early spring bloom, a weaker early winter bloom and subsurface production during the summer. In the western Ionian basin, these two blooms are merged with each other to form a long lasting, gradually evolving winter bloom starting from the begining of January to the end of March. The early spring bloom of the Rhodes gyre, on the other hand, has a shorter lifetime, grows and decays exponentially.

## Acknowledgements

E.Napolitano greatfully acknowledges the support provided by the Universita' di Messina and Istituto Universitario Navale Environmental Marine Sciences Joint Ph.D Program as a visiting fellow at MIT. P. Malanotte-Rizzoli and T. Oguz acknowledge the NSF support OCE-9633145.

## References

Antoine, D., A. Morel, J-M Andre (1995) Algal pigment distribution and primary production in the Eastern Mediterranean as derived from coastal zone color scanner observations. J. Geophys. Res., 100, 16193-16209.

Berland, R.B., A.G. Benzhitski, Z.P. Burlakova, L.V. Georgieva, M.A. Izmestieva, V.I. Kholodov, S.Y. Maestrini (1998) Hydrological structure and particulate matter distribution in Mediterranean seawater during summer. Ocean. Acta, 9, Pelagic Mediterranean Oceanography, 163-177.

Brasseur, P., J.M. Beckers, J.M. Brankart, R. Schoenauen (1996) Seasonal temperature and salinity fields in the Mediterranean Sea: Climatological analyses of a historical data set. Deep-Sea Res.I, 43, 159-192.

Civitarese, G., A. Crise, G. Crispi, R. Mosetti (1996) Circulation effects on nitrogen dynamics in the Ionian Sea. Ocean. Acta, 19, 609-622.

Crise, A., G. Crispi and E. Mauri (1998, in press) A seasonal three-dimensional study of the nitrogen cycle in the Mediterranean Sea. Part I: Model implementation and numerical results. J. Mar. Sys.

Crispi, G., A. Crise, A., E. Mauri (1998, in press) A seasonal three-dimensional study of the nitrogen cycle in the Mediterranean Sea. Part II: Verification of the energy constrained trophic model. J. Mar. Sys.

Doney, S.C., D.M. Glover, R.G. Najjar (1996) A new coupled one dimensional biological-physical model for the upper ocean: applications to the JGOFS Bermuda Time Series Study (BATS) site. Deep-Sea Res., 43, 591-624.

Ediger, D. (1995) Interrelationships among primary production, chlorophyll and environmental conditions in the northern Levantine basin. Ph.D thesis, Middle East Technical University, Institute of Marine Sciences, Erdemli, Icel, Turkey, 187 pp.

Ediger, D. and A. Yilmaz (1996) Characteristics of deep chlorophyll maximum in the northeastern Mediterranean with respect to environmental conditions. J. Mar. Sys., 9, 291-303.

Ediger, D., S. Tugrul, C.S. Polat, A. Yilmaz, I. Salihoglu (1998, in press) Abundance and elemental composition of particulate matter in the upper layer of the northeastern Mediterranean. J. Mar. Syst.

Fasham, M.J.R., H.W. Ducklow, S.M. McKelvie (1990) A nitrogen-based model of plankton dynamics in the oceanic mixed layer. J. Mar. Res., 48, 591-639.

Jassby, A.D. and T. Platt (1976) Mathematical formulation of the relationship between photosynthesis and light for phytoplankton. Limnol. Oceanogr., 21, 540-547.

Levy, M., L. Memery, J-M, Andre (1998) Simulation of primary production and export fluxes in the northwestern Mediterranean Sea. J. Mar. Res., 56, 197-238.

Kuhn, W., and G. Radach (1997) A one dimensional physical-biological model study of the pelagic nitrogen cycling during the spring bloom in the northern North Sea (FLEX'76). J. Mar. Res., 55, 687-734.

Malanotte-Rizzoli, P., B. Manca, M. Ribera, A. theocharis, A. Bergamasco, D. Bregant, G. Budillon, G. Civitarese, D. Georgopoulos, A. Michelato, E. Sansone, P. Scarazzato, E. Souvermezoglou (1997) A synthesis of the Ionian Sea hydrography, circulation and water mass pathways during POEM-Phase I. Prog. Oceanog., 39, 153-204.

May, P.W. (1982) Climatological heat flux estimates of the Mediterranean Sea, Part I: winds and wind stresses. Report 54, NORDA, NSTL Station, 7539529, 56 pp.

McClain, C.R., K. Arrigo, K-S. Tai, D. Turk (1996) Observations and simulations of physical and biological processes at ocean weather station P, 1951-1980. J. Geophys. Res.

McGullicuddy, D.J., J.J. McCarthy, A.R. Robinson (1995) Coupled physical and biological modeling of the spring bloom in the North Atlantic: I. Model formulation and one dimensional bloom processes. Deep-sea Res., 42, 1313-1357.

Oguz, T., H. Ducklow, P. Malanotte-Rizzoli, S. Tugrul, N. Nezlin, U. Unluata (1996) Simulation of annual plankton productivity cycle in the Black Sea by a one-dimensional physical-biological model. J. Geophysical Research, 101, 16585-16599.

Ozsoy, E., A. Hect, U. Unluata, S. Brenner, T. Oguz, J. Bishop, M.A. Latif, Z. Rosentraub (1991) A review of the Levantine Basin circulation and its variability during 1985-1988. Dyn.Atmos. Oceans, 15, 421-456.

Ozsoy, E., A. Hect, U. Unluata, S. Brenner, H.I. Sur, J. Bishop, T. Oguz, Z. Rosentraub, M.A. Latif (1993) A synthesis of the Levantine Basin circulation and hydrography, Deep-Sea Res., 40, 1075-1119.

Rabitti, S., F. Bianchi, A. Boldrin, L.Daros, G. Socal, C. Totti (1994) Particulate matter and phytoplankton in the Ionian Sea. Ocean. Acta, 17, 297-307.

Robinson, A.R., D. J. McGillicuddy, J. Calman, H. W. Ducklow, M.J.R. Fasham, F.E. Hoge, W.G. Leslie, J.J. McCarthy, S. Podewski, D.L. Porter, G. Saure, J.A. Yoder (1993) Mesoscale and upper ocean variabilities during the 1989 JGOFS bloom study. Deep-Sea Res.II, 40, 9-35.

Salihoglu, I., C. Saydam, O. Basturk, K. Yilmaz, D. Gocmen, E. Hatipoglu, A. Yilmaz (1990) Transport and distribution of nutrients and chlorophyll-a by mesoscale eddies in the northeastern Mediterranean. Mar. Chem., 29, 375-390.

Sur, H.I., E. Ozsoy, U.Unluata (1993) Simultaneous deep and intermediate depth convection in the Northern Levantine Sea, winter 1992. Oceanol. Acta, 16, 33-433.

Yilmaz, A., D. ediger, O. Basturk, S. Tugrul (1994) Phytoplankton fluorescence and deep chlorophyll maxima in the northeastern Mediterranean. Oceanol. acta, 17, 69-77.

Yilmaz, A. and S. Tugrul (1998, in press) The effect of cold- and warm-core eddies on the distribution and stoichiometry of dissolved nutrients in the northeastern Mediterranean. J. Mar. Sys.

# PLANKTON CHARACTERISTICS IN THE AEGEAN, IONIAN AND NW LEVANTINE SEAS

I.SIOKOU-FRANGOU, O.GOTSIS-SKRETAS, E.D.CHRISTOU and K.PAGOU
National Centre for Marine Research
Ag.Kosmas, Helliniko, 16604, Athens, Greece

## 1.Introduction

The Eastern Mediterranean Sea, except the N.Aegean Sea and the Adriatic Sea, is characterized by a subtropical climate favourable to thermophilic and halophilic species [1] and therefore its plankton consists mainly of subtropical elements [2]. The Ionian and Levantine Seas offer depths often over 3000 m and have a narrow continental shelf. The main water masses in the Ionian Sea are the Modified Atlantic Water, the Ionian Surface Water, the Levantine Intermediate Water and the Eastern Mediterranean Deep Water [3]. The latter two water masses are present also in the Levantine Sea along with the Surface Levantine Water [4]. The Aegean Sea is divided in two distinct parts: (a) the N.Aegean, which presents an alternation of plateaux and deep troughs and receives Black Sea waters (b) the S.Aegean that is characterized by the Cyclades plateau followed by the deep Cretan basin [5]. The latter communicates with the Ionian and the Levantine Seas through the western and eastern straits of the Cretan Arc respectively.

Several cyclonic and anticyclonic gyres (Figure 1) have been found influencing the distribution of nutrients in the area, which is considered as very oligotrophic [6, 7]. Deep waters of the Levantine Sea are poor in nutrients, since external inputs to the surface waters are limited. Therefore primary production depends on the transportation of nutrients from deep layers through vertical mixing.

Many studies on plankton composition and distribution in the Ionian, Aegean and Levantine Seas were performed in the sixties and seventies [for a review see 8 and 9]. Synoptic cruises were done in these areas since 1985 within international (POEM, MAST) and national projects. Most of these studies have used the same standard methods, facilitating the comparison of data collected in other areas of the Mediterranean Sea. The present paper will attempt an overview based mainly on these data.



*P. Malanotte-Rizzoli and V.N. Eremeev (eds.),*
*The Eastern Mediterranean as a Laboratory Basin for the Assessment of Contrasting Ecosystems*, 205–223.

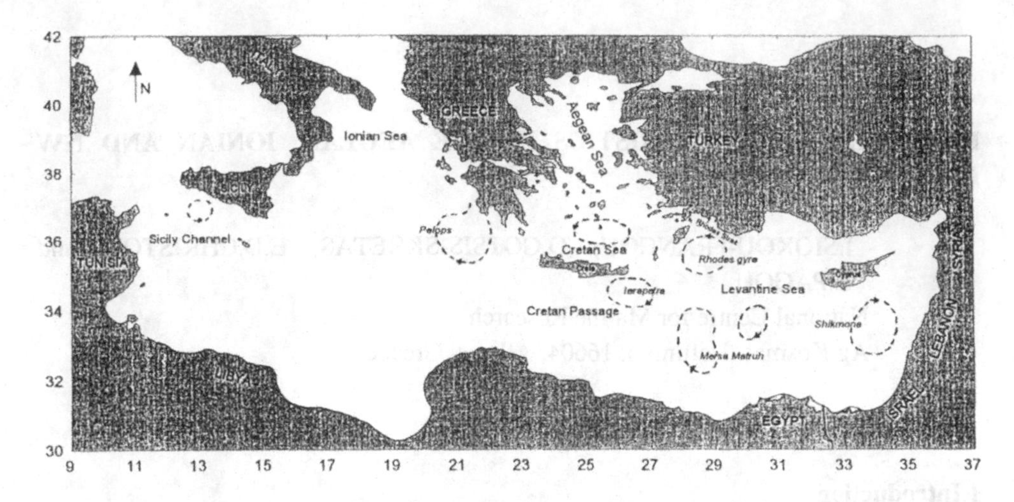


Fig.1. Map of Eastern Mediterranean showing the main gyres [after Mazzocchi *et al.* [10]]

## 2. Phytoplankton

### 2.1. CHLOROPHYLL A

The North Levantine Sea in the Eastern Mediterranean is characterized as one of the most transparrent and less productive seas in the ocean. The results of chlorophyll $\alpha$ concentrations from several studies conducted in the North Levantine Basin, the Aegean and the Ionian Seas (Figure 1) are summarized and discussed. The average over the euphotic zone (down to 100-110 m) chlorophyll $\alpha$ concentrations are usually very low (Table 1). Chlorophyll $\alpha$ variations ranged from 0.09-0.47 mg $m^{-3}$ in the Rhodes cyclonic gyre during 1991-94 [11, 12, 13], from 0.11 to 0.23 mg $m^{-3}$ in the Cretan Passage during 1991-92 [11, 12] and from 0.05-0.28 mg $m^{-3}$ in the Cilician anticyclonic gyre (NE Levantine Basin) [13]. These chlorophyll concentrations in the N.Levantine Sea are comparable to corresponding values from the Cretan Sea (0.13-0.46 mg $m^{-3}$) in 1991-92 [11, 12], the same sea (0.13-0.27 mg $m^{-3}$) during four seasonal samplings in 1994 [14, 15], the eastern Ionian Sea (0.16-0.26 mg $m^{-3}$) in 1988 [16], the offshore Israeli waters (0.06-0.12 mg $m^{-3}$) in 1982-83 [17], the offshore Egyptian waters (0.09-0.79 mg $m^{-3}$) in 1982 [18]. Furthermore, it was reported that the chlorophyll concentrations in the core of Cyprus eddy, during February 1989, scattered in the range of 0.16-0.23 mg $m^{-3}$ with no clear trend at the euphotic zone [19]. Even lower values were recorded in September and November 1989, whereas chlorophyll concentrations were not measurable in the water column below a depth of 200 m.

These low chlorophyll values in the Levantine Basin are due to the extremely low concentrations of nutrients (phosphates and nitrates) especially in the euphotic layer [20, 6, 19]. The high atomic ratio values nitrogen/phosphorus (N:P 21 to 29) reported by Berland *et al.* [21] and N:P of 28-29 reported by Krom *et al.* [22], leads to the conclusion that the eastern Mediterranean Basin is strongly phosphorus limited. On the other hand, the western Mediterranean seems to be richer in nutrients and phytoplankton biomass [23, 24], while Owens *et al* [25] suggested that the western basin appears to be N limited. Berland *et al.* [21] reported that the chlorophyll α was lower in the eastern than in the western Mediterranean Sea during summer 1983. Estrada *et al.* [24] reported average chlorophyll concentrations of 0.32 to 0.45 mg $m^{-3}$ in the layer of 0 - 110 m during seven oceanographic cruises from 1982 to 1987 across the Catalo-Balearic Sea.

TABLE 1. Chlorophyll α concentrations (mg $m^{-3}$) in cyclonic and anticyclonic gyres in the North Levantine Basin and in other Mediterranean seas

| | | Chl α (mg $m^{-3}$) | | DCM (mg $m^{-3}$) | | D-DCM (m) | | |
|---|---|---|---|---|---|---|---|---|
| Area | Time | Range | X | Range | X | Range | X | Ref. |
| Rhodes cyclonic gyre | Oct 91 | 0.06-0.12 | 0.09 | 0.11-0.24 | 0.19 | 50-100 | 75 | 11 |
| Rhodes cyclonic gyre | Mar92 | 0.27-1.05 | 0.47 | 0.42-1.50 | 0.80 | 0-75 | 28 | 12 |
| Rhodes cyclonic gyre | Oct 91 | 0.04-0.16 | 0.07 | 0.07-0.56 | 0.26 | 50-100 | 75 | 13 |
| Rhodes cyclonic gyre | Mar 92 | 0.38-0.49 | 0.45 | 0.36-0.64 | 0.51 | | | 13 |
| Rhodes cyclonic gyre | Jul 93 | 0.28-0.57 | 0.42 | 0.85-1.24 | 1.00 | 55-80 | 67 | 13 |
| Rhodes cyclonic gyre | Mar 94 | 0.20-0.57 | 0.35 | 0.37-1.08 | 0.70 | 40-50 | 45 | 13 |
| Cretan Passage | Oct 91 | 0.08-0.15 | 0.11 | 0.10-0.39 | 0.20 | 50-100 | 75 | 11 |
| Cretan Passage | Mar 92 | 0.08-0.62 | 0.23 | 0.20-0.98 | 0.33 | 0-100 | 58 | 12 |
| Cilician A.G* | Oct 91 | 0.01-0.07 | 0.05 | 0.04-0.20 | 0.11 | 75-130 | 100 | 13 |
| Cilician A.G. | Jul 93 | 0.23-0.24 | 0.22 | 0.71-0.86 | 0.77 | 75-115 | 95 | 13 |
| Cilician A.G. | Mar 92 | 0.19-0.45 | 0.28 | 0.27-0.51 | 0.35 | | | 13 |
| Cilician A.G. | Mar 94 | 0.10-0.15 | 0.12 | 0.20-0.35 | 0.22 | 60-85 | 77 | 13 |
| Cretan Sea | Oct 91 | 0.11-0.15 | 0.13 | 0.20-0.33 | 0.27 | 75 | 75 | 11 |
| Cretan Sea | Mar 92 | 0.27-0.89 | 0.46 | 0.34-1.31 | 0.65 | 25-75 | 50 | 12 |
| Cretan Sea | 1994 | 0.13-0.27 | 0.19 | | | | | 14,15 |
| Eastern Ionian Sea | Mar 88 | 0.15-0.49 | 0.26 | | | 25-75 | 60 | 16 |
| Eastern Ionian Sea | Sep 88 | 0.12-0.25 | 0.16 | | | 50-100 | 75 | 16 |
| Eastern Ionian Sea | Oct 91 | 0.08- | 0.14 | 0.11- | 0.29 | 50-100 | 80 | 11 |

| | | | | | | | | |
|---|---|---|---|---|---|---|---|---|
| | | 0.28 | | 0.79 | | | | |
| E. Medit. (Cyprus eddy) | Febr 89 | | | 0.16-0.25 | | 0-125 | | 19 |
| E. Medit. (Cyprus eddy) | May 89 | | | 0.20-0.35 | | 100 | 100 | 19 |
| E. Medit. (Cyprus eddy) | Sep 89 | | | 0.09-0.13 | | 90-110 | | 19 |
| E. Medit. (Cyprus eddy) | Nov 89 | | | 0.14-0.15 | | 80 | 80 | 19 |
| Offshore Israel | 1982-83 | 0.06-0.12 | | | | | | 17 |
| Offshore Egypt | 1982 | 0.09-0.79 | | | | | | 18 |
| Western Ionian Sea | Oct 91 | | 0.07 | | | | 75 | 27 |
| Eastern Mediterranean | sum. 83 | | | | 0.28 | 90-120 | | 21 |
| Central Mediterranean | sum. 83 | | | | 0.40 | 75-95 | | 21 |
| Western Mediterranean | sum. 83 | | | 0.50-0.85 | | 65-85 | | 21 |
| NW Mediterranean | 1982-87 | 0.32-0.45 | | 0.65-1.45 | | 40-70 | | 24 |

Chl α = averages of water column chlorophyll α
DCM = averages of deep chlorophyll maxima
D-DCM = averages of depth of DCM
X = mathematical averages
Cilician A.G*=Cilician anticyclonic gyre

The presence of a deep chlorophyll maximum (DCM), a common feature in the Mediterranean Sea, has usually been observed below 50 m in the N.Levantine, Aegean and Ionian Seas during summer stratification period. The spatial and vertical distribution of chlorophyll and the formation of DCM in the eastern Mediterranean, apart from the light intensity, is controlled by the euphotic zone nutrient concentrations occurring in cyclonic and anticyclonic gyres. The DCMs usually are formed at shallower depths (D-DCM: 28-75 m) in cyclonic eddy fields (such as the Rhodes cyclone) (Table 1), whereas the depths of the DCMs and the nutricline coincide [13]. The mean concentration of chlorophyll α was 0.47 mg $m^{-3}$ for the 0 to 100 m depth in the Rhodes cyclone in March 1992, with relatively high concentrations ranging from 0.42 to 1.50 mg $m^{-3}$ at the depth of DCM [12]. Corresponding values in July 1993 were 0.42 mg $m^{-3}$ and 0.85-1.24 mg $m^{-3}$ (mean 1.0 mg $m^{-3}$) [13]. At the anticyclonic regions (such as Cilician area) (Table 1) and the Ierapetra anticyclonic gyre in the NW Levantine Basin [12], the DCM was generally located at greater depths (D-DCM: 77-100 m), at the base of the euphotic zone or below it and well above the nutricline, whereas the chlorophyll α concentrations were relatively low compared to those in the Rhodes gyre.

The DCMs are broader in shape and at relative shallower depths in winter-spring conditions [12, 14, 26], than in summer-autumn conditions when the water column is stratified (11, 15, 21, 27]. Generally, there is no consistent trends concerning the vertical distribution of chlorophyll in late winter-early spring, although in certain cases maxima were noted at either 25, 50 or 75 m depth. These inconsistencies in the pattern of the vertical profiles of chlorophyll $\alpha$ indicated a rather unstable environment because of the transitional period (from winter to spring) and the commencement of the thermocline. Chlorophyll concentrations, in all examined areas, showed a clear seasonal pattern with maximum values during late winter-early spring (0.12-0.47 mg $m^{-3}$) and minimum during late summer-early autumn (0.05-0.16 mg $m^{-3}$). The summer thermal stratification situated around 50-100 m [4, 28], apparently prevents the supply of nutrients to the surface waters from the deeper water layers. In winter, vertical mixing processes enrich the euphotic layer with nutrients.

Investigations in the Western Mediterranean basin by Estrada [29] showed that DCM were formed at shallower depths (40-90 m), well below the zone of maximum vertical gradient of the thermocline or the pycnocline. Estrada *et al.* [24] reported maximum chlorophyll concentrations ranging from 0.65 to 1.45 mg $m^{-3}$ at depths ranging from 40 to 70 m across the Catalo-Balearic Sea. Furthermore, Berland *et al* [21] reported that DCM maxima were located at shallower depths (65-80 m) in the western Mediterranean Sea, while they were recorded at 80-120 m in the regions of the eastern and central Mediterranean Sea during summer 1983.

## 2.2 COMMUNITY QUANTITATIVE ASPECT

The results of phytoplankton cell concentrations obtained from several investigations within 1986-92 (POEM cruises) and 1994 (MAST-PELAGOS cruises) in the NW Levantine Basin, the S.Aegean and the E.Ionian Seas are summarized and discussed. During these studies, phytoplankton water samples were collected with NISKIN bottles, fixed with Lugol's solution and, after sedimentation, were examined under an inverted microscope. The total number of diatoms, dinoflagellates, coccolithophores, silicoflagellates and "other groups" such as chlorophyceae, chrysophyceae, prasinophyceae etc. is referred as total phytoplankton. The small phyto-flagellates (diameter of *ca* 3-5 μm) are referred separately as μ-flagellates and can be used as a rough estimation of the small phytoplankton population.

In all sampling seasons (with the exception of spring 1992) total phytoplankton densities were very low ranging from 560 to 29920 cells $l^{-1}$ in the Rhodes area, from 480 to 7000 c $l^{-1}$ in the Cretan Passage and from 240 to 34840 c $l^{-1}$ in the S.Aegean Sea (Table 2). Relatively higher values were observed in the SE Ionian (1560-72090 c $l^{-1}$) and in the N.Aegean Sea (2040-182760 c $l^{-1}$). Phytoplankton cell concentrations, similar to those in the NW Levantine Basin, were found for the offshore Israeli waters during 1981-82 at near surface and at the DCM layer [30]. Similar total phytoplankton densities (9260-68265 c $l^{-1}$) have been detected in the Aegean Sea during five ocenographic cruises from August 1958 to February 1961 [31], in the Petalion Gulf of the central Aegean Sea [32] and in the Ionian Sea [33, 26]. The recorded low cell concentrations in the eastern Mediterranean emphasize the poverty in phytoplankton standing stock which can be attributed mainly to the limiting concentrations of nutrients. On the other hand, higher phytoplankton abundances were recorded in the NW Mediterranean during 1984-86 [23].

TABLE 2. Ranges of total phytoplankton and μ-flagellates cell concentrations in the Eastern Mediterranean.

| Time | Rhodes area | Cretan Passage | South Aegean Sea | East Cretan straits | South-east Ionian Sea | North Aegean | Ref. |
|---|---|---|---|---|---|---|---|
| | | | Total phytoplankton (cells $l^{-1}$) | | | | |
| Mar 86 | 6480-22280 | | 2400-12240 | | 2240-16240 | | 34 |
| Mar 87 | | | 2840-23320 | | 2080-72090 | 2040-10880 | 34 |
| Mar 89 | 560-29920 | 480-7000 | 560-3120 | | | | 35 |
| Mar 92 | 4240-68640 | 2520-6880 | 3120-70080 | | | | 35 |
| Mar 94 | | | 1320-34840 | 7420-20980 | 8760-17600 | | 14 |
| Sep 86 | 2560-25720 | | 1680-9240 | | | 2960-13600 | 34 |
| Sep 87 | 560-7900 | | 1120-25840 | | 4080-10020 | 2120-182760 | 34 |
| Oct 91 | 800-9000 | 720-7000 | 240-2800 | | 1560-6040 | | 11 |
| Sep 94 | | | 3760-11440 | 8760-20040 | | | 15 |
| Jun 94 | | | 3360-15000 | 4280-13080 | 12720-27920 | | 15 |
| | | | μ-Flagellates (cells $l^{-1}$) | | | | |
| Mar 86 | 1120-4160 | | 1240-7880 | | 2080-54000 | | 34 |
| Mar 87 | | | 2640-21760 | | 900-26900 | 2500-11200 | 34 |
| Mar 89 | 3040-506000 | 27500-265000 | 10200-309000 | | | | 35 |
| Mar94 | | | 1700-53300 | 17800-6900 | 16500-28300 | 7600-5900 | 14 |
| Sep 86 | 1000-3200 | | 760-9900 | | | 1560-9680 | 34 |
| Sep 87 | 560-2400 | | 2240-28560 | | 1800-12320 | 3200-2100 | 34 |
| Oct 91 | 7040-6000 | 5760-9280 | 1960-11600 | | | 4400-7500 | 11 |

Phytoplankton abundances presented a seasonal variation with highest values in spring, and lowest values in early autumn (Table 2). As far as the spatial phytoplankton distribution is concerned, relatively higher cell densities were recorded in the E.Ionian than in the S.Aegean Sea, especially during spring 1987, reaching values of 72090 c $l^{-1}$. These higher concentrations may be attributed to the inflow of Adriatic water, richer in nutrients, into the Ionian Sea. From the nutrients and primary production point of view, a great scientific interest presents the Rhodes cyclonic gyre which is a "nutrient oasis" in the N.Levantine Basin. In this area, the nutricline can rise to the base of the euphotic zone so that especially in winter months, nutrients are introduced into the productive zone from the nutricline by vertical mixing [13]. Phytoplankton densities presented a significant increase in spring 1992 at Rhodes cyclonic area (2520 to 70080 c $l^{-1}$) [35] and the presence of large diatoms in high numbers (up to 64000 c $l^{-1}$) in the core of the gyre, suggests nutrient enrichment. The reverse phenomenon occurred at the peripheries and the neighbouring Ierapetra anticyclone, where minimal cell densities (2520-6880 c $l^{-1}$) were observed.

The distribution of phytoplankton densities in the NW Levantine Basin, the Aegean and the Ionian Seas did not present a clear and distinct pattern with depth, although relatively increased phytoplankton values at 50 m and, sometimes, at 75 m depth were observed, especially in summer when the DCM is established. The depth of the maximal phytoplankton cell densities did not always coincide with the depth of chlorophyll maxima and this might merely represent the higher intracellular chlorophyll level due to photoadaptation at lower light intensities [36] not necessarily reflecting greater phytoplankton cell concentrations. Similarly, Kimor *et al.* [30] found no apparent differences between the abundances of the major phytoplankton groups in the DCM and the near surface layer of the offhore Israeli waters during 1981-82, while significant differences between the two layers were determined for chlorophyll $\alpha$ concentration. In the Rhodes area which is influenced by the cyclonic gyre, there was a tendency for higher phytoplankton values at near surface layer during spring.

In the NW Levantine, the Aegean and the Ionian seas, the μ-flagellates densities were always higher than those of phytoplankton (Table 2), whereas their vertical distribution usually was in accordance with that of phytoplankton, not prevailing a distinct pattern among depths. Very high values of μ-flagellates (up to 506000 c $l^{-1}$) were observed in March 1989 in the Rhodes area, the S.Aegean Sea and the Cretan Passage. Therefore, their constant presence and relatively high abundances in these oligotrophic areas classify them among the most important groups of marine plankton in the eastern Mediterranean.

## 2.3 COMMUNITY QUALITATIVE ASPECT

Data obtained during the 1986-87 POEM cruises [34]) showed that diatoms dominated during spring in the Aegean Sea and the Rhodes area, while dinoflagellates were dominant in the Ionian Sea. During summer, dinoflagellates were relatively more abundant in all above cited areas. Coccolithophores also constituted a very important element of the phytoplankton community during spring and late summer, especially in the central and the S.Aegean Seas. During early autumn 1991, the Ionian Sea was characterized by the dominance of coccolithophores and dinoflagellates, whereas few diatoms were recorded. In contrast, diatoms contributed a significant part along with coccolithophores in the Cretan Sea, the Rhodes area and the Cretan Passage [11, 35]. In the Cretan Sea and the Cretan Arc, during 1994 [14, 15], the average seasonal phytoplankton taxa distribution was as follows: diatoms (32.9%) were the dominant group throughout spring, dinoflagellates (53.8%) predominated during summer, coccolithophores (40.8%) prevailed in autumn, whereas "other groups" mainly cryptophyceae and rhodophyceae (31.0%) were the dominant groups in winter. The importance of diatoms and μ-flagellates was also noted by Blasco [32] and Ignatiades [37] for the Petalion and the outer Saronikos Gulf (central Aegean Sea), respectively.

As far as the dominant species is concerned, during spring 1986, the diatoms *Nitzschia closterium* and *Thalassiothrix frauenfeldii* predominated in the S.Aegean Sea and the Rhodes area, while the dinoflagellates *Scrippsiella trochoidea* and *Gymnodinium breve* were more abundant in the Ionian Sea [34]. During spring 1987, the cryptophyceae *Cryptomonas sp.* and the diatom *Chaetoceros affinis* predominated throughout the north and central Aegean Seas. Also, in the N.Aegean Sea the diatoms *Rhizosolenia delicatula* and *Rhizosolenia stolterfothii* were among the dominant species, while in the central Aegean Sea the coccolithophores *Coccolithus pelagicus, Coccolithus fragilis* and *Syracosphaera mediterranea* were abundant. Dinoflagellates predominated in all the studied areas in summer 1986 and were represented mostly by species of the genus *Gymnodinium.* In summer 1987, the most important species in the S.Aegean Sea and the Rhodes area were the cryptophyceae *Cryptomonas sp.*, the dinoflagellate *Exuviella baltica* and *Gymnodinium sp.* and the coccolithophores *C. pelagicus* and *C. fragilis,* while the phytoflagellate *Phaeocystis pouchetii* predominated in the N.Aegean Sea [34].

During autumn 1991 the phytoplankton community in the Ionian Sea was characterized by the dominance of coccolithophores, such as, *Coccolithus sp.* (5-10 μm), *Emilliana huxleyi, Syracosphaera pulchra, Rhabdosphaera tignifer* and the dinoflagellates *S.trochoidea, Gymnodinium spp., Oxytoxum spp.*, whereas very few diatoms were recorded. In contrast, in the Cretan Sea, the Rhodes area and the Cretan Passage, the diatoms *Nitzschia spp., Rhizosolenia fragilissima, Rhizosolenia alata* were relatively abundant. During spring 1992, the predominance of large healthy diatoms such as *C. affinis, T. frauenfeldii, R.fragilissima* and *Nitzschia delicatissima* in the Rhodes cyclonic area suggests nutrient enrichment at this area.

The cluster analysis of samples collected from the Cretan Sea and the Cretan Arc during 1994, did not group them according to area or depth, but groups were separated according to season of sampling. During spring, the diatoms *C. affinis, T. frauenfeldii*, the cryptophyceae *Cryptomonas spp.*, the rhodophyceae *Rhodomonas spp.* and the coccolithophore *Pontosphaera sp.* predominated [14]. This community was replaced during summer by the dinoflagellate *Gymnodinium sp.*, the flagellate *Solenicola setigera* and the diatoms *T. frauenfeldii*, *N. delicatissima* and *R. fragilissima* [15]. In autumn, coccolithophorids predominated with the species *Calyptrosphaera globosa, Pontosphaera sp., Calyptrosphaera superba* and the diatoms *N. delicatissima* and *Nitzschia seriata*, whereas in winter, phytoplankton was characterized by *N. delicatissima, Cryptomonas sp.* and *Cryptomonas huxleyi* [15].

Generally, the species composition of diatoms, dinoflagellates, coccolithophores which were found in the NW Levantine, the S.Aegean and the Ionian Seas, are similar to those observed in the offhore Israeli waters [38, 30], in the Egyptian coast [39], in the central Aegean Sea [37] and in the Ligurian Sea [40]. Most of the above mentioned species are eurythermal of tropical-subtropical character, whereas some of them are nutrient-poor species [41]. The phytoplankton species composition in the eastern Mediterranean show a homogenity, although in certain areas the local hydrographic conditions might affect the community structure.

## 3. Zooplankton

This review is limited to mesozooplankton (collected by a 200μm mesh size net) since the microzooplankton is almost unknown in these areas. On the other hand, studies on the macroplankton are very restricted and performed in the sixties.

### 3.1. QUANTITATIVE ASPECT

According to the current opinion the Eastern Mediterranean is one of the most oligotrophic areas in the world ocean. This oligotrophic character is reflected in zooplankton as low standing stock not only offshore but also in coastal areas.

Most data concern the surface (0-50 m) layer and can present a picture of the spatial distribution of zooplankton quantity. In order to compare the values obtained from different areas of the E.Mediterranean, mean values are given for each area in the Table 3: zooplankton abundance in the N.Aegean Sea is generally higher than that in S.Aegean, Ionian and Levantine Seas. This difference is more pronounced at the eastern part, near the Dardanelles strait, where abundance such as 2000-3000 ind $m^{-3}$ are encountered; these high values should be related to the influence of the Black Sea waters as well as to the topography of the area (extended continental shelf) [42, 43]. A more detailed spatial image was obtained in the 1991-92 cruises when samples were collected along transetcs across cyclonic and anticyclonic gyres (Figure 1). Zooplankton abundance values in the Rhodes area were higher than those detected at all other areas, but the Sicily channel, in autumn 1991 [10] and the highest in spring 1992; they attained 1376 $ind.m^{-3}$ in the 0-50m layer of stations positioned within the core of the Rhodes gyre [44].

These values were higher than those recorded in the same area in spring 1986 [45] and in spring 1988 [46] as well than those found in the Tyrrhenian Sea [47]. The enhanced productivity in this area, was related to the dynamics of the Rhodes gyre [44]. During the cold winter 1992 significant upwelling of deep waters occurred within the core of the gyre [48], resulting in high nutrients and chlorophyll concentrations in the surface layer. (See Souvermezoglou *et al.*, this volume). The significant effect of the gyre upon zooplankton becomes more evident by comparing data collected at stations positioned in the neighbouring anticyclones (Anaximander and Rhodes), where zooplankton abundance was lower than that in the Rhodes cyclone [45, 10, 44].

TABLE 3. Total zooplankton abundance (ind m$^{-3}$) in the 0-50m layer (collected by 200μm mesh size net).

| Season<br>Area | Spring 1988 (1) | Summer 1988 (1) | Autumn 1991 (2) | Spring 1992 (3) |
|---|---|---|---|---|
| N.Aegean | 1125 | 765 | | |
| S.Aegean | 489 | 411 | 108 | 360 |
| Ionian East | 688 | 446 | 181 | |
| West | | | | 595 |
| NW Levantine | 315 | 229 | | |
| Rhodes area | | | 263 | 988 |
| Cretan Passage | | | 192 | 311 |
| Central Levantine | | | 216 | |
| Sicily Channel | | | 579 | |

(1) Siokou et al unpubl. data
(2) 10
(3) 44

Very low number of zooplankters were collected at one station in the SE of Crete (0-100m layer): 85 ind.m$^{-3}$ in January 1987 and 109 ind.m$^{-3}$ in June 1993 [49, 50]. These data are lower than those recorded in the same layer by Pancucci-Papadopoulou *et al.* [45] and Mazzocchi *et al.* [10]. This difference could be due to the use of a 333 μm mesh size which is inefficient for the collection of medium size zooplankters (copepods of the genus *Clausocalanus, Oithona, Paracalanus*), dominant in the Mediterranean Sea. On the contrary Delalo [51] recorded higher values of zooplankton abundance in the 0-200m layer of the entire Levantine Sea (1397 ind.m$^{-3}$) during September, but she used a fine mesh size net (120 μm). Similarly, values reported by Greze [52] and Pavlova [53] for the Ionian and Aegean Seas respectively are higher than those presented in Table 3.

For comparison purposes between different areas of the Mediterranean Sea, zooplankton abundance values integrated in the 0-200m or 0-300m layer are presented in Table 4. Lower values were detected in the Cretan, Levantine and Ionian Seas than in the Balearic Sea for both seasons, whereas they were similar in autumn but higher in spring to those found in the Tyrrhenian Sea.

Although there are no data on the annual cycle of zooplankton in the NW Levantine, the Aegean and Ionian Seas, a seasonality seems to exist with higher values in spring than in late summer-autumn period, in accordance to the seasonal variability of zooplankton in the Mediterranean Sea. Differences become more pronounced in the Rhodes gyre area under strong upwelling conditions.

The decrease of zooplankton abundance with depth is a general pattern in the world ocean, which was also found in the studied areas, although few data for the zone deeper than 300m are few. A sharp reduction of abundance is observed below 100m and virtually all zooplankton was concentrated in the upper 500m. Less than 1 ind.$m^{-3}$ was found below 1000m [45]. The vertical distribution pattern seems to be almost similar in the Aegean, E.Ionian and NW Levantine Seas.

**TABLE 4.** Zooplankton abundance for different Mediterranean areas

| Area | Mesh size | Water column | Season | Abundance (ind.$m^{-3}$) | Reference |
|---|---|---|---|---|---|
| Cretan Sea | 200 | 0-300 | Autumn 1991 | 45 | 10 |
| | | | Spring 1992 | 149 | 44 |
| SE.Ionian | 200 | 0-300 | Autumn 1991 | 62 | 10 |
| W.Ionian | | | Spring 1992 | 250 | 44 |
| Rhodes area | 200 | 0-300 | Autumn 1991 | 66 | 10 |
| | | | Spring 1992 | 221 | 44 |
| Cretan Passage | 200 | 0-300 | Autumn 1991 | 56 | 10 |
| | | | Spring 1992 | 119 | 44 |
| Central Levantine | 200 | 0-300 | Autumn 1991 | 89 | 10 |
| Sicily channel | 200 | 0-300 | Autumn 1991 | 200 | 10 |
| N.Adriatic Sea | 250 | 0-100 | Annual | 500-1000 | 54 |
| S.Adriatic Sea | | 0-500 | | <200 | |
| Balearic Sea | 250 | 0-200 | Spring | 550 | 55 |
| | | | Autumn | 200 | |
| Tyrrhenian Sea | 250 | 0-300 | annual | 83 | 47 |
| | | | spring | 96 | |

However linear regression of log transformed abundances as a function of depth at the same areas showed different slopes. Slopes were similar between Ionian, NW Levantine and Tyrrhenian Seas, whereas they become more abrupt in the Aegean Sea [49, Siokou *et al* unpubl. data). Weikert and Trinkaus [49] concluded that their data evidence a reduced vertical flux in the Levantine Sea when compared to the N.Atlantic.

Nevertheless discrepancies in the general vertical pattern have been detected at a single position or a single sampling period. At the Limnos trough (N.Aegean Sea) the abundance decreased till 500m and was slightly increased deeper due to the abundance of *Calanus helgolandicus.* In the Cretan Passage, Weikert [50] observed strong differences in the vertical pattern between different years and seasons. The standing crop of zooplankton was significantly higher in June 1993 than in January 1987, especially below 1050m. In June 1993 *C.helgolandicus* and *Eucalanus monachus* were found in high concentrations below 600m and down to 4000m, accounting for 70% of total zooplankton, while their abundance was lower (*E.monachus*) or insignificant (*C.helgolandicus*) in January 1987. According to

Weikert [56] these differences were attributed either to the inflow of deep water from the Aegean Sea, where *C.helgolandicus* is indigenous, or to local events of eutrophication of the mixed layer, which induce sporadic phytoplankton blooms and generate habitats where *C.helgolandicus* and *E.monachus* can feed and reproduce.

An important aspect of the zooplankton vertical distribution in the NW Levantine, Ionian and Aegean Seas is the apparent absence of distinct day/night differences [10], a well-known phenomenon in the world ocean. This peculiarity could be due to the paucity of strong mesozooplankton migrants in the Mediterranean intermediate layers [47, 49]. The reduction of strong migrants in Mediterranean waters suggests that copepods play a minimal role in the tranfer of energy to the deep sea and this role can only be achieved by other migrating organisms such as euphausiids and other minor groups [47].

All the above data confirm the oligotrophic character of the Aegean Sea (apart from the northeastern region), the Ionian and the NW Levantine Sea. Nevertheless the existence of the Rhodes gyre in the latter area could temporally differentiate this character by increasing the productivity.

## 3.2 QUALITATIVE ASPECT

Copepods account generally for more than 70% of total zooplankton in the E.Mediterranean Sea. Appendicularians contribute significantly to the total zooplankton numbers representing up to 19% in the upper layers. They are well represented down to the 200m depth [10]. On the contrary, ostracods abundance is low in the upper 50m layer, but it increases with depth contributing <10% to the total community down to 1000m and even more below 1000m [49]. The abundance of cladocera was found low in the E.mediterranean basin, except in NE Aegean Sea where they contributed significantly (up to 40%) [10; Siokou-Frangou *et al* unpubl. data].

Copepods, have been studied down to the species level at all areas. Their number of species increases by the sampling effort in the basin. The most abundant genera in the upper 0-100m layer are *Clausocalanus* and *Oithona.* During the summer-autumn period *Clausocalanus furcatus* and *Oithona plumifera* are dominant in the 0-50m layer in almost entire basin [57, 42, 58], whereas more than one species are dominant in spring [Siokou-Frangou *et al.*, unpubl. data] (Table 5). The 50-100m layer is characterized by the dominance of *Oithona setigera, Clausocalanus paululus, Oncaea media* and *Oncaea mediterranea.* Between 100 and 300m the former two species are abundant, accompanied by *Haloptilus longicornis, Lucicutia flavicornis, Lucicutia gemina* and *Pleuromamma gracilis* [51, 49, 58].

*H.longicornis, Spinocalanus spp., P.gracilis, Pleuromamma abdominalis* are the main copepods between 300 and 700m, whereas *Eucalanus monachus* is dominant below 500m [49, 45]. Generally, deep Mediterraean waters are poor in plankton with respect to those of the Atlantic ocean and characterized by the absence of true bathypelagic species. Instead, Mediterranean deep-sea habitats are populated by a small number of midwater species living at great depths in addition to *Lucicutia longiserrata* which is a true deep-sea calanoid species [59, 49]. The latter species along with *O.setigera, M.minor* and *H.longicornis* are mentioned as the most abundant copepods below 600m in the Ionian and Levantine Seas [59, 60]. In the deep basins of the Aegean Sea two distinct assemblages were found: *Calanus helgolandicus, Oncaea ornata, Chiridius armatus* and *Microcalanus sp.* were abundant in the northern part, whereas *Spinocalanus longicornis, Spinocalanus oligospinosus, E.monachus, Scaphocalanus spp, Temeropia mayumbensis* and *O.setigera* characterized the southern part [60]. Using a fine mesh-size net (50μm) Bottger-Schnack [61] mentioned the dominance of the genus *Oncaea* down to 1850m in the Levantine Sea.

TABLE 5. Abundance of dominant copepod species in percentage of total zooplankton (0-50m layer).

| Spring 1988 (1) | | | | |
|---|---|---|---|---|
| Area | N.Aegean | S.Aegean | E.Ionian | NW Levantine |
| Clausocalanus pergens | 27 | 7.4 | | |
| Clausocalanus jobei | 7.2 | 7.2 | | |
| Clausocalanus spp. | | | 48.5 | 31.9 |
| Oithona plumifera | 8 | 5.9 | 0.3 | 5 |
| Ctenocalanus vanus | 8.6 | 4.9 | 1.2 | 3.6 |

(1) Siokou-Frangou *et al.* (unpubl. data)

| Autumn 1991 (2) | | | | | | |
|---|---|---|---|---|---|---|
| Area | Sicily channel | E.Ionian | Cretan Sea | Cretan Passage | Rhodes area | Central Levantin e |
| Clausocalanus furcatus | 27.5 | 25.6 | 28.4 | 36 | 38 | 21 |
| Clausocalanus paululus | 1.5 | 3.7 | 21 | 4.7 | 12 | 20 |
| Calocalanus pavoninus | 6.7 | 8.6 | 3.3 | 4.7 | 2 | 7 |
| Oithona plumifera | 21.2 | 21 | 7.9 | 10.8 | 8.7 | 8.6 |
| Farranula rostrata | 3.8 | 2.6 | 9.8 | 6.6 | 4.9 | 10.4 |
| Oncaea media | 7.8 | 1.6 | 0.2 | 1.6 | 0.4 | 2.4 |

(2) [58]

With the exception of the Adriatic and the Aegean Seas, there is large communication within the E.Mediterranean Basin from the Sicily channel till the Suez channel. Therefore a strong similarity among the communities of the different regions can be assumed. Although no differences seem to exist in the faunistic aspect, some dissimilarities in the community composition and structure have been found among areas. An almost permanent observation is the distinction of the NE part of the Aegean Sea: in late winter-early spring *C.helgolandicus, Acartia clausi,*

*Centropages typicus, Paracalanus parvus* and *Evadne nordmanni* dominated [43] and during the warm period the cladocerans *Penilia avirostris, Evadne spinifera* and *Evadne tergestina* characterized this area [42]. This distinction, resulting in a mixte neritic-pelagic character, seems to be related to the topography of the area, as well as to the influence of the Black sea waters entering in the Aegean Sea and spreading westwards according to the dominating circulation [42, 43].

Apart from the circulation, mesoscale hydrological structure seems to influence zooplankton composition in the E.Mediterranean Sea. During autumn 1991 dissimilarities appeared in the copepod assemblages in the 50-300m layer and they were possibly related to different hydrological features (cyclonic or anticyclonic gyres) prevailing in the basin, rather than to the geographical position of stations and/or to water masses distribution. The Sicily channel, the Rhodes area and the Cretan Sea, dominated by cyclonic gyres, are characterized by different copepod assemblages when compared to the SE Ionian Sea, the Cretan Passage and the central Levantine Sea where anticyclonic gyres predominate [58]. The influence of the Rhodes gyre upon zooplankton community has been revealed stronger in spring 1992: as mentioned previously, during this period and after the upwelling of deep waters, high values of chlorophyll, diatoms and zooplankton abundance were found in the surface layer. In this area and mainly in the upper 100m layer, zooplankton was highly dominated by *E.monachus* [44]. In spring 1986 and 1988 *E.monachus* was mainly found deeper than 500m [45, 46]. This copepod is considered as a rapidly growing grazer which occupies deep layers during the diapause period and migrates to the surface under upwelling conditions [62]. Although the influence of the Rhodes gyre upon nutrients and chlorophyll concentration and distribution has been found several times, the extreme conditions in 1992 resulted in an upwelling environment. These conditions probably enhanced the zooplankton productivity and especially that of *E.monachus,* resulting in high abundance values. On the other hand the high abundance of *E.monachus* and *C.helgolandicus* in the deep layer of the neighbouring Ierapetra gyre detected in June 1993 [56], could be a secondary result of this upwelling.

### 3.3. BIOGEOGRAPHY

Two thirds of the E.Mediterranean basin are situated S. of 36°N, therefore at relatively low latitudes compared to the Western Mediterranean. Except for the Adriatic and the N.Aegean Sea, this basin is characterized by high temperature and salinity values, attributing a subtropical character. Although both basins are very similar from the faunistic point of view, some differences seem to exist regarding the abundance of the dominant species. During the recent studies in the Gulf of Lion *Clausocalanus spp.* were the most abundant species, followed by *P.parvus, Oithona spp., Oncaea spp., Corycaeus spp., C.typicus, C.helgolandicus* and *Temora stylifera* (63). In the eastern basin, except in the Adriatic and the N.Aegean Sea, *P.parvus, C.helgolandicus* were found rare, whereas the abundance of *C.typicus* and *T.stylifera* was found to be lower.

The faunistic "personality" of the E.Mediterranean lies on its communication with the neighbouring Black and Red Seas; the former has a boreal whereas the latter has a tropical character. Although the communication is limited due to the narrowness of the Dardanelles strait and the Suez channel, the influence of these extreme environments has been detected early. Eight species (*P.parvus, Pontella mediterranea, Centropages kroyeri, Oithona similis, Oithona nana, A.clausi, C.helgolandicus, Anomalocera patersoni*) were considered as Black sea species found in the N.Aegean Sea by Pavlova [53] and Moraitou-Apostolopoulou [64]. The presence of these species in small or even large numbers in other regions of the E.Mediterranean (coastal and offshore), do not support the hypothesis of a true migration of these species from the Black Sea, but rather of an enrichment of the Aegean Sea populations. On the contrary the rare presence of *Pseudocalanus elongatus* in the NE Aegean Sea could support the hypothesis of migration from the Black Sea or even that of the boreal relict species [65]. Among the planktonic species considered as lessepsian migrants few copepods are mentioned mainly to be found in the SE Levantine [8], while *Arietellus pavoninus* was captured in the S.Aegean Sea [66]. The presence of *Calanopia elliptica* and *Centropages furcatus* in the NW Levantine (Siokou *et al* unpubl. data) suggests a spreading of the migrants throughout the entire Levantine Sea.

## 4. Conclusion

The Aegean, Ionian and NW Levantine Seas are oligotrophic areas characterized by subtropical plankton elements, except the NE Aegean Sea, due to its topography and hydrology. Some differences, regarding the chlorophyll vertical distribution, the total plankton abundance and the relative abundance of dominant species, were found between these areas and the Western Mediterranean Sea. The productivity and the community structure seem to be affected by the hydrological features of the basin. Namely the Rhodes cyclonic gyre under extreme meteorological conditions functions as an upwelling area, resulting in enhanced plankton productivity.

### References


1. Furnestin, L. (1979) Aspects of the zoogeography of the Mediterranean plankton, in S.Van der Spoel and A.C. Pierrot-Bults, (eds), *Zoogeography and diversity in plankton*, Bunge Scientific Publ., Utrecht., pp. 191-253.
2. Basescu,, M. (1985) The effects of the geological and physico-chemical factors on the distribution of marine plants and animals in the Mediterranean, in M.Moraitou-Apostolopoulou and V.Kiortsis (eds), *Mediterranean Marine Ecosystems*, Plenum Press, New York and London, pp.195-212.
3. Malanotte-Rizzoli, P., Manca, B.B., Ribera d'Alcala, M., Theocharis, A., Bergamasco, A., Bregant D., Budillon, D., Civitarese, P., Georgopoulos, D., Michelato, A., Sansone, E., Scarazzato, P., and Souvermezoglou, E. (1997) A synthesis of the Ionian Sea hydrography, circulation and water mass pathways during POEM-Phase I, *Prog. Oceanog.* **39**, 153-204.

4. Theocharis, A., Georgopoulos, D., Lascaratos, A., and Nittis, K. (1993) Water masses and circulation in the central region of the Eastern Mediterranean: eastern Ionian, South Aegean and Northwestern Levantine, 1986-1987, *Deep-Sea Res.* Part II Special Issue **40**, 1121-1142.
5. Theocharis, A. and Georgopoulos, D. (1993) Dense water formation over the Samothraki and Limnos plateaux in the North Aegean Sea (Eastern Mediterranean Sea), *Cont. Shelf Res.*, **13 (8/9)**, 919-939.
6. Salihoglu, I., Saydam, C., Basturk, O, Yilmaz, K., Gocmen, D., Hatipoglu, E., and Yilmaz, A. (1990) Transport and distribution of nutrients and chlorophyll-α by mesoscale eddies in the Northeastern Mediterranean, *Mar. Chem.* **29**, 375-390.
7. Yacobi, Y.Z., Zohary, T., Kress, N., Hecht, A., Robarts, R.D., Waiser, M., Wood, A.M., and Li, W.K.W. (1995) Chlorophyll distribution throughout the southeastern Mediterranean in relation to the physical structure of the water mass, *J. Mar. Syst..* **6**, 179-190.
8. Moraitou-Apostolopoulou, M. (1985) Zooplankton communities in the Eastern Mediterranean, in M.Moraitou-Apostolopoulou and V.Kiortsis (eds), *Mediterranean Marine Ecosystems*, Plenum Press, New York and London, pp.303-332.
9. Stergiou, K.I., Christou,E.D., Georgopoulos, D., Zenetos, A. and Souvermezoglou, E. (1997) The Hellenic seas: physics, chemistry, biology and fisheries, *Ocean. Mar. Biol. Ann. Rev*, **35**, 415-538.
10. Mazzocchi, M.G., Christou, E.D., Fragopoulu, N. and Siokou-Frangou, I. (1997) Mesozooplankton distribution from Sicily to Cyprus (Eastern Mediterranean): I. General aspects, *Oceanologica Acta*, **20(3)**, 521-535.
11. Gotsis-Skretas, O., Siokou-Frangou, I., Pagou, K., and Christou, E. (1993) *POEM Data Report 2. POEMBCO91 Cruise, 1 Oct. 91- 15 Nov. 91. POEM - Phase II.* Scientific Research Programme, Greek Contribution, NCMR, Athens, Greece.
12. Gotsis-Skretas, O., Pagou, K., and Asimakopoulou, G. (in press) Distribution of chlorophyll α in the Eastern Mediterranean: Eastern Ionian, South Aegean and Northwest Levantine, *Proceed. POEM-BC workshop, Molitg- les- Bains, July 1996.*
13. Ediger, D. and Yilmaz, A. (1996). Characteristics of deep chlorophyll maximum in the Northeastern Mediterranean with respect to environmental conditions, *J. Marine Systems* **9**, 291-303
14. Gotsis-Skretas, O., Pagou, K., Akepsimaidis, K., and Kougioufas, P. (1995), Phytoplankton, in *CEC/MAST-MTP, sub-project PELAGOS, Hydrodynamics and Biogeochemical fluxes in the Straits of the Cretan Arc (Aegean Sea, Eastern Mediterranean Basin, 1st Annual Progress Report, September 1993 - August 1994,* NCMR, Athens, Greece, pp. 164-185.
15. Gotsis-Skretas, O., Pagou, K., and Asimakopoulou, G. (1996) Seasonal, spatial and vertical variability of chlorophyll α and phytoplankton communities in the straits of the Cretan Arc and the adjacent seas (South Aegean, Ionian and Levantine seas), in *CEC/MAST-MTP, sub-project PELAGOS, Hydrodynamics and Biogeochemical fluxes in the Straits of the Cretan Arc (Aegean Sea, Eastern Mediterranean Basin), 2nd Annual Progress Report, September 1994 - August 1995,* NCMR, Athens, Greece, pp. 315-330.
16. Gotsis-Skretas, O., Pagou, K., Christaki, U., and Akepsimaidis, K. (1993) Distribution of chlorophyll α in the oligotrophic waters of Aegean, Levantine and Ionian seas, in *Proceed. 4th Natl Symp. Oceanogr. Fish.,* pp. 73-76.
17. Berman, T., Townsend, D.W., El Sayed, S.Z., Trees, C.C., and Azov, Y. (1984) Optical transparency, chlorophyll and primary productivity in the Eastern Mediterranean near the Israeli coast, *Oceanologica Acta* **7**, 367-372.
18. Dowidar, N.M. (1984) Phytoplankton biomass and primary productivity of the south-eastern Mediterranean, *Deep-Sea Res.* **31**, 983-1000.

19. Krom, M.D., Brenner, S., Kress, N., Neori, A., and Gordon, L.I. (1993) Nutrient distributions during an annual cycle across a warm-core eddy from the E. Mediterranean Sea, *Deep-Sea Research I* **40**, 805-825.
20. Souvermezoglou, E., Dagre, P., Nakopoulou, H., Psyllidou, R., and Ikonomou, I. (1989) Distribution of nutrients and oxygen in the Eastern Mediterranean Sea, in *Proceed. UNESCO/IOC 2nd POEM Scientific Workshop, Observatorio Geofisico Sperimentale (OGS), Trieste, Italy, 31 May - 4 June 1988*, POEM Sci. Repts **3**, Cambridge Massachusetts, USA, pp. 85-102.
21. Berland, B.R., Benzhitski, A.G., Burlakova, Z.P., Georgieva, L.V. Izmestieva, M.A., Kholodov, V.I., and Maestrini, S.Y. (1988) Conditions hydrologiques estivales en Mediterranee, repartition du phytoplankton et de la matiere organique, *Oceanologica Acta* ,**Special issue No 9,** 163-177.
22. Krom, M.D., Kress, N., Brenner, S., and Gordon, L.I. (1991) Phosphorus limitation of primary productivity in the eastern Mediterranean Sea, *Limnol. Oceanogr.*, **36**(3), 424-432.
23. Estrada, M. (1991) Phytoplankton assemblages across a NW Mediterranean front: changes from winter mixing to spring stratification, in J.D. Ros and N. Prat (eds.), *Homage to Ramon Margalef; or Why there is such pleasure in studying nature, Oecologia aquatica* **10**, pp.157-185.
24. Estrada, M., Marrase, C., Latasa, M., Berdalet, E., Delgado, M., and Riera, T. (1993) Variability of deep chlorophyll maximum characteristics in the Northwestern Mediterranean, *Mar. Ecol. Prog. Ser.* **92**, 289-300.
25. Owens, N.J.P., Rees, A.P., Woodward, M.S., and Mantoura, R.F.C. (1989) Size-fractionated primary production and nitrogen assimilation in the northwest Mediterranean Sea during January 1989, in *First Workshop on the Northwest Mediterranean Sea*. Comm. Eur. Communities Water Pollut. Res. Rep. Paris, pp. 126-135.
26. Yilmaz, A., Ediger, D., Basturk, O., and Tugrul, S. (1994) Phytoplankton fluorescence and deep chlorophyll maxima in the Northeastern Mediterranean, *Oceanologica Acta* **17**, 69-77.
27. Rabitti, S., Bianchi, F., Boldrin, A., Daros, L., Socal, G., and Totti, C. (1994) Particulate matter and phytoplankton in the Ionian Sea, *Oceanologica Acta* **17**, 297-307.
28. Ozsoy, E., Hecht, A., and Unluata, U. (1989) Circulation and hydrography of the Levantine Basin. Results of POEM coordinated experiments 1985/1986, *Prog. Oceanogr.* **22**,125-170.
29. Estrada, M. (1985) Deep phytoplankton and chlorophyll maxima in the Western Mediterranean, in M. Moraitou-Apostolopoulou and V. Kiortsis (eds.), *Mediterranean Marine Ecosystems*, Plenum Press, New York and London, pp. 247-278.
30. Kimor, B., Berman, T., and Schneller, A. (1987) Phytoplankton assemblages in the deep chlorophyll maximum layer off the Mediterranean coast of Israel, *J. Plankt. Res.* **9**, 433-443.
31. Mihailov, A.A. and Denisenko, V.V. (1963) On the phytoplankton of the Aegean Sea, *Trudy Sevastopol. Biol. St.* **17**, 3-12 (in Russian).
32. Blasco, D. (1974) Etude du phytoplancton du Golfe de Petalion (Mer Egee) en Mars 1970. *Rapp. Comm. Int. Mer Medit.* **22**, 65-70.
33. Rouhiainen, M.I. and Georgieva, L.V. (1982) Phytoplankton in the Ionian and Sardinian Seas, *Ekologija Morja* **8**, 24-36.
34. Pagou, K. and Gotsis-Skretas, O. (1989) Plankton communities. I. Phytoplankton, in A. Boussoulengas and A.V. Catsiki (eds.), *Pollution Research and Monitoring Programme in the Aegean and Ionian seas, Report II. 1986-1987*, NCMR, Athens, June 1989, pp. 128-156.
35. Pagou, K., Gotsis-Skretas, O., and Christaki, U. (in press) Phytoplankton communities in the Eastern Mediterranean, *Proceed..POEM-BC workshop, Molitg- les- Bains, July 1996.*

36. Raymont, J.E.G. (Ed.) (1980) *Plankton and Productivity of the Oceans. Vol 1. Phytoplankton,* 2nd ed. Pergamon Press, Oxford.
37. Ignatiades, L. (1969) Annual cycle, species diversity and succession of phytoplankton in lower Saronikos Bay, Aegean Sea, *Mar. Biol.* **3**, 196-200.
38. Lakkis, S. and Lakkis, V.N. (1980) Composition, annual cycle and species diversity of the phytoplankton in Lebanese coastal waters, *J. Plankton Res.* **3**(1), 123-136.
39. Halim,Y., Guergues, S.K., and Saleh, H.H. (1967) Hydrographic conditions and plankton in the South East Mediterranean during the last normal Nile flood (1964), *Int. Revue ges. Hydrobiol.* **52**(3), 401-425.
40. Travers, A. (1965) Microplancton recolte en un point fixe de la mer Ligure (Bouee-Laboratoire du Comexo) pendant l'annee 1964, - *Rec. Trav. St. Mar. End. Bull. 39*, **55**, 11-50.
41. Kimor, B. (1985) Round table on indicator species in marine phytoplankton II. Background presentation, *Rapp. Comm. int. Mer Medit.* **29(9)**, 65-79.
42. Siokou-Frangou, I., Pancucci-Papadopoulou, M.A., and Kouyoufas, P. (1990) Etude de la repartition du zooplancton dans les mers Egee et Ionienne, *Rapp.Comm.int.mer Medit.*, **32,** 221.
43. Siokou-Frangou, I., Pancucci-Papadopoulou,M.A., and Christou, E. (1994a) Sur la repartition du zooplancton superficiel des mers entourant la Grece (printemps 1987), *Biologia Gallo-Hellenica*, **22,** 337-354.
44. Christou, E.D., Siokou-Frangou, I., Mazzocchi, M.G., and Aguzzi, L. (1998) Mesozooplankton abundance in the Eastern Mediterranean during spring 1992, *Rapp.Comm.int.mer Medit.*, **35,** 410.
45. Pancucci-Papadopoulou, M.A., Siokou-Frangou, I., Theocharis, A., and Georgopoulos, D. (1992) Zooplankton vertical distribution in relation to the hydrology in the NW Levantine and the SE Aegean seas, *Oceanologica Acta*, **15,** 365-381.
46. Christou, E.D., Siokou-Frangou, I., Georgopoulos, D., and Theocharis, A. (1990) On the distribution of zooplankton from the Levantine and Ionian seas in relation with the hydrology of the area, in *Proceed.3rd Natl.Symp.Ocean.Fish.*, 495-503.
47. Scotto di Carlo, B., Ianora, A., Fresi, E., and Hure, J. (1984) Vertical zonation patterns for Mediterranean copepods from the surface to 3000m at a fixed station in the Tyrrhenian sea, *J.Plankton Res.*, **6**, 1031-1056.
48. Kontoyannis, H., Theocharis, A., and Kioroglou, S. (1997) Water mass formation in the NW Levantine Basin in 1992 and 1995: similarities and differences in structure characteristics, in *Proceed.5th Natl.Symp.Ocean.Fish.*, 169-172.
49. Weikert, H. and Trinkaus,S. (1990) Vertical mesozooplankton abundance and distribution in the deep Eastern Mediterranean Sea SE of Crete, *J.Plankton Res.*, **12,** 601-628.
50. Weikert, H. (1995) Strong variability of bathypelagic zooplankton at a site in the Levantine Sea- a signal of seasonality in a low-latitude deep-sea? *Rapp.Comm.int.mer Medit,* **34**, 218.
51. Delalo, E.P. (1966) The zooplankton of the eastern Mediterranean (Levantine Sea and Gulf of Syrte), in M.Science (ed) *Investigation of Plankton in the South seas,* **7,** 62-81.
52. Greze, V.N. (1963) Zooplankton of the Ionian Sea. *Okeanol. Issl.*, **9,** 42-59.
53. Pavlova, E. (1966) Composition and distribution of zooplankton in the Aegean Sea, in M.Science (ed) *Investigation of Plankton in the South seas,* **7,** 38-61.
54. Hure, J., Ianora, A., and Scotto di Carlo, B. (1980) Spatial and temporal distribution of copepod communities in the Adriatic Sea, *J.Plankton Res.*, **2(4)**, 295-316.
55. Jansa, X. and Fernandez de Puelles, M.L. (1990) Distribucion espacio-temporal del zooplancton en el mar Balear, *Bol. Inst. Esp. Oceanogr.*, **6**, 107-136.

56. Weikert, H. (1996) Changes in Levantine Deep-Sea Zooplankton, *Proceed. Int.POEM-BC/MTP Symp., Molitg-les Bains, July 1996*, 99-101.
57. Pasteur, R., Berdugo, V. & Kimor, B. (1976) The abundance, composition and seasonal distribution of epizooplankton in coastal and offshore waters of the Eastern mediterranean, *Acta Adriatica*, **18**, 55-80.
58. Siokou-Frangou, I., Christou, E.D., Fragopoulu, N. and Mazzocchi, M.G. (1997) Mesozooplankton distribution from Sicily to Cyprus (Eastern Mediterranean): II Copepod assemblages, *Oceanologica Acta*, **20(3)**, 537-548.
59. Scotto di Carlo,B., Ianora,A., Mazzocchi, M.G. and Scardi, M. (1991) Atlantis II cruise: uniformity of deep copepod assemblages in the Mediterranean Sea. *J.Plankton Res.*, **13(2)**, 263-277.
60. Siokou-Frangou, I., Christou, E. and Pancucci-Papadopoulou, M.A. (1994b) Deep-sea mesozooplankton in the Eastern mediterranean. *7$^{th}$ Deep Sea Biology Symposium, Crete, September 1994.*
61. Bottger-Schnack, R. (1997) Vertical structure of small metazoan plankton, especially non-calanoid copepods. II. Deep Eastern Mediterranean (Levantine Sea), *Oceanologica Acta*, **20(2)**, 399-419.
62. Smith, S.L. (1995) The Arabian Sea: mesozooplankton response to seasonal climate in a tropical ocean. *ICES J. Mar. Sci.*, **52**, 427-438.
63. Kouwenberg, J.H. (1994) Copepod distribution in relation to seasonal hydrographics and spatial structure in the North Western Mediterranean (Golfe du Lion), *Estuar.Coast Shelf Sci.*, **38**, 69-90.
64. Moraitou-Apostolopoulou, M. (1976) Influence de la mer Noire sur la composition de la faune planctonique (Copepodes) de la mer Egee, *Acta Adriatica*, **18(16)**, 271-274.
65. Siokou-Frangou, I. (1985) Sur la presence du copepode calanoide *Pseudocalanus elongatus* (Boeck) en mer Egee du Nord, *Rapp.Comm.int.mer Medit.* **29(9)**, 231-233.
66. Moraitou-Apostolopoulou, M. (1969) Sur la presence en mer Egee d'*Arietellus pavoninus* copepode pelagique cite pour la premiere fois en Mediterranee, *Biologia Gallo-Hellenica*, **2**, 189-191.

# THE EFFECT OF PHYSICAL PROCESSES ON THE DISTRIBUTION OF NUTRIENTS AND OXYGEN IN THE NW LEVANTINE SEA

E. SOUVERMEZOGLOU and E. KRASAKOPOULOU
*National Centre for Marine Research*
*Aghios Kosmas, Ellinikon, 16604, Greece*

Abstract
The extensive study of the nutrient regime in the Eastern Mediterranean since 1985 allowed the detailed description of their spatial and temporal distribution. The combination of the chemical and physical oceanography data revealed the important chemical signature of the physical processes, the circulation features and dynamics. The two main circulation features namely Rhodes cyclone and Ierapetra anticyclone make the NW Levantine a special area to study the interactions between physical and chemical parameters. Thus we have been able to understand the increase of nutrients and plankton biomass during the cold winter 1992 in the Rhodes gyre and confirm the evolution of the Eastern Mediterranean Deep Water from its chemical characteristics.

## 1. Introduction

The Mediterranean Sea has long been known as an impoverished area with low nutrient level, insufficient to support a large biomass [1]. The recent observations confirm the general depletion of nutrients compared with other parts of the world ocean [2]. There is a limited supply to its surface waters from both its lower layers and from the external sources (the Atlantic inflow, riverine discharges and atmospheric input) but the main reason of its poverty is related to its hydrology and circulation as a concentration basin [3].

Until recently, there was sporadic information concerning the nutrients and oxygen distribution in the Eastern Mediterranean Sea. McGill [1] presented a summary of seasonal patterns in oxygen and phosphorus distribution in the Eastern and Western Mediterranen Sea, based on the few chemical measurements carried out during hydrological cruises. The results indicate the low level of biological activity and the oligotrophic character of the Mediterranean Sea. In 1965, the same author [4] compares the available data from the Eastern Mediterranean with those from the Western Mediterranean and Atlantic Ocean. He notes a considerable depletion of nutrients from the western to the eastern part of the Mediterranean Sea.



*P. Malanotte-Rizzoli and V.N. Eremeev (eds.),*
*The Eastern Mediterranean as a Laboratory Basin for the Assessment of Contrasting Ecosystems, 225–240.*

The nutrient regime of the Eastern Mediterranean has been studied extensively in recent years. In the framework of national and international research programs carried out in the Eastern Mediterranean during the past decade 1986-1995 (POEM, POEM-BC, PELAGOS/MTP-I, OTRANTO/MTP-I, Open Sea Oceanography) chemical data were collected simultaneously with the physical data. The analysis of the data sets and the interpretation of the results revealed the existing analogies between the chemical and physical parameters and dynamics in the spatial distributions, the signals corresponding to physical processes and the temporal variability, from seasonal to interannual [5, 6, 7, 8, 9, 10, 11]. The role of the physical processes is more important because the phytoplankton production in the eastern basin can be related to the input of nutrients from the lower layers especially by wintertime vertical mixing.

## 2. Area of study

The Levantine basin is the second largest basin of the Eastern Mediterranean. It is bounded by Asia Minor and the northeast African mainland (Figure 1). The NW Levantine Sea exchanges waters with the Aegean through the Rhodes Strait (sill depth: 350m; width 17 km), Karpathos Strait (sill depth: 850m; width: 43 km), Kassos Strait (sill depth: 1000m; width 67 km) and with the Ionian Sea through the Cretan passage. The major troughs are the Rhodes (4000m ), Antalya (2000-2500m ), Cilicia (1000m ) and Lattakia (1000-1500m) Basins, the Hellenic Trench (3000-3500m) and the Herodotus Abyssal Plain (3000m). The latter two depressions are separated by the Mediterranean Ridge (2500m).

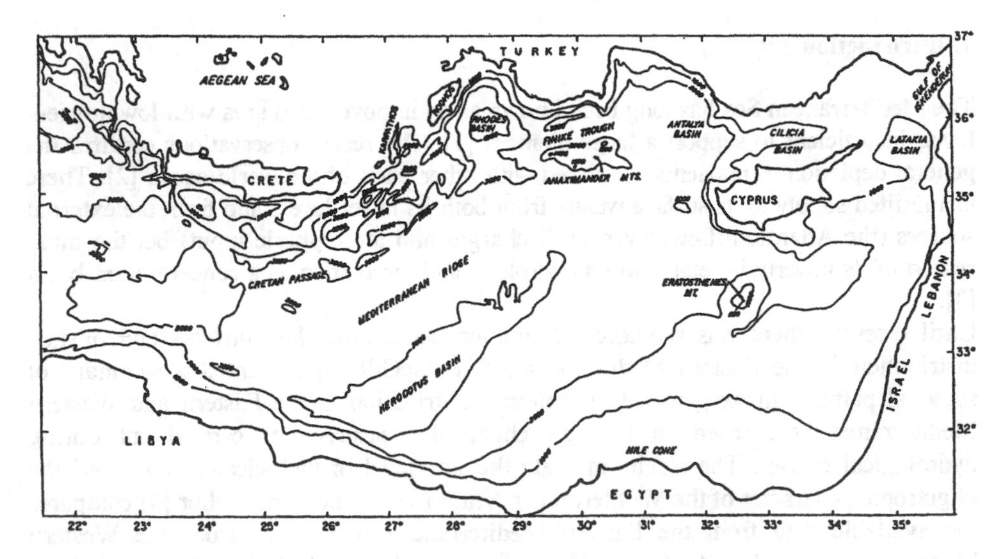


*Figure 1.The bottom topography and geography of the Levantine basin.*

The Levantine Sea was covered from 1986 with coordinated surveys by Greece, Turkey and Israel. Greece covered the NW Levantine Sea (Rhodes gyre area, the eastern passages of the Cretan Arc, the area south east of Crete). Turkey covered the northern Levantine basin and Israel the southern Levantine basin.
The meteorogical and hydrological conditions in the Levantine Sea are often favorable for the convergence and the convective sinking of high salinity waters. The Eastern Mediterranean and specifically the Rhodes gyre is the site of formation of the Levantine Intermediate Water.
The synthesis of the recently collected data allowed to describe the Levantine basin circulation [12,13]. The hydrological features consists of a series of dynamically interacting sub-basin scale eddies (the Rhodes cyclonic, Mersa Matruh anticyclonic and Shikmona anticyclonic gyres) and embedded coherent structures (the Anaximander, Antalya, Cilician and Ierapetra anticyclonic eddies) fed by bifurcating jet flows (the Central Levantine Basin Current and Asia Minor Current, AMC). The general circulation is represented schematically in Figure 2 (from [12] ).

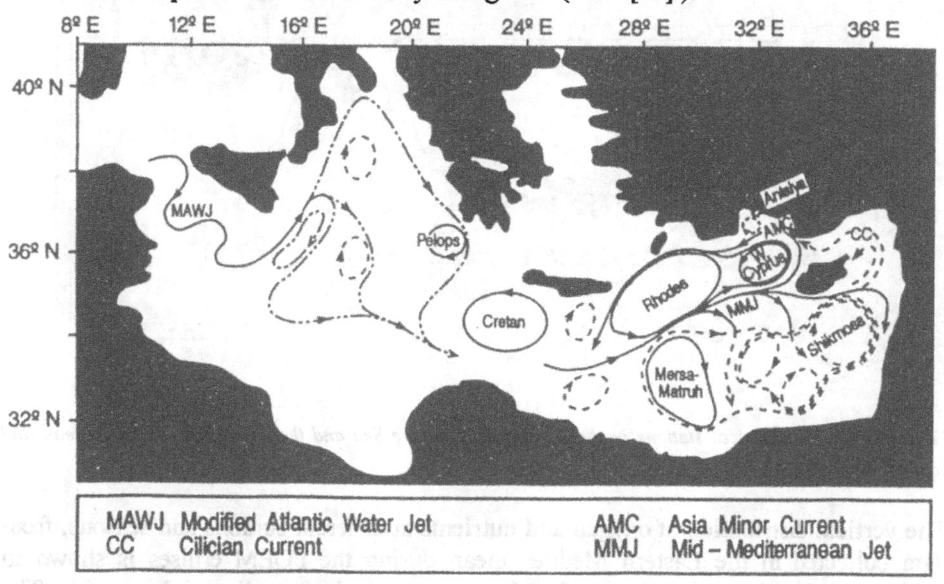


*Figure 2. Schematic representation of the general upper thermocline circulation extended from melding data and dynamics (Robinson and Golnaraghi, [12 ]; Fig.29)*

According to Theocharis *et al* [14] the three existing main water masses in the NW Levantine Sea are: a) the Modified Atlantic Water (MAW), which originates from the Atlantic Ocean, enters in the Eastern Mediterranean through the Sicily channel and is characterised by a salinity minimum, b) the Levantine Intermediate Water (LIW), which is the saltiest water mass of the Eastern Mediterranean and is generated in late winter in several areas of the Levantine Basin and in the South Aegean Sea and c) the Eastern Mediterranean Deep Water (EMDW) which is colder and less saline than LIW. This is produced in the Ionian Sea through the mixing of the transformed LIW with the Adriatic Bottom Water (AdBW) outflowing from Otranto Strait. The layer between

LIW and the EMDW (700-1600m) is considerably uniform (S~38.7, T~13.6 °C) and is called Transitional Mediterranean Water (TMW). Recently during the severe winters of 1992 and 1993 deep water formation and convective overturning in the Rhodes gyre led to the formation of Levantine Deep Water (LDW) [15].

## 3. Results and Discussion

### 3.1. Distribution of nutrients in the NW Levantine Sea

The northern Levantine Sea has been sampled several times since 1986 in the framework of national and multinational research programs. The survey consisted from an extensive grid of stations shown in Figure 3.

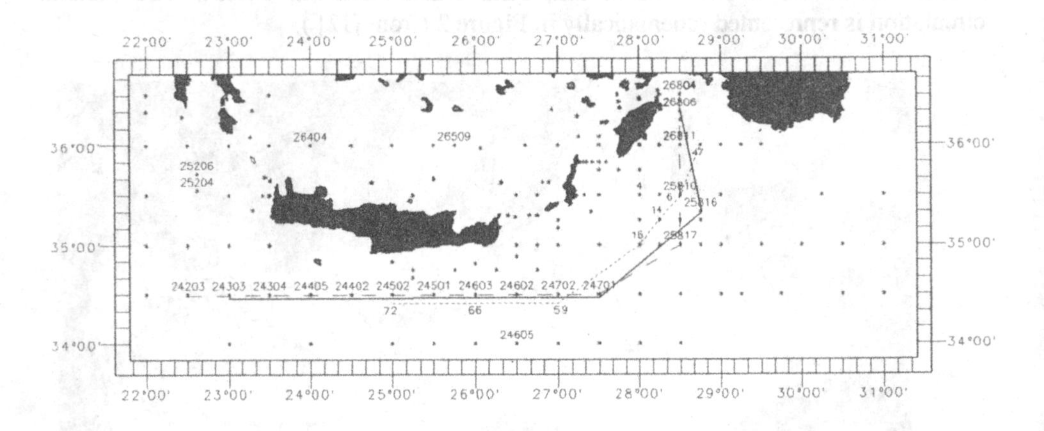


*Figure 3. Grid of chemical stations in the northwest Levantine Sea and the adjacent southern Aegean and eastern Ionian Seas.*

The vertical distribution of oxygen and nutrients at different stations and seasons, from data collected in the Eastern Mediterranean during the POEM cruises is shown in Figure 4. It is evident that the nutrient levels are typical of an oligotrophic region. The nutrient depleted surface layer is separated from the intermediate and deep-water layers by a transitional layer of 100-200m thickness, within which the concentration of nutrients increases rapidly. The concentration of nutrients in the intermediate and deep-water layers is rather constant, increasing in the following order: Aegean< Ionian < Levantine, and of the same order of magnitude as reported by Mc Gill [1, 4]. The oxygen is almost saturated in the surface layer (6ml/l in winter and 4.8 ml/l in summer). We can observe a sharp decrease of oxygen in the transition layer while in the deep water the concentrations are around 4.2 ml/l and are decreasing in the order which the nutrients are increasing: Aegean> Ionian > Levantine.

The northwest Levantine circulation in late summer 1987 is described in details by Theoharis *et al* [14]. The important circulation feature of the Eastern Mediterranean, namely Ierapetra anticyclone, exists during this cruise and presents strong signal mostly in the upper layer. It is found to the west of the Rhodes gyre, just to the southeast of Crete. Figure 5 represents the distribution of oxygen and nutrients along a section south of Crete and east of the eastern straits of the Cretan Arc during late summer 1987. The distributions of oxygen and nutrients present similarities both in the frontal region and the vertical structure. The oxygen and nitrate isoconcentration lines at the station 24603 are depressed about 250m below those found in the adjacent areas. Surface waters up to the depth of 200m (depth of the nutricline) within the gyre are very poor in nutrients ($NO_3 < 1$ μM). The Rhodes gyre is a permanent circulation feature within the Eastern Mediterranean, having a hydrographic structure typical of a cyclonic dome. The depth of the nutricline at the stations 24701 and 25816 is influenced by the Rhodes cyclonic circulation and appears very close to the surface at about 75m. The deep layer (below 1000m) is almost homogeneous. The stations in the Ionian side (24303, 24402) have similar chemical characteristics with the Ionian Sea ($O_2 \sim 4.2$-$4.3$ ml/l, $NO_3 \sim 4.5$-$5.0$ μM). In the Levantine Sea the deep layer is less oxygenated and richer in nutrients ($O_2 \sim 4.1$ ml/l, $NO_3 \sim 5.5$ μM). It is well known from previous investigators that the EMDW is formed in the Northern Ionian Sea and spreads to the south Ionian and Levantine seas. The slight diminution of oxygen and augmentation of nitrate in the EMDW, from Ionian to the NW Levantine basin is related to the decomposition of organic materials during the spreading.

The results obtained during the last years showed that although the Mediterranean in the whole is oligotrophic, locally and temporary high planktonic biomasses can be found. In the cyclonic regions where the nutricline ascends to the base of the euphotic zone, the phytoplankton biomass and the primary production is higher than in the anticyclonic regions where the nutricline is situated at greater depths, limiting the nutrient input to the surface waters during the winter mixing [9]. In order to compare the distribution of hydrological and chemical parameters in a cold and a warm core eddy we chosen the Rhodes cyclone and the Ierapetra anticyclone during different cruises in autumn, winter and spring (Figures 6 and 7). The concentrations of nutrients in the surface waters around the Rhodes gyre (Figure 6) are, for most of the cruises, as poor as in the other areas of the Levantine Sea. The oxycline, the nutricline, the halocline and the thermocline are established near the surface. During the severe winter of 1992 the halocline and the thermocline disappear completely, while the oxycline and nutricline are topped by a very thin layer poor in nutrients and rich in oxygen reaching from the surrounding areas. The nutrients concentrations in the subsurface layer are higher than those measured during the other cruises due to the deep and intense convective mixing. The increase of oxygen, salinity, temperature and the corresponding decrease of nitrate by about 0.35 ml/l, 0.1psu, 0.25 °C and 0.8 μM respectively can be attributed to the deep ventilation of the waters (see also paragraph 3.2). Nevertheless we must take in account the contribution of CDW to the modification of the hydrochemical properties of LDW (see also paragraph 3.3). In the Ierapetra anticyclonic eddy, the oxycline, the nutricline, the halocline and the thermocline are established about 100 meters deeper than in the Rhodes cyclone. The nutricline deepened to about 400m in late winter 1992 while it is situated at about 200m in October 1987, November 1991

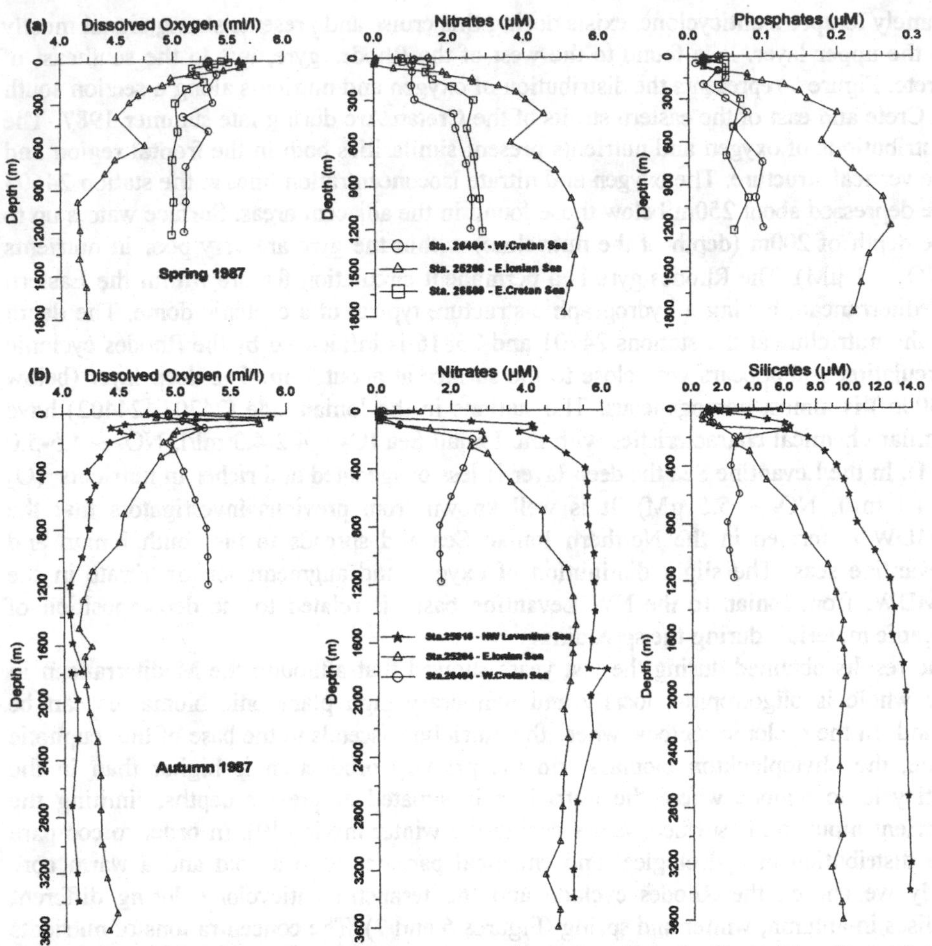


*Figure 4. Vertical distribution of oxygen and nutrients in the northwest Levantine Sea and the adjacent Cretan and eastern Ionian Seas during (a) spring 1987 (b) autumn 1987.*

and April 1995. The increase of oxygen and the corresponding decrease of nitrate below 1000 meters for the period from October 1987 to March 1992 can be attributed to the CDW outflow (see also paragraph 3.3). The evolution of the hydrochemical parameters from 1987 to 1995 due to the outflow of CDW is well reflected in the layer below 1200 meters by an increase of oxygen, salinity and temperature and a decrease of nitrates by about 0.2 ml/l, 0.06 psu, 0.2 °C and 0.6 µM respectively.

The Mediterranean waters, beside their relative poverty in nutrients, are characterised by a nitrate to phosphate atomic ratio different from the open ocean, the Atlantic Ocean in particular. Our results showed that for the NW Levantine Sea the N:P ratio ranges between 23-28, which is much higher than that in the Atlantic Ocean, in conformity with the Redfield's ratio N:P=16:1 [16]. The N:P ratio in the water column of the Levantine Sea vary substantially with depth. Anomalously high values (N:P > 40) were found at the top of the nutricline (Krom *et al* 1991, Yilmaz *et al* 1997). The high values of the N/P ratio in the top of the nutricline appear on the vertical distributions of the ratio of both the Rhodes gyre and the Ierapetra anticyclone during different cruises (Figures 6 and 7). The N:P ratio is rather constant in the layer below 400m and it ranges between 20 and 24. We do not have yet any explanation for the higher values observed in 1995.

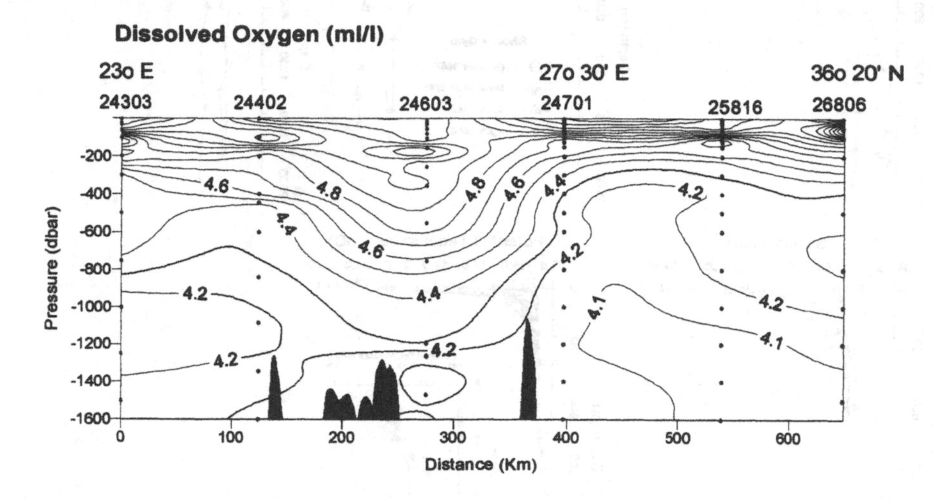


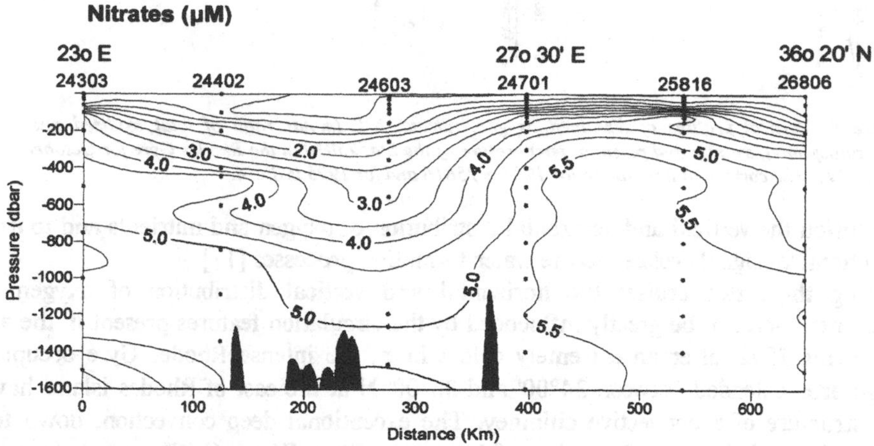


*Figure 5. Vertical distribution of oxygen (ml/l) and nitrate (µM) along a section south of Crete and east of the eastern straits of the Cretan Arc in late summer 1987 (solid line).*

## 3.2. Chemical signals related to the physical processes

Within the frame of POEM-BC program, two surveys in March-92 and April-95, designed to examine the water formation and spreading processes in the Eastern Mediterranean. The complete grid of chemical stations in the area of the eastern straits of the Cretan Arc and the NW Levantine Sea made possible to investigate with a good

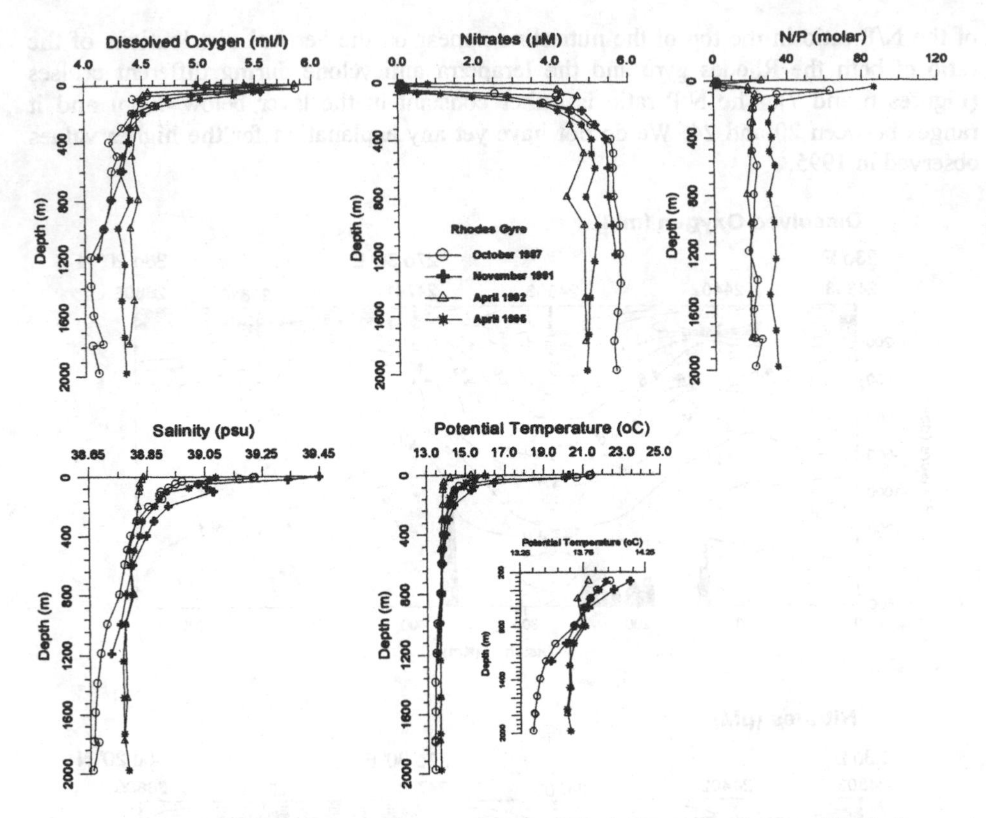


*Figure 6. Vertical profiles of dissolved oxygen, nitrate, N/P (molar ratio of total oxidised nitrogen to orthophosphate), salinity and potential temperature of the Sta. 25810 in the Rhodes Gyre for October 1987 - April 1995. The corresponding station for 1987 is 25816 and for 1995 is Sta. 4.*

resolution the vertical and horizontal distribution of oxygen and nutrients and to detect the chemical signals related to the water formation processes [17].

During these two cruises the horizontal and vertical distribution of oxygen and nutrients appear to be greatly influenced by the circulation features present in the area. In Spring-1992, after an extremely cold winter, the intense Rhodes Gyre occupies a large area extended between 34°00' and 36°00' N at the east of Rhodes island having the structure of a convective chimney. The exceptional deep convection, down to at least 2000m led to the formation of Levantine Deep Water (LDW), contrary to the classical LIW formation in the area. The well mixed core of the chimney had a diameter of about 150 km and was occupied by water with θ~ 13.7 °C, S~38.77 psu and σθ~29.18 [15]. At the 50 dbars level, within the core of the gyre the oxygen minimum is lower than 4.7 ml/l and the nitrate maximum higher than 3.5 μM. The cyclonic gyre extends to the south at 100 dbars level and presents two distinct cores. The lower oxygen ($O_2$<4.5 ml/l) and the higher nitrate concentrations ($NO_3$>4.5 μM) appear in the centre of the northern core (Figure 8). The oxygen varied between 4.4 ml/l and nitrate attained 4.6 to 4.8 μM in the 100-200m layer. In the center of the gyre (station 25810) the whole water column below 200m down to deep layers, is homogenized and

the oxygen varied between 4.3 and 4.6 ml/l (Figure 9), whereas nitrate varied between 4 and 5 μM. These concentrations are considerably higher than those usually found in the area. This important enrichment of the surface layer in nutrients potentially can contribute to the increase of biological production in the region [18] .
To the southeast of Crete within the Ierapetra anticyclone, on the contrary, the oxygen concentrations were higher than 5ml/l and nitrate did not exceed 2.5 μM in the upper layers (Figure 8). It appears that the conditions prevailing in the Ierapetra anticyclone favour the formation of new LIW masses. The lenses of well oxygenated ($O_2$>5.2 ml/l) and poor in nutrients (i.e. $NO_3$<2.0 μM) water detected in the layer between 100 and 200 m (station 24702, Figure 9), are related to the newly formed LIW with θ~ 16 °C, S~39.15 psu and σθ~28.95 [15]. Along the Cretan Passage a succession of local minima and maxima of oxygen in the surface layer (Figure 9), in correspondence with maximum and minimum values of nutrients in the same positions, are detected. The latter is probably associated with the small cyclonic and anticyclonic eddies which are formed by the Mid-Mediterranean Jet (MMJ) that flows and meanders between 33°30' and 35°00' N.

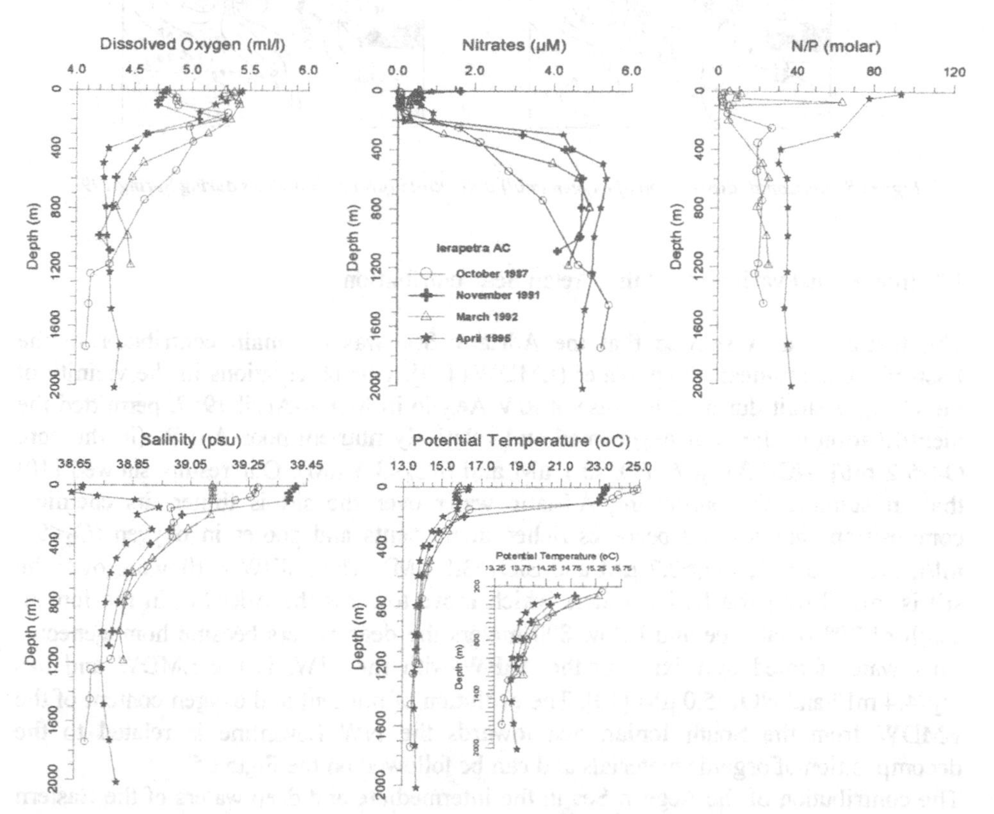


*Figure 7. Vertical profiles of dissolved oxygen, nitrate, N/P (molar ratio of total oxidised nitrogen to orthophosphate), salinity and potential temperature of the Sta. 24605 in the Ierapetra anticyclone for October 1987 - April 1995. The corresponding station for1987 is 24603 and for 1995 is Sta. 59.*

In Spring-95, well oxygenated-poor in nutrients water is observed at the south of Crete, in the northern edge of the Ierapetra anticyclone (34°45' N, 26°15' E) and seems to be associated with newly formed LIW (Figure 10). Similarly, blobs with newly formed water rich in oxygen and poor in nutrients are also detected at other positions along the 34°30' N latitudinal transect (stations 66 and 59, Figure 11). The Rhodes cyclonic gyre dominates the basin to the southeast of Rhodes and Karpathos islands and pushes relatively 'oxygen poor-nutrient rich' ($O_2$<4.4 ml/l; $NO_3$>4.5 μM) deep waters, in the productive layer (Figure 10). In both Spring-92 and 95 cruises, the presence of a small anticyclonic eddy formed by a branch of the Asia Minor Current which enters the Cretan sea through Karpathos strait, affects the distribution of oxygen and nutrients in the area close to the strait (Figures 8 and 10 ).

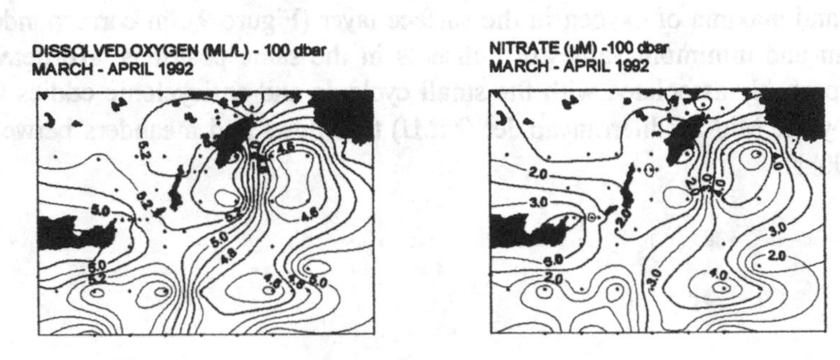


*Figure 8. Horizontal distribution of oxygen (ml/l) and nitrate (μM) at 100 dbars during Spring 1992.*

## 3.3. Interannual variability of the Cretan Sea contribution

The historical data showed that the Adriatic Sea was the main contributor to the Eastern Mediterranean Deep Water (EMDW) [19]. Our observations in the vicinity of the Otranto Strait during the cruise of R/V Aegaio in March-April 1987, permitted the identification of the well oxygenated and relatively nutrient-poor AdBW (in the core $O_2$>5.2 ml/l, $NO_3$<3.0 μM; $PO_4$<0.1 μM and $SiO_4$<3.5 μM). Our results showed [10] that in summer the outflowing Adriatic water over the sill is thiner, its chemical composition changes and becomes richer in nutrients and poorer in oxygen ($O_2$<5.1 ml/l, $NO_3$>3.3 μM, $PO_4$>0.2 μM and $SiO_4$>3.8 μM). The AdBW outflowing over the sill is mixed with the Ionian waters which move towards the Adriatic. In the Ionian, south of 39° N latitude and below 800 meters the deep waters became homogeneous. This water formed by mixing of the AdBW with the LIW, is the EMDW and has $O_2$<4.4 ml/l and $NO_3$>5.0 μM [10]. The evolution of nutrient and oxygen content of the EMDW from the South Ionian Sea towards the NW Levantine is related to the decomposition of organic materials and can be followed on the Figure 5.
The contribution of the Aegean Sea in the intermediate and deep waters of the Eastern Mediterranean has been considered secondary and rather sporadic. It was restricted in the vicinity of the Cretan Arc regions loosing rather quickly its characteristics [14].

Within the period of 1986-1987 the Cretan contribution was identified mainly in the layers below LIW, with nearly deep LIW charactestics, affecting a large area around Cretan Sea. Intermittent outflow of Cretan Deep Water (CDW) is manifested in the same period (late winter and late summer cruises 1987) in form of "high oxygen-low nutrient" ($O_2$>5.0 ml/l, $NO_3$<2.5 μM) patches detected, at a depth of 900m, at the west side of the Cretan Arc, in the Ionian Sea [10]. This patches are also characterised by high salinity and temperature values (S>38.9; T>14.4 °C). The patches of CDW are less pronounced in summer than in winter with lower oxygen and higher nutrient content. Schlitzer *et al* [20] based on CFM-12 (chlorofluoromethane) data from the expedition of F.S. Meteor during the same year, taken together with oxygen and hydrographic measurements, conclude that the AdBW is the only substantial source of deep and bottom water in the Ionian and the Levantine Seas.

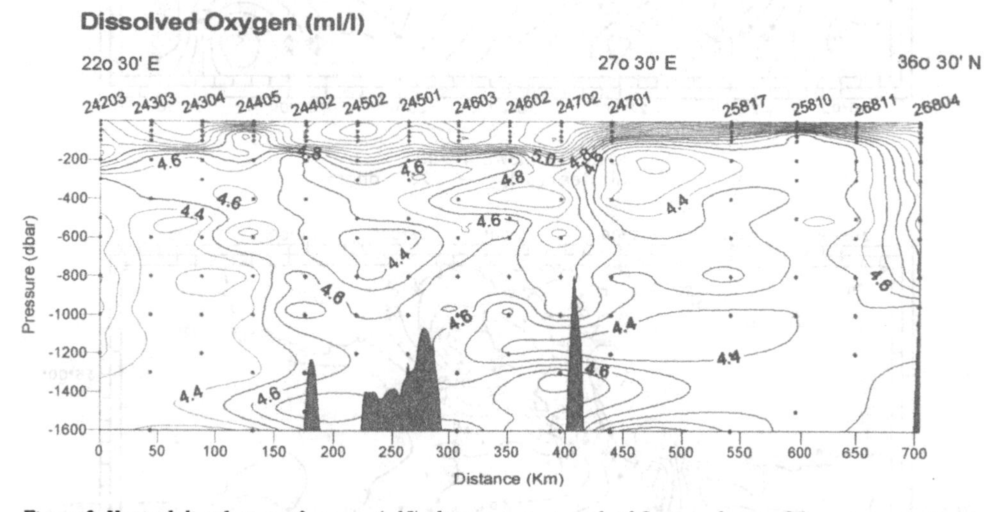


*Figure 9. Vertical distribution of oxygen (ml/l) along a section south of Crete and east of the eastern straits of the Cretan Arc in Spring 1992 (dashed line).*

Since the beginning of this decade a drastic change in the circulation through the straits of the Cretan Arc has been observed [21, 22]. In the years following 1987 the dense waters of Aegean origin added to the deep and bottom sections of the Eastern Mediterranean displace upwards the waters of Adriatic origin with two important consequences on the distribution of oxygen and nutrients [23 ,24] :

a) the important increase of oxygen and decrease of nutrient content in the deep and bottom layers of the Eastern Mediterranean ,

b) the intrusion of "nutrient rich-oxygen poor" Transitional Mediterranean Water (TMW) in the intermediate layers of the Cretan Sea.

This drastic change in the thermohaline circulation is basically induced by the increase of deep water density in the Cretan Sea. The first period of anomaly (1987-1992) is salinity driven (0.1 increase) and the 29.2 isopycnal is raised up to 400m in 1991 and 30m in 1992 [25]. The whole Cretan basin is filled with young and oxygenated ($O_2$~ 5.0) water [26] with densities that reach 29.26. A massive outflow of this water towards

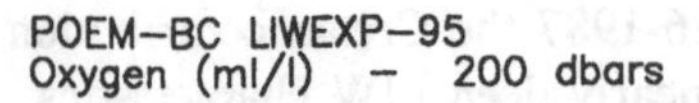


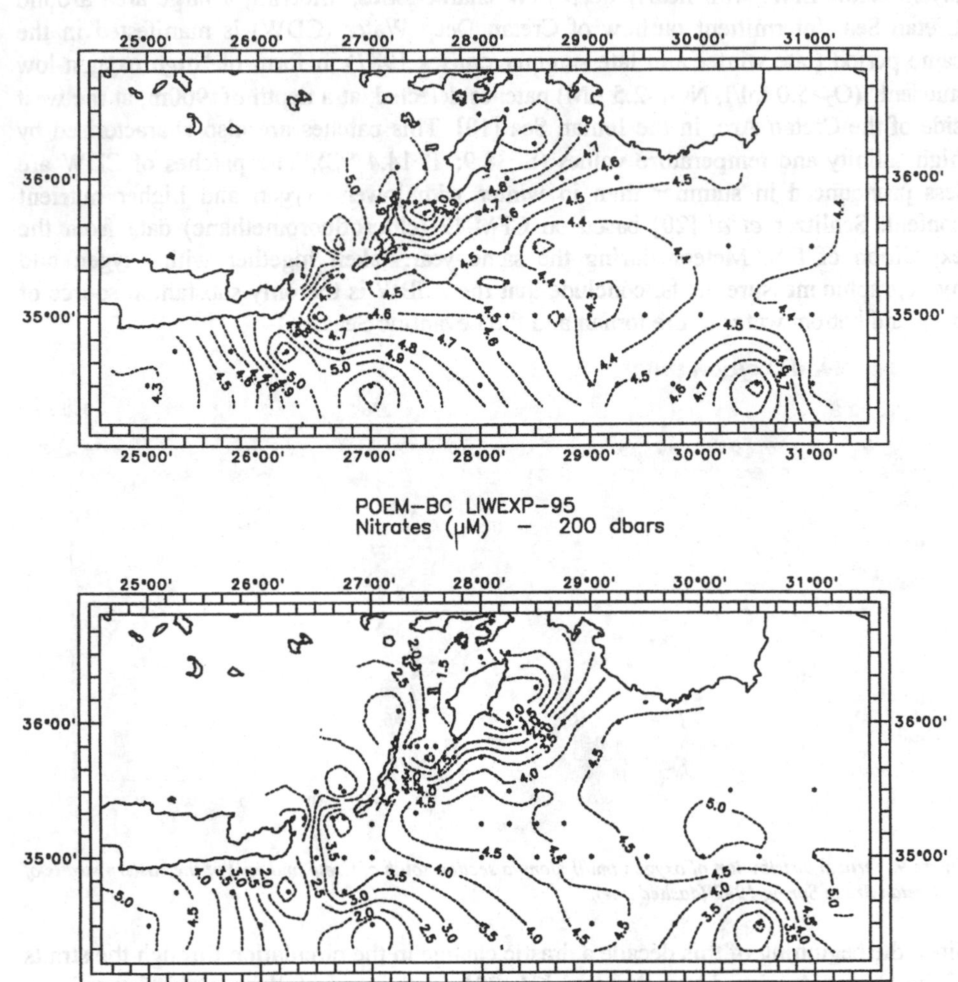


*Figure 10. Horizontal distribution of oxygen (ml/l) and nitrate (μM) at 200 dbars during spring 1995.*

**the Levantine and Ionian Seas mainly through the three deeper straits of the Cretan Arc, namely; Antikythira, Kassos and Karpathos is observed [27, 28] .**

**During 1992, large quantities of outflowing CDW lie beside the DW of Adriatic origin creating important gradients in the deep layer. Patches of oxygen rich-nutrient poor CDW with lateral scale of 100 to 250 kilometers are detected to the south of Crete and eastward of the Eastern Straits of the Cretan Arc (Figure 9).**

**The second period (1992-1995) is driven by temperature (0.3-0.4 °C cooling), while the salinity remains almost in the same high values. This corresponds to an increase of the Cretan water's density that reaches the extreme value of 29.4. By the year of 1995, the**

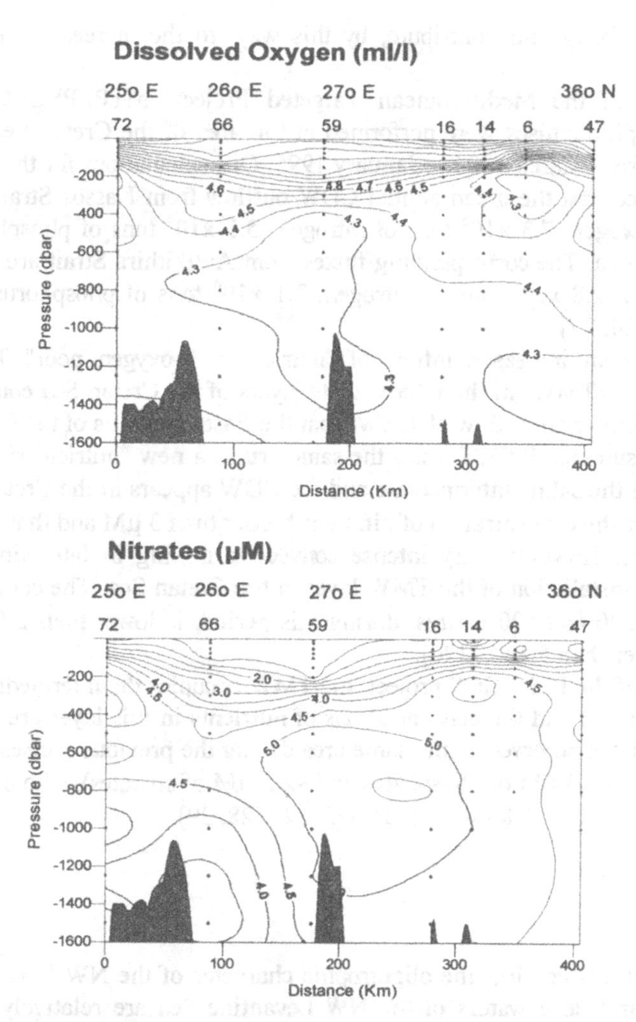


*Figure 11. Horizontal distribution of oxygen (ml/l) and nitrate (μM) along a section south of Crete and east of the eastern straits of the Cretan Arc in Spring 1995 (dotted line).*

**Cretan Sea dense waters have filled the deepest parts of the Ionian and Levantine Seas changing dramatically the properties of the Eastern Mediterranean deep waters.**
**In 1995, the vertical distribution of oxygen and nitrate along the same transect, starting from 25°E longitude (Figure 11), totally differs than that of 1992. The water column below 300 m is homogenised and oxygen concentration varied between 4.3-4.5 ml/l, whereas the nitrate values are in the range of 3.5-5.0 μM (Figure 11). The mean value of oxygen concentration increases from 1987 to 1995 by 0.3 ml/l (13 μM) and that of nitrate concentration decreases by about 1 μM. The old EMDW is lifted up several hundreds of meters enriching with nutrients the intermediate depths of the basin (Figures 5 and 11). We can suppose that in some areas these nutrients can reach the**

euphotic layer by mixing and contribute, by this way, to the increase of biological production.

In the framework of the Mediterranean Targeted Project (MTP)-PELAGOS four seasonal oceanographic cruises were performed in the area of the Cretan Sea and the Cretan Arc straits from March 1994 to January 1995. Our calculations for the period of these cruises, showed that the mean annual CDW outflow from Kassos Strait is about 10.8 x$10^7$ tons of oxygen, 7.3 x$10^5$ tons of nitrogen, 5.5 x$10^4$ tons of phosphorus and 17.9 x$10^5$ tons of silica. The corresponding fluxes from Antikithira Strait are about 4.7 x$10^7$ tons of oxygen, 2.8 x$10^5$ tons of nitrogen, 2.1 x$10^4$ tons of phosphorus and 6.2 x$10^5$ tons of silica [24, 27].

In the same period an increased inflow of "nutrient rich-oxygen poor" Transition Mediterranean Water (TMW) in the intermediate layers of the Cretan Sea compensates the CDW outflow. Important inflow of TMW from the Eastern Straits of the Cretan Arc is observed in late summer 1991. During the same cruise a new "nutrient rich-oxygen poor" layer between the saline intermediate and the CDW appears in the Cretan Sea. In the core of this layer the concentration of nitrate is higher than 3 μM and that of oxygen lower than 4.8 ml/l. However, very intense convective mixing in late winter 1992, prevented from the installation of the TMW layer in the Cretan Sea. The concentration of nitrate between 200 and 600 metres, during this period, is lower than 2.5 μM and that of oxygen higher than 5.1ml/l [26].

During the cruises of the PELAGOS project, the TMW occupies the intermediate layers of the entire Cretan Sea and the concentrations of nutrients in this layer are often two times higher than those observed in the same area during the previous cruises (increase ~2.5 μM of nitrate, ~0.05 μM of phosphates and ~2.5 μM of silicates). The decrease of oxygen in this layer is about 0.8 ml/l (35 μM) [23, 27, 28, 29].

## 4. Conclusions

The recent observations confirm the oligotrophic character of the NW Levantine Sea. The intermediate and deep waters of the NW Levantine Sea are relatively richer in nutrients and poorer in oxygen than the other subbasins of the Eastern Mediterranean. This is related to the circulation and the evolution of the EMDW during its spreading from the Ionian to the Levantine Sea.

In the cyclonic regions the nutricline ascends in the base of the euphotic zone and contribute to the increase of phytoplankton biomass and primary production. During the extremely cold winter of 1992 the formation of LDW within the Rhodes gyre contibute to an important enrichment of the surface layer in nutrients. The tranformation of the Cretan Sea from a secondary and sporadic to a principal source of the EMDW affects dramatically the distribution of oxygen and nutrients in the deep layers of the Eastern Mediterranean. In the same period the upwards displacement of the "old" EMDW resulted to the intrusion of "nutrient rich-oxygen poor" TMW in the intermediate layers of the Cretan Sea.

## References

1. Mc Gill, D.A. (1961) A preliminary study of the oxygen and phosphate distribution in the Mediterranean Sea. *Deep-Sea Res.*, **9** , 259-269.
2. Souvermezoglou E. (1989) Distribution of nutrients and oxygen in the Eastern Mediterranean Sea. Proceedings of the UNESCO/ IOC Second Scientific Workshop, Trieste, Italy, POEM Sci. Repts. 3, Cambridge, Massachusetts, USA, 85-102.
3. Souvermezoglou E. (1988) Comparaison de la distribution et du bilan d'echanges des sels nutritifs et du carbone inorganique en Mediterrannee et en Mer Rouge. *Oceanologica Acta* S. I. **9** , 103-109.
4. Mc Gill D.A. (1965) The relative supplies of phosphate, nitrate and silicate in the Mediterranean Sea. Comm. Int. Explor. Sci. Mer Medit. Rapp. P.V. Reunions, **18**, 734-744.
5. Krom M.D., Kress N., Brenner S., Gordon L.I. (1991) Phosphorus limitation of primary productivity in the Eastern Mediterranean Sea. *Limnology and Oceanography*, **36**, 424-432.
6. Krom M. D., Brenner S., Israilov L., Krumgalz B. (1991) Dissolved nutrients , performed nutrients and calculated elemental ratios in the South-East Mediterranean Sea. *Oceanologica Acta* **14** , 2, 189-194.
7. Krom M. D., Brenner S., Kress N., Neori A. and Gordon I.L. (1992) Nutrient dynamics and new production in a warm-core eddy from the Eastern Mediterranean Sea. *Deep Sea Research* , **39** , 3/4 , 467-480.
8. Krom M. D., Brenner S., Kress N., Neori A. and Gordon I.L. (1993) Nutrient distributions during an annual cycle across a warm-core eddy from the E. Mediterranean Sea. *Deep Sea Research* , **40** , 4 , 805-825.
9. Salihoglou I., Saydam C., Basturk O., Yilmaz K., Gocmen D., Hatipoglu E., Yilmaz A. (1990) Transport and distribution of nutrients and chlorophyll-a by mesoscale eddies in the northeastern Mediterranean. *Mar. Chem.* **29**, 375-390.
10. Souvermezoglou E., Hatzigeorgiou E., Pampidis I., Siapsali K. (1992). Distribution and seasonal variability of nutrients and dissolved oxygen in the northeastern Ionian Sea. *Oceanologica Acta*, **15**, 6 : 585-594.
11. Yilmaz A. and Suleyman T. (1997) The effect of cold- and warm-core eddies on the distribution and stoichiometry of dissolved nutrients in the northeastern Mediterranean. *Journal of Marine Systems*, (in press).
12. Robinson A. R. and Golnaraghi M. (1994) The physical and dynamical oceanography of the Mediterranean sea, in "Ocean processes in climate dynamics: Global and Mediterranean examples". P.Malanotte-Rizzoli and A.R. Robinson, eds., NATO ASI Series C, vol. 419, Kluwer Acad. Publ., Dordrecht, 255-306, 1994.
13. Malanotte-Rizzoli P. , Manca B. , Ribera d' Alcala M., Theocharis A., Bergamasco A., Bregant D., Budillon G., Civitarese G., Georgopoulos D., Korres G., Lascaratos A., Michelato A., Sansone E, Scarazzato P., Souvermezoglou E. (1997) A synthesis of the Ionian Sea hydrography, circulation and water mass pathways during POEM phase I. *Progress in Oceanography* **39**, 153-204.
14. Theoharis A., Gergopoulos D., Lascaratos A., Nittis K. (1993) Water masses circulation in the central region of the Eastern Mediterranean (E. Ionian, S. Aegean and NW Levantine ). *Deep Sea Res.*, II, 40, 6: 1121-1142.
15. Kontoyiannis H., Theocharis A., Kioroglou S. (in press). Formation and early spreading processes during late winter/early spring in the SE Aegean and NW Levantine. Similarities and differences between 1992-1995. Proceedings of International POEM-BC Workshop/Symposium, Molitg les Bains, 3-5 July 1996, France .
16. Redfield A.C., Ketchum B.H., Richards F.A. (1963) The influence of organisms on the composition of sea-water in: The Sea, M.N. Hill, editor, Interscience, New York, **2**, pp 26-77.
17. Souvermezoglou E., Krasakopoulou E., Pavlidou A. (in press).Distribution of oxygen and nutrients in the eastern straits of the Cretan Arc and the NW Levantine Sea during spring (1989-1992-1995). Proceedings of International POEM-BC Workshop/Symposium, Molitg les Bains, 3-5 July 1996, France.
18. Siokou-Frangou I. and Christou E.D., Gotsis-Scretas O., Kontoyiannis H., Krasakopoulou E., Pagou K., Pavlidou A., Souvermezoglou E., Theocharis A. (1997). Impact of physical processes upon chemical and biological properties in the Rhodes gyre area. International Conference: Progress in Oceanography of the Mediterranean Sea, Rome, November 17-19, pp 193-194.
19. Malanotte-Rizzoli P. and Hecht A. (1988) Large scale properties of the Eastern Mediterranean: a review. *Oceanologica Acta*, 11, 4: 323-335.
20. Schlitzer R., Roether W., Oster H., Junghans G.H., Hausmann M., Johannsen H., Michelato A. (1991) Chlorofluoromethane and oxygen in the Eastern Mediterranean . *Deep -Sea Res.*, **38**, 12, 1531-1551.
21. Theoharis A., Georgopoulos D., Karagevrekis P., Iona A., Perivoliotis L., Charalambidis N. (1992) Aegean influence in the deep layers of the Eastern Ionian Sea (October 1991). Rap. com. Int. Mer Med., 33, 235, 12.

22. Roether W., Manca B., Klein B., Bregant D., Georgopoulos D., Beitzel V., Kovacevic V., Luchetta A. (1996) Recent Changes in Eastern Mediterranean Deep Waters. *Science*, 271 :353-359.
23. Souvermezoglou E., Krasakopoulou E., Pavlidou A. (1996) Recent modifications of nutrient and oxygen exchanges throught the straits of the Cretan Arc. Proceedings of T.O.S. International Meeting: The Role of the Ocean in Global Change Research. Amsterdam 8-11 July 1996, 34.
24. Souvermezoglou E., Krasakopoulou E., Pavlidou A., 1997. The impact of the new hydrological regime on the distribution of oxygen and nutrients in the Eastern Mediterranean Sea. International Conference: Progress in Oceanography of the Mediterranean Sea, Rome, November 17-19, pp 357-359.
25. Theocharis A., Kontoyiannis H., Balopoulos E., Georgopoulos D., 1996. Climatological changes in the Cretan Sea (south Aegean Sea) leading to the new hydrological regime of the deep waters in the Eastern Mediterranean. Proceedings of T.O.S. International Meeting : The Role of the Ocean in Global Change Research. Amsterdam 8-11 July 1996, 35.
26. Souvermezoglou E., Krasakopoulou E., Pavlidou A. 1996. Modifications of the nutrients and oxygen exchange regime between the Cretan Sea and the Eastern Mediterranean (1986-1995). Proceedings of International POEM-BC / MTP Symposium, Molitg les Bains, 1-2 July 1996, France, 133-136.
27. Souvermezoglou E., and Pavlidou A., Krasakopoulou E., Georgakopoulou E. , Kontoyiannis H. (1996) Nutrients and oxygen variability and fluxes ; E. Balopoulos Editor. PELAGOS Project. Hydrodynamics and Biogeochemical Fluxes in the Straits of the Cretan Arc (Aegean Sea, Eastern Mediterranean Basin ) Final Report.
28. Souvermezoglou E., Krasakopoulou E. (1998) E volution of oxygen and nutrients deep circulation through the straits of the Cretan Arc: impact on the deep layers of the Eastern Mediterranean Sea. Comm. Int. Explor. Sci. Mer Medit. , Dubrovnic, June 1998. Rapp. du 35e congres de la CIESM. Comite d' Oceanographie Physique.35: 196-197.
29. Balopoulos E. (1997) Hydrodynamics and biogeochemical fluxes in the straits of the Cretan Arc (Aegean Sea, Eastern Mediterranean Basin). Interdisciplinary research in the Mediterranean Sea. A synthesis of scientific results from the Mediterranean targeted project (MTP) phase I 1993-96. European Commission Marine Science and Technology Programme. Edited by E. Lipiatou , 93-125.

# ABUNDANCE AND ELEMENTAL COMPOSITION OF PARTICULATE MATTER IN THE UPPER LAYER OF NORTHEASTERN MEDITERRANEAN

D. EDİGER[1], S. TUĞRUL[1], Ç. S. POLAT[2], A. YILMAZ[1] AND İ. SALİHOĞLU[1]

*1 Middle East Technical University, Institute of Marine Sciences, P.O.Box 28, 33731, Erdemli- İçel/Turkey*

*2 İstanbul University, Institute of Marine Sciences and Management, Müşküle Sok., No 10, Vefa-İstanbul/Turkey*

## ABSTRACT

Suspended particulate (POC, PON, PP) profiles obtained in 1991-1994 indicate the existence of characteristic subsurface maxima near the base of the euphotic zone in the cyclonic Rhodes gyre and its peripherial waters in the Northeastern Mediterranean. Interestingly the N:P of the bulk seston was reasonable during stratification seasons when the surface water was relatively poor in phosphate; but the ratio was unexpectedly low (N:P=6-12) in the late winter of 1992 when the surface layer of Rhodes gyre was occupied with nutrient rich deep waters.

## 1. INTRODUCTION

The eastern Mediterranean is one of the well known region of low productivity over the world due to limited nutrient supply to its surface layer from external and internal sources [1, 2]. The annual rate of phytoplankton production in the basin shows regional fluctuations and has been estimated to range regionally between 16 and 65 $gCm^{-2}$. The seasonality and the magnitude of primary productivity are principally determined by the extent and duration of winter mixing which provides nutrient inputs from intermediate layers to the euphotic zone [3, 4, 5]. Chlorophyll-a (CHL) concentrations, as a simple but rough measure of phytoplankton biomass, range from 0.01-0.6 $\mu gL^{-1}$ in summer to 0.1-1.7 $\mu gL^{-1}$ during the late winter-early spring bloom period [5]. A well-developed deep chlorophyll maximum (DCM) near the base of the euphotic zone is a characteristic feature of the NE Mediterranean throughout almost the whole year [3, 4, 6, 7, 8, 9, 10, 11, 12, 13]. Nevertheless this prominent feature may disappear under severe winter conditions, as experienced in the late winters of 1989 [3, 11] and 1992 [5].

Not unexpectedly, particulate organic matter (seston) content of the eastern Mediterranean surface water is relatively less than in the western basin [9, 14]. During spring-autumn period, when the surface layer is thermally stratified, background seston



*P. Malanotte-Rizzoli and V.N. Eremeev (eds.),*
*The Eastern Mediterranean as a Laboratory Basin for the Assessment of Contrasting Ecosystems*, 241–266.

concentrations of the productive upper layer are principally sustained by the regenerated production. Therefore. non-living particles (detritus) constitute a substantial fraction of the bulk seston content of the euphotic zone [15], as occurred in other oligotrophic seas [16, 17, 18, 19].

The deep water of the NE Mediterranean is relatively rich in dissolved inorganic nutrients ($NO_3$=4-6 μM and $PO_4$=0.15-0.22 μM) but with a relatively high N:P ratio (=26-28) [4] compared to the deep oceanic values (=14-17) [20]. Therefore, nutrient inputs from the deep layer to the surface waters by advective and convective mixing in winter occur with the indicated high ratios [3]. This process most probably leads to phosphorus-limited algal growth in the euphotic zone of the eastern Mediterranean [21. 22]. Moreover. the limited nutrient supply from the deep waters with relatively high N:P ratios is expected to affect the chemical composition of biogenic particles synthesized in the euphotic zone. Unfortunately, no systematic data is available for a sound understanding of seasonal and spatial variability in both the abundance and elemental composition of seston in the northeastern Mediterranean. For this goal, seston samples were collected from the upper layer of the NE Mediterranean in 1991-1994 period. in the scope of the ongoing national oceanographic program. Particulate organic carbon. nitrogen and total particulate phosphorus concentrations of the seston retained on filters have been evaluated on the seasonal and regional basis, together with the hydrographic. optical, chlorophyll and dissolved inorganic nutrient data from the same sites.

## 2. METHODOLOGY

Oceanographic data discussed in this article were obtained during the October-91, March-92. July-93 and March-94 cruises in the NE Mediterranean. The October-91 and July-93 surveys represent the period of seasonal stratification in the surface waters whilst the March data stand for the late winter-early spring conditions. Unfortunately, no data was obtained in summer 1992 and winter 1993. The study area and the positions of sampling stations. located between the longitudes 28°00′-36°00′ E and latitudes 34°00′-36°45′ N, are shown in Figures 1 and 2. respectively.

Water samples for chemical measurements were collected with 5L-Niskin bottles on a Rosette attached to the Sea-Bird CTD probe down to 1000 m. The upper layer from surface to below 1% light depth were sampled for particulate organic carbon (POC), nitrogen (PON). particulate phosphorus (PP) and chlorophyll-a (CHL) analyses. Particulates retained on GF/F filter pads for POC and PON analysis were determined by the conventional dry combustion technique [23], using a Carlo Erba model 1108 CHN analyzer. PP content of the seston was determined by the method of dry combustion + colorimetric measurement [24]. CHL samples. homogenized and extracted into 90% acetone solution. were measured by the standard fluorometric method [25]. using a Hitachi F-3000 Model fluorometer and a commercially available CHL standard (Sigma).

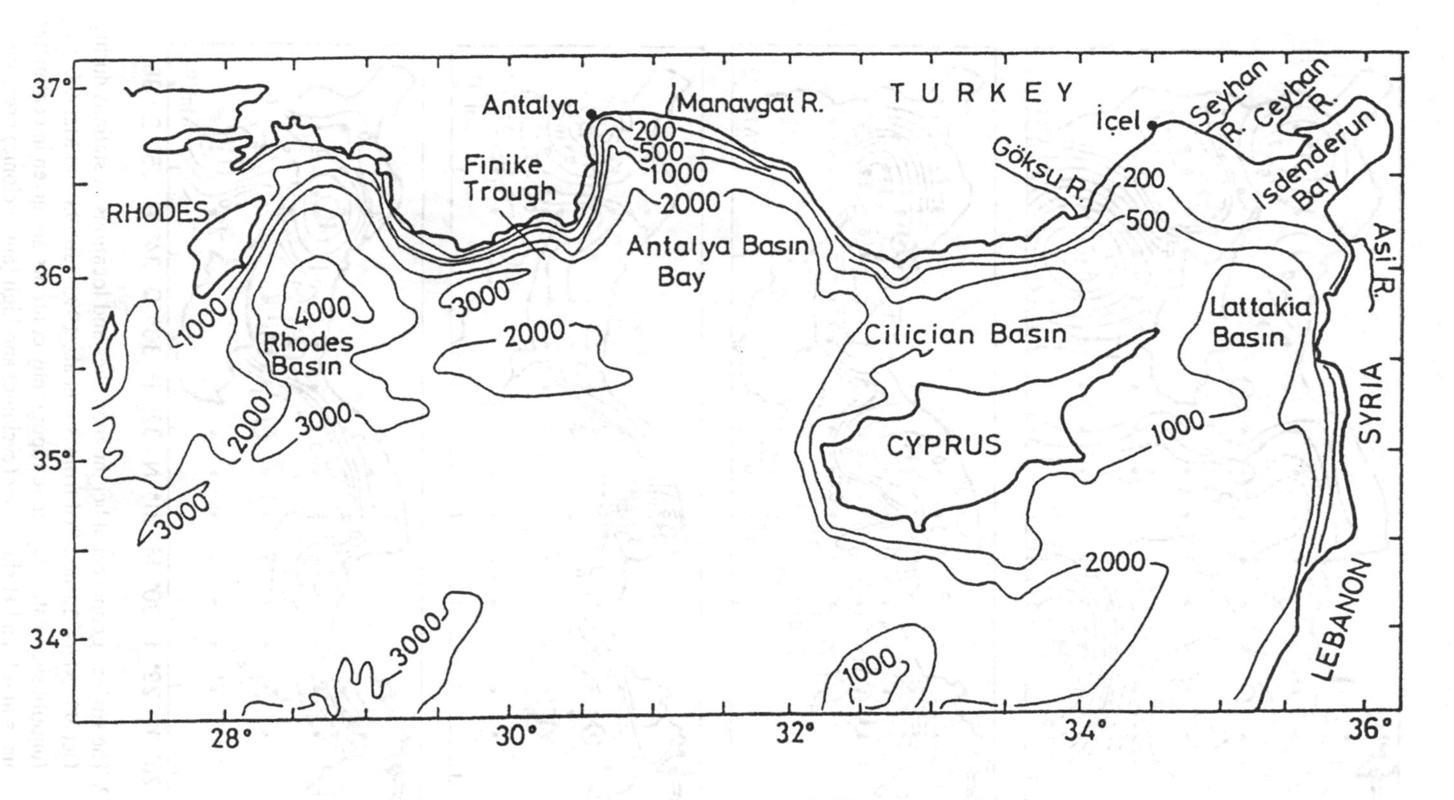


*Figure 1.* Location map and bathymetry of the NE Mediterranean with the nomenclature of major sub-basins

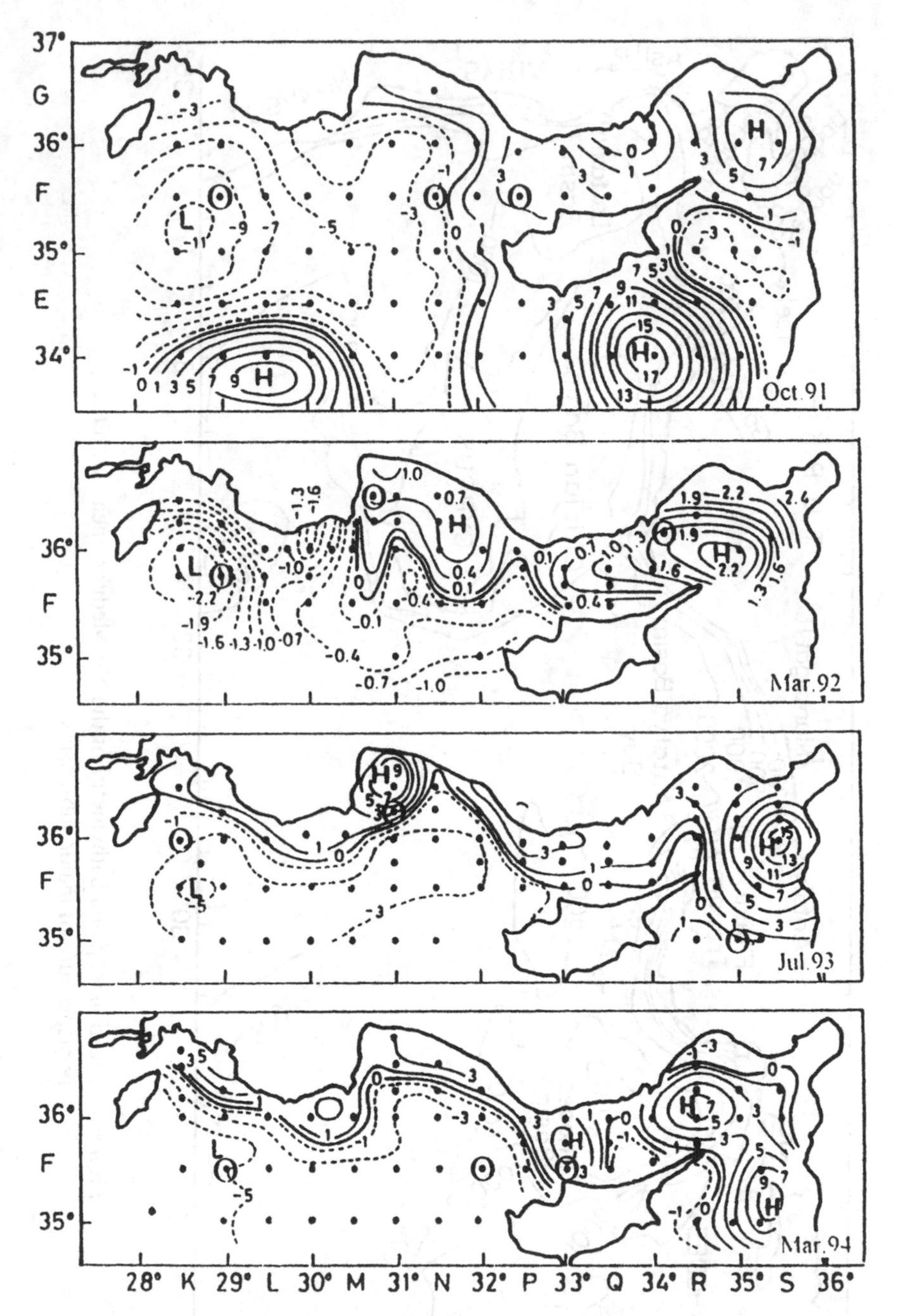


*Figure 2.* The surface geopotential height anomalies and locations of stations during Oct-91, Mar-92 (modified from Sur et al., 1993), Jul-93 and Mar-94 (unpublished data). Dynamic topography contours are given in centimeter units and L and H show low (cyclonic) and high (anticyclonic) pressure areas.

Seawater samples for $NO_3$+$NO_2$ and $PO_4$ were put into 50 mL HDPE bottles and kept frozen until analysis by a Technicon Model two-channel autoanalyser. The analytical methods followed were very similar to those described in [26].

## 3. RESULTS

### 3.1. WATER COLUMN STRUCTURE

The hydrodynamics and hydro-chemistry within the NE Mediterranean Sea display three regions of distinct vertical features. They are the cyclonic Rhodes Gyre (L), the Cilician Basin with a quasi-permanent anticyclonic eddy (H) and fronts+peripheries. As clearly shown in Figure 2. the region off Antalya bay is a typical site for fronts and peripheries between the cyclonic and anticyclonic eddy fields.

The hydrographic properties of the NE Mediterranean upper layer in March 1992 were apparently different from the vertical structure in March 1994 (Fig. 3). Prolonged winter conditions in 1992 permitted the deep water to occupy the entire upper layer of the Rhodes cyclonic region. This process resulted in the formation of a markedly thick, well-mixed (isohaline and isothermal) upper layer in March 1992. For example, typical temperature, salinity and density profiles displayed in Figure 3 demonstrate that the Rhodes upper layer was homogenized thoroughly by advection of the deep water and subsequent convective mixing down to at least 1000 m. The isothermal (~13.8 $^{o}$C) and isohaline (~38.8 ppt) upper layer water possessed slightly higher salinity and temperatures values than the those of the Levantine Deep Water (LDW) determined previously as T=13.6 $^{o}$C and S=38.7 [27, 28], due to mixing with the warmer and saltier surface waters of the Rhodes region under prolonged winter conditions. However, during the period of seasonal stratification LDW may rise up to 50 m in the Rhodes gyre but merely to 100 m in peripheries and 150 m in the Cilician basin (Fig. 3). In other words. the upper layer of the NE Mediterranean is occupied with saltier and warmer waters (thus less dense), separated from LDW by a pycnocline. This quasi-permanent density gradient zone appeared to be situated at relatively shallower depths (50 m) in the Rhodes cyclone in October 1991 and July 1993 (Fig. 3); it was completely destroyed in the core of the Rhodes cyclone during the winter of 1992 [29], as previously experienced in 1987 [30].

In the upper layer of the Cilician Basin. small scale anticyclonic eddies are generally established [27. 28. 31] with the hydrographic structures apparently different from that of the Rhodes cyclone. In summer-autumn period. the Cilician surface water is more saline and warmer (Fig.3), below which a less saline, cooler water (<39 ppt) of Atlantic origin may be seen until winter mixing. A vertically homogeneous water layer at intermediate depths. overlying the LDW during the period of seasonal stratification in the surface water. is called the Levantine Intermediate Water (LIW) and characterised by a temperature of around 15.5 $^{o}$C and salinity of 39.1 [27, 32]. The thickness of LIW layer changes seasonally and regionally. In winter, LIW is mixed thoroughly with the salty surface waters to form a vertically homogenous upper layer.

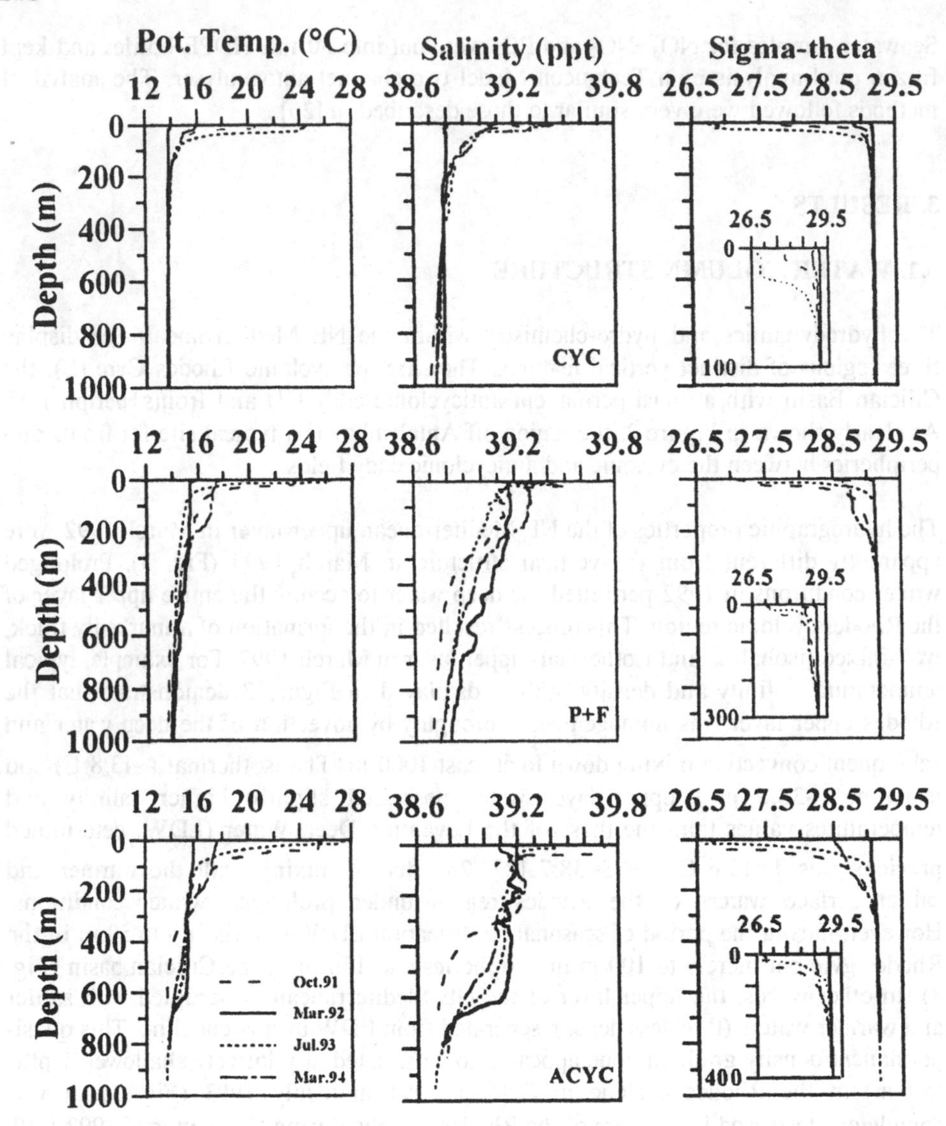


Figure 3. Vertical profiles of hydrographic parameters (after Sur et al., 1993 and unpublished data of the Institute of Marine Marine Sciences, Physical Oceanography Section) for selected stations in the Rhodes Gyre (CYC), Peripherial and Frontal areas (P+F) and Cilician Basin (ACYC) for Oct. 91 - Mar. 94. Station locations are marked in Fig. 2.

As clearly shown in Figure 3, in the frontal zone off the Antalya bay and anticyclonic Cilician region, the upper mixed layer was much thicker (300-600 m) in the winter of 1992 than in the winter of 1994 (200-300 m). A well defined seasonal stratification in the surface water was observed at around 50 m in summer-autumn period (Fig. 3).

### 3.2. WATER TRANSPARENCY

Measurements of irradiance indicate that 1% of surface light intensity (defining the base of the euphotic zone) penetrates to a depth of 70-110 m in the Cilician basin; it is relatively shallow (45-80 m) in the more productive Rhodes cyclonic and Antalya Bay regions. The present data are consistent with those obtained by [33] who determined the 1% light depth to range between 55 and 95 m, with an average value of 80 m for the whole NE Mediterranean. The 1% light depth measured in this study are also depicted on the CHL profiles for a better understanding of the role of basin hydrodynamics on the biochemical and optical properties of the NE Mediterranean.

The downward attenuation coefficient ($K_d$) estimated from the measurements of irradiance in the NE Mediterranean in 1991-1994 was found to range from 0.04 $m^{-1}$ in the anticyclonic region during summer-autumn period, to 0.12 $m^{-1}$ in cyclonic and frontal regions during winter-early spring period with an average value of 0.057 $m^{-1}$.

### 3.3. NUTRIENTS

Basin-wide. long-term studies conducted since 1988 have shown the critical role of convective mixing and advection of deep water in winter on the spatial and seasonal variations of the nutrients in the upper layer of NE Mediterranean [4, 10]. In the Rhodes cyclonic gyre, the nutrient concentrations of the surface water are closely associated with the hydrographic structure. Simply put, in March 1992 when the upper layer was occupied completely by the LDW with the associated chemical properties, vertically uniform nutrient profiles were obtained throughout the water column down to a depth of about 1000 m (Fig. 4). Nevertheless, the nutrient concentrations of the euphotic zone were very similar to the characteristic values of LDW ($PO_4$=0.2 µM, $NO_3$=5.5 µM). In March 1994, the saltier surface water was separated from the LDW by a sharp pycnocline located at 50 m (Fig. 4). indicating less severe winter conditions to be insufficient for the advection of LDW up to the surface layer of the eddy. Thus, nutrient supply to the upper layer from the LDW through the nutricline became very limited; the surface concentrations were measured to as low as about 0.5 µM for $NO_3$ and 0.02 µM for $PO_4$ (Fig. 4). The nutricline strictly coincided with the quasi-permanent pycnocline just located near the base of the euphotic zone in the Rhodes cyclone (Fig. 4). The shallow nutricline permits a partial nutrient supply to the lower depths of the euphotic zone. where the light intensity is always a limiting factor for algal growth at such depths.

In the fronts and periphery off Antalya bay, the nutricline appeared to be located at 75-100 m in summer-autumn period, which permits to a limited nutrient supply to the

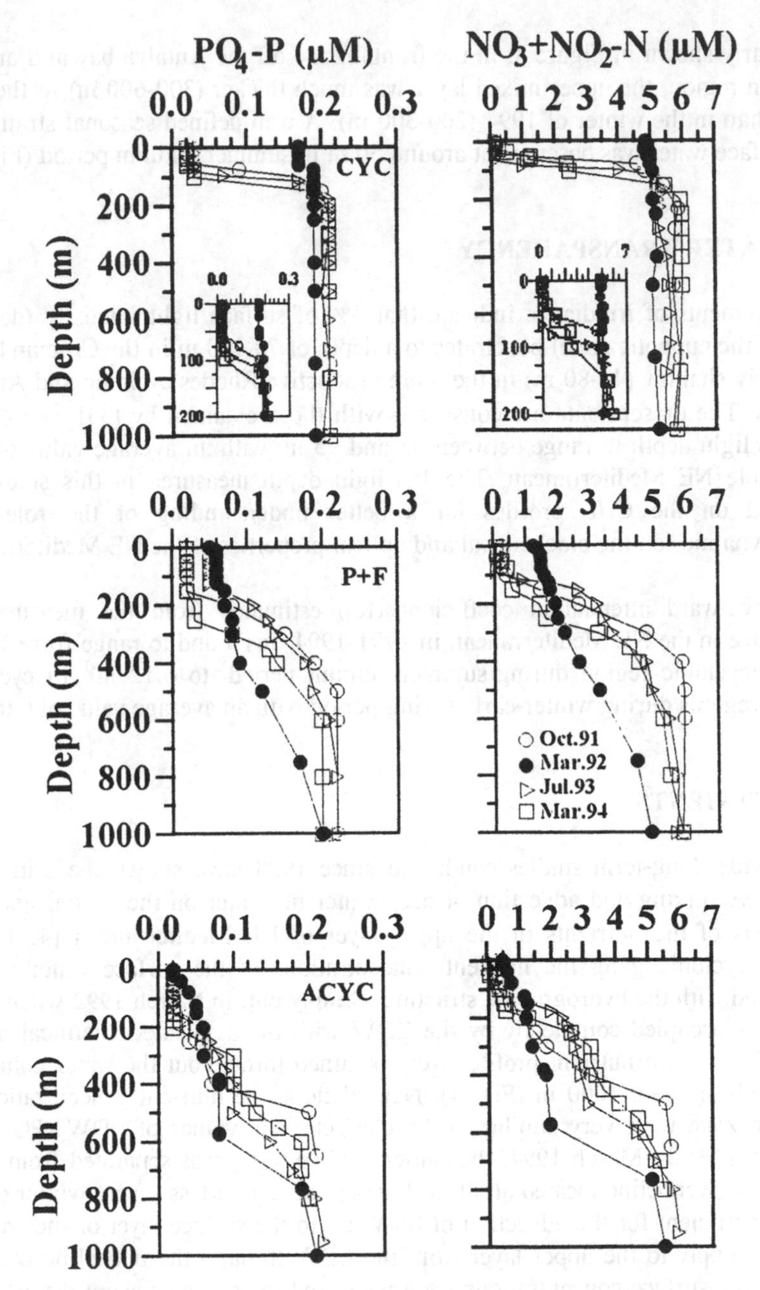


**Figure 4. Vertical profiles of dissolved nutrients for selected stations in the Rhodes Gyre (CYC), Peripherial and Frontal areas (P+F) and Cilician Basin (ACYC) for Oct.91 - Mar. 94. Station locations are marked in Fig. 2.**

lower depths of the euphotic zone (Fig. 4). Under less severe winter conditions, the nutricline depth is slightly modified, as experienced in March 1994 (Fig. 4). However, during the prolonged winter of 1992, the frontal surface waters appeared to be relatively enriched in nutrients by the supply from the nutricline depths; the euphotic zone concentrations increased from 0.02 μM in summer to the levels of 0.05 μM for $PO_4$ and from 0.11 μM to 1.5 μM for $NO_3$. Moreover, the permanent nutricline became broader and moved to a greater depth (500 m) and also the nutrient content of LIW increased apparently in March 1992 relative to those measured in October 1991 (Fig. 4). This supply was provided from LDW through the pycnocline by means of intense convective mixing in winter, which also modified the structure of density gradient between the LIW and LDW (Fig. 3).

In the Cilician basin, where an anticyclonic eddy is formed quasi-permanently, the chemical profiles are also associated with hydrographic structure. As clearly shown in Fig. 4, the saltier surface waters are always poor in nutrients due to establishment of the nutricline at much greater depths than in the Rhodes cyclone. Because the limited nutrient supply from the deeper waters is utilized in photosynthesis the surface concentrations are as low as 0.02 μM for $PO_4$ and 0.2-0.3 μM for $NO_3$ throughout most of the year. Below the euphotic zone, there exists a nutrient-poor aphotic layer which explicitly coincides with the LIW layer (Fig. 4). The nutrient-poor LIW layer may extend down to 300-500 m, as experienced in October 1991 and July 1993 (Fig.4), and to greater depths in the southeastern Mediterranean [3]. The nutrient content of LIW remains almost constant with depth down to the permanent nutricline of the anticyclones; but the concentrations slightly increase from spring to autumn until deep winter mixing, due to the supply of particulate nutrients from the surface layer. In winter, the LIW layer is mixed with the surface waters by convective processes, leading to a net export of nutrients to the surface layer. The nutrient-poor LIW does not appear in the cyclonic region (Figs. 3 and 4). The nutricline below the LIW extend down to the boundary of LDW throughout the basin (Figs. 3 and 4). Although its depth and thickness vary in space and time, the upper and lower boundaries of the nutricline are defined by the 29.00-29.05 and the first appearance of the 29.15 density surfaces, as recently reported [4]. In the Cilician basin the nutricline onset appeared at a relatively shallower depth (200 m) in the winter of 1994 than in the summer-autumn period (Fig. 4).

### 3.4. PARTICULATE ORGANIC MATTER AND CHLOROPHYLL-A IN THE UPPER LAYER OF THE NE MEDITERRANEAN

#### *3.4.1. Rhodes Gyre*

Particulate organic carbon, nitrogen and phosphorus concentrations were relatively low in the surface mixed layer of the cyclonic eddy during the summer-autumn period (Fig. 5a). The highest particulate concentrations were consistently recorded at the depth of the DCM (around 60 m), reaching the levels of 3.16 μM for POC, 0.23 μM for PON and 0.026 μM for PP in October 1991, whilst the corresponding maxima for July 1993 being 6.62, 0.57 and 0.036 μM respectively. In March 1994, the vertical distribution of

POM in the euphotic zone appeared to be consistent with that of CHL-a, with less pronounced maxima observed at shallower depths than in the summer-autumn period. Under the prolonged winter condition of 1992, the Levantine deep waters with their associated chemical properties occupied the surface layer. Efficient convective mixing in the upper layer led to the formation of vertically uniform POM and CHL-a profiles in the core of the gyre (Figs. 5a and 6). Although the entire upper layer of the Rhodes cyclonic region was occupied with the relatively nutrient-rich deep water in March 1992, the highest concentrations of both POM and CHL-a were recorded in the frontal areas off Antalya bay (Fig. 5b and 6 ) as discussed below.

### *3.4.2. Frontal area between the Rhodes eddy and Antalya basin*

Fig. 2 shows the sampling location in March 1992 where the saltier upper layer of the studied site was partly diluted with the deep water by convective mixing down to about 400 m. Consequently, the upper layer was enriched to some extend with dissolved inorganic nutrients as clearly shown in Fig 4. In the region coherent subsurface particulate peaks for March 1992 were recorded at 30-40 m (Fig. 5b) whilst the CHL-a distribution was nearly uniform over the entire euphotic zone with a small increase at the 1 % light depth (Fig. 6). Interestingly, primary productivity measurements at the same location indicated the maximum carbon uptake rate to occur at 35 m [33], coinciding with the particulate maxima and corresponding to the depth of 10 % of the surface irradiance (Fig. 6). In March 1994 when the convective mixing was relatively weak, the subsurface peaks of POC and PON profiles were coincident with the relatively broad DCM established between 40-80 m. On the contrary, the PP profile displayed a slightly increasing trend below 50 m and then retained almost constant down to 90 m, where POC and PON dropped to minimal values. In July 1993, the pronounced particulate maxima well coincided with the DCM formed between 65 and 75 m. Moreover, the POM and CHL-a maxima were as high as those obtained in March 1994. On the other hand, POC and PON data from October 1991 displayed insignificant variation with depth in the upper 150 m. The concentrations obtained below 80 m were similar to July-93 data. At this station the Oct-91 surface CHL-a values were as low as 0.01 $\mu gL^{-1}$ and comparable with the July 1993 measurements. The Oct-91 CHL-a profile displayed a weak and broad peak between 60-120 m whilst the July 1993 peak was formed between 60-80 m.

### *3.4.3. Cilician Basin*

An anticyclonic eddy is formed quasi-permanently in the region (Fig. 2). In March 1992 particulate and CHL-a data displayed vertically almost similar distributions in the euphotic zone; the concentrations of POC and PON decreased slightly at 30-50 m (Fig. 5c) whilst the CHL-a profile displayed a similar subsurface minimum but at shallower depths (20-30m) (Fig. 6) relative to the depth of the particulate minimum. At the base of the euphotic zone an increase was observed in the POC and PON profiles, where the CHL profile displayed a contrary view decreasing markedly from 0.28 $\mu gL^{-1}$ at the surface to 0.14 $\mu gL^{-1}$ at 80 m. Interestingly, the March-92 particulate concentrations were less than or comparable with the October-91 and July-93 values (Fig. 5c) whilst the corresponding CHL-a data displayed an opposite trend in the first 80 m (Fig. 6).

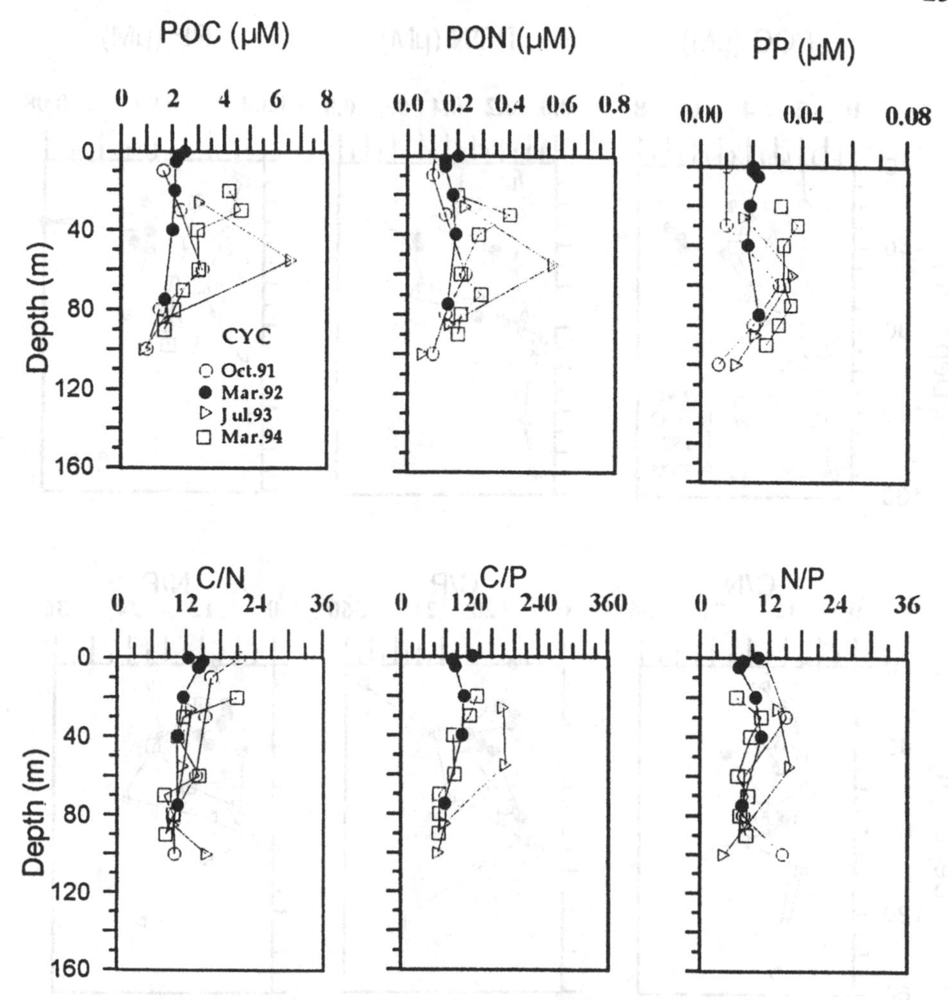


Figure 5a. Vertical profiles of POC, PON and PP and elemental ratios for selected stations in the Rhodes Gyre (CYC) for Oct.91 - Mar.94. Station locations are marked in Fig. 2.

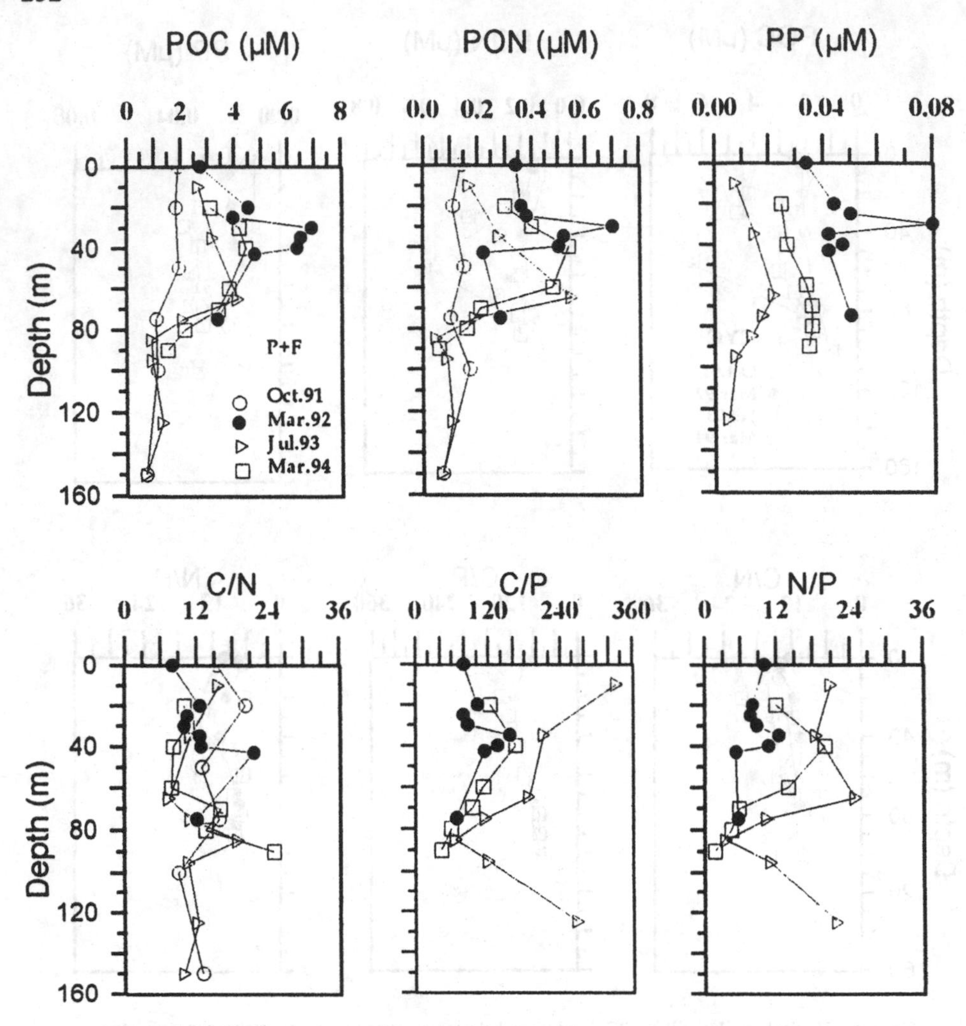


Figure 5b. Vertical profiles of POC, PON and PP and elemental ratios for selected stations in the Peripherial and Frontal areas (P+F) for Oct.91 - Mar.94. Station locations are marked in Fig. 2.

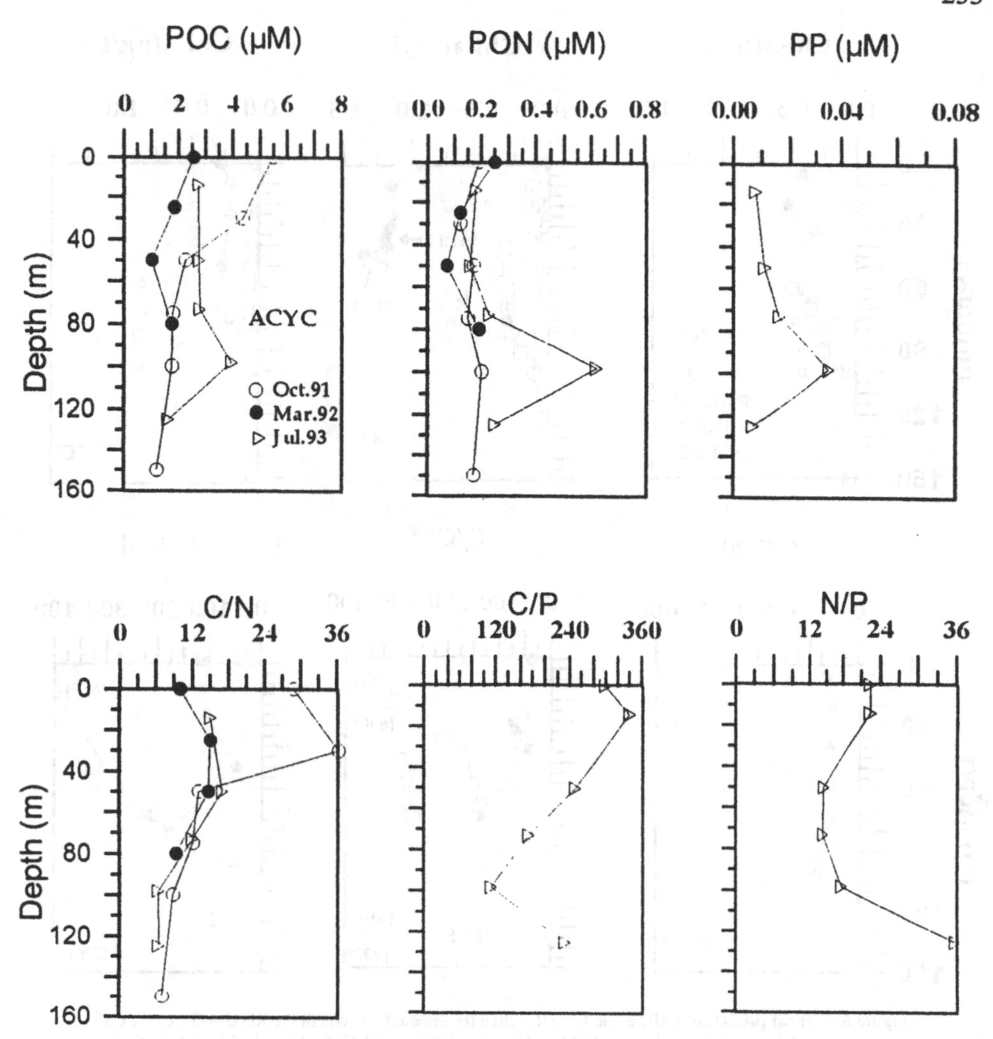


Figure 5c. Vertical profiles of POC, PON and PP and elemental ratios for selected stations in the Cilician Basin (ACYC) for Oct.91 - Mar.94. Station locations are marked in Fig. 2.

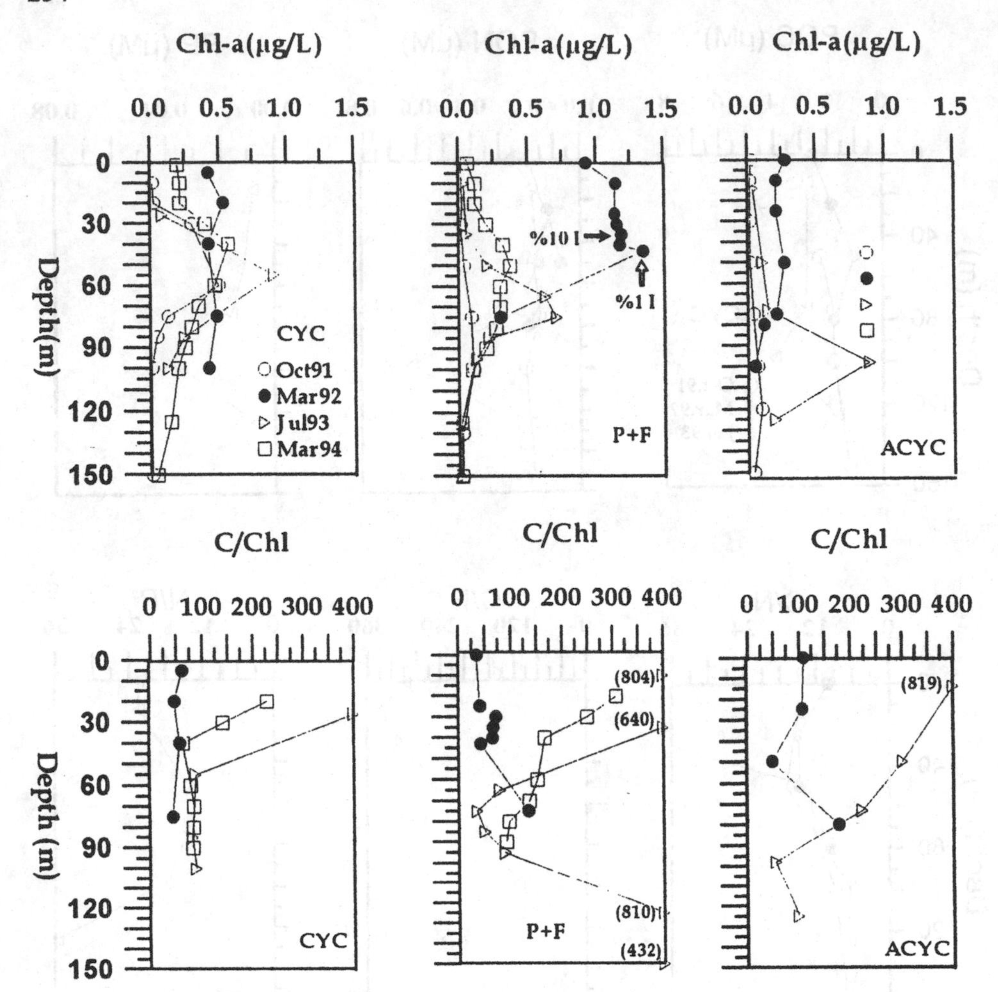


Figure 6. Vertical profiles of CHL-a and C/CHL ratios for selected stations in the Rhodes Gyre (CYC). Peripherial and Frontal areas (P+F) and Cilician Basin (ACYC) for Oct.91 - Mar.94. Station locations are marked in Fig. 2.

Unexpectedly, the highest particulate concentration in the surface water were determined in October 1991 when the CHL content of the euphotic zone being much lower than those of July 1993 and March 1992 (Figs. 5c, 6). The July 1993 particulate profiles in Fig. 5c demonstrate well defined subsurface maxima which coincide with the DCM formed at 100 m, nearly corresponding to the base of the euphotic zone.

### *3.4.4. Average particulate concentrations*

Table 1 shows depth-averaged concentrations of POC, PON and PP for the euphotic zone of hydrodynamically different regions of the NE Mediterranean. It appears that the seasonality in POC and PON content of the cyclonic Rhodes region is higher than in PP values from the same site. In other words, the background concentrations of PP in the cyclonic region is nearly two-fold those for both the frontal water off the Antalya bay and the Cilician basin whereas the means of POC and PON being comparable (Table 1). The average concentrations ranged regionally and seasonally between 1.44 and 3.9 µM for POC, 0.12 and 0.41 µM for PON, and 0.011 and 0.037 µM for PP in 1991-1994. The greatest mean values were estimated in the frontal waters off Antalya Bay in March 1992. However, in July 1993 and March 1994, when vertical mixing was limited and the surface water being poor in dissolved inorganic nutrients, the maximum POC's appeared in the Rhodes region where the nutricline was located immediately below the euphotic zone. In October 1991, the POC concentration was relatively high in the Cilician anticyclonic region due to as yet undefined factors.

Table 1. Average POM concentrations (µM) in the NE Mediterranean

Rhodes Cyclone

| DATE | POC | PON | PP |
|---|---|---|---|
| Oct. 1991 | 1.55±1.48 | 0.13±0.06 | - |
| March 1992 | 2.15±1.20 | 0.16±0.06 | 0.020±0.002 |
| July 1993 | 3.07±2.52 | 0.26±0.21 | 0.020±0.01 |
| March 1994 | 3.70±1.23 | 0.41±0.19 | 0.033±0.01 |

Cilician Basin

| DATE | POC | PON | PP |
|---|---|---|---|
| Oct. 1991 | 2.8±1.71 | 0.16±0.03 | - |
| March 1992 | 1.8±0.62 | 0.16±0.08 | - |
| July 1993 | 2.7±0.82 | 0.29±0.21 | 0.015±0.01 |
| March 1994 | - | - | - |

Off Antalya Bay

| DATE | POC | PON | PP |
|---|---|---|---|
| Oct. 1991 | 1.44±0.51 | 0.12±0.04 | - |
| March 1992 | 3.90±1.43 | 0.29±0.10 | 0.037±0.02 |
| July 1993 | 2.40±1.0 | 0.21±0.13 | 0.011±0.004 |
| March 1994 | 3.45±1.1 | 0.35±0.16 | 0.034±0.005 |

### 3.4.5. Ratios

Fig. 5 displays the vertical variation of the stoichiometric ratios of C, N and P in the bulk seston collected from different depths and locations. Not unexpectedly, the highest C:N ratios (16-36) appeared in October 1991 in the surface mixed layer having very low CHL but higher POC values. In March of 1992 and 1994, and July 1993 the C:N ratio generally varied between 6 and 24 over the euphotic zone; the ratio apparently decreased below 50-60 m of the Cilician basin whereas the vertical variation of particulate ratios being much less pronounced in the Rhodes cyclonic eddy. The particulate ratios from the frontal waters indicated relatively high variation with both depth and season; the C:P and N:P ratios were in the range of 180-340 and 13-21 respectively, in the surface mixed layer. These ratios decreased steadily towards the base of the euphotic zone, indicating faster decay of P-associated organic compounds in the surface mixed layer having very low CHL during summer-autumn period (Fig. 6).

Fig. 6 shows that spatial and seasonal changes in POC:CHL ratio is much greater than in the elemental composition (C:N:P ratios) of the bulk seston. The lowest and vertically almost uniform POC:CHL ratios were recorded in March-92; the ratio was as low as 35 in the cyclonic eddy and in frontal waters, whereas it slightly exceeded 100 in the Cilician basin (Fig. 6). Surprisingly, it exceeded 150-300 in March 1994, especially in the upper 30 m of the euphotic zone of cyclonic and frontal sites. Not unexpectedly, the POC:CHL ratio much increased in less productive summer period. For instance, in October 1991, it was abnormally high, varying between 100 in the cyclone, 2600 in the Cilician basin. Therefore, such large ratios were not depicted in Fig. 6. However, in July-93, the ratio appeared to decrease to levels of 800 in the mixed layer and then to 50 in the lower depths of the euphotic zone .

### 3.4.6. Regression of particulate data

Correlations between particulate data sets (POC, PON and PP) in pairs were examined by applying linear regression analysis. The equations given in Table 2 permit us to estimate molar ratios of C:N:P in the bulk seston during the less and more productive seasons. It appears from the Table that particulate carbon and nutrients (N, P) were highly correlated. However, the correlation coefficients for POC-PP and PON-PP regressions are lower than for the POC-PON regression. The C:N ratio derived from the slope of C-N regression was 9.85 for October 1991, which was apparently higher than the ratios of oceanic plankton (6.7) [34], of the July 1993 (7.2) and of the March 1994 (6.0) but very similar to the March 1992 ratio (9.9). The C:P ratio of the seston was also variable, ranging from 107 in March 1992 to 123 in July 1993. Interestingly, the temporal variation of C:P ratio was less than in the C:N ratio of the bulk seston. The C:P ratios of July 1993 and March 1994 are very similar and higher than an estimate of 107 for March 1992 when the surface water was relatively enriched with dissolved inorganic nutrients by the supply from deep waters (see Fig. 4 ). The N:P ratio of the bulk seston from the nutrient poor surface waters of NE Mediterranean was estimated to be 16.4 and 17.3 for March 1994 and July 1993, respectively. They are very similar to the planktonic ratio of 16 [34]. Unexpectedly, the N:P ratio was anomalously low (11) in March 1992.

Table 2. Linear regression analysis for organic carbon, nitrogen and phosphorus in seston from the euphotic zone of the NE Mediterranean

| Date | Regression | $r^2$ | n | Significance level |
|---|---|---|---|---|
| Oct. 1991 | C=9.85±2.97N+0.26±0.44 | 0.50 | 15 | p<0.01 |
| March 1992 | C=9.9±3.81N+0.30±0.19 | 0.93 | 21 | p<0.001 |
| | C=107±15.7P+0.04±0.43 | 0.75 | 17 | p<0.001 |
| | N=11±1.36P-0.03±0.04 | 0.79 | 17 | p<0.001 |
| July 1993 | C=7.2±0.43N+0.79±0.14 | 0.87 | 43 | p<0.001 |
| | C=123±22.3P+0.49±0.36 | 0.48 | 39 | p<0.001 |
| | N=17.3±2.2P-0.04±0.04 | 0.51 | 39 | p<0.001 |
| March 1994 | C=6.0±1.09N+1.28±0.41 | 0.79 | 10 | p<0.001 |
| | C=118±21P-0.74±0.74 | 0.80 | 10 | p<0.001 |
| | N=16±3.8P-0.22±0.13 | 0.70 | 10 | p<0.003 |

### *3.4.7. Deep Chlorophyll-a Maximum (DCM)*

The formation, maintenance and location of the DCM in the water column are controlled by light attenuation and nutrient availability in the NE Mediterranean as in other seas [35, 36]. The DCM is formed at depths where the light intensity decreases to 0.5-5% of its surface value; in general, it appears at the base of the euphotic zone in the NE Mediterranean [5]. The depth of the DCM fluctuated seasonally from 45 m in the cyclonic region and 100 m in the anticyclonic region during less productive summer-autumn seasons [5]. The eddy fields also affected the depth of the DCM and, thus of the particulate maxima which were located at relatively shallower depths in the cyclonic Rhodes eddy and in the fronts+peripheries. These characteristic features appeared at the base of the euphotic zone in the anticyclonic Cilician region (Figs.5a-c and 6).

In the Rhodes cyclonic region, the depth of the nutricline determines both the thickness of the euphotic zone and the location of DCM throughout most of the year (Fig. 4). The upward diffusion of nutrients was possibly combined with the accumulation of sinking phytoplankton in the lower depth of the euphotic zone, resulting in the establishment and maintenance of the DCM with relatively high CHL-a concentrations, as previously suggested [37] for similar environments. During the cooler winter of 1992, the unusually deep convective mixing all over the NE Mediterranean resulted in relatively high nitrate-based new production estimated using Deep Chlorophyll Maximum method [38] and thus high primary production and very high f-ratio for March 1992 [33].

There appears to be a close correlation between the POM and CHL-a profiles (Figs.5a-c and 6). Apparent increases in the POM concentration at the DCM depths indicate a greater algal biomass within light-limited depth of euphotic zone than in the surface waters. Nevertheless, the co-variance of zooplankton+bacteria+detritus with phytoplankton is very likely to bias the planktonic biomass high. The coincidence of the maxima of POM with CHL generally occurred in the cyclonic eddy and its peripheries, where the pycnocline (thus the nutricline) was situated near the base of the euphotic zone. However, under severe winter conditions, the occurrence of DCM and POM peaks is prevented by intense convective mixing, as occurred in March 1992. A similar hydrodynamic feature was recorded previously in the Southeastern Mediterranean [11].

## 4. DISCUSSION

### 4.1. VERTICAL DISTRIBUTION OF POM

The abundance and vertical distribution of bulk POM in the euphotic zone of the NE Mediterranean appeared to vary by 2-3 times with season and region. The extent of such changes is determined by relative importance of complicated abiotic and biotic processes in the organic carbon and nutrient cycles in the upper layer of the marine ecosystem. In fact, the eastern Mediterranean is a typical region of low productivity and thus of low seston concentration [6, 9, 39] due to limited nutrient supply to the euphotic zone. Relatively low concentrations of POC, ranging between 2 and 5.5 μM in the open sea throughout the year, corroborates the above statement. The average POC's given in Table 1 are comparable to measurements from the Equatorial Pacific [17], from the Southeastern Mediterranean shelf water [9], from open waters of Japan and East China Seas [18] and from the Sargasso Sea near Bermuda [19].

In October 1991, when the nutrient-poor euphotic zone was seasonally stratified and had very low CHL, the surface mixed layer of the Cilician basin was relatively rich in POC compared to that of the cyclonic eddy, leading to anomalously high POC:CHL ratios (100-2600). Such large values (not shown in Fig. 6), indicate detritus-dominated POM pool in the euphotic zone over the greater part of the basin. However, the POC profile increased below the seasonal thermocline in the cyclonic Rhodes region where the nutricline is always situated near the 1% light depth throughout most of the year. This feature admits nutrient influx to the light-limited zone, leading to shade-adapted algal growth in this eddy and its peripheries. Algae in this zone sythesize more pigment relative to organic production [35, 40, 41]. Therefore, POC profiles displayed a less pronounced subsurface POC peak compared to that of the CHL-a (Figs. 5a and 6). Nevertheless, the POC:CHL ratios (60-242) estimated from the measured values within the DCM depths of the eddy were much higher than the cellular Carbon:CHL ratios determined for the light-limited depths of the oligotrophic seas [16, 17, 42], accounting for about 50 % of the total POC measured. In other words, the subsurface POC peak observed in the Rhodes cyclone was most probably dominated together by detritus+bacteria+zooplankton produced *in situ* and by biogenic particles sinking from the upper layer to the top of density gradient zone.

In March 1992, when the water column was well mixed and relatively enriched with nutrients, the lowest POC concentrations were obtained in the cyclonic Rhodes region (Table 1 and Fig. 5a) because strong convective mixing and horizontal flows in the upper layer much exceed the successive growth of algae and thus accumulation of living and non-living seston in the euphotic zone of the cyclonic eddy, other than vertically migrating planktonic species. This physical pressure also modified the characteristic subsurface POM peaks formed quasi-permanently over the basin. Accordingly, the characteristic subsurface maxima of POM appeared in the frontal waters off the Antalya Bay and Cilician basin but not in the cyclonic eddy (Fig. 5a-c). In the front, the POM peak well coincided with the depth of the maximum algal growth rate [33] observed at 10-15 % light depths whereas the CHL distribution being almost homogenous in the well-mixed euphotic zone. The maintenance of the POM peak in the frontal waters may have originated from local increases mainly of zooplankton biomass and partially of detritus+bacteria concentrations. A similar mechanism was previously suggested for the occurrence of subsurface POM peaks in the frontal regions of the Western Mediterranean Sea [43]. In February 1989, the convective mixing was strong enough to mix the water column thoroughly down to 500-600 m in the anticyclonic eddy of the southeastern Mediterranean, resulting in vertically uniform CHL profiles extending far below the photic zone [11]. Unfortunately they had measured neither the bulk seston nor biomass to understand the POM distribution in the photic zone under severe winter conditions. Interestingly, the March-92 CHL-a profile displayed a small subsurface maximum (DCM) at 50 m in the frontal waters, where the particulate concentrations decreased markedly. The maintenance of this weak CHL peak in the well-mixed water column was most probably originated from less grazing pressure on algal biomass produced *in-situ* (with relatively low POC:CHL ratio) in the light-limited zone. However, in March 1994, when the convective mixing was relatively weak and confined to the upper 100 m, the POC and PON peaks were coincident with the relatively broad CHL maximum established between 40-80 m, indicating the algal growth rate to exceed the dilution of the biomass by vertical and horizontal mixing.

The seasonal thermocline developed below the surface mixed layer prevents the nutrient supply to the photic zone from the lower layer of the open sea by vertical mixing. Therefore, in the period of spring-autumn, POM pool in the euphotic zone is principally sustained by the regenerated production, especially in anticyclonic regions where the nutricline is established much below the photic zone. A supplementary conclusion was reached by the observation of the total POC measured in the euphotic zone of the Sargasso Sea was mainly composed of detritus and algal carbon merely made up 10 % of POC [44], assuming a cellular-C:CHL ratio (w/w) of 50 [15]. It should be noted that the bulk seston retained on filters may have contained a considerable quantity of bacteria, depending on sampling depth, location and season because a substantial fraction (50%, on average) of bacteria was recorded to be collected on precombusted GF/F filter pads [45]. On the other hand, the contribution of bacterial carbon may be similar or higher than algal carbon in the oligotrophic seas [44]. This finding strongly suggest that bacterial carbon may constitute from <10% to few 10% of the total POC measured in the euphotic zone of the Levantine basin. Unfortunately no data of total bacteria biomass was available for confirming the considerable contribution of bacteria to the suspended POC pool in the NE Mediterranean upper layer during the less productive seasons. Nevertheless, bacterial carbon may account for a substantial fraction of the POC pool just below 1% light depth [44]. Thus the apparent decrease in the C:N ratio determined in the lower depths of the euphotic zone of the Levantine basin may have been originated partly from the bacterial biomass with relatively low C:N ratios and to some extent from ultraplankton growing in the light-limited zone.

## 4.2. REGRESSION ANALYSIS OF PARTICULATE DATA

The stoichiometry of C, N and P in the bulk seston, derived from the slopes of the linear regressions of particulate data given Table 2, are generally comparable to the Redfield C:N:P ratios of 106:16:1. Nevertheless, the C:N and N:P ratios of the March-92 seston (9.9 and 11, respectively) markedly differ from the Redfield ratio whereas the C:P ratios being very similar. In other words, when the surface waters of NE Mediterranean received sufficient nitrate from deep waters with relatively high N:P ratios (>25) [4], successive algal growth proceeded with apparently low N:P and high C:N ratios compared to the Redfield ratios. Interestingly the uptake rates of C and P by algae being similar in March 1992. Such anomalous C:N and N:P ratios in the plankton-dominated seston (because POC:CHL ratio being as low as 35-175) may have originated in either relatively low uptake rate of nitrate by photoautotrophs or faster decay of nitrogenous organic compounds in the marine ecosystem under physical stresses. Similar anomalous ratios were previously reported for the north Atlantic bloom [46, 47], and for the western coastal area of France [48], and they were attributed to the uptake of more dissolved organic and inorganic carbon constituents relative to nitrogenous constituents by phytoplankton.

Interestingly, changes in the abundance of the bulk seston occurred with an N:P ratio of 16.4-17.3 in the nutrient-poor surface waters of March 1994 and July 1993 periods (see the regressions slope in Table 2). These values are very similar to the conventional planktonic ratio of 16 [34]. This strongly suggests that the uptake of dissolved inorganic nutrients by phytoplankton and decomposition of organic nutrients in POM by bacteria proceeded at similar rates even though the chemical data from the eastern Mediterranean imply phosphorus-limited algal growth. Relatively high C:N ratios in summer-autumn period, when the surface waters were poor in both algal biomass and dissolved inorganic nutrients, were most probably the result of slower decay rate of carbohydrates in suspended organic matter [14].

The intercept of linear regressions of POC vs PON or PP from productive surface waters of the seas could imply the concentration of carbonaceous compounds that do not vary with N- and P-associated organic compounds [14]. The concentration of carbonaceous compounds is very likely to account for a remarkable fraction of the measured POC when the total stock of bulk seston decreases markedly in the euphotic zone. A close correlation was also observed between the magnitudes of intercept and nitrate deficiency in the surface water; a low intercept value is very likely when nutrient concentrations are sufficient not to limit algal growth [14]. Lab-scale studies revealed that the protein content of POM tends to decrease as nitrate became exhausted and carbohydrates and lipids to increase whilst the slope of the regression line was affected slightly [49]. Inspection of the regressions equations in Table 2 indicate that the intercept of POC vs PON regression was relatively low (10 % of POC in Table 1) in March 1992 when the nutrients were not a limiting factor for algal growth, confirming the above suggestion. The intercept, as an index of carbohydrate concentration in POM being low in the euphotic zone, increased apparently in July 1993 and March 1994 when the photic zone was relatively poor in nutrients. Estimates of carbohydrates accounted for about 30 and 36 % of the average POC's for July 1993 and March 1994, respectively. Interestingly, as shown in Table 2, the October-91 intercept of the POC-PON regression line with the apparently high slope (9.85) was relatively low (0.26) although the surface waters being very poor in both algal biomass and dissolved inorganic nutrients. The low intercept strongly suggest that the ratio of carbohydrate to total POC remained almost constant with the increasing concentrations of POM in the euphotic zone.

### 4.3. POC RELATION ON CHLOROPHYLL

The POC content of bulk seston retained on filters includes both living (phyto- and zoo-plankton, bacteria) and nonliving (detritus) organic compounds. Therefore an estimate of algal-C:CHL ratio from the slope of POC-CHL regression is likely biased high whereas the magnitude of pigment-free organic compounds in the bulk seston from the intercept of linear regression may be underestimated if the other components of seston co-vary with the phytoplankton [15]. Therefore, cellular-C:CHL ratios, derived from linear regression of POC on CHL for the NE Mediterranean phytoplankton (Table 3), may have been overestimated either slightly during the bloom period or markedly during the post-bloom period. Consequently, an estimate of phytoplankton biomass from the seasonally varying but vertically invariant C:CHL ratios (the slope of POC vs CHL regression) should be used cautiously in determining the contribution of algal carbon to the POC pool in the euphotic zone of the NE Mediterranean.

Estimates of the POC:CHL ratio appeared to vary seasonally between 45 and 201 for the Levantine euphotic zone (see Table 3). The slopes of the regressions for July 1993 and March 1994 also revealed that the ratio of the DCM zone was apparently higher than the slope of the combined euphotic zone data (see Table 3). Unexpectedly, the regression line of the data from the July-93 light-limited zone and the March-94 euphotic zone had a negative intercept (Table 3). A relatively high ratio and a negative intercept are very likely to appear in the POC-CHL regression when increases in POC much exceed that in CHL from different locations or depths where the POCs reach peak values (see Fig. 2 in [15]).

**Table 3: Regression analysis of POC vs CHL-a in phytoplankton populations in the NE Mediterranean**

| Date | Regression Analysis | $r^2$ | n |
|---|---|---|---|
| Oct.91 | not healthy | - | - |
| Mar.92 | POC=45CHLa+15 | 0.50 | 27 |
| Jul.93 | POC=57CHLa+16 | 0.57 | 45 |
| | *POC=91CHLa-9 | 0.69 | 10 |
| Mar.94 | POC=201CHLa-20 | 0.66 | 15 |
| | *POC=242CHL-38 | 0.99 | 5 |
| Jul.93+Mar.92 | POC=42CHLa+16 | 0.50 | 67 |
| Oct.91+Jul.93+Mar.94 | *POC=60CHLa+15 | 0.50 | 49 |

* data only from the DCM zone including depths of the gradients.

Interestingly the July-93 ratio (57) was merely 25% higher than an estimate of 45 for March 1992 although the March-92 POC:CHL ratios calculated from the measured POC and CHL values at a given depth (Fig. 6) were much lower than in July 1993. The ratios derived from the regressions are not much different from the cellular C:CHL ratios in the mixed layer of other seas [16, 35, 42, 50]. These findings strongly suggest that POM increases over the euphotic zone in July 1993 and March 1992 predominantly determined by the changes in the phytoplankton biomass whereas the heterotrophic+detrital+bacteria carbon (TC) concentration varied independently of the planktonic biomass. However, a contrary view appeared in the regression analysis of March-94 data and measurements within the limited DCM zone in July 1993 and March 1994 (see Table 3). Relatively high POC:CHL ratios estimated from the regressions indicated the TC values co-varied with the algal biomass and predominated the total POC measured in the water column of the euphotic zone.

In October 1991, the POC distribution was almost vertically uniform and relatively high in the surface mixed layer of the periphery and anticyclonic region indicating the detrital material to dominate the POC pool in this nutrient-poor layer. Abnormally high POC:CHL-a ratios in October 1991 represent relatively slow growth rates imposed by nutrient limitation [35]. A similar conclusion for March 1994 period, representing a post-bloom period when CHL concentrations were very low in the nutrient depleted euphotic zone and comparable to the summer-autumn concentrations (Figs. 5a-c and 6).

The ratio estimated from the combined March-92 and July-93 data (including DCM measurements) was as low as 42 for the euphotic zone, apparently lower than a ratio of 60 derived from combined DCM data of 1991-1994. In other words, changes in the bulk seston content of the euphotic zone in March 1992 and July 1993 may have predominantly originated from planktonic biomass and partly from TC with the chemical composition given in Table 2.

Previous studies have shown that the cellular-C:CHL ratio becomes relatively high when algal growth proceeds under light-saturated and nutrient limited conditions; but it decreases markedly in the light-limited depths of the euphotic zone [35, 50]. Moreover, an estimate of algal biomass from the C:CHL ratio derived from the slope of POC-CHL regression is likely to be overestimated due to the possibility of co-variance of TC with the algal biomass [15]. Disregarding such biasing factors, the relatively low POC:CHL ratios estimated for March 1992 and July 1993 indicate that algal carbon accounted for about 10 and 30 % of the average POCs given in Table 1 for these periods, respectively. These values were most probably overestimated to some extent when the lower ratios calculated from cell counts and carbon content of the algal cells [16, 35, 42]. They have also suggested that in the oligotrophic Sargasso Sea the growth rate of phytoplankton must be fast enough to provide sufficient organic carbon for bacterial growth, leading to very high POC:CHL ratio in the bulk seston predominantly composed of detritus and about 10% of algal carbon. A similar conclusion was reached by the estimation of merely 25% of POC measured in the Sargasso Sea in August is provided by bacterial+phytoplankton biomass [19].

## 5. CONCLUSIONS

During the spring-autumn period, when dissolved inorganic nutrients become depleted in the seasonally stratified euphotic zone, phytoplankton biomass is substantially low. Thus, the total of detritus+zooplankton+bacteria dominates the suspended POC pool in the NE Mediterranean upper layer as in other oligotrophic seas [19]. Nevertheless, a close correlation appears between POM and CHL profiles in the cyclonic Rhodes region and its peripheries because the nutricline is always situated near the base of the euphotic zone throughout most of the year. In these sites, the subsurface particulate maxima appear to be consistent with the DCM. The elemental composition (C:N:P ratios) of the bulk seston derived from the regression analysis of particulate data is generally comparable with the Redfield 106:16:1 ratios. However, the C:N ratio of the bulk seston apparently decreases below the seasonal thermocline where irradiance becomes a limiting factor for algal growth. This indicates the suspended organic matter existing in the light-limited deep euphotic zone to be mostly produced *in situ*, rather than the particles sinking from the nutrient-limited surface layer.

Anomalously low N:P ratios (6-12) of the bulk seston appeared in the March 1992, when the euphotic zone was enriched with dissolved inorganic nutrients (and relatively high N:P ratio) and much productive. Such low N:P ratio of the bulk seston strongly suggests that POM is exported to the lower depths with much lower N:P ratios than that of nitrate to reactive phosphate (26-28) in the Levantine deep water (LDW). The profound conflict between the elemental composition of POM produced during spring bloom and nitrate to phosphate (N:P) ratios in the LDW and LIW has led to propose another source for anomalously high nitrate to phosphate ratios observed in such water masses and it has been suggested that both the LDW and LIW had originally very high ratios of biologically labile nitrogen to phosphorus before its sinking to the deep basin during the winter mixing [4]. Relatively fast release of nitrogenous organic compounds from POM of planktonic origin increase DON content of the surface waters during late-winter bloom as observed in the productive waters [51]. They then sink with relatively high concentrations of biologically labile organic nitrogen, yielding higher N:P ratio in the aged intermediate and deep water masses. Nevertheless, this suggestion should be supported by late winter measurements at the origins of LDW and LIW.

*Acknowledgments:* The oceanographic surveys of R/V Bilim were supported by the Turkish Scientific and Technical Research Council (TÜBİTAK) within the frame work of the national oceanographic programme. We express our thanks to various members of our Institute (IMS-METU) as well as the captain and crew of the R/V Bilim for their enduring contribution.

## REFERENCES

1. Bethoux, J.P., (1981) Le phosphore et l'azote en mer Mediterranee, bilans et fertilite potentielle, *Mar. Chem.*, **10**, 141-158.

2. Dugdale, R.C. and Wilkerson F.P., (1988) Nutrient sources and primary production in the eastern Mediterranean. *Oceanol. Acta.* **9**, 179-184.

3. Krom, M.D., Brenner, S., Kress, N., Neori, A., and Gordon L.I., (1992) Nutrient dynamics and new production in a warm core eddy from the Eastern Mediterranean. *Deep-Sea Res.*, **39**, 467-480.

4. Yılmaz, A., and Tuğrul, S., (1997) The effect of cold- and warm-core eddies on the distribution and stoichiometry of dissolved nutrients in the northeastern Mediterranean. *J. Mar. Systems*, (in press)

5. Ediger, D. and Yılmaz, A., (1996) Characteristics of deep chlorophyll maximum in the northeastern Mediterranean with respect to environmental conditions. *J. Mar. Systems*, **9**, 291-303.

6. Berman, T., Townsend, D.W., El Sayed, S.Z., Trees, G.C., and Azov, Y., (1984) Optical transparency, chlorophyll and primary productivity in the Eastern Mediterranean near the Israeli coast. *Oceanol. Acta*, **7**, 367-372.

7. Dowidar, N.M., (1984) Phytoplankton biomass and primary production of the south-eastern Mediterranean. *Deep-Sea Res.*, **31**, 983-1000.

8. Azov, Y., (1986) Seasonal patterns of phytoplankton productivity and abundance in nearshore oligotrophic waters of the Levant basin. *J.Plankton Res.*, **8**, 41-53.

9. Abdel-Moati, A.R., (1990) Particulate organic matter in the subsurface chlorophyll maximum layer of the Southeastern Mediterranean. *Oceanol. Acta*, **13**, 307-315.

10. Salihoğlu, İ., Saydam, C., Baştürk, Ö., Yılmaz, K., Göçmen, D., Hatipoğlu, E., and Yılmaz, A., (1990) Transport and distribution of nutrients and chlorophyll-a by mesoscale eddies in the Northeastern Mediterranean. *Mar. Chem.*, **29**, 375-390.

11. Krom, M.D., Brenner, S., Kress, N., Neori, A., and Gordon L.I., (1993) Nutrient distributions during an annual cycle accross a warm-core eddy from the E. Mediterranean Sea. *Deep Sea Res.*. **40**(4), 805-825.

12. Yılmaz, A., Ediger, D., Baştürk, Ö.. and Tuğrul, S., 1994. Phytoplankton fluorescence and deep chlorophyll maxima in the Northeastern Mediterranean. *Oceanol. Acta*, **17**, 69-77.

13. Yacobi, Y.Z., Zohary, T., Kress, N., Hecht, A., Robarts, R.D., Waiser, M., Wood, A.M., Li, W.K.W., (1995) Chlorophyll distribution throughout the southeastern Mediterranean in relation to the physical structure of the water mass. *J. Mar. Systems*, **6** (3), 179-190.

14. Copin-Montegut, C. and Copin-Montegut, G., (1983) Stochiometry of carbon, nitrogen and phosphorus in marine particulate matter. *Deep Sea Res.*, **30** (1), 31-46.

15. Banse, K. (1977) Determining the carbon to chlorophyll ratio of natural phytoplankton. *Mar. Biol.* **41**, 199-212.

16. Eppley, R.W., Harrison ,W. G., Chisholm ,S. W.and Steward, E., (1977) Particulate organic matter in surface waters off Southern California and its relationship to phytoplankton. *J. Mar. Res.*, **35** (4), 671-695.

17. Eppley, R.W., Chavez, F.P., and Barber, R.T., (1992) Standing stock of particulate carbon and nitrogen in the equatorial Pacific at 150°W. *J. Geophys. Res.*, **97**(C1), 655-661.

18. Chen, C.T.A., Chi, M.L., Being, T.H. and Lei, F.C., (1996) Stoichiometry of carbon, hydrogen, nitrogen, sulfur and oxygen in the particulate matter of the western North Pacific marginal seas. *Mar. Chem.* **54**, 179-190.

19. Caron, D.A., Dam, H.G., Kremer, P., Lessard, E.J., Madin, L.P., Malone, T.C.,.Napp, J.M, Peele, E.R., Roman, M.R. and Youngbluth , M.J., (1995) The contribution of microorganisms to particulate carbon and nitrogen in surface waters of the Sargasso Sea near Bermuda. *Deep Sea Res.* **42**(6), 943-972.

20. Takahashi, T., Broecker, W.S. and Langer, S., (1985) Redfield ratio based on chemical data from isopycnal surfaces. *J. Geophys. Res.*, **90**(C4), 6907-6924.

21. Krom, M.D., Kress, N., and Brenner, S., (1991) Phosphorus limitation of primary productivity in the eastern Mediterranean. *Limnol. Oceanogr.*, **36**, 424-432.

22. Zohary, T and Robarts, R.D., (1996) P-limitation study of the Eastern Mediterranean. Meteor, in C. Hemleben, W. Roether and P. Stoffers (eds.). *Berichte 96-4*. Hamburg, 1996, pp.52-56.

23. Polat, S. Ç. and Tuğrul, S., (1996) Chemical exchange between the Mediterranean and the Black Sea via the Turkish straits, in Briand (ed.), *Dynamics of Mediterranean Straits and Channels*, CIESM Science Series No 2 F. Bulletin Oceanographique Monaco, No special 17, pp.167-186.

24. Karl, D.M., Winn , C.D., Dale, V.W.and Letelier, H.R.(1990) Hawaii Ocean Time-series Program, Field and Laboratory Protocols, September 1990.

25. Holm-Hansen, O., Lorenzen, C.J., Holmes, R.W. and Strickland, Y.M.H.A., (1965) Fluorometric determination of chlorophyll. *J.Cons.Perm. Int. Explor.Mer.*, **30**, 3-15.

26. Grasshoff, K., Ehrhardt, M. and Kremling, K., (1983) Determination of nutrients, in Methods of seawater analysis. 2nd edition, Verlag Chemie GMBH, Weinheim, pp.125-188.

27. Özsoy, E., Hecht, A., and Ünlüata, Ü., (1989) Circulation and hydrography of the Levantine Basin. Results of POEM coordinated experiments 1985-1986. *Prog. Oceanogr.*, **22**, 125-170.

28. Özsoy, E., Hecht, A., Ünlüata, Ü., Brenner, S., Oğuz, T., Bishop, J., Latif, M.A., and Rosentraub, Z., (1991) A review of the Levantine Basin circulation and its variability during 1985-1988. *Dyn. Atmos. Oceans*, **15**, 421-456.

29. Sur, H.I., Özsoy, E., and Ünlüata, Ü., (1993) Simultaneous deep and intermediate depth convection in the Northern Levantine Sea, winter 1992. *Oceanol. Acta*, **16**, 33-43.

30. Gertman, I.F., Ovchinnikov, I.M., and Popov, Y.I., (1990) Deep convection in the Levantine Sea. Rapp.Commn int.Mer medit., **32**, p.172.

31. Özsoy, E., Hecht, A., Ünlüata, Ü., Brenner, S., Sur, H.I., Bishop, J., Oğuz, T., Rosentraub, Z., and Latif, M.A., (1993) A synthesis of the Levantine basin circulation and hydrography, 1985-1990. *Deep-Sea Res.*, **40**(6), 1075-1119.

32. Hecht, A., Pinardi, N., and Robinson, A.R., (1988) Currents, water masses, eddies, and jets in the Mediterranean Levantine basin. *J. Phys. Oceanogr.*, **18**, 1320-1353.

33. Ediger, D., (1995) Interrelationships among primary production chlorophyll and environmental conditions in the Northern Levantine Basin. Ph.D.Thesis, METU, Institute of Marine Sciences, Erdemli, İçel,Turkey,187p.

34. Redfield, A.C., Ketchum, B.H. and Richards, F.H., (1963) The influence of organisms on the composition of sea water. in Hill, M.N. (ed.) The Sea Ideas and Observations, Vol.2. Interscience, New York, pp.27-77.

35. Cullen, J.J., (1982) The deep Chlorophyll maximum: Comparing vertical profiles of chlorophyll-a. *Can. J. Fish. Aquat. Sci.*, **39**, 791-803.

36. Estrada, M., Marrase, C., Latasa ,M., Berdalet ,E., Delgado ,M. and Riera ,T., (1993) Variability of deep chlorophyll maximum characteristics in the northwestern Mediterranean. *Mar.Ecol.Prog.Ser.* **92**, 289-300

37. Kirk, J.T.O., (1983) Light and photosynthesis in aquatic ecosystems. 2nd ed. Cambridge Univ.Press, 508p.

38. Strass, V.H., and Wood, J.D., (1991) New production in the summer revealed by the meridional slope of the deep chlorophyll maximum. *Deep Sea Res.*, **38**, 35-56.

39. Murdoch, W.W. and Onuf, C.P., (1974) The Mediterranean as a system. Part I: Large Ecosystem. *Int. J. Environ. Stud.*, **5**, 275-284.

40. Herbland, A.A. and Voituries, B., (1979) Hydrological structure analysis for estimating the primary production in the tropical Atlantic Ocean. *J. Mar. Res.*, **37**, 87-101.

41. Taguchi, S., Ditullio, R.G.and Laws, E.A., (1988) Physiological characteristics and production of mixed layer and chlorophyll maximum phytoplankton populations in the Caribbean and western Atlantic Ocean. *Deep Sea Res.* **35**(8), 1363-1377.

42. Furuya, K. (1990) Subsurface chlorophyll maximum in the tropical and subtropical western Pacific Ocean: vertical profiles of phytoplankton biomass and its relationship with chlorophyll a and particulate organic carbon. *Mar. Biol.*, **107**, 529-539.

43. Lohrenz, S.E., Wiesenburg, D.A., Depalma, I.P., Johnson, K.S., and Gustafson, D.E., (1988) Interrelationships among primary production, chlorophyll, and environmental conditions in frontal regions of the western Mediterranean Sea. *Deep-Sea Res.*, **35**, 793-810.

44. Li, W.K.W., Dickie, P.M., Irwin, B.D. and Wood, A.M., (1992) Biomass of bacteria, cyanobacteria, prochlorophytes and photosynthetic eukaryotes in the Sargasso Sea. *Deep-Sea Res.* **39**(3/4), 501-519.

45. Lee, S. and Fuhrman, J.A., (1987) Relationships between biovolume and biomass of naturally derived marine bacterioplankton. *Appl. Environ. Microbiol.*, **53**, 1298-1303.

46. Sambrotto, R.N., Savidge. G., Robinson. C., Boyd . P., Takahashi. T., Karl, D.M., Langdon, Chipman, D., Marra, J. and Cadispodi. L,.(1993) Elevated consumption of carbon relative to nitrogen in the surface ocean. *Nature*, **363**, 248-250.

47. Toggweller. J.R., (1993) Carbon overconsumption. *Nature*, **363**. 210-211.

48. Dauchez, S., Queguiner ,B., Treguer, P. and Zeyons, C., (1991) A comparative study of nitrogen and carbon uptake by phytoplankton in a coastal euphotic ecosystem (Bay of Brest. France). *Oceanol. Acta*, **14**, 87-95.

49. Anita, N.J., McAllister. C.D., Parsons. T.R., Stephens, K. and Strickland, J.D.H., (1963) Further measurements of primary production using a large volume plastic sphere, *Lim. Oceanogr.*, 8, 166-183.

50. Li, W.K.W. Zohary, T., Yacobi, Y. Z., Wood, A. M., (1993) Ultraphytoplankton in the eastern Mediterranean Sea: towards deriving phytoplankton biomass from flow cytometric measurements of abundance, fluorescence and light scatter. *Mar.Ecol. Prog. Ser.*, **102**, 79-87.

51. Wafar, M., Corre, P.L.and Birrien, J.L., (1984) Seasonal changes of dissolved organic matter (C,N,P) in permanently well mixed temperate waters. *Lim.Oceanog.* **29**(5), 1127-1132.

# Comparative analysis of sedimentation processes within the Black Sea and Eastern Mediterranean

A. Y. Mitropolsky, S. P. Olshtynsky
Institute of Geological Sciences NAS, Kiev, Ukraine

All cruises of Ukrainian recearch vessels usually begin in the Black and Mediterranean seas. Consequently the Branch of Geology of Oceans and Seas of the Institute of Geological Sciences of the National Academy of Sciences, which is the oldest marine geology scientific centre of Ukraine, have a considerable data base and map archive On the ground of marine geological data obtained in many routes in the Black and Mediterranean seas we have realized the comparative studying of sedimentation processes within the Black Sea and the Levantine Basin of Eastern Mediterranean. Certain results obtained are discussed in this paper. Geochemical aspects of sediment accumulation in the Black and Levantine seas were considered in detail in [1, 2, 3].

Both of these basins may be considered as examples of contrasting ecosystem combinations. They have almost identical area sizes and water mass volumes. The relations of shelf and deep sea zones of both basins are similar. The coastline length and sea shore morphology of the Black Sea are similar those ones of the Levantine Basin in many features (Table 1). Comparative high sedimentation rate is a characteristic of both basins. As well as the Black Sea, the Levantine Basin has a good river, which supplies huge mass of fresh water and drift matter. In both cases the river fan occupies approximately one third of the bottom area.

Surely, there are some difference peculiarities of these basins. They are located in the various climatic zones, and have distinct watersheds and essentially different water exchange relations with neighbour basins. Appreciable diversities of water mass hydrochemistry also take place. In the Black Sea at the depth more 150-200 m the water mass is poisoned with hydrogen sulphide, whereas in the Levantine Basin water is saturated with oxygen from the surface to the bottom and maximum of concentration is in the layer of 200-500 m. Salinity of the Black Sea water forms a half of the value of the Eastern Mediterranean salinity. The Levantine water contains carbon dioxide far



*P. Malanotte-Rizzoli and V.N. Eremeev (eds.),*
*The Eastern Mediterranean as a Laboratory Basin for the Assessment of Contrasting Ecosystems*, 267–280.

less than the Black Sea water because of its low bioproductivity. It has also low content of silica and in its sediments there is less concentration of organic matter. Maps of distribution of various types of sediments are shown on Figures 1, 5.

Besides, such main features of these basins are special characteristics, which depend on basic factors of sedimentogenesis. Let us consider them by order.

The effect of the terrigenous matter supply on the sediment accumulation. It is at once apparent the Nile represents the main source of the suspended terrigenous matter for the Levantine Basin. Its role in sediment forming is the greatest (Table 2). Other sources of the terrigenous matter such as marine abrasion and eolian transport have much less importance. Drifts of this river are spreading through large areas of the Levantine Basin bottom. Before Assuan weir had been built, the Nile was transporting 160-200 million tons per year of suspended material into the sedimentation basin. At present time its drift supplying is estimated to be appximately 30% of that value. Inside the basin the suspended matter is transported by the surface water circulation counter clockwise along Lebanon and Israel shores towards Cyprus. The certain part of drift mass is carried by deep sea streams in the north and north-west directions forming separate lobes of the complicated river fan of the Nile. Terrigenous river silts are enriched with heavy minerals such as monoclinic pyroxenes, amphibole, epidote, rutile, zirconium silicate. Clay minerals of silts are generally represented by montmorillonite. Suspension of Nile water contains a notable amount of iron, titanium, phosporous, organic carbon, chromium, vanadium, copper, and nickel. In geologic scale of time the influence of the river fan of the Nile on the sedimentation processes within the Levantine Basin was especially significant in Messinean.

The hydrodynamic effect on the transport of suspended matter. The surface water circulation in both basins is derected counter clockwise, but deep sea streams may have another directiones particularly upon river fans, near mouthes and submarine canyons. Suspensions are transported by streams through far distans, although a comparatively isolated zone may be arranged inside the circulation area. Just such phenomenon takes place in the Black Sea. There are two halistatic zones where the entering of terrigenous matter is relatively moderate.

The role of the factor of climatic zonation. Basins are arranged in different climatic zones. Southern part of the Levantine sea is effected by arid conditiones, nothern one is placed in the transitional zone. The main part of the Black sea belongs to humid zone. Such climatic zonation impacts mainly the processes of forming and accumulation of carbonates. Maps of distribution of carbonates are demonstrated on Figures 2, 6. The accumulation of chemogeneous carbonates increases in the arid zone, where sediments contain from eight to fifteen times more carbonates than suspensions. The amount of diagenetic carbonates is increasing there. The contents of carbonates in sediments correlates with the contents of aleuritic fraction and becomes above the terrigeneous input. The mineral composition of carbonates is represented in this zone by magnesian calcite.

In the transitional zone the role of organogenic carbonates is increasing, and in the humid zone this source of carbonate becomes predominant.

In the Black sea the total content of calcium carbonate increases from shores towards deep sea regions, in the Levantine basin it decreases from the arid (platform) zone to the transitional (geosyncline) zone. The content of calcite in the both case sharply increases from shores to the upper part of the continental margin. In deep sea regions calcite is predominant. In the Black sea low magnesian calcite prevailes, in the Levantine basin high magnesian calcite is predominant. As far as organic carbonate forms are conerned, in the Black sea coccolites prevail and on the shelf molluscs. In the Levantine basin foraminifers, coccolites and pteropodes are predominant, and on the shelf molluscs and echinoderms. There are also chemogenious and chemogene-diagenetic calcites, ankerites, dolomites, and on the shelf oolites and eolian carbonates.

Organic matter is very important for diagenesis of marine bottom sediments. Maps of distribution of organic carbon in sediments are shown on Figures 3, 7. Comparatively inactive role of organic matter indistribution of macro and micro elements in the upper layer of sediments is caracteristic of the Levantine basin. This fact take place because of deep mineralization of organic matter and its appreciable low content. There is not observed any notable influence on the distribution of such fluent elements as manganese, phosphorus, copper, nickel and molybdenum. Only in sediments with interlayers of sapropel silts redistribution of elements occurs active enough. Therein the content of organic carbonis in the limit 2 - 11%. Such interlayers are notably enriched with molybdenum, copper, nickel, cobalt, uranium, sometimes vanadium and impoverished with manganese. This peculiar properties may be the evidence that in the period of sapropel silt accumulation the sedimentogenesis occured under conditions which were analogous recent ones of the Black sea.

Mean values of concentrations of some elements in bottom sediments are shownen in the Table 3 and on Figures 4, 8 the distribution of lead in sediments is demonstrated. As to peculiarities of the accumulation of trace elements in sediments of Abu Quir Bay it should be noted that the chromium distribution is primarily determined by the suspended matter evacuated from the Nile as well as by the hydrodynamic regime in the bay.

Zirconium presence in the sediments of Abu Quir Bay is due to transportation of coarse sands containing zirconium by the alongshore west currents. Realization of zirconium from the suspended matter broughtby the Nile is of minor importance.

So we may summarise the main characteristics of sedimentation processes in the Levantine basin. Its water is impoverished with dissolved silica as well as suspended one. There is few phytoplankton silica (notable amount of it is observed only in spring and along shores). In the bottom sediment silicon is connected with the terrigeneous fraction. The biogenic factor (food webs) plays the main role in the input of organic matter to deep water layers and bottom sediments. The content of organic carbon in sediments is here from two to three times less than in the Black sea. The appreciable amount of organic carbon is determined only in shell deposits. Such low content of organic carbon in the water and sediments is the reason of poverish of bentic forms of life.

The less concentration of phosphorus and another nutrients in the water timitates the growth of phytoplankton the main base of bioproduction. Comparatively high temperature of the water and the oxygen saturation lead to quick destruction of organic matter.

**Table 1.**
**Some characteristics of the Levantine Basin and the Black Sea**

| Characteristic | Levantine Basin | Black Sea |
|---|---|---|
| Area $km^2$ | 350,000 | 422,000 |
| Volume $km^3$ | 600,000 | 555,000 |
| Salinity g/kg | 42 | 22 |
| Temperature C° | 13 | 8 |

**Table 2.**
**Some characteristics of the Nile, the Danube and the Dnieper**

| Characteristic | Nile | Danube | Dnieper |
|---|---|---|---|
| Length km | 6,500 | | |
| Watershed $km^2$ | 2,867,000 | 817,000 | 500,000 |
| Discharge $km^3$/yr | | 203 | 43 |
| Drifts million t | 160 / 57 | 67.5 | 1.5 |
| River fan $km^2$ | 104,000 | 75,000 | |

**Table 3.**
**Average content of trace element in bottom sediments**
**$10^{-4}$%**

| Region | Cr | Zr | Ti | Mn | Mo | V | Co | Ni | Cu |
|---|---|---|---|---|---|---|---|---|---|
| Black Sea | 130 | 100 | 1100 | 220 | 2.6 | 129 | 11 | 44 | 3.4 |
| Levantine Basin | 50 | 35 | 4500 | 1000 | 1.0 | 78 | 15 | 70 | 36 |
| Delta of the Nile | 80 | 95 | 7200 | 700 | 0.3 | 95 | 10 | 30 | 32 |
| Abu Quir Bay | 120 | 70 | 6800 | 1000 | 0.45 | 97 | 4 | 24 | 29 |

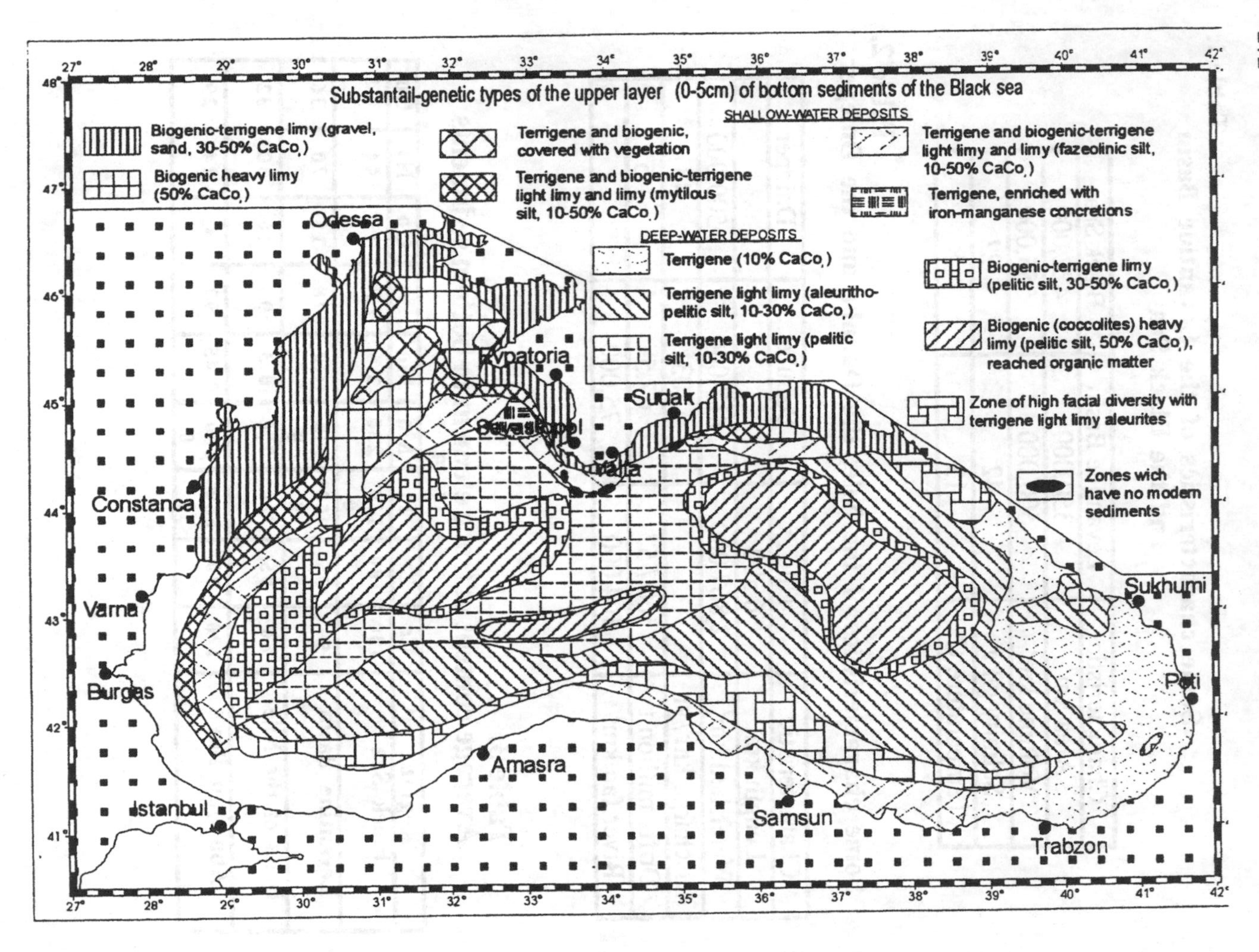
Substantail-genetic types of the upper layer (0-5cm) of bottom sediments of the Black sea
SHALLOW-WATER DEPOSITS
Biogenic-terrigene limy (gravel, sand, 30-50% CaCo$_3$)
Biogenic heavy limy (50% CaCo$_3$)
Terrigene and biogenic, covered with vegetation
Terrigene and biogenic-terrigene light limy and limy (mytilous silt, 10-50% CaCo$_3$)
Terrigene and biogenic-terrigene light limy and limy (fazeolinic silt, 10-50% CaCo$_3$)
Terrigene, enriched with iron-manganese concretions
DEEP-WATER DEPOSITS
Terrigene (10% CaCo$_3$)
Terrigene light limy (aleuritho-pelitic silt, 10-30% CaCo$_3$)
Terrigene light limy (pelitic silt, 10-30% CaCo$_3$)
Biogenic-terrigene limy (pelitic silt, 30-50% CaCo$_3$)
Biogenic (coccolites) heavy limy (pelitic silt, 50% CaCo$_3$), reached organic matter
Zone of high facial diversity with terrigene light limy aleurites
Zones wich have no modern sediments
Odessa
Evpatoria
Sudak
Constanca
Varna
Burgas
Istanbul
Amasra
Samsun
Trabzon
Sukhumi
Poti
27° 28° 29° 30° 31° 32° 33° 34° 35° 36° 37° 38° 39° 40° 41° 42°
48° 47° 46° 45° 44° 43° 42° 41°

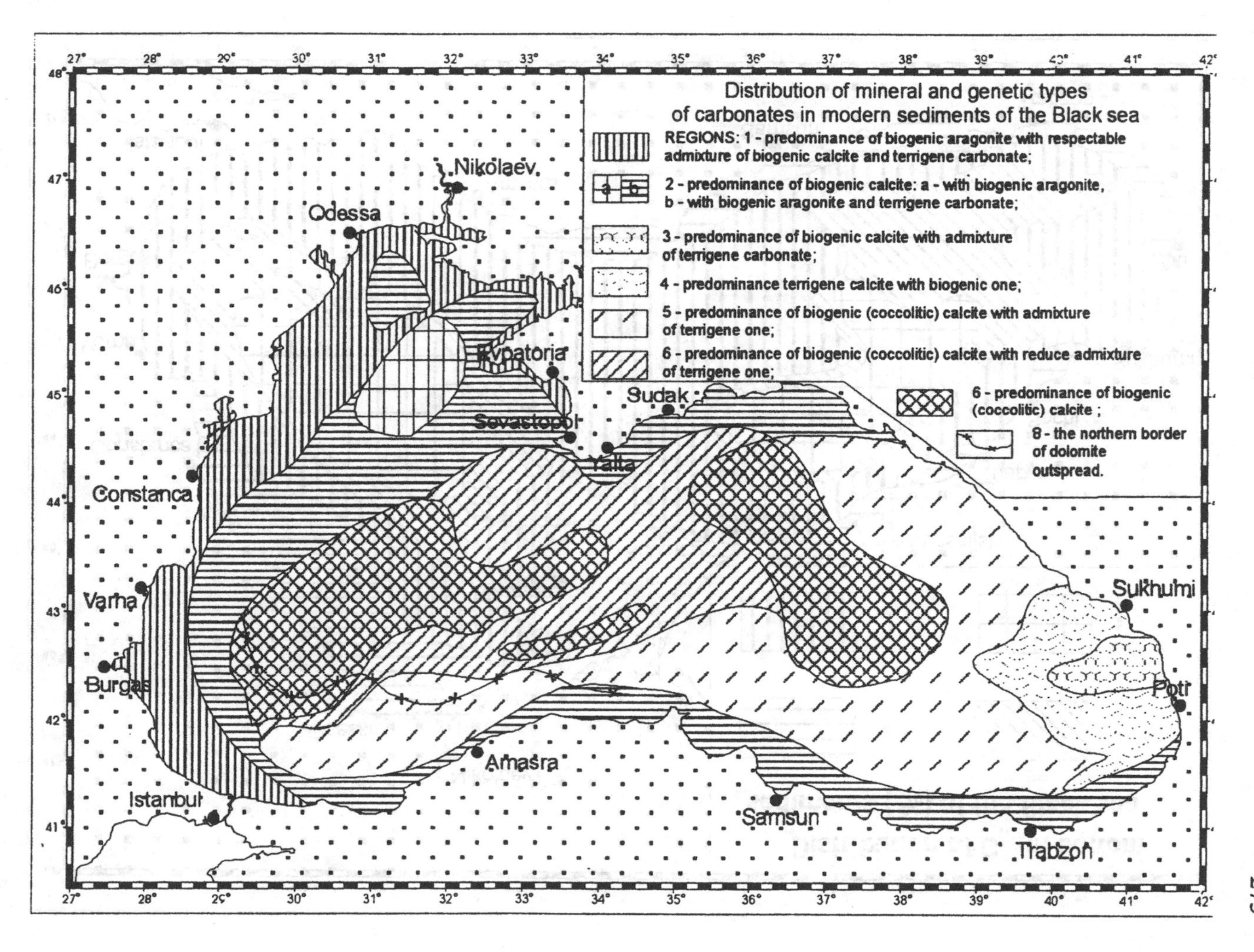
Distribution of mineral and genetic types
of carbonates in modern sediments of the Black sea
REGIONS: 1 - predominance of biogenic aragonite with respectable admixture of biogenic calcite and terrigene carbonate;
a
b
2 - predominance of biogenic calcite: a - with biogenic aragonite, b - with biogenic aragonite and terrigene carbonate;
3 - predominance of biogenic calcite with admixture of terrigene carbonate;
4 - predominance terrigene calcite with biogenic one;
5 - predominance of biogenic (coccolitic) calcite with admixture of terrigene one;
6 - predominance of biogenic (coccolitic) calcite with reduce admixture of terrigene one;
6 - predominance of biogenic (coccolitic) calcite ;
8 - the northern border of dolomite outspread.
Nikolaev.
Odessa
Evpatoria
Sevastopol
Yalta
Sudak
Constanca
Varna
Burgas
Istanbul
Amasra
Samsun
Trabzon
Sukhumi
Poti
27°
28°
29°
30°
31°
32°
33°
34°
35°
36°
37°
38°
39°
40°
41°
42°
48°
47°
46°
45°
44°
43°
42°
41°

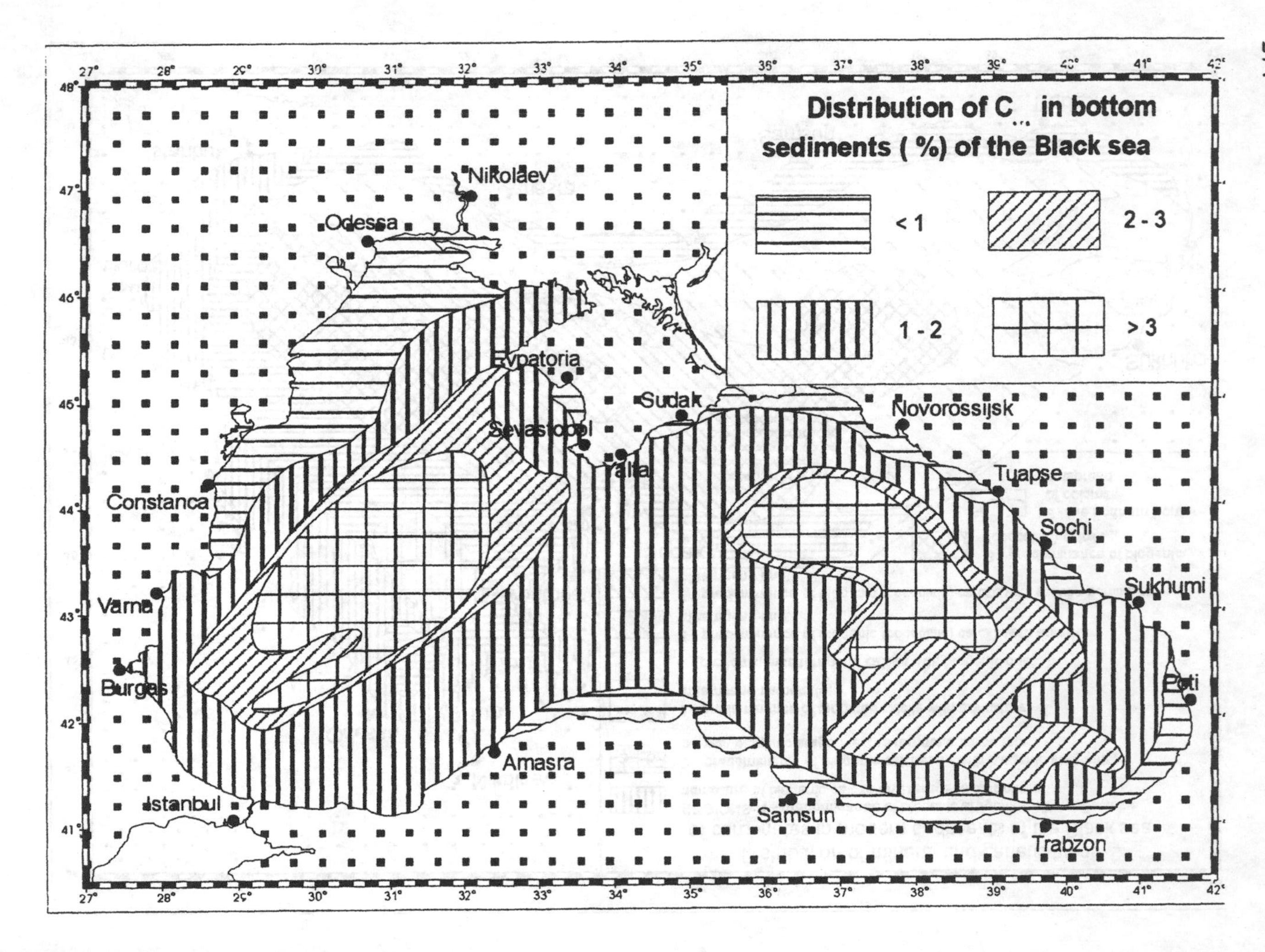
Distribution of C in bottom
sediments ( %) of the Black sea
< 1
2 - 3
1 - 2
> 3
Nikolaev
Odessa
Evpatoria
Sudak
Novorossijsk
Sevastopol
Yalta
Tuapse
Constanca
Sochi
Sukhumi
Varna
Burgas
Poti
Amasra
Istanbul
Samsun
Trabzon

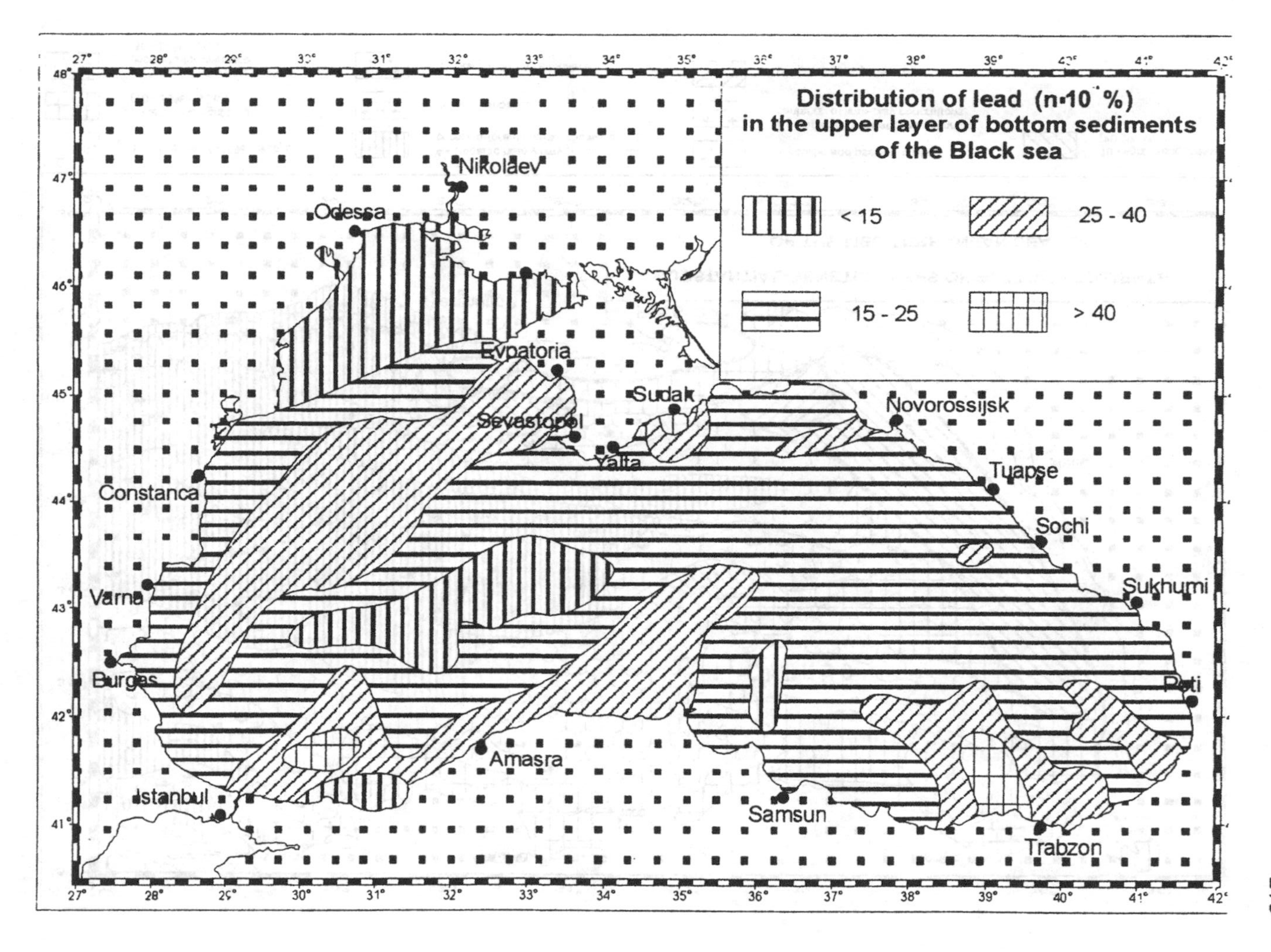

Distribution of lead (n·10 %)
in the upper layer of bottom sediments
of the Black sea
< 15
25 - 40
15 - 25
> 40
Nikolaev
Odessa
Evpatoria
Sudak
Sevastopol
Yalta
Novorossijsk
Tuapse
Sochi
Sukhumi
Poti
Constanca
Varna
Burgas
Amasra
Istanbul
Samsun
Trabzon

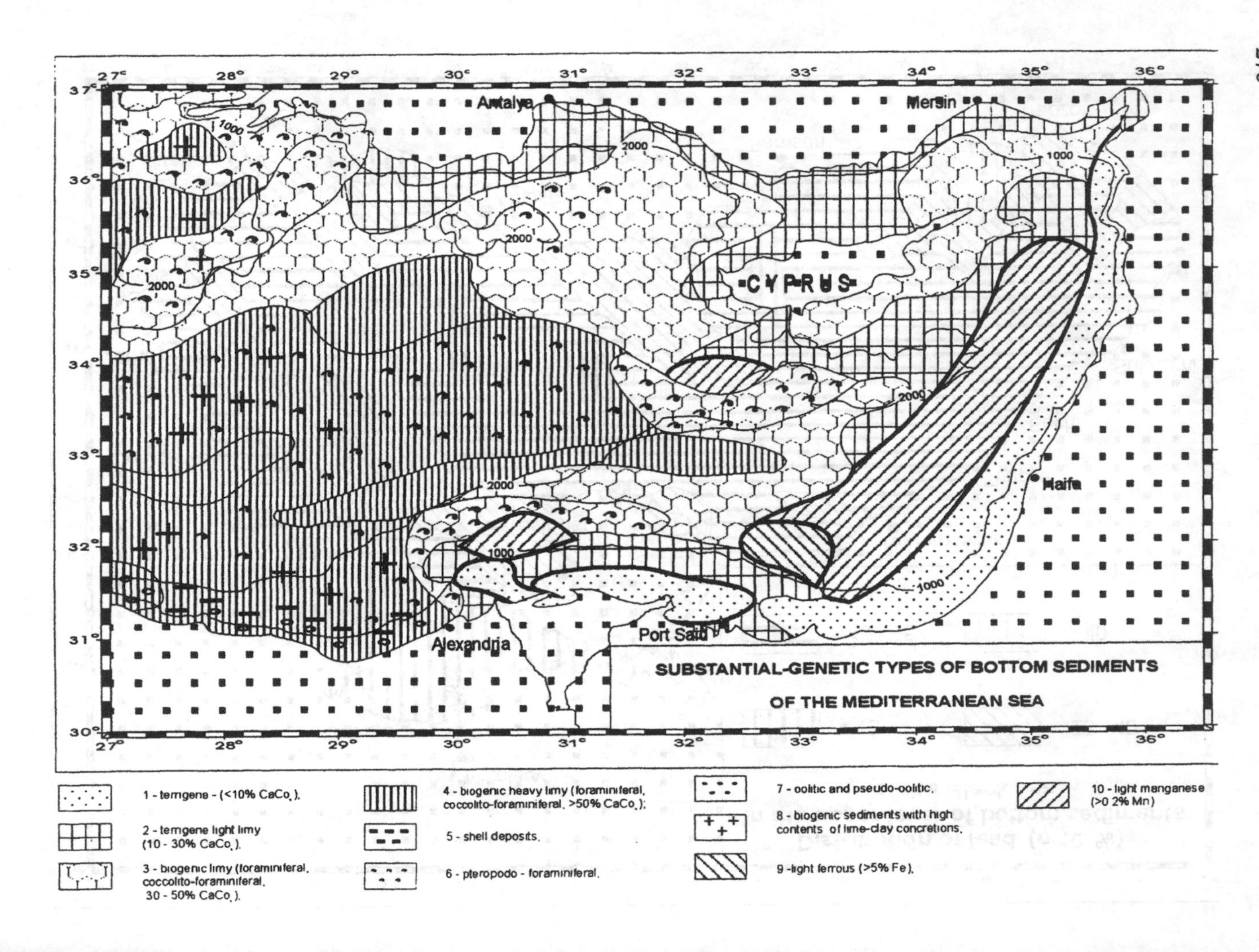
SUBSTANTIAL-GENETIC TYPES OF BOTTOM SEDIMENTS
OF THE MEDITERRANEAN SEA
Antalya
Mersin
CYPRUS
Haifa
Port Said
Alexandria
27°
28°
29°
30°
31°
32°
33°
34°
35°
36°
37°
1000
2000
1 - terrigene - (<10% CaCo$_3$).
2 - terrigene light limy (10 - 30% CaCo$_3$).
3 - biogenic limy (foraminiferal, coccolito-foraminiferal, 30 - 50% CaCo$_3$).
4 - biogenic heavy limy (foraminiferal, coccolito-foraminiferal, >50% CaCo$_3$);
5 - shell deposits.
6 - pteropodo - foraminiferal.
7 - oolitic and pseudo-oolitic.
8 - biogenic sediments with high contents of lime-clay concretions.
9 - light ferrous (>5% Fe).
10 - light manganese (>0 2% Mn)

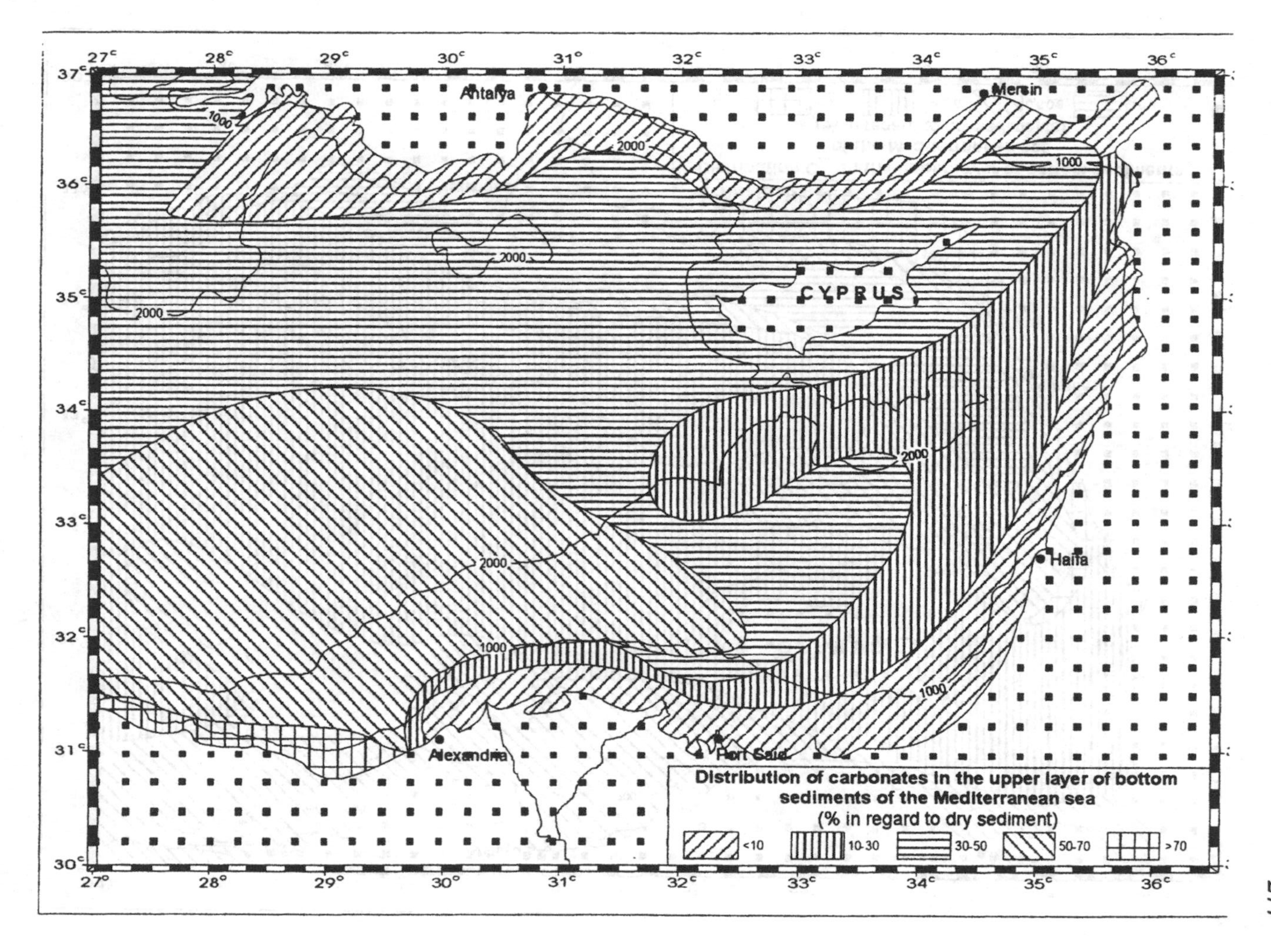

Distribution of carbonates in the upper layer of bottom sediments of the Mediterranean sea
(% in regard to dry sediment)
<10
10-30
30-50
50-70
>70
CYPRUS
Antalya
Mersin
Haifa
Port Said
Alexandria
1000
2000
27°
28°
29°
30°
31°
32°
33°
34°
35°
36°
37°

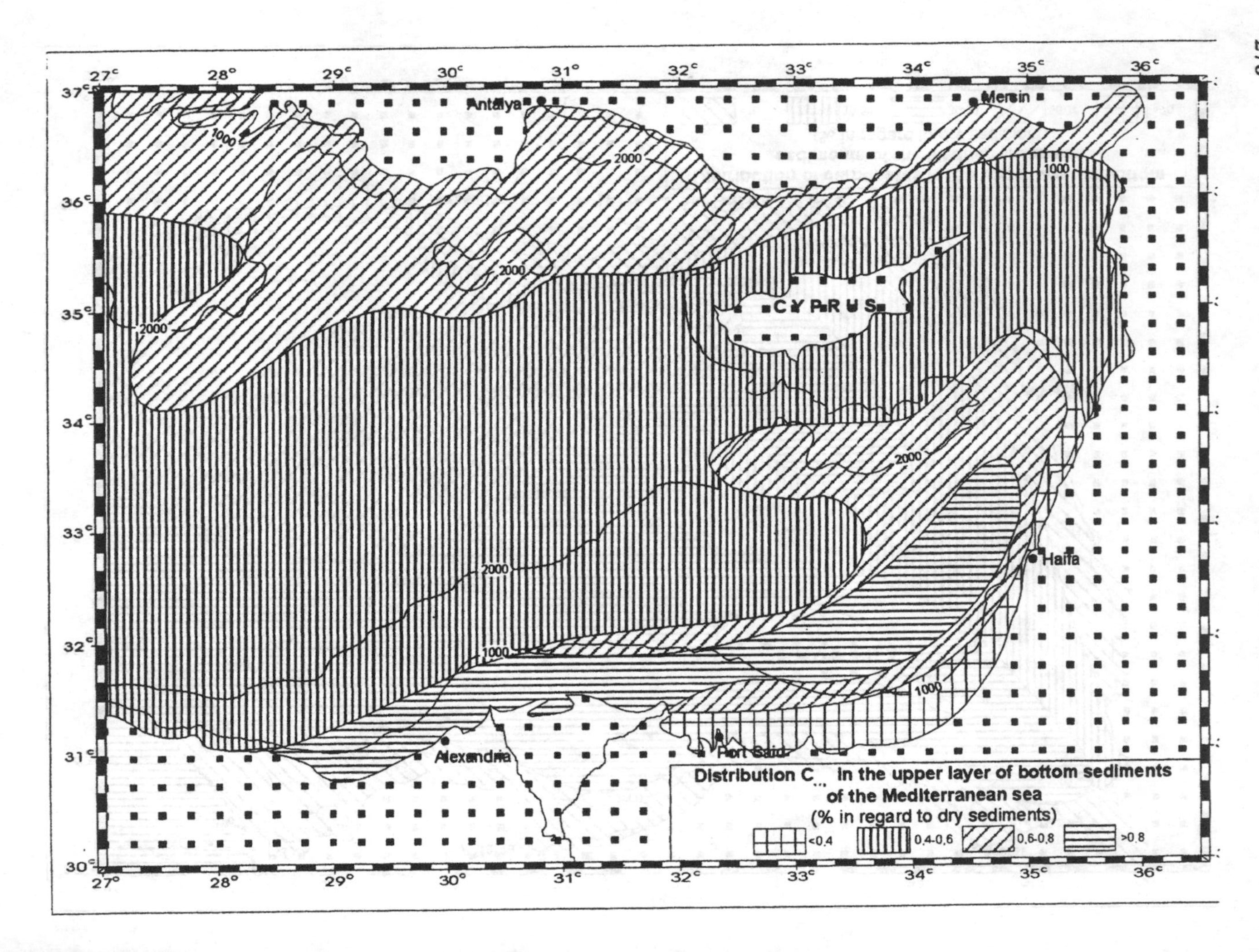
Distribution C... in the upper layer of bottom sediments
of the Mediterranean sea
(% in regard to dry sediments)
<0,4
0,4-0,6
0,6-0,8
>0,8
Antalya
Mersin
CYPRUS
Haifa
Alexandria
Port Said
1000
2000
27°
28°
29°
30°
31°
32°
33°
34°
35°
36°
37°

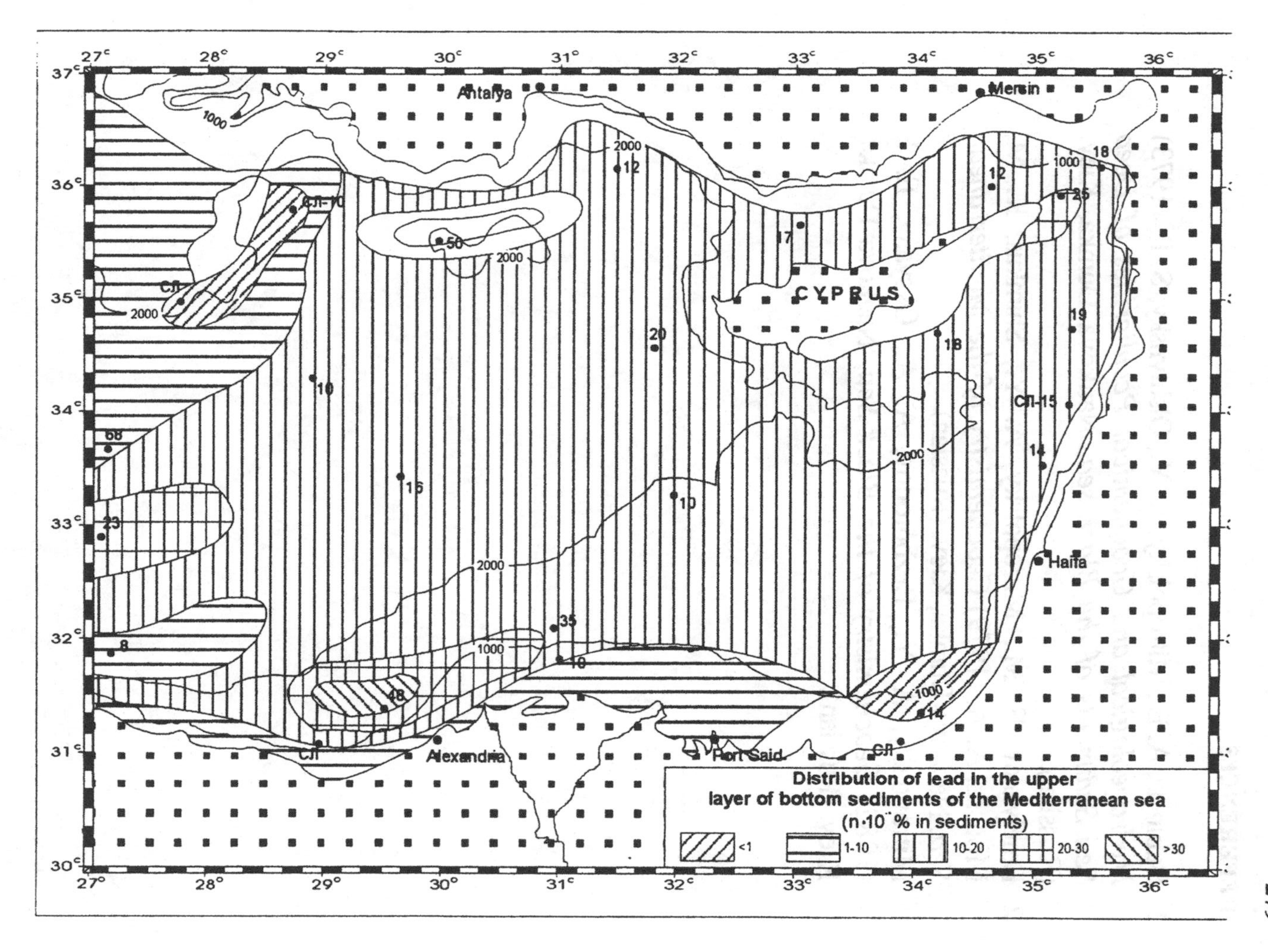
Antalya
Mersin
CYPRUS
Haifa
Alexandria
Port Said
Distribution of lead in the upper
layer of bottom sediments of the Mediterranean sea
% in sediments)
<1
1-10
10-20
20-30
>30

**REFERENCES**

1. Babinetz, A. E., Mitropolsky, A. Y., Olshtynsky, S. P. (1973) *Hydrogeological and Geochemical Peculiarities of Deep Sea Sediments of the Black sea*, Naukova Dumka, Kiev. (Russian).
2. Emelyanov, E. M., Mitropolsky, A. Y., Shimkus, K. M, Moussa, A. A. (1979) *Ceochemistry of the Mediterranean Sea*, Naukova Dumka, Kiev. (Russian).
3. Mitropolsky, A. Y., Bezborodov, A. A., Ovsyany, E. I. (1982) *Geochemistry of the Black Sea*, Naukova Dumka, Kiev. (Russian).

# SENSITIVITY TO GLOBAL CHANGE IN TEMPERATE EURO-ASIAN SEAS (THE MEDITERRANEAN, BLACK SEA AND CASPIAN SEA): A REVIEW

EMIN ÖZSOY
*Institute of Marine Sciences, Middle East Technical University, P. O. Box 28, Erdemli, İçel 33731 Turkey*

**Abstract.** Common features of the three relatively isolated Euro-Asian Seas are reviewed to evaluate their sensitivity to climatic or anthropogenic change. The projection of the effects of Global Change occur through physical linkages, mediated by global, basin-scale and meso-scale processes, which need to be better unterstood for better forecasts. Prominent interannual / interdecadal signals and large scale controls are evident, and in some cases the changes are of a magnitude detected for the first time in the history of modern observations.

## 1. Introduction

The Levantine Sea, Black Sea, and the Caspian Sea are the remotest, climatically coupled, progresively isolated interior Seas of the Euro-Asian continent. All three Seas are neighbored by high mountain chains, vast continental flatlands, deserts and fertile lands, in a transition between the Atlantic and Indian Oceans. Ocean-atmosphere-land interactions in this environment of contrasting marine and continental climates, complex land and sea bottom topography, and energetic mid-latitude atmospheric motions make the region prone to extremes, and result in pigment patterns with marked regional differences (Figure 1). As a result, the feedbacks to the global system could be disproportionately large in comparison to the size of the region.

If any property is common among these three Seas, it is their sensitivity to Global Change. Inland seas, with their smaller inertia, respond faster to climatic forcing compared to the global oceans. For the same reason, they are more sensitive to environmental degradation, with the Euro-Asian Seas



*P. Malanotte-Rizzoli and V.N. Eremeev (eds.),*
*The Eastern Mediterranean as a Laboratory Basin for the Assessment of Contrasting Ecosystems,* 281–300.

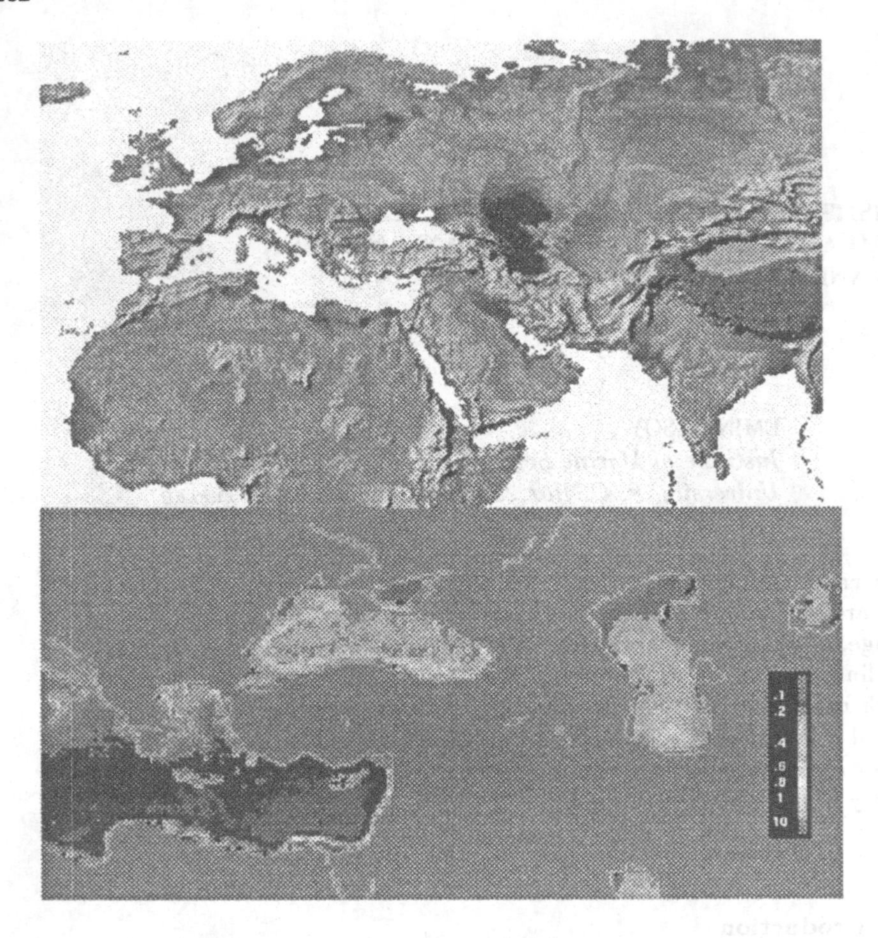


*Figure 1.* Top: Topography of Euro-Asian continent. The Caspian Sea, currently ~28$m$ below sea level, shown in blue, bottom: CZCS average pigment concentration ($mg/m^3$) in the eastern Mediterranean, Black Sea and Caspian Sea regions. Data after NASA/GSFC

being among the most troubled waters of the world ocean (IOC, 1993). Land use / cover and subsequent hydrological changes in the adjacent lands lead to desertification and scarcity of water (Moreno and Oechel, 1995; Jeftic, 1992, 1996; Glanz and Zonn, 1997), amplified by cultural and socio-economic contrasts in the region.

At present the causal relationships explaining the evolution of the ocean-atmosphere system projected onto the region are not well established. Often the changes are mediated through sub-basin and meso-scale processes, and are therefore difficult to be identified.

## 2. Large-Scale Controls

The region is one of the foremost areas of the world where interannual and long term climatic variability is predominant. In the Mediterranean region such variability is well known (*e.g.* Garrett *et al.*, 1992; Robinson *et al.*, 1993; Malanotte-Rizzoli and Robinson 1994; Jeftic et al., 1992, 1996). The Mediterranean variability appears correlated with global teleconnection patterns, and coupled with the Indian Monsoon system and El Nino / Southern Oscillation (ENSO) (Ward, 1996). For example, good correlation has been found between heavy rain and snow in Israel during the last 100 years and ENSO (Alpert and Reisin, 1986). Similarly, global versus regional climate interaction has been emphasized in the case of the Caspian Sea (Rodionov, 1994) as well as in the Black Sea (Polonsky *et al.*, 1997).

Weather in Europe, extending well into Eurasia, is to a large extent determined by conditions in the North Atlantic, and in particular the North Atlantic Oscillation (NAO) quantified by the normalized anomaly of the sea level pressure difference between the Azores and Iceland (Hurrell, 1995). Severe weather in Europe occurs when the NAO index is positive. The NAO accounts for about one third of the hemispheric interannual variance, and accounts for surface temperature changes as well as evaporation-precipitation anomalies in the European and the Eastern Mediterranean regions (Hurrell, 1995, 1996; Marshall, 1997). It has been linked to sea level changes in the Caspian Sea (Rodionov, 1994), to Danube river runoff directly influencing Black Sea hydrology (Polonsky *et al.*, 1997), as well as to surface winter temperatures, precipitation and river runoff in Turkey (Cullen, 1998).

Large scale control is well expressed in long range atmospheric transport patterns, suggested by the simultaneous occurence (Li *et al.*, 1996; Andreae, 1996) and parallel dependence on the interannual NAO patterns (Moulin *et al.*, 1997) of the transport of aerosol dust from the Sahara desert into the Mediterranean and tropical Atlantic regions. An exceptional case studied during the first half of April 1994 (Özsoy *et al.*, 1998; Kubilay *et al.*, 1998), has illustrated the role of large scale controls. Atmospheric blocking in the Atlantic had triggered upper air jet interactions and meridional circulations on a hemispherical scale (Figure 2a-c). These interactions resulted in large scale subsidence and cyclogenesis resulting in an anomalous pattern of dust simultaneously transported towards the subtropical Atlantic and the eastern Mediterranean regions, leading to maximum dust concentrations in 30 years of measurements in Bermuda and similarly high values in Erdemli (Özsoy, *et al.*, 1998). It is most interesting that the average sea level pressure was characterized with the typical dipole pattern of the NAO (Fig 2d), with a corresponding high index value of ~4, suggesting significant

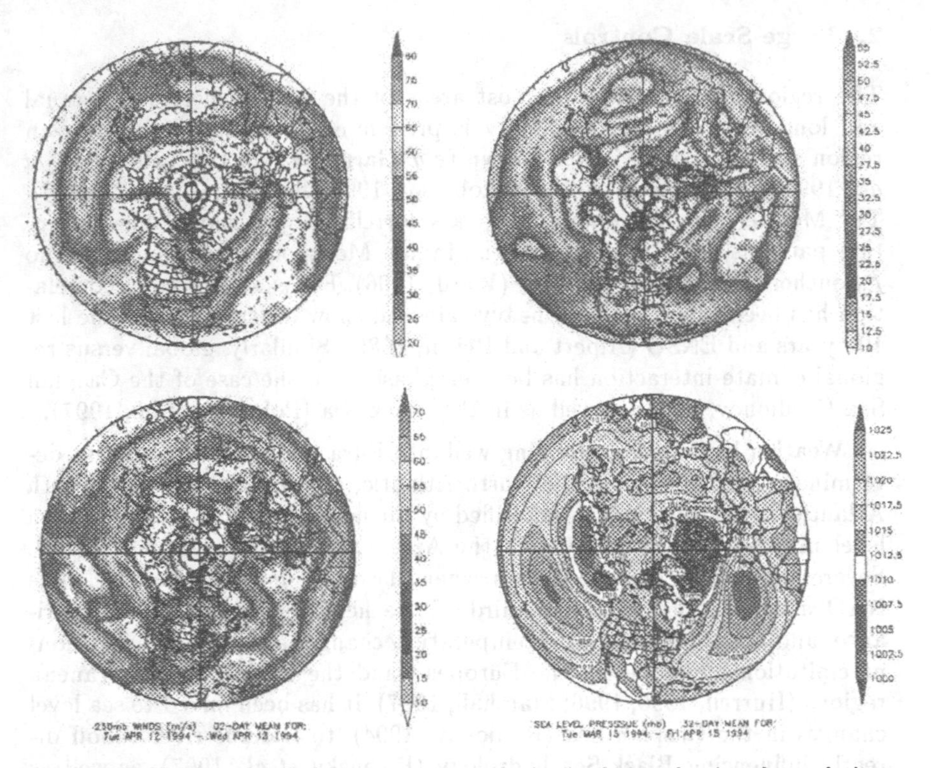


*Figure 2.* **Northern hemisphere upper atmospheric circulation showing the evolution of the polar and subtropical jets. 250*hPa* wind speed (shading) and direction (vectors) based on two day averages for (a) March 31 - April 1, (c) April 4 - 5, 1994, (c) April 12 - 13, 1994 and (d) the monthly average sea level pressure for the period March 15 - April 15, 1994. The source for the data is NOAA Climate Diagnostic Center (plotting page web adress http://www.cdc.noaa.gov/HistData/**

links between high index NAO circulation and the anomalous dust event. In addition to the links with NAO, the Atlantic dust flux also appears well correlated with the African drought and the ENSO (Prospero and Nees, 1986).

## 3. Similarities in Regional Cooling Patterns

There is a significant degree of synchronism displayed between the Levantine, Black and Caspian Seas, in terms of the air and sea surface temperatures displayed in Figure 3. This is because of the proximity of the three regions, but also a result of the possible large scale controls discussed above, and by Özsoy and Ünlüata (1997). In the same Figure, comparisons are also made with the NAO and SO indices and with solar transmission (an indicator of volcanic dust in the atmosphere). Some cold years appear

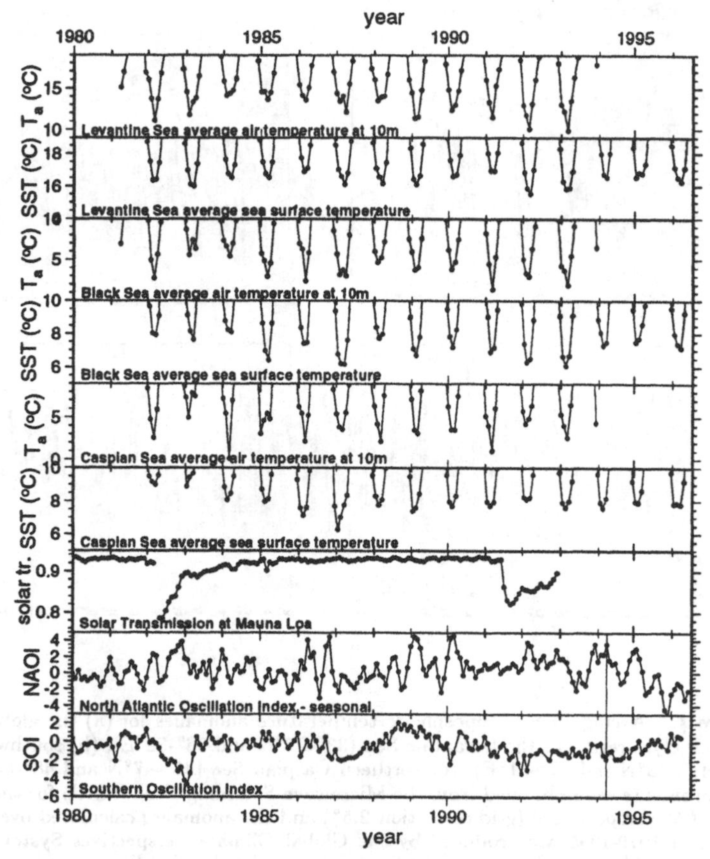


*Figure 3.* Time series of average air (ECMWF/ERA re-analyses, 6 hr forecast temperature at 10*m* height) and sea surface temperatures (ESA ERS1/ATSR derived monthly averages) for the Levantine, Black and Caspian Seas, Solar transmission at Mauna Loa, and the climatic indices NAOI (seasonal averages) and SOI, for the last two decades.

linked with negative values of the SO Index in Figure 3 (*e.g.* 1982-83, 1986-87, the 1990's), often cited as ENSO years (*e.g.* Meyers and O'Brien, 1995). Similarly, some years (1983, 1986-87, 1989-90, 1992-93) are characterized by high NAO indices.

Relatively cooler winters are detected in the years 1982-83, 1985, 1987, 1989, 1991 and 1992-93 in Figure 3. Some of the cold years correspond to well known cases of convection and deep water formation in the regional seas (*e.g.* from recent data *in 1987*: dense water intrusion into the Marmara Sea from the Aegean, Beşiktepe *et al.*, 1993; deep water formation in the Rhodes Gyre region, Gertman et al., 1990; main pycnocline erosion in the Black Sea, unpublished data; *in 1989*: extensive LIW formation in

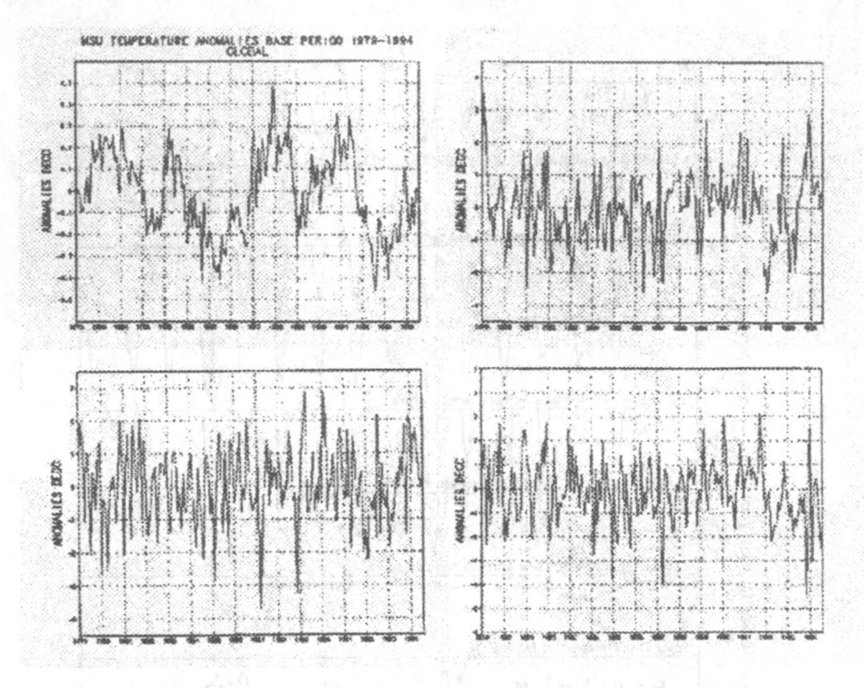

*Figure 4.* Average lower tropospheric temperature anomalies for (a) the globe, (b) the Rhodes Gyre region of the Levantine Sea (33°-37°N and 26°-32°E), (b) southwest Black Sea (41°-43°N and 27°-31°E), (c) northern Caspian Sea (45°-47°N and 48°-53°E). The measurements were obtained from the Microwave Sounding Unit (MSU) for the 0-5.6$km$ layer of the troposphere (grid resolution 2.5°) and the anomalies calculated over the base period of 1979-1995 are produced by the Global Climate Perspectives System (GCPS) at the http://www.ncdc.noaa.gov/onlineprod/prod.html web address.

northern Levantine Sea (Özsoy *et al.*, 1993); *in 1992 and 1993*: Black Sea Cold Intermediate Water (CIW) formation and pycnocline erosion, Ivanov *et al.*, 1997a,b; simultaneous intermediate and deep water formation in the Rhodes Gyre region of the Levantine Sea; Sur *et al.*, 1992, Özsoy *et al.*, 1993). There are also some surprises: the year 1987 is one of the coolest years in all three seas, but there is no corresponding decrease of air temperature in the Caspian Sea. Secondly, while the air temperature displays various anomalous years in the Caspian Sea, the sea surface temperature does not respond to it very effectively, except the year 1987, unlike the pattern observed for the other two seas.

Cooling in the lower troposphere (Figure 4a) occurred globally in 1992-1993 (Spencer and Christy, 1992), following other cooling periods of 1982, 1985-86, 1989 in the last two decades. Significant drops of Temperature drops of 1-2°C occurred in the entire region extending from the Rhodes Gyre of the Levantine Sea to the southern Black Sea and the northern Caspian Sea (Figures 4b-d). An event of global climatic significance (Fiocco *et al.*, 1996) the eruption of Mount Pinatubo volcano in June 1991, resulting in stratospheric warmings (Angell, 1993), decreased solar energy inputs (Dutton and Christy, 1992; Dutton, 1994) and anomalous temperatures (Halpert *et al.*, 1993; Boden *et al.*, 1994) in the entire northern hemisphere in 1992-93. Anomalous cold temperatures appeared in the Middle East in very similar spatial patterns during the winters of 1992 and 1993, covering the Black Sea, eastern Mediterranean, and African regions in both years (Özsoy and Ünlüata, 1997). In Turkey, the winter of 1992 was the coldest in the last 60 years (Türkeş *et al.*, 1995), and in Israel, it was the coldest in the last 46 years (Genin *et al.*, 1996).

## 4. Surface Fluxes

To study the effects of climate variability in the three seas, the loss terms of the surface fluxes are computed from uniform quality, decadal atmospheric re-analysis data obtained from the ECMWF at 1° resolution (mean sea level pressure, $10m$ wind, $2m$ atmospheric and dew point temperatures and cloudiness, produced every 6 $hr$ intervals from $6hr$ global forecasts), and monthly average sea surface temperature based on ERS1/ATSR satellite data for the period 1981-1994. The air-sea fluxes are calculated by a method of iteratively reconstructing the atmospheric variables at $10m$ height from ECMWF supplied fields based on Monin-Obukhov turbulent boundary layer theory, then using bulk formulae to compute the wind-stress, moisture flux, sensible ($q_s$) and latent ($q_l$) heat fluxes (Launiaien and Vihma, 1990; Vihma, 1995) as well as the longwave radiation ($q_b$) fluxes (Bignami *et al.*, 1995). The values are then averaged over the sea domain and a month to produce the values in Figure 5, where the standard deviations are also marked.

The comparison of fluxes in the three different seas shows coincidence between the active periods of Black Sea and Levantine Sea, and a lesser degree of sychronism between them and the Caspian Sea. On the other hand, the Black Sea and Caspian Sea momentum fluxes are larger, more seasonal and more variable than the Levantine Sea, where the larger events come in interannual pulses. The sensible heat flux increases from the Levantine Sea to the Black Sea, reaching a maximum in the Caspian Sea.

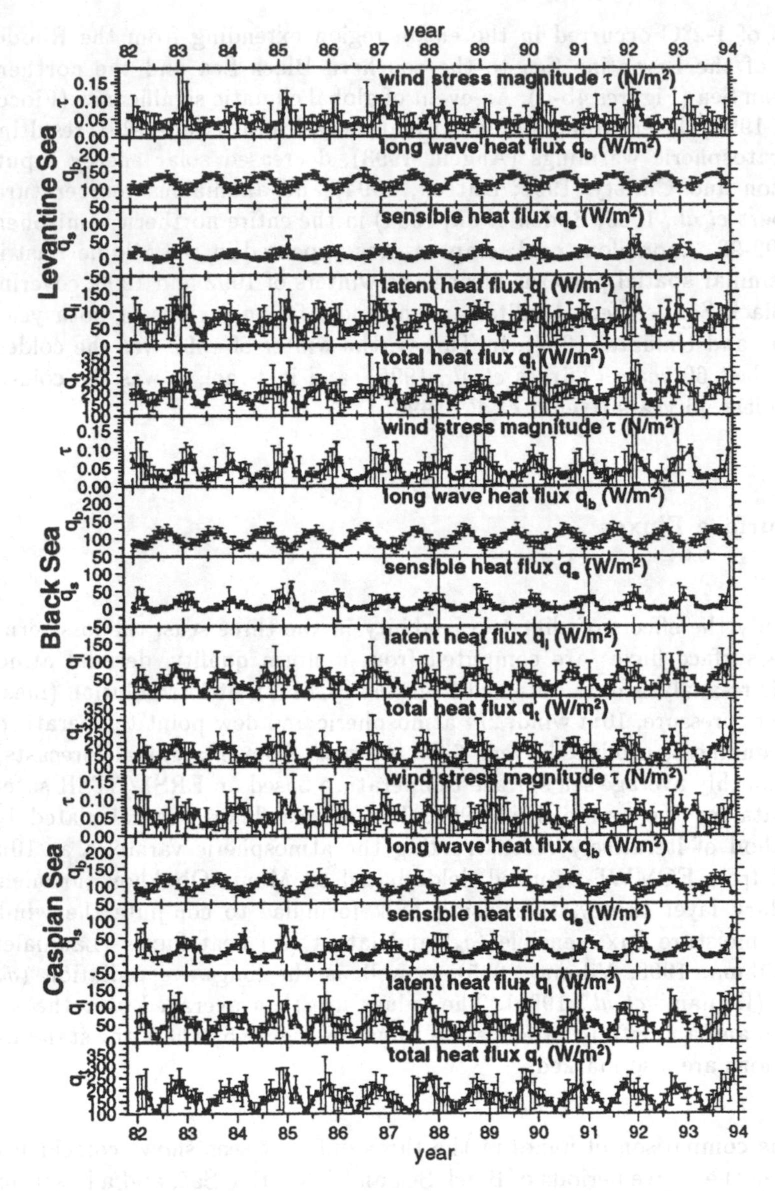


*Figure 5.* Wind stress and heat flux components for the Levantine, Black and Caspian Seas computed from the ECMWF re-analysis, 6 hourly forecast fields. The values are averaged over each basin and over one month periods. The error bars denote one $\sigma$ standard deviation.

## 5. Changes in the Eastern Mediterranean

### 5.1. THERMOHALINE CIRCULATION

The mean residence time varies considerably from ~100 years for the Mediterranean and shorter for the Caspian, to ~2000 for the Black Sea. The

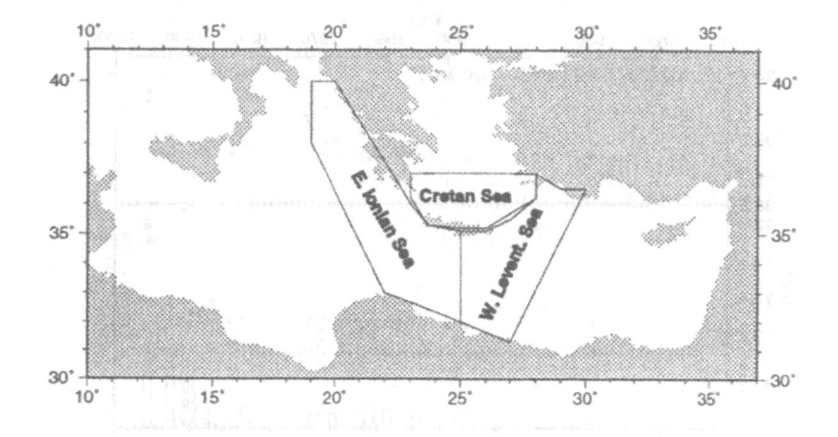


*Figure 6.* Areas chosen for deep water analyses in the western Levantine, eastern Ionian and Cretan Seas

Mediterranean 'conveyor belt' (*i.e.* the 3-D thermohaline circulation, partly connected to the North Atlantic through the Gibraltar Strait), could have global impact, by regulating water masses and overturning in the North Atlantic (Reid *et al.*, 1994; Hecht *et al.*, 1996), with a potential, yet speculative contribution (Johnson, 1997a,b) to the triggering of paleoclimatic transitions (Broecker *et al.*, 1985).

The conceptual schemes of Mediterranean thermohaline circulations (*e.g.* Robinson and Golnaraghi, 1994) have been drastically changed by the discovery of deep water formation (a) through dense water outflow from the Aegean Sea, and (b) simultaneously with LIW in the Rhodes Gyre core. It is now evident that the 'conveyor belt' circulation in the entire Mediterranean is undergoing changes. Increases in the deep water temperature and nutrients (Bethoux *et al.*, 1989; Bethoux, 1989, 1993) appear coupled to the annual deep convection patterns (Gascard, 1991; Madec and Crepon, 1991; Send *et al.*, 1996) in the western Mediterranean.

The real surprise has recently come from the eastern Mediterranean, where deep water was found to form in the center of the permanent Rhodes Gyre, simultaneously with LIW on its periphery (Gertman *et al.*, 1990; Sur *et al.*, 1992; Özsoy *et al.*, 1993) with a recurrence interval of a few years depending on cooling. Furthermore, a climatically induced switching in the closed cell 'conveyor belt' is now evident. The classical scheme of deep water renewal by dense water (fresh, cold) outflow from the Adriatic Sea (Roether and Schlitzer, 1991; Schlitzer *et al.*, 1991; Roether *et al.*, 1994), have been replaced by the dense water (salty, warm) outflow from the Aegean Sea (Roether *et al.*, 1996), starting in the early 1990's. The event has been detected for first time since the beginning of oceanographic observations,

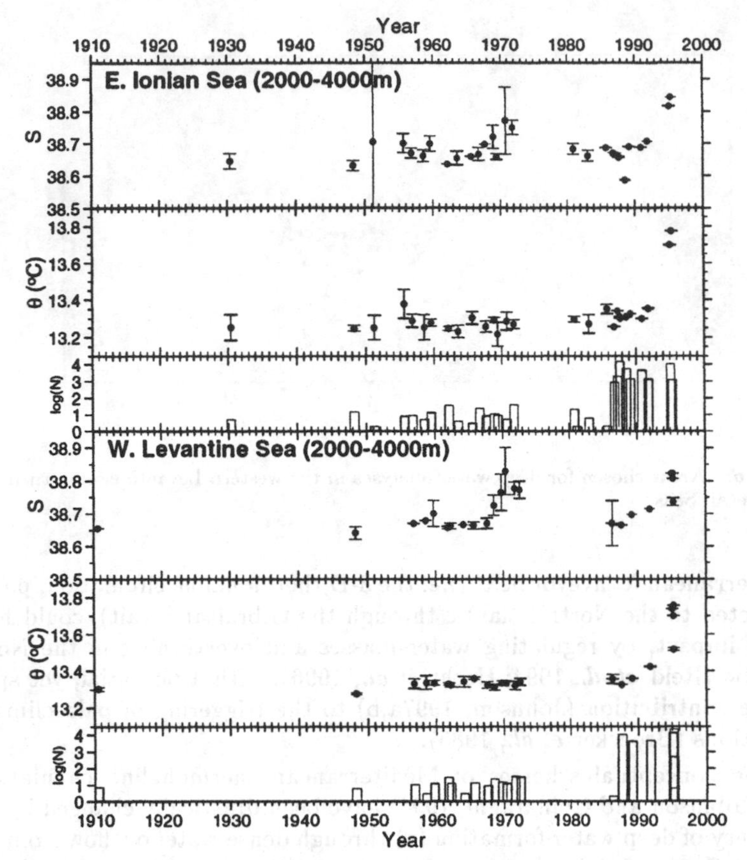


*Figure 7.* **Average salinity, potential temperature, and log number of observations in the depth interval of 2000-4000*m* in the western Levantine, eastern Ionian Sea areas. The error bars denote standard deviation. The averages are obtained from individual data sets contained within the combined MODB / POEM data and grouped into 1 year intervals falling within the specified depth range.**

although intermediate depth ($\sim$ 1,000-1500$m$) Aegean intrusions of lesser magnitude were well known (Roether and Schlitzer, 1991).

The changes in deep water first became evident when an unusually warm, saline water mass below 1000$m$ was detected south of the Island of Crete during summer 1993 (Heike *et al.*, 1994; A. Yılmaz, pers. comm.). Anomalous heat fluxes were detected in the deep sediments of Ionian Sea deep brine lakes in 1993-94, capped by a double diffusive interface in contact with the new deep water (Della Vedova *et al.*, 1995), suggesting a fresh transient event. Although an exact date for the Aegean dense water outflow can not be established, observations suggest a strong pulse in 1992-93, superposed on a background trend starting in the late 1980's.

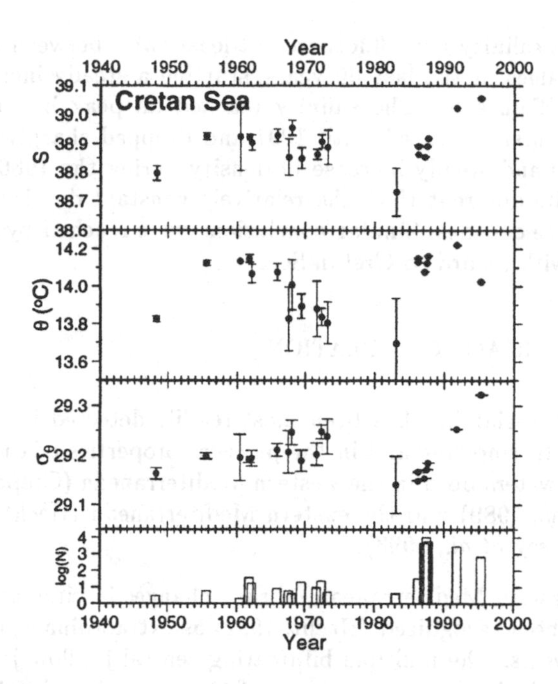


*Figure 8.* Average salinity, potential temperature, $\sigma_\theta$ density and log number of observations in the depth interval of 1000-2000$m$ in the Cretan Sea. The error bars denote standard deviation. The averages are obtained from individual data sets contained within the combined MODB / POEM data and grouped into 1 year intervals falling within the specified depth range.

An analysis of the MODB historical data base (Brasseur *et al.*, 1996) combined with the last 10 years of Physical Oceanography of the Eastern Mediterranean (POEM) data was made by Özsoy and Latif (1996) to address the effect of Aegean outflow in the observed changes in deep water. For this purpose, the three regions adjacent to the Aegean - Eastern Mediterranean junction (Figure 6) were selected.

The average properties in the depth range of 2000-4000$m$, with the 95 % confidence limits are shown in Figure 7 the western Levantine and eastern Ionian boxes (Figure 6), together with the number of observations in each annual cluster. The vertical resolution, accuracy and coverage of the measurements increased in the second half of 1980's (POEM program). If we disregard the large salinity changes associated with measurement drift in the 1960's and 1970's, and trust the relatively more accurate temperature measurements, an increase in deep water temperature (and possibly salinity) is evident in both regions in the late 1980's, topped by a rapid increase during 1992-95.

In the Cretan Sea deep water (maximum depth ~2000$m$), both temperature and salinity have fluctuated at least twice between minimum and maximum values in the last 60 years, starting a steady increase after the mid 1980's (Figure 8). The salinity reached its peak in 1995, while the temperature first increased until 1991 and dropped sharply in 1991-95 to yield a rapid and steady increase in density during the 1980's and 1990's, which is quite different than the relatively constant density values of the previous three decades. Similar conclusions were reached by Theocharis *et al.*, (1996) with regard to Cretan Sea.

## 5.2. NEAR-SURFACE CIRCULATION

Upper ocean variability has been most readily detected in the circulation of the main thermocline and in the physical properties of the intermediate and surface waters both in the western Mediterranean (Crepon *et al.*, 1989; Barnier *et al.*, 1989) and the eastern Mediterranean (Hecht, 1992; Sur, *et al.*, 1992; Özsoy *et al.*, 1993).

In the eastern Mediterranean, abrupt changes in circulation and water masses has been recognized (Hecht, 1992) as extraordinary, multi-annually recurrent events. The multiple bifurcating central jet flow joining the Asia Minor jet cyclonically east and west of Cyprus and anticyclonically along the eastern coast of Levant in 1985-86 has abruptly changed to a better organized, cyclonic flow around Cyprus in the 1988-1990 period, coincident with the Rhodes Gyre deep convection in 1987, the massive penetration of low salinity modified Atlantic water into the northern Levantine, and the disappearence of a coherent anticyclone in the Gulf of Antalya the same year and its re-appearence in 1990 (Özsoy *et al.*, 1991, 1993). The surface salinity steadily increased from 39.1-39.3 in 1985-86 up to 39.5 in 1989 and 39.7 in 1990. During the same period changes occurred in in the pattern of formation and maintenance of LIW. The LIW trapped in the Antalya anticyclonic eddy in 1985-86 disappeared together with the eddy itself in 1987. LIW with increasingly higher salinities was observed in the winter periods of the following years, reaching from ~39.0 in 1989 to ~39.2 in 1989-90 (Özsoy *et al.*, 1993). An increasing trend was evident in the Shikmona Gyre core salinity, temperature and density during the 1988-1994 period, with abrupt changes in the winters of 1990 and 1992 (Brenner, 1996). Salinity increases in the entire region, together with circulation changes leading to the entry of salt water into the Aegean, and blocking of LIW to spread westward by multi-centered anticyclonic region in the Southern Levantine, have been shown by Malanotte-Rizzoli *et al.*, (1998) for the POEM survey of October 1991. The changes imply a salt redistribution pattern favoring the creation of Aegean dense water outflow.

## 6. Hydrological Cycles and Sea Level

The sea-level, besides being a good indicator of climatic fluctuations, is a sensitive measure of hydrometeorological driving factors in enclosed and semi-enclosed seas. In the Black Sea, sea-level responds non-isostatically to atmospheric pressure and the total water budget, which are both highly variable on interannual and seasonal time scales (Sur *et al.*, 1994), strictly controlled by the variable inputs of large rivers and the dynamical constraints imposed the Turkish Straits (Özsoy *et al.*, 1998a). Simple models used to understand the time dependent response has had limited success to produce and explain sea-level variations of large amplitude in this non-tidal Sea, unless special effects of wind-setup on the hydraulics of the Bosphorus are included (Ducet and Le Traon, 1998).

In the Caspian Sea, sea-level changes depend on the regional hydrometeorological regime linked to global climate (Radionov, 1994). The sea level changes with climatic and anthropogenic components are of great economic and environmental importance for the surrounding countries. Interestingly, the sea-level change influences the residence time of the deep waters, and therefore has a direct bearing on the health of the Caspian Sea. Abrupt changes in sea-level have occurred twice since 1830's, as well as earlier in history, as the fate of Khazars occupying in its shores in the 10th century stand witness. The sea level has first dropped from a -25 m in 1930's down to -29 m by 1978, and has risen to the present -27 m after 1977. Hydrogen sulfide has been detected in deep waters prior to 1930 when sea-level was high, as a result of insufficient ventilation, limited by the decreased volume of dense water formed on the ice-covered, shallow northern Caspian shelf (average depth $\sim 2m$) under the influence of the large, variable inputs of the Volga river (Kosarev and Yablonskaya, 1994).

## 7. Marine Ecosystems

The marine ecosystems of interior seas are especially vulnerable, as significant changes in nutrient supplies are taking place in their confined waters. In the Black Sea, and the Caspian Sea, supplied by large rivers, eutrophication is leading to losses of habitats, species diversity, and consequent economical value (Zaitsev and Mamayev, 1997, Kosarev and Yablonskaya, 1994). The eastern Mediterranean system may be undergoing change related to the Nile river (*e.g.* Turley, 1997). Ecosystem disasters in even smaller, confined waters of the the Aral Sea, Kara Boğaz Göl, and Azov Sea, are better recognized. The introduction of extraneous species represent anthropogenic effects, either by filling a niche as in the case of the Black Sea, or through Lessepsian migration from the Red Sea, in the case of the eastern Mediteranean (Por, 1978; Galil, 1993; Gücü *et al.*, 1994;

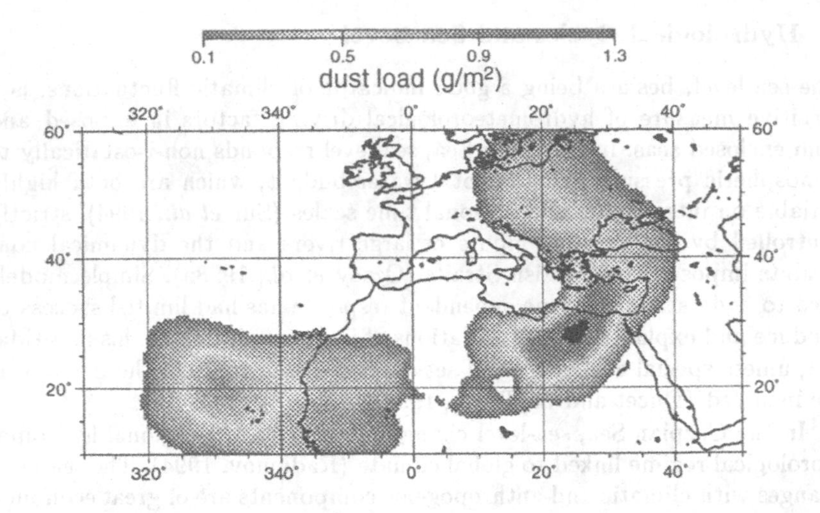


*Figure 9.* **Model derived tropospheric dust load ($g/m^2$) at 18:00 on April 6, 1994.**

Kıdeyş and Gücü, 1995). On the other hand, the natural variabilities are frequently of the same magnitude as anthropogenic effects, which makes diagnosis difficult.

The basic ingredients and biological machinery of the marine ecosystems are often not sufficiently resolved (*e.g.* the sub-oxic zone in the Black Sea, Oğuz, *et al.*, *this volume*; or the roles of P-limitation and cyanobacteria in the eastern Mediterranean, Zohary and Robarts, 1996; Li *et al.*, 1993). In addition to these biochemical factors, dynamical features are of first order importance for the ecosystem. In the Black Sea, upwelling, coastal flows and river supply play important roles (Sur *et al.*, 1994, 1996; Özsoy and Ünlüata, 1997). In the eastern Mediterranean, where the surface nutrients are low, nutrient upwelling at fronts (Özsoy *et al.*, 1993), and nutrient supply in the river mouths and coasts appear significant. Ecosystem changes associated with an uplift of the nutricline following the Aegean dense water outflow are suspected in the eastern Mediterranean (Roether *et al.*, 1996), and are perhaps associated with changes in deep zooplankton (Weikert, 1996). The strong cooling and mixing in 1992-93 produced a massive algal bloom near the Rhodes Gyre (Yılmaz *et al.*, 1996; Ediger and Yılmaz, 1996), and in Eilat, where it lead to the destruction of coral reefs (Genin *et al.*, 1996). These events coincided with the enhanced pycnocline erosion (Ivanov *et al.*, 1997a,b), followed by a massive bloom (Vladimirov *et al.*, 1997) in the following summer of 1992 in the Black Sea. The bloom in the Black Sea apparently produced atmospheric non-sea-salt sulfate aerosols of marine biological origin swept southwards and detected in the Erdemli measurements (Özsoy *et al.*, 1998c).

## 8. Aerosol Dust

Aerosol desert dust is important in the climate system, primarily because of its effects on heat budgets (Charlson and Heintzenberg, 1995), especially in semi-enclosed seas (Gilman and Garrett, 1994), on the heterogeneous chemistry of the tropospheric and greenhouse gases (Dentener *et al.*, 1996), as well as on the biogeochemical cycles in the marine environment (Duce *et al.*, 1991).

Dust transport is an important climatic process which is particularly active in the Mediterranean area (Guerzoni and Chester, 1996), with an increasing incidence in recent years associated with drought and desertification. The role of large scale controls in establishing conditions for transport, such as in the spectacular April 1994 case, are outlined in Section 2. The dust load for this unique case of simultaneous transport (Andrae, 1996; Moulin *et al.*, 1997) in a shallow layer across the Atlantic and within a rapidly developing cyclone towards the eastern Mediterranean and Black Sea regions, resulting in anomalously high concentrations in Barbados and Erdemli (Li *et al.*, 1996, Özsoy *et al.*, 1998b; Kubilay *et al.*, 1998) is shown in Figure 9.

Sahara dust is known to be the main source contributing to the formation of the fertile 'red soils' of the eastern Mediterranean. It is hypothesized, based on satellite observations and concurrent dust measurements (unpublished material, IMS-METU) that the transport and deposition on sea surface of eroded dust from the Sahara and Arabian deserts has an impact on short term, episodic phytoplankton blooms in the eastern Mediterranean. There is often a striking coincidence of 'high reflectivity' from the sea surface typically associated with *E. huxleyi* blooms, and the incursions of dust, supported by the coincidence of the fertilization events with high fluxes of carbonates in sediment trap obsetvations. However, the working hypothesis has to be better defined and tested with detailed observations.

## 9. ACKNOWLEDGEMENTS

This study benefited from a number of activities, most notably the POEM and POEM-BC, international cooperative programs, the NATO SfS Projects TU-FISHERIES, TU-BLACK SEA and TU-REMOSENS projects, as well as other work carried out by the IMS-METU. Special thanks are for the Turkish General Directorate of Meteorology for providing access to the ECMWF data on a collaborative basis.

## References

1. Alpert, P. and T. Reisin (1986). An Early Winter Polar Air Mass Penetration to the Eastern Mediterranean, *Mon. Wea. Rev.*, **114**, 1411-1418.
2. Andreae, M. O. (1996). Raising Dust in the Greenhouse, *Nature*, **380**, 389-390.
3. Angell, J. K. (1993). Comparison of stratospheric warming following Agung, El Chichon and Pinatubo volcanic eruptions, *Geophys. Res. Lett.*, **20**, 715-718.
4. Barnier, B., M. Crepon, and C. Le Provost (1989). Horizontal Ocean Circulation Forced by Deep Water Formation. Part II: A Quasi-Geostrophic Simulation, *J. Phys. Oceanogr.*, **19**, 1781- 1793.
5. Beşiktepe, Ş., Özsoy, E., and Ü. Ünlüata (1993). Filling of the Marmara Sea by the Dardanelles Lower Layer Inflow, Deep- Sea Res., 40(9), 1815-1838.
6. Bethoux, J. P. (1989). Oxygen Consumption, New Production, Vertical Advection and Environmental Evolution in the Mediterranean Sea, *Deep-Sea Res.*, **36**, 769-781.
7. Bethoux, J. P. (1993). The Mediterranean Sea, a Biogechemistry Model Marked by Climate and Environment, in: N. F. R. Della Croce (editor), *Symposium Mediterranean Seas 2000*, University of Genoa, Santa Margherita Ligure, pp. 351-372.
8. Bethoux, J. P., Gentili, B., Raunet, J. and D. Talliez (1989). Warming Trend in the Western Mediterranean Deep Water, *Nature*, 347, 660-662.
9. Bignami, F., S. Marullo, R. Santoleri and M. E. Schiano (1995). Longwave Radiation Budget in the Mediterranean Sea, *J. Geophys Res.*, **100**(C2), 2501-2514.
10. Brenner, S. (1996). The Relationship Between Atmospheric Forcing and the Shikmona Gyre, International POEM-BC/MTP Symposium, Molitg les Bains, France, 1-2 July 1996, pp. 87-89.
11. Charlson, R., J. and J. Heintzenberg (1995). *Aerosol Forcing of Climate*, Wiley, 416 pp.
12. Crepon, M., M. Boukthir, B. Barnier, and F. Aikman (1989). Horizontal Ocean Circulatiom Forced by Deep Water Formation. Part I: An Analytical Study, *J. Phys. Oceanogr.*, **19**, 1781-1793.
13. Cullen, H. (1998). North Atlantic Influence on Middle Eastern Climate and Water Supply, *Climatic Change* (submitted).
14. Della Vedova, B., Foucher, J.-P., Pellis, G., Harmegnies, F., and the MEDRIFF Consortium (1995). Heat Flow Measurements in the Mediterranean Ridge Indicate Transient Processes of Heat Transfer between the Sediments and the Water Column (MAST II - MEDRIFF Project), *Rapp. Comm. Int Mer Medit.*, **34**, 100.
15. Dentener, F. J., G. R. Carmichael, Y. Zhang, J. Lelieveld and P. J. Crutzen (1996). Role of mineral aerosol as a reactive surface in the global troposphere, *J. Geophys. Res.*, **101**, 22,869- 22,889, 1996.
16. Duce *et al.* (1991) The atmospheric input of trace species to the world ocean, *Global Biogeochem. Cycles*, **5**, 193-256, 1991.
17. Ducet, N. and P. Y. Le Traon (1998). Response of the Black Sea Mean Level to Atmospheric Pressure and Wind Forcing, (submitted).
18. Dutton, E. G. and J. R. Christy (1992). Solar Radiative Forcing at Selected Locations and Evidence for Global Lower Tropospheric Cooling Following the Eruptions of El Chichon and Pinatubo, *Geophys. Res. Lett.*, **19**, 2323-2316.
19. Dutton, E. G. (1994). Atmospheric Solar Transmission at Mauna Loa, In: T. A. Boden, D. P. Kaiser, R. J. Sepanski, and F. W. Stoss (editors), Trends 93: A Compendium of Data on Global Change, ORNL/CDIAC-65, Carbon Dioxide Information Analysis Center, Oak Ridge, Tenn., USA.
20. Ediger, D. and A. Yılmaz (1996). Characteristics of Deep Chlorophyll Maximum in the Northeastern Mediterranean with respect to Environmental Conditions, *J. Mar. Sys.*, **9**, 291-303.
21. Fiocco, G., D. Fua and G. Visconti (1996). *The Mount Pinatubo Eruption, Effects on the Atmosphere and Climate*, NATO ASI Series, Kluwer Acad. Publ., Dordrecht, 310 p.

22. Galil, B. S. (1993). Lessepsian Migration. New Findings on the Foremost Anthropogenic change in the Levant Basin Fauna, in: N. F. R. Della Croce (editor), *Symposium Mediterranean Seas 2000*, University of Genoa, Santa Margherita Ligure, pp. 307-323.
23. Garrett, C., Outerbridge, R. and K. Thompson (1992). Interannual Variability in Mediterranean Heat and Buoyancy Fluxes, *J. Climatology.*
24. Gascard, J. C. (1991). Open Ocean Convection and Deep Water Formation in the Mediterranean, Labrador, Greenland and Weddell Seas, in: P. C. Chu and J. C. Gascard (editors) *Deep Convection and Deep Water Formation in the Oceans*, Elsevier, Amsterdam, pp. 241-265.
25. Genin, A., Lazar, B. and S. Brenner (1995). Vertical Mixing and Coral Death in the Red Sea following the Eruption of Mount Pinatubo, *Nature*, **377**, 507-510.
26. Gertman, I. F., I. M. Ovchinnikov and Y. I. Popov (1990). Deep convection in the Levantine Sea, *Rapp. Comm. Mer Medit.*, **32**, 172.
27. Gilman, C. and Garrett C., Heat flux parameterizations for the Mediterranean Sea: The role of atmospheric aerosols and constraints from the water budget, J. Geophys. Res., **99**, 5119-5134, 1994.
28. Glantz, M. H. and I. S. Zonn (1997). *The Scientific, Environmental and Political Issues in the Circum-Caspian Region*, NATO ASI Series, 2, Environment - vol. 29, Kluwer Academic Publishers, Dordrecht.
29. Gücü, A. C., F. Bingel, D. Avşar and N. Uysal (1994) Distribution and Occurrence of Red Sea Fish at the Turkish Mediterranean Coast-Northern Cilician Basin. *Acta Adriat.*, **34**(1/2): 103- 113.
30. Guerzoni, S. and R. Chester (1996). *The Impact of Desert Dust Across the Mediterranean*, Kluwer Academic Publishers, Dordrecht, 389pp.
31. Halpert, M. S., Ropelewski, C. F., Karl, T. R., Angell, J. K., Stowe, L. L., Heim, R. R. Jr., Miller, A. J., and D. R. Rodenhuis (1993). 1992 Brings Return to Moderate Global Temperatures, *EOS*, **74**(28), September 21, 1993, 436-439.
32. Hecht, A. (1992). Abrupt Changes in the Characteristics of Atlantic and Levantine Intermediate Waters in the Southeastern Levantine Basin, *Oceanol. Acta*, 15, 25-42.
33. Hecht, M. W., W. R. Holland, V. Artale and N. Pinardi (1996). North Atlantic Model Sensitivity to Mediterranean Waters (unpublished manuscript).
34. Heike, W., Halbach, P., Türkay, M. and H. Weikert (1994). Meteorberichte 94-3, Mittelmeer 1993, Cruise No. 25, 12 May - 20 August 1993, (Cruise Report), (Section 5.2.7).
35. Hurrell, J. W. (1995). Decadal Trends in the North Atlantic Oscillation: Regional Temperatures and Precipitation. *Science*, **269**, 676-679.
36. Hurrell, J. W. (1996). Influence of Variations in Extratropical Wintertime Teleconnections on Northern Hemisphere Temperatures, *Geophys. Res. Lett.*, **23**, 665-668.
37. IOC (1993). Priorities for Contaminants in Specific Marine Areas, International Oceanographic Commission, Electronic Library, Information Document INF - 923, Paris, 1 March 1993.
38. Ivanov, L., Ş. Beşiktepe and E. Özsoy (1997a). The Black Sea Cold Intermediate Layer, in: E. Özsoy and A. Mikaelyan (editors), *Sensitivity to Change: Black Sea, Baltic Sea and North Sea*, NATO ASI Series (Partnership Sub-series, Environment, 27), Kluwer Academic Publishers, Dordrecht, pp. 253-264.
39. Ivanov, L., , Ş. Beşiktepe and E. Özsoy (1997b). Physial Oceanography Variability in the Black Sea Pycnocline, in: E. Özsoy and A. Mikaelyan (editors), *Sensitivity to Change: Black Sea, Baltic Sea and North Sea*, NATO ASI Series (Partnership Sub-series, Environment, 27), Kluwer Academic Publishers, Dordrecht, pp. 265-274.
40. Jeftic, L., J. D. Milliman and G. Sestini (1992). *Climatic Change and the Mediterranean*, UNEP, Edward Arnold, London, 673pp.
41. Jeftic, L., S. Keckes and J. C. Pernetta (1996). *Climatic Change and the Mediterranean*, Volume 2, UNEP, Arnold, London, 564pp.

42. Johnson, R. G. (1997a). Ice Age Initiation by an Ocean-Atmosphere Circulation Change in the Labrador Sea, *Earth Planet. Sci. Lett.*, **148**, 367-379.
43. Johnson, R. G. (1997b). Climate Control Requires a Dam at the Strait of Gibraltar, *EOS*, July 8, 1997.
44. Kıdeyş, A. E. and A. C. Gücü (1995). Rhopilema nomadica: a Poisonous Indo-Pacific Scyphomedusan New to the Mediterranean Coast of Turkey, *Israel J. of Zoology*, **41**(4): 615- 617.
45. Kosarev, A. N. and E. A. Yablonskaya (1994). *The Caspian Sea*, Backhuys Publishers, Haague, 259 pp.
46. Kubilay, N., E. Özsoy, S. Nickovic, I. Tegen and C. Saydam (1998). A Hemispheric Dust Storm Affecting the Atlantic and Mediterranean (April 1994): Analyses, Modelling, Ground-Based Measurements and Satellite Observations, (submitted).
47. Launiaien, J. and T. Vihma (1990). Derivation of Turbulent Surface Fluxes - An Iterative Flux-Profile Method Allowing Arbitrary Observing Heights, *Environmental Software*, **5**(3), 113-124.
48. Li, W. K. W., T. Zohary, Y. Z. Yacobi and A. M. Wood (1993). Ultraphytoplankton in the Eastern Mediterranean Sea: Towards Deriving Phytoplankton Biomass from Flow Cytometric Measurements of Abundance, Fluorescence and Light Scatter, it Marine Ecology Progress Series, **102**, 79-87.
49. Li, X., H. Maring, D. Savoie, K. Voss and J. M. Prospero (1996). Dominance of Mineral Dust in Aerosol Light Scattering in the North Atlantic Trade Winds, *Nature*, **380**, 416-419.
50. Malanotte-Rizzoli, P. and A. R. Robinson (editors) (1994). *Ocean Processes in Climate Dynamics: Global and Mediterranean Examples*, NATO ASI Series C: Mathematical and Phycical Sciences, vol. 419, Kluwer Acad. Publ., Dordrecht, 437 pp.
51. Malanotte-Rizzoli, P., B. B. Manca, M. Ribera d'Alcala, A. Theocharis, S. Brenner, G. Budillon and E. Özsoy (1998). The Eastern Mediterranean in the 80s and in the 90s: The Big Transition in the Intermediate and Deep Circulations (submitted).
52. Madec, G. and M. Crepon (1991). Thermohaline-Driven Deep Water Formation in the Northwestern Mediterranean Sea, in: P. C. Chu and J. C. Gascard (editors) *Deep Convection and Deep Water Formation in the Oceans*, Elsevier, Amsterdam, pp. 241-265.
53. Marshall, J. and Y. Kushnir (1997). A 'white paper' on Atlantic Climate Variability, an unpublished working document.
54. Meyers, S. D. and J. J. O'Brien (1995). Pacific Ocean Influences Atmospheric Carbon Dioxide, *EOS*, **76**(52), 533-537.
55. Moreno, J. M. and Oechel, W., Eds. (1995). *Global Change and Mediterranean Ecosystems*, Springer Verlag, Ecological Studies, **117**, New York, N.Y.
56. Moulin, C., C. E. Lambert, F. Dulac and U. Dayan (1997). Control of atmospheric export of dust from North Africa by the North Atlantic oscillation. *Nature*, **387**, 691-694.
57. Özsoy, E. (1981). *On the Atmospheric Factors Affecting the Levantine Sea*, European Center for Medium Range Weather Forecasts, Reading, U.K., Technical Report No. 25, 30 pp.
58. Özsoy, E., A. Hecht, Ü. Ünlüata, S. Brenner, T. Oğuz, J. Bishop, M. A. Latif and Z. Rozentraub (1991). A Review of the Levantine Basin Circulation and its Variability during 1985-1988, *Dyn. Atmos. Oceans*, **15**, 421-456.
59. Özsoy, E., Hecht, A., Ünlüata, Ü., Brenner, S., Sur, H. İ., Bishop, J., Latif, M. A., Rozentraub, Z. and T. Oğuz (1993). A Synthesis of the Levantine Basin Circulation and Hydrography, 1985-1990, *Deep-Sea Res.*, **40**, 1075-1119.
60. Özsoy, E. and M. A. Latif (1996). Climate Variability in the Eastern Mediterranean and the Great Aegean Outflow Anomaly, International POEM-BC/MTP Symposium, Molitg les Bains, France, 1-2 July 1996, pp. 69-86.
61. Özsoy, E. and Ü. Ünlüata (1997). Oceanography of the Black Sea: A Review of

Some Recent Results, *Earth Sci. Rev.*, **42**(4), 231-272.

62. Özsoy, E., Latif, M. A., Beşiktepe, Ş., Çetin, N., Gregg, N. Belokopytov, V., Goryachkin, Y. and V. Diaconu (1998a). The Bosphorus Strait: Exchange Fluxes, Currents and Sea-Level Changes, L. Ivanov and T. Oğuz, editors, Proceedings of NATO TU-Black Sea Project - Ecosystem Modeling as a Management Tool for the Black Sea : Symposium on Scientific Results, Crimea, (Ukraine), June 15-19, 1997.
63. Özsoy, E., N. Kubilay, S. Nickovic and C. Saydam (1998b). A Hemispheric Dust Storm in April 1994: Observations, Modelling and Analyses, XXIIIrd NATO/CCMS International Technical Meeting on Air Pollution, Varna, 28 Sep. - 2 Oct. 1998. (submitted for publication).
64. Özsoy, T., C. Saydam, N. Kubilay, O. B. Nalçacı and İ. Salihoğlu (1998c). Aerosol Nitrate and Non-Sea-Salt Sulfate Over the Eastern Mediterranean, XXIIIrd NATO/CCMS International Technical Meeting on Air Pollution, Varna, 28 Sep. - 2 Oct. 1998. (submitted for publication).
65. Polonsky, A., E. Voskresenskaya and V. Belokopytov, Variability of the Northwestern Black Sea Hydrography and River Discharges as Part of Global Ocean-Atmosphere Fluctuations, in: E. Özsoy and A. Mikaelyan (editors), *Sensitivity to Change: Black Sea, Baltic Sea and North Sea*, NATO ASI Series (Partnership Sub-series 2, Environment, 27), Kluwer Academic Publishers, Dordrecht, 536 pp., 1997.
66. Por, F. D. (1978): Lessepsian Migration - The influx of Red Sea Biota into the Mediterranean by way of the Suez Canal, Springer Verlag, 215 pp.
67. Reid, J. L. (1994). On the Total Geostrophic Circulation of the North Atlantic Ocean: Flow Patterns, Tracers and Transports, *Prog. Oceanogr.*, **33**, 1-92.
68. Reiter, E. R. *Handbook for Forecasters in the Mediterranean; Weather Phenomena of the Mediterranean Basin*; Part 1: General Description of the Meteorological Processes, Environmental Prediction Research Facility, Naval Postgraduate School, Monterey, California, Technical Paper No. 5-75, 344pp, 1979.
69. Robinson, A. R., Garrett, C. J., Malanotte-Rizzoli, P., Manabe, S., Philander, S. G., Pinardi, N., Roether, W., Schott, F. A., and J. Shukla (1993). Mediterranean and Global Ocean Climate Dynamics, *EOS*, **74**(44), November 2, 1993, 506-507.
70. Robinson, A. R. and M. Golnaraghi (1994). The Physical and Dynamical Oceanography of the Mediterranean Sea, in: Malanotte-Rizzoli P. and A. R. Robinson (editors), *Ocean Processes in Climate Dynamics: Global and Mediterranean Examples*, NATO ASI Series C: Mathematical and Physical Sciences, vol. 419, Kluwer Academic Publishers, Dordrecht, pp. 255-306
71. Rodionov, S. N., *Global and Regional Climate Interaction: The Caspian Sea Experience*, Water Science and Technology Library, v. 11, Kluwer Academic Publishers, Dordrecht, 256 pp., 1994.
72. Roether, W. and R. Schlitzer (1991). Eastern Mediterranean Deep Water Renewal on the Basis of Chlorofluoromethane and Tritium Data, *Dyn. Atmos. Oceans*, **15**, 333-354, 1991.
73. Roether, W. V. M. Roussenov and R. Well (1994). A Tracer Study of the Thermohaline Circulation of the Eastern Mediterranean, in: P. Mlanaotte-Rizzoli and A. R. Robinson (editors), *Ocean Processes in Climate Dynamics: Global and Mediterranean Examples*, NATO ASI Series C: Mathematical and Physical Sciences, vol. 419, Kluwer Academic Publishers, Dordrecht, pp. 371-394
74. Roether, W., Manca, B., Klein, B., Bregant, D., Georgopoulos D., Beitzel V., Kovacevic, V., and A. Luchetta (1996). Recent Changes in Eastern Mediterranean Deep Waters, *Science*, **271**, 333- 335.
75. Schlitzer, R., Roether, W., Oster, H., Junghans, H.-G., Hausmann, M., Johannsen H. and A. Michelato (1991). Chlorofluoromethane and Oxygen in the Eastern Mediterranean, *Deep-Sea Res.*, **38**, 1351-1551.
76. Send, U., J. Font and C. Mertens (1996). Recent Observation Indicates Convection's Role in Deep Water Circulation, *EOS*, **77**(7), February 13, 1996, 61-65.

77. Spencer, R. W. and J. R. Christy (1992). Precision and Radiosonde Validation of Satellite Gridpoint Temperature Anomalies. Part II: A Tropospheric Retrieval and Trends During 1979-1990, *J. Climate*, **5**, 858-866.
78. Sur, H. I., Özsoy, E., and Ü. Ünlüata (1992). Simultaneous Deep and Intermediate Depth Convection in the Northern Levantine Sea, Winter 1992, *Oceanol. Acta*, **16**, 33-43.
79. Sur, H. İ., Özsoy, E. and Ü. Ünlüata, (1994). Boundary Current Instabilities, Upwelling, Shelf Mixing and Eutrophication Processes In The Black Sea, *Prog. Oceanog.*, **33**, 249-302.
Sur, H. İ., Y. P. Ilyin, E. Özsoy and Ü. Ünlüata, (1996). The Impacts of Continental Shelf / Deep Water Interactions in the Black Sea, *J. Mar. Sys.*, **7**, 293-320.
80. Theocharis, A., H. Kontoyannis, S. Kioroglou and V. Papadopoulos(1996), International POEM-BC/MTP Symposium, Molitg les Bains, France, 1-2 July 1996, pp. 58-60.
81. Türkeş, M., Sümer, U., and G. Kılıç, 1995. Variations and Trends in Annual Mean Air Temperatures in Turkey With Respect To Climatic Variability, International Journal of Climatology, 15, 557-569.
82. Turley, C. M. (1997). The Changing Mediterranean Sea - A Sensitive Ecosystem ?, Abstracts Volume, International Conference on the Progress in Oceanography of the Mediterranean Sea, EC-MAST Programme, Rome, November 17-19, 1997.
83. Vihma, T. (1995). Atmosphere-Surface Interactions over Polar Oceans and Heteregenous Surfaces, *Finnish Marine Res.* **264**, 3-41.
84. Vladimirov, V. L., V. I. Mankovsky, M. V. Solov'ev and A. V. Mishonov (1997). Seasonal and Long-term Variability of the Black Sea Optical Parameters, in: E. Özsoy and A. Mikaelyan (editors), *Sensitivity to Change: Black Sea, Baltic Sea and North Sea*, NATO ASI Series (Partnership Sub-series 2, Environment, 27), Kluwer Academic Publishers, Dordrecht, 536 pp., 1997.
85. Wallace, J. M. and M. L. Blackmon, Observations of Low-frequency Atmospheric Variability, in: B. J. Hoskins, and R. P. Pearce, *Large-Scale Dynamical Processes in the Atmosphere*, Academic Press, 397 pp., 1983.
86. Ward, N. (1995). Local and Remote Climate Variability Associated with Mediterranean Sea-Surface Temperature Anomalies, European Research Conference on Mediterranean Forecasting, La Londe les Maures, France, 21-26 October 1995.
87. Weikert, H. (1996). Changes in Levantine Deep Sea Zooplankton, International POEM-BC/MTP Symposium, Molitg les Bains, France, 1-2 July 1996, pp. 99-101.
88. Yılmaz A., D. Ediger, Ç. Polat, S. Tuğrul, and İ. Salihoğlu (1996). The Effect of Cold and Warm Core Eddies on the Distribution and Stoichiometry of Dissolved Nutrients and Enhancement of Primary Production and Changes in Elemental Composition of Phytoplankton in the Eastern Mediterranean, International POEM-BC/MTP Symposium, Molitg les Bains, France, 1-2 July 1996, pp. 122-127.
89. Zaitsev, Yu. and V. Mamaev (1997). Marine Biological Diversity in the Black Sea, A Study of Change and Decline, Black Sea Environmental Series, Black Sea Environmental Series, Vol. 3, GEF Black Sea Environmental Programme, United Nations Publications, New York, 208 pp.
90. Zohary, T. and R. Robarts (1996). Experimental Study of Microbial P-limitation in the Eastern Mediterranean, International POEM-BC/MTP Symposium, Molitg les Bains, France, 1-2 July 1996, p. 128.

# SOME ASPECTS OF THE WATER EXCHANGE THROUGH THE DARDANELLES STRAIT

V.N. EREMEEV[1], N.M. STASHCHUK[1], V.I.VLASENKO[1],
V.A. IVANOV[1], O. USLU[2]
[1]*Marine Hydrophysical Institute, NASU*
*2, Kapitanskay St., Sevastopol, 335000, Ukraine.*
[2] *Institute of Marine Sciences and Technology*
*1884/8, Sokak N010, 35340 Inciralti-Izmir, Turkey.*

## 1. Introduction

This paper focuses on describing the hydrographic characteristics of the Black Sea brackish waters flowing through the Dardanelles to the highly saline Athos basin of the Aegean Sea. This region of the Aegean Sea is located between the areas with markedly different physical and geographical conditions, where an intense interaction and transformation of several water masses within a relatively small area takes place. The Aegean Sea connects to the Black Sea by the Dardanelles/ Marmara Sea/ Bosphorus system at the east and joins the Levantine Sea through several passages and straits located between the islands Crete and Rhodes (Fig.1). The Aegean Sea has very irregular coastlines and bottom topography with many islands. It is appears to have the most complicated hydrometeorological regime among the Mediterranean Sea basins. It is divided into three parts isolated from one to another by islands and sills: Athos Basin, Chios Basin and Crete Basin. The water exchange within the surface layer of these parts of the sea is restricted because of the narrow and rather shallow straits between the islands. Important The knowledge on the water formation and spreading characteristics are of crucial importance for the Aegean Sea ecology.

The hydrological regime of the Black Sea is quite different than that of the Mediterranean. Its water balance is characterized by excess of river discharge and precipitation over the evaporation, while in the case of the Mediterranean, the annual rate of evaporation exceeds the total precipitation and river runoffs. Therefore, throughout the year, the Black Sea level remains above the Mediterranean one.

Brackish Black Sea waters (BSW) outflowing from the Dardanelles, as a rule, form a well-pronounced plume traceable over a large range in the adjacent zone. As the density of Black Sea waters is much less than that of the Aegean Sea, and the Black Sea water flows continuously from the strait to the mixing zone, the plume's free surface rises above the level of ambient waters. This causes the plume to spread over the Aegean Sea waters, thereby reducing thickness. Horizontal turbulent diffusion and turbulent mixing through the lower interface between brackish waters and salty waters hampers the propagation of Black Sea waters in the Athos basin and contribute to the generation of sharp frontal boundaries along the plume's perimeter.



*P. Malanotte-Rizzoli and V.N. Eremeev (eds.),*
*The Eastern Mediterranean as a Laboratory Basin for the Assessment of Contrasting Ecosystems*, 301–312.

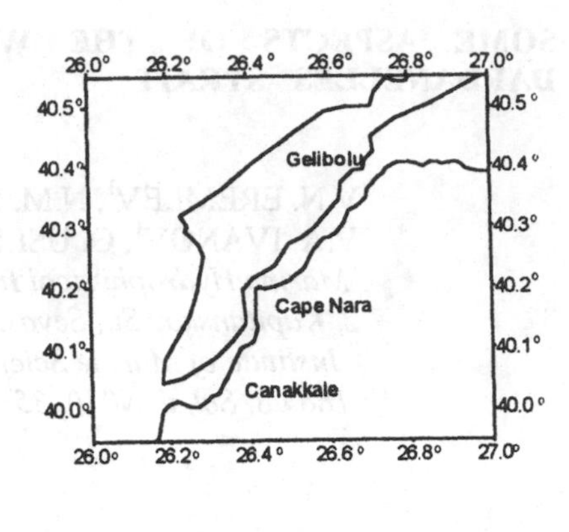


*Figure 1.* **Map of the Aegean Sea with three sub-basins: Athos, Chios and Crete Basins; Plan view of the Dardanelles Strait.**

The mechanism responsible for the frontogenesis of such a plume front are the horizontal pressure gradients, resulting, basically, from the inclination of the surface of brackish water layer and the opposite inclination of the front between the plume waters and ambient ones. As long as the two water masses interact, the front and the frontal zone will exist. The front may change its position due to fluctuations of the intensity of brackish water source. The front's location can be also impacted by tides and drift currents.

The Dardanelles Strait is 120 km in length, ranging in width from 1296 m to 18520 m, and with maximum depth of 106 m (Fig.1). A two layer currents system exists in the Dardanelles. The maximum speed of the upper current is observed at the surface, being in excess of 100 cm/s and more, and swiftly decreasing with depth. Below 20 m depth, the strait is filled with highly-saline Aegean Sea waters, which are transported by the countercurrent to the Marmara Sea, and then through the Bosporus to the Black Sea. The speed undercurrent is on the order of 10 cm/s.

In summer, Black Sea waters exiting the Dardanelles have a salinity of about 26-28 ppt [4]. Off the exit their salinity increases up to 36-37ppt, on the flow is separated into several branches [7]. The low salinity waters are observed generally along the northern and western shores of the sea (33-34 and 36-38ppt respectively). Simultaneously, the salinity dramatically increases as one moves from 25° E toward to the Turkish coast (up to 38.8-39.0 ppt). In winter, the reduction of the inflow of Black Sea waters and an intensive convective mixing caused by significant cooling of the air lead to the general increase of surface waters salinity (up to 35.0-38.9 ppt).

The results of investigations based on in situ observations and mathematical simulations of structure and evolution of the frontal zone occurring in the North Aegean Sea area were published in [6]. The present paper describes the satellite NOAA-AVHRR thermal data showing the distribution of the Black Sea water inflow in the Aegean Sea. This is followed by description of the in-situ data obtained by flow system for the surface layer on the route along Marmara Sea\ Dardanelles\ Aegean Sea during 1995.

## 2. Analysis of Satellite Data

The BSW, which is characterized by low salinity and low density, is confined in the surface layer of the North Aegean Sea. Because the Black Sea water is colder than the surrounding waters (a difference of up to 3-5$^0$ C), the surface water temperature could be utilized as a «tracer» for describing the distribution of the BSW in the North Aegean Sea.

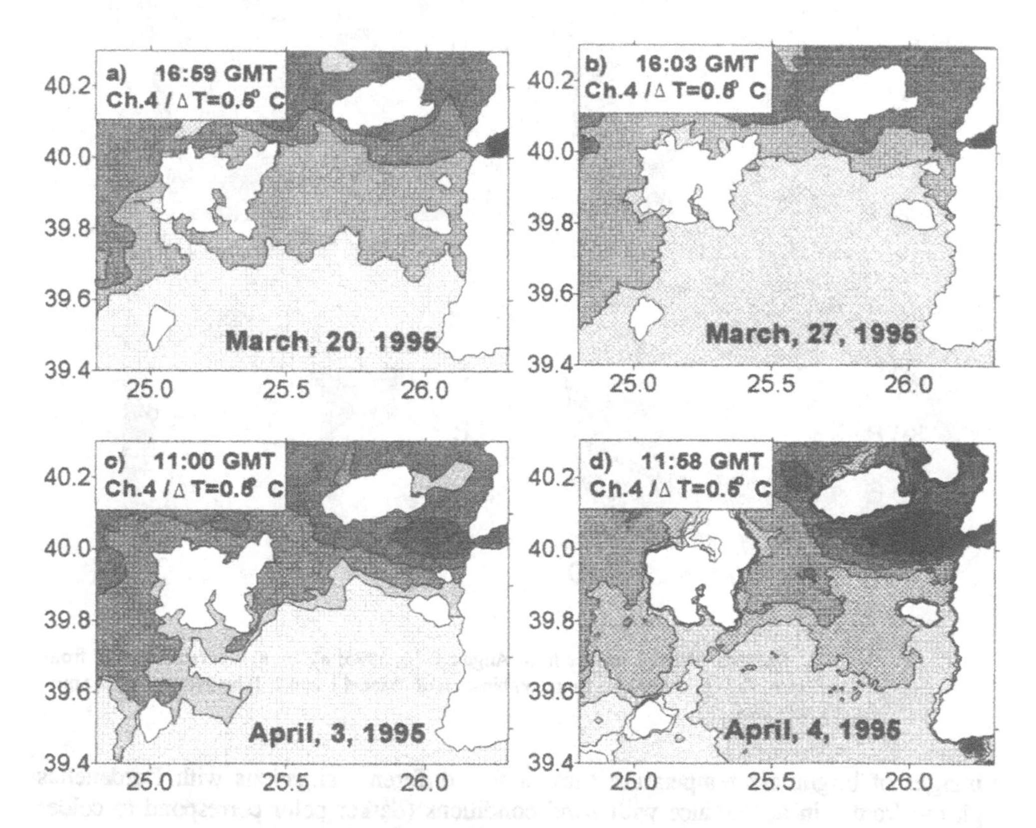


*Figure 2.* Satellite infrared AVHRR images (NOAA-14) of brightness temperature for periods of March, 20, 1995 (a): March, 27, 1995 (b); April, 3, 1995 (c ); April, 4, 1995 (d). Darker shades indicate colder brightness temperature.

The typical example of the surface Black Sea waters distribution in the Aegean Sea for a cold season is presented in Figure 2. The inflowing to the Aegean Sea BSW is much colder than surrounding one and can be readily detected with the use of surface images in winter. The series of satellite (NOAA-14, Ch-4) AVHRR infrared

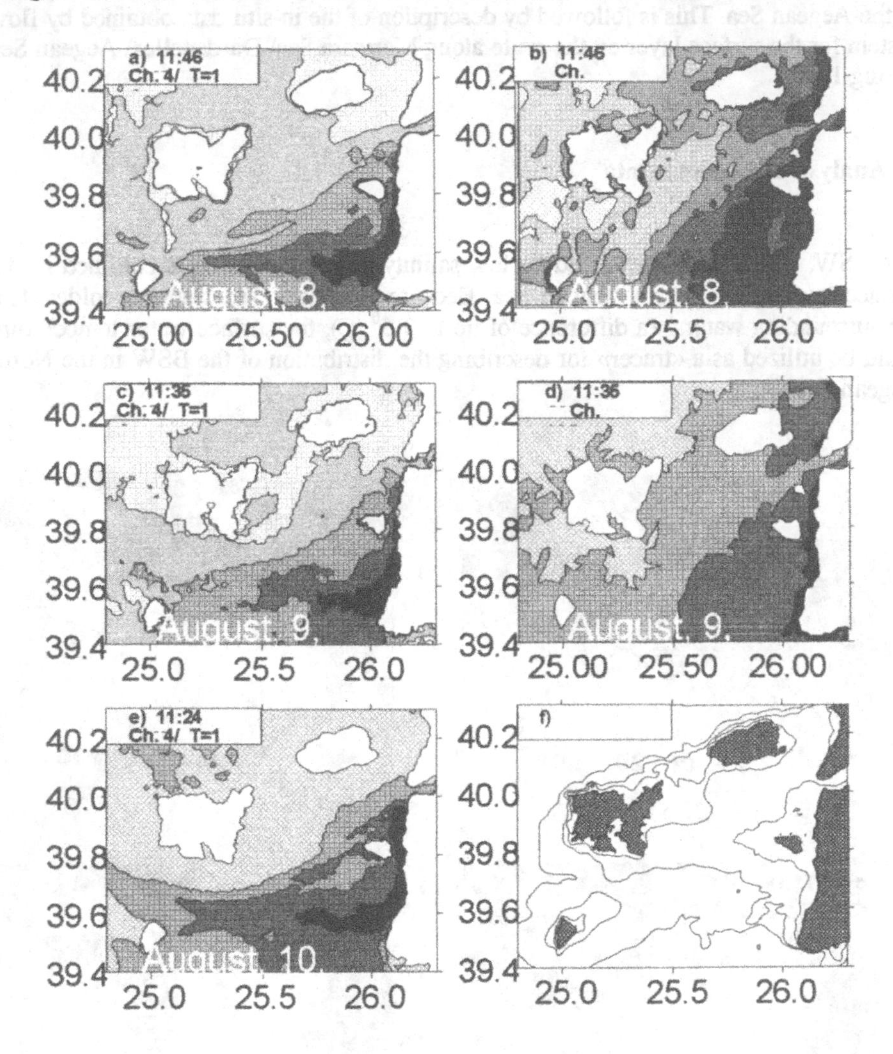


*Figure 3.* Satellite AVHRR images from August 8-10, 1995; a), c), e) - infrared images from channel 4; b),d) - optical images from combination of channel 1 and 2; f) bathymetry of the pre-Dardanelles area.

images of brightness temperature show a few different situations with Dardanelles plume front in accordance with wind conditions (darker color correspond to colder water). For the image from March 20, 1995 the blockage for the Black Sea water inside the Dardanelles is clearly seen. This is due to strong southeasterly winds prevailing the region during this period. Such situation of the absence of the front in the pre-Dardanelles area observed in cruise of R/V «Professor Kolesnikov» during December, 1993. Another situation is presented at March, 27, 1995 image when the cold Black Sea waters spread northward by the wind action. The calm situation took place during April, 3 and 4. So, a few typical positions of cold water front is clearly seen from these scenes.

Another situation occurred in the summer season. The summer images are much more difficult to analyze from the point of view of the Black Sea water distribution. Satellite infrared AVHRR images (NOAA-14) of temperature for period August 8-10, 1995 is presented at the left column in Figure 3. Darker shades indicate colder brightness temperature. Images of visible channel (combination of channel 1 with channel 2) for August 8 and 9, 1995 is shown in the right panels of the Figure 3. Because the difference between the temperatures of inflow BSW and the ambient Aegean Sea waters in summer is negligibly small the position of the thermohaline front can not be defined clearly from the presented images. From these summer scenes (August, 8,9,10) only the development of upwelling processes at the Asia Minor area is clearly seen and verified by the direct measurements. Moreover, even optical images do not put any clarity to BSW dynamics.

## 3. Analysis of Surface Temperature and Salinity

Analysis of the sea surface images indicated that it is difficult to define the position and configuration of the outer boundary of the BSW front in summer season, when the difference between the temperatures of inflowing and ambient waters is small. In winter, as this difference, as a rule, is larger the sea surface images can be useful and provide good results.

In this connection it is worth to pay attention to one circumstance. The usually accepted idea on the two-layer current structure of the in the Dardanelles leads to the assumption that the BSW pass through the Dardanelles in the upper thin 10-20 m layer and mix with the Aegean Sea waters in the pre-Dardanelles area of the Athos Basin. If it is true, the Dardanelles front should be identified rather easy by the use of the sea surface satellite images when the difference between the Aegean and Black Sea waters equal to several degrees. But the latest in-situ measurements in the Dardanelles have shown that the mixing of the Aegean and Black Sea waters involves more complicated physical processes. The mixing process takes place not only in the zone of the direct frontal contact of two water masses in the Aegean Sea but also in the strait on the way of BSW through the Dardanelles. So, the temperature of inflowing from the Dardanelles waters is not equal to the water temperature in the Black and Marmara seas, but has intermediate value. This interesting fact makes the procedure of the sea surface satellite image analysis more difficult and less applicable. This is illustrated below by means of continuous high-resolution records (space step is equal to approximately 100 m) of the surface water temperature and salinity in Bosporus/ Marmara/ Dardanelles/ Aegean system perform in the spring and summer, 1995.

Figure 4 shows values of temperature and salinity measured at 5 m depth along the way from the Marmara Sea to the Aegean Sea. The measurements begin in the

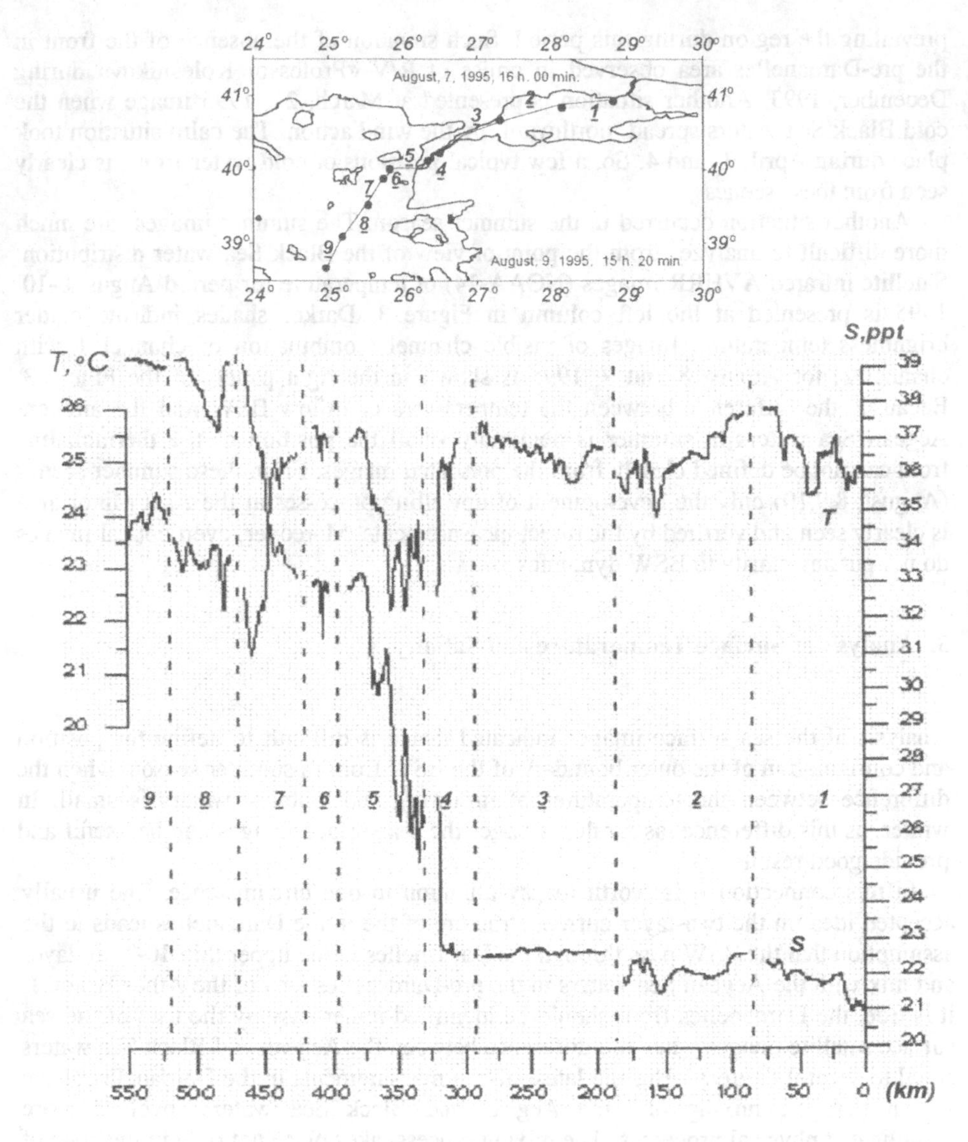


*Figure 4.* **Values of surface temperature and salinity measured along way from the Marmara Sea to the Aegean Sea from August, 7-9, 1995.**

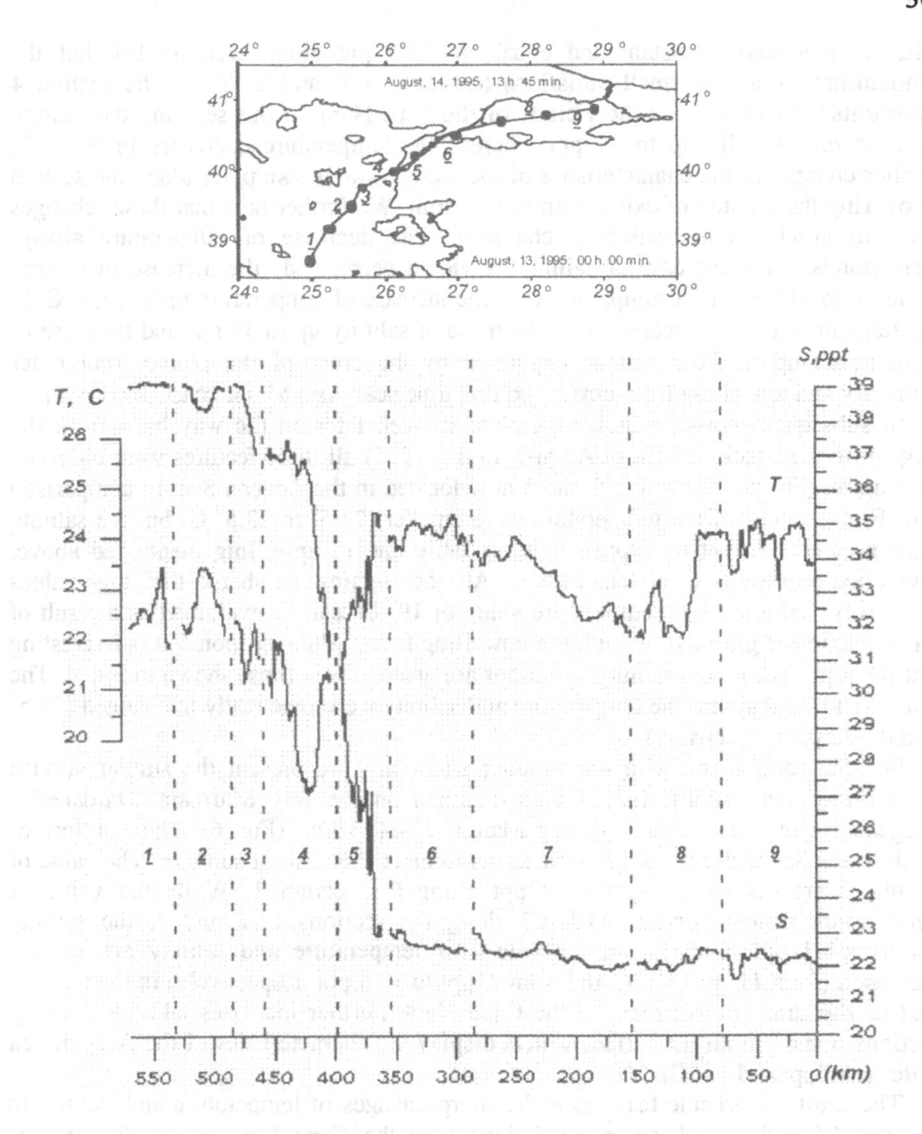


*Figure 5.* **Values of surface temperature and salinity measured along way from the Marmara Sea to the Aegean Sea from August, 13-14, 1995.**

**Marmara Sea in August, 7, 1995 in 16.00h. and terminate in the Aegean Sea in August, 9, 1995 in 15h. 20 min. The thin line corresponds to the temperature (left scale) and the thick line to the salinity (right scale). Sections 1 and 2 are placed in the Marmara Sea where the salinity and temperature have values from 21 ppt to 22 ppt and $24^0$ to $25.5^0$ C. The section 3 starts in Marmara Sea and covers the northern part of the Dardanelles to the north of the Cape Nara. It is interesting to note that salinity**

value is practically constant and equals to 22.5 ppt along sections 1-3, but the temperature value has small variations between 24.5$^0$ and 25.5$^0$ C. The section 4 represents the variations in the vicinity of the Cape Nara. In this section, the salinity increase dramatically up to 27 ppt whereas the temperature decreases to 22.5$^0$ C. Further changes in the characteristics of the surface water take place along the section 5 covering the vicinity of exit region of the strait. We further note that these changes have an synchronous oscillating character. The decrease of temperature always corresponds to the increase of salinity or vice a verse, and the increase in salinity value up to 33.5 ppt is accompanied with the increase of temperature up to 25.5$^0$ C in the Aegean Sea. In the section 7 the increase of salinity up to 39 ppt and decrease of temperature up to 21$^0$C can be explained by the cross of the plume front outer boundary and upwelling front developed this time near Asia Minor coast (see Fig.3).

In subsequent survey which took place a week later on the way back from the Aegean to the Black Sea (from August, 13-14, 1995) all these features were observed once again (Fig. 5). Section 1, 2 and 3 now located in the Aegean Sea. In comparison with Fig.4, values of the temperature were smaller (21.5$^0$ to 23.5$^0$ C) but the salinity were greater (37.5ppt to 39ppt). It is probably due to upwelling mentioned above, developed near the coast of Asia Minor. At the section number 4 the high values of salinity and very low temperature value of 19$^0$ C can be explained as a result of counteraction of plume front with the upwelling front. Along section 5 it is interesting that the temperature and salinity variations are analogous to those shown in Fig. 4. The rest of the Dardanelles the temperature and salinity were practically unchanged (24.5$^0$ C and 22.5ppt respectively).

For the comparison with the summer situation, we present the similar surface temperature and salinity distributions obtained on the way Marmara \Dardanelles \Aegean for the cold season during March, 22-24, 1995 (Fig. 6). The inflow of fresh Black Sea water in cold season seams to be is less than in summer. The value of salinity increases from 24ppt to 27ppt along the section 1. While the value of temperature almost constant (10$^0$ C) along the sections 1, 2 and 3, the salinity undergoes 1 ppt. Once again the values of temperature and salinity are sharply increased (from 11$^0$ to 13.5$^0$ C and from 27ppt to 32.5 ppt, respectively) in the western part of the strait, downstream of the Cape Nara. Further increases take place along sections 6 and 7 until the surface waters display the characteristics of the Aegean Sea water (38.5ppt and 16$^0$ C).

The most remarkable features is the sharp changes of temperature and salinity in Figures 4,5 and 6 with same oscillations near the Cape Nara where the strait is narrowed and bended almost 90 degrees. This indicates a mixing of inflowing fresh water with outflowing Aegean waters taking place in this region. It was also shown in [2] that the Dardanelles exercises efficient hydraulic controls on the upper-layer flow, due to contraction at Nara Pass and abrupt expansion of the width at its Aegean exit. The decrease of surface temperature in summer and its increase in winter in the region of the Canakkale strait deals with that the part of lower layer flow entrains into the upper layer. This implies that a part of the Aegean Sea water flows back to the Aegean Sea as a part of the surface layer flow.

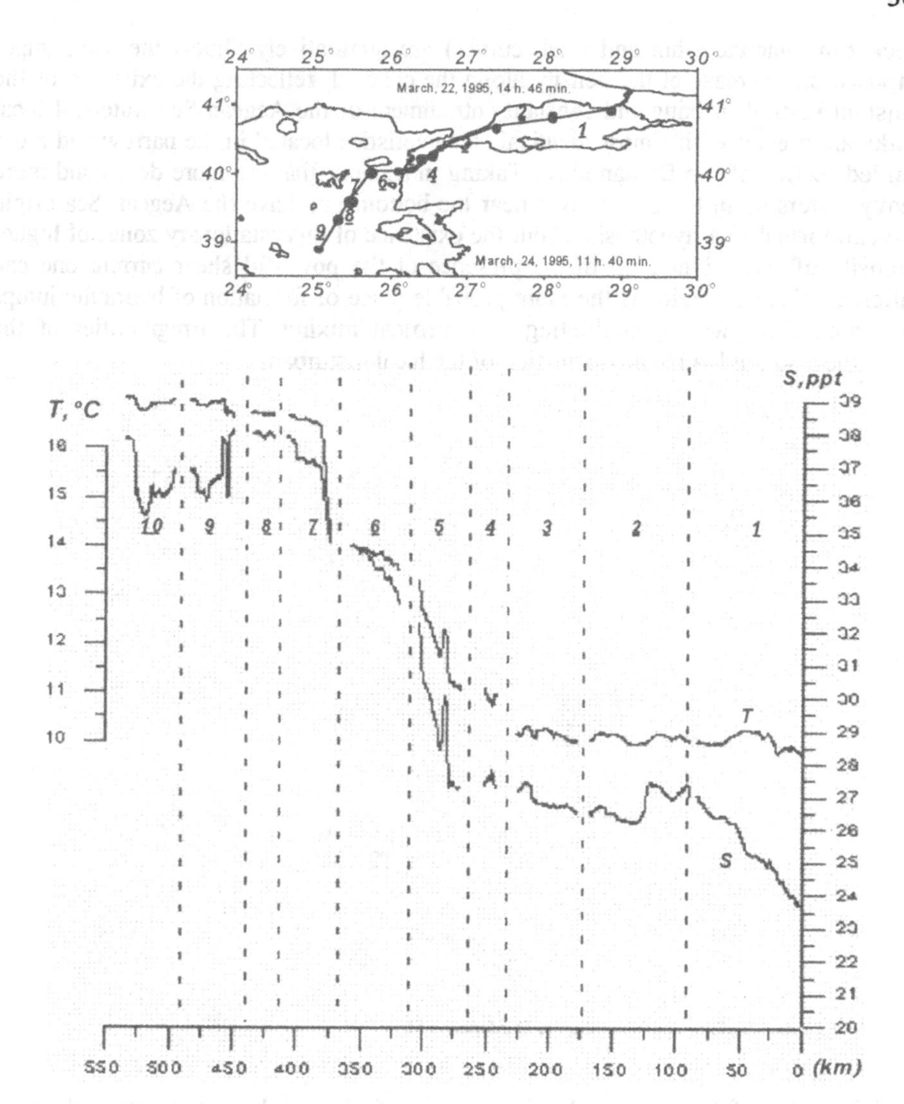


*Figure 6.* **Values of surface temperature and salinity measured along way from the Marmara Sea the Aegean Sea from March, 22-24, 1995.**

For Bosphorus it was shown in [1] by means of numerical modeling that basically two-layer flow structure involves an intermediate entrainment layer in which there is a system of recirculation that may be expected to play a prominent role in the exchange of water properties between the surface and bottom layers.

For three measurements described above the resulting $\sigma_t$ density variations along the Dardanelles are shown in Figure 7. Density profiles recorded during August, 7-9, 1995 are shown in Figure 7 by thin line. The thick line is the density obtained for August, 13-14, 1995 and the dotted line was calculated for sections 3, 4 and 5 for March, 22-24, 1995 (Fig.6). It is clearly seen that the two summer curves obtained a week time interval (thin and thick curves) are qualitatively almost the same super imposed on increase of the density along the channel, reflecting the existence of the constant vertical mixing and constant entrainment of the Aegean Sea waters. 4 local peaks are present with almost identical characteristics located in the narrow and more bended section of the Dardanelles . Taking in account that the more dense and more heavy waters lie in the lower layer near the bottom and have the Aegean Sea origin one can formulate a hypothesis about the existence of quasistationary zones of higher intensity of vertical mixing. In the presence of the powerful shear current one can anticipate that this region is the more probable place of formation of hydraulic jumps and internal lee waves, facilitating the vertical mixing. The irregularities of the coastline also can lead to the formation of the local disturbances.

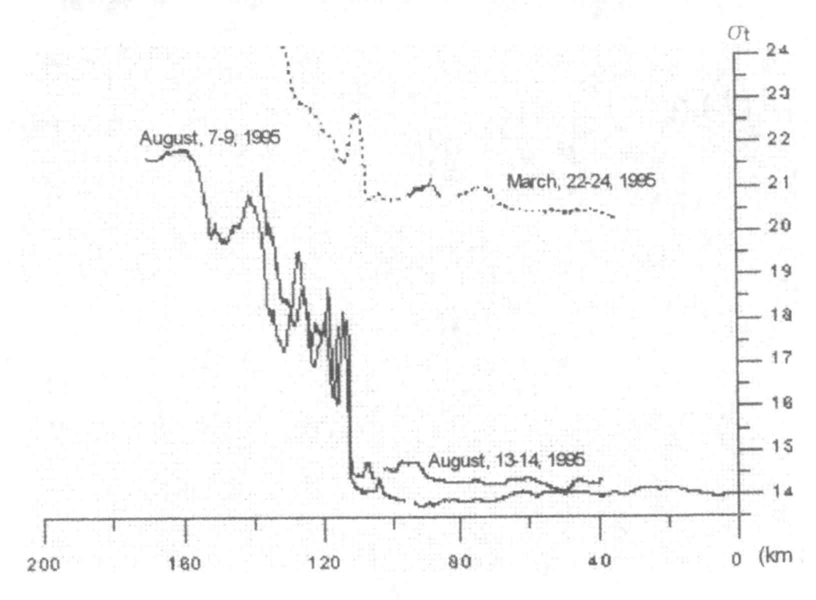


*Figure 7.* **Surface density along the Dardanelles Strait.**

The value of the surface density in winter is larger than analogous values in summer season. Nevertheless the density jump in winter season occurs also in the same place and this fact can be the additional confirmation of this hypothesis.

Presented results obtained in the Dardanelles Strait illustrate several features of flow that may be related to the internal hydraulics adjustment of the exchange flow as was shown in Bosphorus Strait by Oguz *et al.* [3]. Their results were based on fine resolution (CTD data ~1 m in the vertical and ~1-2 km in the horizontal) hydrographic measurements and two-layered model experiments. The model computations show that almost half of the lower layer flow coming into the region from the eastern Marmara basin is lost by entrainment into the upper layer. Consequently, the upper layer salinity prescribed as 18 ppt at the Black Sea termination of the strait increases typically to 23-25 ppt at southern exit region. The lower layer salinity of 38 ppt at the Marmara end of the strait, on the other hand, may decrease up to 33 ppt at the northern sill region showing that the Mediterranean effluent is diluted to a considerable extent before it joins into the western Black Sea.

## 4. Summary and Conclusions

One of the important questions of a study of eastern Mediterranean is the research of the dynamics of the Aegean Sea waters and their interaction with adjoining basins. Besides other factors (such as evaporation, precipitation, river discharge, the interaction with Levantine Basin). The Black Sea inflow through the Dardanelles Strait plays a major role in the Aegean Sea circulation. The level of the Black Sea always is higher than a level of the Aegean sea. This circumstance is the main reason that the permanent inflow of brackish BSW exists in the highly saline Athos Basin of the Aegean sea. Thus the annual discharge of inflowing water according to [4] is equal to 350 $km^3$ per year. The largest inflow takes place in summer season, the least in winter.

The presence of many islands and complex bottom relief are the basic factors for the water exchange between various basins of the Aegean Sea occurs mainly restricted to the upper layers of the sea. The more light BSW inflows through the Dardanelles in the upper 10-20-meter layer and spreads on rather large distance from the strait . It flows along the northern part of the Aegean Sea, gradually mixing up with the ambient waters. The thermohaline front arises in the immediate proximity to a strait in a zone of contact of the Black Sea and Aegean Sea waters. As the Black Sea waters become less dense, it is concentrated in a thin upper surface layer, which is exposed more to atmospheric processes. The location and shape of the front are therefore largely determined by meteorological conditions prior to the observations. It implies that forecasting of a hydrological situation in the pre-Dardanelles area is rather difficult task.

The position of the brackish waters of the Black Sea origin can be traced by the satellite images. However, such analysis can be carried out only under condition of large contrast of temperatures (a few degrees). For a winter season this condition as a rule is valid. The less salty BSW in a winter season is also characterized by lower temperature and thus easily can be identified on the satellite images. The analysis of the winter images of a sea surface of the pre-Dardanelles area of the Aegean Sea has allowed to describe some probable typical hydrological situations under various meteorological conditions - from the complete lock-out of the front in a strait during a strong counter wind, up to a real position of the front with a quiet weather.

This approach however is less applicable for summer season as the difference of temperatures of inflowing and ambient waters becomes insignificant. Since the temperature contrast between the Marmara and Aegean Seas are much less as compared to the winter month. An analysis of the pump casting data, which have been carried out on the course of a vessel, has shown, that salinity and temperature of surface water gradually increase in the strait in the direction to its exit. The presence of the powerful surface currents in upper 20-meter layer directed to the Aegean Sea (up to 1m/sec) and the countercurrent in the near-bottom layer creates conditions of entrainment of salty waters into the upper layer, mixing and returning back to the Aegean Sea. By the results of in-situ measurements, it was revealed a few local quasistable mixing zones in the strait. They are located in the region of narrowest and most curved portion of the Dardanelles.

It is interesting that analysis of only surface temperature and salinity done by pumping system in the Dardanelles had shown that submaximal exchange exists in this strait. Such withdrawal was done early in [5] from more complex analysis of CTD data (~1 m in the vertical and ~1-2 km in the horizontal ).

## 5. Acknowledgments

Authors wish to thank Dr. S.V. Stanichny (MHI) for supplying us with the satellite data. Thanks are due to Prof. T. Oguz of the Institute of Marine Sciences, Turkey, for his fruitful comments.

## 6. References

1. Johns, B. and Oguz, T. (1990) The modelling of the flow of water through the Bosphorus, *Dynamics of Atmospheres and Oceans*, **14**, 229-258.
2. Oguz, T. and Sur, I., (1989) A two-layer model of water exchange through the Dardanelles Strait, *Oceanologica Acta*,**1**, 23-31.
3. Oguz, T., Ozsoy E., Latif, M.A., Sur, H.I. and Unluata, U., (1990) Modeling of hydraulically controlled exchange in the Bosphorus Strait, *J. Phys. Oceanogr.*, 7, 945-965.
4. Ovchinnikov, I.M., Plakhin, E.A., Moskalenko, L.V., Neglyad, K.V., Osadchiy, A.S., Fefoseyev, A.F., Krivosheya, V.G. and Voytova, K.V. (1976) *Hydrology of the Mediterranean Sea,* Gidrometeoizdat, Leningrad.
5. Unluata, U.,Oguz, T., Latif, M.A., Ozsoy, E. (1990) On the physical oceanography of the Turkish straits, in L.J. Pratt (ed.), *The Phys Oceanography of Sea Straits*, Kluwer Academic Publishers, Dordrecht, pp.25-60.
6. Vlasenko, V.I., Stashchuk, N.M., Ivanov, V.A., Nikolaenko, E.G., Uslu, O. and Benli H., (1996) Influence of the water exchange through the Dardanelles on the thermohaline structure of the Aegean Sea, in F. Briand (ed.), *Dynamics of Mediterranean Straits and Channels*, CIESM Science Series, No 2, Monaco, pp. 147-165.
7. Zodiatis, G. (1994) Advection of the Black Sea ware in the North Aegean Sea.- *The Global Atmosphere and Ocean Systems*, **1**, 41-60.

# BIO-OPTICAL CHARACTERISTICS OF THE AEGEAN SEA RETRIEVED FROM SATELLITE OCEAN COLOR DATA

V.I.BURENKOV, O.V.KOPELEVICH, S.V.SHEBERSTOV,
S.V.ERSHOVA, M.A.EVDOSHENKO
P.P.Shirshov Institute of Oceanology Russian Academy of Sciences
Nakhimovski prospect 36, Moscow 117851, Russia

## 1. Introduction.

The specific goal of this work is to demonstrate the possibilities of new ocean color satellite sensors such as the Sea-viewing Wide Field-of-view Sensor (SeaWiFS) for studying processes in the Eastern Mediterranean, with the Aegean Sea as an example. The Aegean Sea appears as an appropriate basin for study of contrasting ecosystems, because it is subjected to intrusion of water masses of different origin with pronounced contrast in bioproductivity: the eutrophic water from the Black Sea, and the oligotrophic water from the Levantine Basin.

SeaWiFS gives a nearly comprehensive global view of the oceans every two days. The ocean bio-optical parameters derived routinely from SeaWiFS data include chlorophyll *a* concentration; "CZCS-type" pigment concentration (the summation of chlorophyll *a* and phaeophytin concentrations): and the diffuse attenuation coefficient at 490 nm, *K(490)* [1,2]. SeaWiFS data have potential for the derivation of other seawater optical characteristics such as the absorption coefficient with separation of contributions arising from yellow substance and phytoplankton pigments, and the backscattering coefficient. These are important quantities for calculation of light propagation in water bodies, both solar radiation and laser pulses, and they can be also used to study various processes of suspended and dissolved matter transformation as well the ocean dynamics in the upper ocean. Methods to retrieve these quantities have been already considered [3-6]. In this work the bio-optical characteristics retrieved from SeaWiFS data in September-October, 1997 have been mapped for the Aegean Sea by using algorithms developed by the authors.

## 2. SeaWiFS Data.

The SeaWiFS project produces LAC and GAC data products where the terms "LAC" and "GAC" refer to the spatial resolution and mean respectively "Local Area Coverage" and "Global Area Coverage". LAC products have a spatial resolution of 1.1 km at nadir (directly beneath the satellite), and pixels are contiguous; on board SeaWiFS only limited amount of LAC data are recorded but such kind of data are broadcast to High Resolution



*P. Malanotte-Rizzoli and V.N. Eremeev (eds.),*
*The Eastern Mediterranean as a Laboratory Basin for the Assessment of Contrasting Ecosystems,* 313–326.

Picture Transmission (HRPT) ground stations situated throughout the world. GAC products are continuously created onboard the satellite by selecting every fourth pixel on every fourth scan line so the pixels are spaced at 4.4 km intervals. The level-1A products include the row image data recorded onboard the satellite and all instrument and spacecraft telemetry data together with appended instrument calibration and navigation data; these data allow to calculate calibrated radiances measured at the top of the atmosphere with reference to time and geographical coordinates. Level-2 products are derived from the level-1A radiance data with no remapping, so level-2 data correspond to the original pixel position. Before computing level-2 data, pixels are eliminated if they contain cloud, sun glint, or other abnormalities; for the accepted pixels the atmospheric correction algorithm is applied to derive the normalized water-leaving radiances, and, by means of the bio-optical algorithms, in-water characteristics.

The SeaWiFS data used were searched, viewed and ordered via the World Wide Web from the Goddard Space Flight Center Distributed Active Archive Center (GSFC DAAC). These data included the GAC and LAC data recorded on the satellite's tape recorder and downlinked to the ground station in Wallops, USA as well as the LAC data captured by the HRPT ground station in Dundee, Scotland (the LAC data for the Aegean Sea were available only from this HRPT station in September-October, 1997). When the level-1A LAC data are used, the normalized water-leaving radiances are calculated with the SeaDAS software package (SeaDAS - SeaWiFS Data Analysis System) specially developed by the SeaWiFS Project for the processing, display, analysis, and quality control of all SeaWiFS data products and kindly made available by NASA.

## 3. The models and algorithms.

The starting data for calculation of bio-optical characteristics are the values of the normalized water-leaving radiances, $L_{WN}(\lambda_i)$, in the 5 spectral bands of SeaWiFS with $\lambda_i$ equal to 412, 443, 490, 510, and 555 nm; the $L_{WN}$ values are independent of viewing and solar geometry and optical characteristics of the atmosphere [7]. These values are related to the values of the ocean irradiance reflectance $R(\lambda)=E_u/E_d$ just beneath the sea surface ($E_u$ and $E_d$ are respectively upwelling and downwelling irradiances just beneath the sea surface ) by the following formula [7]

$$L_{WN}(\lambda) = F_o\, M\, R(\lambda)/Q, \tag{1}$$

where $F_o$ is the extraterrestrial solar irradiance, $Q$ is the ratio of upwelling irradiance to upwelling radiance towards zenith. $M$ is defined by

$$M = \frac{(1-r_L)(1-r_E)}{m^2\,(1-rR)} \tag{2}$$

where $r_L$ , $r_E$ are the Fresnel reflectances of the sea surface for upwelling radiance and downwelling irradiance, $r$ is the water-air reflectance for upwelling irradiance, and $m$ is the seawater refraction index. The values of $r_L$, $r_E$ and $r$ are accepted to be equal to 0.021,

0.04, and 0.48 respectively [7]. Formulae (1)-(2) can be modified to obtain the spectral radiance reflectance $\rho(\lambda)=\pi L_u/E_d$ ($L_u$ is the upwelling radiance just beneath the surface), which is used in the bio-optical algorithm. The values of $\rho(\lambda)$ and $R(\lambda)$ are linked by a simple relationship: $\rho(\lambda)=\pi R(\lambda)/Q$; the value of $Q/\pi$ is accepted to be equal to 1.2 [8].

The value of $\rho(\lambda)$ obtained from (1)-(2) can be directly related to the absorption $a(\lambda)$ and backscattering $b_b(\lambda)$ coefficients by the following formula [7]

$$\rho(\lambda) = \pi( l_1 X + l_2 X^2), \tag{3}$$

where $l_1$=0.0949, $l_2$=0.0794, and

$$X = b_b/(b_b + a). \tag{4}$$

Values of $b_b$ and $a$ can be retrieved from (4) with values of $X$ obtained from (1)-(3) and represented as

$$a(\lambda) = a_w(\lambda) + a_{ph}(\lambda) + a_g(\lambda), \tag{5}$$

$$b_b(\lambda) = b_{bw}(\lambda) + b_{bp}(\lambda), \tag{6}$$

where $a_w(\lambda)$ and $b_{bw}(\lambda)$ are absorption and backscattering coefficients of pure sea water (with their values in accordance with Pope, Fry [9] and Shifrin [10]), $a_{ph}(\lambda)$ is the spectral absorption of phytoplankton pigments, $a_g(\lambda)$ is the absorption by chromophoric dissolved organic matter (CDOM), as it is called "yellow substance" ("gelbstoff"), and $b_{bp}(\lambda)$ is the backscattering of suspended particles. The absorption of phytoplankton pigments $a_{ph}(\lambda) = C_{chl}\, f_{ph}(\lambda)$, where $f_{ph}(\lambda)$ is specific absorption of phytoplankton pigments, is represented by the power function

$$f_{ph}(\lambda) = A(\lambda) C_{chl}^{-B(\lambda)}, \tag{7}$$

where $A$ and $B$ are positive wavelength-dependent parameters [11]. Spectral dependence of the CDOM absorption $a_g(\lambda)$ is assumed in the usual form $a_g(\lambda)=a_g exp(-s(\lambda-440))$, where $a_g$ is value of the CDOM absorption at $\lambda$= 440 nm, $s$ is a parameter which value was accepted to be equal to 0.017 $nm^{-1}$. The spectral dependence of suspended particles backscattering coefficient has the traditional form $b_{bp}(\lambda)=b_{bp}(\lambda/550)^{-n}$, where $b_{bp}$ is the particle backscattering coefficient at the reference wavelength 550 nm. The value of parameter $n$ is determined from the following relationship [3]

$$n = -1.09 + 3.05\, \rho(443)/\rho(490). \tag{8}$$

The unknown parameters $C_{chl}$, $a_g$ and $b_{bp}$ are determined from the system of five nonlinear equations (3) using the least square technique. Because of nonlinear dependence of phytoplankton absorption on chlorophyll concentration, a number of iterations must be done to obtain a solution of this system with the proper accuracy. These iterations increase the computation burden drastically with dealing with great amount of space ocean color sensors data. To reduce the computational time, the shape of the phytoplankton specific absorption curve is assumed to be independent of the chlorophyll concentration, more precisely it is calculated for a chlorophyll concentration of 0.5 $mg/m^3$. Further calculations showed that this assumption does not increase errors in the retrieval of bio-optical parameters.

Along with producing new level 2 products ($a_g(440)$, $a_{ph}(440)$, $b_b(550)$, and $C_{chl}$ and $K_d(490)$ with the authors' algorithms) corresponding the original pixel position, level 3 data products for all bio-optical variables are derived by "binning" level 2 data into 4 km grid cells (contrary to the usual SeaWiFS 9 km grid cells) and averaging them spatially and temporally within every cell over September and October of 1997. Although September is generally taken as a summer month whereas October as an autumn one, there was found no pronounced distinction between these months; so we also averaged the data over these two months. Both the level 2 and level 3 data are mapped on a Mercator projection with the computer code developed for PC by the authors; the level 2 images are limited to the part of the satellite swath encompassing the Aegean Sea. Examples of the mapped data are given in Plates 1-3.

## 4. Results.

Plates 1-3 display spatial distributions of the values of $a_g(440)$, $b_b(550)$, and $C_{chl}$ derived from SeaWiFS GAC data over the Aegean Sea. These quantities were chosen as they relate directly to three important seawater components: color organic matter (yellow substance), particulate matter, and photosynthetic phytoplankton. Spatial distributions of concentrations of these components depend on disposal and characteristics of the sources producing these components as well as on processes of their transfer from the sources and their transformation. The most important sources in the Aegean Sea, namely, the inflow of the Black Sea water through the Dardanelles, the photosynthetic primary production, and the run-off of rivers should be considered, as well as, to a lesser extent, the coast erosion, volcanoes, and the inflow of aerosol particles. As for the primary production, it should be noted that its values are usually very low in the most part of the Mediterranean Sea due to impoverished nutrient concentration. The primary production is enhanced only where nutrient concentration within the surface layer increases due to such phenomena as local upwellings, river run-off, and, regarding the Aegean Sea, the inflow of the Black Sea water through the Dardanelles. All the three most important sources mentioned above act mainly in the northern part of of the Aegean Sea, and their effect can be distinguished in Plates 1-3. As to the transfer processes in the Aegean Sea, one of them is an advection due to the movement of water masses. The general circulation in the Aegean Sea is cyclonical with two basic currents. One of them, associated with the Black Sea water, coming through the Dardanelles, moves west to the Gulf of Salonika, and then proceeds southward along the coast of Greece; the other current, opposite in direction and associated with the Levantine water, coming through the south-eastern straits, moves north along the coast of Turkey. These currents are best displayed by the distribution of the $b_b(550)$ values derived from the level 2 data on September 27 (Plate 1); they are not as well distinguished in the mean distribution of the $b_b(550)$ values (Plate 2). This can probably be explained that the circulation in the Aegean Sea is strongly affected by winds and, although the general circulation is rather stable, the meridional main streams can disintegrate forming both cyclonic and anticyclonic eddies according to the wind conditions [12].

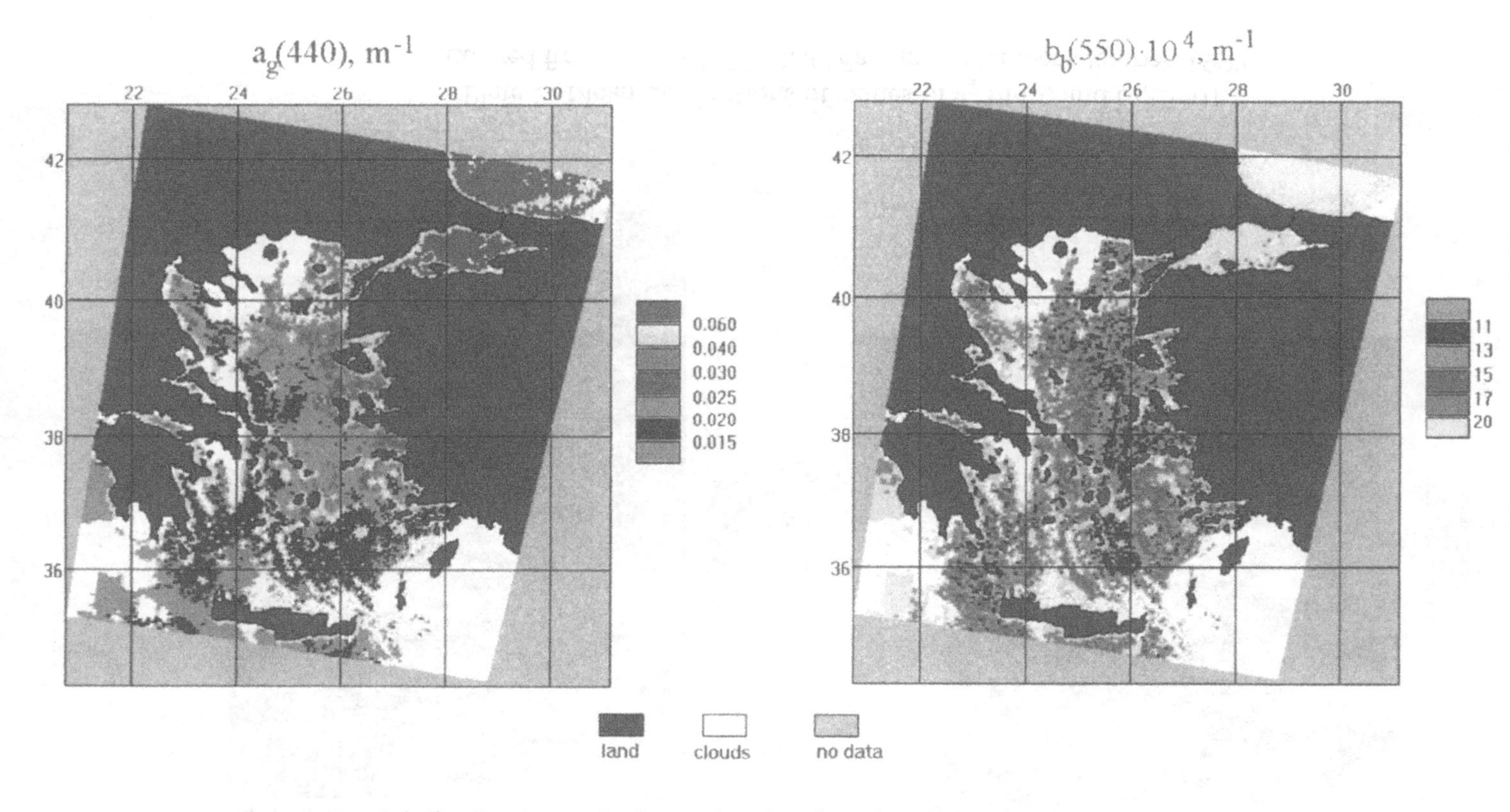


Plate 1. Spatial distributions of values of $a_g(440)$ and $b_b(550)$ in the Aegean Sea on September 27, 1997 (derived from the SeaWIFS Level 2 GAC data)

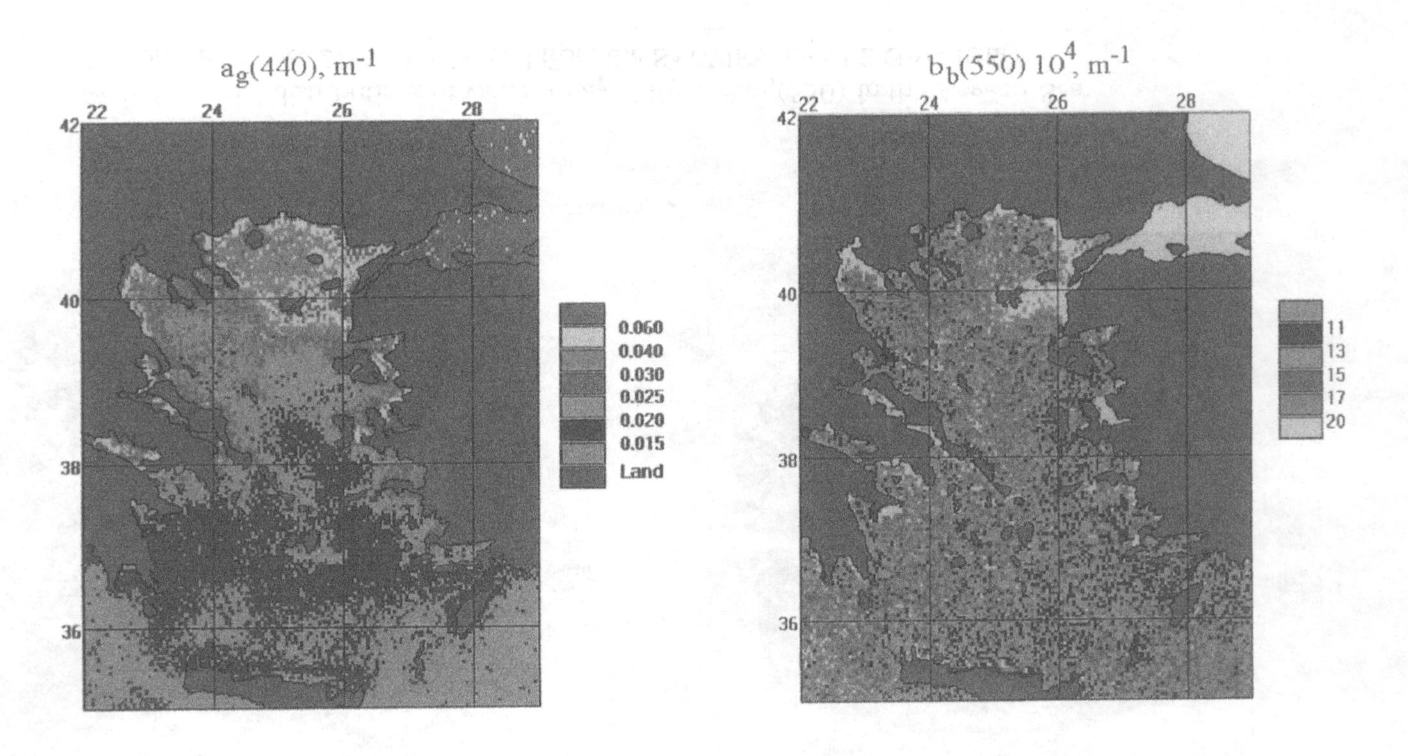


Plate *2*. Mean distributions of values of $a_g(440)$ and $b_b(550)$ derived from all SeaWIFS GAC data in September-October 1997

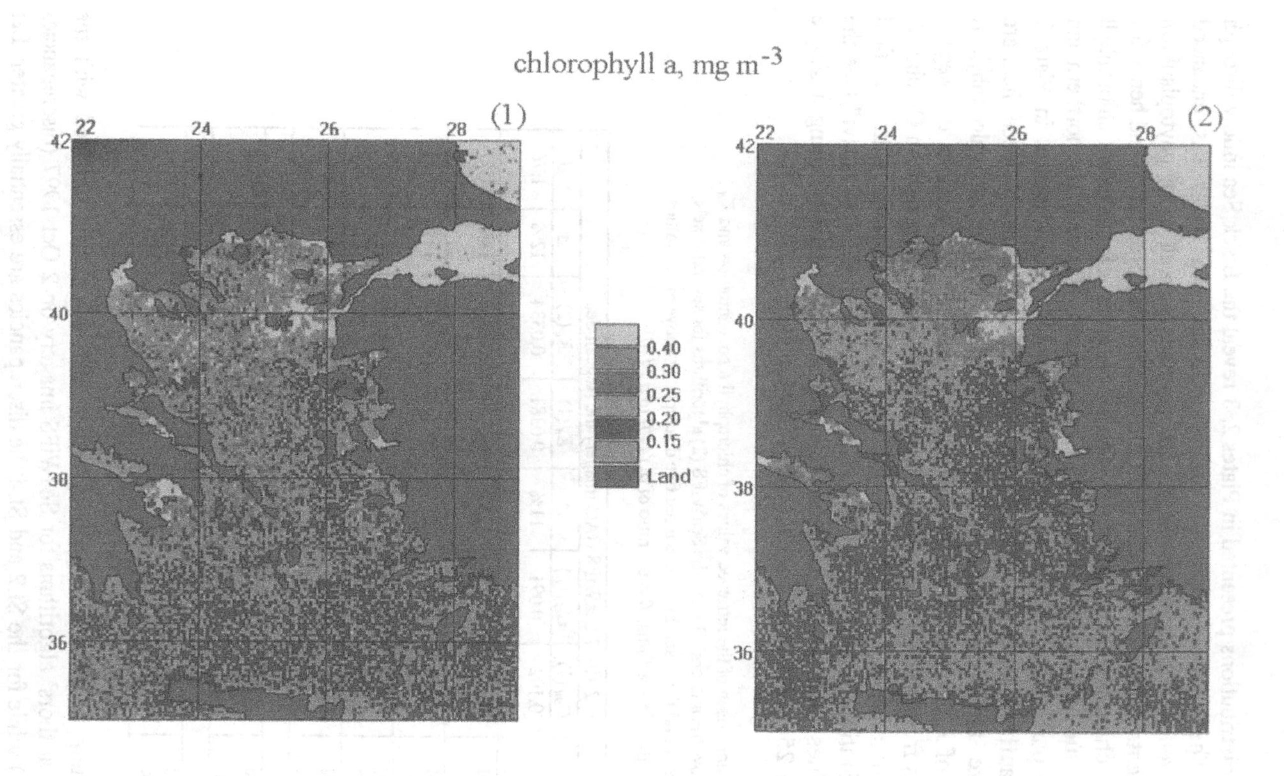


Plate 3. Mean distributions of chlorophyll values retrieved by the authors'(1) and SeaWIFS (2) algorithms. September - October, 1997

All mean distributions presented in Plates 2, 3 reveal the Black Sea inflow through the Dardanelles. The reason is that the inflowing Black Sea water has enhanced concentrations of yellow substance and suspended particles as well as of phytoplankton pigments and nutrients; the latter is beneficial for growing primary production and hence for further increasing chlorophyll concentration. The other areas of enhanced chlorophyll concentrations are due to the run-off of rivers Marica, Struma, Vardar in the northern and the north-western parts of the Aegean Sea. Comparing chlorophyll images in Plate 3 derived using the authors' and SeaWiFS algorithms, some discrepancies between them are noted. To analyze these discrepancies, we compared the values of chlorophyll *a* concentration and of $K_d(490)$ retrieved by both algorithms for several pixels in the Aegean and the Black Seas from SeaWiFS level 2 GAC data. They include the data on October 2, 8, and 11; the data on October 8 and 11 were contemporaneous with the field measurements at Stations 2 and 4 during the validation cruise of R/V Akvanavt" (see the next section). Values of $K_d$ by the authors' algorithm are calculated by using a simple relationship: $K_d = 1.25(a+b_b)$ based on the work of Gordon [13].

Table 1. Comparison of the retrieved values of chlorophyll a concentration and $K_d$ calculated by the authors' (1) and the SeaWiFS (2) algorithms for several pixels in the Aegean and the Black Seas; $\Delta$ is a relative discrepancy between the values retrieved by these algorithms; $\beta$ is the ratio of $a_g(440)$ to $a_p(440)$.

| | 2.10.97 SeaWiFS GAC imagery the Aegean Sea. | | | | | | |
|---|---|---|---|---|---|---|---|
| | $C_{chl}$ (1) | $C_{chl}$ (2) | $\Delta$ | $K_d$ (1) | $K_d$ (2) | $\Delta$ | $\beta$ |
| 36.98°N 24.01°E | 0.102 | 0.091 | 11% | 0.0381 | 0.0334 | 12% | 1.76 |
| 37.02°N 26.01°E | 0.118 | 0.119 | 1% | 0.0403 | 0.0350 | 13% | 1.77 |
| 38.01°N 26.01°E | 0.184 | 0.160 | 13% | 0.0437 | 0.0376 | 14% | 1.31 |
| 38.00°N 24.99°E | 0.256 | 0.231 | 10% | 0.0493 | 0.0434 | 12% | 1.32 |
| 39.99°N 24.99°E | 0.479 | 0.388 | 17% | 0.0642 | 0.0592 | 8% | 1.40 |
| | St.4 in the Aegean Sea | | | | | | |
| 39.29°N 25.12°E | 0.128 | 0.184 | 44% | 0.0479 | 0.0434 | 10% | 2.98 |
| | St.2 in the Black Sea | | | | | | |
| 42.96°N 35.59°E | 0.491 | 1.06 | 138% | 0.100 | 0.125 | 25% | 3.71 |

As seen from Table 1, there is a good agreement between results obtained with the SeaWiFS and the authors' algorithms for SeaWiFS imagery on 2 Oct 1997 (discrepancies are less than 20%) while for the St.2 and St.4 the discrepancies are essentially greater. Let us call attention to the differences in the values of $\beta$ which are less than 2 for all points on 2 Oct 1997, but are 3-4 for the St.2 and St.4. Thus the discrepancies appear for the waters with relative enhanced concentrations of yellow substance. Note the values of $K_d(490)$ agree quite satisfactorily in even such cases.

**5. Validation of the algorithms using contemporaneous field measurements.**

Validation of the algorithms was performed during the scientific cruise of RV "Akvanavt" in the Black and Aegean Seas, October 6-24, 1997. The field studies, carried out contemporaneously with measurements by satellite ocean color sensors SeaWiFS and MOS-IRS, included:

- measurements of spectral upwelling radiances and surface irradiances by deck and floating radiometers;
- measurements of vertical profiles of nadir upwelling radiance and downwelling irradiance by means of a submersible radiometer MER-2040 (these measurements were performed by specialists from the Polish Institute of Oceanology, Sopot);
- determination of atmospheric spectral optical thicknesses by means of a sun photometer;
- measurements of vertical profiles of seawater beam attenuation by a submersible transmissometer;
- sampling and conservation of water samples for determination of phytoplankton pigment concentration.

Five drift stations were carried out during the cruise, two of them in the Aegean Sea on October 11 and 16, but unfortunately SeaWiFS was out of service on October 16. The station 4 on October 11 was located southward of Lemnos; the measurements under SeaWiFS were carried out at 11:09.55 GMT; the coordinates at this moment were 39.32 N, 25.12 E; the zenith and azimuth angles of observation were respectively 37.7° and 250.5°; the solar zenith and azimuth angles 48.8° and 201.5°. The weather was cloudless but with a southern wind of about 10 m/s and a wavy sea. Figure 1 shows the vertical distribution of the water temperature and beam attenuation coefficient at 530 nm, *c(530)*, at this station.

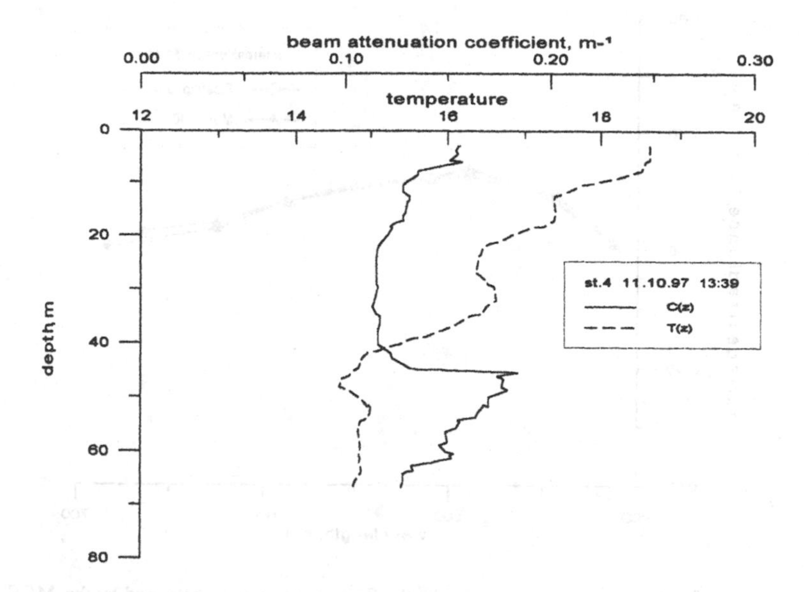


Fig.1. Vertical profiles of the beam attenuation coefficient and temperature at the St.4.

As seen, there was rather clear water in the surface layer - the values of *c(530)* were in the range from 0.11 to 0.16 $m^{-1}$; the Secchi depth was 25 m but probably it was underestimated due to bad conditions of observation. Chlorophyll *a* concentration was 0.09 $mg/m^3$ at 0 and 10 m, 0.08 $mg/m^3$ at 20 m. The values of aerosol optical thickness, $\tau_a$, at 8 spectral bands of the sun photometer are given in Table 2:

Table 2. The values of aerosol optical thickness $\tau_a$ at St.4 in the Aegean Sea.

| $\lambda$, nm | 400 | 442 | 492 | 520 | 557 | 678 | 778 | 862 |
|---|---|---|---|---|---|---|---|---|
| $\tau_a$ | 0.313 | 0.274 | 0.270 | 0.244 | 0.218 | 0.182 | 0.174 | 0.142 |

Measurements of upwelling radiance and downwelling irradiance in the spectral range 390-700 nm with spectral resolution 2.5 nm were carried out by means of a new floating spectroradiometer constructed in the P.P.Shirshov Institute of Oceanology Russian Academy of Sciences (SIO RAS). This instrument measures upwelling radiance just beneath the sea surface and downwelling irradiance just above the sea surface at a distance about 50 m from ship, so the difficulties connected with influence of sun glint, correction for light reflected by the sea surface, influence of ship shadow are avoided.

Absolute radiometric calibration of the radiance and irradiance channels was performed in the laboratory before the expedition. Accuracy in measurements of the upwelling radiance and downwelling irradiance is about 5%. Optical measurements were also carried out (only in the Black Sea) by the instrument MER-2040 (Biospherical Instrument Corporation). Examples of intercalibration of these instruments are shown in Fig. 2, 3. The discrepancy between values of $L_u$ and $E_d$ measured by these two instruments was generally in limits of 5%.

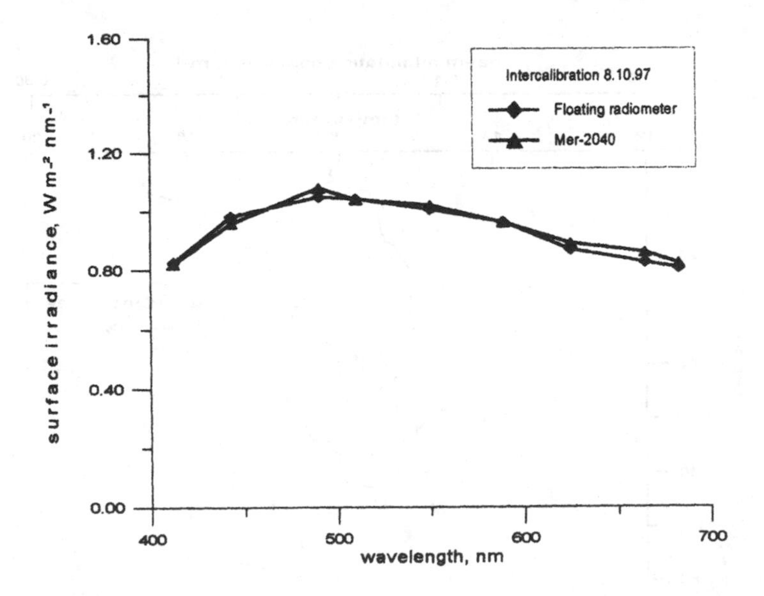


Fig.2. Comparison of surface irradiances measured by the floating spectroradiometer and by the MER-2040 at the St.2 (GMT 10:30).

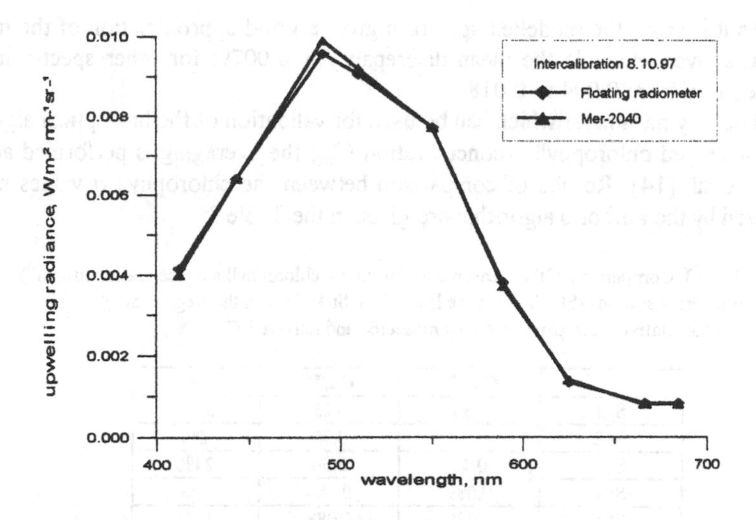


Fig.3. Comparison of upwelling radiances measured by the floating spectroradiometer and by the MER-2040 at the St.2 (GMT 10:30).

Values of $L_u$ and $E_d$ measured by the floating spectoradiometer at different stations in the Black and Aegean Seas are used to obtain values of spectral reflectance $\rho(\lambda)$ for validation of the bio-optical algorithm described above. The bio-optical parameters were retrieved both with the five SeaWiFS spectral channels and with 17 wavelengths in the spectral range 410-570 nm (step10 nm). The values of the retrieved parameters do not differ significantly for these two cases, so next the bio-optical algorithm is only applied to the SeaWiFS spectral channels. Fig. 4 shows an example of comparison between measured and modelled reflectance spectra.

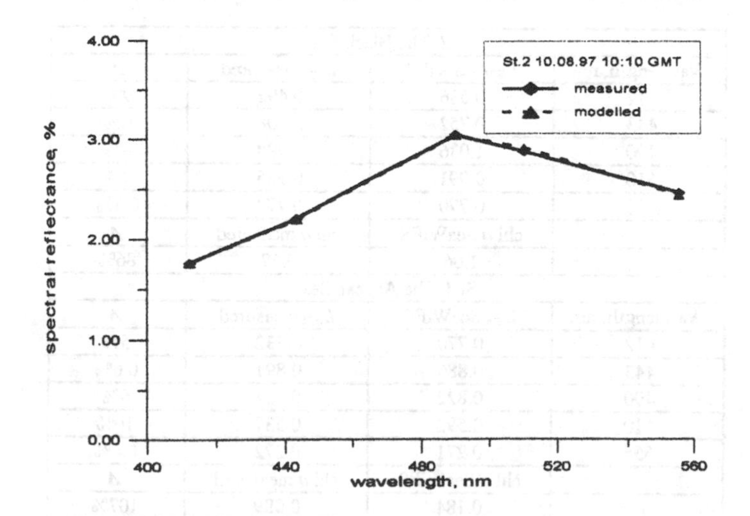


Fig.4. Comparison of measured and modelled reflectance spectra for the St.2.

As it is seen, the modelled spectrum gives a good approximation of the measured one. For the given example the mean discrepancy is 0.0079; for other spectra measured these values vary from 0.004 to 0.018.

The only parameter which can be used for validation of the bio-optical algorithm is the depth-averaged chlorophyll *a* concentration $C_{chl}$; the averaging is performed according to Gordon et al. [14]. Results of comparison between the chlorophyll *a* values measured and retrieved by the author's algorithm are given in the Table 3.

Table 3. Comparison of the measured and retrieved chlorophyll a concentration (mg $m^{-3}$) at different stations (St.1-3 are in the Black Sea, St.4, 5 are in the Aegean Sea). $\Delta$ is the relative discrepancy between measured and retrieved $C_{chl}$.

| | $C_{chl}^{meas}$ | $C_{chl}^{retr}$ | $\Delta$ |
|---|---|---|---|
| St.1 | 0.35 | 0.23 | 33% |
| St.2 | 0.57 | 0.56 | 2% |
| St.3 | 0.45 | 0.56 | 24% |
| St.4 | 0.089 | 0.086 | 3% |
| St.5 | 0.076 | 0.088 | 16% |

As it is seen, the agreement between measured and retrieved chlorophyll *a* concentrations is quite satisfactory.

The SeaWiFS level-2 products were evaluated by concurrent sub-satellite measurements at two stations under appropriate weather conditions. One of these stations was located in the Black Sea, the other in the Aegean Sea. Data of sub-satellite measurements of the upwelling radiance and downwelling irradiance were transformed into water-leaving radiances for 5 SeaWiFS spectral channels. Results of comparison between values of $L_{WN}(\lambda_i)$ and *Chl* derived from the SeaWiFS satellite data and data of *in situ* measurements are presented in Table 4.

Table 4. Comparison of the SeaWiFS water-leaving radiances (W $m^{-2}$ $\mu m^{-1}$ $sr^{-1}$) and chlorophyll *a* concentration with sub-satellite measurements for two stations; $\Delta$ is the relative discrepancy between the satellite and *in situ* data.

| St.2. The Black Sea. | | | |
|---|---|---|---|
| Wavelength, nm | $L_{WN}$ SeaWiFS | $L_{WN}$ measured | $\Delta$ |
| 412 | 0.536 | 0.492 | 9% |
| 443 | 0.752 | 0.708 | 6% |
| 490 | 1.036 | 0.998 | 4% |
| 510 | 0.991 | 0.925 | 7% |
| 555 | 0.770 | 0.772 | 0.3% |
| | chl *a* SeaWiFS | chl *a* measured | $\Delta$ |
| | 1.06 | 0.57 | 86% |
| St.4. The Aegean Sea. | | | |
| Wavelength, nm | $L_{WN}$ SeaWiFS | $L_{WN}$ measured | $\Delta$ |
| 412 | 0.770 | 0.732 | 5% |
| 443 | 0.886 | 0.891 | 0.6% |
| 490 | 0.822 | 0.784 | 6% |
| 510 | 0.592 | 0.537 | 10% |
| 555 | 0.271 | 0.272 | 0.4% |
| | chl *a* SeaWiFS | chl *a* measured | $\Delta$ |
| | 0.184 | 0.089 | 107% |

Table 4 shows that the agreement between satellite and in situ data is quite satisfactory for water-leaving radiances and much worse for chlorophyll a concentration. Discrepancies between water-leaving radiance data can have various causes: errors in measurements; due to atmospheric correction algorithm; differences between nadir radiances (for field measurements) and radiances in the direction of the satellite sensor. Discrepancies in chlorophyll *a* concentration can also be due to the bio-optical algorithm used in the SeaWiFS data processing. This algorithm is mainly applicable to Case-1 waters for which there is a definite correlation between phytoplankton pigments and CDOM absorption; therefore, the SeaWiFS algorithm can give significant errors for the Case-2 waters with predominance of absorption by yellow substance. Such cases are shown in the previous section. Comparison of the data given in Table 3 and 4 shows essentially better agreement between retrieved and observed chlorophyll *a* concentrations for the bio-optical algorithm proposed in this paper. This analytical algorithm does not assume any correlation between contributions in absorption arising from phytoplankton pigments and from yellow substance as well as with the particle backscattering and can be applied for different water types. As it is given in Table 1, the authors' algorithm derives chlorophyll values for St. 2 and 4 equal to 0.49 and 0.128 which are much closer to the measured values of 0.57 and 0.089 than derived by SeaWiFS algorithm (1.06 and 0.184).

Of course the amount of experimental data is not enough for making final conclusions, and further studies are required.

This work was supported by the funds from the Russian Ministry of Sciences and Technologies under the project "Satellite Ocean Color Scanner SeaWiFS" and from the Contract NAS15-10110 between NASA and RSA. The authors are grateful to the GSFC DAAC for providing the SeaWiFS data, and to Robert Frouin for editing suggestions.

## 6. References

1. Campbell, J.W., Blaisdell, J.M., and Darzi, M. (1995) Level -3 Sea WiFS data products: spatial and temporal binning algorithms, *NASA Technical Memorandum 104566*, **32**, 1-73.

2. Acker, J.G. (1997) Sea WiFS data available at the GSFC DAAC, *Backscatter*, **8**, 8-14.

3. Carder, K.L., Lee, Z, Hawes, S., and Chen, F.R. (1995) Optical model of ocean remote sensing: Application to ocean color algorithm development, *COSPAR Colloquium Space Remote Sensing of Subtropical Oceans*, 15A3-1-13A3-6.

4. Weidemann, A.D., Stavn, R.H., Zaneveld, J.R.V., and Wilcox, M.R. (1995) Error in predicting hydrosol backscattering from remotely sensed reflectance, *J. Geophys. Res.*, **100**, 13,163-13,177.

5. Roesler, C.S., and Perry, M.G. (1995) In situ phytoplankton absorption, fluorescence emission, and particulate backscattering spectra determined from reflectance, *J. Geophys. Res.*, **100**, 13,279-13,294.

6. Lee, Z., Carder, K.L., Peacock, T.G., Davis, C.O., and Mueller, J.L. (1996) Method to derive ocean absorption coefficients from remote-sensing reflectance, *Appl. Opt.*, **35**, 453-462.

7. Gordon, H.R., Brown, O.B., Evans, R.H., Brown, J.W., Smith, R.S., and Clark, D.K. (1988) A semianalytic radiance model of ocean color, *J. Geophys. Res.*, **93**, 10,909-10,924.

8. Kopelevich, O.V., Sheberstov, S.V., and Farroukhchine, R.Kh. (1998) Application of the statistical method of atmospheric correction of satellite ocean color data in the coast upwelling area, *Issled. Zemli iz Kosmosa*, 113-122.

9. Pope, R.M., and Fry, E.S. (1997) Absorption spectrum (380-700 nm) of pure water. II. Integrating cavity measurements, *Appl. Opt.*, **36**, 8710-8723.

10. Shifrin, K.S.(1988) *Physical optics of ocean water*, AIP Translation Series, Amer. Inst. Phys., New York.

11. Bricaud, A., Babin, M., Morel, A., and Claustre, H. (1995) Variability in the chlorophyll-specific absorption coefficients of natural phytoplankton: Analysis and parameterization, *J. Geophys. Res.*, **100**, 13,321-13,332.

12. Ovchinnikov, I.M. (1976) The currents in the straits and seas of the Meditterranean Basin, in V.A.Burkov (ed.), *Gydrologiya Sredizemnogo morya*, Gidrometeoizdat, Leningrad, pp.342-344.

13. Gordon, H.R. (1989) Can the Lambert - Beer low be applied to the diffuse attenuation coefficient of ocean water?, *Limnol. Oceanogr.*, **34**, 1389-1409.

14. Gordon, H.R., Clark, D.K., Brown, J.W., Brown, O.B., Evans, R.H., and Broenkow, W.W. (1983) Phytoplankton pigment concentrations in the Middle Atlantic Bight: comparison of ship determinations and CZCS estimates, *Appl. Opt.*, **22**, 20-36.

# A MATHEMATICAL FRAMEWORK FOR ZOOPLANKTON POPULATION DYNAMICS MODELING: INDIVIDUAL-BASED APPROACH

Y.M. PLIS and A.Y. PLIS
*Ukrainian Scientific and Research Institute of Ecological Problems*
*6 Bakulina St., Kharkiv 310166, Ukraine*

**Abstract.** A mathematical framework for plankton population temporal-spatial variability and assessment is presented. The applied approach is based on equations describing the transport of plankton population numerical density and biomass, as well as aging and changing body' components of individual plankton organisms. Equations for mean and variances dynamics of population organisms is also presented. A curvilinear mesh is used to improve the computational efficiency of the numerical algorithm, which is based on the method of component-by-component splitting.

## 1. Introduction

Plankton are important and abundant organisms in an aquatic ecosystem and primary consumer of plant materials and detritus. Like fishes, zooplankton are integrators of contaminants in aquatic environment, so are effective indicators of ecological stress. Plankton population dynamics is an important biological component of the forcing over the marine ecosystem. Collecting field data about plankton population numerical density and biomass temporal-spatial distributions is a very expensive and time consuming procedure. At the same time mathematical models prove to be relatively inexpensive and effective tools for obtaining information about temporal-spatial variability of the key physical, chemical and biological components of aquatic ecosystems.

At present time most food web models do not take individual properties of organisms into account and aggregate them on the populational level, mostly by scalar biomass characteristics. In ecotoxicology, however, this approach has not proved to be successful because an impact of any pollutant occurs at the level of the individual, not at the population level. Thus intra-population resolution is needed if one is to estimate the ability of a population, consisting of individuals of differing ages and physiological characteristics, to withstand severe environmental stress by pollutants impact.

Individual-based models can greatly increase our understanding of the system



*P. Malanotte-Rizzoli and V.N. Eremeev (eds.),*
*The Eastern Mediterranean as a Laboratory Basin for the Assessment of Contrasting Ecosystems*, 327–333.

under study because they allow us to include a description of the actual mechanisms and processes that determine the vital rates of the different classes of individuals that make up the interacting populations.

Our modeling approach applies equations describing transport of population characteristics (number of organisms, integral lipid content, integral accumulated toxicant content and etc.) in the water body space, as well as aging and changing physiological characteristics of individual organisms. In general case, it takes into account following physical and biological processes: hydrodynamics, advective-diffusive transport, chemical and biological transformation of pollutants, bioaccumulation, individual organism's life-history and population dynamics, lethal and sublethal impacts of toxicants.

The approach is enhanced by taking into account not only mean characteristics of aquatic populations but also the variances of their distributions. Although knowledge of a population's mean characteristic is important information for environmental managers, it is often more important to know what the condition of the tails of the population is. Consequently, this approach should increase the utility of the developed model for assessing the ecological risks posed by toxicants or other physical and chemical alterations in aquatic communities.

## 2. Basic Equations

### 2.1. ZOOPLANKTON POPULATION DYNAMICS

Independent variables of our zooplankton population model consist of time $t$, spatial coordinates $\mathbf{x}=(x_1,x_2,x_3)$, physiological properties of an individual plankton organism $\boldsymbol{m}=(m_1,m_2,...,m_n)$, and its age $a$. Following Cloutman and Cloutman, (1994), we introduce a distribution function $P(t,\mathbf{x},\boldsymbol{m},a)$ such that $Pd\mathbf{x}d\boldsymbol{m}da$ is the number of individuals of the plankton population in the volume of a linear space of our independent variables between $\mathbf{x}$, $\boldsymbol{m}$, $a$ and $\mathbf{x}+d\mathbf{x}$, $\boldsymbol{m}+d\boldsymbol{m}$, $a+da$.

The transport $P$ can be described by the classical advection-dispersion equation combined with the equation of a physiologically structured population dynamics

$$\frac{\partial P}{\partial t} + LP + \Lambda P = -DP + Q_P,$$
$$LP = \nabla_\alpha (v^\alpha P - D^{\alpha\beta} \nabla_\beta P), \quad \Lambda P = \frac{\partial P}{\partial a} + \sum_{j=1}^{n} \frac{\partial}{\partial m_j}\left(\frac{dm_j}{dt} P\right), \tag{1}$$

where $D$ and $Q_P$ are, respectively, a mortality rate, and rate of external source of plankton; $\nabla_\alpha$ is the covariant derivative; $v^\alpha$ is the contravariant component of a velocity of currents; $D^{\alpha\beta}$ is the tensor of coefficients of the turbulent diffusion; $m_j$ are elements of the vector $\boldsymbol{m}$; $n$ is the number of components of the $\boldsymbol{m}$.

The term $LP$ describes the transport along independent spatial variables $\mathbf{x}$. The term $\Lambda P$ describes the transport along $n$ independent physiological variables and age $a$.

The birth process for the population is described by a boundary condition

$$P(t,\mathbf{x},\mathbf{m},a=0) = \int_0^A \int_{\mathbf{m}} \beta(t,\mathbf{x},\mathbf{m},\gamma)\, P(t,\mathbf{x},\mathbf{m},\gamma)\, d\mathbf{m}\, d\gamma\,, \tag{2}$$

where $A$ is a limiting age of a plankton organism; $\beta(t,x,m,\gamma)$ is a birth function that represents the number of neonates born to an individual of age $\gamma$ and mass-characteristics $\boldsymbol{m}$ at time $t$.

## 2.2 . ZOOPLANKTON INDIVIDUAL MODEL

To specify the equation (1) we use a mathematical model describing the age dynamics of a plankton individual. In general, such dynamics can be described by a system of nonlinear equations for elements $m_j$ of the vector $\boldsymbol{m}$

$$\partial m_j / \partial a = F_j(\mathbf{m}, t, a), \tag{3}$$

where $F_j$ is a rate function, describing the input and output flows of mass-characteristics $m_j$.

## 2.3. MOMENTS OF MASS-DISTRIBUTIONS

To avoid solving the equations (1) for all individual organisms composing the population we integrate the equation (1) over $\boldsymbol{m}$. Assuming in accordance with
Metz and Diekmann (1986) that

$$\int_{\mathbf{m}} \Lambda P\, d\mathbf{m} = 0, \tag{4}$$

we obtain the equation for calculations of the population numerical density of an age-structured plankton population as

$$\partial N/\partial t + \partial N/\partial a + L N = -DN + Q, \tag{5}$$

where $N = \int_{\mathbf{m}} P\, d\mathbf{m}$, $Q = \int_{\mathbf{m}} Q_P\, d\mathbf{m}$.

In an effort to preserve the information about mass diversity between individual organisms with the same age we consider mass characteristics $\boldsymbol{m}$ as a set of random variables. Their diversity originates from the fact that individual organisms with different age and corresponding mass characteristics have offspring of which mass characteristics differ also.

Let $m_j$ be one of the components of the vector $\mathbf{m}$ for an individual organism at age $a$, which was born by an individual at age γ, and $p$ is the probability of this event. For each $m_j$ we can introduce following expressions for the mean $\bar{m}$ and variance $\sigma^2$:

$$\bar{m}(t,x,a) = \int_0^A p(t,x,a,\gamma)\, m_j(t,x,a,\gamma)\, d\gamma,$$
$$\sigma^2(t,x,a) = \int_0^A p(t,x,a,\gamma)[\bar{m}(t,x,a) - m_j(t,x,a,\gamma)]^2\, d\gamma. \tag{6}$$

The probability $p$ describes the mass distribution between individual organisms of the same age. At the time $t$ and age $a$=0 this function is

$$p(t,x,0,\gamma) = \beta(t,x,\gamma)[\int_0^A \beta(t,x,v)N(t,x,v)dv]^{-1}. \tag{7}$$

As result of the assumption that the mortality $D$ is the same for individual organisms of the same age we can obtain a following condition of the invariance of $p$ with the age $a$

$$p(t+a,x,a,\gamma) = p(t+a,x,0,\gamma). \tag{8}$$

## 2.4. ADVECTION-DIFFUSION TRANSPORT

Using the assumption about ability of zooplankton to vertical migrations the transport term $LP$ of equation (1) was written using curvilinear orthogonal in the horizontal plane coordinates $(\xi,\eta)$ (Plis, 1992)

$$LP = \frac{1}{Jh}[\frac{\partial}{\partial\xi}(JhS^{\xi}) + \frac{\partial}{\partial\eta}(JhS^{\eta}) - \frac{\partial}{\partial\xi}(hD_L\frac{\partial P}{\partial\xi}) - \frac{\partial}{\partial\eta}(hD_L\frac{\partial P}{\partial\eta})],$$
$$S^{\xi} = \int_0^h v^{\xi}\, dz, \quad S^{\eta} = \int_0^h v^{\eta}\, dz, \tag{9}$$

where $J = \partial(x,y)/\partial(\xi,\eta)$ is the Jacobian of the transformation from Cartesian coordinates $(x, y)$ to curvilinear coordinates $(\xi,\eta)$; $h$ is a depth; $v^{\xi}$ and $v^{\eta}$ are the contravariant components of current velocity; $D_L$ is the horizontal coefficient of turbulent diffusion; $S^{\xi}$ and $S^{\eta}$ are components of the full stream vector.

This approach permits the water body to be represented in the transformed coordinates as a simple rectangular shape and to be approximated by a grid with the optimal number of nodes for numerical approximation at an acceptable level of accuracy. It was employed to improve the efficiency of a numerical algorithm in regard to computational memory for the essentially multivariable population model.

## 3. Algorithm of Solution

### 3.1 . OPERATOR SPLITTING METHOD

Component-by-Component operator splitting method has been applied as an approximate technique to integrate equation (5) with advection-diffusion term (9) over an arbitrary time interval $\Delta t$. This integration can be written as

$$N_{t+\Delta t} - N_t = \int_t^{t+\Delta t} LNdt - \int_t^{t+\Delta t} \frac{\partial N}{\partial a} dt + \int_t^{t+\Delta t} (-DN + Q)\, dt \qquad (10)$$

The integration (10) is split into three stages. Each of the stages uses solutions of the previous stage as initial conditions. The same procedure should be applied to integration of equations describing dynamics and spatial distribution of population's mass-characteristic $M_j = Nm_j$. The attractive feature of operator splitting procedure is that each stage can be solved using a different numerical technique that is specially suited to achieve high accuracy for each integral in (10).

The finite difference approximation of the first integral in (10) was obtained using the QUICK scheme, originally developed by Leonard (1979). At moment $t = t + \Delta t$ in any $(i,k)$-th cell of the grid, approximating the water body, the solution was obtained as the solution of the following system of three-diagonal algebraic equations

$$\sum_{l=-1}^{1} \Psi^{\xi}{}_{ik} N_{ik} (t + \Delta t, a + \Delta a) = \Phi^{\xi}{}_{ik} (t, a), \quad \sum_{k=-1}^{1} \Psi^{\eta}{}_{ik} N_{ik} (t + \Delta t, a + \Delta a) = \Phi^{\eta}{}_{ik} (t, a),$$

with coefficients $\Psi$ and $\Phi$, describing the properties of the numerical scheme.

The second integral in (10) is solved by implementing the 'coherent' cohort model analytical solution (Brewer, 1989). The third integral is solved by a second-order explicit Runge-Kutta method. For plankton population structured by $K$ age groups we should apply the same procedure for each of these groups.

### 3.2. DYNAMICS OF $m$ AND $\sigma^2$

The algorithm of numerical calculations of age-dynamics and transport of moments of plankton populational characteristics was initially proposed by Plis and Barber (1995).

Let $m'(\mathbf{m}, t, a)$ be a solution of equation (3) for any $m_j$ at the initial condition $m'(\mathbf{m}, t-a, a) = m(t-a, a)$,

so, that

$$m(\mathbf{m}, t, a, \gamma) = m'(\mathbf{m}, t, a) + \varepsilon\,(\mathbf{m}, t, a, \gamma), \qquad (11)$$

where $\varepsilon$ supposed to be small comparatively with $m'$.

Applying the Euler-Cauchy approximation method to equation (3) and using the Teylor-series expansion of $F(m, a)$ about $m = m'$ we obtain with the order of magnitude $O(\varepsilon^2)$ that

$$m(t+\Delta a, a+\Delta a) = m'(t+\Delta a, a+\Delta a),$$

$$\sigma^2(t+\Delta a, a+\Delta a) = \prod_{i=0}^{k} g^2(t-a+i\Delta a)\,\sigma^2(t-\Delta a, 0), \qquad (12)$$

where

$k\Delta a = a$, $g(a) = 1 + (\partial F/\partial \mathbf{m})_{m=m'(a)}$.

### 3.3. SPATIAL TRANSPORT OF $\mathbf{m}$ AND $\sigma^2$

The transport of mean $E$ and variance $Var$ of the population mass-characteristic $M=Nm_j$ at any $(i,j)$-th cell of the computational grid can be calculated as

$$E(M_{i,j}) = \sum_{k=-1}^{1}\sum_{m=-1}^{1}\phi_{i+k,j+m}m_{i+k,j+m},$$

$$Var(M_{i,j}) = \sum_{s=-1}^{1}\sum_{l=-1}^{1}\sum_{k=-1}^{1}\sum_{m=-1}^{1}\phi_{i+s,j+l}\phi_{i+k,j+m}\rho_{i+s,j+l,i+k,j+m}\sigma_{i+s,j+l}\sigma_{i+k,j+m}, \qquad (13)$$

where $\rho$ is a covariance coefficient; $\phi$ are the coefficients describing the hydrological conditions of a simulation scenario and properties of the numerical scheme.

Let us assume that in each cell of our spatial grid the individual mass-characteristics $m$ associated with the organisms of the same age are statistically dependent ($\rho = 1$), but they are statistically independent ($\rho = 0$) relative to the same variables in neighboring cells. So that at any moment $t$ after the transport was accomplished, the mean and variance of $m$ are

$$m_{i,j}(t,a) = \frac{E(M_{i,j}(t,a))}{N_{i,j}(t,a)}, \quad \sigma^2_{i,j}(t,a) = \frac{Var(M^2_{i,j}(t,a))}{N^2_{i,j}(t,a)}. \qquad (14)$$

## 4. Application and Conclusions

The presented approach was applied to zooplankton population modeling in the upper portion of Lake Hartwell (South Carolina, USA). The case of PCBs impacting zooplankton population was investigated. To develop the model of an individual zooplanktoner we applied the mathematical model describing the life history and bioenergetics of an individual daphnid (Hallam et al., 1990). An individual organism was represented as consisting of three body components: protein, lipid and accumulated toxicant.

The simulations gave us the information about tendencies of seasonal dynamics of spatial distributions of zooplankton populational and individual biomass characteristics, numerical densities of plankton organisms and their age-distributions.

Application of the approach provide, we believe, both diagnostic and prognostic information beyond that available from traditional approaches to the evaluation of ecological effects. The mathematical framework should be effective practical tools for evaluations of ecological risk of plankton populations, and can produce much valuable information at a high level of.

## 5. References

1. Brewer, J.W. (1989) Spreadsheets, PC's, and the finite-difference solutions for ecological distribution, *Ecological Modeling*, 47,65-83.
2. Cloutman, D.G and Cloutman L.D. (1994) A unified mathematical framework for population dynamics modeling, *Ecological Modeling*, 71,131-160.
3. Hallam, T.G., Lassiter R.R., Li J., and Suarez L.A. (1990) Modelling individuals employing an integrated energy response: application to *Daphnia*, *Ecology*, 71(3),938-954.
4. Leonard, B.P. (1979) A stable and accurate modelling procedure based on quadratic upstream modelling. *Computational Methods in Applied Mechanical Engineering*, 19,59-98.
5. Metz, J.A.J. and O. Diekmann (1986) The dynamics of physiologically structured populations, in *Lecture Notes in Biomathematics*, Springer-Verlag, Berlin, Vol.68.
6. Plis, Y.M. (1992) An approach to calculating wind-driven currents and transport of substances in unstratified water bodies using curvilinear coordinates, *Water Resources Research*,28(1), 83-88.
7. Plis Y.M. and Barber M.C. (1997) Patterns of response of zooplankton populations to toxicants: A modeling study, in *Fish Physiology, Toxicology and Water Quality*, EPA/600/R-97/098, pp.245-251.

# SPATIAL AND SEASONAL VARIABILITY OF WATER AND BIOGEOCHEMICAL FLUXES IN THE ADRIATIC SEA

M. GACIC**, G. CIVITARESE*, L. URSELLA**
*** Osservatorio Geofisico Sperimentale, P. O. Box 2011, I-34016 Trieste (Italy)*
** CNR - Istituto Sperimentale Talassografico, V. le R. Gessi 2, I-34123 Trieste (Italy)*

***Abstract***

*The intensive basin-wide hydrographic sampling and current measurement programme within one-year period (1995/1996) in the Adriatic Sea made it possible to calculate estimates of the southward fluxes of water, suspended and dissolved matter at selected transects. From presented results it was shown that about 75% of the water entering the Adriatic through the Strait of Otranto recirculates within the South Adriatic Gyre. The contribution of the Northern Adriatic in terms of the water represents only 4 to 5% of the total volume of water exchanged through the Strait of Otranto. Consequently, the influence of the Northern Adriatic shallow area in determining bio-chemical properties of the deep South Adriatic Pit is rather small. This conclusion is in a good agreement with the recently obtained evidences that the Adriatic Sea as a whole acts as a mineralization basin with respect to the Ionian Sea.*

## 1. Introduction

The Adriatic Sea can be divided into three distinct sub-basins on the basis of their oceanographic characteristics and bathymetric features (Fig. 1). The northern part is a shallow shelf area with a gently sloping bottom down to the 100 m isobath at its southern limit. The central part encounters the Middle Adriatic Pit; a circular 270 m deep area delimited by the 170 m deep Palagruza Sill to the south. Finally, the South Adriatic Pit is roughly of a circular shape and the deepest portion of the Adriatic Sea (maximum depth about 1200 m). It is delimited by the 750 m deep Sill of Otranto in the south and by narrow smooth shelves along the coasts.
The freshwater discharge in the Adriatic is rather high (up to 5700 $m^3$/sec), and more than 50% of this discharge is concentrated in the northern shallow part.
The circulation in the Adriatic is forced by the longitudinal pressure gradient caused by the freshwater contributions to the northern shallow part as well as by the differential



*P. Malanotte-Rizzoli and V.N. Eremeev (eds.),*
*The Eastern Mediterranean as a Laboratory Basin for the Assessment of Contrasting Ecosystems*, 335–357.

cooling/heating of the water column [1] during winter/summer (the heat losses/gains in the northern part of the Adriatic are larger than in the Central and South Adriatic). During the winter, dense water is formed in both the northern shelf area and in the South Adriatic Pit. The Northern Adriatic shelf area is a source of two water masses, one relatively fresh in the surface layer due to the influence of the riverine inflow, and the other cold and dense formed in the winter, which occupies the bottom layer. The southward spreading of these two water masses, occurs in the form of a narrow swift surface coastal current along the Italian shore with a width of the order of 10 kilometers, and in the form of a bottom density driven current, respectively. The much larger compensating inflow occurs along the eastern coast.

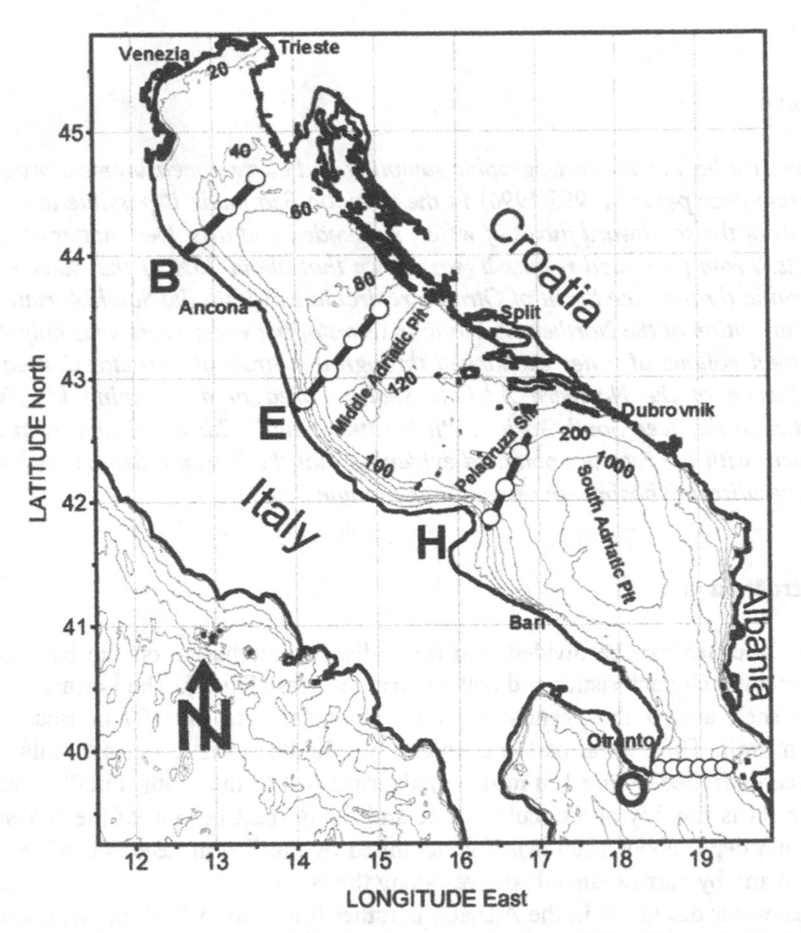


*Figure 1.* **Bathymetric map of the Adriatic Sea. Depths are given in meters. CTD and biogeochemical sampling transects are marked by lines, while the mooring positions are marked by open circles.**

This transversal asymmetry is possible since the width of the Adriatic Sea is about 200 km, i.e. an order of magnitude larger than the maximum value of the internal Rossby radius of deformation [2, 3].

One of the objectives of the project PRISMA-1 (Programma di RIcerca e Sperimentazione per il Mare Adriatico) was to assess longitudinal fluxes of water, and suspended and dissolved matter across four transects delimiting the three Adriatic sub-basins and in the Strait of Otranto - the communicating inlet between the Adriatic and Ionian Seas (Fig. 1).

In this paper, the estimated fluxes of water, and suspended and dissolved matter based on the measurements carried out during the period May 1995 - February 1996 for the summer (stratified water column) and winter (homogeneous water column) will be presented and discussed. Furthermore, the aim of the present research is to assess the importance of the Northern Adriatic eutrophic area in the functioning of the entire Adriatic as a unique physical and biogeochemical system. Since the coverage of the selected transects by measurements was not complete (the measurements were carried out up to the Croatian territorial waters), only estimates of the southward transport along the western shore could have been given.

The paper is organized as follows: in section 2, a brief description of the experimental design is given; section 3 contains a short discussion of the vertical distributions of thermohaline properties, and of some chemical parameters like dissolved oxygen and nutrients, as well as chlorophyll a; in section 4, the vertical distributions of the longitudinal current components and their seasonal variability are described; section 5 gives the methodology of flux calculations and includes a discussion on the estimates of dissolved and suspended matter fluxes; finally, section 6 contains conclusions.

## 2. Data sets and experimental design

Four seasonal cruises with the *R/V Urania* were carried out in May, August, and October 1995 and February 1996. The spatial distribution of the sampling locations is given in the Fig. 1. The following parameters were sampled: temperature and salinity with a SeaBird CTD probe and dissolved oxygen, dissolved inorganic nitrogen (DIN), phosphates, silicates, chlorophyll a and total suspended matter (TSM) determined by standard chemical methods [4, 5]. Current measurements were carried out using moored current meters at selected locations (see Fig. 1 for instrument locations). Independently, but simultaneously, an intensive current measurement and hydrological sampling program was undertaken in the Strait of Otranto [6, 7]. In addition, a basin-wide survey with a ship-borne ADCP was carried out on a continuous basis from May 1995 until February 1996 with the *R/V Thetis*.

## 3. Vertical distribution of physical and biogeochemical properties

The phenomenological description of the spatio-temporal variability of the thermohaline and biogeochemical parameters will be focused on the comparison of the functioning of the shallow northern area influenced heavily by the freshwater discharge and vertical convection, with that of the deeper Adriatic areas which is mainly influenced by the horizontal advection of open-sea waters. In this context, attention will be mainly paid to the northernmost transect B located in an area 50 m deep and to transect H (see Fig. 1 for transect locations).

Vertical distributions of the water temperature reveal a complete vertical homogenization of the water column along transect B only in winter (Fig. 2a, b and c). In the salinity field, the vertical homogenization is evident in all seasons excepting summer. A coastal layer is prominent in the salinity and density fields in all seasons. Its width, however varies in time reaching a minimum in winter when it does not exceed 10 km, due to the smaller horizontal length scales in this season (yearly minimum of the baroclinic radius of deformation).

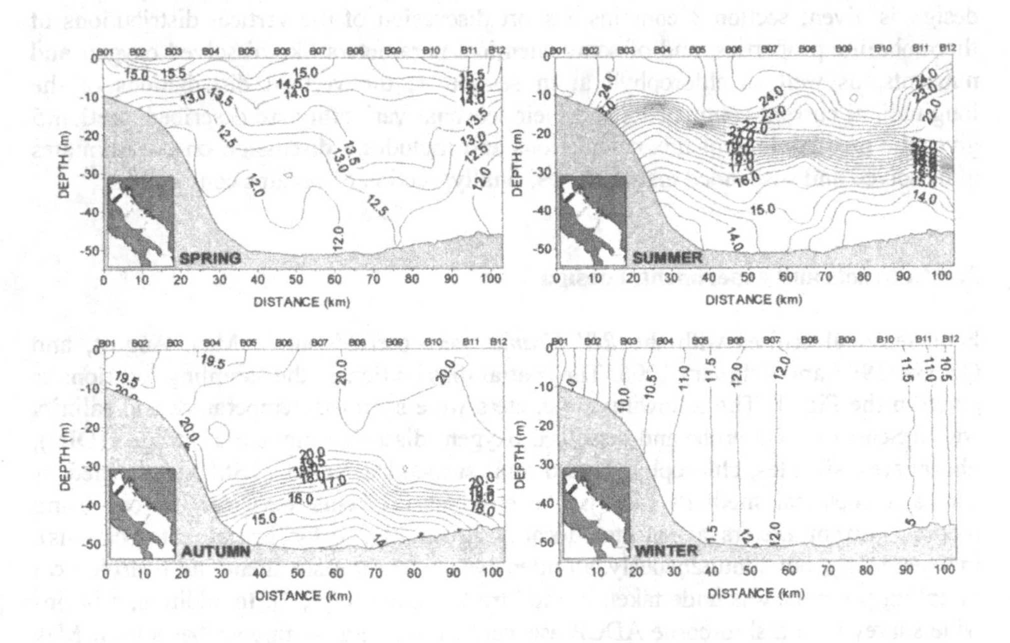


*Figure 2a.* Vertical distributions of temperature (C) for transect B.

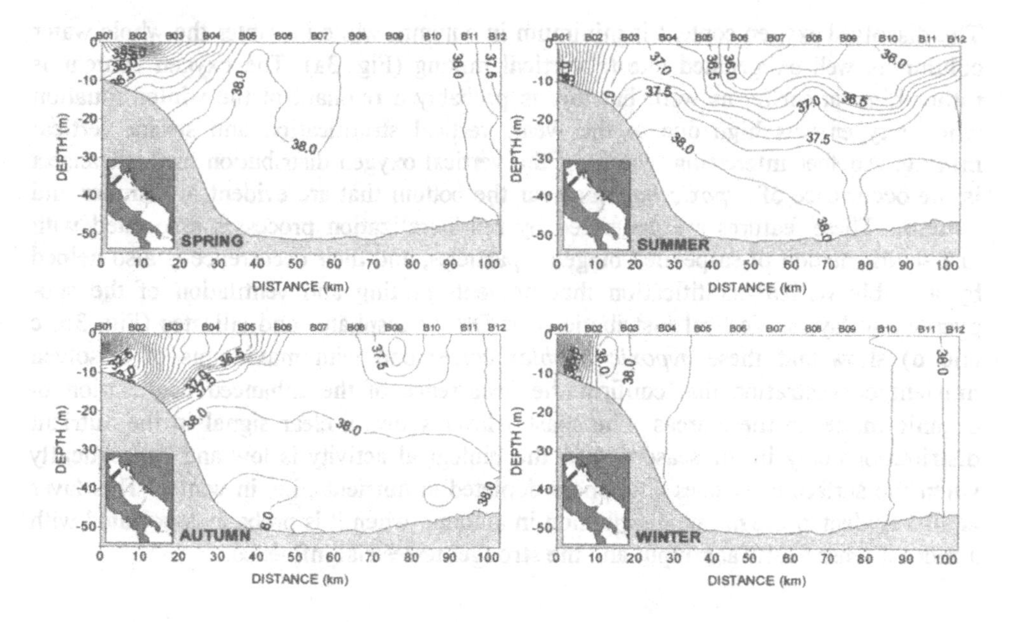


*Figure 2b.* **Vertical distributions of salinity for transect B.**

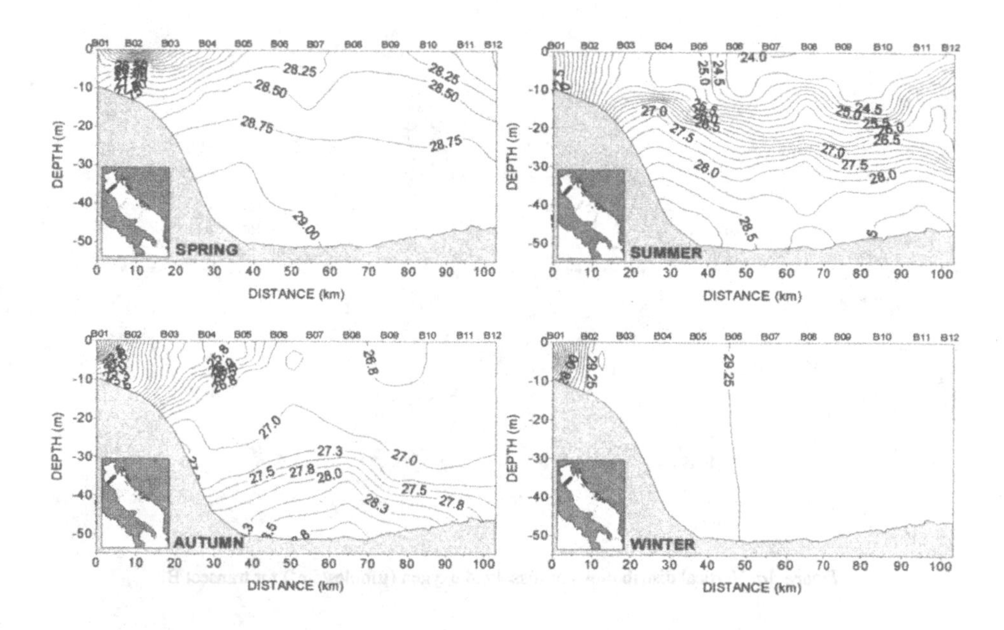


*Figure 2c.* **Vertical distributions of density (kg/m$^3$) for transect B.**

The dissolved oxygen content is minimum in autumn, while in winter the whole water column is well oxygenated due to vertical mixing (Fig. 3a). The oxygen content is rather high in spring as well, but this is probably a remnant of the winter situation when oxygen was high due to the weak vertical stratification and strong vertical mixing. Another interesting feature of the vertical oxygen distribution in this transect is the occurrence of *hypoxic bubbles* near the bottom that are evident in summer and autumn. These features are generated by remineralization processes associated with fast sedimentation of suspended biogenic particles, and their occurrence is also helped by a stable vertical stratification that prevents mixing and ventilation of the sub-pycnocline layers. Vertical distributions of DIN, phosphates and silicates (Fig. 3b, c and d) show that these *hypoxic bubbles* correspond with maximums of dissolved nutrient concentration that confirm the occurrence of the enhanced degradation of organic matter in these areas. The coastal layer shows a clear signal in the nutrient distributions only in the season when the biological activity is low and consequently when the surface layer does not appear depleted in nutrients, i.e. in winter. This layer is also evident in the nitrate distribution in autumn, when it is probably associated with a more intense freshwater input and the stronger terrestrial influence.

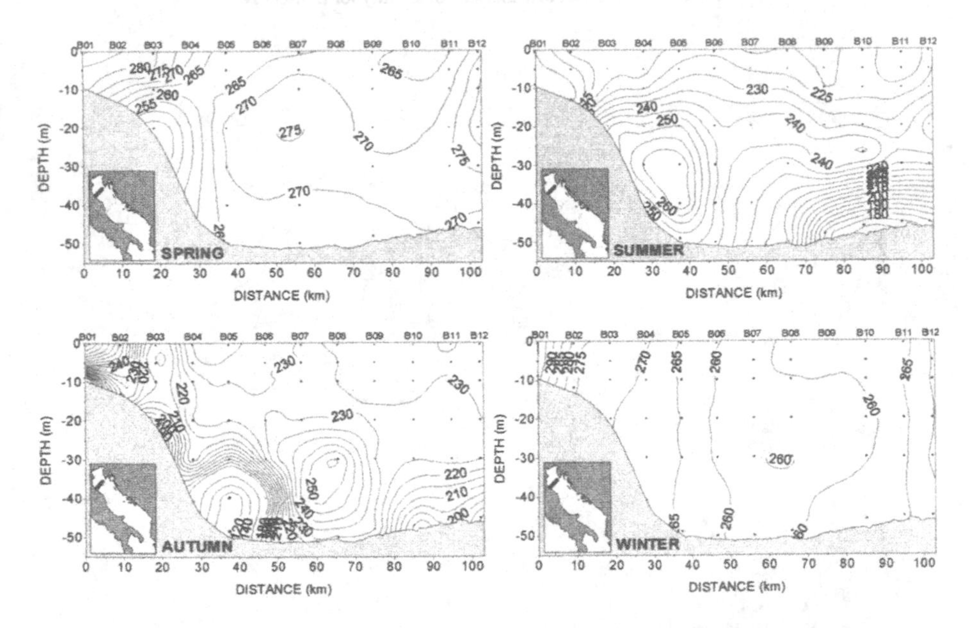


*Figure 3a.* Vertical distributions of dissolved oxygen (µmoles/dm$^3$) for transect B.

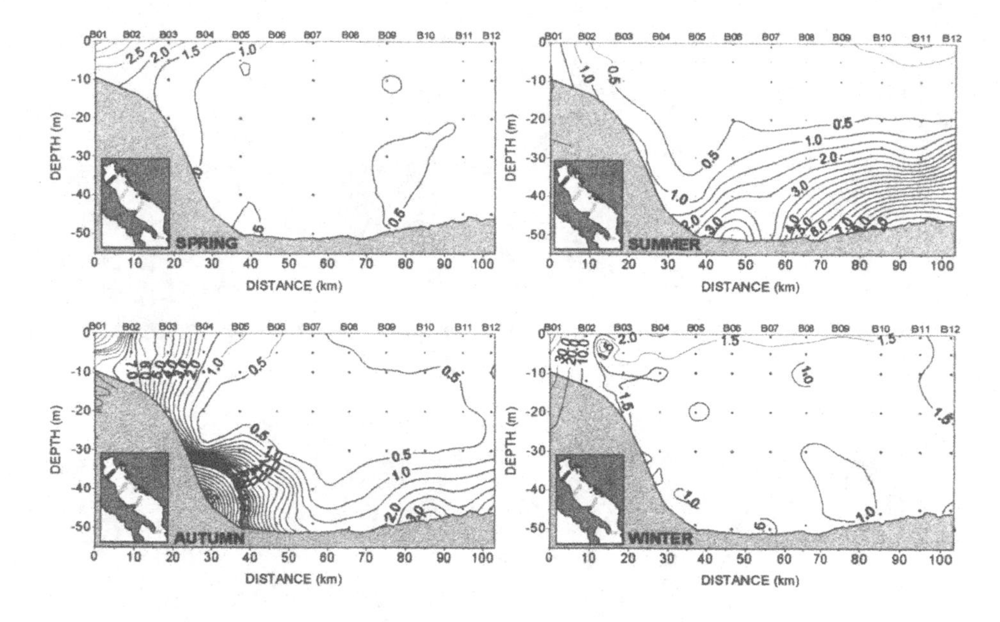


*Figure 3b.* **Vertical distributions of DIN (μmoles/dm³) for transect B.**

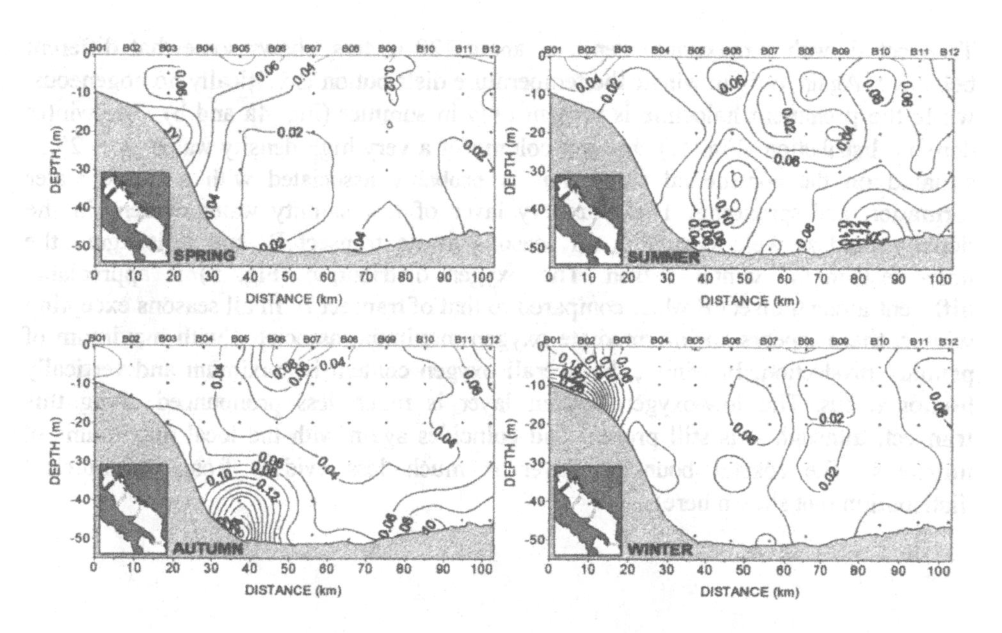


*Figure 3c.* **Vertical distributions of phosphate (μmoles/dm³) for transect B.**

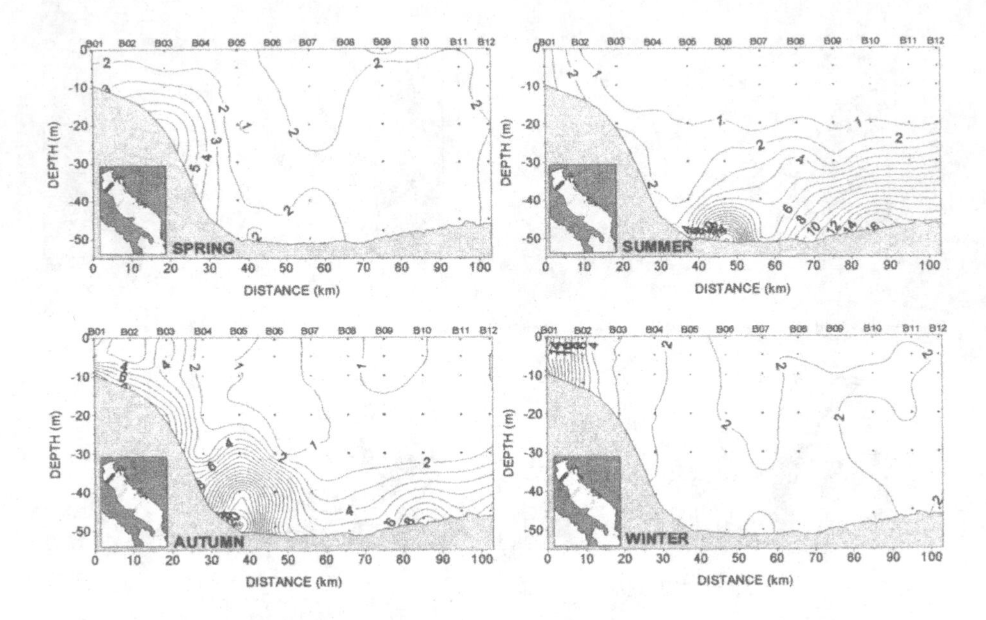


*Figure 3d.* Vertical distributions of silicate (μmoles/dm$^3$) for transect B.

Transect E, with a maximum depth of about 120 meters, shows somewhat different behavior. Again, only in winter the temperature distribution is vertically homogeneous, while the prominent halocline is present only in summer (Fig. 4a and b). The winter density distribution (Fig. 4c) shows a column of a very high density water ($\gamma_\theta > 29.3$) situated on the continental slope that is probably associated with a dense water formation and spreading. The boundary layer of low salinity water evident in the density field as well, occurs in all seasons as on transect B, and it is again the narrowest in the winter season. The oxygen distribution (Fig. 5) is appreciably different along transect E when compared to that of transect B. In all seasons excepting winter, there appears an intermediate oxygen maximum associated with maximum of primary production. In winter, the overall oxygen content is maximum and vertically homogeneous. The low-oxygen bottom layer is much less pronounced along this transect, although it is still present and coincides again with the local maximums of nutrients. The coastal boundary layer is much less evident from the nutrient distribution (not shown here).

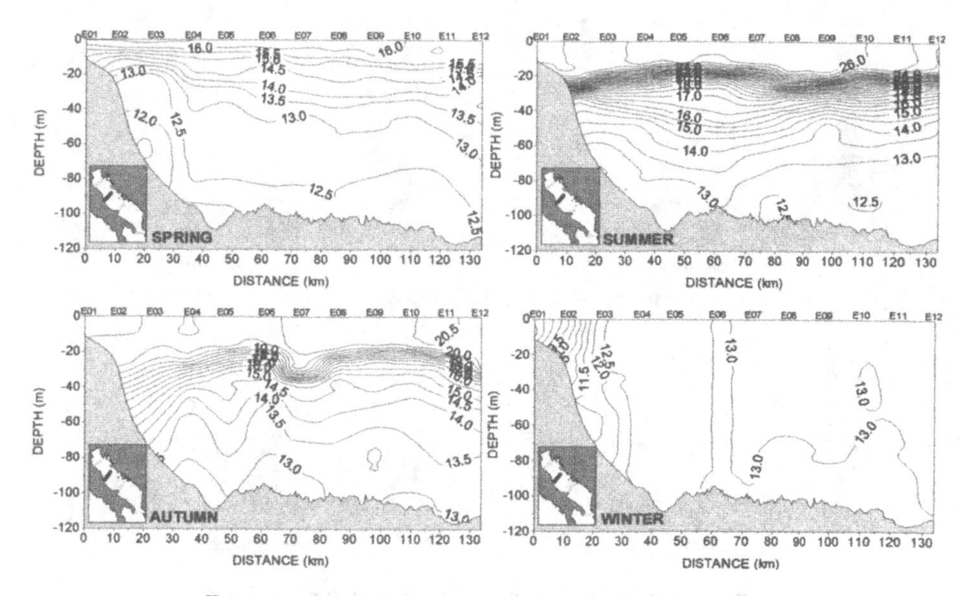


***Figure 4a.*** **Vertical distributions of temperature (C) for transect E.**

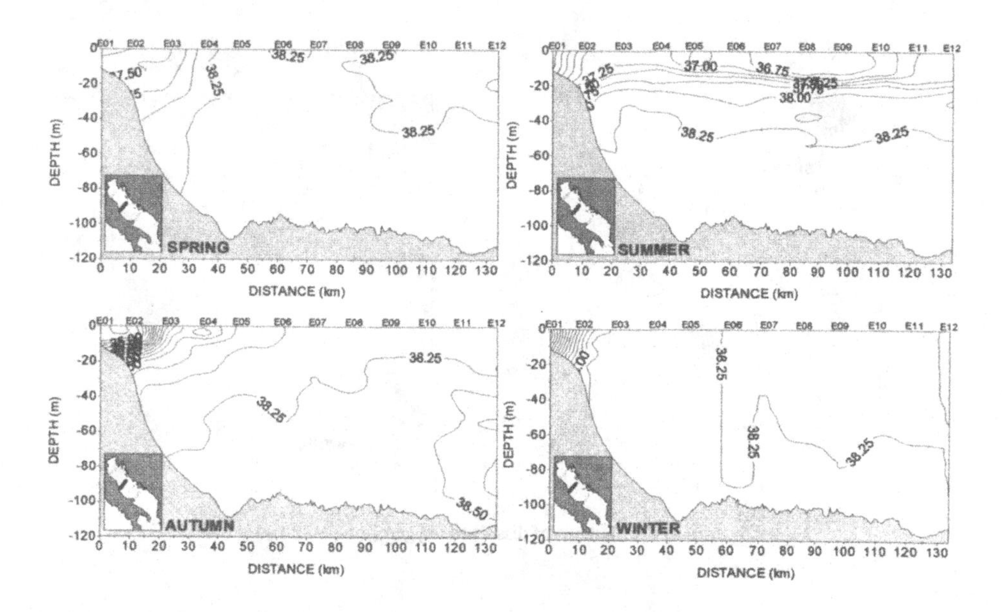


***Figure 4b.*** **Vertical distributions of salinity for transect E.**

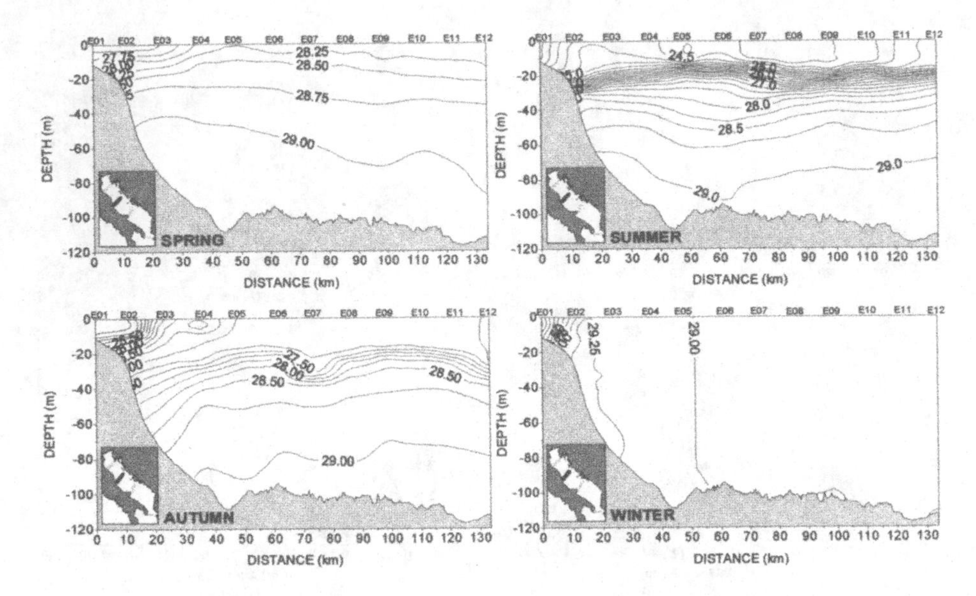


*Figure 4c.* **Vertical distributions of density (kg/m³) for transect E.**

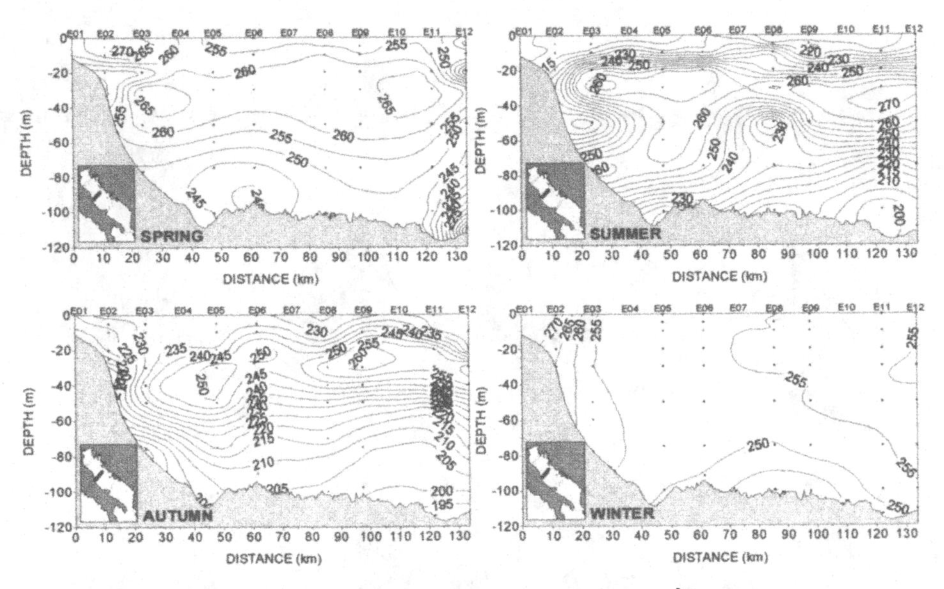


*Figure 5.* **Vertical distributions of dissolved oxygen (µmoles/dm³) for transect E.**

Deeper water areas (transect H with a maximum depth of about 230 m) display important influences of the horizontal advection processes in sub-pycnocline layers. Temperature and salinity transects (Fig. 6a and b) reveal a signal associated with the cold ($T<12.5^0$ C) and relatively fresh ($S<38.4$) vein at the continental shelf break at a depth of about 100 m.

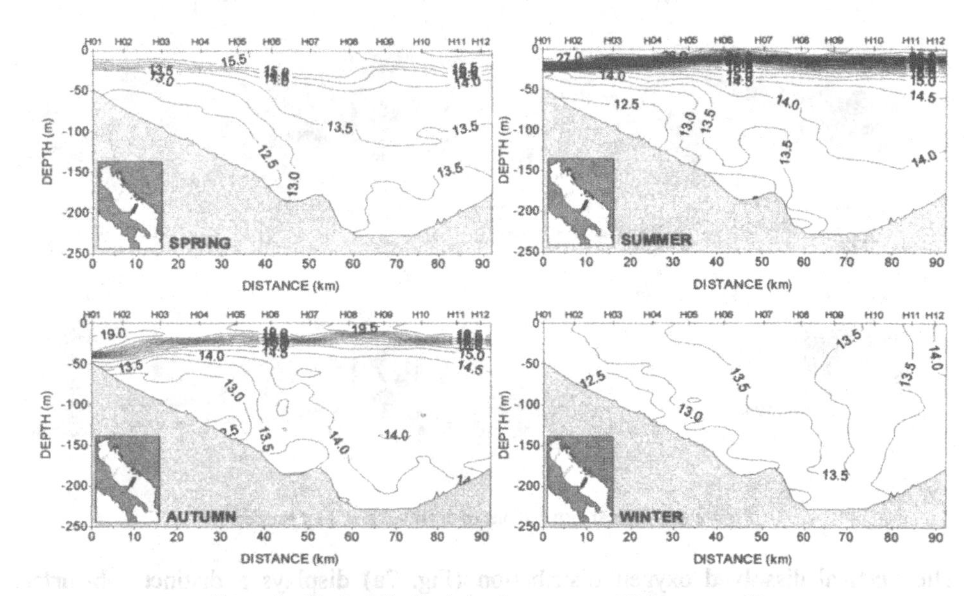


*Figure 6a.* **Vertical distributions of temperature (C) for transect H.**

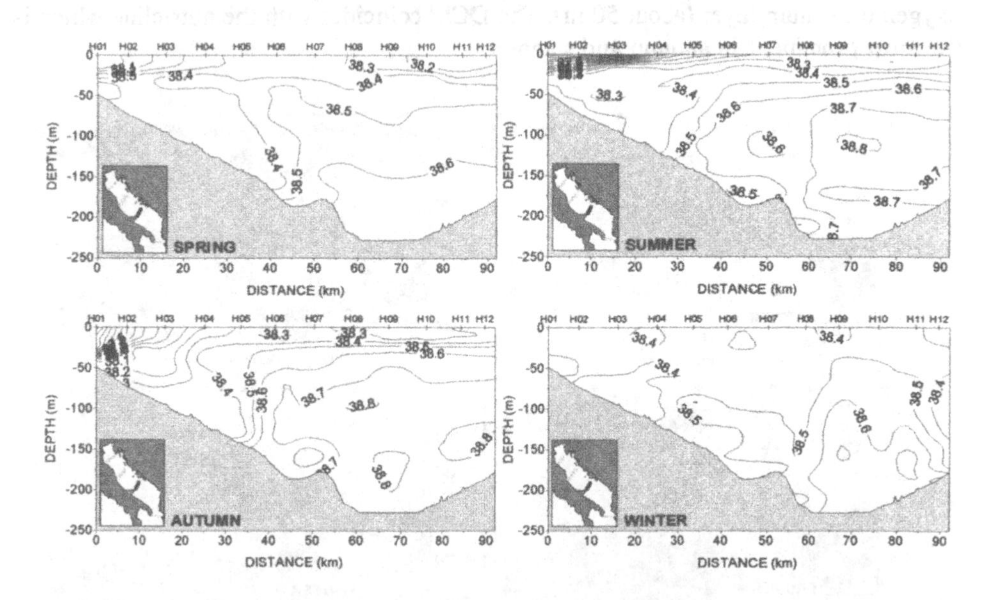


*Figure 6b.* **Vertical distributions of salinity for transect H.**

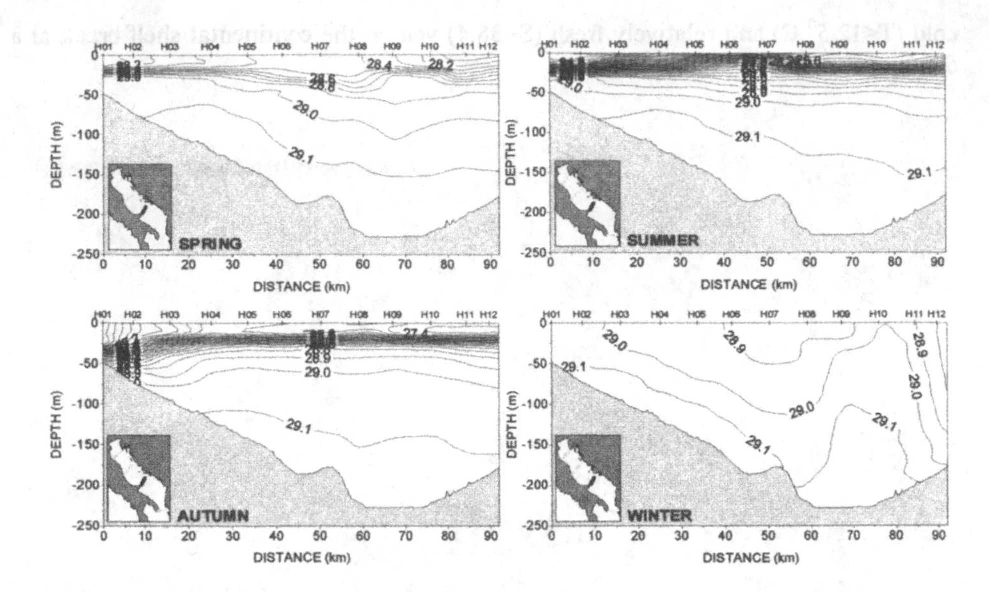


*Figure 6c.* Vertical distributions of density (kg/m$^3$) for transect H.

The vertical dissolved oxygen distribution (Fig. 7a) displays a distinct sub-surface maximum at a depth of about 25 m in all seasons excepting winter, that is associated with a Deep Chlorophyll Maximum (DCM) layer (Fig. 7e) situated slightly below the oxygen maximum layer (about 50 m). The DCM coincides with the nutricline which is the most prominent in autumn and summer.

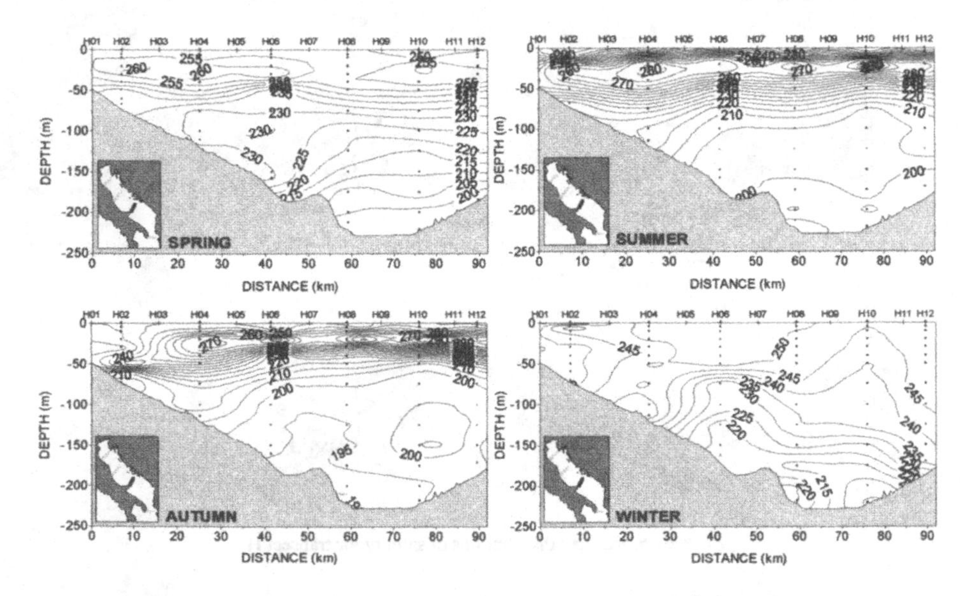


*Figure 7a.* Vertical distributions of dissolved oxygen (μmoles/dm$^3$) for transect H.

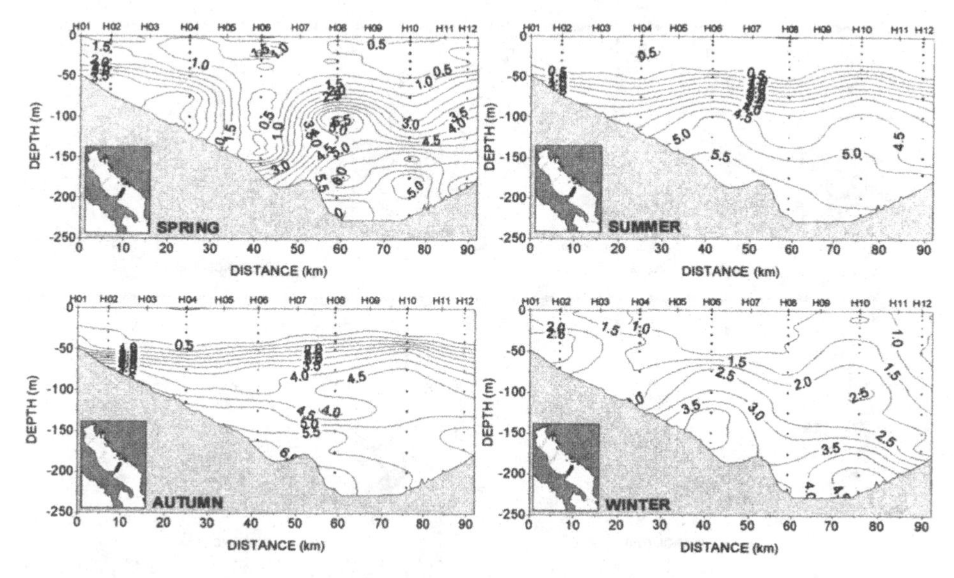


*Figure 7b.* **Vertical distributions of DIN (μmoles/dm³) for transect H.**

**The sub-pycnocline pattern in the dissolved oxygen distribution is similar to that of the distributions of DIN and phosphate (Fig. 7b and c), but is different from the silicate distribution (Fig. 7d). The maximum concentrations of silicate in the deeper layers of transect H are the result of the advection of the diatom-dominated waters from the Northern Adriatic.**

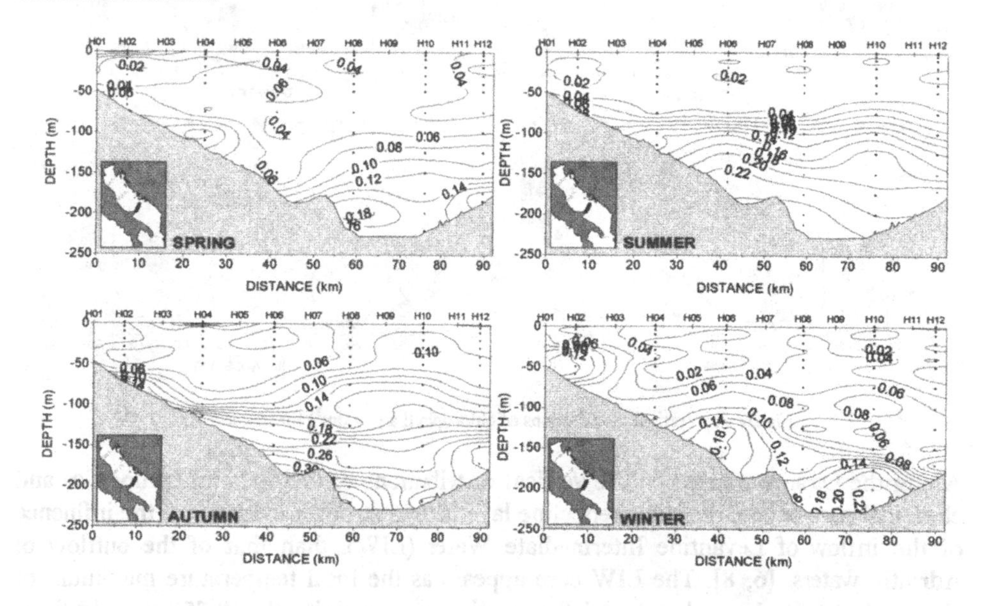


*Figure 7c.* **Vertical distributions of phosphate (μmoles/dm³) for transect H.**

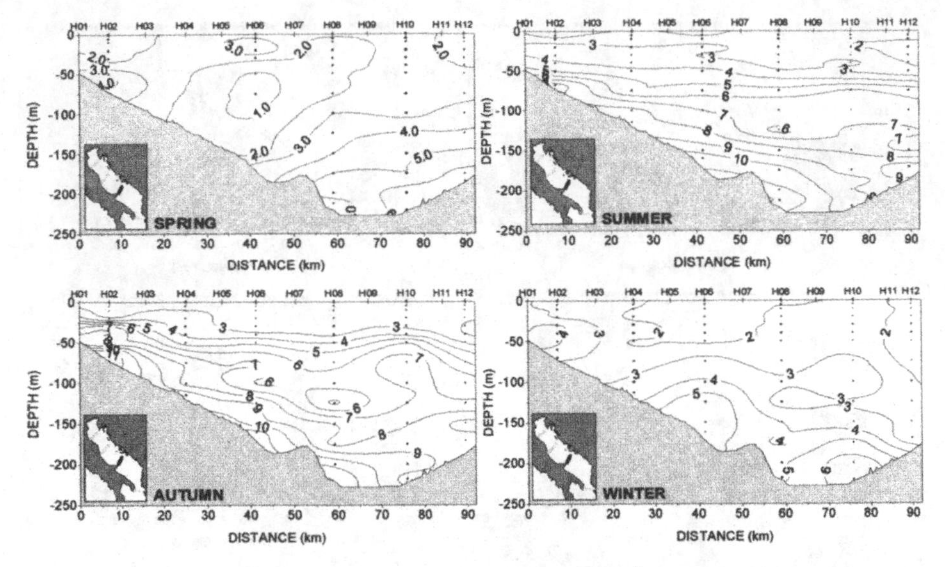


*Figure 7d.* **Vertical distributions of silicate (µmoles/dm$^3$) for transect H.**

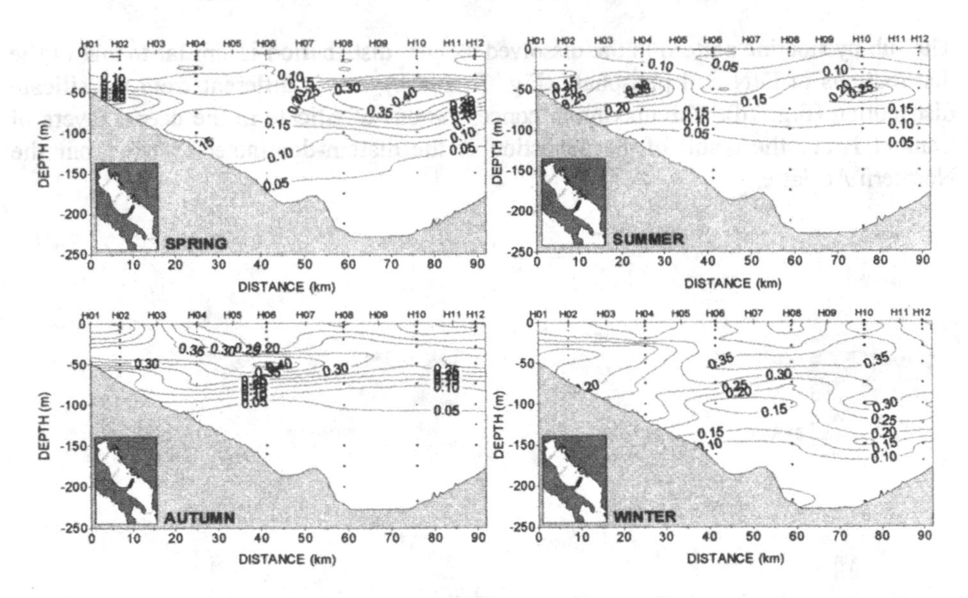


*Figure 7e.* **Vertical distributions of chlorophyll a (µg/dm$^3$) for transect H.**

Along the Otranto transect in the vertical distributions of thermohaline properties and chemical parameters, the sub-pycnocline layer shows stronger evidence of the influence of the inflow of Levantine Intermediate Water (LIW), than that of the outflow of Adriatic waters, [6, 8]. The LIW core appears as the local temperature maximum of about 14 C (not shown here) and the maximum of salinity (S>38.75) at a depth of about 300 m (Fig. 8).

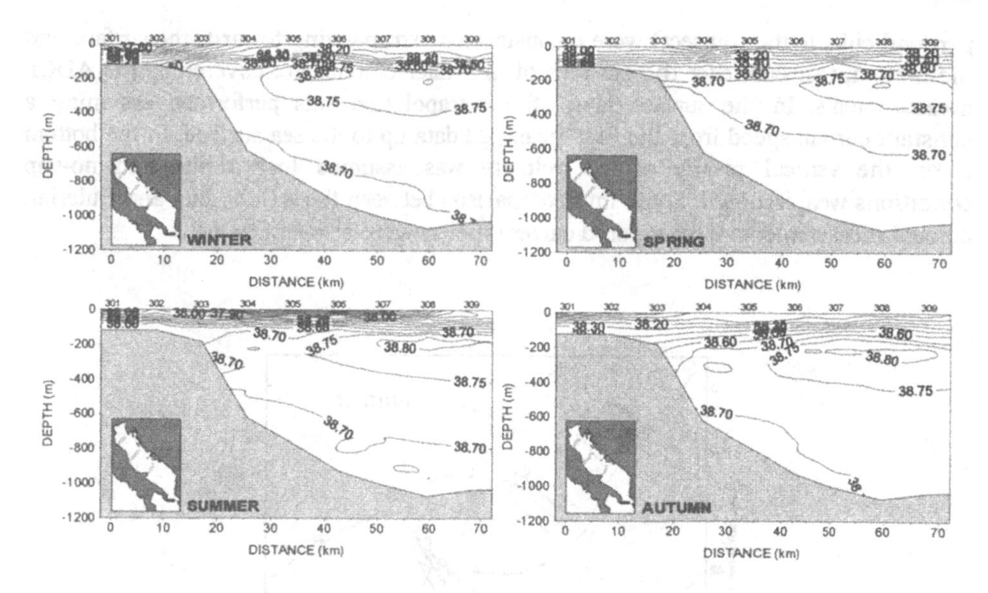


*Figure 8.* Vertical distributions of salinity for transect O (Strait of Otranto).

The tongue of LIW is pressed against the eastern coast but reaches almost the western continental slope. The seasonal thermocline is absent only in winter while the halocline is present throughout the year. The coastal boundary layer is evident in the salinity field only in autumn and winter. In addition, in winter the coastal boundary layer is characterized by the presence of the less saline water that is at the same time colder than the adjacent off-shore waters. The surface nutrient-depleted layer is present in all seasons and the intermediate layer of the maximum nutrient concentrations and the minimum dissolved oxygen content coincides in this area with the LIW core. The exception are silicates for which the maximum layer occurs below the LIW depth [8].

## 4. Longitudinal current component distributions

The entire period of the ADCP measurements was divided into two time-segments; the stratified season (May - September, 1995), designated as "summer", and the season characterized by a vertically homogeneous water column (October, 1995 - February 1996) denoted henceforth as "winter", which were then analyzed separately (the winter and summer spatial coverage of the ADCP measurements are shown in Fig. 9).

The tidal signal was eliminated with the method developed by Candela *et al.* [9]. Afterwards, the detided current field was expressed as a sum of the steady component and a time-varying one. The latter one will be called the "residual" current field in the rest of the paper and contains the time variability on all time scales from the tidal periods up to the entire record length. Some details on the data analysis are presented in a separate paper [10]. Vertical transects of a longshore current (current component perpendicular to the transect) were reconstructed extrapolating towards the surface and the bottom, to account for the portions of the water column not covered by the ADCP measurements. In the surface layer, the extrapolation was performed assuming a constant current speed from the first measured data up to the sea surface. In the bottom layer, the vertical profile of the velocity was assumed logarithmic and no-slip conditions were imposed. Some intercomparison between the ADCP data and Eulerian current measurements from moored current meters were also carried out.

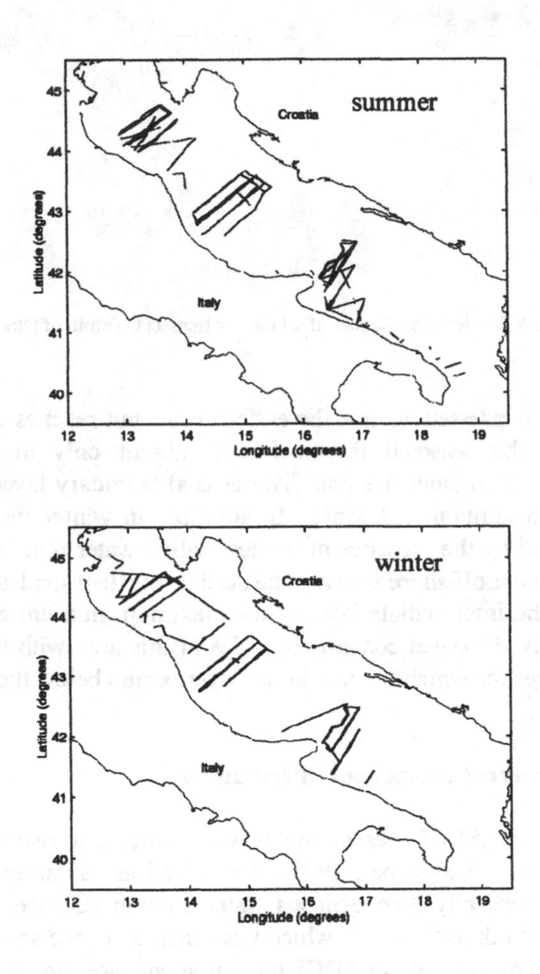


*Figure 9.* Ship-borne ADCP measurements for summer and winter. For the definition of the two seasons, see text.

Steady longshore current components for winter and summer for locations where moored current meter data were available were compared with the averaged low-pass moored currentmeter data. The results of the intercomparison were found to be satisfactory [11] considering the fact that the two measurement methods are completely independent, and also considering that the data processing and methods of calculations of the steady current field are very different.

The residual (time-dependent) current field reveals a strong small-scale variability in both time and space (Fig. 10). Typical horizontal scales are of the order of ten kilometers, while in the vertical during the stratified season, the typical scales are of the order of ten meters. During the winter season, these meso-scale structures extend from the surface to the bottom, i.e. the vertical shear is very small. The signal associated with the coastal boundary layer is clearly present in the steady current field in both summer and winter seasons but very rarely in the residual current field. In fact, the southward coastal current was noted in the residual current field only on one occasion subsequent to a strong Po River discharge pulse (Fig. 10).

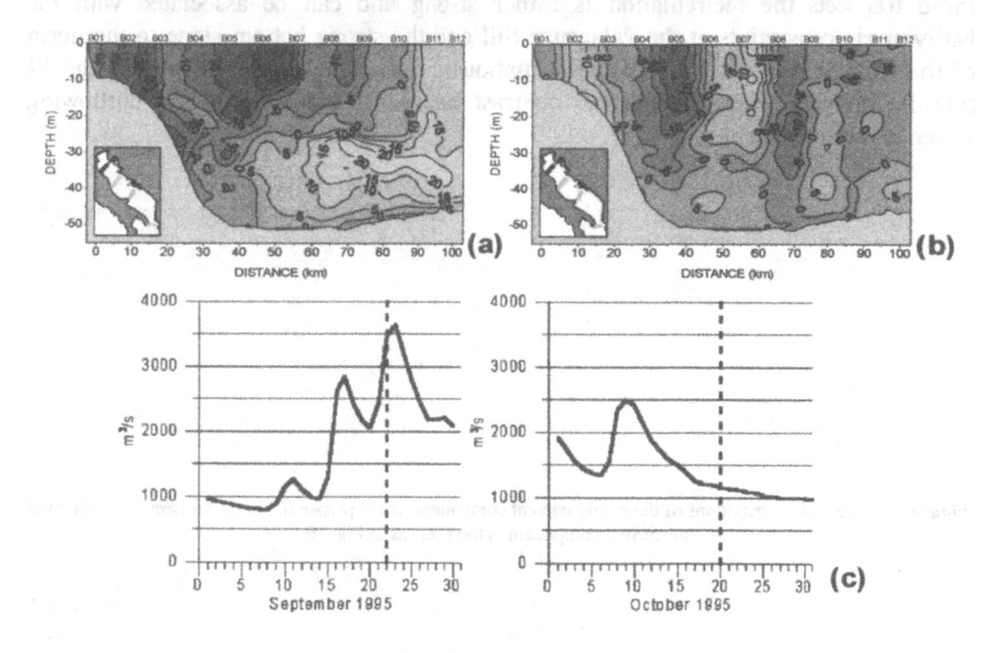


*Figure 10.* **Vertical distributions of the residual current component (cm/s) perpendicular to the transect B for 22 September 1995 (a) and 20 October 1995 (b). The current is defined negative if directed out of the paper. The corresponding Po river discharge rate is depicted below (c). Vertical dashed line indicates the ADCP measurement date.**

Comparison of summer and winter steady currents (Fig. 11a, b, c and d) reveal for all transects a more pronounced horizontal shear (cyclonic shear) in the upper thermocline layer during the winter than in summer. This is mainly due to the strengthening of the coastal boundary current during the winter. From the Strait of Otranto where there is a complete coverage by current measurements of the entire transect, an increase in both inflowing and outflowing surface and intermediate currents in the winter season is evident. On the other hand, the Adriatic Deep Water (ADW) outflow in the Strait of Otranto shows weaker variability from one season to the other.

Typically, the entire water column along transects is characterized by the cyclonic shear decreasing from the surface to the bottom. However, along the transect E to the north of the Middle Adriatic Pit, the shear changes sign from the cyclonic in the surface layer to the anticyclonic in the bottom layer. This may be associated with the remnants of the anticyclonic gyre whose presence was evidenced over the Middle Adriatic Pit during one part of the year [12], that is in spring. It is also interesting to note that an appreciable part of the transect area is occupied by the inflowing current only along the transect B. Along transects E and H, major portions of the area, if not all of it, are occupied by the southward current. Since, with the ADCP surveys, only the minor eastern part of these transects remained uncovered, it means that along both these transects the recirculation is rather strong and can be associated with the bathymetric constraints at the Palagruza Sill and the strong bottom slope to the north of the Middle Adriatic Pit. This recirculation is noticeable especially well in the IR satellite imagery when the thermal contrast between the inflowing and outflowing water is very prominent.

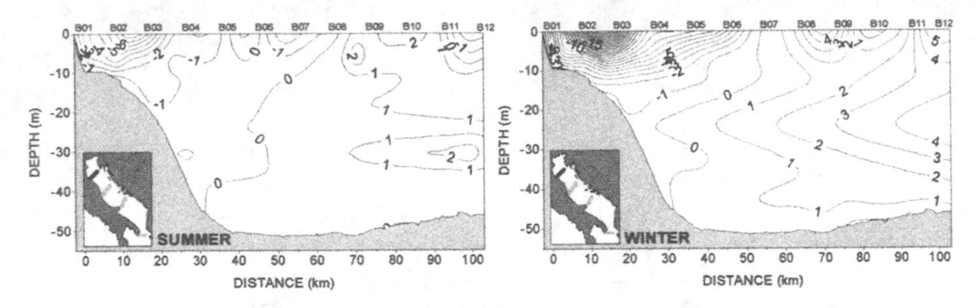


*Figure 11a.* Vertical distributions of the steady current component (cm/s) perpendicular to transect B. The sign of the current component is the same as in Fig. 10.

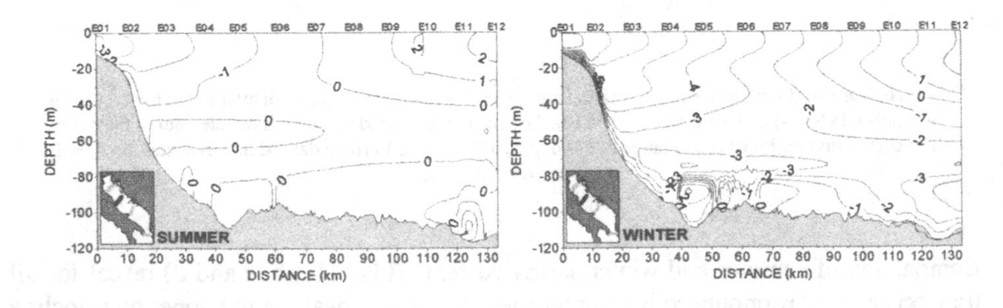


*Figure 11b.* Vertical distributions of the steady current component (cm/s) perpendicular to transect E. The sign of the current component is the same as in Fig. 10.

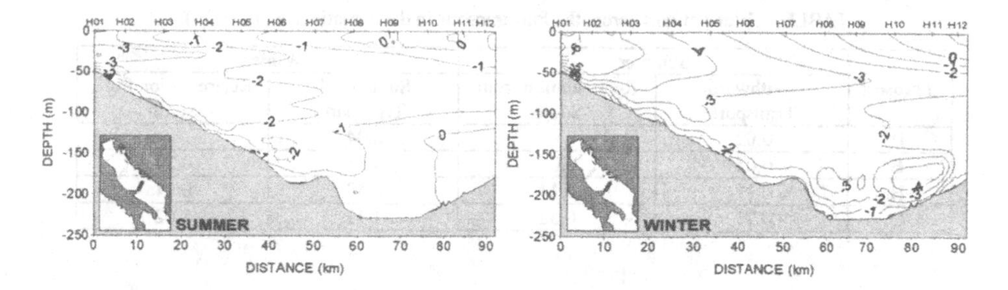


*Figure 11c.* **Vertical distributions of the steady current component (cm/s) perpendicular to transect H. The sign of the current component is the same as in Fig. 10.**

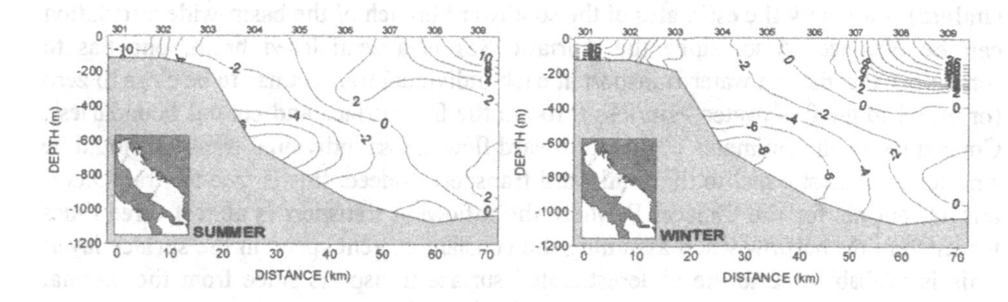


*Figure 11d.* **Vertical distributions of the steady current component (cm/s) perpendicular to transect O (Otranto Strait). The sign of the current component is the same as in Fig. 10.**

## 5. Flux calculations

Steady southward fluxes of water, and dissolved and particulate matter for the winter and summer seasons were estimated using the steady longshore current component and the average distributions of biogeochemical parameters. The mean "summer" distribution of biogeochemical parameters was defined as an average of the spring (May, 1995) and summer (August, 1995) cruise data, while the "winter" distribution was defined as an average of the autumn (October, 1995) and winter (February, 1996) cruise data. First, for each transect and for each data set, an interpolation on a regular grid was performed using the kriging method. A grid of 200 X 200 points was reconstructed. The interpolation procedure was carried out using the PC software SURFER 6.0. Subsequently, by averaging, the summer and winter interpolated fields were obtained. These average fields were then used for the calculations of water, dissolved and suspended matter fluxes. For what concerns the Strait of Otranto, fluxes were obtained re-calculating on a six-months time scale the results from Civitarese *et al.* [8].

TABLE 1. Water fluxes across the four transects in the Adriatic Sea ($10^6$ $m^3/s$).

| Transect | Summer | | Winter | |
|---|---|---|---|---|
| | Southward Transport | Recirculation from South | Southward Transport | Recirculation from South |
| B | 0.02 | - | 0.04 | - |
| E | 0.05 | 0.03 | 0.28 | 0.24 |
| H | 0.19 | 0.14 | 0.39 | 0.11 |
| O | 0.71 | 0.52 | 1.41 | 1.02 |

Analyzing the estimated water fluxes (Tab. 1), it is important to remember that all of the transects except the one at Otranto are only partially covered with current measurements. Since the measurements were carried out starting from the western (Italian) coast, only the estimates of the southward branch of the basin-wide circulation can be obtained. Also, since the Adriatic Sea is a semiclosed basin, one has to remember that the net water transport at each individual transect has to be close to zero (or equal to the freshwater gain/loss through the free surface and coastal boundaries). Consequently, the estimates of the northward flow across individual transect should be smaller or at most equal to the southward transport. Indeed, this is true for transects E and H, but not for the transect B where the inflowing transport is almost three times larger than the outflow when assuming the constant current speed in the surface layer. This is probably due to the underestimated surface transports since from the thermal wind relationship in areas with maximum horizontal density gradients, we obtained a surface velocity of up to 15 cm/s while the constant velocity assumption resulted in surface currents of up to 5 cm/s. In addition, the relative importance of this surface layer increases in shallow water since it becomes an important fraction of the total transect area. Thus at the transect B, the assumption of the vertically constant current speed can introduce an appreciable error in estimates of the southward transport. Also, the missing part of the transect between the westernmost measurement point and the coast can appreciably contribute to the error in the transport estimates. Therefore, we have reconstructed the surface layer velocities at the transect B calculating the geostrophic shear from the horizontal density distribution as obtained from the hydrographic campaigns. Then the surface current was allowed to go to zero logarithmically from the westernmost measurement location towards the coast. After that, at each location the vertical logarithmic profile with no-slip conditions was assumed. The obtained water fluxes then are still unbalanced but the difference between the two is now much smaller than it was before with the assumption of the constant current speed in the surface layer.

The obtained values of the fluxes reveal stronger longitudinal flows during winter than during summer. This difference is especially prominent along transect E since the re-circulation rate to the north of the Mid-Adriatic Pit in winter is almost one order of magnitude larger than in summer. On average, during both summer and winter, about 25% of the volume that enters the Adriatic through the Strait Of Otranto passes over the Palagruza Sill to the Middle and Northern Adriatic. This means that 75% of the water entering the Adriatic remains trapped in the South Adriatic Pit. As said earlier, the varying influence of the bottom topography under different stratification conditions on the re-circulation is most evident along transect E, whereas along transect H the re-circulation rate changes by only small amount from summer to winter. If we consider the volume of water passing through transect B as a contribution of the Northern Adriatic, it follows that it adds to the South Adriatic Pit a volume of water that is very small: during winter, it represents only 4% of the total water volume exchanged through the Strait of Otranto, while during summer its contribution is only 3% of the water exchanged with the Ionian Sea.

In Table 2, the estimates of biogeochemical southward fluxes through four investigated transects are summarized. Either in summer or in winter the fluxes increase from north to south, owing not only to an obvious increase in the water transport, but also to the intrinsic biogeochemical characteristics of the Adriatic sub-basins. The relative increase in the amount of the exchanged inorganic nutrients inside the Adriatic, that is excluding the Strait of Otranto transect, is maximum passing from the transect E to the transect H in summer.

TABLE 2. Biogeochemical transports across the four transects in the Adriatic Sea (units for DIN, Phosphate and Silicate in $10^6$ moles, for Chl a in tons, for TSM in $10^3$ tons).

| Transect | Summer | | | | | Winter | | | | |
|---|---|---|---|---|---|---|---|---|---|---|
| | DIN | Phosphate | Silicate | Chl a | TSM | DIN | Phosphate | Silicate | Chl a | TSM |
| B | 303 | 13 | 485 | 192 | 309 | 3516 | 48 | 2350 | 995 | 2633 |
| E | 650 | 24 | 1455 | 860 | 1150 | 10390 | 303 | 14490 | 920 | 3720 |
| H | 7140 | 456 | 11760 | 420 | 690 | 14780 | 588 | 25780 | 1260 | 3770 |
| O | 51250 | 2110 | 79700 | - | - | 85780 | 3150 | 127900 | - | - |

This increase can be explained in terms of the relation of the total transect depth and the surface nutrient depleted layer thickness. The two are almost the same at transect B while at E and H the water depth is larger than the thickness of the surface nutrient depleted layer. In winter, the relative increase in the nutrient exchange rate is maximum passing from the transect B to transect E (two to five times), while the flux increase from the transects H to E is reduced. On the contrary, in summer the Chlorophyll a and TSM southward fluxes, after an increase from transect B to transect E, decrease by about 50% from E to H. This decrease is due to the presence of the South Adriatic water that recirculates across the transect H, and which is characterized by a phytoplankton biomass lower than in the Middle and Northern Adriatic areas. In winter, the southward increase in Chlorophyll a and TSM fluxes is reduced since the longitudinal phytoplankton biomass variations are not so pronounced as in summer. In any case, transect H in correspondence to the abrupt change of the bottom depth (from 170 to 1200 m) appears as a discontinuity zone separating the two areas of the Adriatic: the northern part, relatively shallow and productive, and the oligotrophic deep water areas of the South Adriatic Pit.

## 6. Conclusions

Estimates of the southward water fluxes have indicated that the Northern Adriatic adds to the South Adriatic Pit only 3-4% of the total water volume exchanged through the Strait of Otranto. Also, it was shown that at the Palagruza Sill, the northern border of the South Adriatic Pit, about 75% of the water entering through the Strait of Otranto recirculates. Thus, the Palagruza Sill represents a discontinuity area separating the relatively shallow and productive northern shelf from the oligotrophic South Adriatic Pit.

Water and biogeochemical fluxes estimates at the Strait of Otranto, on the other hand, made it possible to assess the relative importance of the northern productive area in determining the overall biogeochemical physionomy of the Adriatic Sea as a unique water body. These calculations, in fact, show that the Adriatic Sea through the Strait of Otranto exports inorganic nutrients and imports particulate organic carbon and nitrogen. This means that the Adriatic Sea acts as a mineralization site in which the oligotrophy is more marked than in the basin with which exchanges the biogeochemicals. Furthermore, this suggests that the Northern Adriatic, even though being an organic matter source, does not have an important influence on the whole Adriatic biogeochemical physionomy.

In interpreting these results one should however bear in mind that they are obtained on the basis of a single year realization (1995/96), and on the other hand, it is well known that the year-to-year variations of the oceanographic conditions in the Adriatic are rather large [12, 13]. Consequently, the longitudinal fluxes, as well as the exchange with the Ionian Sea are subject to strong interannual variations. Thus, one should assess whether these conclusions are obtained for extreme conditions of the longitudinal exchange or they are representative of the average conditions. As mentioned earlier, the most important driving force of the longitudinal fluxes is the pressure gradient which is to a large extent dependent on the buoyancy input and, during the winter, on the magnitude of the surface heat losses and subsequent dense water formation. Thus, climatic conditions of the specific year have to be compared to the long-term climate variations.

***Acknowledgments.*** This research was supported by the Italian Ministry of University, Scientific Research and Technology, the National Council of Research and the Central Institute for Applied Marine Research within the project PRISMA1 (Programma di RIcerca e Sperimentazione per il Mare Adriatico). The data set originated from:

*Istituto di Ricerca sulla Pesca Marittima (IRPEM)*, resp. A. Artegiani;
*Istituto Talassografico di Trieste (ITT)*, resp. D. Bregant;
*Osservatorio Geofisico Sperimentale (OGS)*, resp. E. Accerboni;
*Laboratorio di Biologia Marina (LBM)*, resp. S. Fonda-Umani;
*Istituto Centrale per le Ricerce Applicate al Mare (ICRAM)*, resp. M. Giani;
*Istituto di Biologia Marina (IBM)*, resp. P. Franco;
*Istituto di Ricerca Sulle Acque (IRSA)*, resp. A. Puddu.

The dataset was kindly put at our disposal by the Scientific Coordinator of the PRISMA1 project, R. Pagnotta. We express our thanks to S. Ghergo for technical assistance in data recovery. Useful comments by T. Hopkins to an early draft of this paper were greatly appreciated. Finally, it is almost impossible to name all the technicians, oceanographic vessel crew members and scientists which gave a great contribution in collecting and analysing this data set.

## 7. References


1. Orlic, M., Gacic M. and LaViolette, P.E. (1992) The currents and circulation of the Adriatic Sea, *Oceanologica Acta* **15**, 2, 109-124.
2. Paschini, E., Artegiani, A. and Pinardi, N. (1993) The mesoscale eddy field of the middle Adriatic Sea, *Deep-Sea Research* **40**, 1365-1377.
3. Bergamasco, A., Gacic, M., Boscolo, R. and Umgiesser, G. (1996) Winter oceanographic conditions and water mass balance in the Norrthern Adriatic (February 1993), *Journal of Marine Systems* **7**, 67-94.
4. Strickland, J.D.H. and Parsons, T.R. (1968) *A practical handbook of seawater analysis,* Fisheries Research Board of Canada, Ottawa.
5. Grasshoff, K., Ehrardt, M., Kremling, K. (Eds.) (1983) *Methods of seawater analysis,* Verlag Chemie, Weinheim, 419 pp..
6. Gacic, M., Kovacevic, V., Manca, B., Papageorgiou, E., Poulain, P.-M., Scarazzato, P. and Vetrano, A. (1996) Thermohaline properties and circulation in the Strait of Otranto, in F. Briand (ed*.), Dynamics of Mediterranean Straits and Channels*, Bulletin de l'Institut Oceanographique, Monaco, n. special 17, CIESM Science Series **2**, 117-145.
7. Civitarese, G., Gacic, M., Vetrano, A., Boldrin, A, Bregant, D., Rabitti, S. and Souvermezoglou E. (in press) Biogeochemical fluxes through the Strait of Otranto (Eastern Mediterranean), *Continental Shelf Research.*
8. Civitarese, G., Boldrin, A., Bregant, D., Cozzi, S., De Lazzari, A., Gacic, M., Kovacevic, V., Krasakopoulou, E., Rabitti, S., Souvermezoglou, E. and Vetrano, A. (in press) Nutrients and particulate matter dynamics and exchanges in the Otranto Strait. *EC Ecosystems Research Report Series - The Adriatic Sea.*
9. Candela, J., Beardslay, R.C. and Limeburger, R. (1992) Separation of tidal and sub-tidal currents in ship-mounted Acoustic Doppler Current Profiler observations, *Journal of Geophysical Research* **97**, C1, 769-788.
10. Ursella, L., Accerboni, E., Gacic, M. and Mosetti, R. (1998) Basin-wide use of a ship-mounted ADCP gives a new picture of the Adriatic Sea circulation, *Rapp. Comm. Mer Medit.,* 35th CIESM Congress Proceedings, 202-203.
11. Ursella, L. and Gacic, M. (1998) Studio della circolazione dell'Adriatico per mezzo di profilatore acustico ad effetto Doppler (ADCP), in F. Crisciani (ed.), *Tecnologie marine: attualità e prospettive,* CNR - Istituto Talassografico di Trieste, 119-148.
12. Gacic, M., Marullo, S., Santoleri, R. and Bergamasco, A. (1997) Analysis of the seasonal and interannual variability of the sea surface temperature field in the Adriatic Sea from AVHRR data (1984-1992), *Journal of Geophysical Research,* **102**, C10, 22937-22946.
13. Buljan, M. and Zore-Armanda, M. (1976) Oceanographical properties of the Adriatic Sea, in H. Barnes (ed.), *Oceanography and Marine Biology Annual Review,* Aberdeen University Press, **14**, 11-98.

# Mediterranean Sea trophic characteristics interpreted through three-dimensional coupled ecohydrodynamical models

A. Crise, G. Crispi and C. Solidoro
Osservatorio Geofisico Sperimentale, P.O. Box 2011 34016 Trieste, Italy
E-mail: acrise@ogs.trieste.it, gcrispi@ogs.trieste.it, csolidoro@ogs.trieste.it

## Abstract

The spatial variability of the Mediterranean ecosystem in response to the variability in nutrient supply and demand is addressed through basin-wide aggregated ecological hydrodynamical coupled models. A discussion of a multi-nutrient two-phytoplankton ecosystem submodel is presented together with a more aggregated implementation including dissolved inorganic nitrogen, phytoplankton and detritus. This modelling effort gives an interpretation of the nutrients gradients in terms of the effects of the interplay between both biological and dynamical processes. The general oligotrophy of the Mediterranean Sea is in principle explained by the inverse estuarine circulation. Three elements in particular seem to be relevant in creating the North-South and East-West gradients: the different physiography of the two subbasins, the detrital fall-out from the fertile zone and the cyclonic and anticyclonic wind driven structures. The vertical fluxes of biogenic material and the nutrient exchanges at the bottom of the euphotic zone seem to play an important role in modifying the N:P ratio at depth along the east-west axis.

## Introduction

The Mediterranean has long been recognized as one of the largest macronutrients-depleted areas in the world with a trophic concentration too low to sustain relevant biomass concentrations (McGill, 1961). The general oligotrophic regime is explained by the inverse estuarine circulation: the estimates of nitrate fluxes at Gibraltar Strait agree on a net loss (inflow in the superficial layer minus outflow in the bottom one) ranging from 1.25 Mtons/year (Sarmiento et al., 1988) to 3.11 Mtons/year (Béthoux, 1979), compensated by



*P. Malanotte-Rizzoli and V.N. Eremeev (eds.),*
*The Eastern Mediterranean as a Laboratory Basin for the Assessment of Contrasting Ecosystems*, 359–381.

natural and anthropogenic sources (such as river runoff, atmospheric inputs, and nitrogen fixation). This general picture alone is not sufficient to explain the well known east-west increasing trophic gradient, which is present both in the surface and in depth. This paper focuses on how these gradients arise from the interactions between general circulation processes and the biogeochemical cycles. The interplay between these phenomena has been investigated by means of a 3D coupled ecohydrodynamical model. In it the cycles of nitrogen has been simulated through a very simplified schematization of its major functional compartments. The rationale in aggregating the overwhelming complexity of a marine system in a lumped variable deterministic approach is that a model is, in any case, a simplification of reality. This implies that the first aim of a model, in our opinion, is not a detailed forecasting of a multitude of parameters, but rather to capture the major features of the ecosystem machinery, so gaining a keener insight on the driving mechanisms , and learning something about processes and large scale ecological responses of the ecosystem to different scenarios of environmental conditions.

## Trophic gradient(s) and dynamical processes of the Mediterranean Sea.

The Mediterranean peculiarity is mainly derived from the different dynamics active in the Eastern and Western Mediterranean (henceforth E.Med and W.Med). This difference can be explained in terms of several factors that can be summarized in the table 1.

| Process | Western Mediterranean | Eastern Mediterranean |
|---|---|---|
| Biochemical exchanges | Atlantic Ocean and E.Med | Marginal seas and W.Med |
| Deep water formation | Within the basin | Mostly outside the basin |
| Coastal upwellings | Intense in the northern sector | Scarce |
| Euphotic Layer | Shallow | Deep |
| Nutricline | Shallow | Deep |
| Buoyancy content | Low | High |

Tab.1 Typical features of the Eastern and Western Mediterranean Sea in relation to the oligotrophic regime.

All the typical features reported in this table for the eastern subbasin can induce directly or indirectly a more severe oligotrophy in the upper layer compared with the western basin, thus directly influencing primary production and ultimately affecting the energy availability to the upper trophic levels.

The exchanges with other basins lead to reduced import of nutrient in the E.med because the main source of nutrients in the upper ocean, due to Surface Atlantic Water, is located in W.Med while in the E.Med the inflowing current coming from W.Med is practically nutrient-depleted (Rabitti et al., 1994). On the other hand, the exchanges with regional seas do not modify the general unbalance in nutrient sources. The influence of the Adriatic Sea seems to have reduced importance in nutrient exchanges with the Ionian Sea through Otranto Strait (Civitarese et al.,1998) while Cretan Sea, being the most impoverished nutrient area of the E.Med, cannot be assumed as a source of nutrients for the rest of the eastern basin, even though only preliminary fluxes estimates are now-a-day available.

Mediterranean **deep water formation** sites are in the W.Med the Gulf of Lions (Schott et al.,1996; Rhein, 1995; Schott and Leaman, 1991) and in the Adriatic Sea (Ovchinnikov et al., 1985; Roether and Schlitzer, 1991) and in recent years also in the Aegean basin (Roether et al., 1996) in the E.Med. The convective mixing is an important source of nutrients in particular because of overturning of water column and the final mixing of its physical and biochemical properties, What is relevant to the nutrient budget in the surface layer is that the nutrient inport is local, i.e, is effective only where convection takes place, while deep water spreading can ultimately affect the whole subbasin. The dense water formation then directly fertilizes the upper layer of the Gulf of Lions, while no counterpart is present for the E.Med.

The influence of the wind driven **coastal upwellings** has recognized to be one of the most effective processes in nutrient supply in the North Western Mediterranean as shown by Millot and Wald (1981) using remotely sensed images and simulated by Pinazo et al. (1996). Conversely no evidence for upwellings in infrared images is presented by LeVourch et al. (1992) in the E.Med.

In this basin, moreover, the low chlorophyll concentration and the reduced influence of rivers discharge on the open ocean water allows the light to penetrate in depth giving way to **a deep euphotic layer** (Berman et al., 1984) while in the W.med the transparency is lower and the euphotic zone is much shallower.

In the E.Med the **nutricline** erosion by primary producers creates a deep nutrient-depleted layer where regenerated production dominates during stratification season and turbulent entrainment is difficult not only during summer, when the onset of a well developed seasonal pycnocline decouples the mixed layer from the biomass maximum, but also in winter.

During the mixing period, in W.Med the shallower nutricline and energetic wind forcing allow a ubiquitous increase of the chlorophyll concentration in the surface layer. In contrast, the deeper nutricline present in the Eastern Mediterranean can be affected by mixing only when dense water formation and convection processes take place, typically at the end of the winter.

This is confirmed by the analysis of the **buoyancy contents** calculated by MODB Med5 gridded data set (fig.1) in Autumn and in Winter. The maps are produced by assuming reference levels which roughly correspond to the nutricline depths, shallower in the western basin (80m) and deeper in the eastern one (160m): in Autumn, the buoyancy supply necessary to homogenize the whole water column till the nutricline is almost twice in the eastern than in the western basin.

Another interesting aspect of the buoyancy content above the nutricline is that, during the stratification season, cyclonic areas tend to decrease the buoyancy while the opposite is true in the anticyclones. The prevailing cyclonic circulation in the northern side of the Mediterranean contributes to the shoaling of nutricline along the northern coasts and, relatively speaking, this holds also for Rhodes and Cretan Gyres. Cyclonic gyres then act as a preconditioning factor in winter mixing as shown in buoyancy content typical of Winter situation in both basins. The analysis of the Winter buoyancy content shows that in the

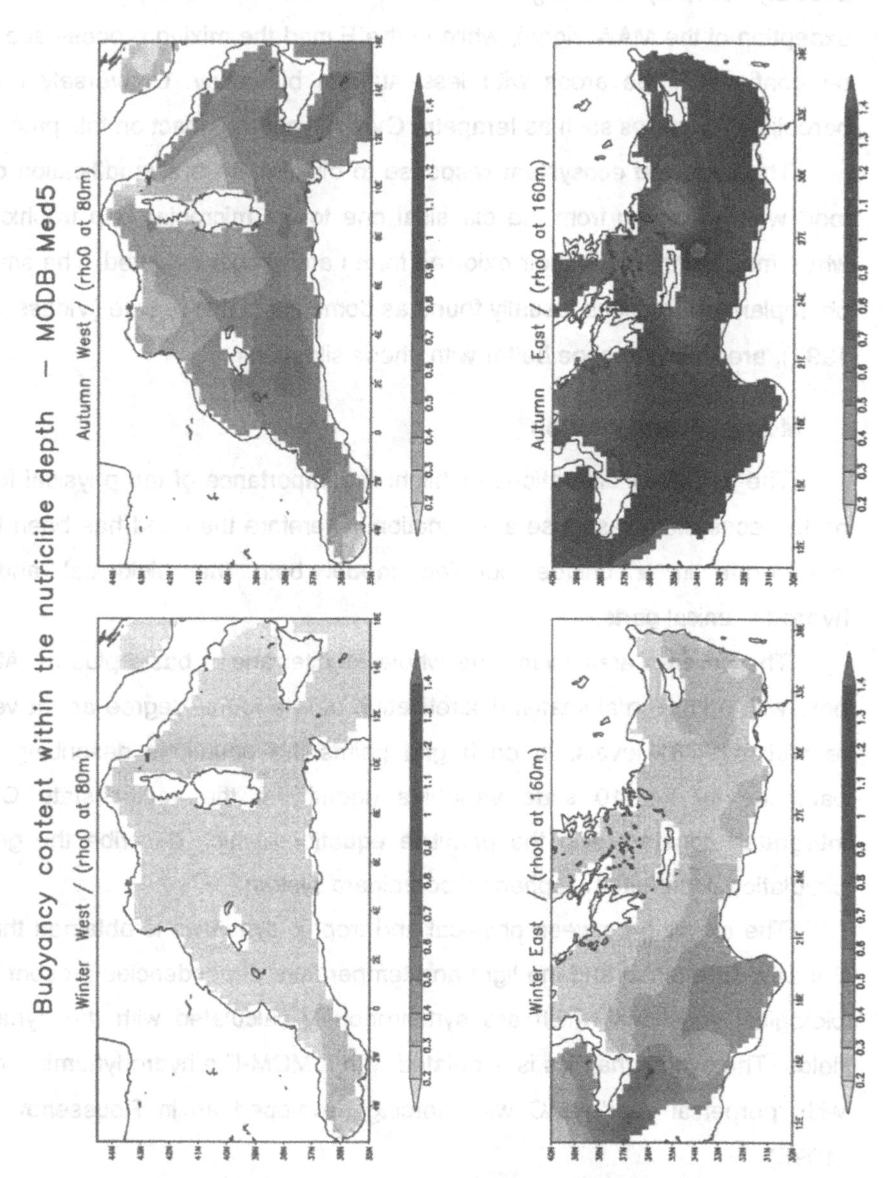


**Fig.1.** Buoyancies contents (m²/s²) estimated in Western and Eastern Mediterranean by the MODB Med5 gridded data sets in Autumn and Winter

whole W.Med the water column is homogenized untill the nutricline (and possibly deeper) inducing a nutrients entrainment everywhere with the exception of the MAW signal, while in the E.med the mixing process seems to be confined in the areas with less Autumn buoyancy. Conversely marked baroclinic structures such as Ierapetra Gyre have less impact on this process.

The expected ecosystem response to oligotrophy is a modification of the food web, switching from the classical one to the microbial loop trophic path when macronutrients in their oxidized forms are almost depleted. The smallest phytoplankton fractions, usually found as dominant in this regime (Vidussi et al., 1997), are known to cope better with these situations.

**Model Implementation**

The above considerations highlight the importance of the physical forcing on the ecosystem response and function. Therefore the need has been felt to incorporate in a unique coupled model both the biological and the hydrodynamical parts.

The covered area spans the whole Mediterranean basin plus an Atlantic box, with a horizontal spatial discretization of one fourth degree and a vertical resolution of 31 levels. In each grid points the equations describing mass balances for the 10 state variables (identifying the vector state C) are integrated, together with the primitive equations which describe the general circulation pattern in the spherical coordinate system.

The coupling between physical and trophic dynamics is obtained through the advective terms and the light and temperature dependencies present in the biological equations which are synchronously calculated with the dynamical fields. The hydrodynamics is simulated with a MOM-like hydrodynamical model with 'perpetual year' NMC wind forcing developed as in Roussenov et al. (1995).

The three-dimensional formulation is needed in order to appraise the influence of horizontal transport, upwelling/downwelling processes and vertical mixing on the nutrient and possibly on phytoplankton cycling and distribution.

The biological processes are conceptualized by a mass flow scheme (fig.2)

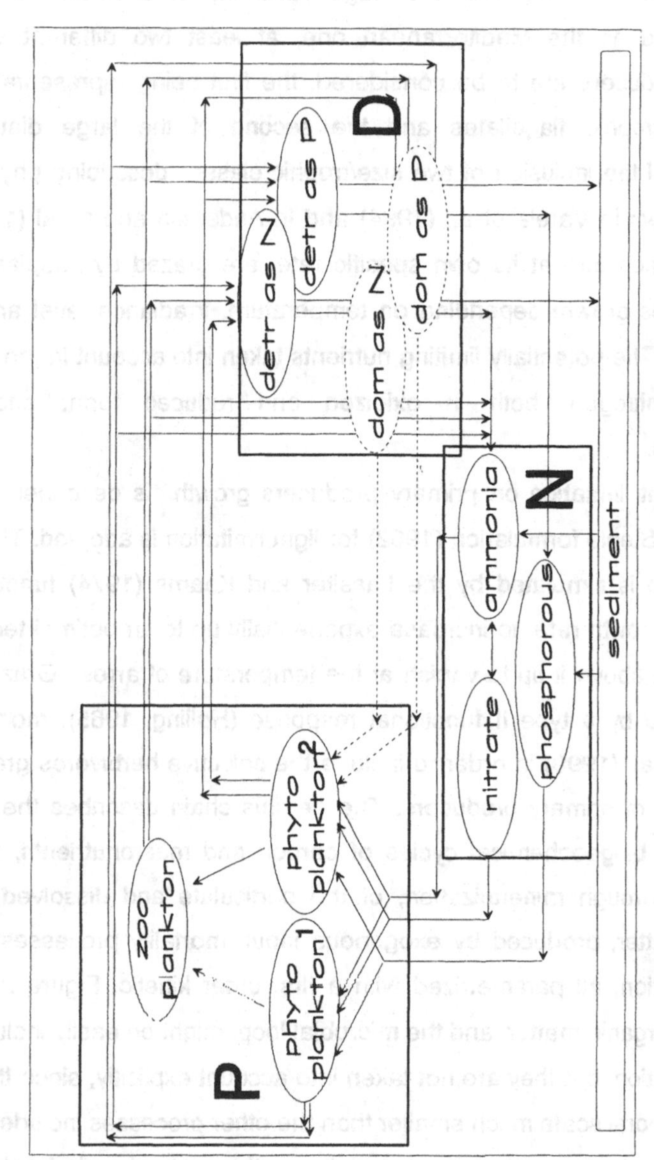


**Fig.2. Conceptualized biological cycle. Mass flows between major functional compartments are depicted. NPD further aggregation is indicated as well.**

with the objective to describe and explain the N:P ratio anomalies and the primary production of the pelagic ecosystem in the Mediterranean Sea.

In order to cope with the huge variability of a basin as large and differentiated as the Mediterranean one, at least two different species of primary producers are to be considered, the first being representative of the small autotrophic flagellates and the second of the large diatoms. The relevance of the inclusion of two size/trophic classes describing phytoplankton pool is shown in Varela et al. (1994) and in Andersen and Nival (1988). Both fractions, each one at its own specific rate, are grazed by zooplankton, and both species growth depending on temperature, irradiance level and nutrient availability. The potentially limiting nutrients taken into account in the model are inorganic nitrogen, both in oxidized and reduced form, and reactive phosphorus.

Nutrient limitation on primary producers growth is described by Monod kinetic and Steele formulation (1962) for light limitation is adopted. The effect of temperature is simulated by the Lassiter and Kearns (1974) function, which allows the growth rate to increase exponentially up to an optimal temperature, and decline above it up to vanish at the temperature of arrest. Grazing activity is described by a type II functional response (Holling, 1965), modified as in Fasham et al. (1990) in order to include the selective herbivores grazing upon two groups of primary producers. The detritus chain describes the remaining part of the biogeochemical cycles of carbon and macronutrients, that is the recycling, through mineralization, of the particulate and dissolved non-living organic matter, produced by exogenous input, mortality processes, excretion and exudation, all parametrized with a first order kinetic. Figure 2 shows as dissolved organic matter, and the microbial loop, might be easily included in our schematization, but they are not taken into account explicitly, since they have a spatio-temporal scale much smaller than the other processes included.

Light and temperature vary according with seasonal and night/day cycles, and the exponential light absorbtion with depth takes into account also self-shading effects. At Gibraltar Strait biological tracers are allowed to be exchanged with an academic ocean where all tracers are relaxed to prescribed concentrations using nudging, in order to properly condition the surface inflow.

Biomass nitrogen- and phosphorus-equivalents are assumed to be constant and follow the standard Redfield ratio, with the exception of the P and N detrital contents, which are allowed to remineralize with different speeds.

The dynamics of Dissolved Oxygen is also reproduced, since this variable, besides being frequently sampled, is an aggregated index of the quality of a water body.

The 10 variables submodel has been tested on a 1-D water column dynamics, driven by the same energetic input, light intensity at the surface and temperature derived by the NPD 3D simulation in the Ligurian Sea. In particular, the seasonal evolution of temperature profiles, being computed with the full 3D model, takes into account also the advection.

The nominal trajectory of model correctly reproduces the formation of the well known Deep Chlorophyll Maximum, as can be seen in Fig. 3 In fact, the dynamic of the primary production is first triggered by light and temperature and therefore the model shows a bloom of diatoms in the early spring, followed by the bloom of micro-flagellates, which reach their maximum productivity at higher level of light intensity and temperature. Such blooms cause a rapid depletion of nutrients, which is more pronounced near the surface, where light intensity is higher. Grazing activity starts affecting the phytoplanktonic stocks toward the end of the spring.

A local sensitivity analysis, based on the linearization of the trajectory of the model, has been performed on the 1D model, indicating that many parameters, if slightly changed, have similar influence on model output of chlorophyll and nutrients. (Solidoro et al., 1998). This points out once more the uselessness of a too detailed biological description as long as experimental data are restricted to currently available informations at basin scale.

As indicated in figure 2, the model can be further simplified, by the aggregation of functionally similar state variables, up to a representation taking into account a generic nutrient (N), a unique planktonic pool (P) and a detritus (D). This NPD model is described in details in Crise et al. (1998).

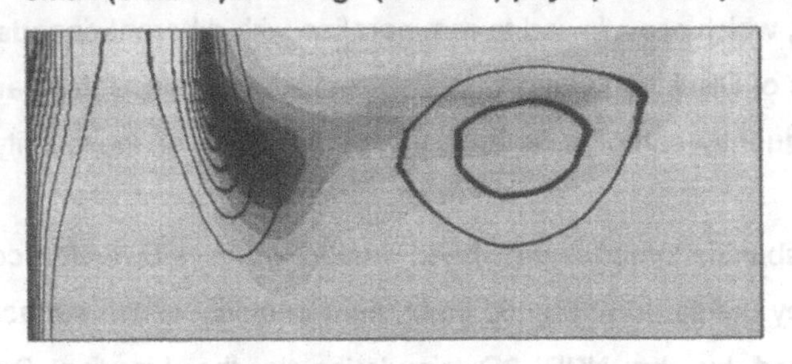


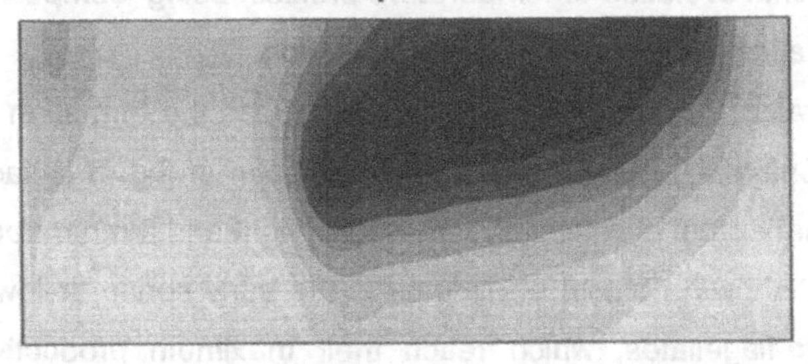

**Fig.3. Time evolution of the densities of phyto (2a) and zooplanktonic (2b) pools, and of the concentration of nutrients (2c), as computed by the 1d vertical model.**

**Results and discussion**

The trophic gradients in Mediterranean seem to be a permanent feature, thus the climatological distribution of the NPD model results, obtained by averaging 24 months, retains the signal. In fact, NPD simulations exhibit a pronounced ***east-west positive gradient of the meridionally integrated nutrients*** (figure 4a), which is experimentally well known and, in our model, induced mainly by the different physical, and physically driven, processes in the two basins.

In the fig.4a the dissolved inorganic nitrogen is compared, meridionally averaged, against nitrates experimental data obtained and processed as in Crise et al. (submitted), both averaged in the upper 200m over two years. The model estimates are qualitatively and quantitatively representative of the signal present in the data even if an underestimation is clearly evident in correspondence of the Rhodes Gyre area. The large error bars associated to the data averages are connected to the seasonal variability of the nutrient cycle in the upper layer.

The model analysis suggests that the skewness of nutrient contents is connected to the Gibraltar contribution in the upper layer, which accounts for the largest part of nutrients inputs in the Mediterranean. The Atlantic Surface Water,flowing along the African coast, reaches the Levantine Basin after some bifurcations and modifications along its path, carrying memory of its biogeochemical content, which is less and less discriminable because of the vertical leakage induced by the biological fallout: carbon (and nutrients) assimilation by primary producers creates, in the superficial layer, new organic matter that, during the multiple transformations within the food web, is partially released in detrital form, giving way to a net downward flux.

Few relevant terrestrial sources are present along the African coast, thus preserving the eastward decreasing nutrient and biomass contents. This mainly explains why in the southern Mediterranean the gradient is well preserved.

The Alboran Sea plays a key role in trapping the nutrient enriched surface waters of Atlantic origin. The inorganic nitrogen entering through the Gibraltar

**Fig.4.Two years average of zonal (a) and meridional (b) averages in the first 200 m layer: total nitrogen (white dot) and phytoplankton (black dot) expressed in mmolN/m³.**

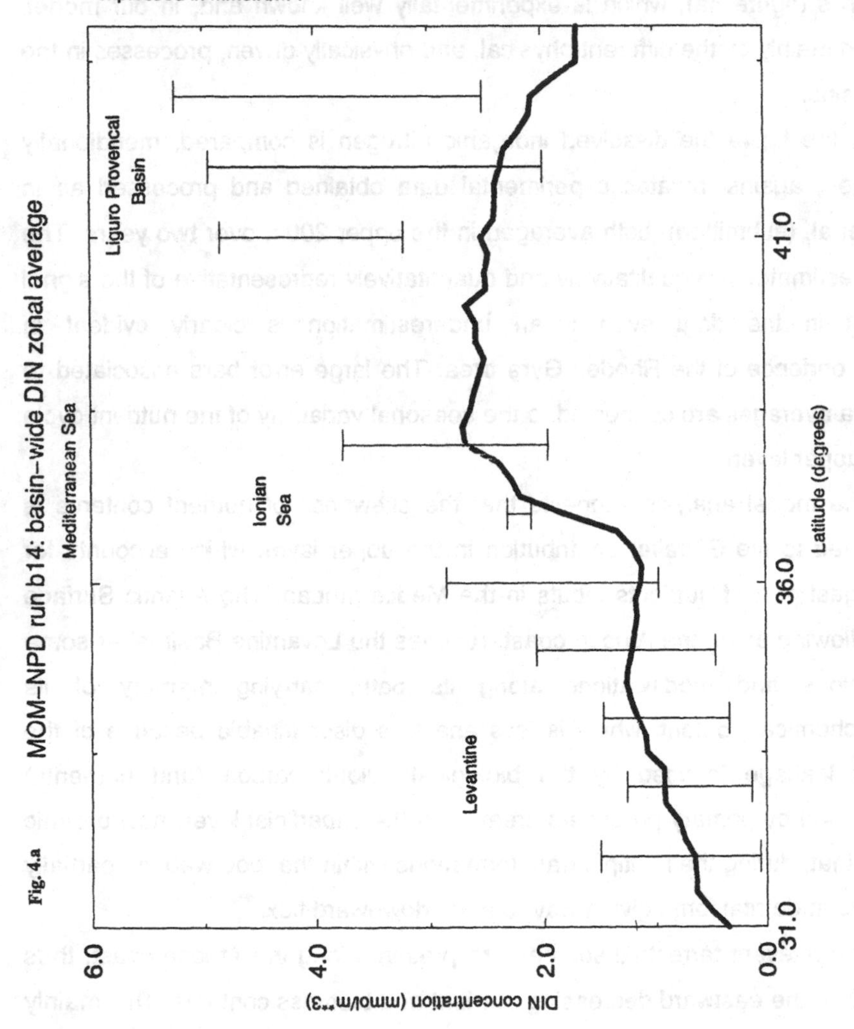

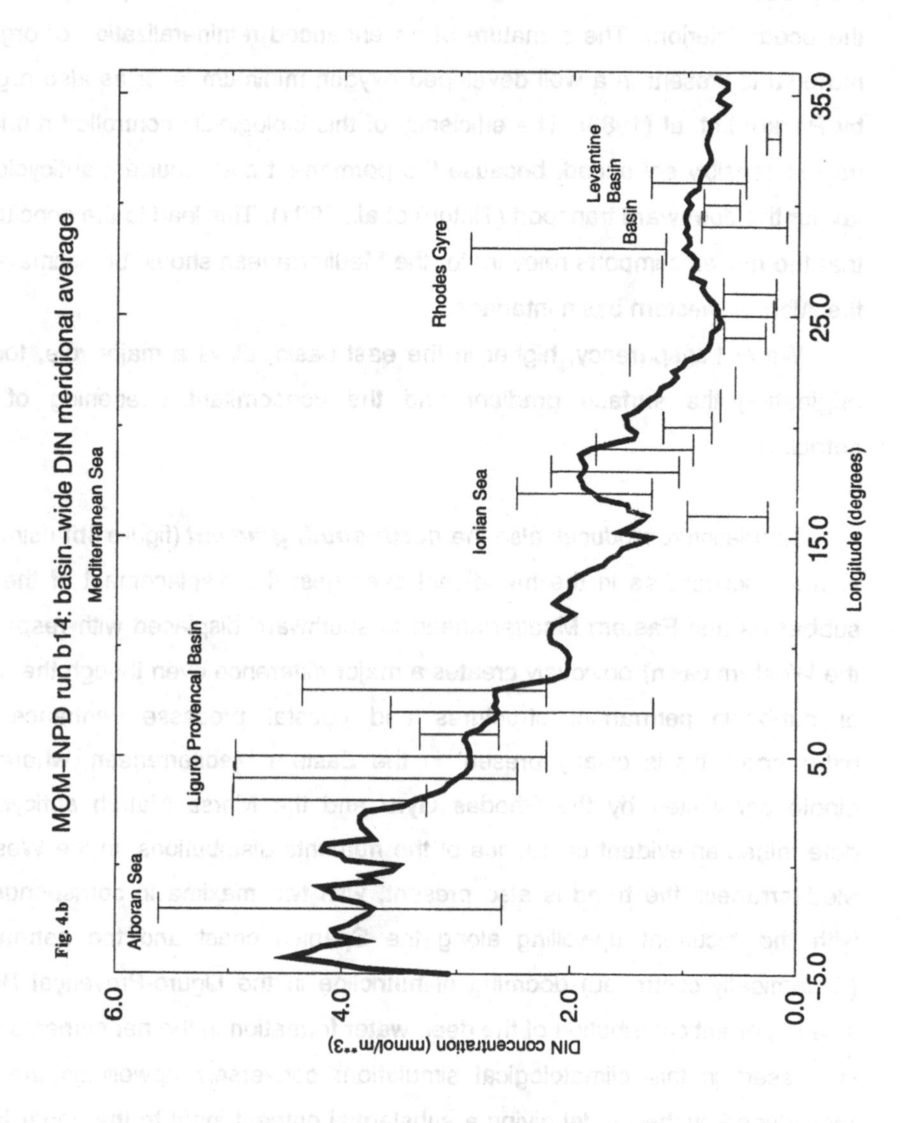
MOM–NPD run b14: basin–wide DIN meridional average
Mediterranean Sea
Alboran Sea
Liguro Provencal Basin
Ionian Sea
Rhodes Gyre
Levantine Basin
Basin
DIN concentration (mmol/m**3)
0.0
2.0
4.0
6.0
–5.0
5.0
15.0
25.0
35.0
Longitude (degrees)

Fig. 4.b

Strait is indeed quickly incorporated by autotrophic activity in this basin, where the biological pump creates, on average, an export of biogenic material outside the productive zone which contributes to maintain the vertical trophic gradient in the ocean interiors. The signature of an enhanced remineralization of organic material is present in a well developed oxygen minimum layer as also argued by Packard et. al (1988). The efficiency of this biologically controlled nutrients trap is sensibly enhanced, because the permanent and recurrent anticyclones favour the downward transport (Tintoré et al., 1991). This lead to the conclusion that the nutrients imports relevant for the Mediterranean should be estimated at the Alboran-western basin interface.

Water transparency, higher in the east basin, plays a major role, too, in originating the surface gradient and the concomitant deepening of the nutricline.

Simulation reproduces also the ***north-south gradient*** (figure 4b) using the same procedure as in the meridional averages: the displacement of the two subbasins (the Eastern Mediterranean is southward displaced with respect to the Western basin) obviously creates a major difference even though the effect of subbasin permanent structures and coastal processes enhance this difference. This is clearly present in the Eastern Mediterranean, where the dipole constituted by the Rhodes Gyre and the Mersa Matruh anticyclone determines an evident unbalance of the nutrients distributions. In the Western Mediterranean the trend is also present, with two maxima in corrispondence with the recurrent upwelling along the Spanish coast and the permanent (dynamically controlled) dooming of nutricline in the Liguro-Provençal Basin. The important contribution of the deep water formation in the net nutrients uplift is missed in this climatological simulation: conversely upwellings are well reproduced by the model giving a substantial nutrient input to the upper layer, evident in particular along the Liguro-Provençal basin. On top of this dynamically adjusted distribution other surface effects contribute to modulate in space and time the nutrients availability and the phytoplankton standing crop.

Ekman pumping, evaluated as in McClain and Firestone (1993) by using 1980-1988 NMC montly mean winds, shows a pronounced meridional gradient during winter and spring, while is neutral during the rest of the year (figure 5). The vertical velocity, positive in the northern area and negative in the southern one, contributes to maintain the meridional gradient.

The relationship between nutrients and barotropic velocity in function of depth is used in order to correlate higher nutrients concentration with cyclonic gyres structures.

Figure 6 depicts the relation between the stremfunction and the DIN at 30m, and 160m depth, where *R* and *L* identify respectively the dot clusters relative to respectively Rhodes Gyre and the Gulf of Lyon Gyre. In the proximity of the pycnocline where the biological uptake is active the cyclonically induced displacement is low (fig. 6a) and virtually no correlation is noticeable, whereas below the euphotic zone the stronger the cyclone the higher the nitrogen concentration as shown by fig. 6b (Crise et. al., 1998).

The effect on the nutrients availability on phytoplankton distribution is shown in figure 7 where April model simulation (fig.7a) and April CZCS level 3 image (fig. 7b) are compared. Model result is able to reproduce the major features of satellite data over quite a large range of value (the logarithmic scale goes from 0 to 7 mg/m3) showing also the presence of a WE basin-wide gradient and a well detectable NS tilt in surface pigment conentration. Further details on this comparison can be found in Crispi et al. (1998).

## Conclusions

More ecologically detailed conceptualizations than the NPD aggregated level are required if more details (i.e. competition among different producers, or multinutrient limitation) are to be investigated. As an example figure 8 shows the spatial distibution of the N/P ratio, as simulated by the ecological model in its 10 state variable version. Experimental data are known to be qualitatively in agreement to this picture (Spencer, 1983: Krom et al., 1991), but no quantitative comparison are here presented, since a full discussion of the

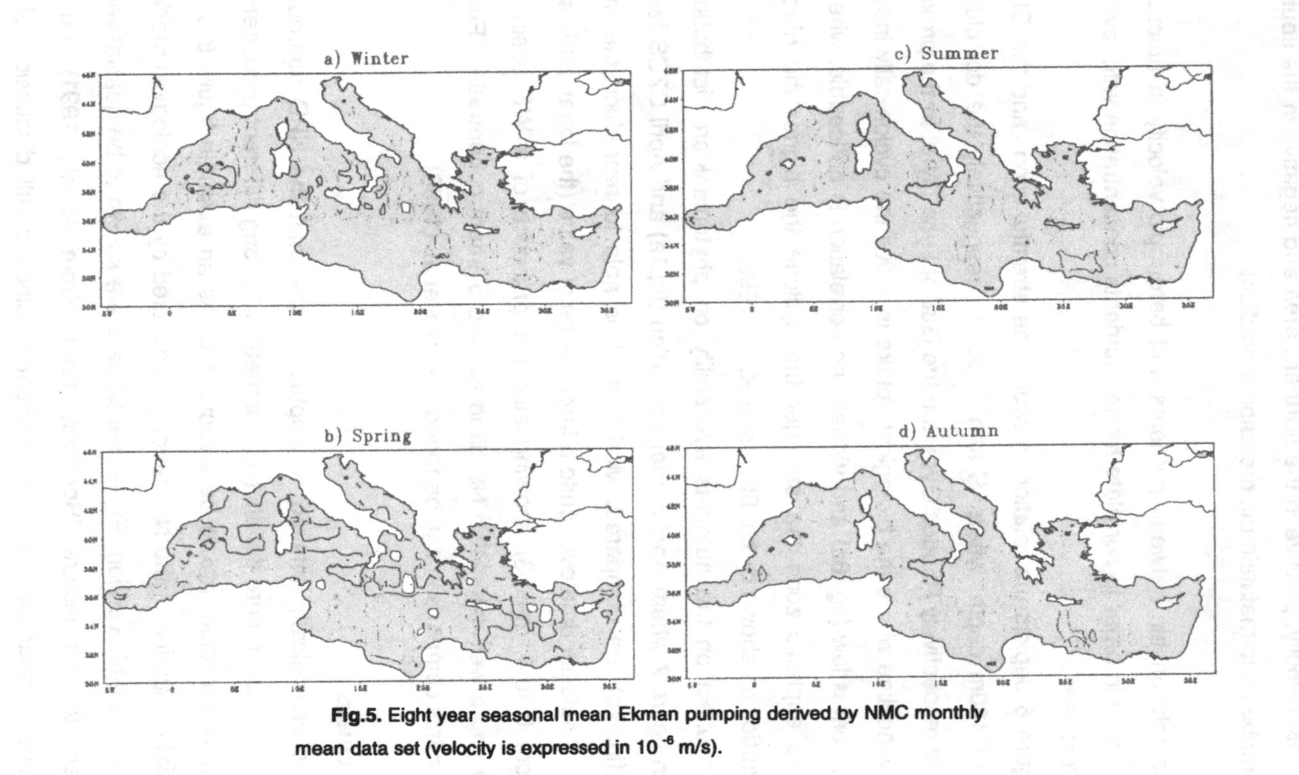


**Fig.5.** Eight year seasonal mean Ekman pumping derived by NMC monthly mean data set (velocity is expressed in $10^{-6}$ m/s).

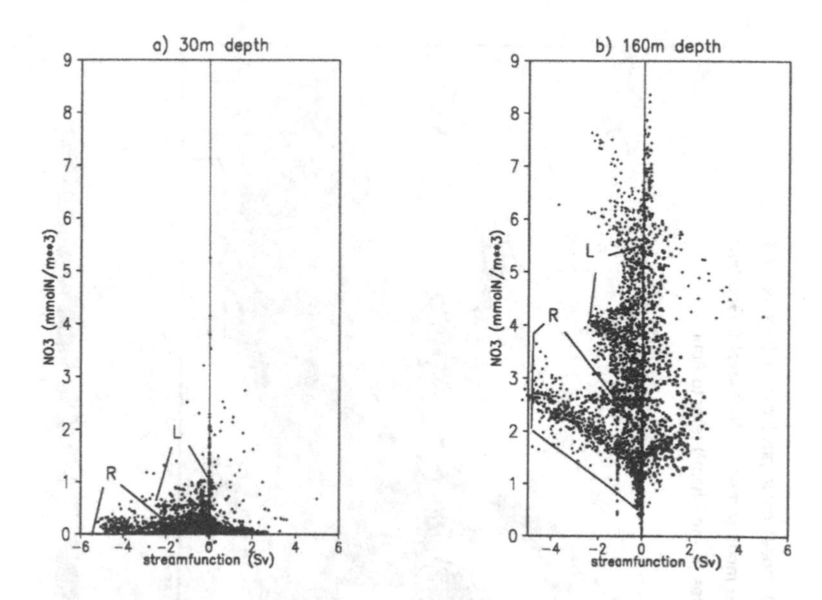


**Fig.6.** Nitrates (mmolN/m³) vs streamfunction (Sv) scatter plots at 30m and 160m depth as predicted by MOM-NPD model (R = Rhodes Gyre, yellow dots; L= Gulf of Lion, blue dots).

**Fig.7.** Comparison between NPD model chlorophyll concentrations and surface pigments as obtained from the monthly mean CZCS level 3 Goddard Space Flight Center data set. Both plates (in mg Chl/m³) refer to April.

Fig. 7.a

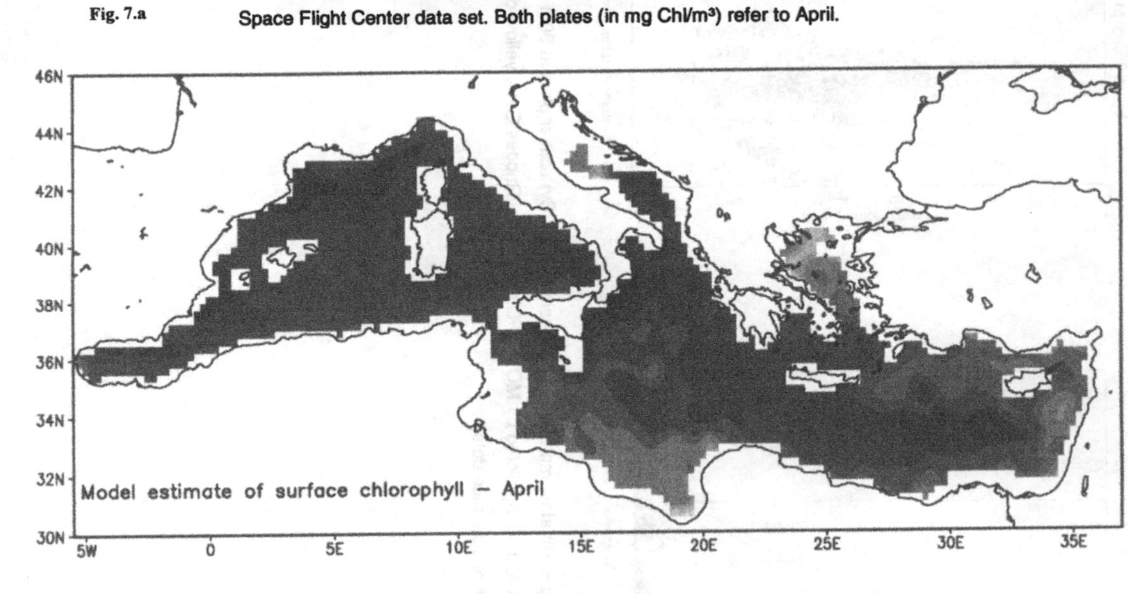


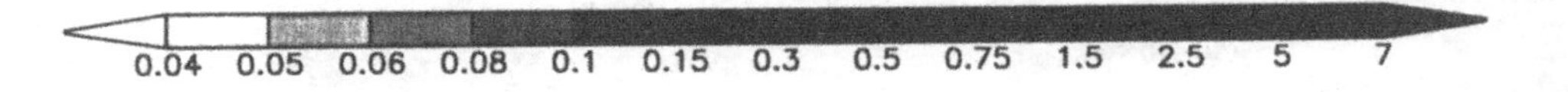

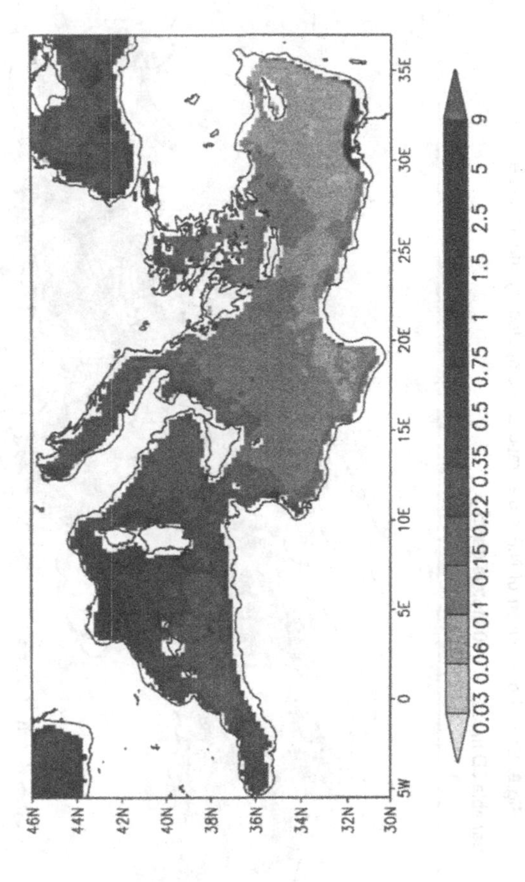
CZCS lev.3 Pigments concentration (Mg/m**3) – April
46N
44N
42N
40N
38N
36N
34N
32N
30N
5W
0
5E
10E
15E
20E
25E
30E
35E
0.03 0.06 0.1 0.15 0.22 0.35 0.5 0.75 1 1.5 2.5 5 9

Fig. 7.b

**Fig.8.** Spatial distribution of N:P molar ratio as computed by the 10 state variable 3D model, at 400 meters depth.

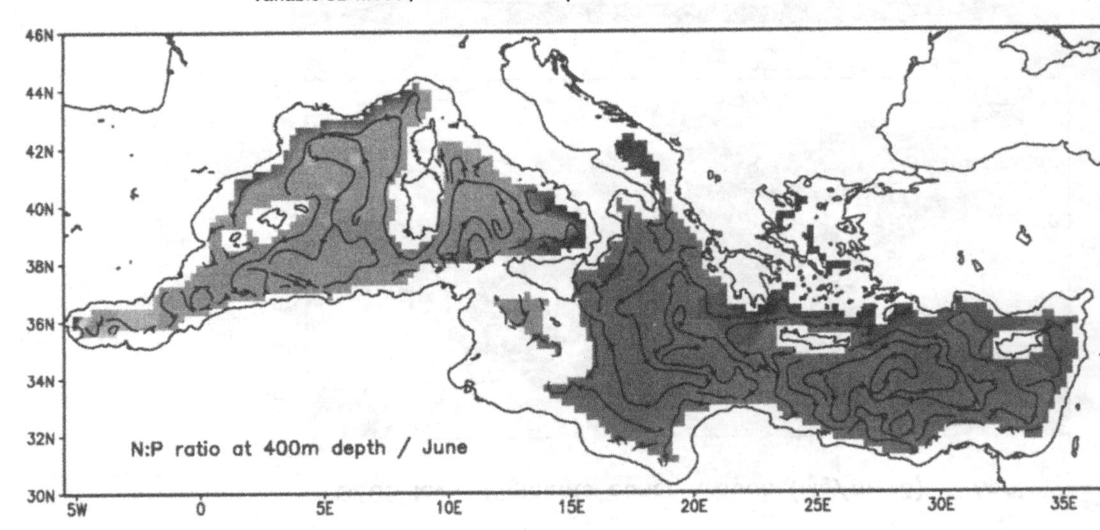

results of this model is beyond our present purpose.

The paper here presented offers an interpretation of the trophic gradients of the Mediterranean, and suggests a possible explanation of the different oligotophic regimes and ecosystem response in the East and West Mediterranean, based on an integrated approach to physical-biochemical processes. Our results indicate that aggregated models can be considered a proper tool for the study of phenomena going on over a basin-wide spatial scale and a seasonal temporal one.

**Acknowledgments** This work was partially supported by EU MAST MATER project (contract n. MAS3-CT96-0051). The authors thank Valentina Mosetti and Isabella Tomini for their valuable work in preparing experimental data and remotely sensed images.

All the maps included in this paper are obtained with the public domain software GrADS.

**References**

Andersen, V. and P. Nival, 1988. Modele d'ecosisteme pelagique des eaux cotieres de la Mer Ligure. Oceanol. Acta, Special Issue 9, 211-217.

Berman, T., D.W. Townsend, S.Z. El Sayed, C.C. Trees and Y. Azov, 1984. Optical transparency, chlorophyll and primary productivity in the Eastern Mediterranean near the Israeli coast. Oceanol. Acta, 7, 3, 367-372.

Béthoux, J.P., 1979. Budgets of the Mediterranean Sea, their dependence on the local climate and on the characteristics of the Atlantic waters. Oceanol. Acta, 2, 157-163.

Civitarese, G., M. Gacic, A.Vetrano, A.Boldrin, D.Bregant, S.Rabitti and E. Souvermetzoglou, (1998). Biogeochemical fluxes through the Strait of Otranto (Eastern Mediterranean). Cont. Shelf Res., in press.

Crise, A., G. Crispi and E. Mauri, 1998. Seasonal three-dimensional study of the nitrogen cycle in the Mediterranean Sea. Part I. Model implementation and numerical results. J. Mar. Sys, in press.

Crispi G., A. Crise and E. Mauri, 1998. Seasonal three-dimensional study of the nitrogen cycle in the Mediterranean Sea. Part II.Verification of the energy constrained trophic model. J. Mar Sys., in press.

Crise, A., J.I. Allen, J. Baretta, G. Crispi, R. Mosetti and C. Solidoro. The Mediterranean pelagic ecosystem response to physical forcing. Progr. Oceanogr., submitted.

Fasham M. J. R., H.W. Ducklow and S.M. MacKelvie, 1990. A nitrogen based model of plankton dynamics in the oceanic mixed layer. J. Mar. Res. 48, 591-639.

Holling, C. S. 1965. The functional response of predator to prey density. Mem. Entomol. Soc. Can. 45, 5-60

Krom, M.D., N. Kress, S. Brenner and L.I. Gordon, 1991. Phosphorus limitation of primary productivity in the eastern Mediterranean Sea. Limnol. Oceanogr. 36,3 424-432.

Lassiter R.R. and D.K. Kearns, 1974. Phytoplankton population changes and nutrient fluctations in a simple aquatic ecosystem model. In: Modelling the eutrophication process, edited by E.J. Middlebrookers D.H., Falkenberger and T. E. Maloney. Ann Arbour Science,131-188.

LeVourch J., C. Millot, N. Castagné, P. Le Borgne et J.P Orly, 1992. Atlas of Thermal Fronts of the Mediterranean Sea Derived From Satellite Imaginery, Musèe Océanographique Monaco,146pp.

McGill, D.A., 1961. A preliminary study of oxygen and phosphate distributions in the Mediterranean Sea. Deep-Sea Res., 8, 259-269.

McClain, R. and J. Firestone, 1993. An Investigation of Ekman Upwelling in the North Atlantic. J. Geoph. Res., 98, C7,12,327-12,339.

Millot, C. and L. Wald, 1981. Upwellings in the Gulf of Lions.In. Coastal upwelling. edited by F.A. Francis. AGU, 529pp.

Ovchinnikov, I.M., V.I. Zats, V.G. Krivosheya and A.I. Udodov,1985. Formation of deep Eastern Mediterranean waters in the Adriatic Sea. Oceanology, 25, 704-707.

Packard, T.T., H.J. Minas, B. Coste, R. Martinez, M.C. Bonin, J. Gostan, P. Garfield, J. Christensen, Q. Dortsch, M. Minas, G. Copin-Montegut and C. Copin-Montegut, 1988. Formation of the Alboran oxygen minimum zone. Deep-Sea Res., 35, 7 ,1111-1118.

Pinazo, C., P. Marsaleix, B. Millet, C. Estournel and R. Vehil, 1996. Spatial and temporal variability of phytoplankton biomass in upwelling areas of the north-western Mediterranean: a coupled physical and biogeochemical modelling approach. J. Mar. Sys.,7, 161-191.

Rabitti, S., F. Bianchi, A. Boldrin, L. Daros, G. Socal and C. Totti, 1994. Particulate matter and phytoplankton in the Ionian Sea. Oceanol. Acta, 17, 3, 297-307.

Rhein, M., 1995. Deep water formation in the western Mediterranean. J. Geophys. Res., 100,(C4), 6943-6959.

Roether, W., B.B. Manca, B. Klein, D.Bregant, D. Georgopulos, V. Beitzel, V. Kovacevic and A. Luchetta, 1996. Recent changes in the Eastern Mediterranean deep water. Science, 271, 333-335.

Roether, W. and R. Schlitzer, 1991. Eastern Mediterranean deep water renewal on the basis of chlorofluoromethane and tritium data. Dyn. Atm. Oc., 15, 333-354.

Roussenov, V.,E. Stanev., V. Artale and N. Pinardi, 1995. A seasonal model of the Mediterranean Sea general circulation. J. Geoph. Res., 100, C7, 13, 515-13, 538.

Sarmiento, J.L., T. Herbert and J.R. Toggweiler, 1988. Mediterranean nutrient balance and episodes of anoxia. Global Biogeoch. Cycles, 2,4,427-444.

Schott F. and K.D. Leaman, 1991. Observations with moored acoustic Doppler current profilers in the convection regime in the Golfe du Lion. J. Phys. Oceanogr., 21, 558-574.

Schott, F., M. Visbeck, U. Send, J. Fischer and L. Stramma, 1996. Observations of Deep Convection in the Gulf of Lions. Northern Mediterranean, during the Winter of 1991/92. J. Phys. Oceanogr., 26, 4, 505-524.

Solidoro, C., A. Crise, G. Crispi and R. Pastres, 1998. Local sensitivity analysis of Mediterranean Sea trophic chains. In: Second International Symposium of Sensitivity Analysis of Model Output Proceedings edited by K.Chan, S. Tarantola and F. Campolongo, 277-280.

Spencer, D. (ed.) 1983. GEOSECS, Indian Ocean expedition. V. 5-6.GPO.

Steele, J.H., 1962. Environmental control of photosynthesis in the sea. Limnol. Oceanogr., 7,137-150.

Tintoré, J., D. Gomis and S. Alonso, 1991. Mesoscale dynamics and vertical motion in the Alboran Sea. J. Phys. Oceanogr., 21, 6, 811-822.

Varela, R.A., A. Cruzado and J. Tintoré, 1994. A simulation analysis of various biological and physical factors influencing the deep-chlorophyll maximum structure in the oligotrophic seas. J. Mar.Sys., 5, 143-157.

Vidussi, F., H. Claustre, J-C. Marty, A. Luchetta and B. Manca, 1997. Phytoplankton distribution in relation to the physical structures of the Eastern Mediterranean Sea. In: International Conference on Progress in Oceanography of Mediterranean Sea, Abstracts Volume, 209-210.

# SOUTH ADRIATIC ECOSYSTEM: INTERACTION WITH THE MEDITERRANEAN SEA

I. MARASOVIĆ (1), D. VILIČIĆ (2) & Ž. NINČEVIĆ (1)
*Institute of Oceanography and Fisheries, 21000 Split, P.O.Box. 500, Croatia (1)*
*Faculty of Science, University of Zagreb, 10000 Zagreb, Rooseveltov trg 6, Croatia (2)*

## 1. Abstract

The causes of irregular increase of primary production in the southern Adriatic waters are discussed. Increase of primary production is related to the periods of intensified inflows of the Mediterranean waters into the Adriatic Sea carrying higher nutrient quantities. Our results confirm that the increase of primary organic production is related to intensified inflow of the Mediterranean water into the Adriatic. However, our hypothesis differs to a certain extent from the earlier ones. We assume that the upwelling, reported south of Palagruža Sill which is provoked by the intensified inflow of the Mediterranean water, causes enrichment of South Adriatic waters by "autochtonous" South Adriatic nutrients.

## 2. Introduction

The Adriatic Sea is a part of the Mediterranean Sea and linked to the eastern Mediterranean by the Strait of Otrant. The oceanographic properties of the Adriatic Sea are determined by its geographical location and morphology, climatic conditions of the area, fresh water inflow from rivers and water exchange with the Mediterranean. The length of the Adriatic Sea is about 800 km and its maximum width is 200 km. Geomorphologically it can be divided into three parts: the northern, the middle and the southern Adriatic. The northern Adriatic is very shallow, with an average depth of 30 to 40 m and maximum of 70 m. It is strongly influenced by the northern Italian rivers, especially the Po River. The middle Adriatic is much deeper, reaching 280 m in the Jabuka Pit. It is separated from the southern Adriatic by the 170 m deep Palagruža Sill. The southern Adriatic is considerably deeper. Its average depth is 900 m, and its deepest part the 1300-m-deep Adriatic Pit. Trough the Otrant Strait it is connected with the Mediterranean Sea.
The earliest research of the Adriatic Sea dates from the past century. However, systematic and regular measurements in the middle and southern Adriatic began in the 1950s. Temperature, salinity, transparency, oxygen and phosphate measurements were done on a monthly basis [1-3]. In the 1960s, primary production was included in the measurements performed in the middle Adriatic [4-6]. On the basis of standard oceanographic parameters, Buljan [7] reported that the advection of the Mediterranean waters was an important factor, which caused increased productivity in the Adriatic Sea. During such periodical "ingressions" [8], the Mediterranean waters, relatively rich in nutrients, are carried into the Adriatic, increasing productivity of oligotrophic middle and south Adriatic waters. Buljan & Zore-Armanda [2] supposed that the increase of productivity occurs primarily due to phosphorus inflow from the Eastern Mediterranean. Buljan & Zore-Armanda [9] also observed an increase in temperature and salinity on an annual scale, which coincided in time with such "ingresions".
Some regularity in the year-to-year production fluctuations was observed and related to the strong advection from the Mediterranean to the Adriatic [10]. A change in the phytoplankton species composition was observed [11], as well as greater biomass and changed species composition in zooplankton communities [12]. These authors found that stronger advection coincided with higher primary production, higher zooplankton biomass and changes in species composition.
The most important feature of the Mediterranean waters advecting into the Adriatic (in the intermediate layer) is their high salinity (chlorinity >21.3 parts per thousand according to Buljan & Zore-Armanda [2]). This high salinity is characteristic of the Levantine basin, which has one of the highest salinities of all the world seas (>39 psu)[13, 14]. The temperature of Levantine waters is higher than that of the Adriatic waters, so that 'ingressions' are reflected upon the temperature as well [15]. Relating these phenomena to certain climatic factors, Zore-Armanda [15-18] stated that the most important factor enhancing the water exchange between the two basins is the horizontal pressure gradient over the eastern Mediterranean.
A large number of studies up to the 1970s showed that the intensity of water exchange between the Adriatic Sea and the Ionian Sea was the most important factor of long-term production fluctuations in the middle and southern Adriatic [19]. The discharge of the Po River affected production considerably only in the shallow northern Adriatic, while the Mediterranean "ingressions" were of no significance there [20, 21].



*P. Malanotte-Rizzoli and V.N. Eremeev (eds.),*
*The Eastern Mediterranean as a Laboratory Basin for the Assessment of Contrasting Ecosystems,* 383–405.

### 3. Sampling and Methods of Analyses

The present study is based on data from four cruises in the middle and southern Adriatic Sea (Fig.1) during April 1986, April 1987, April 1992. The cruises were performed by R/V "A. Mohorovicic" and R/V "ACADEMIC STRAHOV" as the part of the POEM programme.
Phytoplankton was sampled by 5-1 Niskin bottles. Samples were preserved in a 2% neutralized formaldehyde solution. Cell counts were obtained by the inverted microscope method [22]. Samples of 25 ml were analysed microscopically after 48 h. The phytoplankton cells with a maximum length between 2 and 20 μm were designated as nanoplankton, and cells longer than 20 μm as microplankton. The microplankton cells were counted under magnifications of 200x and 80x. For the smaller, more abundant microplankton cells transacts across the central part of the counting chamber base-plate were made at 200x. Nanoplankton cells were counted in 20-30 randomly selected fields of vision along the counting chamber base-plate, under the magnification of 320x. To avoid miscounting, only easily identifiable nanoplankton cells were counted. Inclusion of a particle in the counts depended on the presence of a cell wall and granular appearance of the cell contents. The precision of the counting method was about ± 10%. Assuming that the biomass (fresh weight) was equal to the total cell volume, the latter was calculated according to Smayda [23]. Cell density and cell sizes were determined simultaneously in each sample from measurements obtained using the inverted microscope. Cell volumes of various species were determined according to cell models (geometrical bodies) constructed by means of light microscopy (or scanning electron microscopy) microphotographs and drawings [24]. As many cells as possible were measured to ensure that the sample mean was close to the population mean. Unfortunately, it was only possible to measure a small number of the less common species, which resulted in their volumes being necessarily approximate.
For chlorophyll *a* determinations, 2 litres of seawater were filtered through a Whatman GF/F glass filter, made basic with magnesium carbonate, and chlorophyll *a* concentrations subsequently determined in the laboratory using a Turner Model 112 fluorometer following the method of Strickland & Parsons [25]. Size fractionation of the phytoplankton samples for chlorophyll *a* measuring was done on 10 μm for nanoplankton and 2 μm for picoplankton.
Salinity was determined using an Autolab –MK-IV inductive salinometer and controlled by argentometric titration. Temperature was measured by Richter-Wiese reversing thermometers.

### 4. Results

For the first time, results of studies conducted in the southern Adriatic in the spring of 1987, suggested that this particular region may also have exceptionally high productivity levels during some seasons of the year. Due to the inadequacy and irregularity of past research and studies, which were performed in the southern Adriatic, the most of the studies considered this area as one of very low productivity.
The chlorophyll *a* concentration data obtained in 1987 show that concentrations are unusually high for the open sea of the southern Adriatic (Fig.2); moreover, they exceed those levels found in the open waters of the middle and northern Adriatic. This is particularly noticeable in the central part of the Dubrovnik-Bari profile (Stations 13, 14), located in the South Adriatic Pit. The relationship between microplankton and nanoplankton fractions of phytoplankton is shown in Fig. 3 and is calculated on the basis of cell volume. It can also be concluded that the microplankton fraction is significantly more abundant than the nanoplankton fraction (the picoplankton fraction was not measured in 1987). The increased quantity of microplankton in relation to nanoplankton is a common characteristic of regions with high productivity.
Similar studies were carried out also in April 1986. At that time the concentrations of chlorophyll *a* were considerably lower in the same area (Fig.4). The quantities of microplankton in relation to nanoplankton were also considerably lower (Fig. 5) than those found in the same time period in 1987.
In addition, pronounced difference was found with respect to the species diversity of phytoplankton. In 1987, the species diversity was significantly higher in comparison with that found in 1986, more specifically, diatoms were best represented in the phytoplankton community.
Research was also conducted in the southern Adriatic and Otrant in 1992. Sampling was done at seven profiles and included a large number of stations. During that period, a very high productivity level was recorded throughout the southern Adriatic; particularly, at the profile south of the Palagruža Sill (profile A). Concentration values of chlorophyll *a* were high at the profile south of the Palagruža Sill, in the South Adriatic Pit area, as well as at stations near the Albanian coast and the western stations, north of the Otrant Strait (Figs. 6,7,8,9,10,11,12). The highest biomass was recorded on eastern stations between Palagruža Sill and Jabuka Pit in all layers in the water column (Figs. 13,14,15).

At the time of these studies, the general pattern of the microplanktonic fraction being characteristic of productive regions was confirmed, since the largest fraction of microplankton was found in the most productive regions. Nanoplankton organisms were poorly represented, while the presence of picoplankton was even poorer (Fig.16). Diatoms were especially well represented in the microplankton fraction (Fig.17), which is also characteristic of regions with naturally high productivity.
With respect to the vertical distribution of phytoplankton biomass, a regularity, which was noted earlier in the Mid-Adriatic waters, reoccurs [26]. As a rule, maximum concentrations of chlorophyll *a* always occur in the subsurface layer between depths of 30 and 75 m ( Figs.13,14,15). The DCM (Deep Chlorophyll Maximum) is more obvious, the higher the concentrations of chlorophyll *a*. Different authors attribute the development of DCM to a variety of causes, most often, it is explained as the adaptation of phytoplankton to the reduction of light in the deeper layers (after such an account, it is not the number of phytoplankton cells which increases but the quantity of chlorophyll *a*). In such a case DCM depth mainly doesn't coincide with the depth of the highest density of phytoplankton cells. Research performed in the southern Adriatic is, to some extent, inconsistent with the preceding interpretation since, the maximum cell abundance and maximum biomass (calculated from the cell volume), as well as the DCM depth were present at the same level at the same time (Figs.18,19). As can be seen in Fig.2 and 3, all maxima are significantly more pronounced in 1987, during the year with higher productivity than in 1986.

## 5. Discussion

Buljan [27] recorded that the periodic increase in productivity levels in the southern Adriatic coincided with increases in salinity and temperature of the seawater. He attributed this phenomenon to an increased inflow of the Mediterranean waters into the Adriatic and refers to it as "Adriatic ingressions". According to Buljan, the Mediterranean water is saltier, warmer and richer in nutrients, which directly reflects the increase in productivity. Pucher-Petković [4, 5] shows that an increase in productivity is also related to changes in phytoplankton composition. Pucher-Petković et al. [11] and Marasović et al. [28] concluded that an increase in productivity with a time lag of two to three years results in the increased catch of small blue fish in the Adriatic waters. Zore-Armanda [18] explains the "Adriatic ingressions" phenomenon through the changes of various climatic factors; that is, that pressure differences between the northern and southern Adriatic provide the driving mechanism for advection of the Mediterranean waters into the Adriatic.
Most of the reported results were obtained during the studies of the open middle Adriatic, while the southern Adriatic waters were studied only for physico-chemical properties ( T, S, nutrients). On the basis of nutrient increase in the southern and middle Adriatic and the higher primary production and phytoplankton cell density in the middle Adriatic, Buljan assumed that the primary production increase in the southern Adriatic, as well, at the time of "ingression". However, it was not confirmed through biological parameters all until 1987. It could not even be assumed that primary production could reach so high values in the southern Adriatic. In April 1987, the first phytoplankton bloom was recorded from the waters of the southern Adriatic, observed by Viličić et al. [29] as phytoplankton cell numbers, and Marasović et al. [30] as a chlorophyll *a* concentration. Results of our study show that the concentration values of chlorophyll *a* in the middle of the South Adriatic Pit exceededsurpassed 3 mg chl *a* $m^{-3}$ . Similar studies were performed in the same area in April 1986, but results of the analysis of the phytoplankton communities and the concentrations of chlorophyll *a* were significantly lower than those obtained in 1987.
Moreover, in April 1986, the highest concentrations of chlorophyll *a* were about 0.5 mg chl *a* $m^{-3}$; as well, the number of phytoplankton cells was one order magnitude lower than that recorded in April 1987. The analysis of the physico-chemical properties of the seawater showed the differences in the maximum water salinity between these two years: in 1987, salinity in the middle of the South Adriatic Pit exceeded 39.0‰, while in 1986, the highest value was 38.60‰ (Fig.20). All the above mentioned results show that the inflow of the Mediterranean waters into the Adriatic was much stronger in 1987 than in 1986. The analysis of weather conditions in 1986 and 1987, pointed to the fact that there was twice as much precipitation during the period from November 1985 to April 1986 than in the same period in 1986-1987.
Frequency of wind directions for 1.5 months prior to the sampling period show that in 1986, the south wind prevailed (SE wind – Scirroco), while in 1987, the north wind prevailed (NE – Bora) (Fig.21).
Similar weather conditions prevailed both in the spring 1987 and 1992. In addition, in April 1992, extremely high concentrations of chl *a* $m^{-3}$ were recorded from the waters of the southern Adriatic. Particularly high concentrations were recorded from the southern side of the Palagruza Sill and the middle of the South Adriatic Pit. High productivity of this region is also characterized by large quantities of microplankton in comparison to nanoplankton and picoplankton. Moreover, the areas with high productivity, as well as the most productive seasons are, as a rule, characterized by higher microplankton

fraction of phytoplankton. A large proportion of diatoms in the phytoplankton communities is also common in regions of high productivity, something which was fairly evident in 1987 and 1992. The DCM (Deep Chlorophyll Maximum) was pronounced in 1987 as well as in 1992. It is, as a rule, always present in the open sea of the middle and southern Adriatic [26]. The DCM is most often found between 30 and 75 m depth; and particularly above 50 m during the most productive period; and in deeper layers in the less productive period. This is probably due to the transparency of the water column, which is reduced as a result of increased cell density.

All the above mentioned results in this paper confirm earlier observations which showed that higher productivity in the south Adriatic waters is related to the increased inflow of the Mediterranean waters into the Adriatic. However, our hypothesis on the origin of nutrients differs to some extent. Buljan [27], as well as many other authors, believe that the Mediterranean waters transport higher quantity of nutrients into the Adriatic, which in turn, increases productivity. Our results suggest another possibility, which could be more credible due to the fact that Ionian Sea and Eastern Mediterranean waters are extremely poor in nutrients and have low levels of productivity [31-35]. These waters cannot affect an increase in productivity of the South Adriatic waters by enrichment through "alochtonous" nutrients, but they can cause "upwelling" due to their strong inflow.

"Upwelling" or mixing, causes the transport of "autochtonous" South Adriatic nutrients from deeper bottom layers into the subsurface layers. When the inflow of water is particularly strong, the south Adriatic water expands across the Palagruza Sill *all* the way to the northern Adriatic [36, 29]. This is evident due to the flora and fauna common to the southern Adriatic occasionally found in large quantities in waters of the northern Adriatic. Still, the largest mass of the Mediterranean water, entering the Adriatic, remains in its southern part not crossing the Palagruža Sill [37]. In fact, the incoming Mediterranean currenthindered by the natural barrier, the Palagruža Sill, changes it direction towards the Italian coast along which it goes back to Otrant. Shift of the Mediterranean water at the Palagruža Sill is a possible causeor vertical mixing or upwelling. The results from 1987 are indicative of a strong upwelling in the southern Adriatic, since considerable quantities of diatoms, particularly of *Chaetoceros* genus, were recorded from deep aphotic layers (100 – 500 m). At the same time, some copepod species were recorded from the same layer, which are otherwise normally found in the surface layers [29]. The very fact that the area below the Palagruža Sill and South Adriatic Pit is most productive in the entire southern Adriatic, also indicated that the increase in productivity is due to the south Adriatic upwelling. In addition the southern Adriatic parts show considerably lower productivity, even though the Mediterranean water quantities are almost the same or even greater there than in the South Adriatic Pit. Therefore, we believe that the South Adriatic upwelling is the main cause of the higher production in the southern Adriatic enhanced by an intensified penetration of the Mediterranean water.

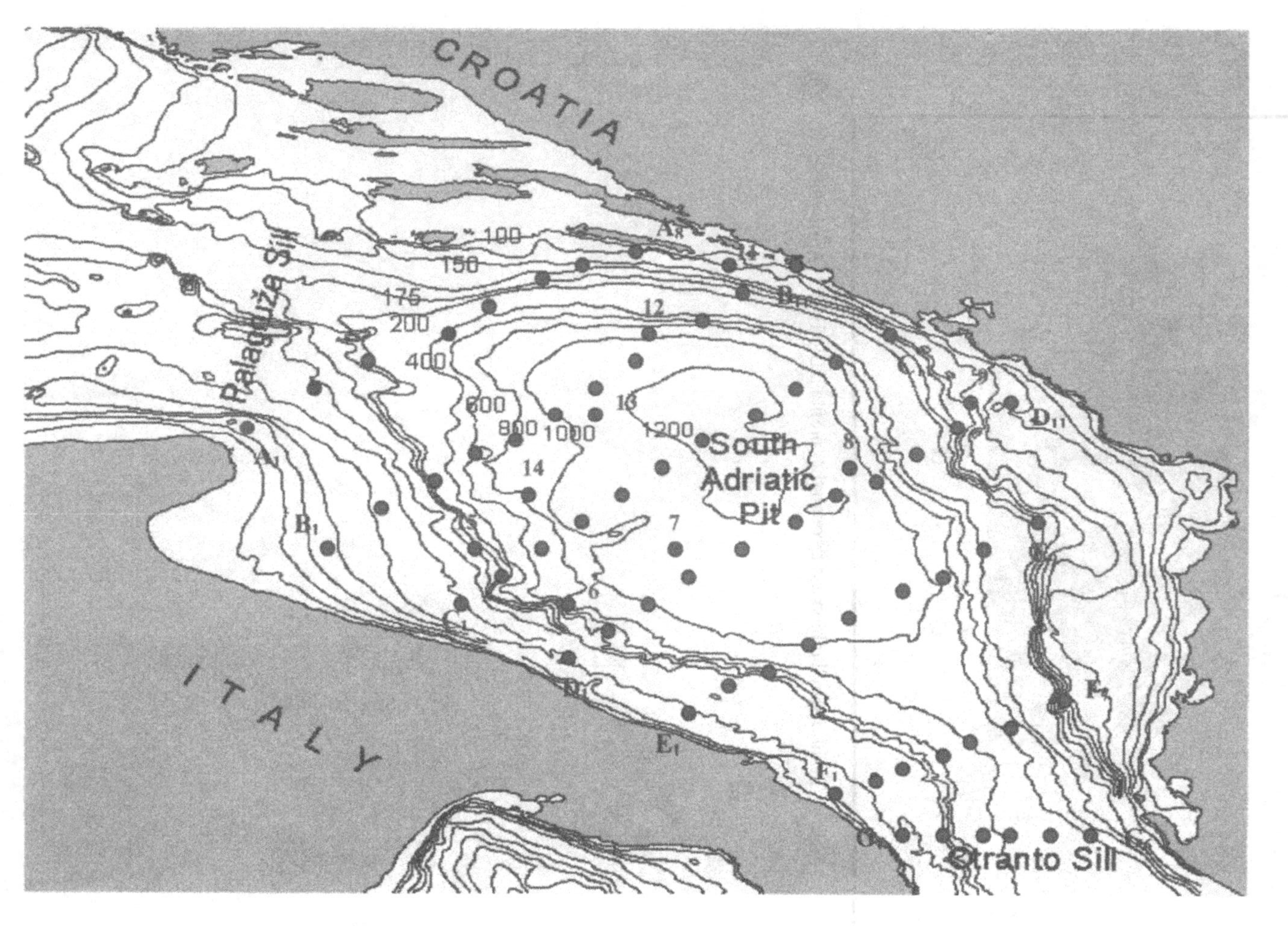


Figure 1. Map of research area and transects of sampling stations

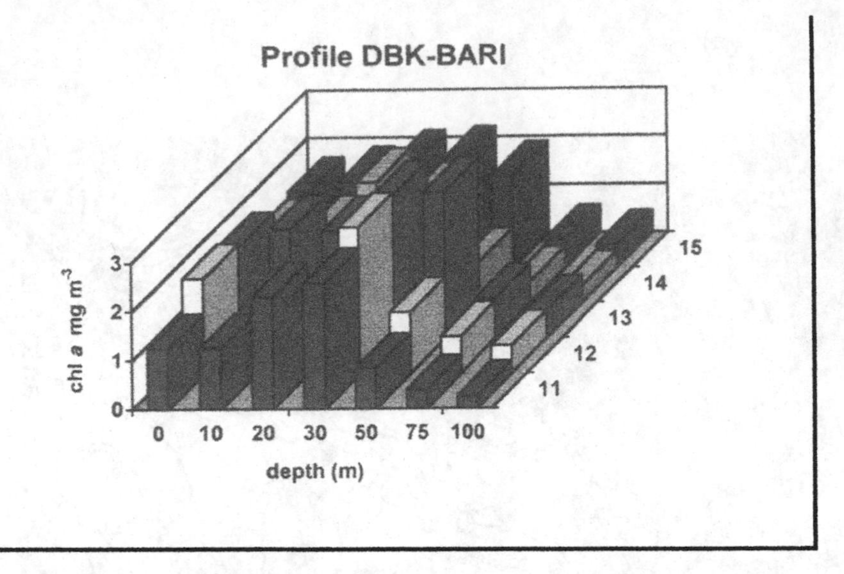


Figure 2. Chlorophyll *a* biomass through the profile DBK-BARI (April 1987)

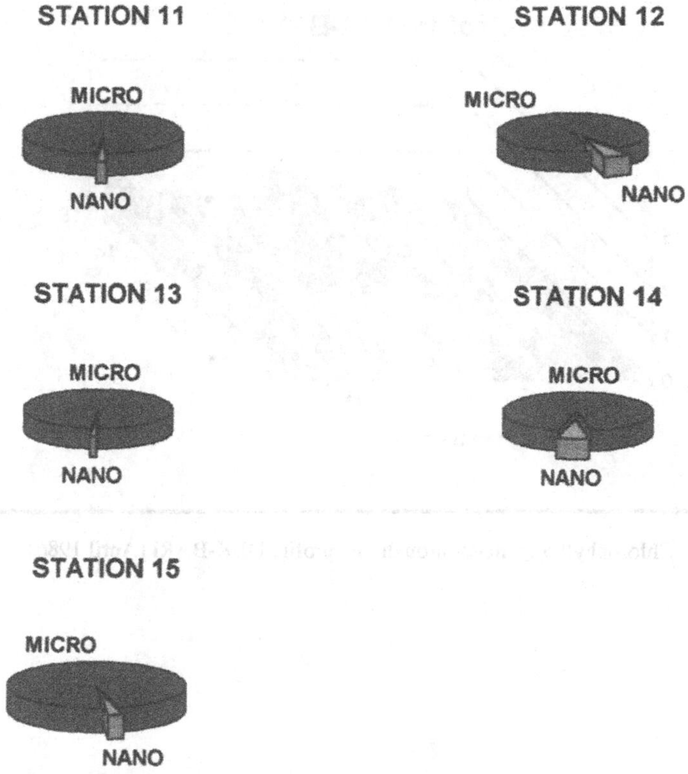

STATION 11

MICRO

NANO

STATION 12

MICRO

NANO

STATION 13

MICRO

NANO

STATION 14

MICRO

NANO

STATION 15

MICRO

NANO

Figure 3. Relation between microplankton and nanoplankton through the profile DBK-BARI in April 1987

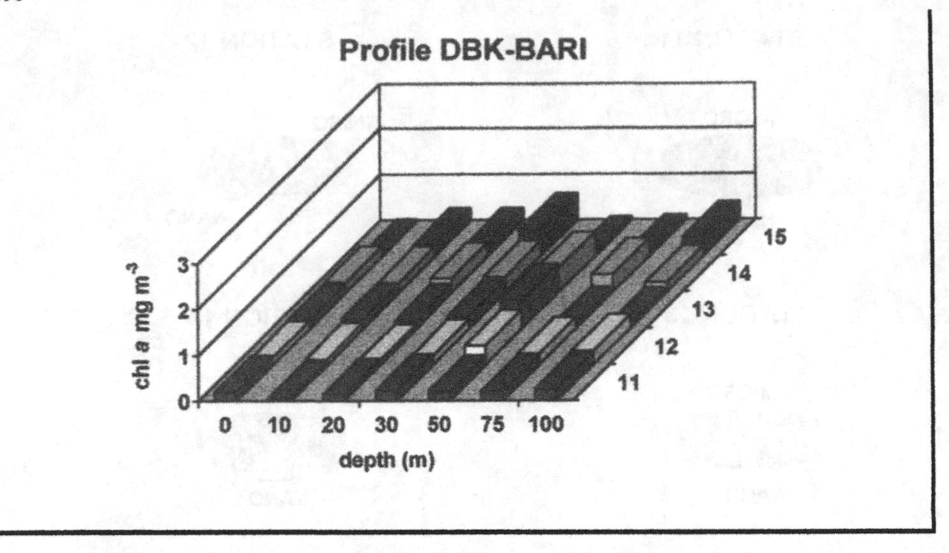


Figure 4. Chlorophyll *a* biomass through the profile DBK-BARI (April 1986)

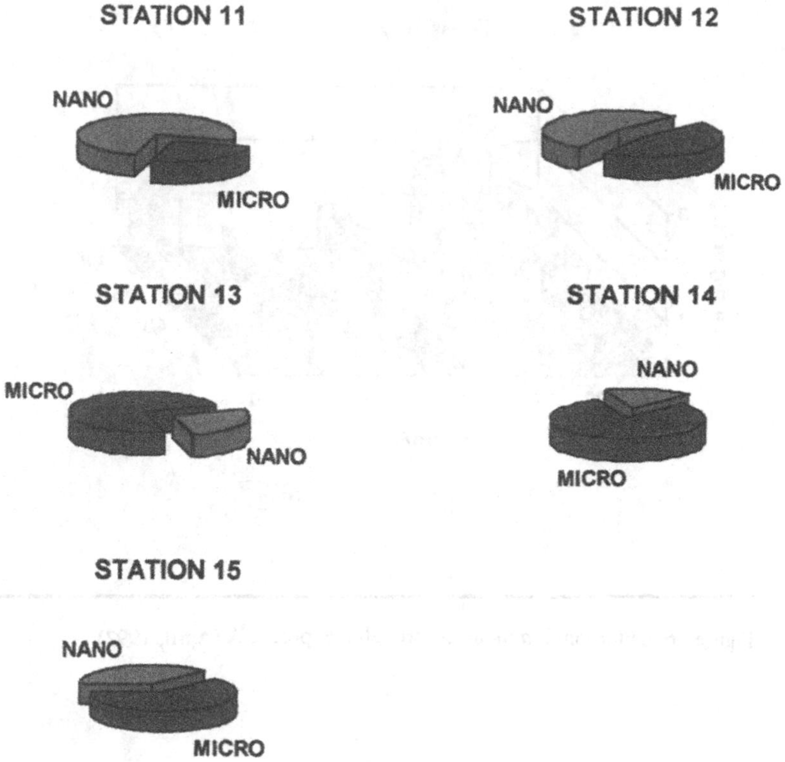

STATION 11

NANO

MICRO

STATION 12

NANO

MICRO

STATION 13

MICRO

NANO

STATION 14

NANO

MICRO

STATION 15

NANO

MICRO

Figure 5. Relation between microplankton and nanoplankton through the profile DBK-BARI in April 1986

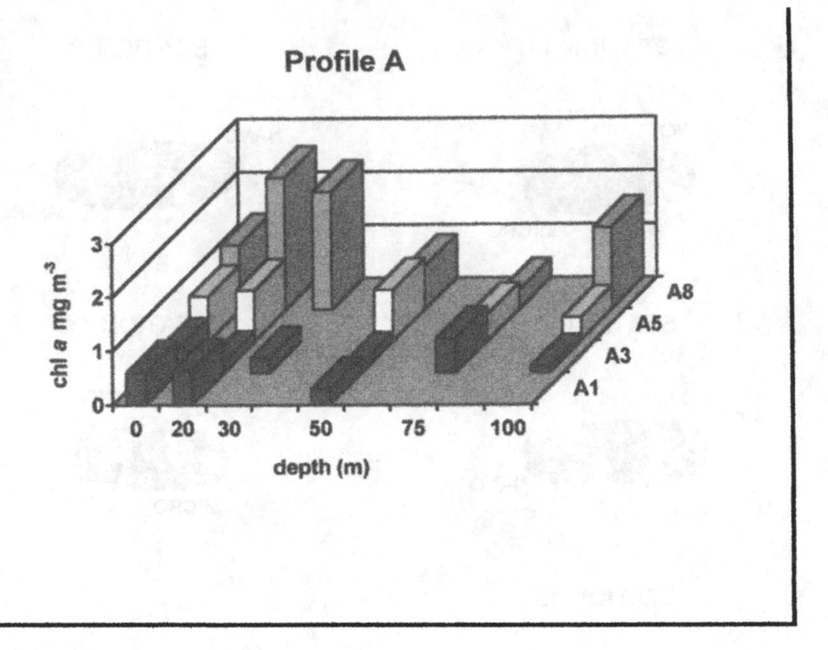


Figure 6. Chlorophyll *a* biomass through the profile A (April, 1992)

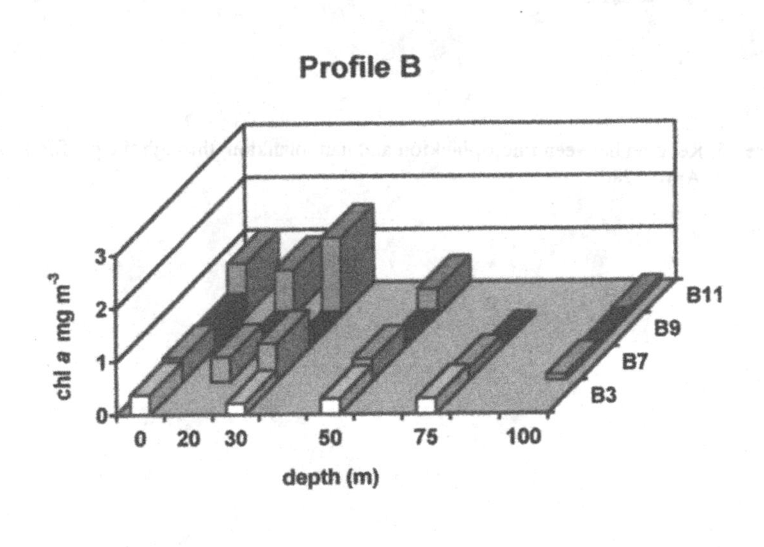


Figure 7. Chlorophyll *a* biomass through the profile B (April, 1992)

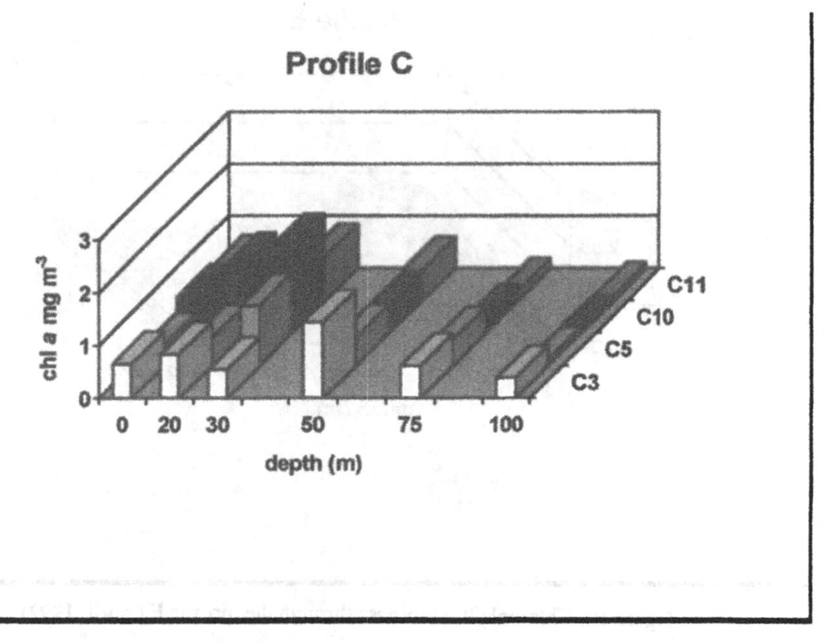


Figure 8. Chlorophyll *a* biomass through the profile C (April, 1992)

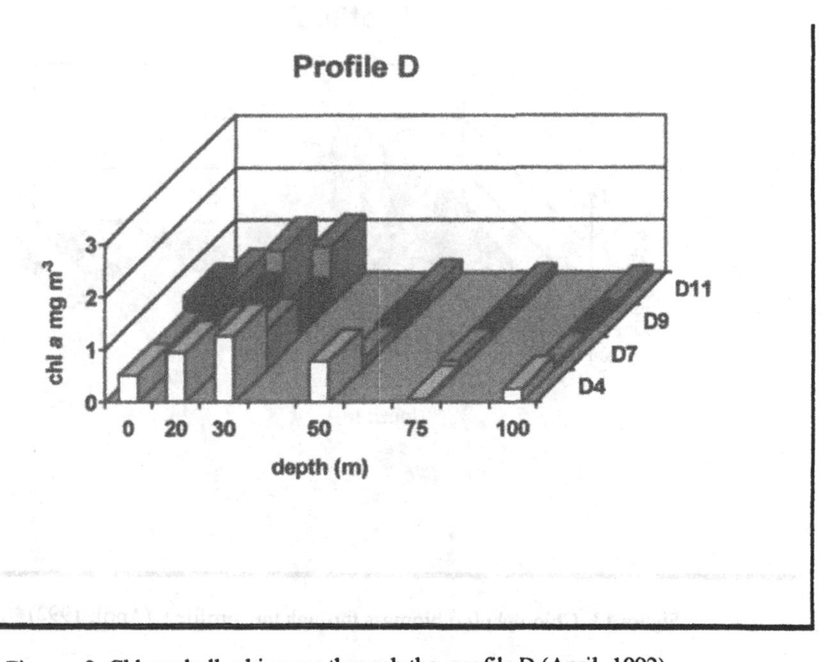


Figure 9. Chlorophyll *a* biomass through the profile D (April, 1992)

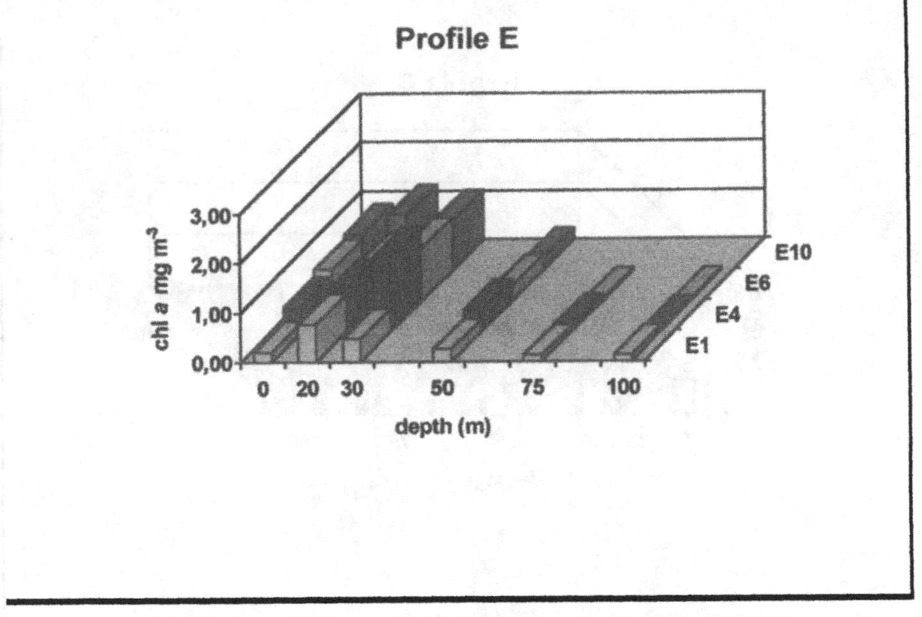


Figure 10. Chlorophyll *a* biomass through the profile E (April, 1992)

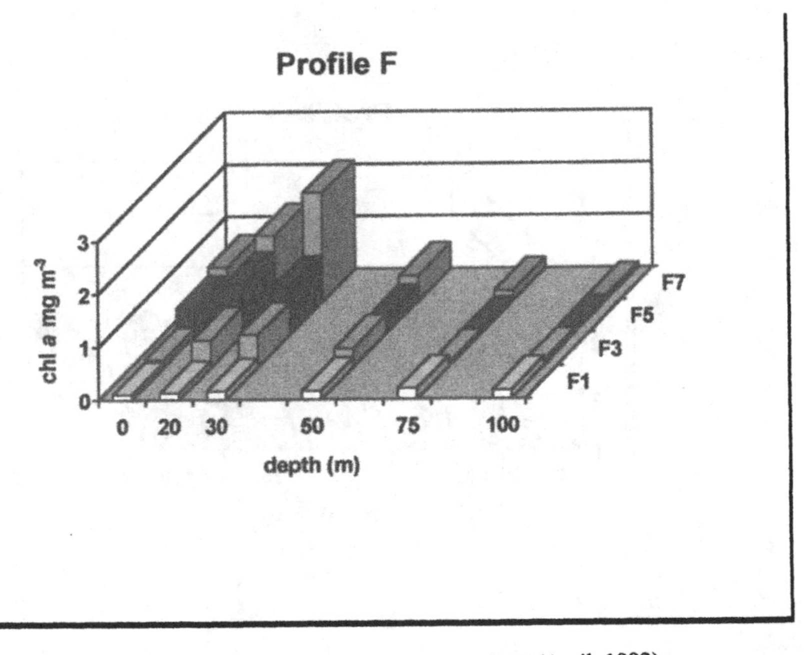


Figure 11. Chlorophyll *a* biomass through the profile F (April, 1992)

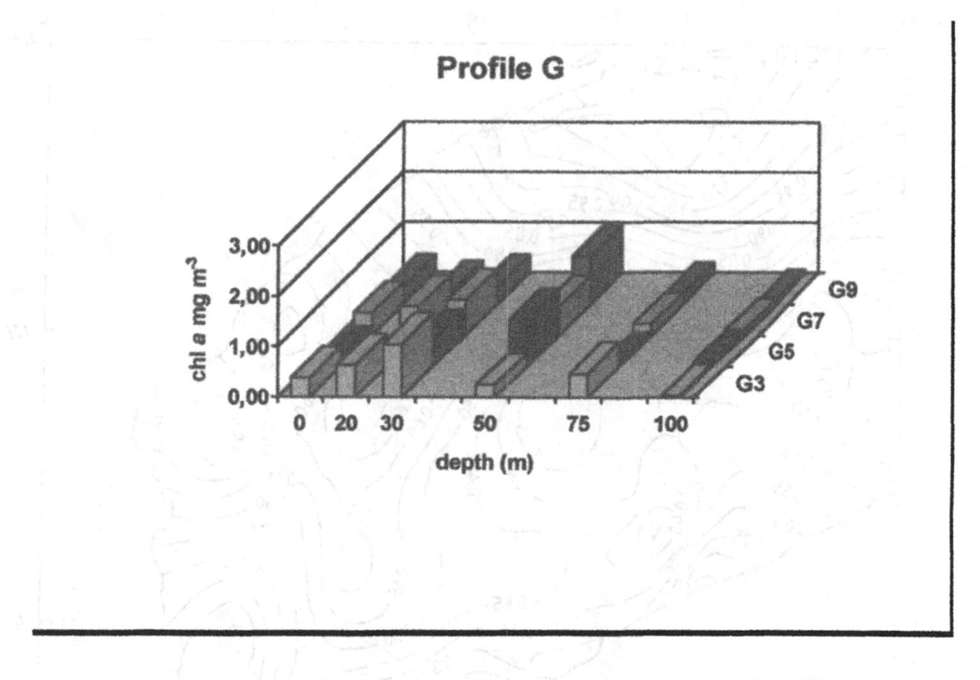


Figure 12. Chlorophyll *a* biomass through the profile G (April, 1992)

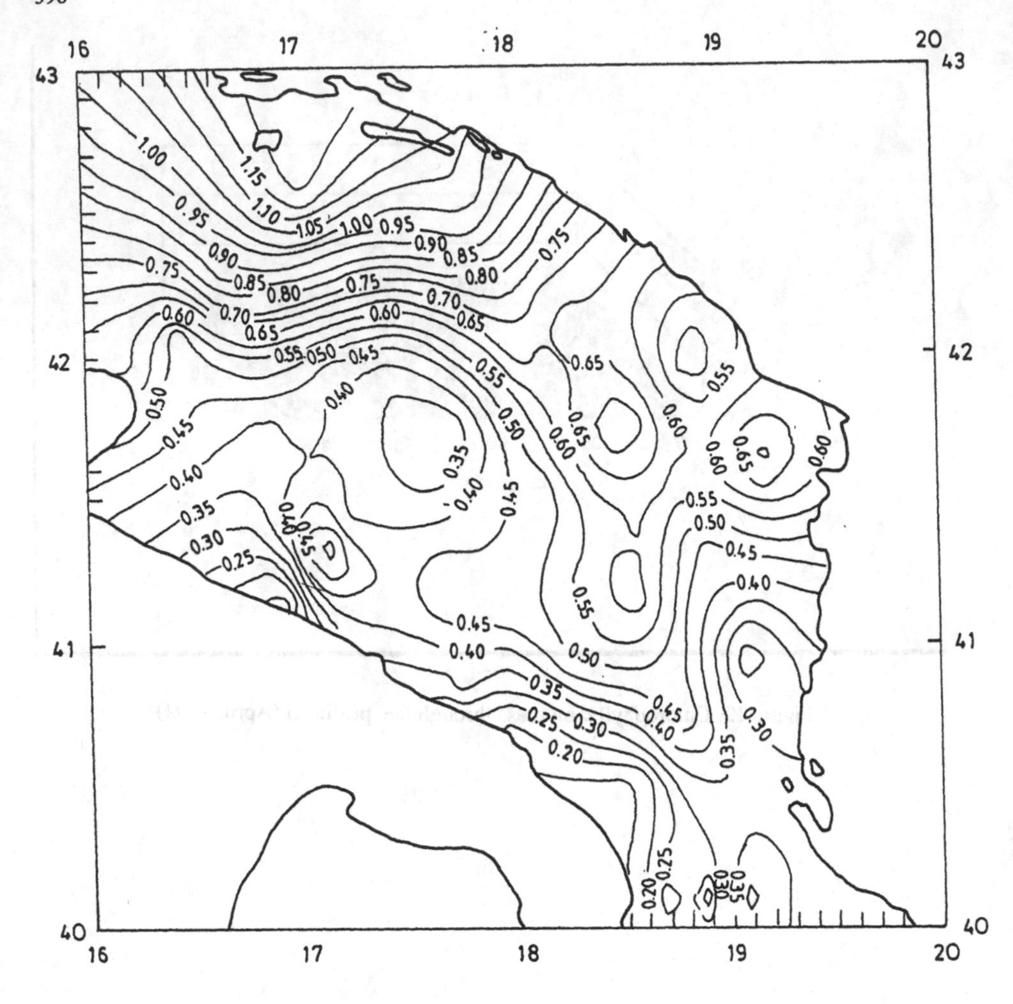


Figure 13. Distribution of the clorophyll *a* biomass in the South Adriatic on the surface (0 m) (April 1992)

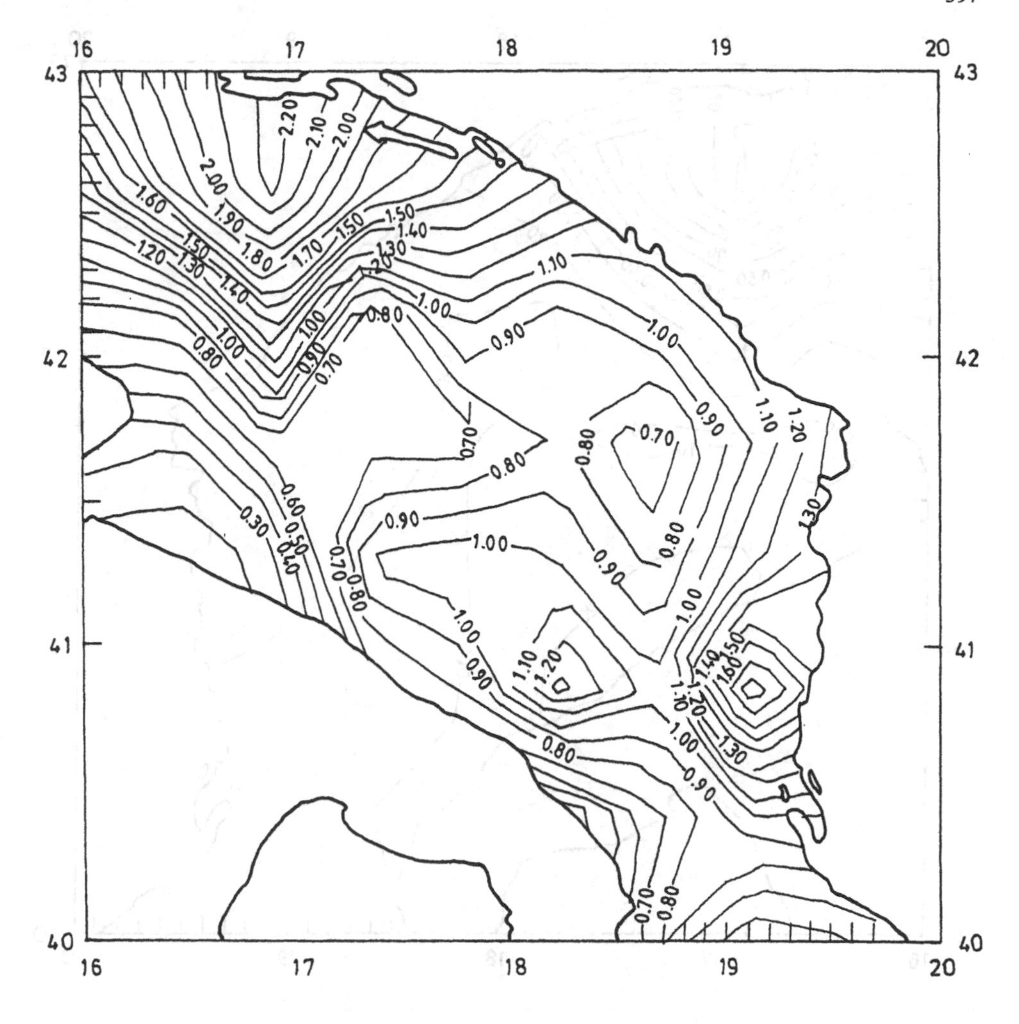


Figure 14. Distribution of the clorophyll *a* biomass in the South Adriatic in the subsurface layer (30 m) (April 1992)

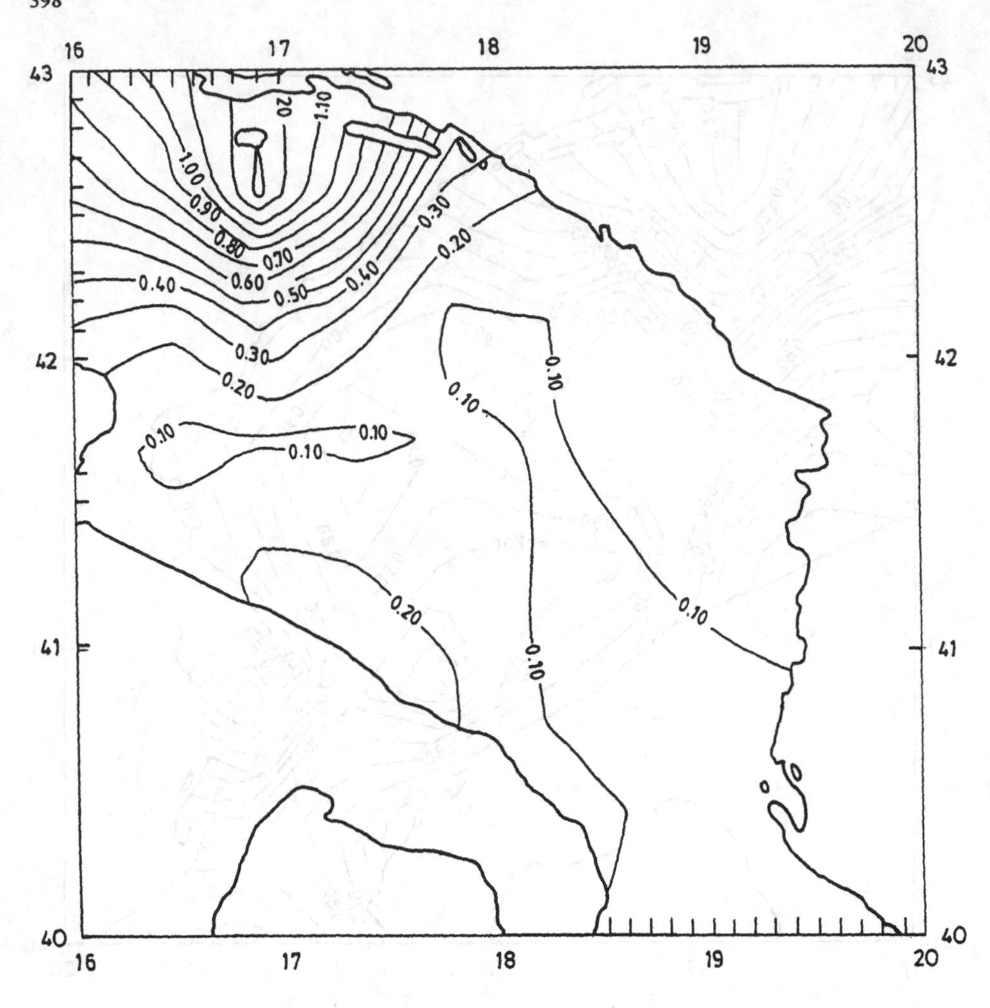


Figure 15. Distribution of the clorophyll *a* biomass in the South Adriatic in the subsurface layer 100 m depth (April 1992)

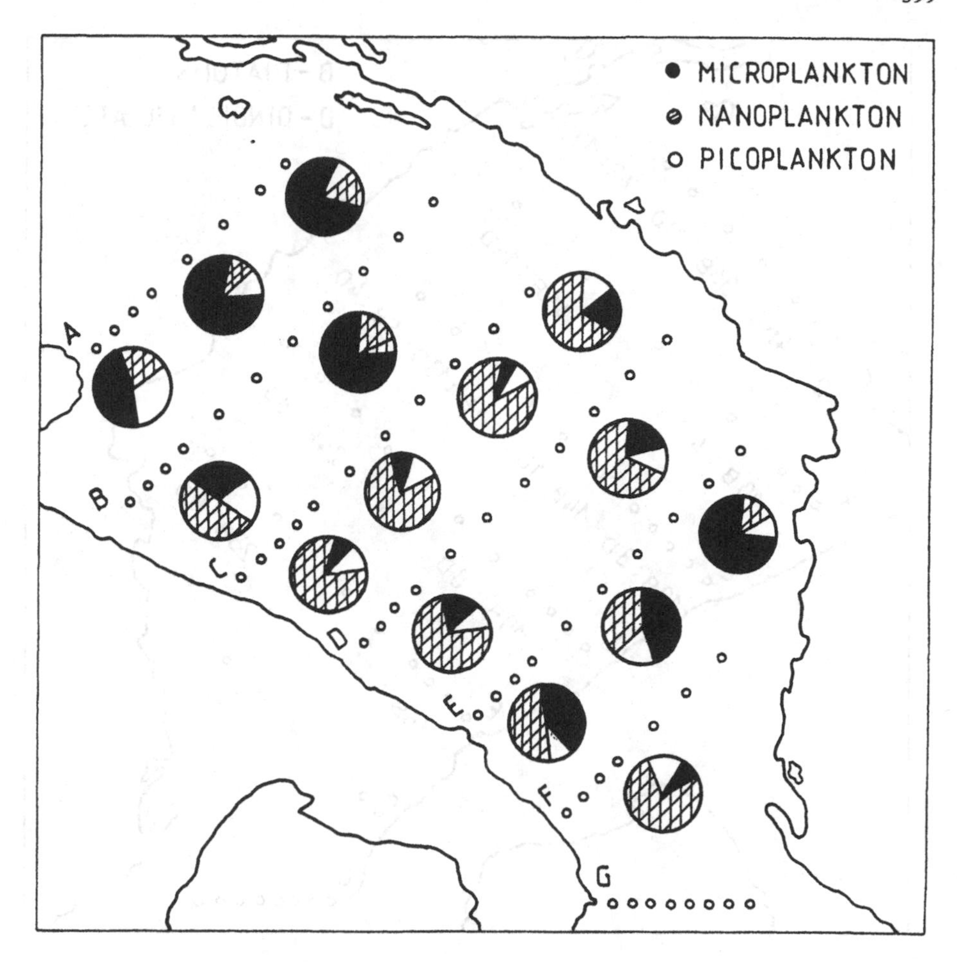


Figure 16. Relation between microplankton, nanoplankton and picoplankton in the phytoplankton community during the April 1992

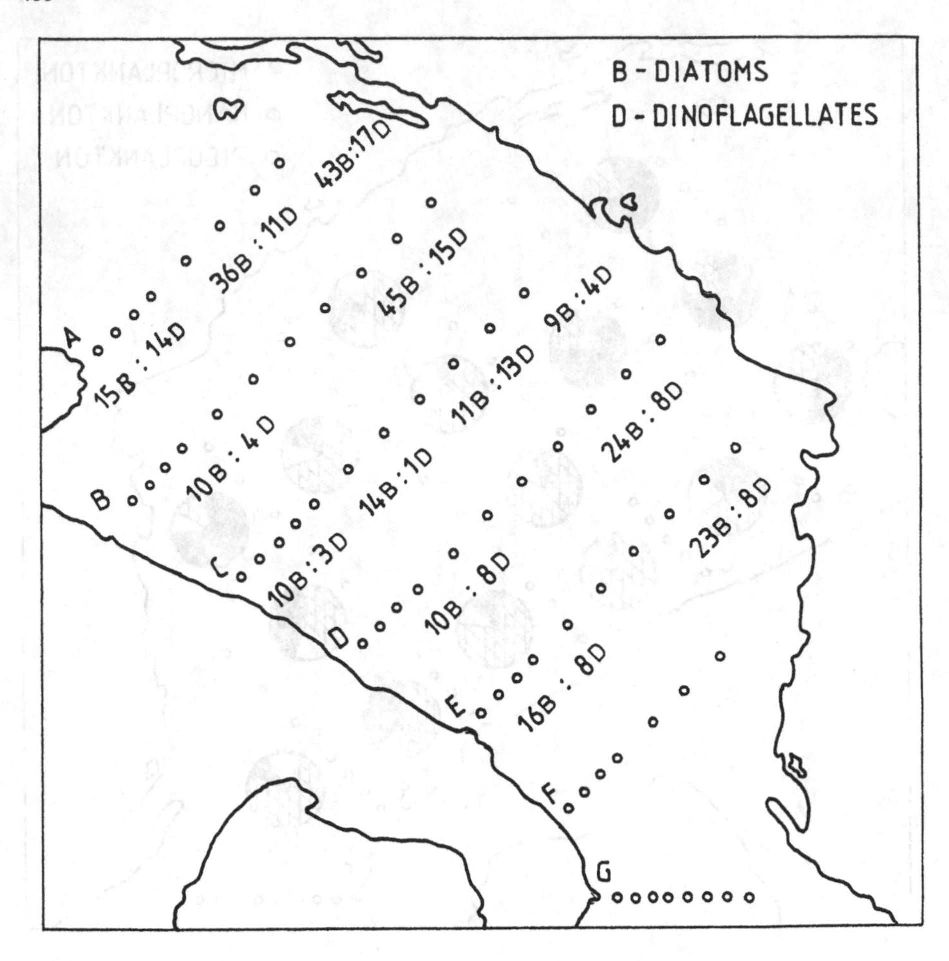


Figure 17. Relation between number of diatom and dinoflagellate species in South Adriatic (April 1992)

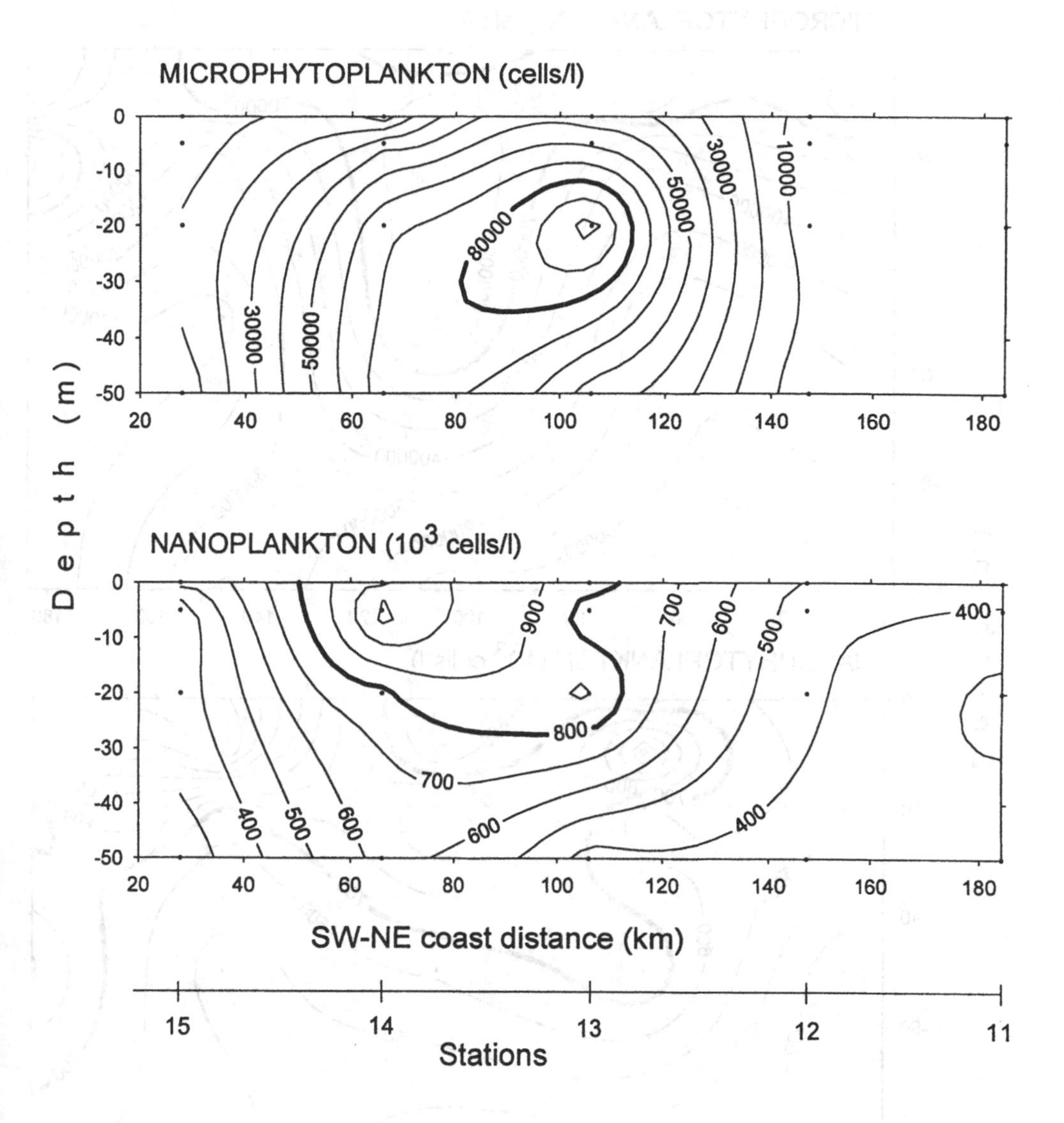


Figure 18. Distribution of the microphytoplankton and nanoplankton cells in the South Adriatic (April 1986)

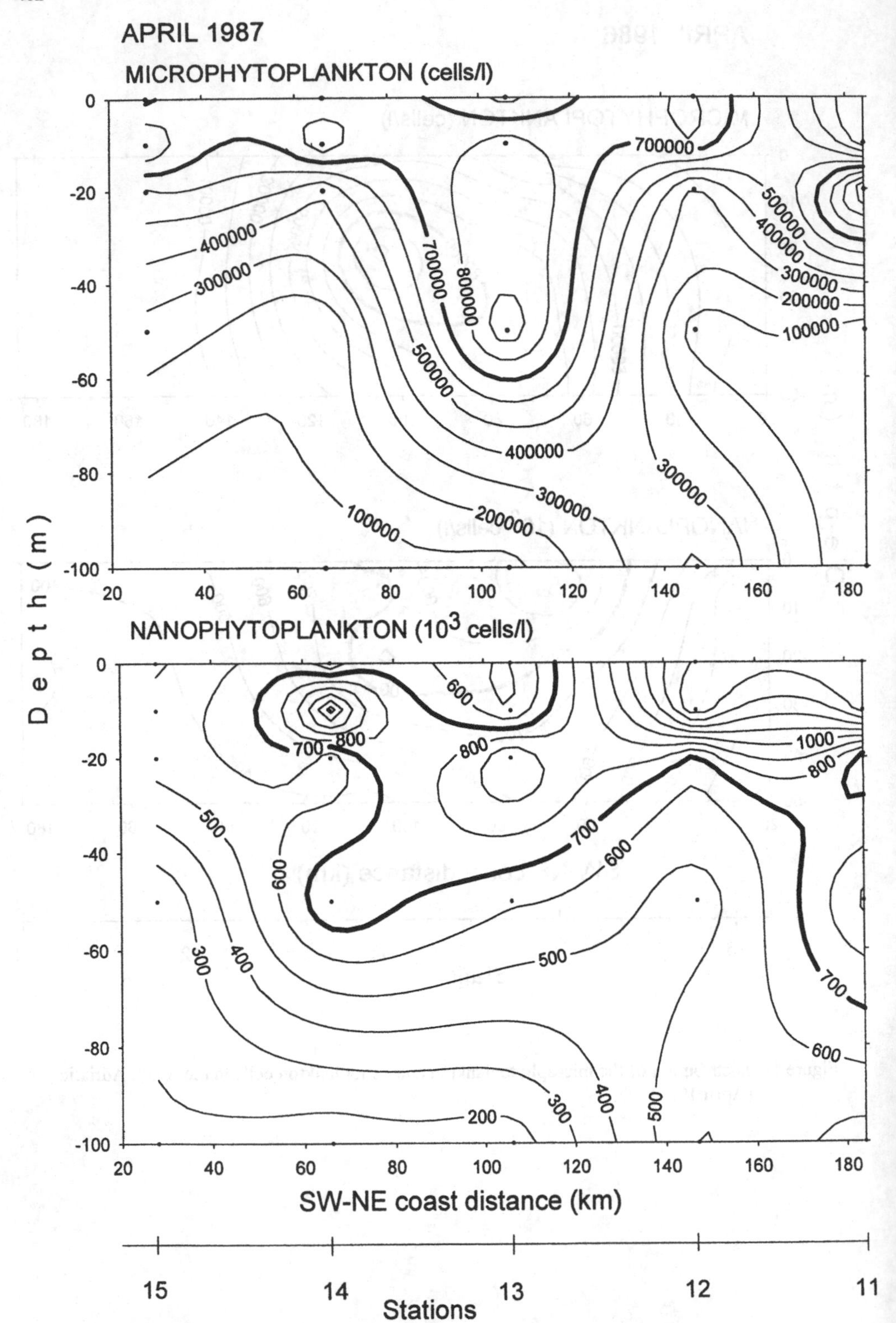


Figure 19. Distribution of the microphytoplankton and nanoplankton cells in the South Adriatic (April 1987)

SALINITY

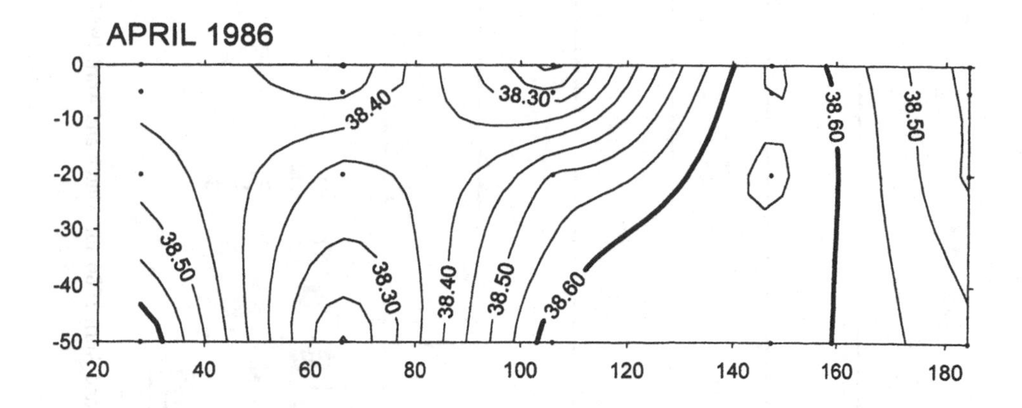


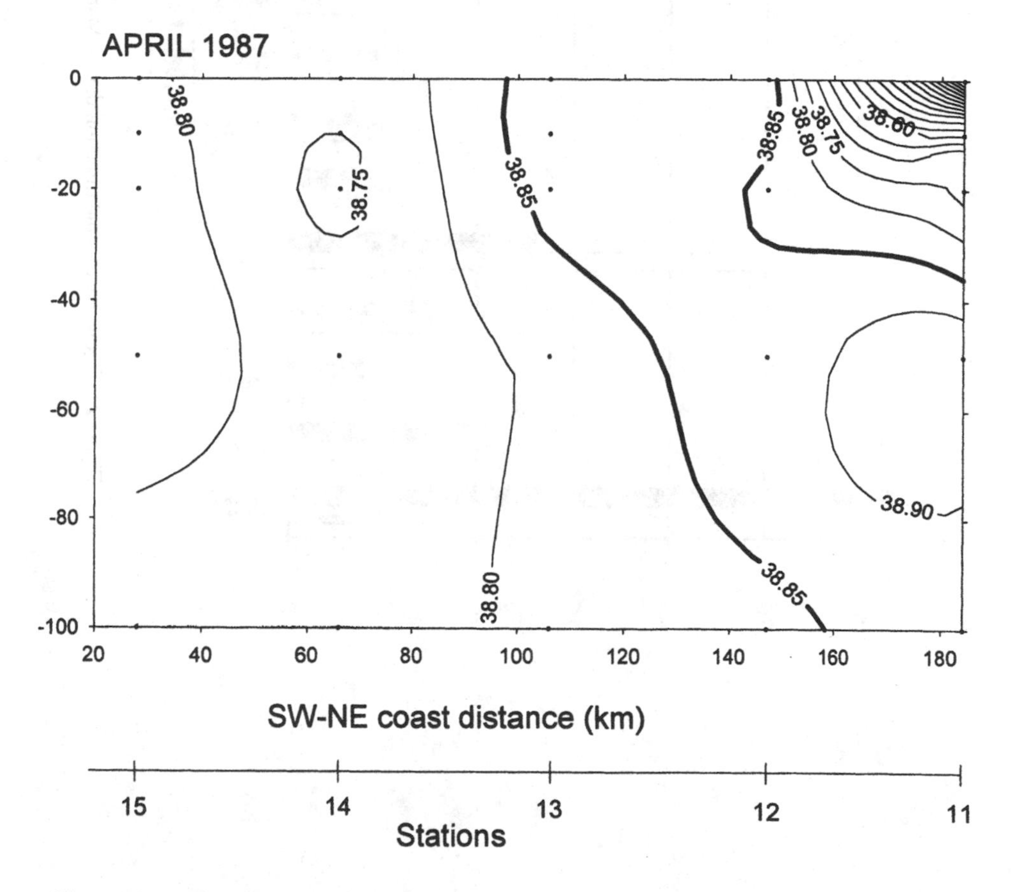


Figure 20. Vertical distribution of salinity on transects DBK-Bari (April 1986 and April 1987)

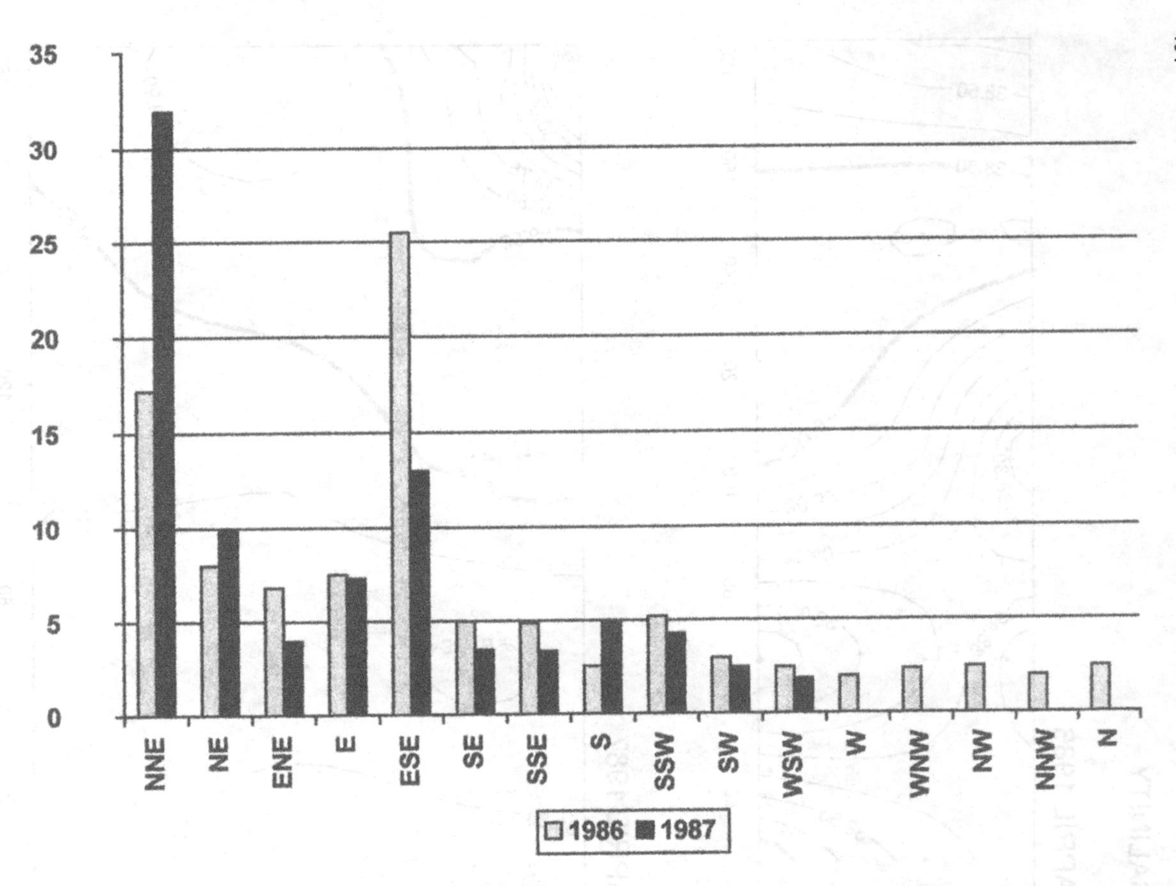


Figure 21. Frequency of wind direction during March and April 1986 and March and April 1987

## 6. References


1. Buljan, M. and Zore-Armanda, M. (1966) Hydrographic data of the Adriatic Sea collected in the period from 1952-1964, *Acta Adriatica* **12**, 1- 438.
2. Buljan, M. and Zore-Armanda, M. (1979) Hydrographic properties of the Adriatic Sea in the period from 1965 through 1970, *Acta Adriatica* **20**, 1- 362.
3. Zore-Armanda, M., Bone M., Dadić, V., Morović, M., Ratković, D., Stojanoski, L. and Vukadin, I. (1991) Hydrographic properties of the Adriatic Sea in the period from 1971 through 1983, *Acta Adriatica* **31**, 5- 547.
4. Pucher-Petković, T. (1971) Recherches sur la production primaire et la densite des populations du phytoplancton en Adriatique moyenne (1962-1967), *Rapp. Comm. int. Mer Medit.* **20**, 339-343.
5. Pucher-Petković, T. (1973) Recherches preliminaire sur la photosynthese du nanoplancton et du micloplancton danc les eaux de l'Adriatique moyenne, *Rapp. Comm. int. Mer Medit* **21**, 445- 448.
6. Pucher-Petković, T., Marasović, I., Vukadin, I. and Stojanoski, L.(1978) Time series of productivity parameters indicating eutrophication of the Middle Adriatic waters, in J.F. Caddy and M. Savini (eds.), *Fifth technical consultation of stock assessment in the Adriatic*, GFSM, Rome, pp. 41-50.
7. Buljan, M. (1968) Fluktuacije oceanografskih svojstava srednjeg Jadrana u razdoblju od 1962-1967, *Pomorski zbornik* **6**, 845-865 (in Croatian).
8. Buljan, M. (1957) Fluctuation of temperature in the open Adriatic, *Acta Adriatica* **8**, 1-26.
9. Buljan, M. and Zore-Armanda, M. (1976) Oceanographical properties of the Adriatic Sea, *Oceanogr. Mar. Biol. Ann. Rev.* **14**, 11-98.
10. Pucher-Petković, T. (1970) Sezonske i višegodišnje fluktuacije primarne produkcije u srednjem Jadranu, *Pomorski zbornik* **8**, 847-856 (in Croatian).
11. Pucher-Petković, T., Zore-Armanda, M. and Kačić, I. (1971) Primary and secondary production of the Middle Adriatic in relation to climatic factors, *Thalassia Jugoslavica* **7**, 301- 311.
12. Vučetić, T. (1970) Fluktuacije zooplanktona na srednjem Jadranu, *Pomorski zbornik* **8**, 867- 881 (in Croatian).
13. Tziperman, E. and Malanotte-Rizzoli, P. (1991) The climatological seasonal circulation of the Mediterranean Sea, *J. Marine Research* **49**, 411- 434.
14. Morcos, S. (1972) Sources of the Mediterranean Intermediate Water in the Levantine Sea, in A.L. Gordon (ed.), *Studies in physical oceanography II*, Gordon & Breach Science Publishers, New York, pp. 185- 206.
15. Zore-Armanda, M. (1963) Les Masses d'eau de la mer Adriatique, *Acta Adriatica* **10**, 1- 94.
16. Zore-Armanda, M. (1969) Origine possible des fluctuations de la salinite de l'eau de la mer Adriatique, *Rapp. Comm. int. Mer Medit* **19**, 719- 792.
17. Zore-Armanda, M. (1969) Water exchange between the Adriatic and the Easte Mediterranean, *Deep-Sea Research* **16**, 171- 168.
18. Zore-Armanda, M. (1971) Influence of the long-term changes in the oceanographic/meteorological conditions in the North Atlantic on the Mediterannean, Proceedings of the "The Ocean World", Joint oceanographic Assembly (Tokio 1970), 151-154.
19. .Pucher-Petković, T. and Zore-Armanda, M. (1973) Essai d'evaluation et prognostique de la production an fonction des facteurs du millieu dans l'Adriatique, *Acta Adriatica* **15**, 1-39.
20. Revelante, N. and Gilmartin, M. (1977) The effects of the northen Italian rivers and eastern Mediterranean ingressions on the phytoplankton of the Adriatic Sea, *Hydrobiologija* **56**, 229- 240.
21. Gilmartin, M. and Revelante, N. (1983) The phytoplankton of the Adriatic Sea. Standing stock and priymary production, *Thalassia Yugoslavica* **19**, 173 –178.
22. Utermohl, H. (1958) Zur Vervollkommnung der quantitativen Phytoplankton methodik, *Mitt. Int. Verein. Theor. Angevw. Limnol.* **9**, 1 – 38.
23. Smayda, T. J. (1978) From phytoplankters to biomass, in A. Sournia (ed. ), *Phytoplankton manual*, UNESCO, Paris, pp. 273 – 279.
24. Viličić, D. (1985) An examination of cell volume in dominant phytoplankton species of the middle and southern Adriatic waters, Int. Rev. Ges. Hydrobiol. **70**, 117 – 132.
25. Strickland, J. D. H. and Parsons, T. R. (1972) A Practical Handbook of Seawater Analyses, *Fish. Res. Bd. Can. Bull.* **167**, 1 – 310.
26. Marasović, I. and Pucher – Petković, T. (1988) Deep Chlorophyll Maximum in the Adriatic, *Rapp. Comm. int. Mer Medit.* **31**, 226.
27. Buljan, M. (1953) The nutrient salts in the Adriatic waters, *Acta Adriatica* **5**, 225 – 237.
28. Marasović, I., Pucher – Petković, T. and Alegria – Hernandez, V. (1988) Phytoplankton productivity of the Adriatic Sea in relation to pelagic fisheries, N*otes* **72**, 1 – 8.
29. Viličić, D., Vučak, Z., Škrivanić, A. and Gržetić, Z. (1989) Phytoplankton Blooms in the Oligotrophic Open South Adriatic Waters, *Marine Chemistry* **28**, 89 – 107.
30. Marasović, I., Gačić, M., Pucher-Petković, T. and Gržetić, Z. (1990) An analysis of subsurface phytoplankton blooms in relation to nutrient content in the southern Adriatic waters, IV POEM Scientific Workshop, Apstracts of presentation, August 27 – September 1 1990, Venice, Italy.
31. Becacos – Kontos, T. (1977) Primary production and environmental factors in an oligotrophic biome in the Aegean Sea, *Marine Biology* **42**, 93 – 98.
32. Berman, T., Townsend, D. W., El Sayed, S. Z., Trees, C. C. and Azov, Y. (1984) Optical transparency, chlorophyll and primary productivity in the eastern Mediterraean near the Israeli coast, *Oceanologica Acta* **7**, 367 – 371.
33. Azov, Y. (1986) Seasonal patterns of phytoplankton productivity and abundance in near shore oligotrophic waters of the Levant Basin (Mediterranean), *J. Plankton Research* **8**, 41 – 53.
34. Kimor, B. and Wood, E. J. F. (1975) A plankton study in the eastern Mediterranean Sea, *Marine Biology* **29**, 321 – 333.
35. Kimor, B., Berman, T. and Schneller, A. (1987) Phytoplankton assemblages in the deep chlorophyll maximum layers of the Mediterranean coast of Israel, *J. Plankton Research* **9**, 433 – 443.,
36. Kršinić, F. (1998) Vertical distribution of protozoan and microcopepod communities in the South Adriatic Pit, *J. Plankton Research* **20**, 1033 – 1060.
37. Gačić, M., Civitarese, G. and Ursella, L. (1998) This issue

# THE BIOLUMINESCENCE FIELD AS AN INDICATOR OF THE SPATIAL STRUCTURE AND PHYSIOLOGICAL STATE OF THE PLANKTONIC COMMUNITY AT THE MEDITERRANEAN SEA BASIN.

Tokarev Yu.N. [1], E.P. Bitukov[1], R.Williams[2], V.I. Vasilenko[1], S.A. Piontkovski[1], B.G. Sokolov[1].
[1]Institute of Biology of the Southern Seas 335011 Sevastopol, Ukraine,
[2]Plymouth Marine Laboratory, Prospect Place, Plymouth PL1 3DH, UK.

ABSTRACT The purpose of this work is to show the possibilities of using the characteristics of the bioluminescent fields to monitor the marine planktonic communities. The data bank contains 3500 vertical casts of bioluminescent potential and near 1000 samples of the planktonic organisms obtained at 500 oceanographic stations executed in 21 expedition to the Mediterranean Sea basin in 1970-1995. Studies were carried in different seasons in the Mediterranean and the Black Seas where different trophic conditions and considerably different species composition and abundance occurred in the phytoplankton. There are of course differences between abundance of various dinoflagellates and an intensity of measurement of bioluminescence, but the main features of these appeared to be similar, the intensity of bioluminescence increased in proportion to their number and physiological state.

Seasonal changes are also well developed and mostly in the Black Sea. Two intensive periods of bioluminescence were recorded, one in May-June and the other one, more intensive, in October-November. The bioluminescent potential was achieving 1.4 $10^{-2}$ microwatt $cm^{-2}$ $l^{-1}$, which exceeded minimum numbers in February 500 times. Seasonal cycles are weakly developed in the oligotrophic regions of the Mediterranean Sea: differences of the bioluminescent potential between summer and winter periods achieve 3.5 times only.

Macroscale trends of bioluminescence changes within the Mediterranean basin are comparable with that of the plankton spatial distribution. General trend of bioluminescence increase, from the Aegean Sea towards the west and from Algerian coast to Spain, was evaluated on the scale of the whole basin. Bioluminescence in the central part of the Black Sea is 3 times weaker than that in the Alboran Sea. However, it is one order higher than that in the central part of the Mediterranean Sea.

As we have stated there are regions of «clean» and «polluted» waters in both neritic and open sea waters. For example, in the Mediterranean Sea the sentral regions of the Alboran, Ionian and Aegean Seas, and the central part of the western gyre and Karadag marine reserve of the Black Sea were all considered to have relatively good water quality while the regions of the intensive shipping (straits, southern part of the Ionian Sea, near-Bosphorus region, Black Sea north-western part) and densely inhabited riparian zones (Crimean southern coast) were considered to be regions of «ecological risk». There are considerable differences between the parameters of the regression lines from the studied regions. For example, to obtain the same bioluminescence intensity in the 0-100 m layer, the quantity of dinoflagellates in polluted regions of the Mediterranean Sea would have been twice as abundant as those from «clean» waters. Similarly the numbers of bioluminescent algae in the Black Sea would have increased 3-4 times to obtain the same bioluminescent intensity. As the anthropogenic pressure in the Black Sea is higer, due to the population density in the river drainage areas and to the low water exchange the differences are more apparent. It can be presumed, that an organism's metabolic state (and hence its bioluminescence) could serve as an indicator for the levels of pollution in the environment.

Key words: bioluminescence, plankton, spatial structure, anthropogenic pollution



*P. Malanotte-Rizzoli and V.N. Eremeev (eds.),*
*The Eastern Mediterranean as a Laboratory Basin for the Assessment of Contrasting Ecosystems,* 407–416.

The purpose of this work is to show the possibilities of using the characteristics of the bioluminescent fields to monitor the marine planktonic communities. The characteristics of bioluminescence of marine organisms are used in fisheries around the world [3,12], in studies of the macro- and mesoscale plankton communities spatial structure [9,20], in taxonomy [13], in investigations of the temporal plankton variability [8]. In laboratory experiments we have shown the high sensitivity of bioluminescent systems of marine organisms, of different taxa, to the influence of different physical and chemical fields of anthropogenic origin [6,18]. Those approaches were initiated in the studies of planktonic communities of the Mediterranean basin seas.

MATERIALS and METHODS

The instrumental measurements of bioluminescence in the Mediterranean basin were started by the Laboratory of Biophysical Ecology, IBSS in 1965 [2,5]. The submersible bathyphotometers, created in the laboratory have been used [4,19]. Together with bioluminescence measurements, temperature of the studied layer, speed of drift of the vessel, force and direction of wind were taken. Normally, measurements of bioluminescence were carried out 2 hours after sunset. This permitted the exclusion of the influence of daylight on the rhythm of light emission of plankton bioluminescents and their vertical migration.

Phytoplankton were sampled with a rosette of 5-l water bottles, fixed 1 m higer than the bathyphotometer, from intense bioluminescence layers and standard horizons. Zooplankton sampling was carried out in the upper 20m layer, using a Juday Net (36 cm mouth diameter, mesh size 116 micrometers). The total plankton biomass (wet weight) caught by net sampling was determined back in the laboratory and certain of the samples were processed to species level.

The data bank contains 3500 vertical casts of bioluminescent potential and near 1000 samples of the planktonic organisms obtained at 500 oceanographic stations executed in 21 expedition to the Mediterranean Sea basin in 1970-1995.

RESULTS and DISCUSSION.

The well developed vertical stratification of bioluminescent field might be noted as its typical feature on the scale of 1-100m (Fig.1,2). The minimal thickness of layers consist from 3 to 7 m. The degree of development of such a stratification increases in waters with enhanced biological productivity. However, almost in all cases, one might notes the upper low-bioluminescent layer (8-15 m thick), the quazistacionaric layer of the enhanced bioluminescence in the thermocline (Fig.1,2).

General trend of bioluminescence increase, from the Aegean Sea towards the west (Fig.3), was evaluated on the scale of the whole basin. Maximal numbers of bioluminescence were noted within zones where the Atlantic and Mediterranean water masses do interact, within cyclonic eddies and the divergent zones. For instance, bioluminescence recorded within this local dynamic zones exceed its average number in the Ionic Sea 30 times. Bioluminescence in the central part of

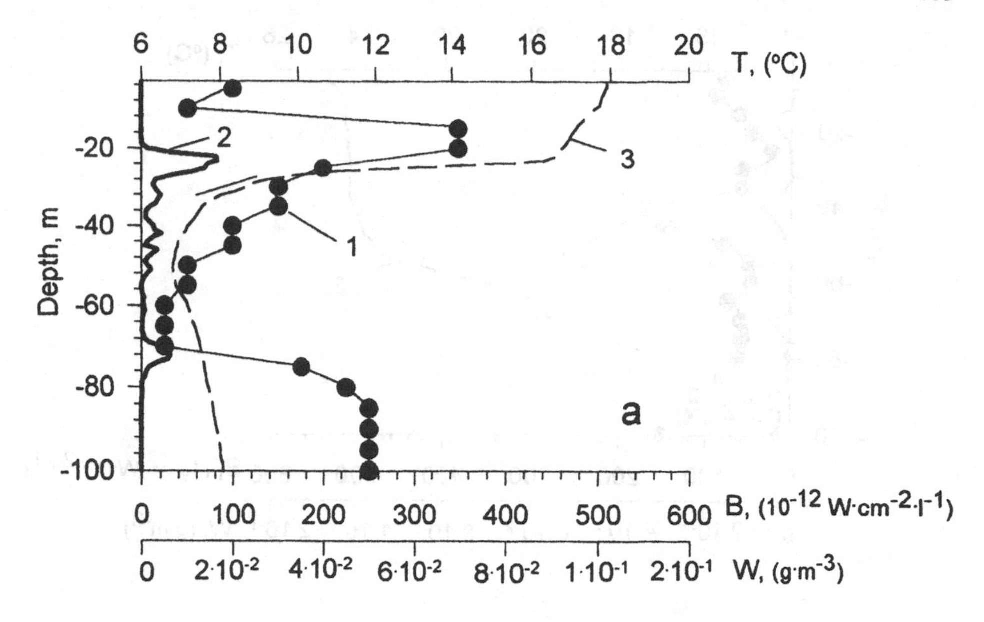


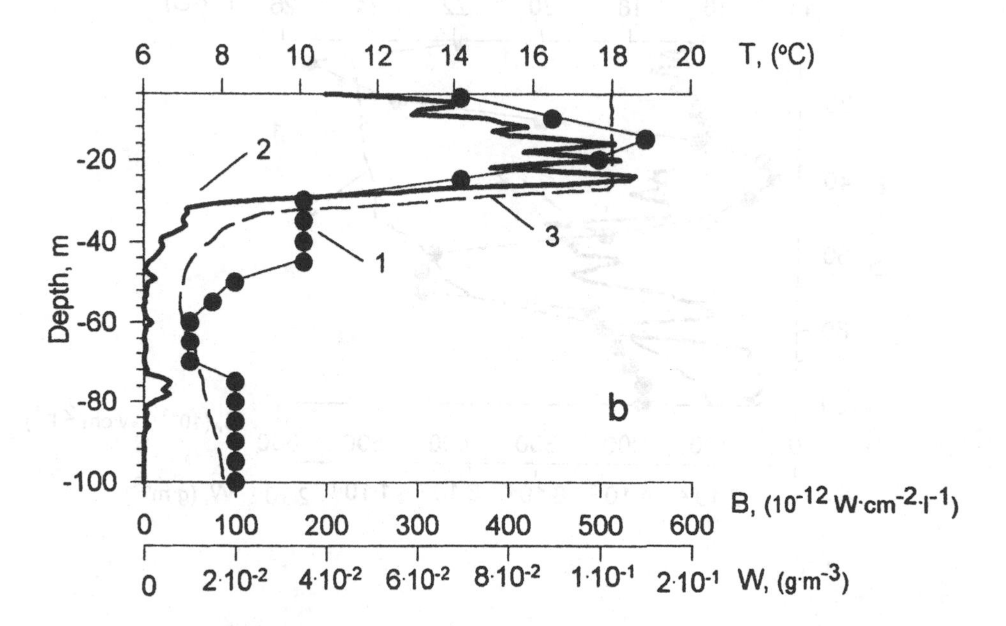


Fig. 1. The typical vertical structure of plankton biomass (1), bioluminescence (2) and temperature (3) in the Black Sea photic layer at day (a) and night (b) time

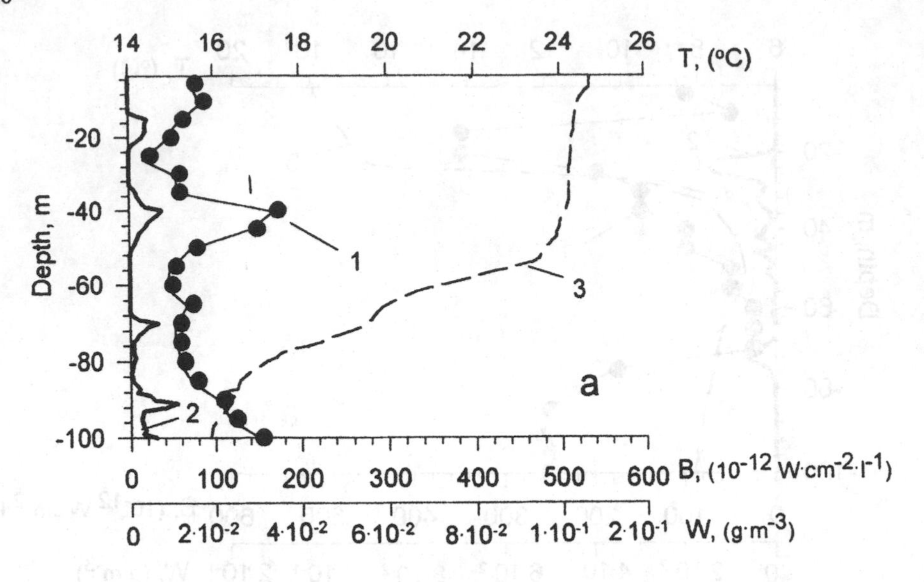


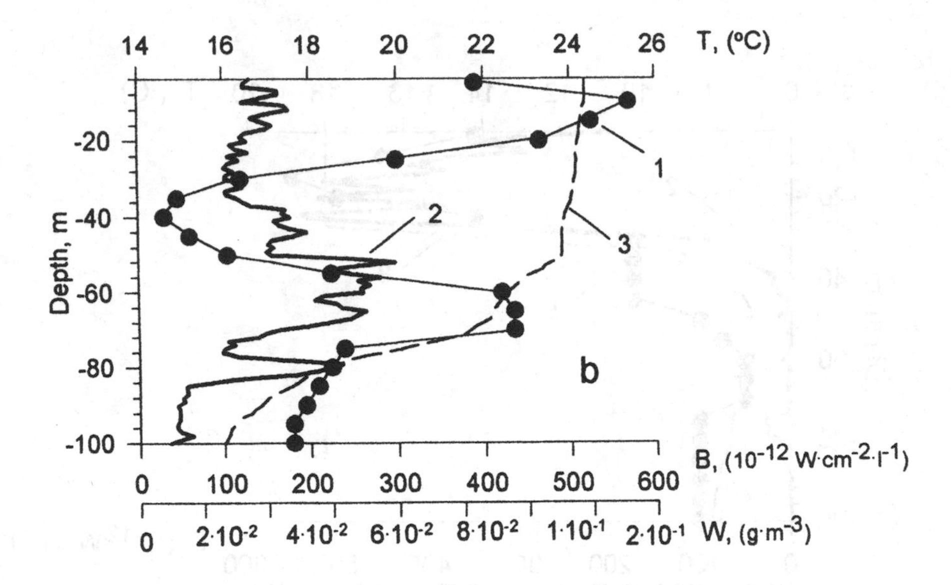


Fig. 2. The typical vertical structure of plankton biomass (1), bioluminescence (2) and temperature (3) in the Ionian Sea photic layer at day (a) and night (b) time

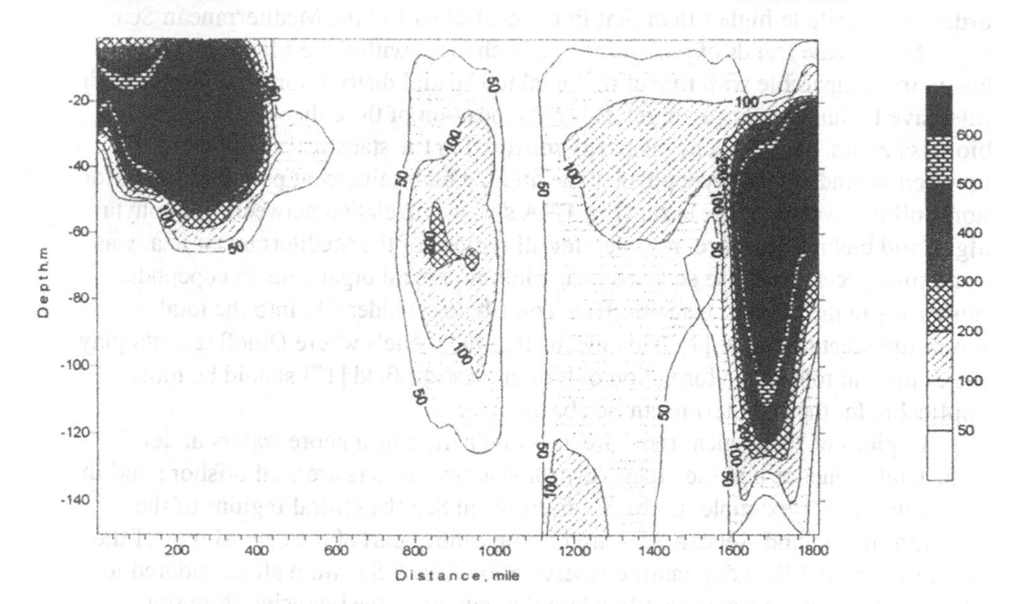


**Fig.3 General trend of bioluminescence from the Black Sea to the Alboran Sea.**

the Black Sea is 3 times weaker than that in the Alboran Sea. However, it is one order of magnitude higher than that in the central part of the Mediterranean Sea.

Macroscale trends of bioluminescence changes within the Mediterranean basin are comparable with that of the plankton spatial distribution [10]. Zones of the intensive bioluminescence are related to the position of the enhanced plankton biomass zones. This tendency is partly mirrored in the statistical relationship between abundance of Dynophyta algae and the bioluminescent potential found for non-polluted waters of the basin (Fig.4). A strong correlation between Dynophyta algae and bioluminescence, revealed for all regions of the Mediterranean Sea, was quite unexpected, because such a strong bioluminescent organisms as copepods, numerous in the plankton, should have contribute considerably into the total bioluminescent potential [7]. This means that the models where Dinoflagellata play the dominant role in the formation of bioluminescent field [17] should be more applicable for the Mediterranean Sea basin.

Regions of "ecological risk" are as a rule neritic near shore waters under industrial influence and the "relatively satisfactory" waters are well offshore and in the open sea. For example, in the Mediterranean Sea the sentral regions of the Alboran, Ionian and Aegean Seas, and in the central part of the central part of the western gyre and Karadag marine reserve of the Black Sea were all considered to have relatively good water quality while the regions of the intensive shipping (straits, southern part of the Ionian Sea, near-Bosphorus region, Black Sea north-western part) and densely inhabited riparian zones (i.e., Crimean southern coast) were considered to be regions of "ecological risk". Evidence for this simple classification is present universally in the literature [1,11,14,16] for the neritic zones of the world's oceans and the Mediterranean Sea, Table 1.

Data from such areas of our studied regions are given in Table 2 in the form of the squares of the correlation coefficients and the values of the regression coefficients. In column "a" regression parameters are given for comparatively clean waters, in column "b" - those in polluted waters. In column "c" these coefficients change when the two curves from each region are united. There are considerable differences between the parameters of the regression lines from the studied regions. For example, to obtain the same bioluminescence intensity in the 0-100 m layer, the quantity of dinoflagellates in polluted regions of the Mediterranean Sea would have to be twice as abundant as those from "clean" waters. Similarly the numbers of bioluminescent algae in the Black Sea would have to increase 3-4 times to obtain the same bioluminescent intensity (Fig.4).

Indeed, resistance of any organisms to environmental influence depends on the metabolic state of the intracellular systems. How quickly the organisms are able to adapt to anthropogenic pollutants is determined primarly by the energetic needs of the organism [15]. It can be presumed, that an organism's metabolic state (and hence its bioluminescence) could serve as an indicator for the levels of pollutants in the environment. These conclusions have been repeatedly confirmed in our

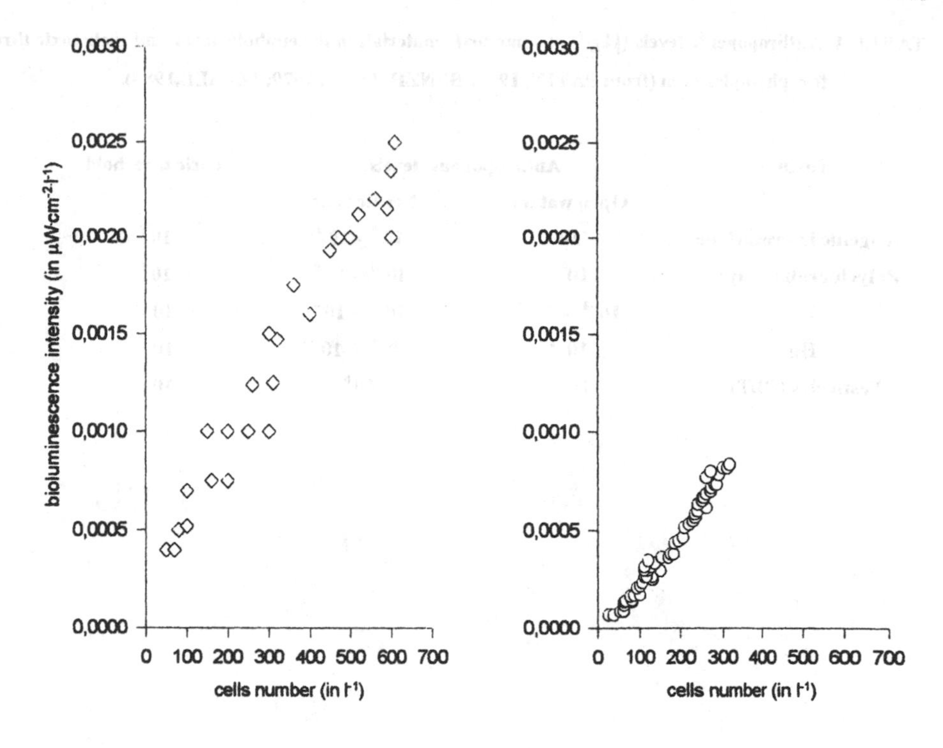


**Fig.4. Regressions between bioluminescence intensity and numbers of dinophyte algae in the upper 100-m.**

**TABLE 1. Anthropogenic levels (μg/l) of some toxic materials in the euphotic layer and their toxic threshold for phytoplankton (from PATIN, 1977; BENZHITSKY, 1979; IZRAEL,1984).**

| Toxin | Anthropogenic levels | | Toxic threshold |
|---|---|---|---|
| | Open waters | Neritic zones | |
| Petrogenic hydrocarbons | 10 | $10^2 - 10^4$ | $10^2$ |
| Polychlorobiphenyls | $10^{-2}$ | $10^{-1} - 10^0$ | $10^1$ |
| Pb | $10^{-1} - 10^0$ | $10^1 - 10^4$ | $10^{-1}$ |
| Hg | $10^{-2}$ | $10^{-1} - 10^0$ | $10^{-1}$ |
| Pesticides (DDT) | $10^{-2}$ | $10^0$ | $10^{-1}$ |

**TABLE 2. Regression equations between bioluminescence intensity and number of the dinophyte algae. Explanation in text.**

| Parameter | Mediterranean Sea | | | Black Sea | | |
|---|---|---|---|---|---|---|
| | «a» | «b» | «c» | «a» | «b» | «c» |
| n | 58 | 21 | 79 | 29 | 20 | 49 |
| a | 9.2E-06 | 4.3E-06 | 9.3E-06 | 1.9E-05 | 1.1E-05 | 5.3E-05 |
| b | 0.2E-03 | 0.1E-03 | 0.4E-03 | 0.1E-02 | 0.3E-02 | 0.3E-02 |
| $r^2$ | 0.983 | 0.853 | 0.830 | 0.950 | 0.864 | 0.284 |

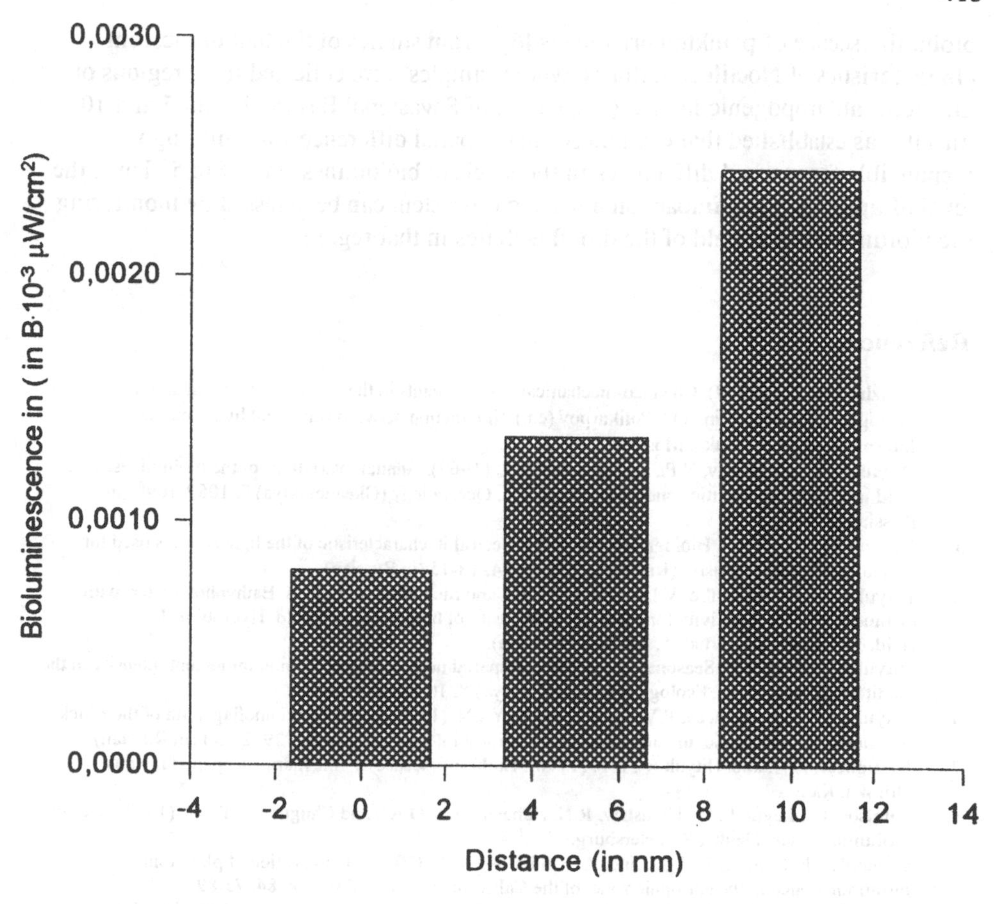


**Fig.5 Reaction of the *Noctiluca miliaris* to electric stimulation from regions within Sevastopol Bay.**

bioluminescence of plankton organisms [6]. From studies of the bioluminescent characteristics of Noctiluca miliaris, where samples were collected from regions of different anthropogenic impact (i.e., centre of Sevastopol Bay and from 5 and 10 nm) it was established that even these small spatial differences in sampling were responsible for marked differences in the levels of bioluminescence, Fig.5. Thus, the level of anthropogenic impact on a marine ecosystem can be assessed by monitoring the bioluminescence field of the dinoflagellates in that region.

## References


1. Benzhitsky, A.G. (1979). Chemico-mechanical contaminants in the ocean environment and their ecological significants, in G.G. Polikarpov (ed.), Interaction between water and living matter. Intern.Symposium, Nauka, Moscow 2, pp. 31-33.
2. Bityukov, E.P., Rybasov, V.P., and Shaida, V.G. (1967). Annual variations of the bioluminescence field intensity in the neritic zone of the Black sea, Oceanology (Okeanologiya) 7, 1089-1099 (in Russian).
3. Bityukov , E.P. (1967). Biological basis of the spectral it characteristic of the light sources used for catching fish, Fish Indusrty (Ribnoe hozyaistvo) 4, 13-15 (in Russian).
4. Bityukov , E.P., Vasilenko, V.I. Tokarev, Yu.N., and Shaida ,V.G. (1969). Bathyphotometer with distance-switched sensitivity for estimating intensity of bioluminescent field, Hydrobiol. J. (Hidrobiologicheskii Jurnal) 5, 82-86 (in Russian).
5. Bityukov, E.P. (1982). Seasonal variability and spatial non- informity of bioluminescent intensity in the mediterranean, Marine Ecology (Ecologiya morya) 8, 10-20 (in Russian).
6. Bityukov, E.P., Evstigneev, P.V. , and Tokarev ,Yu.N. (1993). Luminous Dinoflagellata of the Black Sea under anthropogenic impact, Hidrobiol J (Hidrobiologicheskii Jurnal) 29, 27-34 (in Russian).
7. Evstigneev, P.V., and Bityukov , E.P. (1990). Bioluminescence of the marine Copepoda, Naukova Dumka, Kiev.
8. Gitelson, I.I., Levin, L.A. , Utyushev, R.N. , Cherepanov, O.A. , and Chugunov, Yu.N. (1992). Ocean bioluminescence, Nauka, S.- Petersburg.
9. Greenblatt, P.R., Feng, D.F. , Zirino, A. , and Losee, J.R. (1984). Observation of planktonic bioluminescence in the eutrophic zone of the California Current, Mar.Biol. 84, 75-82.
10. Greze, V.N. (1989). Pelagial of the Mediterranean Sea as ecological system, Naukova Dumka, Kiev.
11. Israel,Yu.A. (1984). Ecology and control of the state of the natural environment, Hydrometeoizdat, Moscow.
12. Makiguchi, N, Arita, M. , and Asai, Y. (1980). Application of a luminous bacterium to fish-attracting purpose, Bull.Jap.Soc.Sci. Fish.Nissuishi. 46, 1307-1312.
13. Nealson, K.H., Arneson, A.C. , and Huber, M.E. (1986). Identification of marine organisms using kineticand spectral properties of their bioluminescence, Mar. Biol. 91, 77-83.
14. Patin, S.A. (1977). Chemical pollution and its inflluence on hydrobionts, in M.E. Vinogradov (ed.), Ocean Biology. Biological production of the Ocean, Nauka, Moscow, pp. 322-331.
15. Schmidt-Nielsen, K. (1982). Animal physiology. Adaptation and envirionment, Mir, Moscow.
16. Stepanov, V.N., Zhivitsky, A.V. , and Berdnikov, L.A. (1992). Economical aspects of protection of the Black Sea against pollution, in Assesssment of land-based sources of marine pollution in the seas adjacent to the C.I.S. Abst. Intern. Conf., Sevastopol 1, pp. 13-15
17. Tett, P.B. (1971). The relation between dinoflagellates and the bioluminescence of sea water, J.Mar.Biol.Ass.UK 51, 183-206.
18. Tokarev, Yu.N. (1976). Action of gamma-radiation on bioluminescence of Noctiluca miliaris, Radiobiology (Radiobiologiya) 16, 131-134 (in Russian).
19. Vasilenko, V.I., Bityukov, E.P., Sokolov, B.G., and Tokarev, Yu.N. (1997). Hydrobiophysical device "SALPA" of the Institute of Biology of the Southern Seas used for bioluminescent investigation of the upper layers of the ocean, in J.W.Hastings, L.J., Kriska and P.E. Stanley (eds.), Bioluminescence and Chemiluminescence. Melecular reporting with photons. / Proceedings of 9th Intern.Symp., Woods Hole, J.Willey & Sons, Chichester, pp. 549-552.
20. Vinogradov, M.E., Gitelson, I.I. , and Sorokin, Yu.N. (1970). The vertical structure of a pelagic comminity in the tropic ocean, Mar.Biol. 16, 187-194.

# SATELLITE OBSERVATIONS OF SIMILAR CIRCULATION FEATURES IN SEMI-ENCLOSED BASINS OF THE EASTERN MEDITERRANEAN WITH RESPECT TO MARINE ECOSYSTEMS INVESTIGATIONS

Y.P. ILYIN, E.M. LEMESHKO, S.V. STANICHNY, D.M. SOLOVIEV
*Marine Hydrophysical Institute of the Ukrainian Academy of Sciences,*
*2, Kapitanska St., 335000 Sevastopol, Ukraine*

## 1. Introduction

Contrast marine ecosystems are the convenient objects for the observations from space in visible and thermal bands of radiance. Ecosystem eutrophication and pollution levels can be recognised as well as the circulation features and water exchange within the studied area. Most contrast ecosystems exist in semi-enclosed basins of the Mediterranean Sea – Adriatic, Aegean and Black seas. Strong natural and antropogenic spatial differences in water properties are maintained in these seas allowing the wide application of the satellite ecological monitoring facilities. Complex international research and monitoring projects for these regions may be developed by means integration of field data collection and numerical ecosystem modelling with the satellite imagery receiving and analysis based on the Marine Hydrophysical Institute (MHI) receiving station and specialists' team.

Present communication is addressed to characterise abilities of MHI in satellite remote sensing with special attention to the applications in the Eastern Mediterranean contrast ecosystem studies. Recent examples of the satellite data analysis are considered in connection with some similar features of semi-enclosed basins listed above.

## 2. MHI facilities for satellite data receiving and processing

MHI is a leading oceanographic institute in Ukraine with about 20-year experience in ocean remote sensing and it has got hardware and software for receiving and processing of satellite information. The short description of the available satellite information in visible and thermal bands is contained in Table 1. Receiving station region overlaps territory from Spain to Mid Asia. MHI has satellite data archives for the Black Sea and the East Mediterranean region for the last 7 years (HRPT-information) and 10 years (APT - data).

Special self-made software gives the opportunity to process satellite images and meteorological information for different problems of modern oceanology. The typical chain of procedures for the AVHRR image processing is receiving - calibration - multichannel calculations (sea surface temperature, suspended matter) - geographical positioning - map calculation - land masking - data extraction or addition.



*P. Malanotte-Rizzoli and V.N. Eremeev (eds.),*
*The Eastern Mediterranean as a Laboratory Basin for the Assessment of Contrasting Ecosystems,* 417–422.

TABLE 1. Visible and infrared (IR) satellite data available on the MHI receiving station

| Mode | Satellite | Instrument & chann. amount | Channels numbers | Resolution, km |
|---|---|---|---|---|
| 1.7 Ghz HRPT | NOAA | AVHRR, 5 | 1,2: Vis.; 3-5: IR | 1.1 at nadir point |
| " | SeaStar | SeaWiFS, 8 | 1-8: Vis. | " |
| 137 Mhz APT | NOAA | AVHRR, 2 | 1(d) or 3(n) , 4 | 4 at any point |
| " | Meteor, Sich | MSU-M, 1 | 1: Vis. | 1.2 at any point |

## 3. The Eastern Mediterranean seas as the objects for satellite observations

Informational properties of satellite images of the ocean surface are determined by manifestation of ecological and dynamical phenomena in optical and thermal tracer distribution. For the visible band such tracers are suspended matter and phytoplankton (chlorophyll-like) pigments and for the thermal infrared (IR) band – spatial fluctuations of sea surface temperature (SST). Most oceanological studies made by the satellite images have been executed for the regions of shelf, boundary currents, upwellings where respectively high background contrasts of SST are maintained all-the-year-round or seasonally, or sources of optically active constituents exist (e.g., river inflow, algae bloom).

Thermal IR and visible images of the sea surface have different informational capacity for the different regions and seasons. For example, in the moderate latitudes of a spring-summer season, SST gradients related to water dynamics phenomena can be significantly suppressed due to intensive surface heating and upper mixed layer development. For the same latitudes and season, visible manifestations of water circulation are the most distinguishable after the river floods when high amount of suspended matter inputs in the sea, or during the intensive phytoplankton blooms. Thus thermal and visible satellite data can mutually supplement one another for the long-time regional ecological studies or monitoring.

All these reasonings are fair for the East Mediterranean semi-enclosed seas. They are situated in the moderate latitudes, having along-meridian extension 5-6 degrees of latitude. That provides them with enough gradients of solar radiation fluxes supporting the background meridianal gradients of SST (Table 2). In addition, there are warm water inputs in southern parts of the seas through the straits, while cold sources exist in their northern parts: a winter shelf cooling and a cold riverine water inflow (the input through the Dardanelles Strait for the Aegean Sea). The Fresh water input in the north parts of the seas and the saline water inflow through the southern straits together with meridianal gradients of evaporation maintain strong background contrasts of surface water salinity (see Table 2). Besides, northern fresh-water sources are also the sources of high amount of suspended matter, nutrients and pollutants forming not only specific ecological situation but also the contrasts in optical water properties.

General surface water circulation patterns are similar for three basins and have cyclonic character on basin scale (rim currents) with several sub-basin gyres [7,8]. So, in a wintertime, the cold low-saline turbid water moves from northern shelf zones and river (strait) mouths along the western shores in southward direction. The warm high-saline transparent water generally expands from the southern straits to the north along the eastern shores. In summer, the manifestation of circulation patterns in water properties is

TABLE 2. Comparison of some hydrologic parameters for the semi-enclosed seas surface water

| Basin name | Latitudes, deg. | Temperature, Cels. | | Salinity, ppt | |
|---|---|---|---|---|---|
| | | Summer | Winter | Summer | Winter |
| the Adriatic Sea | 40-46 | 22-24 | 8-14 | 33.0-38.25 | 36.0-38.5 |
| the Aegean Sea | 36-41 | 24-26 | 12-16 | 30.0-39.25 | 36.0-39.2 |
| the Black Sea | 41-47 | 22-26 | 2-8 | 12.0-18.4 | 16.0-18.4 |

more complex because of seasonal heating of northern shelf-originated and riverine water. In that case, western along-shore currents carry warm low-saline and turbid water to the south. These general circulation patterns are disturbed by the diverse mesoscale processes (coastal upwellings, eddies, meanders, mushroom-like structures), especially in summer time when the rim currents weaken [10]. Mesoscale features of water circulation play an important role in the interaction and exchange between shallow (coastal) and deep see as well as in the renewing of polluted coastal ecosystems [9].

Thus, the Adriatic, the Aegean and the Black seas have many common features, such as similar climatic conditions, high degree of isolation from the other regions of the Mediterranean Sea, the desalination of surface water at the north and the cyclonic character of general circulation. In addition to that, the hydrological regime of the Aegean Sea is complicated by geo-morphological conditions, because this basin has extremely complex bottom relief and coastline, including a lot of islands. In spite of some differences, all of the three basins are very similar as objects for satellite observations.

## 4. Examples of regional studies

Numerous studies were executed in last decades by the integration of satellite images with in-situ data or/and modelling results. As the examples, investigations can be referred concerning the Black Sea large- and meso-scale water circulation features [3,4,7,9,10], the North Adriatic Sea water dynamics [5] and the North Aegean Sea flow phenomena [2,6]. Some recent results reflecting the connection of water circulation features with ecological processes are presented below.

The analysis of five-year AVHRR data has allowed to detect the interesting feature connected with algae (coccolithophorids) bloom in the Black Sea. Coccolithophorids play an important part in carbon cycle, and their optical properties permit to detect the bloom processes in visible channels of the AVHRR scanner.

Bloom processes depend on the hydrological conditions and nutrient concentration in subsurface layer. The main nutrient sources in the Black Sea are river inflows basically concentrated in the north-western part of the sea. The processes of nutrient redistribution are determined by a particular dynamic situation.

On the basis of satellite data analysis for a winter period in the NW part of the sea two typical hydrological situations are allocated. In the first case, the cold jet of the southward current is formed along the western coast as the result of river inflow, shallow water cooling with favour north-eastern winds or without a strong wind effect. It blocks the nutrient propagation from a coastal zone to the central part of pond. In the second case, off-shore advective flows are induced, cross-stream gradients of SST are washed out and the exchange with central part of the sea is executed under the influence of strong westerlies. Both of these situations are presented on Figure 1.

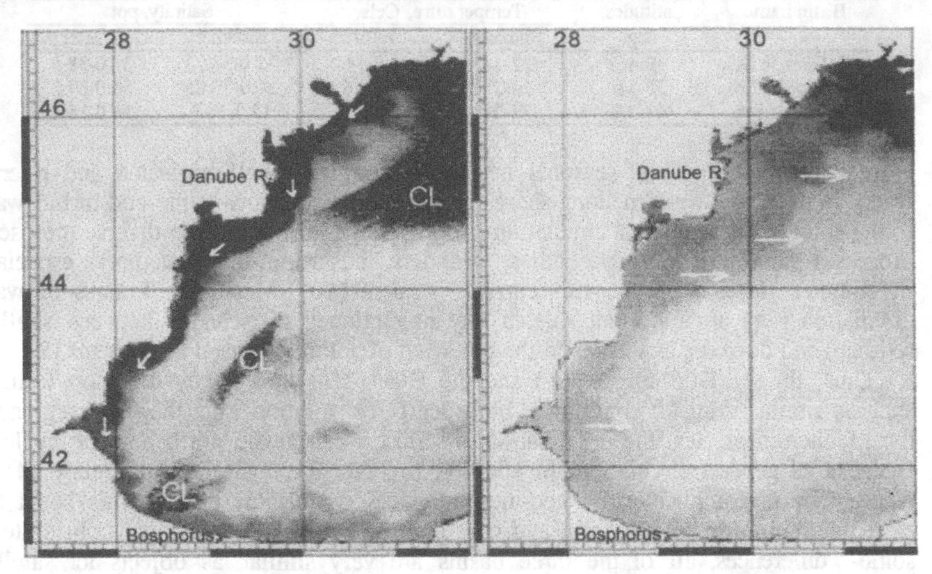


Figure 1. NOAA AVHRR channel 4 thermal images of the Western Black Sea for February 2, 1996 (left) and February 21, 1998 (right). Dark tones correspond to cold water and light tones - to warm water. Land is masked by white color. White arrows indicate prevalent movements of cold (rich in nutrients) water at the western shores. CL denotes cloudiness.

As the result of nutrient redistribution in the first case, bloom processes in June and August weaken and for the second one the intensive coccolithophorid bloom is observed (Figure 2). Thus the analysis of current hydrological situation in a winter period permits to predict the summer coccolithophorid bloom in the Black Sea. Besides, a

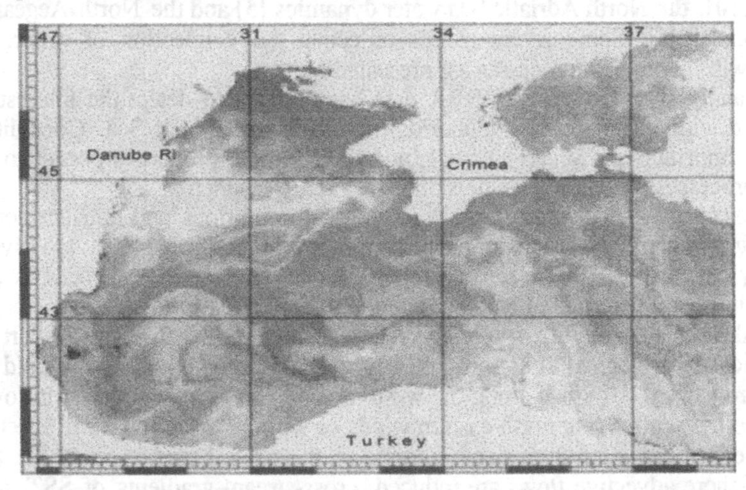


Figure 2. NOAA AVHRR channel 1 visible image of the Black Sea for June 26, 1993. Lighter tones correspond to higher concentrations of suspended matter. Land is masked by white color. See text for description.

large amount of algae is a good tracer for the visualisation of water circulation features on visible satellite images. Here we note the clear manifestation of the two-core structure of anticyclonic gyre west to the Crimea ("Sevastopol eddy") described by Sur and Ilyin in [10].

Phenomena analogous to one described above can occur not only in the Black Sea but also in other semi-enclosed basins because of their similarities having considered in Section 3. Figure 3 presents the winter thermal AVHRR images showing the water circulation pattern in the Adriatic Sea (left) and in the Aegean Sea (right). The cold water from northern shelf regions modified by the fresh inflows of rivers (low-saline flow of the Dardanelles Strait in case of the Aegean Sea) moves cyclonically along the western (and northern in the Aegean Sea) shores towards the south. Instead, the warm water moves to the north along the eastern shores inflowing from neighbour south basins. Schematically, white (cold stream) and black (warm stream) arrows indicate these water movements on the Figure 3.

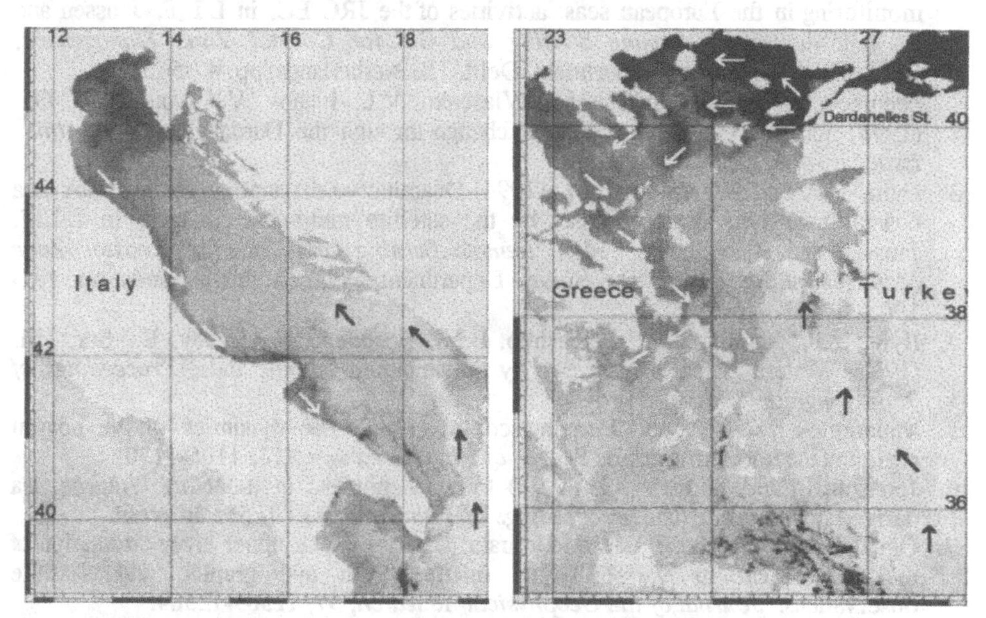


Figure 3. NOAA AVHRR channel 4 thermal images of the Adriatic Sea, February 27, 1998 (left) and the Aegean Sea, February 20, 1998 (right). Tones as for Figure 1. See text for the description.

## 5. Problems for prospective applications of satellite information

Basing on the MHI experience and facilities, a number of items can be specified with respect to the Eastern Mediterranean ecosystem studies by means the satellite data application.

Firstly, there are meso-scale phenomena of the surface water circulation and their relations with marine ecological processes. Moreover, satellite data are the most important part in the observations of such processes due to the typical temporal (~ 1-10 day) and spatial (~ 1-10 km) scales of meso-scale dynamical structures - eddies, upwellings, frontal zone displacements, horizontal wave processes etc. Especially, it concerns with the Aegean Sea circulation because of its complexity induced by the shoreline and islands.

Secondly, quantitative approaches can be developed and applied in regional ecological studies and monitoring, such the estimation of surface currents and coastal trapped waves parameters, basing on the time series of satellite images [3,6].

Thirdly, the modern bio-optical (and ecological) state and tendencies within marine ecosystems or whole basins can be evaluated by means the precise multi-channel scanner SeaWiFS data in comparison with data of CZCS obtained 15 years ago [1].

## 6. References


1. Barale, V. and Schlittenhardt, P. (1994) Remote sensing and coastal zone monitoring in the European seas: activities of the JRC EC, in L.L.F. Janssen and R. Allewijn (eds.), *Remote Sensing and GIS for Coastal Zone Management*, Rijkswaterstaat, Survey Department, Delft, The Netherlands, pp. 47-56.
2. Eremeev, V.N., Stashchuk, N.M., Vlasenko, V.I., Ivanov, V.A. and Uslu, O. (1998) Some aspects of the water exchange through the Dardanelles Strait, *this issue*.
3. Ilyin, Y.P. and Lemeshko, E.M. (1994) Quantitative description of the Black Sea coastal zone dynamical processes by the satellite multi-time images, in L.L.F. Janssen and R. Allewijn (eds.), *Remote Sensing and GIS for Coastal Zone Management*, Rijkswaterstaat, Survey Department, Delft, The Netherlands, pp. 181-190.
4. Ilyin, Y.P., Ivanov, V.A., Lemeshko, E.M., Besiktepe, S., Ozsoy, E., Sur, H.I. (1998) Western Black Sea currents by the ship and satellite data, *Proceedings of 35th Congress CIESM*, in press.
5. Malanotte-Rizzoli, P. and Bergamasco, A. (1983) The dynamics of the coastal region of the northern Adriatic Sea, *J. of Phys. Oceanogr.*, **13**, 1105-1130.
6. Jonsson, L. and Zodiatis, G. (1998) Flow phenomena in the North Aegean Sea derived from satellite data, *Proceedings of 35th Congress CIESM*, in press.
7. Oguz ,T., La Violette, P.E. and Unluata, U. (1992) The upper layer circulation of the Black Sea: its variability as inferred from hydrographic and satellite observations, *Journal of the Geophysical Research*, **97**, 12569-12584.
8. Ovchinnikov, I.M., Plakhin, E.A., Moskalenko, L.V., Neglyad, K.V., Osadchiy, A.S., Fedoseyev, A.F., Krivosheya, V.G. and Voytova, K.V. (1976) *Hydrology of the Mediterranean Sea*, Hydrometeoizdat, Leningrad.
9. Sur, H.I., Ozsoy, E, Ilyin, Y.P. and Unluata, U. (1996) Coastal/Deep Ocean Interactions in the Black Sea and their Ecological/Environmental Impacts, *Journal of Marine Systems*, **7**, 293-320.
10. Sur, H.I. and Ilyin, Y.P. (1997) Evolution of satellite derived mesoscale thermal patterns in the Black Sea, *Progress in Oceanography*, **39**, 109-151.

# FLORA AND BOTTOM VEGETATION ON THE DEEP-WATER BANKS OF THE MEDITERRANEAN SEA

N.A.Milchakova, I.K.Evstigneeva, I.N.Tankovskaya
*Institute of Biology of the Southern Seas (IBSS),Sevastopol 335011, Crimea, Ukraine. Phone: (0692)52-41-10, Fax: (0692) 59 28 13, E-mail: milcha @ibss.iuf.net*

## Abstract

177 species of algae and 1 seagrass were identified for ten deep-sea banks in the Mediterranean Sea (Aegean Sea, Tunisian Strait and Strait of Malta), red algae making up more than half of the number. Bottom vegetation includes nein types of communities which occupy depths from 20 down to 115 m, and have polydominant, mosaic, multistratum structure. Algae flora of the banks is temperate to warm-water, while half of the species are common to both the Mediterranean and the Black Sea.

## Introduction

Conservation of the existing biodiversity is of high priority for modern science. Algal flora of the Mediterranean Sea numbers about 1200 species (Boudouresque et Verlaque, 1978; Kalugina-Gutnik, 1981; Riedl, 1983). Some localities display remarkably high floristic diversity, among them are deep-sea banks. However, until recently the knowledge about algal flora of the banks was regrettably scarce (Furnari, 1984). We endeavoured to fill this gap by making a description of the species and phytogeographical composition and the occurrence of algae on some banks of the Mediterranean Sea.

## Material and methods

Preparing this report, we used materials collected during research cruises



*P. Malanotte-Rizzoli and V.N. Eremeev (eds.),*
*The Eastern Mediterranean as a Laboratory Basin for the Assessment of Contrasting Ecosystems,* 423–430.

to the Mediterranean in 1979-1986 (property of IBSS). Algae for the collections were taken in the during the 87, 88, 89 and 90-th research cruises of the R/V Academician Kovalevsky (spring and autumn time). Collection sites were the banks of Pantelleria, Lampedusa, Avventura, Tolbot and Herd (Tunisian Strait and Strait of Malta) and Johnston,Mancell, Seaney, Stocks and Brooker(Aegean Sea). The description of bottom vegetation and flora is based on the analysis of data collected at 88 stations (22 in the Aegean Sea, 56 in the Tunisian Strait and 10 in the Strait of Malta). Phytobenthos of deep-water banks of the Aegean Sea and of the Tunisian Strait has been described in details also. Samplings were made with the bottom grab "Okean-50", a beam trawl and Seegsby trawl.

Total number of algal species found there are 177; of them 34 green, 47 brown, 96 red, and one seagrass. All areas rich with bottom vegetation are very similar in the amount of species (73-109) that 41-62% of the total number of described algae.

Prof. A.A.Kalugina-Gutnik headed the work during the cruises to the Mediterranean Sea; we dedicate this paper to her blessed memory.

**Description of the taxonomic composition.**

***Aegean Sea.***

115 species of algae were descriped: 21 green, 32 brown and 62 red. Bottom vegetation of the banks occupies depths from 37 down to 115. Owing to high water transparency red algae occur even 195 m deep (Palmophyllum crassum, Peyssonnelia rubra and Pseudolithophyllum expansum).

There dominant associations were identified and described in the vegetation cover. All investigated phytocenoses had polydominant structure and distributed in a mosaic pattern. Compared with sandy and sandy-silty areas of the sea floor, hard substratum displayed the greatest species richness, especially at Johnston bank. Seasonal variations in the structure of phytocenoses are the most pronounced down to 50 m depth, being smoother at the lower boundary of occurence.

In spring the prevailing component are perennial brown and red algae like Sargassum hornschuchii, Codium bursa, Cystoseira corniculata, Phyllophora nervosa and Vidalia volubilis. In summer, annual brown algae, in particular Arthrocladia villosa, become dominant. On the banks of Brooker, Stocks and Seaney Phyllophora nervosa and Peyssonnelia squamaria are the basic component of vegetation cover.

Association of Cystoseira - Phyllophora nervosa - Vidalia volubilis. The incorporated communities inhabit the depths 37 down to 50 m and concentrate mostly on hard bottom substatum. The association comprises 64 algal species; in addition to the three basic algae typical representatives are genera Sargassum, Dictyopteris, Nereia and Fauchea.

Association Phyllophora nervosa - Peyssonnelia squamaria occupies sandy substrate at 50-90 m depth. The species composition is represented by 32 species, most typical are Palmophyllum crassum, Udotea petiolata, Acrodiscus vidovichii, Schysimenia dubyi. In late summer brown algae (Striaria attenuata, Nereia filiformis, Dictyota dichotoma) dominate.

Association Halopteris filicina - Callymenia microphylla grows on sandy-silty substrate at 100-195 m depths. It includes 9 species. In addition to the main algal species, also numerous are Zonaria tournifortii, Phyllophora nervosa and Schysimenia dubyi.

Special features of the communities were the multistratum polydominant structure and the lack of epiphytic synusiae.

Over 50% of algae are common for the Black and Aegean seas (Kalugina-Gutnik, 1975). However, the immigration of species has been reported: Phyllophora brodiae, Laurencia paniculata and L. parvipapillosa have been identified in the Mediterranean first, while the overuse of rivers may cause euryhaline and stenothermic algae to migrate into the Black Sea where they would occupy vacant environmental niches identical to Mediterranean. The most probable environmental niches for the coming aliens are deep-water areas near the Crimean and Caucasian shores. In these circumstances, studies of bottom plants of the eastern and central Mediterranean acquire special significance (Kalugina-Gutnik, 1981).

***Tunisian Strait***

Prof. Kalugina-Gutnik identified 159 algal species in the study area: 28 green, 45 brown and 85 red algae (Table 1). Flora of the deep-water banks differs considerably from that of near-shore waters of the Tunisian-Sicillian region (Kalugina-Gutnik, 1981, 1983).

*Lampedusa bank* is a vast submarine plato with flat bottom and insignificantly varying depths. Algal species numbered 109 (23 green, 35 brown and 51 red algae). The vegetation cover is relatively uniform, being composed of three associations. Brown and red algae prevail. Spring is the season of the dominance of perennial algae like Cystoseira fucoides, Vidalia volubilis, Gracilaria verrucosa. In late summer large brown algae Arthrocladia villosa, Dictyota dichotoma, Dictyopteris polypodioides, Sporochnus pedunculatus grow high above the thicket of perennials contributing to the multistratum structure of polydominant communities.

TABLE 1. The number of algal species collected in the Mediterranean Sea (numerator – numbers, denominator - %) in 1979-1980 (R/V 87, 89, 90 cruises Ak. Kovalevsky)

| Phylum | Aegean Sea | Banks | | | | Total |
|---|---|---|---|---|---|---|
| | | Lampedusa | Pantelleria | Avventura and Tolbot | Herd | |
| Chlorophyta | 21<br>64.7 | 22<br>64.7 | 14<br>41.2 | 13<br>38.2 | 5<br>14.7 | 34<br>19.2 |
| Phaeophyta | 32<br>68.1 | 35<br>74.5 | 31<br>65.9 | 28<br>59.6 | 4<br>8.5 | 47<br>26.6 |
| Rhodophyta | 62<br>65.6 | 52<br>54.2 | 53<br>55.2 | 67<br>69.8 | 5<br>5.2 | 96<br>54.2 |
| Total | 115<br>66.1 | 109<br>61.6 | 98<br>55.3 | 108<br>61.0 | 14<br>7.9 | 177<br>100 |
| R<br>P | 2.0 | 1.5 | 1.7 | 2.4 | — | 2.0 |

Cystoseira fucoides + C. corniculata - Vidalia volubilis - Gracilaria verrucosa dominate the associations, its phytocenoses occupy 50 to 70 m depths. The frequency of occurrence of Halimeda tuna, Udotea petiolata, Polysiphonia squamaria and Rytiphlaea tinctoria is about 100%. In this association was found to include 89 species of algae altogether.

Association of Arthrocladia villosa - Udotea petiolata + Zonaria tournefortii were found down to 70 m depth and comprised 36 species, constant species were Caulerpa prolifera, Acetabularia mediterranea,

Schizymenia dubyi. Vegetative cover was poor at that locality.

Association of Halopteris filicina + Callymenia microphylla + Peyssonnelia squamaria inhabit the depth from 70 down to 115 m. Seven species were identified as composing the algal communities. In addition to dominant species, Valonia macrophyza, Rytiphlaea tinctoria were occasionally seen.

*Pantelleria bank.* Algal species found in the locality numbered 98: 15 green, 32 brown and 51 red algae. The thicket of macrophytes concentrated mostly in the depths from 20 to 115 m. Local vegetation cover displays a multistratum structure, the lower stratum being composed of Peyssonnelia squamaria and Zanardinia prototypus. At the depth of 20 m the dense thicket of Posidonia oceanica with occasional Vidalia volubilis, Sphaerococcus coronopifolius, Sargassum hornschuchii and Cystoseira discors grow. In spring large and small varieties of red algae prevail and in autumn brown. The abundance of both attached and non-attached varieties of Ulva rigida, thallomes of which are found as deep as 105 m depths, is a special feature of the bank. Algal flora develops the greatest richness at the depth 20 to 65 m.

Five algal associations were defined. At the 20-30 m depths the community of Laminaria rodriguezi - Sargassum linifolium - Cystoseira, which includes 54 species is located. Some bottom sites shelter phytocenoses of Posidonia oceanica - Vidalia volubilis, the number of contributing species being 11. 47 algal species form the community of Arthrocladia villosa - Cystoseira at the depths 45-60 m; there are multistratum, mosaic and polydominant in structure.

The communities of Cystoseira - Polysiphonia elongata - Ulva rigida grow deeper (60-80 m), comprise 51 species, the vegetation cover is thin and patchy. The deepest (80-115 m) site of Pantelleria bank is where phytocenoses composing Halopteris filicina + Callymenia microphylla grow; 14 species of algae make poorly developed cover.

*Avventura bank* is an extensive plato with flat seafloor and the depth varying from 40 to 85 m. Vegetation cover is thick and diverse, and composed of 14 green, 28 brown and 65 red algae, 107 species altogether. Phytocenoses are polydominant and multistratum. Three algal associations are similar to those on Pantelleria bank; red and brown algae Cystoseira, Sargassum, Arthrocladia, Sporochnus, Halopteris and Callymenia dominate, showing a remarkable diversity of size and the shape of thallomes.

Association of Laminaria rodriguezi - Sargassum linifolium - Cystoseira display the greatest species diversity. Local algal flora manifests distinct seasonal changes.

## Phytogeographical composition of algal flora

Mediterranean algal flora contributes notably to the species composition of adjacent seas. The Adriatic Sea is the centre of speciation. Algal flora on the investigated deep-water banks of the Aegean Sea and Tunisian Strait is much like that of the Black Sea (Kalugina-Gutnik, 1981).

Correlation between the aforementioned algal groups (phytogeographical division) as specified for the all deep-water banks of is shown in Table 2. Lower boreal species dominate algal flora of the deep-water banks of the Aegean Sea. The share of subarctic to boreal and boreal to tropical species is also significant. Arctis to boreal, cosmopolitan and tropical algae are only minor (Table 2).

Lower boreal species dominate algal flora of the banks Pantelleria and Avventura; subarctic to boreal are prevailing in waters of Lampedusa bank (Tunisian Strait). The share of endemic algae varies insignificantly, and is the largest on Avventura bank where endemic and arctic-to-boreal species are more abundant in comparison with other banks under study.

## Conclusions

1. 177 algal species and one seagrass were collected and described on 10 deep-water banks of Aegean and Mediterranean Seas. The greatest species diversity was found on the banks of Johnston, Avventura and Lampedusa. Floristic composition does not very considerably between the banks.

2. Maximums of algal species concentrate at the depth 20 to 70 m. The vegetation is represented by phytocenoses of ten types, their structure is polydominant, multistratum and mosaic. Deep-water brown and red algae

TABLE 2. Phytogeographical composition of the algal flora in deep-water banks of Mediterranean Sea (numerator – numbers , denominator - %).

| Phytogeographical division (algal group) | Location | | | | | Total |
|---|---|---|---|---|---|---|
| | Aegean Sea | Tunisian and Malta Strait, banks | | | | |
| | | Lampedusa | Pantelleria | Avventura and Tolbot | Herd | |
| Arctic to boreal | 3* / 2.6 | 1 / 0.9 | 2 / 2.0 | 3 / 2.8 | — | 4 / 2.3 |
| Subarctic to boreal | 23 / 20.0 | 30 / 27.5 | 21 / 21.2 | 22 / 20.4 | 3 / 21.4 | 43 / 24.3 |
| Lower boreal | 30 / 26.1 | 25 / 22.9 | 30 / 30.3 | 34 / 31.4 | 3 / 21.4 | 43 / 24.3 |
| Boreal to tropical | 25 / 21.7 | 24 / 22.0 | 19 / 19.2 | 21 / 19.4 | 4 / 28.6 | 33 / 18.6 |
| Subtropical | 18 / 15.6 | 14 / 12.8 | 12 / 12.1 | 12 / 11.1 | 3 / 21.4 | 22 / 12.4 |
| Tropical | 4 / 3.5 | 4 / 3.7 | 5 / 5.0 | 6 / 5.5 | — | 7 / 4.0 |
| Cosmopolitans | 3 / 2.6 | 4 / 3.7 | 2 / 2.0 | 2 / 1.8 | 1 / 7.1 | 5 / 2.8 |
| Endemics | 9 / 7.8 | 7 / 6.4 | 7 / 7.1 | 8 / 8.3 | — | 16 / 9.0 |
| Total | 115 / 65.0 | 109 / 61.6 | 98 / 55.9 | 108 / 61.0 | 14 / 7.9 | 177 / 100.0 |

prevail, among green algae Siphonales spp. were the most abundant.

3. The algal flora is temperate to warm-water. Taxonomic composition is special, with over half of the algal species being from the Black Sea.

4. Conservation of the biodiversity in unpolluted model areas of the Mediterranean is an essential element at conserving the marine plants richness in the adjacent seas and water areas.

## References

1. Boudouresque, C.F. et Verlaque M. (1978) V• g• tation marine de la Corse (M• diterran• e) I. Documents pour la flore des algues, *Bot. Mar.*, 21, 265-275.

2. Furnari, G. (1984) The benthic marine algae of southern Italy, floristic and geobotanic considerations, *Webbia*, 38, pp. 349-369.

3. Kalugina-Gutnik, A.A. (1975) *Phytobenthos of the Black Sea*, Naukova dumka, Kiev, 247 p. (in Russian).

4. Kalugina-Gutnik, A.A. (1981) The phytogeographical characteristic of algoflora of Mediterranean Sea, *Resume in Journess Etud. System. et Biogeogr. Mediterranean*, Cagliari, 13-14 oct., Monaco, 1980, p. 97.

5. Kalugina-Gutnik, A.A. (1983) Comparative description of the flora and vegetation on some [deep-sea] banks of the eastern and central Mediterranean Sea, *Abstracts of the VII-th Congress of the All-Union Botanical Society*, Nauka Publishing House, Leningrad, pp. 84-85.

6. Riedl, R. (1983) *Fauna und Flora des Mettelmeeres*, Verlag Paul Parey, Hamburg and Berlin, 836 p.

# ANTHROPOGENIC EUTROPHICATION OF NEAR-SHORE WATERS AND MACROALGAL BIODIVERSITY IN THE LEVANTINE SEA

O.V.MAXIMOVA, N.V.KUCHERUK

*P.P.Shirshov Institute of Oceanology Russian Academy of Science, Moscow, Russia.*

The material was sampled during the 27-th cruise of r/v "Vityaz" in October – November 1993 in the region of Latakia (Syria). About 30 qualitative samples of macroalgae were collected in 5 coastal points at the depth 0,1–2,0 meters (Figure 1). Those points differed by the level of oil pollution and anthropogenic eutrophication.

1. Point **A** - in the central part of the port of Latakia, at the base of El-Borge Castle. Eutrophication + very intensive oil pollution. Sheltered site.
2. Point **B** – Port entrance. High level of pollution. Semi-exposed site.
3. Point **C** - to the north from the port, at the cape in the Fishermen Harbor. Eutrophication + oil pollution. Semi-exposed locality.
4. Point **D** – to the south from the port, near the breakwater preventing the oil pollution of this locality. The eutrophication due to vicinity of Latakia town is rather high. Exposed coast with intensive water movement.
5. Point **E** - El-Kuobban Bay near the Marine Research Center. Very clear water with minimal anthropogenic influence. Exposed coast with high wave activity.

The distance between points **D** and **E** is approximately 10 km. We identified the macroalgae species using [1, 2, 7, 8].

All the localities in the port of Latakia or in its vicinity (**A,B,C**) were deprived of brown algae (*Phaeophycophyta*) which are known to be oligosaprobic group [3]. In the most polluted point **A** the only macroalga obtained was polysaprobic *Enteromorpha intestinalis (Chlorophycophyta).* In the point **B** three species were sampled: *Corallina*



*P. Malanotte-Rizzoli and V.N. Eremeev (eds.),*
*The Eastern Mediterranean as a Laboratory Basin for the Assessment of Contrasting Ecosystems,* 431–435.

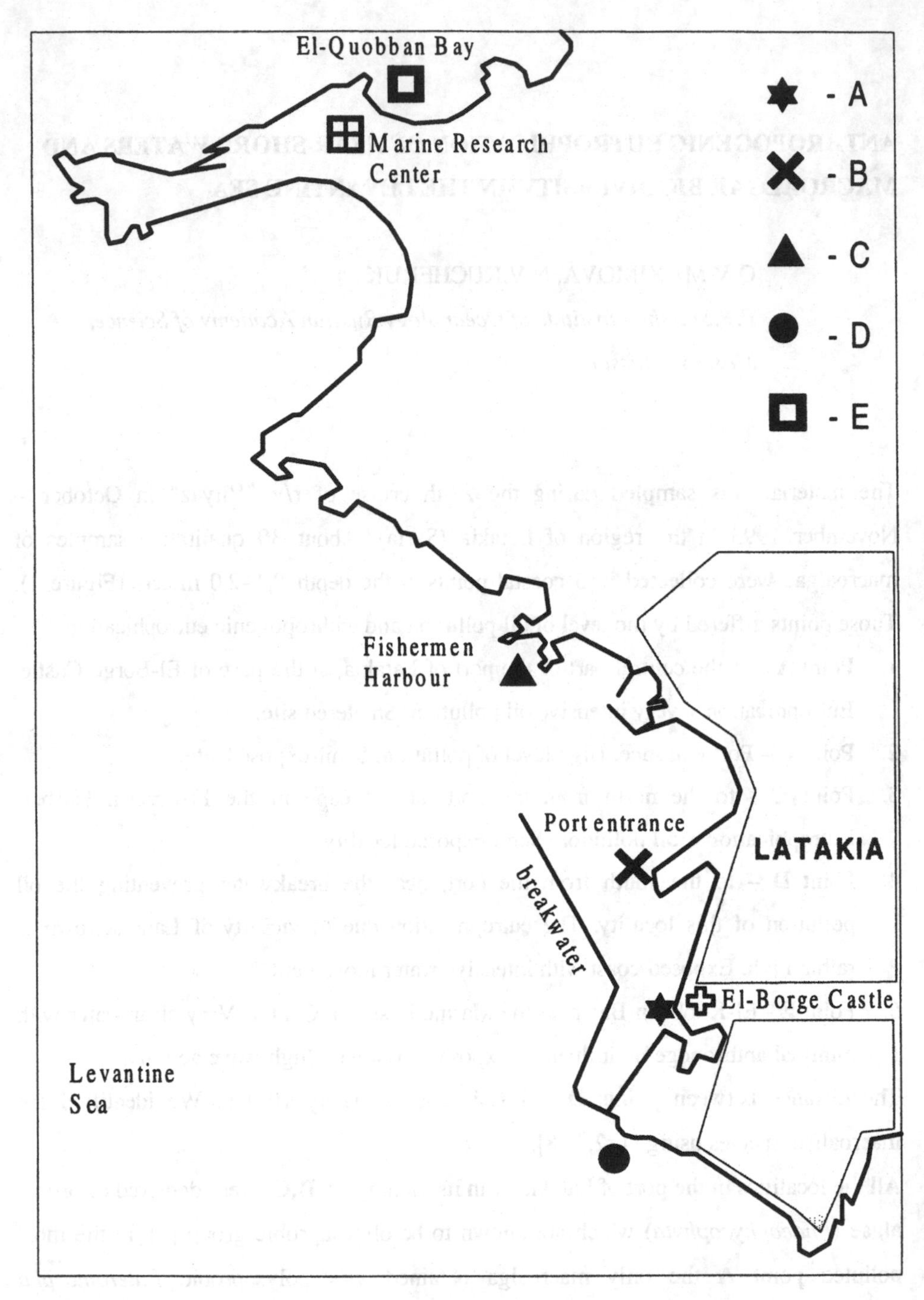


Figure 1. Scheme of macroalgae sampling points along Latakia coast.

*mediterranea* with epiphytic *Polysiphonia denudata (Rhodophycophyta)* and *Cladophora coelothrix (Chlorophycophyta)*. In the point **C** five species were collected: *E.intestinalis*, mesosaprobic *Ulva lactuca*, *Codium vermilara (Chlorophycophyta)*, *C.mediterranea* and oligosaprobic *Jania rubens (Rhodophycophyta)* (Figures 2, 3).

Points **D** and **E** were similar by the environmental conditions, the only difference was in the eutrophication level. In mostly clear point **E** 23 macroalgae species (5 *Chlorophycophyta*, 9 *Phaeophycophyta*, 9 *Rhodophycophyta*), 1 species of *Cyanophyta* (*Oscillatoria sp.*, a component of algo-bacterial mats) and 2 species of marine phanerogams (*Cymodocea nodosa* and *Halophyla stipulacea*) were sampled.

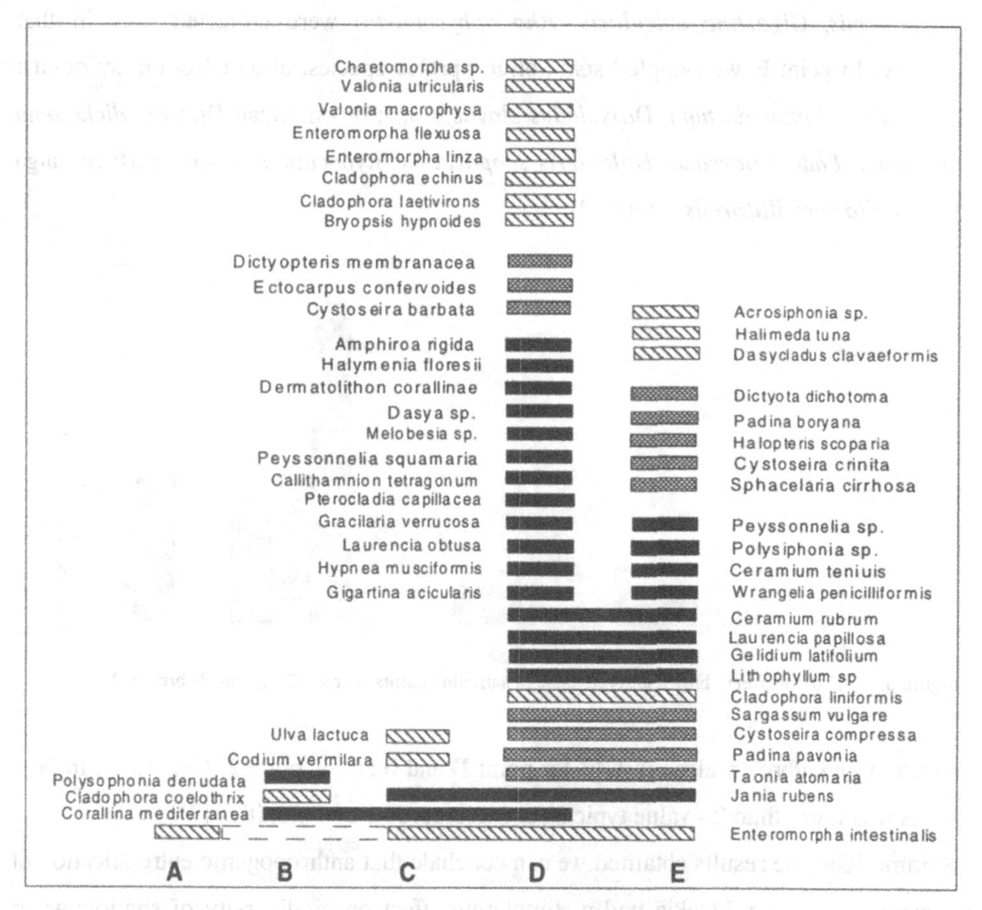


Figure 2. Macroalgae species composition in sampling points.

- green algae - brown algae - red algae

The most algae-rich point was **D**: 36 species (10 *Chlorophycophyta*, 7 *Phaeophycophyta*, 18 *Rhodophycophyta*, 1 *Cyanophyta)*. Points **D** and **E** had some common species: 2 green algae (*E.intestinalis, Cladophora liniformis*); 4 brown algae (*Taonia atomaria, Padina pavonia, Sargassum vulgare, Cystoseira compressa*); 5 red algae (*Gelidium latifolium, Jania rubens, Ceramium rubrum, Laurencia papillosa, Lithophyllum sp.)*. Only *C.rubrum* and *E.intestinalis* are the polysaprobic algae, all the others are meso- or oligosaprobic ones [3]. In point **D** number of mesosaprobic species is higher (22) than in point **E** (10); *Enteromorpha linza, E.flexuosa, Cladophora laetivirons (Chlorophycophyta), Gracilaria verrucosa, Laurencia obtusa, Hypnea musciformis, Gigartina acicularis (Rhodophycophyta)* were collected only in that locality. In point **E** we sampled some oligosaprobic species, absent from other points: green algae *Halimeda tuna, Dasycladus clavaeformis*; brown algae *Dictyota dichotoma f.implexa, Padina boryana, Halopteris scoparia (= Stypocaulon scoparium);* red alga *Wrangelia penicilliformis* (Figure 2).

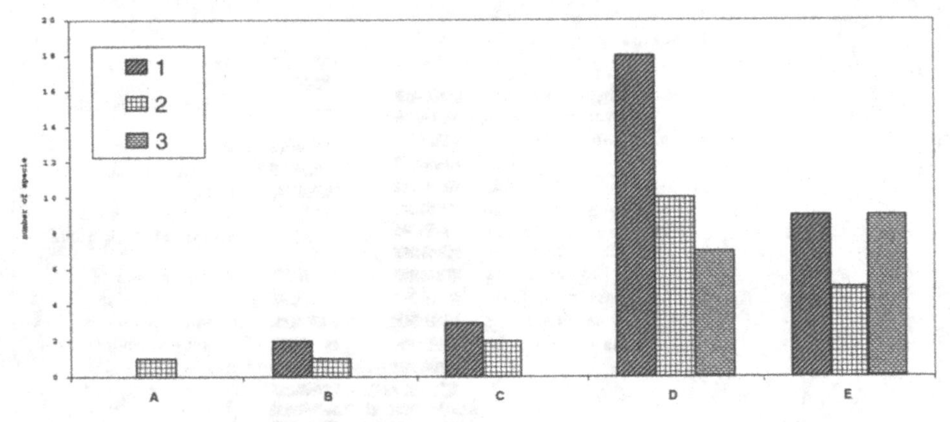


Figure 3. Ratio of the green, brown and red algae in sampling points. 1- red , 2 - green, 3- brown algae.

Ratio of green/brown algae is 1,43 for point **D** and 0,56 for point **E** (Figure 3); in both cases it is lower than 2 - value typical for mesosaprobic communities [3].

Summarizing the results obtained we can conclude that anthropogenic eutrophication of coastal waters near Latakia had a stimulating effect on biodiversity of shallow-water macroalgal communities. In oligotrophic waters of the Levantine Sea only the regions of town sewage effluents had the values of biogenous concentrations as high as 0,8-0,9 μg-at/l $NO_3$ and 0,1-0,3 μg-at/l $PO_4$ [4]. These values are just normal for natural development of most macroalgal species. The oil pollution suppressed the macroalgal vegetation: the minimal biodiversity characterized the communities of the Latakia port. The temporary stimulating effect of eutrophication would change into the effect of macroflora degradation while the increasing of antropogeneous stress and especially in the case of cumulative influence of oil and biogenic pollution. Such situation is observed for many coastal regions of World Ocean and for Mediterranean in particular [1, 6]. One of the most dramatic examples is the fate of the unique Zernov's Phyllophora Meadow in the N-W of Black Sea: it had been formed due to the biogenic flux of great rivers (Danube, Dnieper, Dniester), and it is degrading nowadays because of anthropogenic hyper-eutrophication of these rivers [3 , 5].

**References**

1. Aleem , A.A. (1993) *Marine Algae of Alexandria,* Alexandria.
2. Delepine, R., Boudouresque, C.-F., Orestano, C., Noailles M.-C., Asensi, A. (1987) Algues et autres vegetaux marins, in W.Fisher, M.Schneider, M.-L.Bauchot (eds.), *Mediterranee et Mer Noir. Vegetaux et Invertebres Vol.1.*, FAO, CEE, Rome, pp. 1-136.
3. Kalugina-Gutnik, A.A. (1975) *Phytobenthos of the Black Sea,* Naukova Dumka, Kiev. (In Russian).
4. Lukashov, Yu.F. (1993) Report of hydrochemical group, Report on scientific investigations in 27 cruise of r/v "Vityaz", IO RAS, Moscow. (In Russian).
5. Maximova, O.V., Kucheruk, N.V. (1993) Eco-morphological plasticity of black-sea Phyllophora nervosa and the problem of Zernov's Phyllophora meadows existence, in V.B.Vozzhinskaya and E.P.Turpaeva (eds.), *Biology of the Black Sea Agarophytes: Phyllophora nervosa (D.C.) Grev.,* IO RAS, Moscow. (In Russian).
6. Pérès, J.M. (1967) The Mediterranean Benthos, *Oceanogr. Mar. Biol. Ann. Rev.* **5**, 449-533.
7. Riedl, R. (1983) *Fauna und Flora des Mettelmeeres,*: Verlag Paul Parey, Hamburg and Berlin.
8. Zinova, A.D. (1967) *Key to the Green, Brown and Red Algae in the Southern Seas of the USSR,* Nauka, Moscow-Leningrad.

# DEEP-SEA CALCAREOUS *RHODOPHYCOPHYTA* COMMUNITIES IN THE LEVANTINE SEA

O.V.MAXIMOVA[1], SAKER FAYES [2]

[1] *P.P.Shirshov Institute of Oceanology Russian Academy of Science, Moscow, Russia.*

[2] *Higer Institute of Applied Sciences and Technology, Damascus, Syria.*

This study is based on materials from two cruises of r/v "Vityaz": N 24 (1992) and N 27 (1993). In total 13 qualitative samples were collected at 6 stations on southern syrian shelf (depth 25-90 metres) (Figure 1) by means of trawling (Seegsby trawl) and dredging (grabb “Ocean-0,1”) (Table 1). The community discovered there reminds the "*Lithophyllum* trotoir", "trotoir à *Tenarea*" or "coralligenous bank" [3], well known for the Western Mediterranean. J.M. Pérès [3] mentioned the same structures in the Eastern Mediterranean "covering very large areas – on thousands of square metres" (p.500). The square of discovered by our expeditions "coralligenous meadow" is about some **tens of square kilometres.** Such formation is being described for the first time. The second pecularity of this biocenosis is its location above silty substratum. The thickness of "coralligenous" layer is no more than 10 cm.

Table 1. Material obtained in 24$^{th}$ and 27$^{th}$ cruises of r/v "Vityaz".

| N of station | Depth, m | Number of samples |
|---|---|---|
| 3639 | 90 | 1 (1 grabb) |
| 3640 | 50-63 | 2 (1 grabb and 1 trawl) |
| 3641 | 25 | 1 (1 grabb) |
| 3870 | 78-90 | 1 (united from 5 grabbs) |
| 3871 | 45 | 2 (1 trawl and united from 5 grabbs) |
| 4006 | 40-60 | 6 (5 grabbs and 1 trawl) |



*P. Malanotte-Rizzoli and V.N. Eremeev (eds.),*
***The Eastern Mediterranean as a Laboratory Basin for the Assessment of Contrasting Ecosystems*, 437–440.**

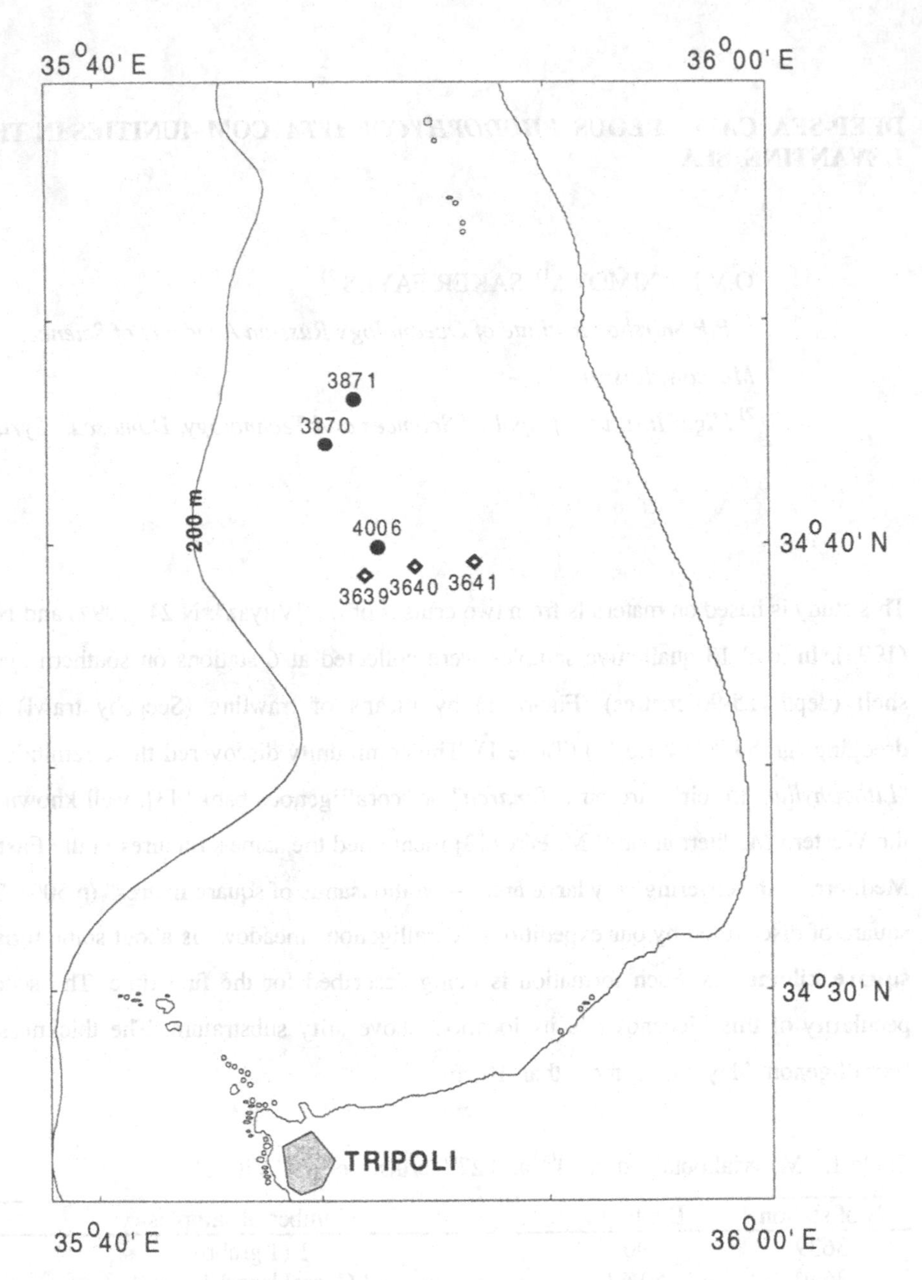


Figure 1. Map of stations with "coralligenous bank" macroalgal community.

◇ - 24 cruise r/v "Vityaz" ● - 27 cruise r/v "Vityaz"

This community is very algae-rich, especially for deep-water environment: 35 species of macroalgae (7 *Chlorophycophyta*, 5 *Phaeophycophyta*, 23 *Rhodophycophyta*). We must mention, that this list is not complete: some species from the collection are not identificated yet because of lack of special guides for the region. Identification was carried out using [1, 2, 5, 6, 7].

The main role in forming of the community belongs to 2 groups of macroalgae: 1) calcareous red algae *(Rhodophycophyta: Crypyonemiales: Corallinaceae) Lithophyllum racemus, Pseudolithophyllum expansum, Lithothamnion sp., Neogoniolithon sp., Amphiroa rigida* and some others, and 2) siphonous green algae *(Chlorophycophyta: Caulerpales = Siphonales) Caulerpa scalpelliformis* and *Udotea petiolata*. The first species is one of Lessepsian migrants, mentioned in the Eastern Mediterranean for the first time in 1930 [4].We consider siphonous macroalgae to be the pioneer species of this community. *Caulerpa* and *Udotea* can occupy soft substrata and they have long ranched rhizoms and numerous rhizoids, extracting slippery substance of polysaccharide nature. The particles of dead thalli of calcareous algae (first of all *A.rigida*) and exoskeletan material (fragments of shells, carapaces) of benthic animals, little stones, grains of sand stick to this mucilage, forming the cover around the creeping parts of the thalli. Crustacean red algae and other species with prostrate calcified and non-calcified thalli (such as brown *Zanardinia prototypus* or red *Hildenbrandtia prototypus* and *Peyssonnelia sp.*) begin to develop above these covers together with different sessile benthic animals (bryozoans, sponges, hydroids, ascidians, polychaetes etc). Thus the second layer of the cover forms – the thick crust, developing into calcareous concretions, upon which the vegetation of algae with vertical soft thalli begins. Among them the most common species are: *Codium bursa, C.decoricatum, Anadyomene stellata, Valonia utricularis (Chlorophycophyta), Gelidium latifolium, Pterosiphonia pennata, Polysiphonia (fructiculosa?), Halymenia ulvoidea, Kallimenia microphylla, Botriocladia boergesenii, Griffithsia opuntioides (Rhodophycophyta).* No seagrasses neither large brown algae, common for this type of benthic community in Western Mediterranean [3] were sampled.

The rhizoms of *Caulerpa* and *Udotea* die under the crustacean cover, but the young parts of the thalli grow through the holes in the crust, the plants continue their vegetation, and the community develops. During our field activities we have observed all the described stages of this development.

**References**


1. Aleem, A.A. (1993) *Marine Algae of Alexandria,* Alexandria.
2. Littler, D.S., Littler, M.M., Bucher, K.E., Norris, J.N. (1989) *Marine Plants of the Caribbean,* Smithsonian Institution Press, Washington, D.C.
3. Pérès, J.M. (1967) The Mediterranean Benthos, *Oceanogr. Mar. Biol. Ann. Rev.* **5**, 449-533.
4. Por, F.D. (1978) *Lessepsian migration,* Springer Verlag, Berlin
5. Riedl, R. (1983) *Fauna und Flora des Mettelmeeres,*: Verlag Paul Parey, Hamburg and Berlin.
6. Tseng, C.K.(ed.) (1983) *Common Seaweeds of China,* Science Press, Beijing, China.
7. Zinova, A.D. (1967) *Key to the Green, Brown and Red Algae in the Southern Seas of the USSR,* Nauka, Moscow-Leningrad.

# INTERMEDIATE WATER FORMATION IN THE LEVANTINE SEA : THE RESPONSE TO INTERANNUAL VARIABILITY OF ATMOSPHERIC FORCING

K.NITTIS
*National Centre for Marine Research, Aghios Kosmas, 16604 Hellenikon, Athens Greece*
A.LASCARATOS
*University of Athens, Dep. Of Applied Physics, University Campus Build. PHYS-V, 15784 Athens Greece*

## 1. Introduction

The Levantine Intermediate Water is the characteristic water mass that occupies the intermediate layers (200-400) of the Mediterranean sea. It is characterized by high salinity and temperature and it is known to be formed in the eastern most part of the Mediterranean sea, the Levantine basin. From this area it starts spreading to the rest of the basin, passing through the straits of Sicily to the western Mediterranean and finally through the strait of Gibraltar to the Atlantic Ocean. Its high salinity is a major preconditioning factor for the formation deep water in the southern Adriatic (source of Eastern Mediterranean Deep Water) and the Gulf of Lions (source of Western Mediterranean Deep Water).

Different areas have been proposed in that past as possible LIW formation sites. The Rhodes cyclonic gyre, a permanent general circulation feature of the area [1] is suggested in most cases [2][3][4] as the main formation area. There are, nevertheless, indications of intermediate water formation in the north Levantine [5], the SE Aegean [6] or the south Levantine [7]. This variety of areas is a first strong indication that the LIW formation process undergoes a significant interannual variability.

In this work, we attempt to simulate this inter-annual variability of LIW formation using a high resolution numerical model initialized and forced by realistic data. Similar numerical experiments, forced by climatological monthly means have shown that LIW is formed in the Rhodes Gyre with an annual



*P. Malanotte-Rizzoli and V.N. Eremeev (eds.),*
*The Eastern Mediterranean as a Laboratory Basin for the Assessment of Contrasting Ecosystems,* 441–446.

formation rate of 1.2 Sv [8]. The experiments presented here reveal the significant inter-annual variability of surface forcing and the resulting variability of LIW formation characteristics. They also show, the importance of synoptic scale extreme atmospheric events compared to the low frequency mean cooling.

## 2. The numerical model

Our numerical model of the Levantine sea, is based on the Princeton Ocean Model (POM), a 3-D primitive equation ocean model designed for both open ocean and coastal sea studies [9]. An eddy resolving grid of 5.5 km is used in the horizontal (first internal rossby radius is 10-14 km). In the vertical, 30 sigma-layers are used with logarithmic distribution near the surface. The model

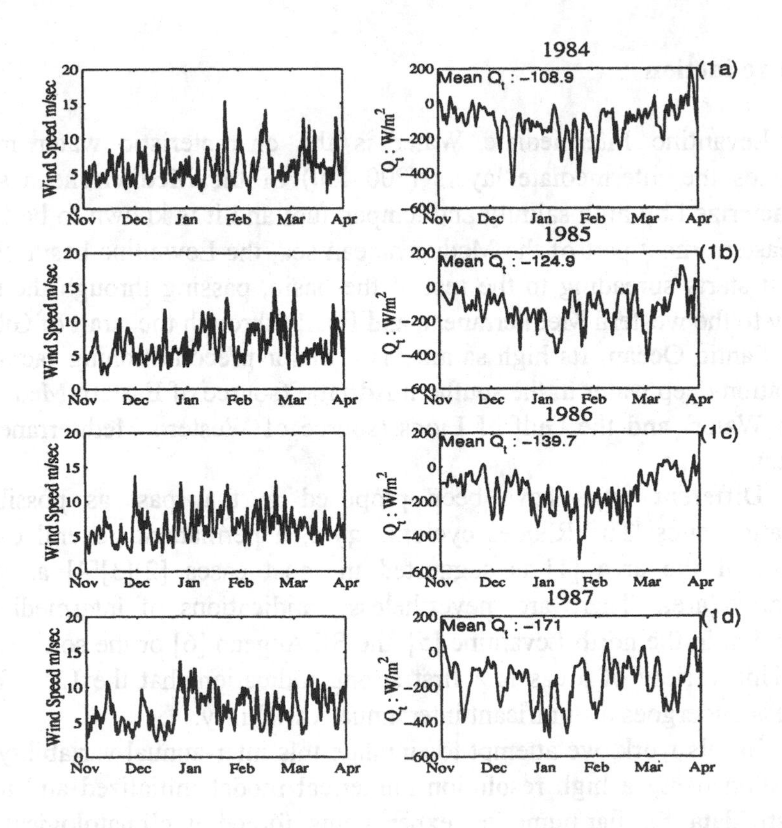


Figure 1 : Wind speed and total heat flux time series

is initialized by the MED-4 data set [10]; the same data base is used for the computation of seasonal T and S profiles along the open boundary (western boundary).

The surface boundary conditions are the fluxes of heat, fresh water (evaporation - precipitation) and momentum (wind stress). All of them are computed at each time step using the sea surface temperature computed by the model at the previous time step and atmospheric parameters taken from different data sets [8]. In this study, we use the 12hr NMC analysis from the period 1980-1988. This data set includes wind speed $W$, air temperature $T_a$ and relative humidity $R_h$. Two more parameters are needed for the calculation of surface fluxes : the cloud cover $C$ and the precipitation $P$. The first one is taken from the COADS 2x2 monthly mean data set for the same period, while precipitation is derived from monthly climatological data [11].

## 3. Results and Discussion

From the 1980-88 period, 4 characteristic winters are chosen as case studies : a mild winter (1984) two typical (1985-1986) and one heavy winter (1987). The inter-annual variability of winter (November - March) heat loss in the Levantine is found to exceed 35 $W/m^2$ : it ranges between 108 $W/m^2$ (1984) and 171 (1987) with a mean value of 135 $W/m^2$ (fig 1a-d). The total budget of each winter is mainly controlled by the characteristics of the short time scale events : the winter of 1987 is characterized as a very cold one, due to the increased frequency, duration and intensity of these events. On the contrary, the winter of 1985 is "typical", or rather "mild", because despite their large number, the synoptic events were relatively weak.

In figure 2 we present maps of the 29.05 isopycnal depth for each of the four winters. The snapshots are from the end of February, the usual period of maximum formation rate [8]. The density of 29.05 is characteristic of LIW in the Eastern Mediterranean and it's depth is a good indicator of the formation site : in the region where this depth becomes zero we have ventilation of the specific isopycnal and formation of new water. During 1984, the year of minimum heat loss, only 0.64 Sv (almost half of the climatological value) of LIW was formed. The area of intermediate water formation was restricted in the center of the Rhodes cyclonic gyre (fig 2a).

During the next year (1985) the winter mean heat loss was 125 $W/m^2$ (i.e. less than the climatological value of 135 $W/m^2$) but we had the maximum LIW formation rate (1.38 Sv). This is attributed to the strong cooling event during the first days of March (fig 1b). Due to the pre-existing weak stratification, this storm event (although not exceptionally strong) was able to create a more than 350m deep mixed layer and form large amounts of LIW. This example indicates that not only the intensity but also the timing of the synoptic events control their effect on the formation of LIW.

During the winter of 1986 we observe outcropping of 29.05 not only in the Rhodes gyre but in the southern parts of the Levantine as well. This is related to a number of atmospheric depressions that crossed the basin along the coast of Africa. This southward shift of their route, created a strong north-south gradient of heat loss with maximum values to the south. Although the area of deep mixing is now extended (2-3 times larger than 1985) the computed LIW formation rate is as low as 0.67 Sv because the water masses in the south

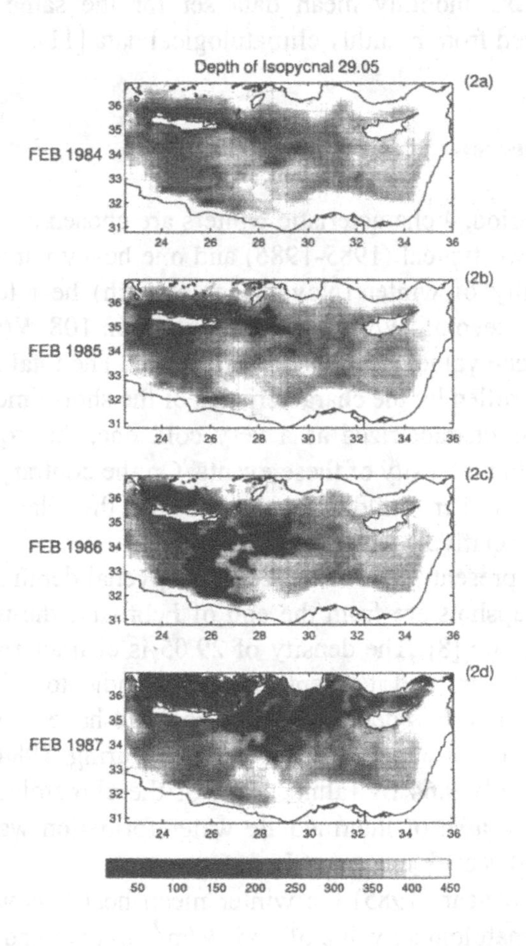


Figure 2: Depth of isopycnal 29.05 during late February

Levantine do not have LIW characteristics (they are fresher and cooler). This year is a typical example of the controlling role that the spatial variability of heat flux can play.

Finally, the winter of 1987 is characterized by large scale deep mixing and formation of intermediate water all over the north Levantine sea (fig 2d). Three very strong events during the beginning of January lead to early LIW formation during these days in the center of the Rhodes gyre. In the following weeks, the water in the center of the gyre becomes even denser (29.16) and leads to convection down to 1000m depth. During the same period, we have mixing down to intermediate depths (200-300m) in the periphery of the gyre and along the coast of Asia Minor. We, therefore, have during this year simultaneous deep and intermediate water formation in the north Levantine. These results are in agreement both with observations from this year [12] or from years with similar conditions (winter of 1992 [5]) as well as with previous numerical experiments with 1-D mixed layer model [4].

## 4. Conclusions

The results of the numerical experiments presented here, suggest that the inter-annual variability of meteorological conditions over the Levantine can strongly affect the characteristics (formation rate, site, duration) of the LIW formation process. The Rhodes cyclonic gyre is the area where LIW is usually formed under typical atmospheric conditions. During exceptionally cold winters, other regions in the north Levantine can also be sources of intermediate water with simultaneous deep water formation in the center of the Rhodes gyre (example of 1987). Mixing down to intermediate depths can be observed even in the south Levantine, but the characteristics of the water mass formed in this region are outside the typical LIW T&S ranges (example of 1986). All these departures from the "mean climatological state" are related to very strong variability of synoptic scale atmospheric events : their intensity, frequency, duration and route, determine the final characteristics of the formation process each year.

## 5. Acknowledgments

This work has been carried out in the framework of the MTP-I/ MERMAIDS-II research project. We acknowledge the support of the European Commission's Marine Science and Technology Program under contract MAS2-CT93-0055.

**6. References**


1. Robinson A.R., M. Golnaraghi, W.G. Leslie, A. Artegiani, A. Hecht, E. Lassoni, A. Michelato, E. Sansone, A. Theocharis and U. Unluata, 1991 : "The Eastern Mediterranean general circulation : features, structure and variability", Dyn. Atmos. Oceans, 15, 215-240.
2. Lacombe, H., and Tchernia, P., 1960 : Quelques traits generaux de l'Hydrologie Mediterrane, Cahiers Oceanographiques 12, 527- 547.
3. Ovchinnikov I.M., 1984 : The formation of Intermediate Water in the Mediterranean , Oceanology, 24, 168-173.
4. Lascaratos A., R.Williams and E.Tragou, 1993 : A Mixed Layer Study of the formation of Levantine Intermediate Water, JGR Vol.98, No.C8, 739-749
5. Sur H.I., E.Ozsoy and U.Unluata, 1992 : Simultaneous deep and intermediate convection in the Northern Levantine sea, winter 1992, Oceanologica Acta, 16, 33-43
6. Wust G., 1961 : On the vertical circulation of the Mediterranean Sea, J. Geophys. Res., 66(10), 3261-3271
7. Morcos S.A., 1972 : Sources of Mediterranean Intermediate Water in the Levantine Sea, Studies in Phys. Oceanog., Vol.2, A.L.Gordon (ed.), Gordon & Breach, New York, pp.185-206
8. Lascaratos A. and K.Nittis, 1998 : A high resolution 3-D numerical study of Intermediate water formation in the Levantine Sea, Jour. Geophysical Res., in press
9. Blumberg A. F. and G. L. Mellor, 1987: A description of a three-dimensional coastal ocean circulation model. Three Dimensional Coastal Ocean Circulation Models, Coastal Estuarine Sci., 4, edited by N.S. Heaps, pp 1-16, AGU, Washington D.C. 1987.
10. Brasseur P., Beckers J.M., J.M.Brankart and R. Schoenauen, 1996 : Seasonal temperature and salinity fields in the Mediterranean Sea : climatological analyses of a historical data set, Deep Sea Res., 43(2), 159-192.
11. Jaeger, L., 1976 :Monatskarten des Niedersschlags fur die gange Erde, Ber. Dtsch Wetterdienstes, Band 18, No. 139.
12. Gertman I.F., I.M.Ovchinnikov and Y.I.Popov, 1990 : Deep convection in the Levantine Sea, Rapp. Comm. Int. Medit., 32,1

# SEA SURFACE TEMPERATURE FIELDS AND AIR-SEA FLUXES IN THE RHODES GYRE DURING LIWEX'95 EXPERIMENT

A.LASCARATOS
*University of Athens, Dep. Of Applied Physics, University Campus Build. PHYS-V, 15784 Athens Greece*
K.NITTIS and A.THEOCHARIS
*National Centre for Marine Research, Aghios Kosmas, 16604 Hellenikon, Athens Greece*

## 1. Introduction

During the period 18-December-1994 to 20-April-1995, a multinational collaborative experiment was carried out in the NW Levantine Sea with main objective to study the Levantine Intermediate Water (LIW) formation process. During this POEM (Physical Oceanography of Eastern Mediterranean) coordinated experiment, the Rhodes Gyre area and the adjacent seas were visited by 5 research vessels that carried out intensive hydrological observations of high spatial resolution. The experiment was designed to study the three phases of the LIW formation process : a) the preconditioning, b) the formation and c) the spreading phase.

The LIW is known to be formed in the Rhodes cyclonic gyre [1][2]. During the whole formation process, but especially during the first two phases (preconditioning and convection) air-sea buoyancy fluxes are the main driving mechanism. In this paper, we investigate the temporal variability of sea surface temperature fields and air-sea heat fluxes in the Rhodes gyre area, throughout the LIWEX'95 experiment period. SST data are taken from NOAA AVHRR weekly averaged images from December 94 to April 95. For the same period we use ECMWF analysis data of wind components (*U* and *V*), air temperature $T_a$ , relative humidity $R_h$ and cloud coverage *C*. Using appropriate bulk formulae we compute the various components of the heat budget. We then study the mixed layer depth during the period of LIWEX'95 and we make simple calculations on the relation between the mixed layer depth and the air-sea buoyancy fluxes.



*P. Malanotte-Rizzoli and V.N. Eremeev (eds.),*
*The Eastern Mediterranean as a Laboratory Basin for the Assessment of Contrasting Ecosystems,* 447–452.

January 1995

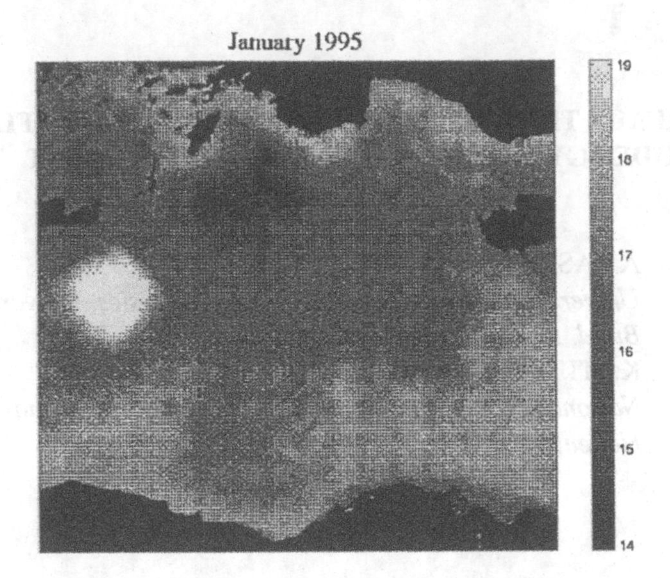


Figure 1 : Weekly SST field from AVHRR image

## 2. Variability of SST fields

The sea surface temperature fields in the NW Levantine are dominated by two anomalies of opposite sign : the extended Rhodes cyclonic gyre with low temperature signal to the east and the smaller but very energetic Ierapetra anticyclone to the west with the maximum surface temperature of the whole East Med (figure 1). The first one is a permanent feature of the Mediterranean sea general circulation, while the second is rather recurrent since it disappears during certain months of the year, or for even longer periods [3]. The Rhodes cold cyclonic gyre is bounded to the north by the warm Asia Minor current that runs along the coast carrying to the west saline and warm waters of Levantine origin, and to the south by the multi-lobe system of the Mersa-Matruh anticyclone. This pattern of SST remains unchanged during the whole period of the experiment although the horizontal gradients tend to weaken gradually mainly due to decrease of temperature in the "warm" areas of the domain.

Using the weekly SST fields, the temperature in the center of the Rhodes gyre was computed for the whole December-April period (figure 2). The temperature undergoes a continuous decrease from 15.2°C at the beginning of December to 14.3°C at mid February. It then increases gradually up to 15.5°C in the first week of April followed by an abrupt and substantial rise up to 17.2°C by the end of the month. This sudden increase is attributed to the build up of a thin and light surface layer that will gradually evolve to a strong subsurface seasonal thermocline.

## 3. Air-Sea buoyancy fluxes

From the weekly values of SST and the high frequency atmospheric data over the Rhode gyre area, we computed the air-sea heat fluxes during the five months period. The total heat flux $Q_t$ is the sum of the various heat loss components :

$$Q_t = Q_s - (Q_b + Q_s + Q_h) \quad (1)$$

where $Q_b$ is the long-wave back radiation, $Q_s$ the sensible heat, $Q_h$ the latent heat flux and $Q_s$ is the gain (solar insolation). The calculation of these components is done through appropriate bulk formulas that use the different atmospheric parameters (wind speed, air temperature, air humidity and cloud cover) and the sea surface temperature. For this applications we have used the May formula for the long-wave radiation and the Kondo bulk formula for the latent and sensible heat fluxes, as proposed by [4] and [5]. The resulting time series of total heat flux is presented in figure 3. From December to approximately mid February the center of the gyre looses heat to the atmosphere, then after a month of almost zero $Q_t$, the budget becomes positive. The time series of heat flux is dominated by strong heat loss events (300-700 $W/m^2$) of 2-7 days duration, while during the rest of the period heat fluxes are generally low (less than 100 $W/m^2$) either negative or positive. All the strong events are related to intrusion of cold and dry air masses of polar origin over the Levantine. They usually accompany storm events (cyclones) that cross the area during winter period with a typical frequency of twice a month. The generally small values of $Q_t$ (close to zero) outside the synoptic episodes, are

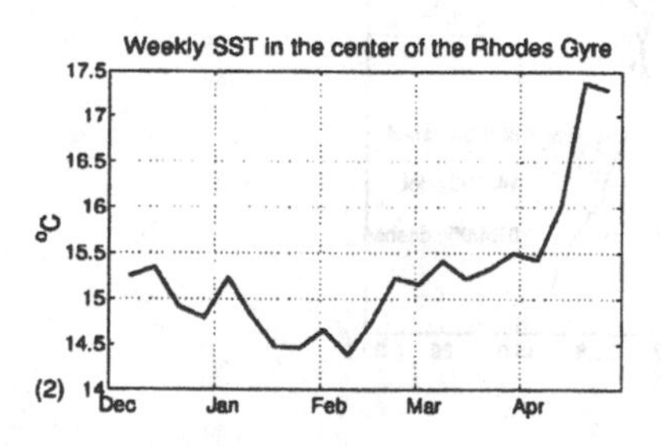


Figure 2 : Weekly mean SST in the center of the gyre

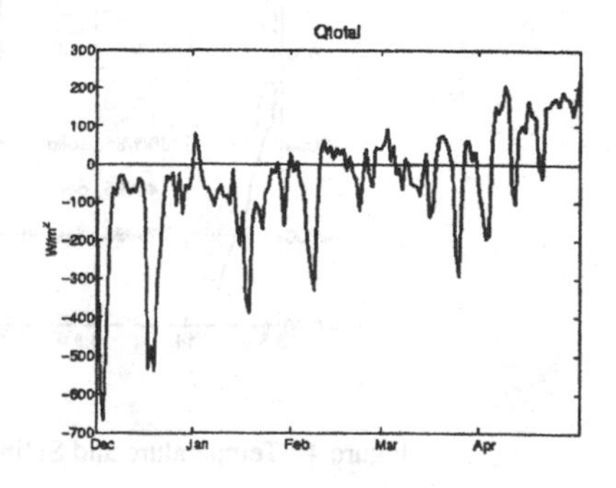


Figure 3 : Time series of total heat flux

related to the low SST's in the center of the gyre which reduce both latent and sensible heat exchanges. In areas outside the gyre, such as the Asia Minor current, where SST's are considerably higher, the heat losses are higher during both 'normal periods' and storms. For this reason the SST variations are far more significant in these areas than in the center of the gyre : in the Ierapetra gyre, the difference between December and March is more than 3.5°C while this difference is only 1°C in the Rhodes gyre.

The time series of heat loss indicates that during that year we had an "early" winter. In fact, the most severe storms and the maximum cooling period was during December instead of January or February. A second remark is that the time at which the net heat budget of the ocean switches from negative to positive values, is shifted earlier by a whole month compared to the climatologicaly expected middle of March. Both the early winter start and early passage to spring are clearly reflected in the SST's of the Rhodes gyre. Indeed in December, the temperature at the center of the gyre is already as low as 15.2°C, a value usually encountered at the end of the cooling period [2]. This can be interpreted as an early indication of deep, instead of intermediate, water formation in the center of the gyre for that winter.

This is indeed verified by the hydrological observations. Temperature and salinity profiles from the center of the Rhodes Gyre (figure 4) indicate that by the end of January convection has already penetrated at intermediate layers. The mixed layer temperature is as low as 14.15°C, much lower than the typical LIW temperature. Although a mixed layer is not clearly defined from that profile, the vertical structure indicates mixing down to more than 500m

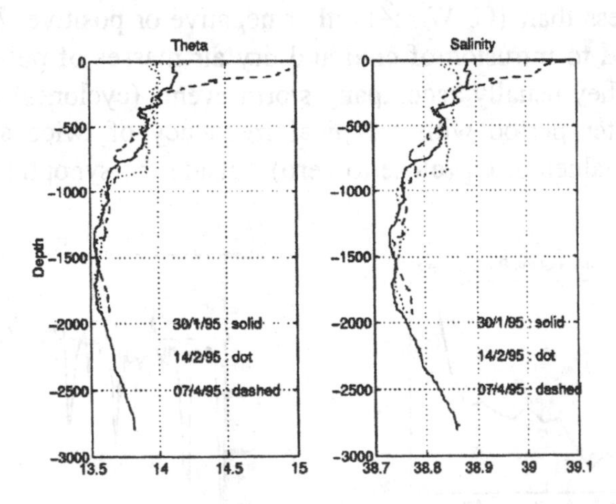


Figure 4 : Temperature and Salinity profiles in the center of the gyre

probably during a previous storm event. In two weeks period (14 February) water mass properties are well mixed down to 900m with temperature less than 14.0$^{o}$C which means formation of water with density much higher that typical LIW. By the beginning of April, the upper 200m are completely re-stratified, in total agreement with the time series of total heat flux that indicates an early heating period.

Assuming one dimensional process, the depth of mixed layer can be computed from the formula [6]:

$$dh/dt = B/(N^2H) \quad (2)$$

where $B=c_w ga Q_t$. This equation was solved numerically using the $N^2$ profile of January 30$^{th}$ and the heat loss time series. Starting the numerical integration from the end of January, we computed a mixed layer depth equal to 730m for February 14$^{th}$. That estimate is very close to the ~900m value indicated by the data. This means that simple mixed layer physics is a good first approximation of intermediate water formation process. The fact that the final mixed layer depth is underestimated could be either due to the omission of the fresh water contribution to buoyancy flux or an indicator that other, more effective, processes such as convective overturning also contribute to the formation of LIW.

## 4. Conclusions

The surface temperature field in the NW Levantine during the LIWEX'95 experiment period is characterized by the presence of the cold-warm dipole of the Rhodes cyclone and the Ierapetra anticyclone. Apart of the periods of extreme meteorological events, heat losses are minimum in the center of the Rhodes gyre due to the permanent low SST values in that area. The time series of air-sea fluxes indicates that the winter of 1994-95 was characterized by a shift of the cooling period by about one month earlier. This is reflected in the time series of SST in the center of the gyre that reach their minimum value as early as the beginning of February instead of March. It is, furthermore, the reason for the early deep mixing (almost 1000m) observed by the end of January and mid February. Finally, simple calculations using characteristic profiles from the center of the gyre and air-sea buoyancy fluxes, indicate that mixed layer dynamics can describe, to a first approximation, deep mixing and water formation in the Rhodes gyre.

**5. References**

1. Ovchinnikov I.M., 1984 : The formation of Intermediate Water in the Mediterranean, Oceanology, 24,168-173.
2. Lascaratos A., R.Williams and E.Tragou, 1993 : A Mixed Layer Study of the formation of Levantine Intermediate Water, JGR Vol.98, No.C8, 739-749
3. Lascratos A and S.Tsandilas, 1998 : Seasonal cycle of a coherent eddy in the Eastern Mediterranean Sea, submitted
4. Castellari S., N. Pinardi and K.D.Leaman, 1998 : A model study of air-sea interaction in the Mediterranean Sea, Jour. of Marine Systems, in press
5. Lascaratos A. and K.Nittis 1998 : "A high resolution 3-D numerical study of intermediate water formation in the Levantine Sea", JGR, in press
6. Turner J.S., 1973 : Buoyancy effects in fluids, Cambridge University Press, 368 pages

# INTERANNUAL VARIABILITY OF THE CIRCULATION AND HYDROGRAPHY IN THE EASTERN MEDITERRANEAN (1986-1995)

ALEXANDER THEOCHARIS and HARILAOS KONTOYIANNIS
*National Centre for Marine Research*
*Aghios Kosmas, Hellinikon 16604, Athens, Greece*

ABSTRACT. The eastern Mediterranean Sea, one of the poorly known sea regions before the '80s, became the object of very intensive research since 1985. The synthesis of the results based on the data sets that were collected in the frame of international (POEM, POEM-BC, EU/MAST/MTP) and national research programs revealed significant information about the circulation, hydrography and their variability in this Basin during the decade 1986-95. The emerged three scales of motion, the basin, sub-basin and mesoscale, interacting within the general circulation pattern, as well as the hydrological regime present interannual to interdecadal variability. This variability plays important role in all upper and lower thermocline processes, as intermediate and deep water mass formation, transformation, spreading, modification of water mass pathways and vertical structure as well as basins' interaction

## 1. Introduction

The eastern Mediterranean has been widely recognized as an ideal basin for studying fundamental oceanic processes of global importance. During the decade 1986-1995, international (POEM, EU/MAST, EU/MTP) and national research programs were carried out in all the sea-regions of this Basin. Therefore, comprehensive data sets were collected in the frame of basin-wide, multiship general surveys and intensive regional process studies. The eastern Mediterranean is a semi-enclosed Sea characterized by a complex coastline and rugged bottom topography. It consists of four major Seas, the Ionian, the Levantine, the Adriatic and the Aegean (Fig. 1). It communicates to the West with the western Mediterranean through the Strait of Sicily (sill depth ~430m), where a two-layer system occurs for the water mass (and salt) exchange: (i) The Modified Atlantic Water (MAW), the relatively less saline water confined in the upper layer, flows eastwards and (ii) the saline Eastern Mediterranean Intermediate Water, the so called Levantine Intermediate Water (LIW), westwards, within the lower layer [1]. Occasionally, some deeper waters are also transported westwards. The Eastern Mediterranean is also connected to the Northeast with the Black Sea through the Strait of Dardanelles that joins the Aegean with the Marmara Sea. At this point it receives a considerable amount of less saline water that mainly affects the hydrology of the



*P. Malanotte-Rizzoli and V.N. Eremeev (eds.),*
*The Eastern Mediterranean as a Laboratory Basin for the Assessment of Contrasting Ecosystems,* 453–464.

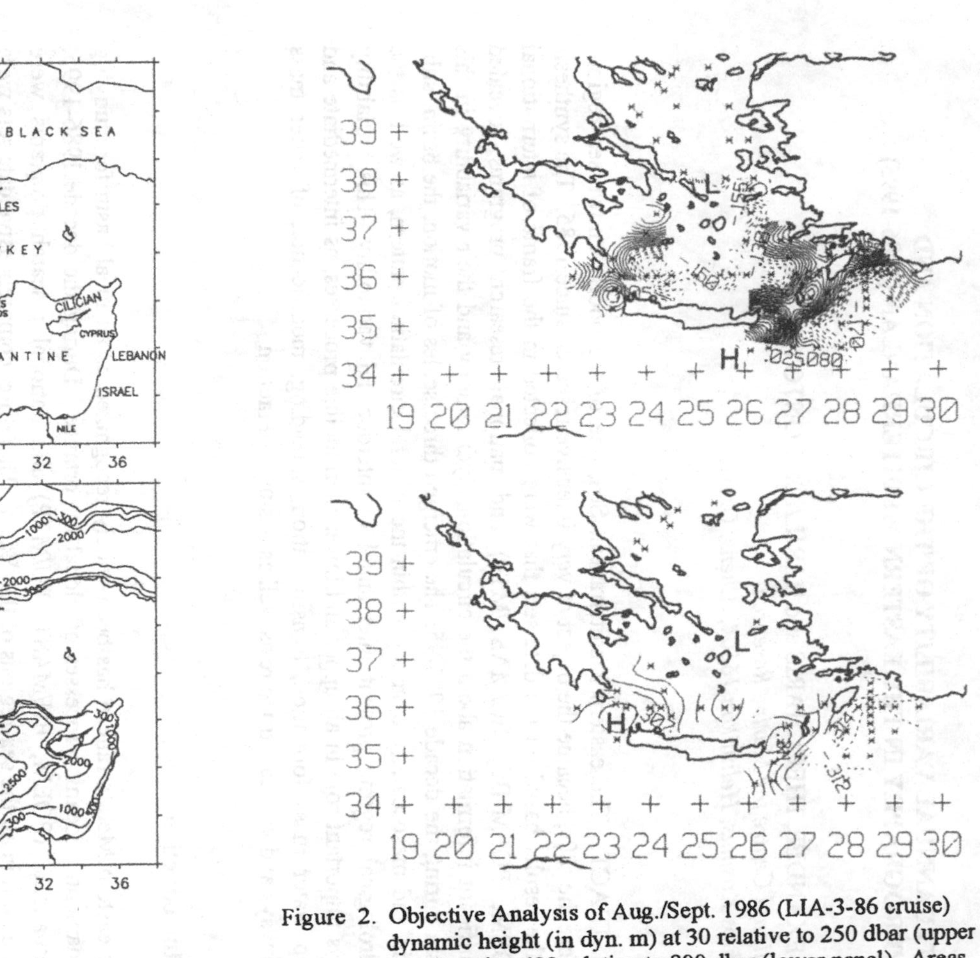


Figure 1. Map of the Eastern Mediterranean Sea (upper panel) Bathymetric chart (lower panel).

Figure 2. Objective Analysis of Aug./Sept. 1986 (LIA-3-86 cruise) dynamic height (in dyn. m) at 30 relative to 250 dbar (upper panel) and at 400 relative to 800 dbar (lower panel). Areas with error greater than 60% are blanked.

Aegean Sea [7, 12]. The Aegean Sea exchanges waters with the Ionian and Levantine Basins through a series of six straits, namely the Cretan Arc Straits, with sill depths varying from 150 to 1000m [13]. Two types of Intermediate Water are formed: the Levantine Intermediate Water (LIW) mainly in the Levantine Basin and the Cretan Intermediate Water (CIW) in the South Aegean. The Eastern Mediterranean Deep Water (EMDW) lies below 1000-1500m down to the bottom in the Ionian and Levantine Basins and has its origin in the Adriatic and the Aegean Seas. In between the above mentioned water masses the so called transitional waters lie in depths ~500-1200m. The pathways and spreading of all water masses are modified by the multiscaled complex and variable in space and time circulation that consists of elements spanning over three predominant spatial scales interacting within the general circulation pattern: (i) the two basin scale thermohaline cells, (ii) the upper thermocline sub-basin gyres driven by wind and thermohaline forcing and (iii) the most energetic embedded mesoscale eddies interconnected and fed by currents and jets [3, 6, 13, 17]. The variability of the circulation structures is manifested in terms of the number, shift and water content of their centers, intensification/relaxation, boundary oscillations, meanders and branching, all of which evolve in different time scales from seasonal to interannual, not yet all of them definitely understood [8]. We present observational evidence of this variability that influences important physical processes of global relevance, as water mass formation, transformation, spreading and mixing, and basins' interactions.

## 2. The sub-basin and mesoscale circulation structures

During the period 1986-1995 sixteen cruises were carried out in the central region of the eastern Mediterranean that comprises the eastern Ionian, the Cretan Passage, the NW Levantine and the south Aegean Seas. A succession of permanent and recurrent cyclones and anticyclones occupies the Mediterranean area outside the Aegean Sea along the Cretan Arc. The deep, strongly barotropic, anticyclonic Pelops gyre and the Cretan Cyclone, confined to the upper thermocline, dominate in the eastern Ionian, while the permanent Rhodes cyclonic gyre in the NW Levantine. The eastern part of the Cretan Passage is occupied by an intense anticyclonic eddy, namely the Ierapetra Anticyclone, that mostly presents very strong surface signal [13]. The Asia Minor Current (AMC) meanders westwards along the northern and western boundary of Rhodes gyre transporting the warm and saline waters of Levantine origin. It bifurcates at the eastern Cretan Arc Straits, thus providing salt in the Aegean Sea. Finally, the Mid-Mediterranean Jet (MMJ) flows eastward along the Cretan Passage carrying MAW. The time scales, seasonal to interannual, of the existing variability are not yet totally resolved. The gyres modify the pathways of the surface and intermediate water masses.

Strong variability appears in the sub-basin scale structures (Figs. 2-8):

(a) Pelops gyre, one of the most important quasi permanent and strong structures, is surface intensified in December 1986 (Fig. 3) and October 1991 [4], while its surface signature is weak in April 1986, April 1987 (Fig. 4) and September 1987 [13].

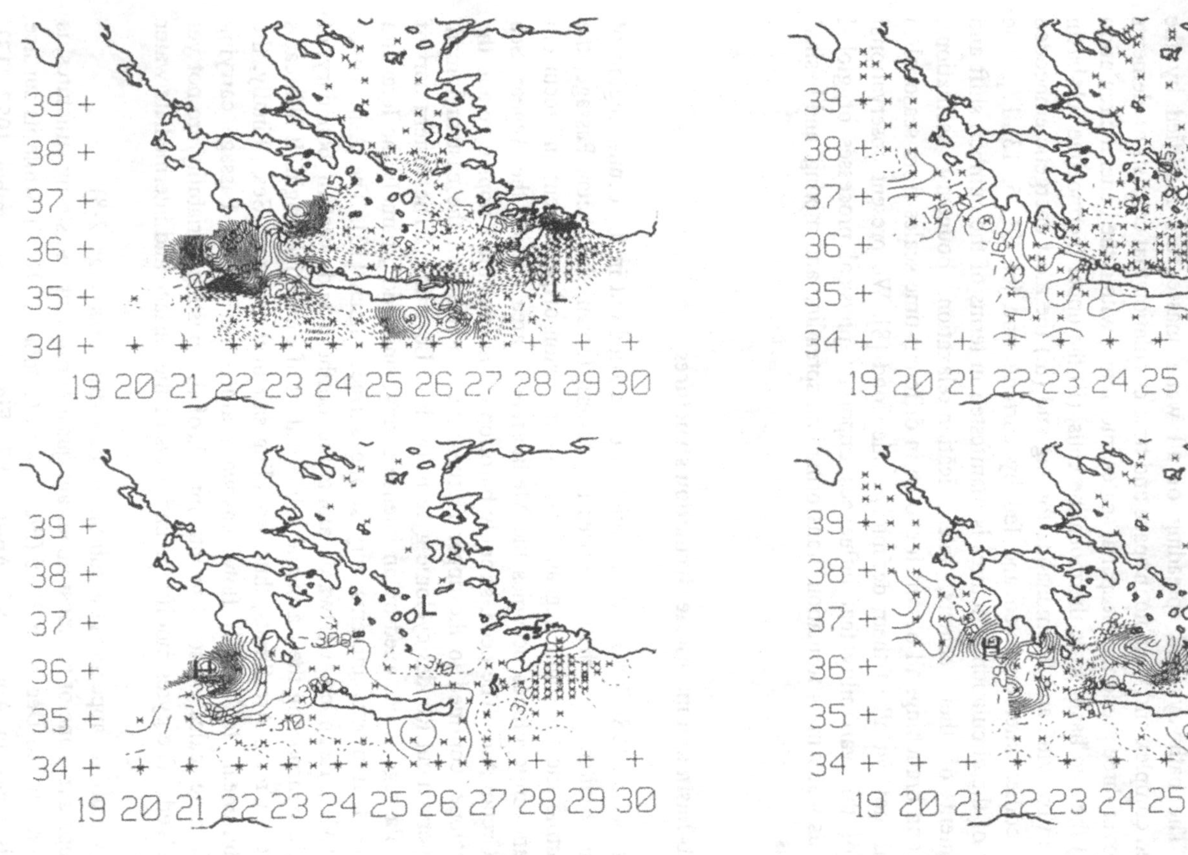


Figure 3. Objective Analysis of Nov./Dec. 1986 (POEM-III-86 cruise) dynamic height (in dyn. m) at 30 relative to 250 dbar (upper panel) and at 400 relative to 800 dbar (lower panel). Areas with error greater than 60% are blanked.

Figure 4. Objective Analysis of Mar./Apr.1987 (POEM-IV-87 cruise) dynamic height (in dyn. m) at 30 relative to 250 dbar (upper panel) and at 400 relative to 800 dbar (lower panel). Areas with error greater than 60% are blanked.

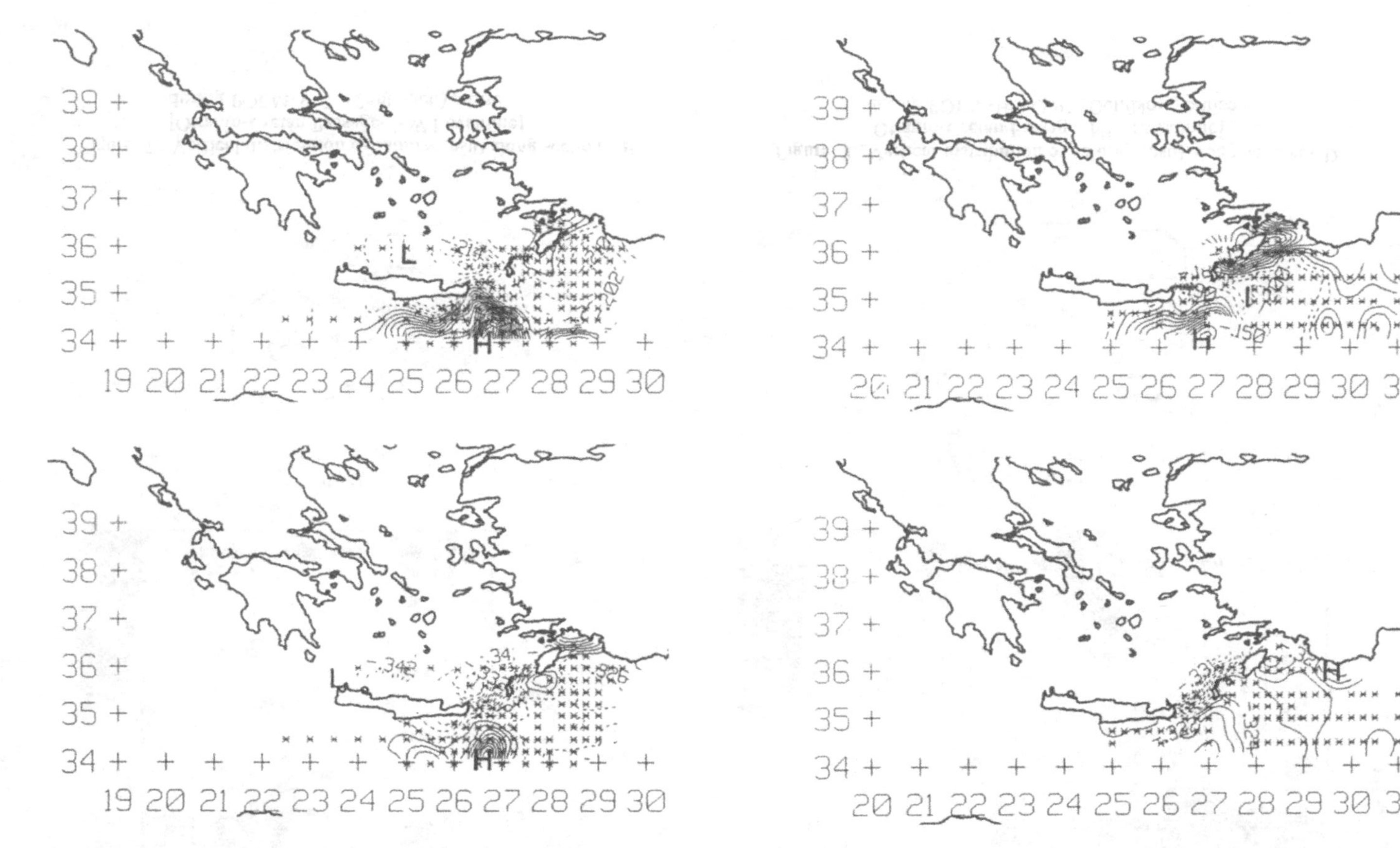


Figure 5. Objective Analysis of Mar./Apr. 1992 (POEM-BCO-92 cruise) dynamic height (in dyn. m) at 30 relative to 250 dbar (upper panel) and at 400 relative to 800 dbar (lower panel). Areas with error greater than 60% are blanked.

Figure 6. Objective Analysis of April 1995 (LIWEX'95 cruise) dynamic height (in dyn. m) at 30 relative to 250 dbar (upper panel) and at 400 relative to 800 dbar (lower panel). Areas with error greater than 60% are blanked.

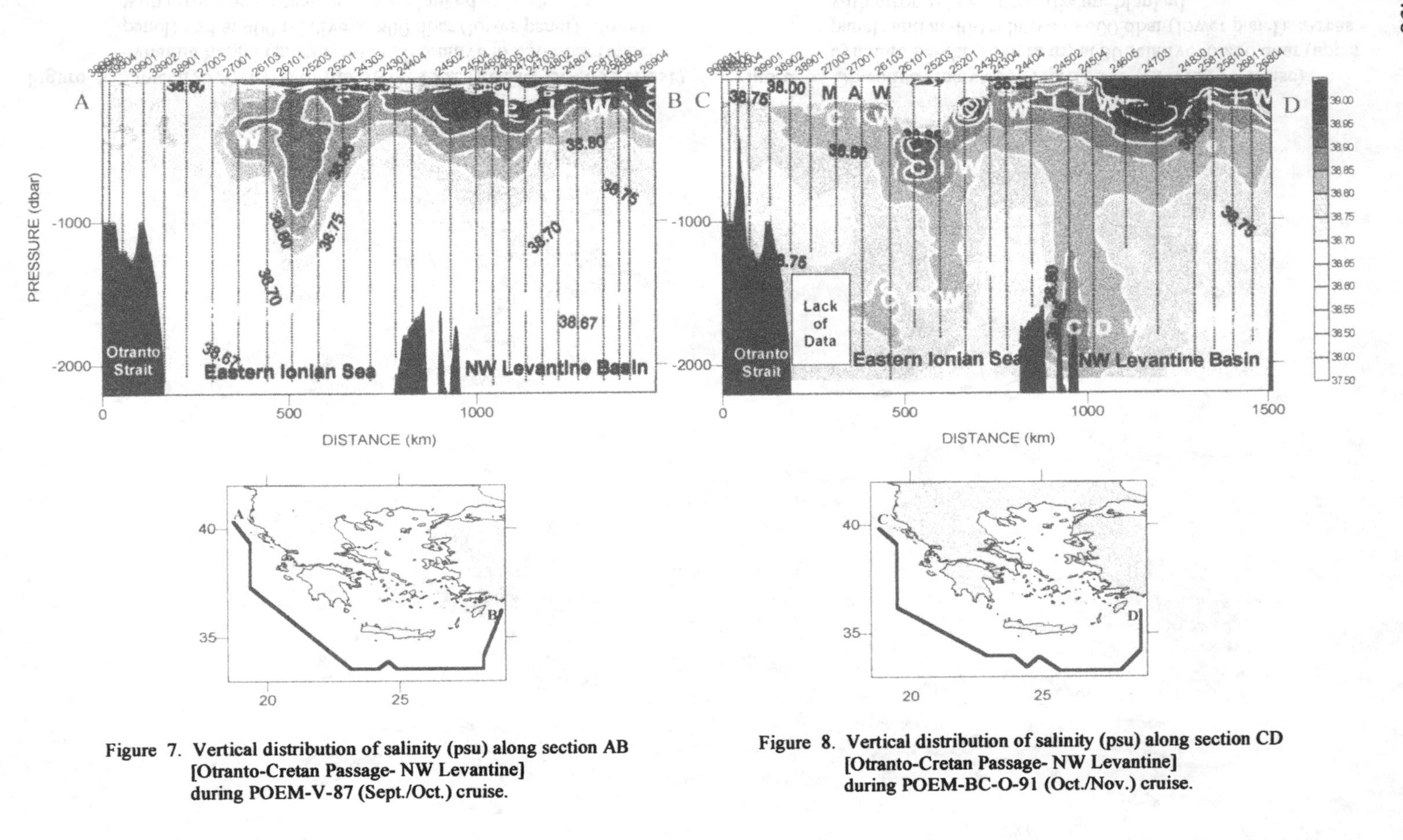


Figure 7. Vertical distribution of salinity (psu) along section AB [Otranto-Cretan Passage- NW Levantine] during POEM-V-87 (Sept./Oct.) cruise.

Figure 8. Vertical distribution of salinity (psu) along section CD [Otranto-Cretan Passage- NW Levantine] during POEM-BC-O-91 (Oct./Nov.) cruise.

However, it is always subsurface intensified and penetrates barotropically down to 800-1500 m (Figs. 7, 8). It has one or two centers that are shifted in space, i.e., in April 1986 both centers are located in the vicinity of the Peloponesian coast [13], while in April 1987 (Fig. 4) there is only one centre. In other periods, regardless the number of centers, the gyre is shifted away from the coasts. In September 1991, Pelops is encircled by a major meander of the Mid-Ionian Jet (MIJ), that flows meridionally along the 19° E [4]. Its size varies from 200 to 300 km in "diameter", even though its shape and orientation present also variability. It is worth mentioning that its water mass content varies in time, i.e., in April 1986 and 1987 traces of MAW are detected at its periphery, in September 1987 MAW cannot intrude in the core of this gyre, but did flow around (Fig. 7), while on the contrary during December 1986 it enters the core and in October 1991 fresh MAW occupied a thick layer within its core (Fig. 8)

(b) the Ierapetra Anticyclone, one of the strongest upper thermocline circulation features presents seasonal to interannual variability regarding its dimensions, intensity, position, number of centers and even appearance [5]: It is absent in March 1986, at least at its known position, and presents an obvious signal in November 1986 with two lobes, a strong surface signature in September 1986 (Fig. 2) and 1987 [13], and it can also be partially detected, due to few data available, in September 1989 very near to Kassos Strait (not shown). Finally, this Anticyclone became very strong and important after 1990. It appears larger and stronger in October 1991 (300 instead of 150-200 km in diameter), in April 1992 and April 1995 (Figs. 5, 6) considerably contributing to the significant change of the circulation pattern in the Levantine Basin in the upper and intermediate layers, e.g., the blocking of the traditional pathway of LIW towards the Ionian [4]. MAW flows eastwards along its periphery intruding also in its core. Depending on its distance from Kassos Strait, it plays different roles in the water mass transport, e.g., in September 1987, when it was located near the southeast tip of Crete, waters exiting from the Cretan Sea are threading around it [13], while in October 1991 low salinity waters, i.e. MAW or Ionian Surface Water (ISW), are entrained from the Cretan Passage by the Anticyclone to enter the Cretan Sea through Kassos Strait [4].

(c) The Cretan Cyclone presents also variability in its shape and dimensions. During autumn 1986 (Fig. 3) and 1987 its surface signal was strong with a diameter of about 300 km, while in 1991 was restricted down to 150 km towards the southwest tip of Crete. It plays significant role in bringing back to the eastern Ionian Sea the MAW or ISW. It also facilitates by this way the MAW intrusion in the Aegean through the western Cretan Arc Straits.

(d) The Rhodes cyclonic gyre always occupies the NW Levantine Basin and considerably preconditions the formation of LIW and/or the Levantine Deep Water (LDW) in this region [2, 10]. Since 1992 its core is affected by deep dense waters of Aegean origin [15]. It has a rather east-west orientation and can be characterized as a multilobe gyre with a 300-400 km diameter. Its westernmost extension can reach the west tip of Crete, as in April 1986 [13]. It is surrounded by anticyclonic gyres such as the Anaximander, Ierapetra, Marsa Matruh and some mesoscale eddies that develop in the vicinity of the eastern Cretan Arc Straits (Rhodes, Karpathos, Kassos). It presents boundary oscillations. Its southern extensions cover crescent-shape areas surrounding several anticyclonic gyres and eddies to its south, as the Ierapetra and Mersa Matruh. In

April 1995, it is connected with the west Cyprus cyclonic gyre forming a large cyclonic flow region with two lobes expanded over the entire north Levantine Basin.

(e) The Asia Minor Current meanders westwards along the southern coasts of Asia Minor carrying the warm and most saline waters from the East. Branches of AMC intrude into the SE Aegean through the eastern Cretan Arc Straits. This branching occurs almost always at Rhodes Strait. A strong branch at Kassos Strait has identified only once, in April 1986, while at Karpathos it seems to occur more often [13, 14, 15]. This is the mechanism of providing salt in the Aegean Sea, enhancing the dense water formation processes in the Cretan Sea and the surrounding shelf areas.

(f) The Mid-Mediterranean Jet flows eastwards carrying the MAW to the Levantine Basin. It presents spatial variability as it was clearly detected along the Cretan Passage only in April 1986 [8, 13]. However, some branches of it as well as its continuation in the Levantine Basin, the so called Central Levantine Basin Current (CLBC) [4, 6], were identified during other periods (1985, 1987, 1991, 1995). This is an evidence that it has been shifted southwards in the Cretan Passage.

In the south Aegean a complex pattern exists consisting of successive cyclonic and anticyclonic eddies and meandering currents of permanent, recurrent or transient nature. It is worth mentioning that the circulation pattern in the south Aegean presents a seasonal evolution in the period 1986-87 [13], while during 1994-95 it became quasi stable [15]. The most persistent feature since 1986 is a large cyclonic gyre that is centered at Mirtoan Sea and covers also the western Cretan Sea and the adjacent Kiklades Plateau. It exhibits boundary oscillation and shift of centre. Its southward extension occasionally reaches the western tip of Crete and incorporates the mesoscale cyclonic eddy that develops at the Strait area. Its diameter can reach 200km. The branches of the AMC meander in the south Aegean interior westwards and/or deflected northwards in the eastern Aegean presenting significant shifts. On the other hand the main current flows eastwards among the Cretan eddies (October 1987). In the recent years a persistent dipole covers the central and eastern Cretan Sea showing also boundary oscillation, but not seasonal variability during 1994-95 [15].

## 3. The thermohaline circulation

The basin scale thermohaline circulation is composed by two cells: the internal closed deep vertical cell, the eastern Mediterranean "conveyor belt" comprising the Ionian and Levantine Basins and the external open cell related to the upper and intermediate water mass exchange between the Eastern and Western Mediterranean Basins and the Atlantic Ocean.

The first cell since the existence of observations was driven by the deep water produced in the south Adriatic. The Eastern Mediterranean Deep Water (EMDW) of Adriatic origin occupied the deep and bottom layers of the entire eastern Mediterranean (Fig. 7). Since the beginning of this century the Aegean Sea was considered to be a sporadic source of dense waters that could be traced outside the Cretan Arc Straits at intermediate and/or deeper layers, down to 800-1200 m [14].

The second cell concerns (i) the inflow of MAW from Sicily Strait that spreads throughout the entire Basin within the surface layer and (ii) the formation of the LIW (200-600 dbar) that flows towards the Western Basin and finally to the Atlantic Ocean. Another LIW branch is directed northward in the eastern Ionian (Fig. 7) following the west coasts of Greece [3, 16] and finally enters the Adriatic Sea, where it participates in the formation of the Adriatic Deep Water.

The significant changes of the south Aegean water mass characteristics during the last 8-9 years [14] have considerably influenced both thermohaline cells (Figs. 7, 8). The origin and pathways of all the water masses at intermediate and deep layers were changed. The traditional westward pathway of LIW was interrupted in 1991 by the new type of Cretan Intermediate Water (CIW) that exits from the western Cretan Arc Straits, circulates anticyclonically around Pelops gyre (Fig. 8) and spreads into the eastern Ionian [4]. Moreover, the Cretan Sea in the south Aegean was filled with the new very dense water mass ($\sigma_\theta$~29.2-29.4), namely the Cretan Dense Water (CDW) (Fig. 10), thus becoming the source of the relatively warmer and more saline Eastern Mediterranean Deep and Bottom Water [9, 14]. Its massive outflow was first observed through the Cretan Arc Straits in 1991 when it occupied the deep layers of the adjacent Mediterranean regions down to 2-2.5 km (Fig. 8) [11]. Then it reached the bottom of the Basin in 1995 pushing upwards several hundreds meters the EMDW of Adriatic origin [9]. The latter is found shallow enough in the vicinity of the Cretan Arc Straits through which they intrude into the Aegean, compensating the deep CDW outflow [15]. Therefore, the water mass thermohaline properties and structure of the south Aegean have undergone significant shift. The weak stratification of the '80s that favoured the open ocean deep convection, e.g. 1987, 1992, has been replaced by strong density gradients that confine the mixing process in the upper 250m (Figs. 9, 10). This climatic shift in the Aegean is reflected in the deep outflow through the Cretan Arc Straits during the crucial period 1986-1994. Moreover, the variability at the Straits concerns the evolution of the water masses in the entire water column. the deep transport rates (0.1-1.1 Sv) and the behaviour of each of the straits [14]. It is worth noting that the reanalysis of the historical data revealed two more, similar but weaker, events occurred within the last 50 years in the Aegean, thus indicating an interdecadal variability.

## 4. Summary

Several national and international research programs that carried out within the period 1986-1995 in the Eastern Mediterranean have contributed significantly to our better knowledge and understanding of this Basin's complex system. We display the interannual up to interdecadal variability of the circulation elements and the hydrography of the Basin's central region. The synthesis of the results obtained during this period revealed and confirmed the multiscaled character of the circulation that consists of three scales interacting in the general pattern. The sub-basin and the mesoscale features exhibit considerable spatial variability manifested as shift of gyres, eddies and currents, number of vortices' centers, boundary oscillations, branching along with modifications of the water mass content. These structural elements modify the

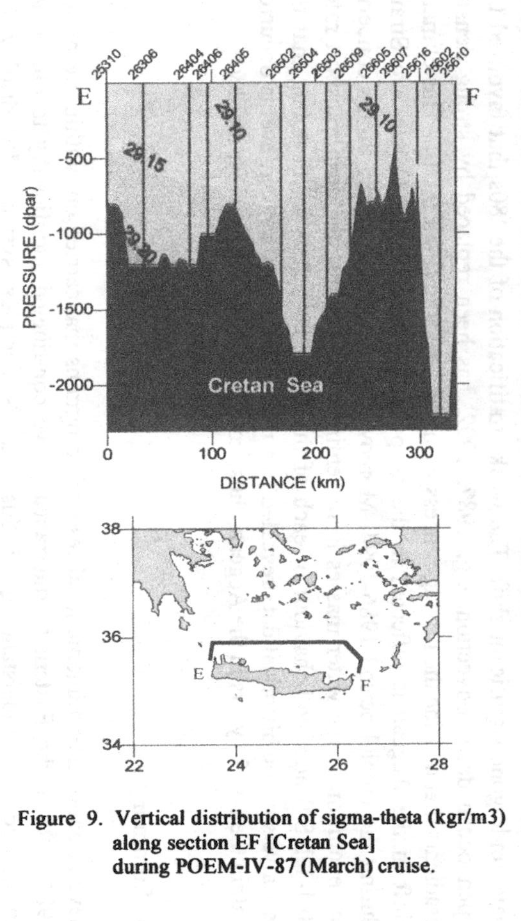


Figure 9. Vertical distribution of sigma-theta (kgr/m3) along section EF [Cretan Sea] during POEM-IV-87 (March) cruise.

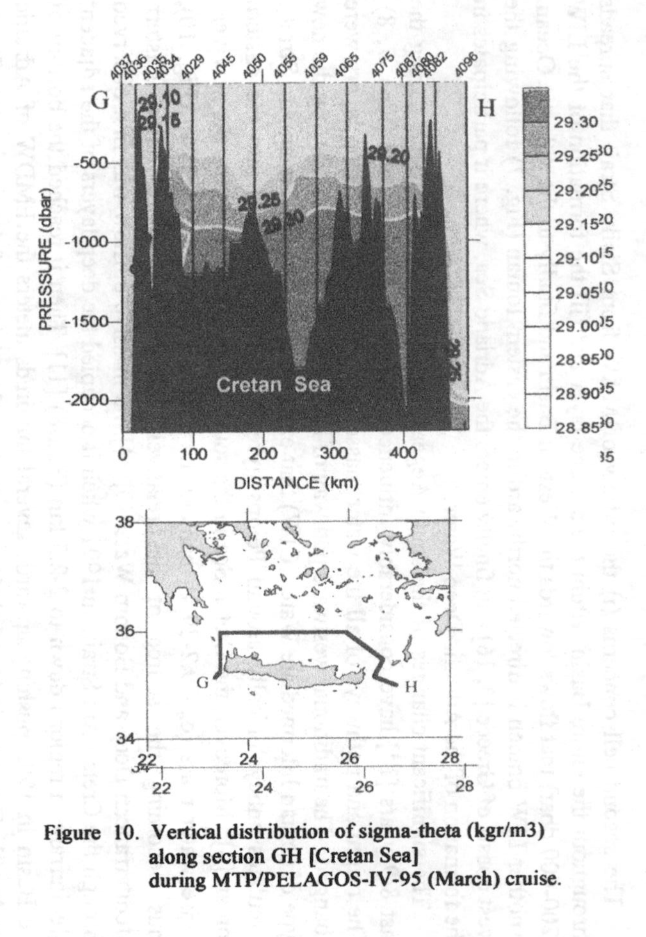


Figure 10. Vertical distribution of sigma-theta (kgr/m3) along section GH [Cretan Sea] during MTP/PELAGOS-IV-95 (March) cruise.

water mass pathways and the flows through the straits. Furthermore, the basin scale thermohaline cells have experienced a unique "climatic shift" since the beginning of observations in the Mediterranean. All three scales of motion influence important processes, as water mass formation, transformation, spreading and basins' interactions. The recent changes are of global relevance due to the important role of the Mediterranean in the World Ocean. Their further implications in the adjacent seas and in the biogeochemical components of the system are mostly unknown and depend on their duration and intensity. Therefore, research and monitoring are needed for the entire eastern Mediterranean and adjacent regions in order to follow the evolution and study the phases that probably lead to a new equilibrium of the system.

## Acknowledgments

The used data sets have been obtained in the frame of (a) the National program of the National Centre for Marine Research (NCMR) "Open Sea Oceanography" and the international POEM and POEM-BC programs and (b) the European Union 's Mediterranean Targeted Project (MTP) PELAGOS under the contract MAS2-CT93-0059. We acknowledge the Hellenic Secretariat of Research and Technology and the European Union 's Marine Science and Technology (MAST) Program and the NCMR for their financial support respectively. We thank the Intergovernmental Oceanographic Commission (IOC) of UNESCO under the auspices of which the POEM program was carried out. Sincere thanks are also due to the officers and the crew of R/V AEGAEO.

## References

1. Astraldi M., G. P. Gasparini, S. Sparnocchia, M. Moretti and E. Sansone (1996). The characteristics of the water masses and the water transport in the Sicily Strait at long time scales. In: *Dynamics of Mediterranean Straits and Channels*, F. Briand ed., 95-116.
2. Malanotte-Rizolli P., A. R. Robinson, W. Roether, B. Manca, , A. Bergamasco, S. Brenner, G. Civitarese, D. Georgopoulos, P. J. Haley, S. Kioroglou, H. Kontoyannis, N. Kress, M. A. Latif, W. G. Leslie, E. Ozsoy, M. Ribera D' Alcala, I. Salihoglu, E. Sansone, A. Theocharis (1996). Experiment in Eastern Mediterranean Probes Origin of Deep Water Masses. *EOS*, 77,32, 305,311.
3. Malanotte-Rizolli P., B. Manca, M. Ribera D' Alcala, A. Theocharis, A. Bergamasco, D. Bregant, G. Boudillon, G. Civitarese, D. Georgopoulos, A. Michelato, E. Sansone, P. Scarazzato, E. Souvermezoglou (1997). A Synthesis of the Ionian Sea Hydrography, Circulation and Water Mass Pathways during POEM Phase-I. *Prog. in Oceanog.*, 39, 153-204.
4. Malanotte-Rizolli P., B. Manca, M. Ribera D' Alcala, A. Theocharis, S. Brenner, G. Boudillon, E. Ozsoy (1998). The eastern Mediterranean in the '80s and in the '90s: The transition in the intermediate and deep circulation, in preparation..
5. Laskaratos A. and S. Tsandilas (1998). Seasonal cycle of a coherent eddy in the Eastern Mediterranean Sea, submitted.
6. Ozsoy E., A. Hecht, U. Unluata, S. Brenner, H. I. Sur, J. Bishop, M. A. Latif, Z. Rosentraub and T. Oguz (1993). A synthesis of the Levantine Basin circulation and hydrography, 1985-1990. *Deep Sea Research II*, 40, 6, 1075-1119.
7. Poulos S. E., P. G. Drakopoulos, M. B. Collins (1997). Seasonal variability in sea surface oceanographic conditions in the Aegean Sea (Eastern Mediterranean): an overview. *Journal of Marine Systems*, 13, 225-244.

8. Robinson A.R., M. Colnaraghi, W.G. Leslie, A. Artegianni, A. Hecht, E. Lazzoni, A. Michelato, E. Sansone, A. Theocharis, U. Unluata (1991). The Eastern Mediterranean general circulation: features, structure and variability. *Dynamics of Atmospheres and Oceans*, 15, 215-240.
9. Roether W, B.Manca, B.Klein, D.Bregant, D.Georgopoulos, V.Beitzel, V.Kovacecic and A.Lucetta (1996). Recent changes in the Eastern Mediterranean Deep Waters, *Science*, 271, 333-335
10. Sur H. I., E. Ozsoy, U. Unluata (1992). Simultaneous deep and intermediate convection in the Northern Levantine Sea, winter 1992. *Oceanologica Acta*, 16, 33-43.
11. Theocharis A., D.Georgopoulos, P.Karagevrekis, A.Iona, L.Perivoliotis and N.Charalambidis (1992). Aegean influence in the deep layers of the eastern Ionian sea. *Rapp. Comm. int. Mer Medit*, 33, 235
12. Theocharis A.and D. Georgopoulos (1993). Dense water formation over the Samothraki and Limnos Plateaux in the north Aegean Sea (Eastern Mediterranean Sea). Continental Shelf Research 13, 8/9, 919-939.
13. Theocharis A., D. Georgopoulos, A. Lascaratos, K. Nittis (1993). Water masses and circulation in the central region of the Eastern Mediterranean (E. Ionian, S. Aegean and NW. Levantine). *Deep Sea Research*, II, 40, (6), 1121-1142.
14. Theocharis A, E. Papageorgiou, H. Kontoyannis, K. Nittis and E. Balopoulos (1998). The Evolution of the Aegean Water's Influence in the Deep Thermohaline Circulation of the Eastern Mediterranean (1986-1995). *Rapp. Comm. int. Mer Medit.* 35 (1), 200-201.
15. Theocharis A., E. Balopoulos, S. Kioroglou, H. Kontoyannis and A. Iona (1998). A synthesis of the circulation and hydrography of the South Aegean Sea and the Straits of the Cretan Arc (March 1994-January 1995). *Prog. in Oceanog.* (accepted).
16. Theodorou A. J., D. Georgopoulos and A. Theocharis (1990). On the contribution of Levantine Intermediate Water to the hydrographic structure of the northeastern Ionian Sea. *Thalassographica* 13, suppl. 2, 57-66.
17. The POEM group (1992). General circulation of the Eastern Mediterranean. *Earth Science Reviews*, 32, 285-309.

# STRUCTURES AND CHARACTERISTICS OF NEWLY FORMED WATER MASSES IN THE NW LEVANTINE DURING 1986, 1992, 1995

HARILAOS KONTOYIANNIS, ALEXANDER THEOCHARIS,
KOSTAS NITTIS
*National Center for Marine Research*
*Aghios Kosmas, Hellinikon, 16604 Athens, Greece*

ABSTRACT. In early spring of 1986, 1992 and 1995, newly formed deep water masses were observed in the cyclonic Rhodes Gyre of the Eastern Mediterranean at depths reaching ~1000 m in 1995 and exceeding 1000 m in 1986 and 2000 m in 1992. Levantine Intermediate Water (LIW) formation was observed at localized source areas south of the East Cretan Straits at the periphery or near the center of the Ierapetra Anticyclone. Adittional massive LIW formation was observed in 1992 in anticyclonic circulation structures north of Cyprus. The lateral scales of the newly formed water masses in cyclonic structures appear roughly proportional to the penetration depth of the convection, so that a large lateral scale indicating a massive production would be associated with a deep rather than an intermediate formation.

## 1. Introduction

The Levantine Sea of the Eastern Mediterannean (Figure 1) has been historically considered as the source region of the Levantine Intermediate Water (LIW) which is the main contributor to the saline Mediterannean tongue spread into the Atlantic. LIW formation has been suggested to occur in the Northern Levantine, in the broad area between Rhodes and Cyprus [1], [2]. Nielsen [3] first noted the region near Rhodes, while Moskalenko and Ovchinikov [4] have indicated that the area between Turkey and Cyprus contributes large amounts of LIW. Morkos [5], on the other hand, concluded that there are possibly three different source regions: a) around Rhodes, b) to the north of Cyprus, and c) a secondary one in the southern Levantine. Observational evidence after 1985 has confirmed the earlier observations on LIW formation mentioned above, and has also revealed that deep water formation has occured in the cyclonic Rhodes Gyre to the southeast of Rhodes [6], [7].

In April of 1986, 1992, and 1995, three CTD surveys were conducted by the R/V Aegaeo in the south Aegean and northwestern Levantine in the framework of multinational collaborative programs [8]. We examine these three hydrographic datasets with focus on the newly formed water masses. Specifically, we seek to identify nearly-homogenized newly-formed water blobs rich in salt and oxygen indicating recent



*P. Malanotte-Rizzoli and V.N. Eremeev (eds.),*
*The Eastern Mediterranean as a Laboratory Basin for the Assessment of Contrasting Ecosystems,* 465–473.

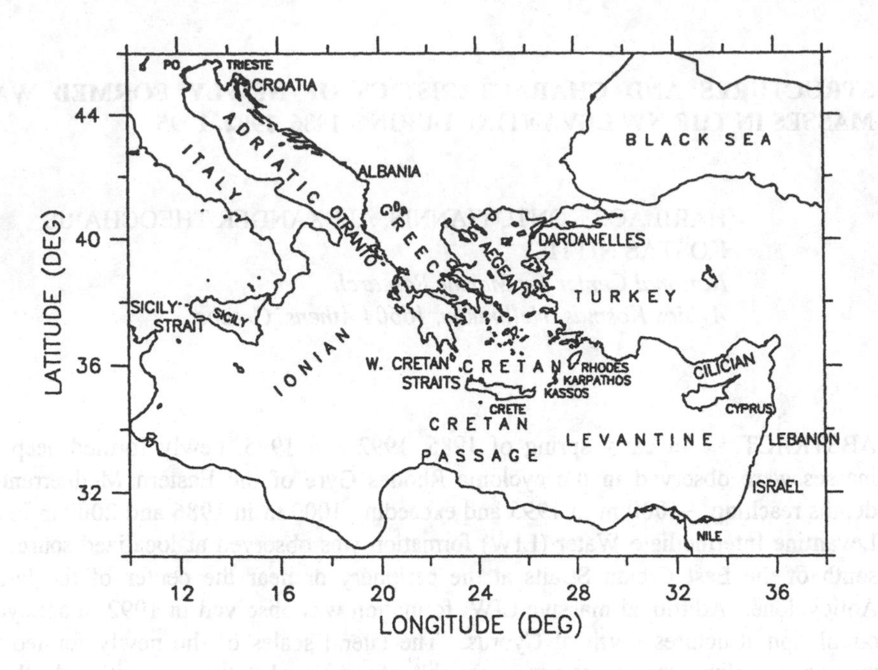


Figure 1. The Eastern Mediterannean.

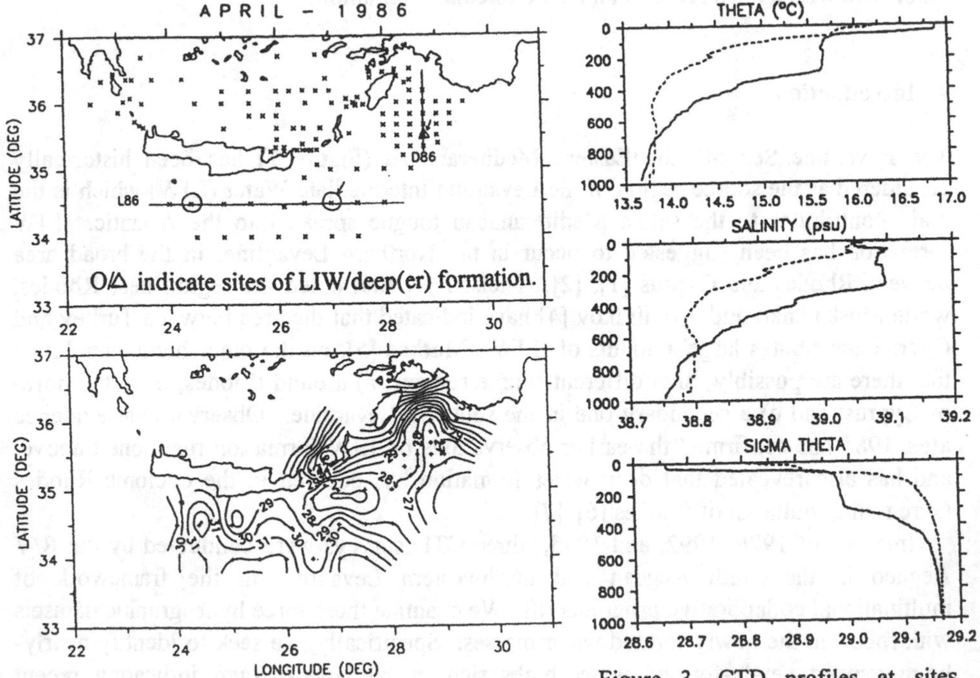


Figure 2. Stations grid (upper) and dynamic height field (dyn cm) at 50 m wrt 500 m (lower) during April 1886.

Figure 3. CTD profiles at sites shown with arrows in fig. 2. Solid/dashed for O/Δ.

formation, and we report their location with respect to the local circulation features, their T/S characteristics and their (vertical/horizontal) scales.

## 2. Formation Sites

Figures 2, 4, 6 show the station grids, the local circulation features, and the formation sites where we observed homogenizations or very weak vertical gradients in the CTD profiles. The circles refer to homogenizations in the upper 300 m and the triangles refer to homogenizations to deeper layers down to ~1000 m in 1986 and 1995 and exceeding 2000 m in 1992. It appears that during 1986, 1992, and 1995, the northern part of the cyclonic Rhodes Gyre, shown to the southeast of the Rhodes Island, favors mostly water formation at depths deeper than the intermediate (300-400 m) depths, whereas the formation of LIW occurs in the area south of the East Cretan Straits. In 1986 cores of LIW were found in anticyclonic meanders of the Mid Mediterannean Jet, while in 1992 and 1995 the newly formed LIW blobs were found in the Ierapetra Anticyclone. We should mention that in 1995 the LIW formation occurs at the periphery of the Ierapetra Anticyclone in a region of cyclonic local circulation due to meandering of the Ierapetra Anticyclone. In this year, the area south of the East Cretan Straits is influenced by higher salinities due to outflow of Cretan Water in the upper 200 m and favors the formation of intermediate saline water masses [9].

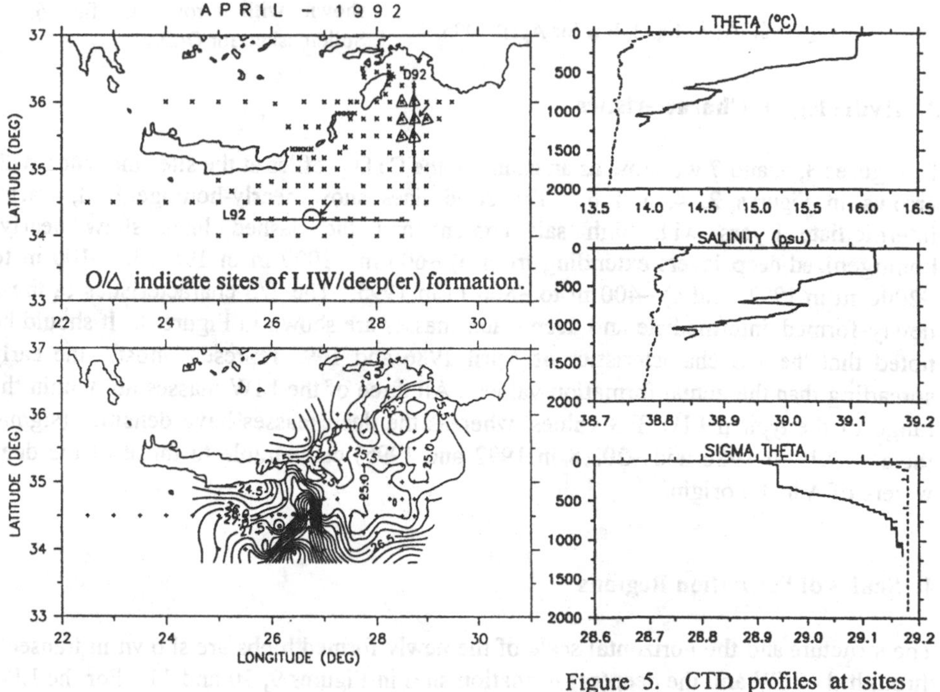


Figure 4. As in fig. 2, but for April 1992.

Figure 5. CTD profiles at sites shown with arrows in fig. 4. Solid/dashed for O/Δ.

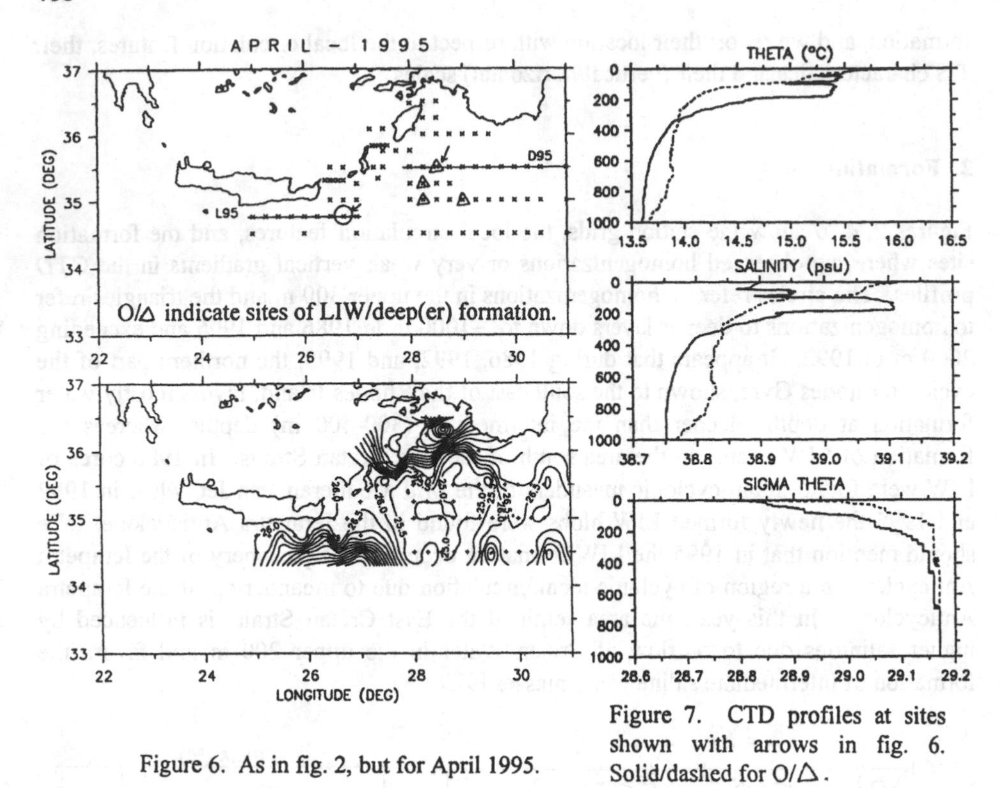


Figure 6. As in fig. 2, but for April 1995.

Figure 7. CTD profiles at sites shown with arrows in fig. 6. Solid/dashed for O/Δ.

## 3. Hydrological Characteristics

In Figures 3, 5, and 7 we show as an example the CTD profiles at the sites indicated with arrows in Figures 2, 4, and 6. The solid lines show nearly-homogenized, warm, intermediate layers with high salt content and the dashed lines show nearly-homogenized deep layers extending from a) ~600 m ~1000 m in 1986, b) ~100 m to ~2000 m in 1992, and c) ~400 m to ~800 m in 1995. The T/S characteristics of these newly-formed intermediate and deep water masses are shown in Figure 8. It should be noted that the T/S characteristics of April 1986 and 1995 represent mostly the early spreading than the actual formation values. All three of the LIW masses are within the range of the typical LIW T/S values, whereas the deep masses have densities (sigma-theta ~29.16 in 1986 and ~29.18 in 1992 and 1995) comparable to those of the deep waters of Adriatic origin.

## 4. Scales of Formation Regions

The structure and the horizontal scale of the newly formed blobs are shown in transects through the LIW and the deep(er) formation sites in Figures 9, 10 and 11. For the LIW water formation we show the structure of the salinity field along the transects L86, L92,

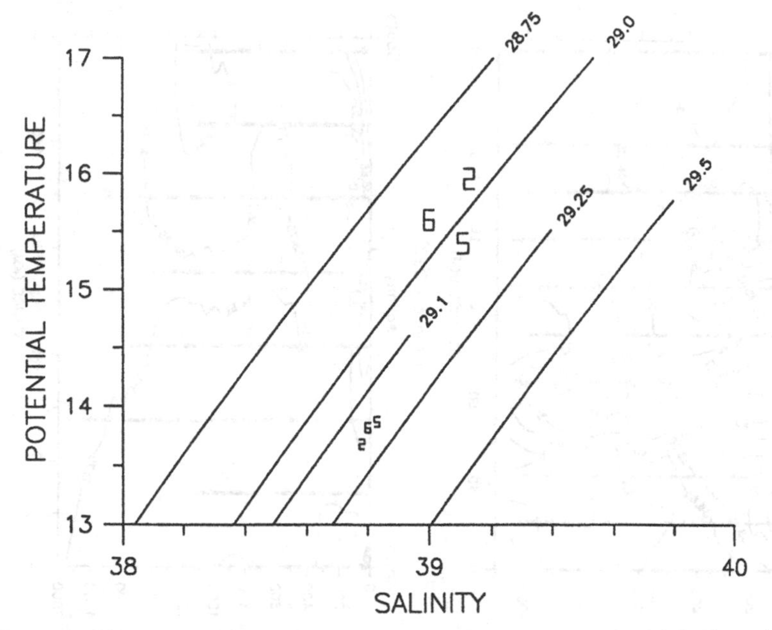


Figure 8. T/S characteristics of newly formed water masses. Large/small numbers for LIW/deep(er) characteristics; 6, 2, 5 for 1986, 1992, 1995 respectively.

and L95 whereas the corresponding structure for the deep water formation is shown along transects D86, D92, and D95 (transect locations in Figures 2,4,6). For the LIW and deep formation in 1995 we also show the CTD dissolved oxygen distributions along L95 and D95. The dissolved oxygen distributions through the deep formation sites in 1986, and 1992 are based only on bottle data. These distributions, presented in [10] and [11], support our water formation observations based on salinity distributions. The local maxima in salinity in the upper 300 m indicate the newly formed pools. The horizontal dimension of the LIW blobs ranges from ~80 km in 1986, when the homogenization reached the depth of 300 m, to ~30-40 km in 1995, when the corresponding vertical scale was ~200 m. Deep maxima in salinity and oxygen are identified in 1986 [10] and 1995 in the northern part of the Rhodes Gyre indicating a possible recent deep formation. The vertical/horizontal scale of these blobls is about 400 m/60 km. In 1992, however, when the hydrographic cruise took place during the actual deep formation, the entire dome of the Rhodes Gyre with a lateral scale of ~150 km is homogenized down to depths exceeding 2000 m. During the same year, a massive production of LIW in the region to the east of the Rhodes Gyre and north of Cyprus has been reported by Sur et al. [7]. The newly-formed saline water mass extended down to ~300-400 m and had a latitudinal extent of ~400 km stretching through neighbouring anticyclonic circulation structures. The existing evidence indicates that in most of the cases the LIW newly formed pools are found near or within anticyclonic circulation structures and are rather localized with smaller lateral scales, whereas the cyclonic Rhodes Gyre favors deeper convection. The lateral scales of the newly formed water masses in cyclonic structures appear nearly proportional to the penetration depth of the convection.

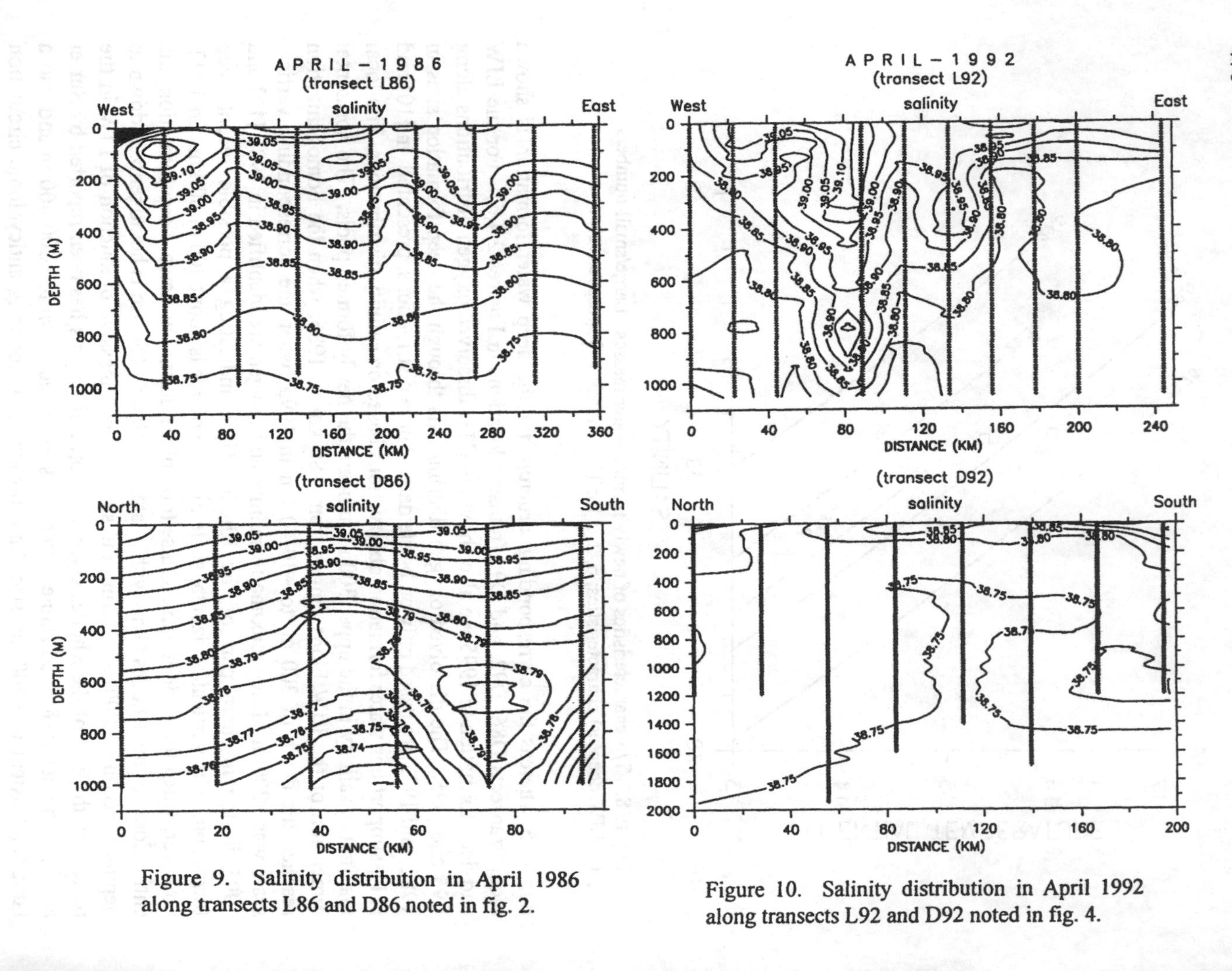


Figure 9. Salinity distribution in April 1986 along transects L86 and D86 noted in fig. 2.

Figure 10. Salinity distribution in April 1992 along transects L92 and D92 noted in fig. 4.

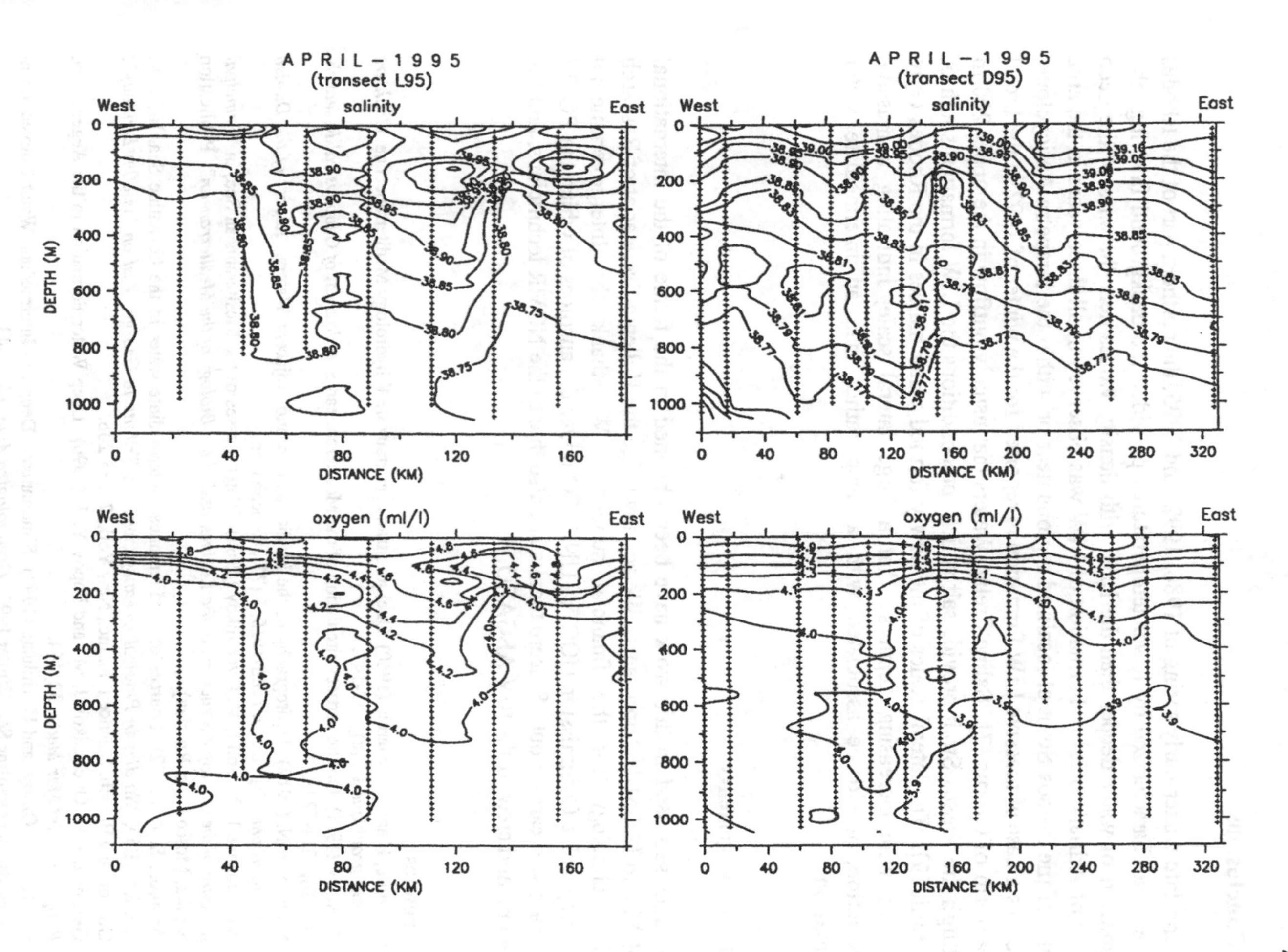


Figure 11. Salinity and oxygen distributions in April 1995, transects L95 and D95 noted in fig. 6.

## 5. Conclusions

During late winter-early spring of 1986, 1992, and 1995, the northern part of the Rhodes Gyre area characterized by lower temperatures provided the background favoring the formation of water deeper than the LIW with density values near the ones of the deep water of Adriatic origin. Formation of LIW was observed in all three years in the area south of the Kassos Strait at localized regions near or within the Ierapetra Anticyclone, whereas a case of massive LIW formation is reported for the winter of 1992 in the broad area north of Cyprus [7]. Saline Cretan Waters that usually outflow in the upper 200 m through the Kassos Strait provide salt which preconditions the LIW formation south of the Stait [9]. The lateral scales of the newly formed water masses in the Rhodes Gyre increase with increasing depth so that a large lateral scale, indicating a massive production, would be associated with a deep rather than an intermediate water formation.

## Acknowledgements

The data sets used in this work have been obtained in the frame of the international POEM and POEM-BC programs. We acknowledge the Hellenic Secretariat of Research and Technology for the financial support. We thank the Intergovernmental Oceanographic Commission (IOC) of UNESCO under the auspices of which the POEM program was carried out. Sincere thanks are also due to the NCMR technicians, and the officers and crew of the R/V AEGAEO.

## References

1. Lacombe, H and P. Tchernia (1960) Quelques traits generaux de l' hydrology Mediterraneenne. *Cahhiers Oceanographique* **12 (8)**, 527-547.
2. Wust, G. (1961) On the vertical circulation of the Mediterannean Sea. *Journal of Geophysical Research,* **66 (10),** 3261-3271.
3. Nielsen, J. N. (1912) Hydrography of the Mediterannean and Adjacent Waters. *Report of the Danish Oceanographic Expedition 1908-1910,* **1 (77),** Copenhagen.
4. Moskalenko L. V. and I. M. Ovchinikov (1965) Water masses of the Mesiterannean Sea, in "*Principal Account of the Geologic Structure, of the Regime, and of the Biology of the Mediterranean*". Publication Nauka, Moscow, (in Russian).
5. Morcos, S. A. (1972) Sources of Mediterannean intermediate water in the Levantine Sea. In A. L. Gordon (Ed.), "*Studies in Physical Oceanography. A Tribute to George Wust on his 80th Birthday*", Gordon and Breach, Science Pub. Inc., New York, **2**, 185-206.
6. Gertman, J. E., Ovchinikov J. M., and Popov Y. J. (1990) Deep Water Formation in the Aegean Sea. *Rapp Comm Int Mer Medit,* **32,** 164.
7. Sur, H. I., E. Ozsoy, and U. Unluata (1992) Simultaneous Deep and Intermediate Water Convection in the Northern Levantine Sea, Winter 1992. *Oceanologica Acta,* **16,** 33-43.
8. The POEM group (1992) General Circulation of the Eastern Mediterannean. *Earth Science Reviews,* **32,** 285-309.
9. Theocharis, A. and H. Kontoyiannis (1995*) Levantine Intermediate Water Experiment 1995, R/V AEGAEO.* Survey Technical Report, National Centre for Marine Research, Ag. Kosmas, Hellinikon, 16604 Athens, Greece.

10. Theocharis, A and D. Georgopoulos (1989) LIW Distribution and Vertical Structure in the Northwest Levantine,in *"POEM Scientific Report #3"*, Harvard Universtity, Cambridge, Massachusetts.
11. Souvermezoglou, E and E. Krasakopoulou (1999) The effects of physical processes on the distribution of nutrients and oxygen in the NW Leventine Sea, *This volume*.

# EUTROPHIC SIGNALS IN THE CHEMISTRY OF THE BOTTOM LAYER OF THE NORTHERN ADRIATIC SEA

E. SOUVERMEZOGLOU and E. KRASAKOPOULOU
*National Centre for Marine Research*
*Aghios Kosmas, Ellinikon, 16604, GREECE*

## 1. Introduction

The Northern Adriatic Sea is the largest semi enclosed continental shelf area in the Mediterranean Sea. It receives approximately one third of the Mediterranean runoff and is one of the most productive regions in the Mediterranean Sea [1]. It plays also a vital role in the Mediterranean circulation by producing, during the winter, the deep water for the Eastern Mediterranean, the North Adriatic Bottom Water (NAdBW).
Important information on the eutrophication of the northern Adriatic can be extracted by monitoring the distribution of oxygen, nutrient and related parameters in its bottom layer. In the framework of the ELNA (Eutrophic Limits of the Northern Adriatic Sea) program six oceanographic cruises were performed from May 1993 to November 1994 in the northern Adriatic Sea. The range of nutrients concentration and of related parameters of the layer a few meters above the bottom of the shallow northern Adriatic Sea are used in order to define the areas with: near anoxic conditions; prevailing bottom photosynthesis; nutrient regeneration; ratio of denitrification [2].
The Jabuka Pit, a continental shelf depression, extends laterally across the Adriatic from Pescara to Sibenik and acts as a collection basin for the NAdBW. Historical data show that the Jabuka Pit bottom water is not annually renewed when the volume and/or the density of the newly formed NAdBW is insufficient to displace the resident bottom water [3]. However it is renewed at least every second year. During the ELNA cruises the deep waters of Jabuka Pit remained practically isolated after the strong flushing occurred in late winter 1993 [4]. Important information on the carbon and nutrient recycling can be extracted through the monitoring of the chemical properties evolution of this trapped water mass for almost two years in the deeper parts of Jabuka Pit.[5].

## 2. Methods

The survey consisted from an extended grid of stations in the northern and central Adriatic Sea (Figure 1), visited almost seasonally (May-June 1993, July 1993, January 1994, March 1994, July-August 1994 and November 1994). During these cruises



*P. Malanotte-Rizzoli and V.N. Eremeev (eds.),*
*The Eastern Mediterranean as a Laboratory Basin for the Assessment of Contrasting Ecosystems*, 475–481.

nutrients and oxygen samples were collected from several depths, while the dissolved inorganic carbon ($TCO_2$) sampling was restricted to the bottom layer with the justification that the bottom signal of this parameter would give the best indication of the carbon retention in the system. The sampling procedures and the analysis methods are described in detail in ELNA's Data Reports [6].

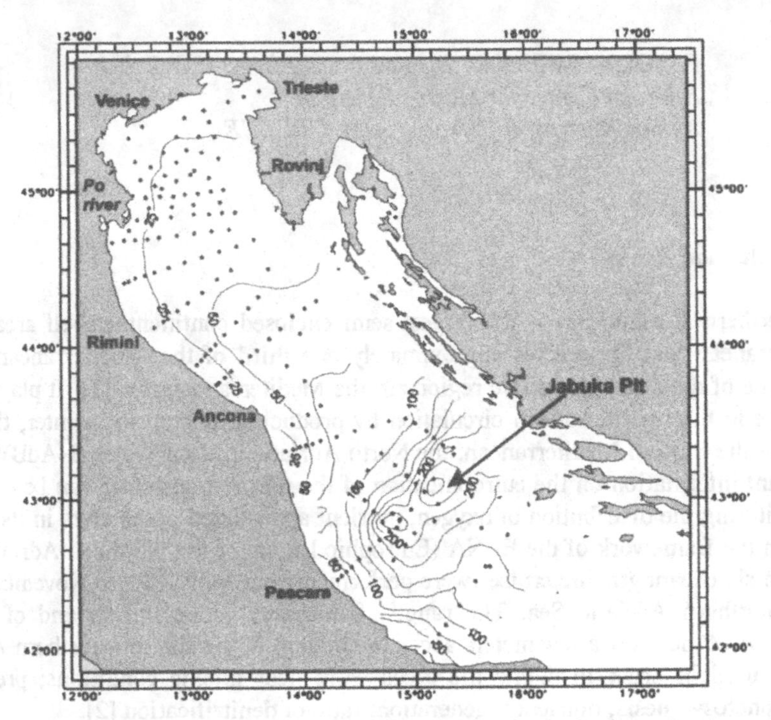


*Figure 1. Bathymetric map and stations positions in the northern and central Adriatic Sea during the cruises of the ELNA project.*

## 3. Results and Discussion

### 3.1. Distribution of nutrients and of related parameters in the bottom layer

The analysis of the contouring plots of the northern Adriatic Sea (north of 43°30' N) designed using the concentration of nutrients and other related parameters, in the layer just above the bottom reveals the existence of three different areas: i) nutrient poor areas with prevailing photosynthetic production - **'a'**, ii) nutrient poor areas with negligible photosynthetic production - **'b'** and iii) nutrient rich areas related to the Po river discharge - **'c'**. The phosphate, nitrate and silicate concentrations of 0.05μM, 1μM and 3.5μM respectively, were set as limits to distinct nutrient-rich and -poor areas. Evidence of prevailing photosynthetic activity is based on oxygen supersaturation or negative values of Apparent Oxygen Utilisation (AOU).

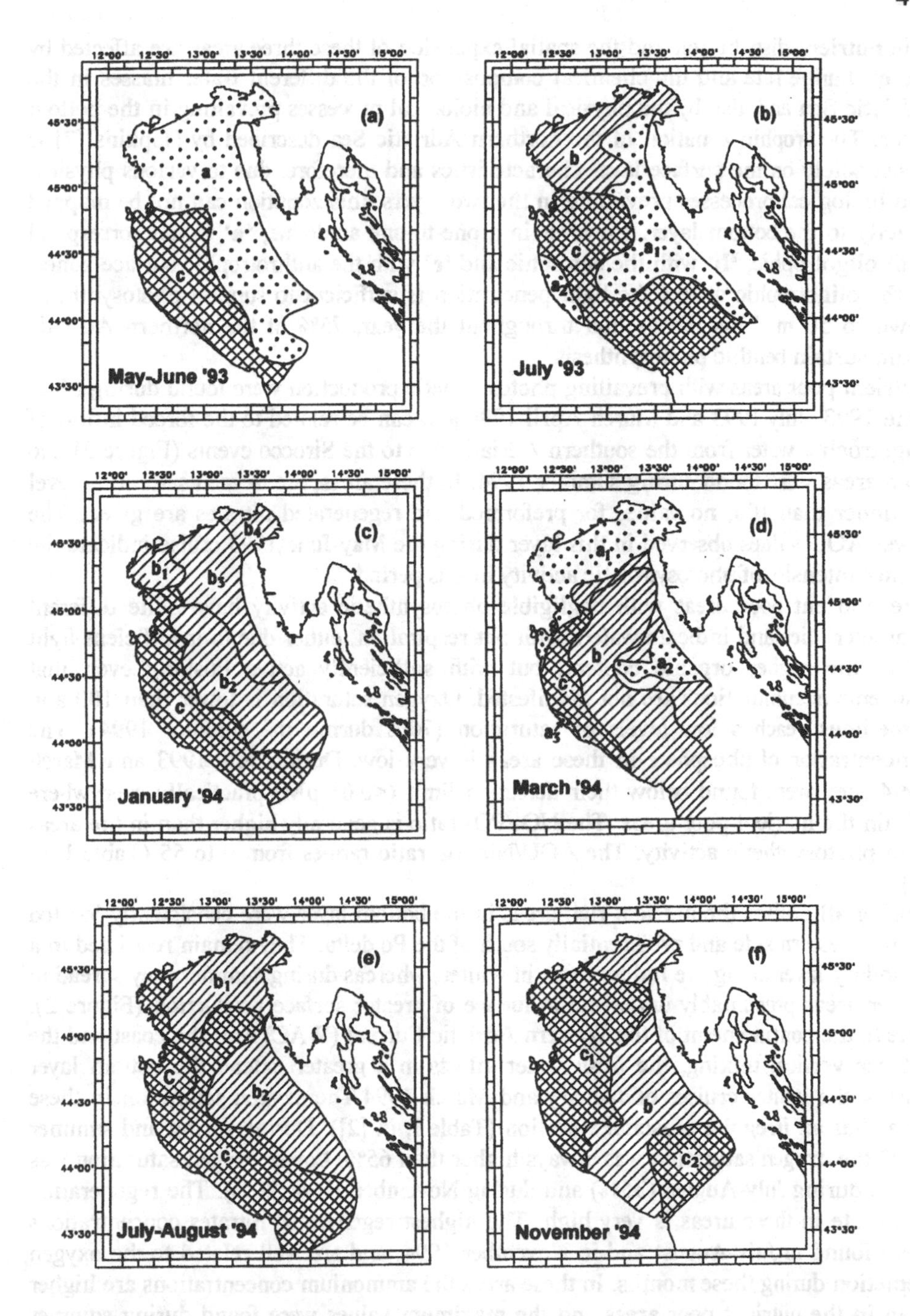


***Figure 2. Spatial distribution of the different areas occurring in the bottom layer of the northern Adriatic Sea during (a) May '93, (b) July '93, (c) January '94, (d) March '94, (e) July-August '94 and (f) November '94.***

The nutrient distribution and the spatial expansion of these three areas are affected by the exchange rate and the chemical composition of the different water masses in the Adriatic Sea and also by the physical and biological processes prevailing in the bottom layer. The trophic zonation of the northern Adriatic Sea described by Hopkins [7] is based solely on the surface layer characteristics and therefore, due to various physical and biological processes prevailing in the two layers, this zonation can not be mapped exactly to the bottom layer. However in a one-to-one analogue, **'a'** would correspond with oligotrophic, **'b'** with the eutrophic and **'c'** with the anthrotrophic surface zones. In the oligotrophic waters the light penetration is sufficient to support photosynthesis down to 50 m. This means that throughout the year, 75% of the northern Adriatic could sustain benthic photosynthesis.

Nutrient poor areas with prevailing photosynthetic production were found during May-June 1993, July 1993 and March-April 1994 and can be related to the forced inflow of oligotrophic water from the southern Adriatic due to the Sirocco events (Figure 2). No such areas were found during summer 1994. In these areas, the oxygen saturation level is higher than 100, no values for preformed and regenerated nitrates are given. The lower AOU values observed in this layer during the May-June 1993 cruise indicate the higher intensity of photosynthetic activity in this period.

The nutrient poor areas with negligible photosynthetic activity have quite different characteristics and indicate waters with net respiration, either due to insufficient light or to the higher organic content, but with sufficiently active flushing even that nutrients accumulations are not manifested. Oxygen saturation is lower than 100 and sometimes reaches low levels of saturation (70% during the summer 1994). The concentration of phosphate in these areas is very low. During July 1993 and March 1994 they were found below their detection limit ($<0.03$ µM) practically everywhere within the nutrient poor areas. The $NO_3/PO_4$ ratio is generally higher than in the areas with photosynthetic activity. The AOU/Ninorg. ratio ranges from 0 to 55 (Table I, in [2]).

During all cruises the nutrient rich areas in the bottom layer were consistently located on the western side and preferentially south of the Po delta. They remain restricted in a boundary layer along the Italian coast in winter, whereas during summer they spread in larger areas presumably under the influence of greater surface production (Figure 2). Due to the confinememt of the Western Adriatic Current (WAC) near the coast and the intense vertical mixing, the fresh water affects in a greater extend the bottom layer during the winter cruises of January and March 1994.The oxygen saturation in these areas has an irregular seasonal variation (Table I, in [2]). During spring and summer 1993 the oxygen saturation was always higher than 65%. Lower oxygen saturation was found during July-August (35%) and during November 1994 (20%). The regeneration of nitrate in these areas is very high. The highest regenerated nitrates concentrations were found in July-August and in November 1994, and are well related to the oxygen depletion during these months. In these areas the ammonium concentrations are higher than in the nutrient poor areas and the maximum values were found during summer (Table I, in [2]). The $NO_3/PO_4$ ratio rises with respect to the nitrate concentration in January 1994. It remains at very high levels during the March 1994 and July-August

1994 cruises, and starts to decrease in the November 1994 cruise. Our results show that the AOU/Ninorg. regenerative ratio at the bottom layer of the northern Adriatic, is generally higher than the stoichiometric ratio for nitrogen regeneration in the ocean (AOU/N=8.625; [8]). The maximum values were found in July-August 1994 (AOU/Ninorg: 70-150) near the Po river plume and in the western side of the Rimini transect. During this period the fresh water discharge and the water exchange rates between the northern Adriatic and the southern parts of the Adriatic is reduced. These very high ratios indicate that a significant part of the regenerated nitrate is subsequently denitrified and lost from the ecosystem in the form of nitrogen gases, resulting in an apparent surplus of oxygen consumption. These results are in accordance with those previously stated [9], remarking that denitrification can occur in the northern Adriatic at least in summer in the sediments which are quite reduced immediately below the surface.

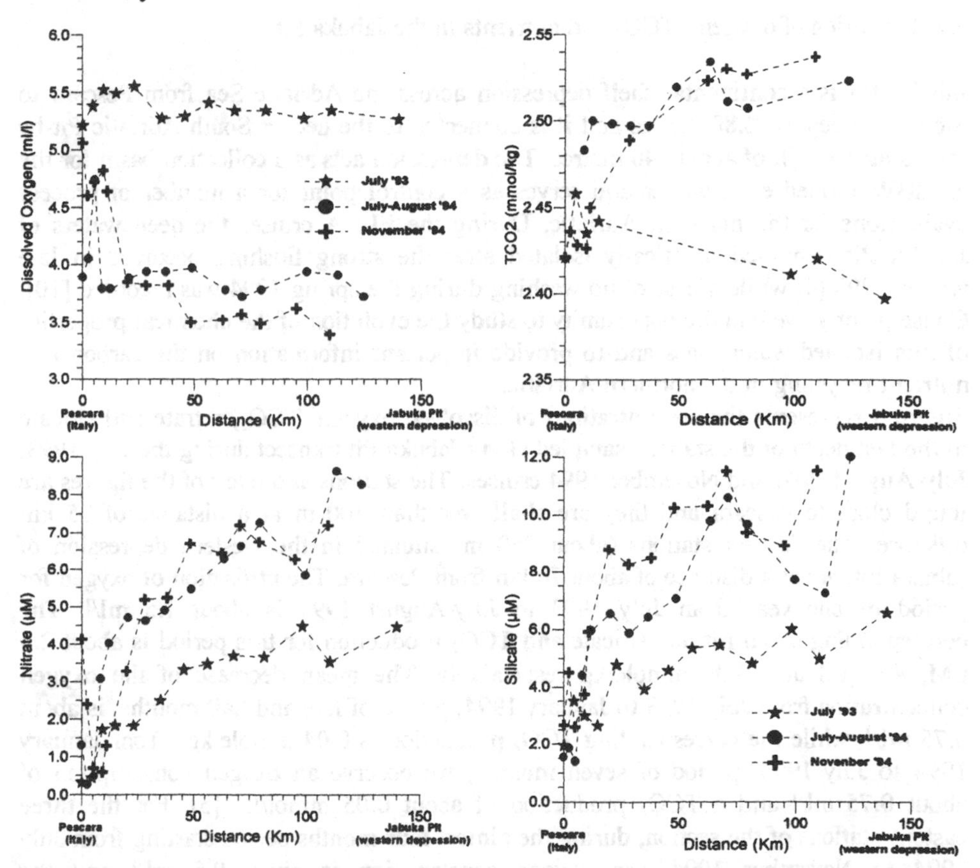


*Figure 3. Comparative distributions of dissolved oxygen, dissolved inorganic carbon ($TCO_2$), nitrate and silicate at the last sampling depth of the stations along a transect from Pescara to the western depression of Jabuka Pit.*

In the frame of the ELNA project the carbon dioxide system is measured systematically for the first time in the northern Adriatic Sea. The bottom signal of this parameter gives the best indication of carbon retention in the system. The decomposition of the organic material, imported directly from the Po and the minor rivers and produced *in situ* in the northern Adriatic results in an increase of the concentration of the dissolved inorganic carbon ($TCO_2$) particularly in the bottom layer. There exists an inverse correlation of $TCO_2$ with oxygen but a direct correlation with nutrients in the bottom layer. The water mass close to the Italian coast is richer in $TCO_2$ and poorer in oxygen. The maximum values of $TCO_2$ near the Po River are about 2.55 mmole/kg in July 1993, 2.58 mmole/kg in July-August 1994 and 2.60 mmole/kg in November 1994. The corresponding minimum concentrations oxygen near the Po River are 2.93 ml/l in July 1993, 1.15 ml/l in July-August 1994 and 0.99 ml/l in November 1994 [5].

### 3.2. Evolution of oxygen , $TCO_2$ and nutrients in the Jabuka Pit

Jabuka Pit is a continental shelf depression across the Adriatic Sea from Pescara to Sibenik to depths <280 metres and it is connected to the deeper South Adriatic Pit by the Palagruza Sill of about 140 metres. The depression acts as a collection basin for the NAdBW formed each winter and serves as a control point for a number of process evaluations for the northern Adriatic. During the ELNA cruises the deep waters of Jabuka Pit remained practically isolated after the strong flushing occurred in late winter 1993 [4] while a case of no washing during the spring 1994 was recorded [10]. Consequently, we had the opportunity to study the evolution of the chemical properties of this isolated water mass and to provide important information on the carbon and nutrient recycling in the northern Adriatic.
Figure 3 represents the concentrations of dissolved oxygen, $TCO_2$, nitrate and silicate in the last depth of the stations sampled along Jabuka Pit transect during the July 1993, July-August 1994 and November 1994 cruises. The stations at the left of the figures are found close to Pescara and they are shallower than 100 m at a distance of 25 km offshore. The deepest stations (about 250 m) situated in the western depression of Jabuka Pit, are at a distance of about 70 km from Pescara. The utilisation of oxygen for period of one year from July 1993 to July-August 1994 is about 1.5 ml/l. The corresponding mean nitrate, silicate and $TCO_2$ production for this period is about 2.5 μM, 4.0 μM and 0.09 mmole/kg respectively. The mean decrease of the oxygen concentration from July 1993 to January 1994, period of five and half months, is about 0.75 ml/l, while the corresponding $TCO_2$ production is 0.04 mmole/kg. From January 1994 to July 1994, period of seven months, we observe an oxygen consumption of about 0.75 ml/l and a $TCO_2$ production of about 0.05 mmol/kg [5]. For the three eastern stations of the section, during the almost four months period starting from July 1994 to November 1994, the oxygen consumption is about 0.5 ml/l and the corresponding $TCO_2$ production 0.02 mmol/kg.
The oxygen consumption to the $TCO_2$ regeneration ratio calculated for these different periods ranges between 0.65 and 1.08 and is lower than the theoretical Redfield ratio ($\delta O_2:\delta CO_2 = 1.3$). In contrary, the rate of oxygen consumption to nitrate and silicate

regeneration calculated for the period from July 1993 to July 1994 ($\delta O_2:\delta NO_3 \sim 26$ and $\delta O_2:\delta SiO_2 \sim 16$) is much higher than the theoretical Redfield ratio ($\delta O_2:\delta NO_3 = 8.6$) and the Richards ratio ($\delta O_2:\delta SiO_2 = 9.2$; [11]). For a constant amount of oxygen consumption there is more carbon and less nutrient regeneration than the predicted from the theoretical Redfield model. The explanation must be related to the composition of the organic matter which is remineralised. In fact Sambrotto *et al* [12] presented time series from both coastal and open ocean which demonstrate that net organic carbon production greatly exceeded that predicted from the consumption of nitrate and the Redfield C:N ratio.

## 4. Conclusions

The six cruises performed from May 1993 to November 1994 allowed us to define and to follow the spatial evolution of three different areas in the near bottom layer of the shallow northern Adriatic Sea. The distinction between these areas was based on nutrients concentrations; further analysis showed that various biochemical conditions prevail in the three different areas. The nutrient rich areas are the most disturbed. The oxygen saturation has irregular seasonal variation and during periods of stratification lets down to the 20% level. The $NO_3/PO_4$ and AOU/Ninorg. ratios greatly exceed the theoretical values. The isolation for about two years of the newly formed NAdBW in the Jabuka Pit permitted to extract important information on the carbon and nutrient recycling.

## References


1. Sournia A. (1973) La production primaire planctonique en Mediterranee. Essai de mise a jour. Bull. Etud. commun. Medit., 5, 128pp.
2. Souvermezoglou E., Krasakopoulou E., Battilotti M. (1998) Seasonal variation of nutrients and nutrient ratios in the bottom layer of the Northern Adriatic Sea. In: Hopkins, Artegiani, Cauwet, Degobbis, Malej (editors) *Ecosystems Research Report Series -The Adriatic Sea*, EU/Environment, Brussels (in press).
3. Artegiani A., Azzolini R., Salusti E. (1989) On the dense water in the Adriatic Sea. *Oceanologica Acta*, 12 (2), 151-160.
4. So.Pro.Mar. (1994) ELNA First Annual Progress Report. EC/Environment, Brussels.
5. Souvermezoglou E., Krasakopoulou E. (1998) Sources and sinks of Dissolved Inorganic Carbon ($TCO_2$) in the northern and central Adriatic Sea In: Hopkins, Artegiani, Cauwet, Degobbis, Malej (editors) *Ecosystems Research Report Series -The Adriatic Sea*, EU/Environment, Brussels (in press).
6. So.Pro.Mar. (1995) ELNA - SPM Data Report 7, EC/Environment, Brussels.
7. Hopkins, T.S. (1996) ELNA Final Scientific Report. EC/Environment , Brussels.
8. Redfield A.C., Ketchum B.H., Richards F.A. (1963) The influence of organisms on the composition of seawater. in : The sea, ideas and observations, vol. 2, edited by A. R. Robinson. Springer Verlag, Berlin, 3-15.
9. Giordani P. and Angiolini L. (1983) Chemical parameters characterizing the sedimentary environment in a NW Adriatic coastal area. *Estur. Coast. Shelf Sci.*, 17, 159-167.
10. So.Pro.Mar. (1995) ELNA Second Annual Progress Report. EC/Environment, Brussels.
11. Richards F.A. (1958) Dissolved silicate and related properties of some western North Atlantic and Caribbean waters. *J. Mar. Res.*, 17 :449-465.
12. Sambrotto R.N., Savidge G., Robinson C., Boyd P., Takahashi T., Karl D.M., Langdon C., Chipman D., Marra J., Codispoti L. (1993) Elevated consumption of carbon relative to nitrogen in the surface ocean. *Nature*, 363, pp. 248-250.

# WATER MASSES AND TRANSPORTS BETWEEN THE AEGEAN AND LEVANTINE BASINS DURING LIWEX '95

B. Manca[1], A. Theocharis[2], S. Brenner[3], H. Kontoyiannis[2], and E. Sansone[4]

*1 Osservatorio Geofisico Sperimentale, P.O.Box 2011, 34016 Trieste, Italy*

*2 National Centre for Marine Research, Aghios Kosmas, Hellinikon 16604, Athens, Greece*

*3 Israel Oceanographic Limnological Research, Haifa 31080, Israel*

*4 Istituto Universitario Navale, Via Acton 38, 80133 Napoli, Italy*

**Abstract.** Within the framework of the POEM-BC program, a multinational collaborative experiment LIWEX '95 in the Levantine basin was undertaken in winter-spring 1995, focussing on the different phases of the process of formation of the Levantine Intermediate Water (LIW). This paper presents the results obtained pertaining to the water exchange between the Aegean and Levantine basins through the eastern straits of the Cretan Arc, where transport rates are also been calculated. In addition, modifications in the thermohaline properties of the deep waters related to the climatic changes in the characteristics of the open upper and closed deep thermohaline cells of the Eastern Mediterranean that began during late '80s, are evidenced.

## 1. Introduction

Coordinated experiments in the Eastern Mediterranean by the international P.O.E.M. (Physical Oceanography of the Eastern Mediterranean) program have revealed the existence of three predominant scales of motion interacting within the general circulation pattern: the basin scale, the sub-basin scale gyres, composed alternatively of permanent or quasi-permanent cyclonic and anticyclonic structures, and the most energetic mesoscale features that interact with the sub-basin scale circulation [1,2].

During the latter half of the '80s, the thermohaline basin-wide circulation presented a relatively unchanging overall general picture. However, very strong differences have started to occur within the POEM-BC (the POEM enlarged to the Biological and Chemical studies) in the beginning of the '90s in the intermediate and deep layers that complicate the deep circulation [3,4].

A more detailed and complex circulation pattern,involving the sub-basin and mesoscale features, has emerged from the investigations carried out during the Levantine Intermediate Water Experiment in 1995 (LIWEX '95) in the Levantine basin



*P. Malanotte-Rizzoli and V.N. Eremeev (eds.),*
*The Eastern Mediterranean as a Laboratory Basin for the Assessment of Contrasting Ecosystems*, 483–494.

proper. A series of multiship surveys were conducted within the framework of a multinational collaborative effort in the period December 28, 1994 to April 20, 1995 [5]. The surveys were specifically designed to focus on the formation of the Levantine Intermediate Water (LIW), the most important water mass of the entire Mediterranean, covering the three major phases of this process: (i) the preconditioning phase in January 1995 (with the German R/V Meteor); (ii) the mixing and convection phase in February 1995 (using the Turkish R/V Bilim); (iii) the spreading and transformation phase of the newly formed water mass (utilising the Greek R/V Aegaeo, the Italian R/V Urania and the Israeli R/V Shikmona simultaneously) in March-April '95. The fieldwork consisted of CTD stations where vertical profiles of temperature, salinity and dissolved oxygen were obtained, and water samples were collected for bio-chemical determinations.

Here, we focus on results obtained during the preconditioning and spreading phases of the experiment. The upper thermocline flow pattern in the Levantine basin and the water exchanges between the Aegean Sea and western Levantine basin are investigated. The baroclinic long-term current component driven by the internal pressure gradient between the two adjoining basins has been estimated as well, contributing to a better understanding of their regional oceanography during the LIWEX '95.

## 2. Circulation pattern during the preconditioning phase of LIW experiment

The dynamic height anomaly at the surface with reference to the 800 dbar level (Fig. 1), obtained during the preconditioning phase of the experiment on January 1995, basically shows the main features of the upper thermocline general circulation in the Levantine basin (for a general picture of the overall circulation see for instance the paper of Robinson et al.[1]). From the examination of the data at R/V Meteor station network, marked with dots in Fig. 1, a detailed identification of the sub-basin scale structures cannot be made. However, the cyclonic circulation centred over the topographic depression Southeast of Rhodes island (i.e. the Rhodes gyre) encompasses a broad area extending from 28° to 30° E and from 34.5° to 36° N. The cyclonic eddy located west of Cyprus, observed in the same location during previous POEM surveys, is connected to the Rhodes gyre by the Central Levantine Basin Current [6], which bifurcating southwest of Cyprus closes the broad cyclonic circulation area in the Northern Levantine. The cyclonic circulation extends longitudinally from the Eastern Cretan Arc Straits to the west of Cyprus and is bordered to the north by the westward-meandering Asia Minor Current (AMC) which flows along the Turkish coastline, where the Antalya anticyclone may be also recognised, and to the south by the anticyclonic complex presumably incorporating the Ierapetra, located southeast of Crete, the Mersa-Matruh and the Shikmona gyres located along the Egyptian coast.

An understanding of the major dynamical features driving the water exchange between the Levantine and Ionian basins through the Cretan Passage is obtained from the vertical distributions of potential temperature (θ), salinity and density ($\sigma_\theta$) along

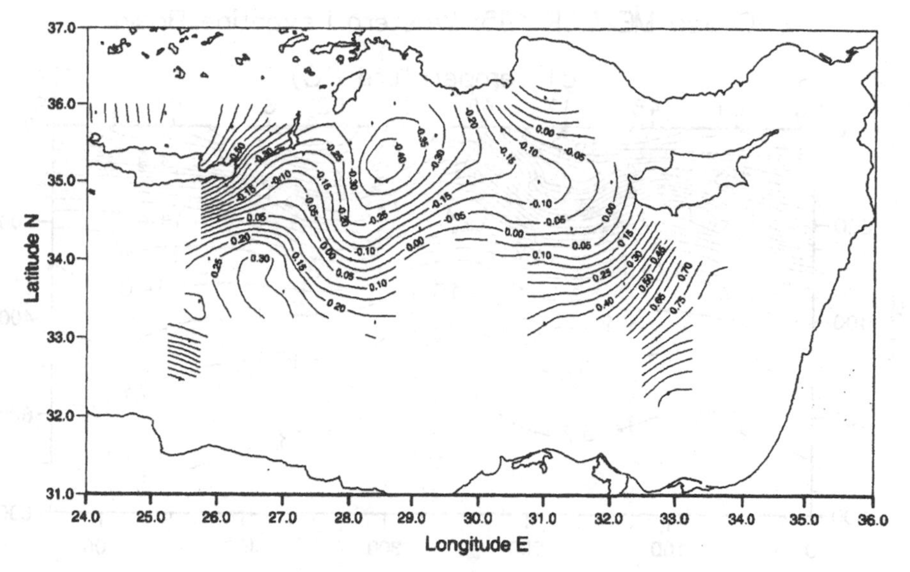


**Figure 1. Surface dynamic height (in $m^2/s^2$) with reference to 800 dbar level, during the preconditioning phase of LIWEX '95, January 1995, R/V Meteor. The network of CTD stations is marked by dots.**

the westernmost transect occupied by the R/V Meteor in the Levantine basin (Fig.2).

The upper panels present in greater detail the first 800 m of the water column along the transect. The 29.10 kg/m$^3$ isopycnal (Fig. 2c) may be considered to roughly approximate the base of the pycnocline, which is conditioned mainly by the temperature field. The pycnocline rises from 300 dbar at the southernmost station (left side of the panels) to less than 100 dbar at the northern station (right side), where it borders the western rim of the Rhodes gyre. The behaviour of the pycnocline is most indicative of the general west-east flow pattern in the upper layer. The deepening of the isopycnals at the left side of the section is mostly due to the salinity front which causes the Mid Mediterranean Jet (MMJ) transporting light and less saline Modified Atlantic Water (MAW) ($\theta$>15.5 °C, S<38.85,$\sigma_\theta$<29.0 kg/m$^3$). It occupies the southernmost portion of the section in the upper 150 m of the water column. Superimposed on the general west-east flow pattern, some sub-basin and mesoscale features are also evident in the oscillations of the isopycnals. The Ierapetra anticyclonic eddy, which penetrates barotropically down to 500 dbar, is detected at station 47 located in the middle of the Cretan Passage, while the doming of the isopycnals at the stations 46 and 45, located in proximity to the eastern tip of Crete, shows that in this area a cyclonic eddy prevails. The features revealed in the northern part recirculate

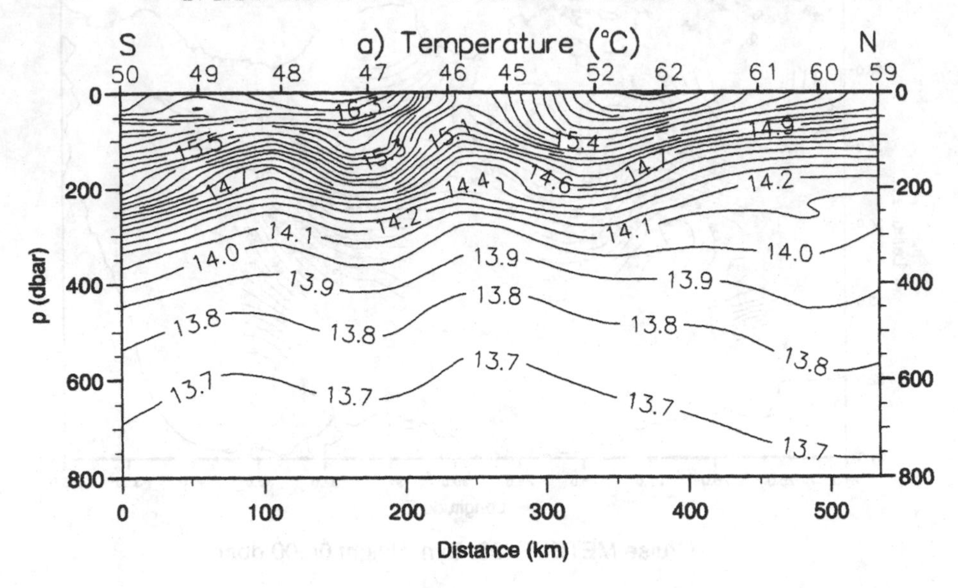


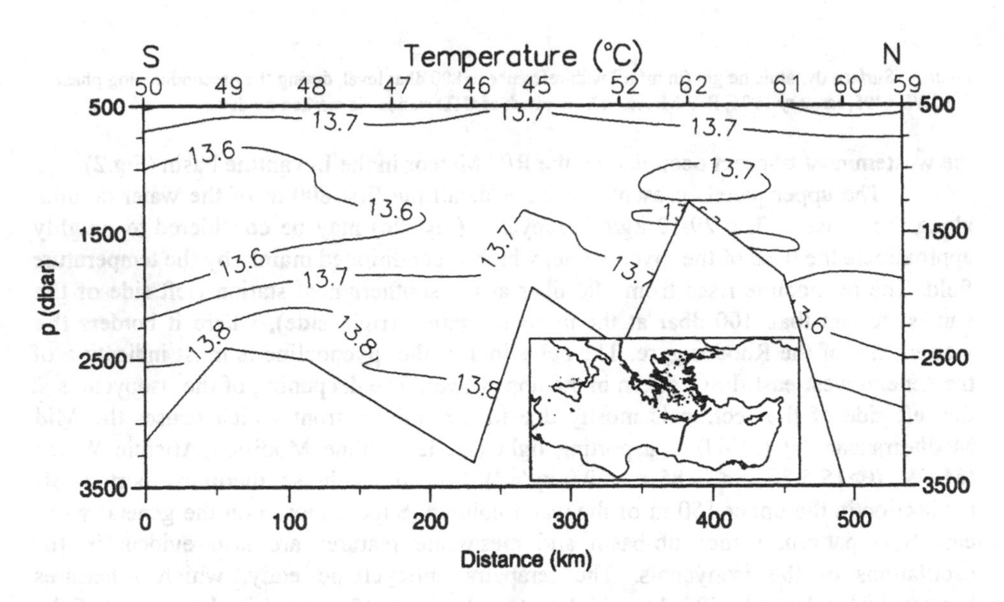


**Figure 2. Vertical distributions of (a) potential temperature, (b) salinity, and (c) potential density down to 800 dbar (upper panels) and from 500 dbar to the bottom (lower panels), along the westernmost section of the R/V Meteor cruise in the Levantine basin, in January 1995. The hydrological stations along the section are indicated at the top of the panels.**

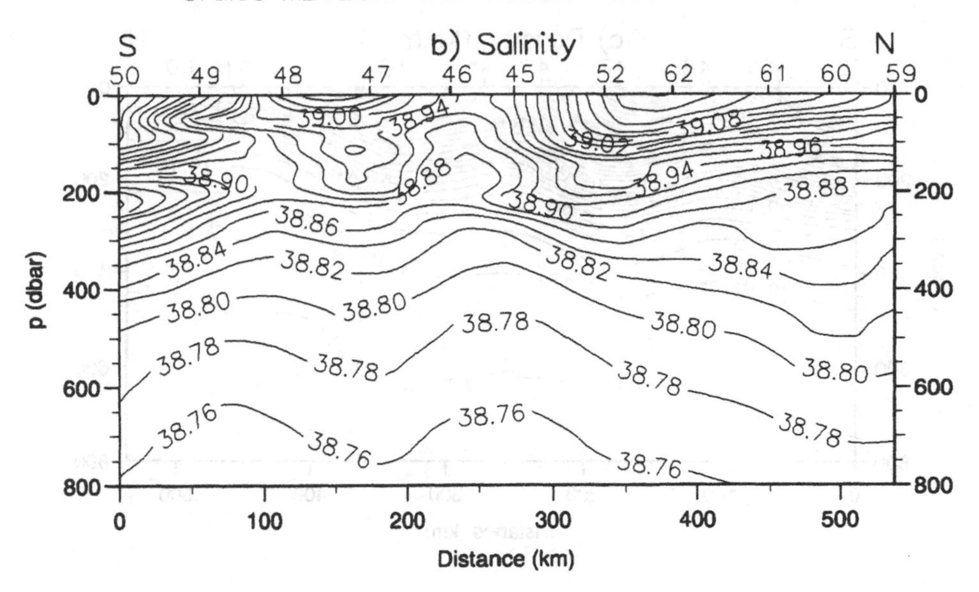


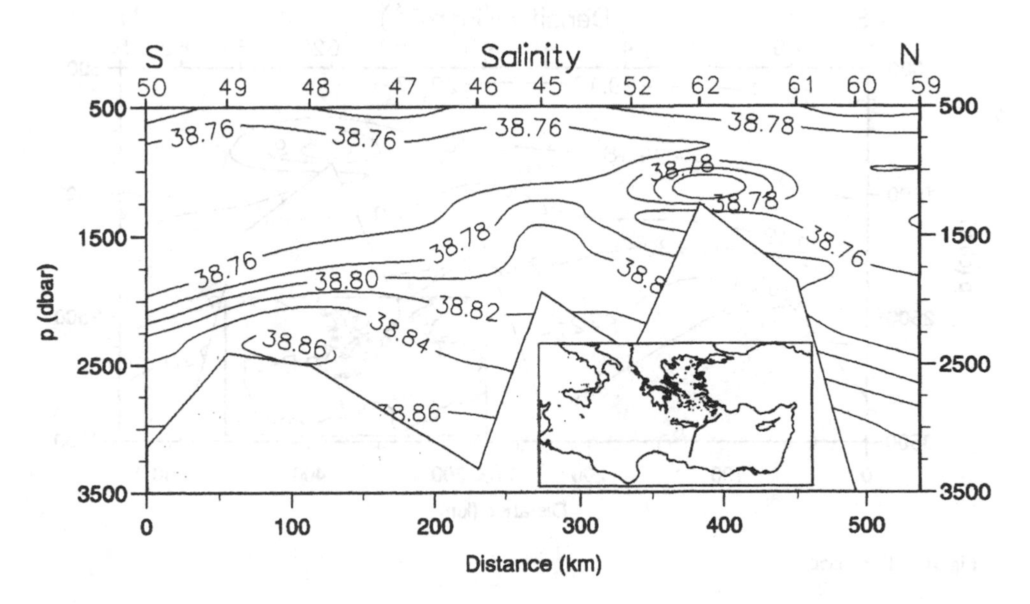


**Figure 2. Continued.**

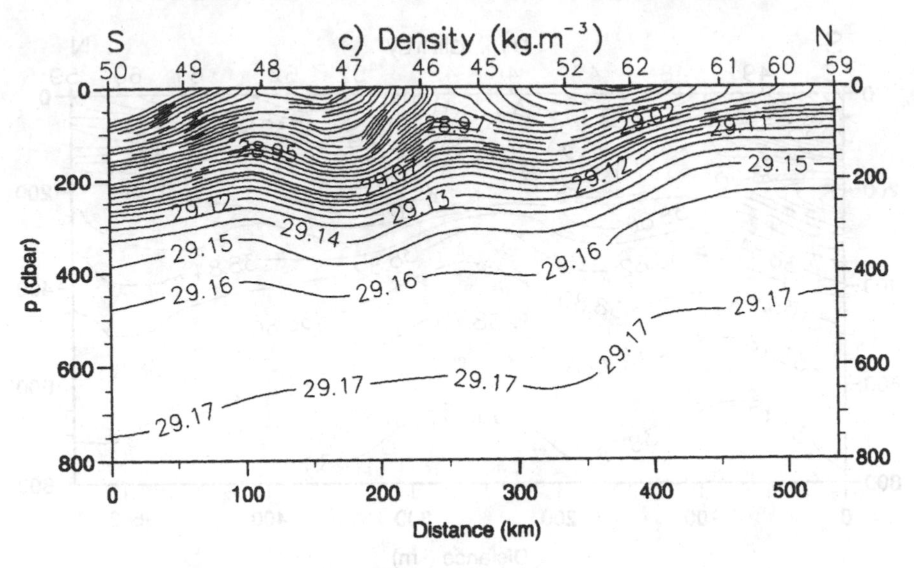


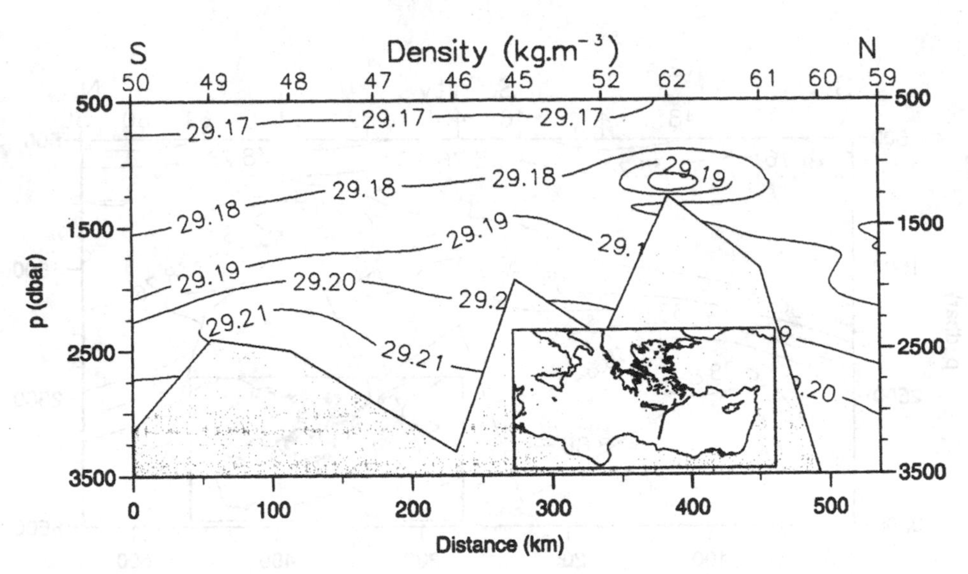


Figure 2. Continued

very salty and warm Levantine Surface Water(LSW) ($\theta$>15.5 °C, S>39.0), advected in in this region by the AMC along the western border of the Rhodes gyre.

In the intermediate layer, the LIW signature ($\theta\cong$15 °C, S>38.9) is restricted to the southern part of the section (Fig. 2b), with properties that differ from those of the salty water downwelled in the centre of the Ierapetra anticyclone ($\theta$>15.5 °C, S>38.94). In the deepest part of the transect (lower panels) patches of warmer, saltier and denser water of Aegean origin ($\theta \cong 13.7$-$13.8$, $S \cong 38.78$-$38.86$, $\sigma_\theta \cong 29.18$-$29.22$ kg/m$^3$) are detected both close to the Straits (station 62) and near the African coast (station 49). This water flowing out from the straits of the Cretan Arc (Kassos and Karpathos straits) sinks immediately to the bottom layer, displacing the colder and fresher Eastern Mediterranean Deep Water (EMDW) of Adriatic origin ($\theta < 13.7$ °C, $S < 38.76$) upwards. It can also be noted that the signature typical of the colder EMDW reaches levels, shallow enough to be incorporated into the flow pattern that intrudes into the Aegean at intermediate levels compensating the outflow at the bottom [7,8].

## 3. Upper thermocline flow patterns and transports across the Cretan Arc Straits

The co-ordinated multiship survey, conducted in March-April 1995 incorporating the most comprehensive station grid spaced broadly every 1/4° (in latitude and longitude), was able to capture the sub-basin and, to some extent, the mesoscale features.

The figures 3a-c depict the vertical distributions of potential temperature ($\theta$), salinity and density ($\sigma_\theta$) in the upper 500 m of the water column, along the section close to the Eastern Cretan Arc Straits and outlined in Fig. 5. They evidence the intense dynamics and complexity of the regional hydrography. The main features are: (i) the presence of a consistent volume of warmer and saltier LSW ($S > 39.04$; $\theta > 16.0$ °C) at the right end of the transect; (ii) the presence of MAW ($S < 38.92$) at the left end of the transect; (iii) the wide central dome bordering the Rhodes gyre, with the convective chimney indicated by the outcropping of 29.06 kg/m$^3$ isopycnal; (iv) the homogeneous water ($\theta \cong 14.9$ °C; $S \cong 39.0$) at the centre of the doming structure which has reached 100 m of depth; (v) the patches of surface ventilated high salinity water ($S > 39.02$) which slide along the doming isopycnals of the Rhodes gyre; (vi) the oscillations of the isopycnals on both sides of the central doming suggesting baroclinic instabilities of a wavelength of about 75 km.

The upper thermocline water exchange rates across the Cretan Straits have been based upon geostrophic velocities calculated from the transversal pressure gradients (Fig. 4a-c). A reference level of 500 dbar was used in the dynamical computation. We have adopted the convention that positive flow enters the page, so the relative positive geostrophic velocities at the right hand of the transect (Fig. 4a) correspond to the westward flow pattern of the AMC transporting Levantine waters in the upper 300 m at a rate of about 2.1 Sv, while the maximum speed exceeded 12-14 cm/s. This flow weakens substantially at the 300 m depth-level, permitting as a reasonable approximation, the assumption of this horizon as a surface of no motion for further calculations in the interior of the straits. The return flow, detected at the northernmost stations, is due to the anticyclonic eddy close to Rhodes island (Fig. 3c). The flow pattern bifurcates in correspondence to the Karpathos strait. In fact, a transport of 0.6 Sv was calculated for the water entering the Karpathos strait (Fig. 4b), which implies a residual transport of 1.55 Sv flowing along the rim of the Rhodes gyre. A cyclonic mesoscale eddy is detected in the Kassos Strait (Fig. 4c). Here the dynamics

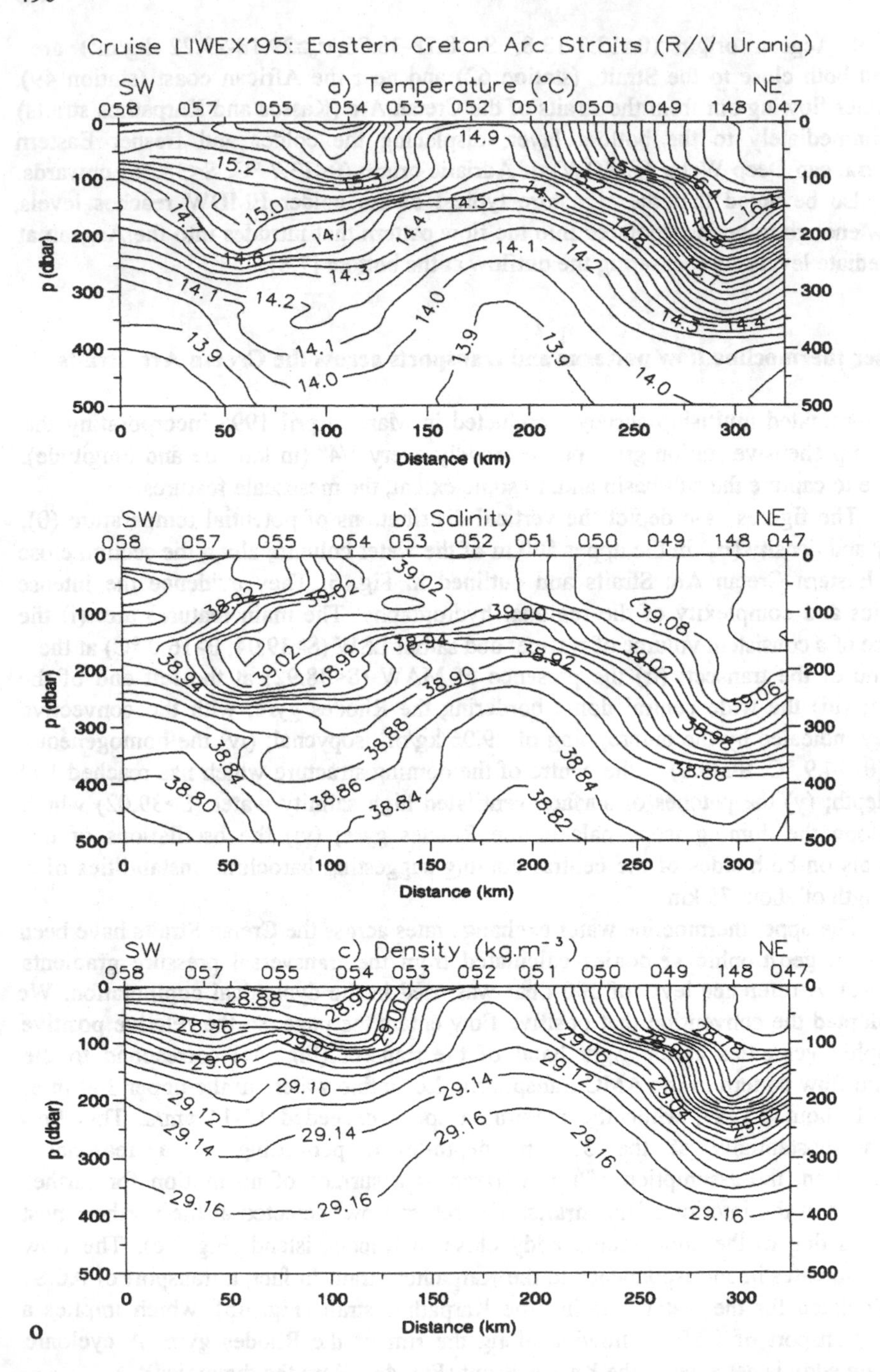


O

**Figure 3. Vertical distributions of (a) potential temperature, (b) salinity, and (c) potential density down to 500 dbar, along the south-north cross-section (south to the left) flanking the Eastern Cretan Arc Straits, and outlined in Fig. 5, during the spreading phase of the survey LIWEX '95 (March-April 1995).**

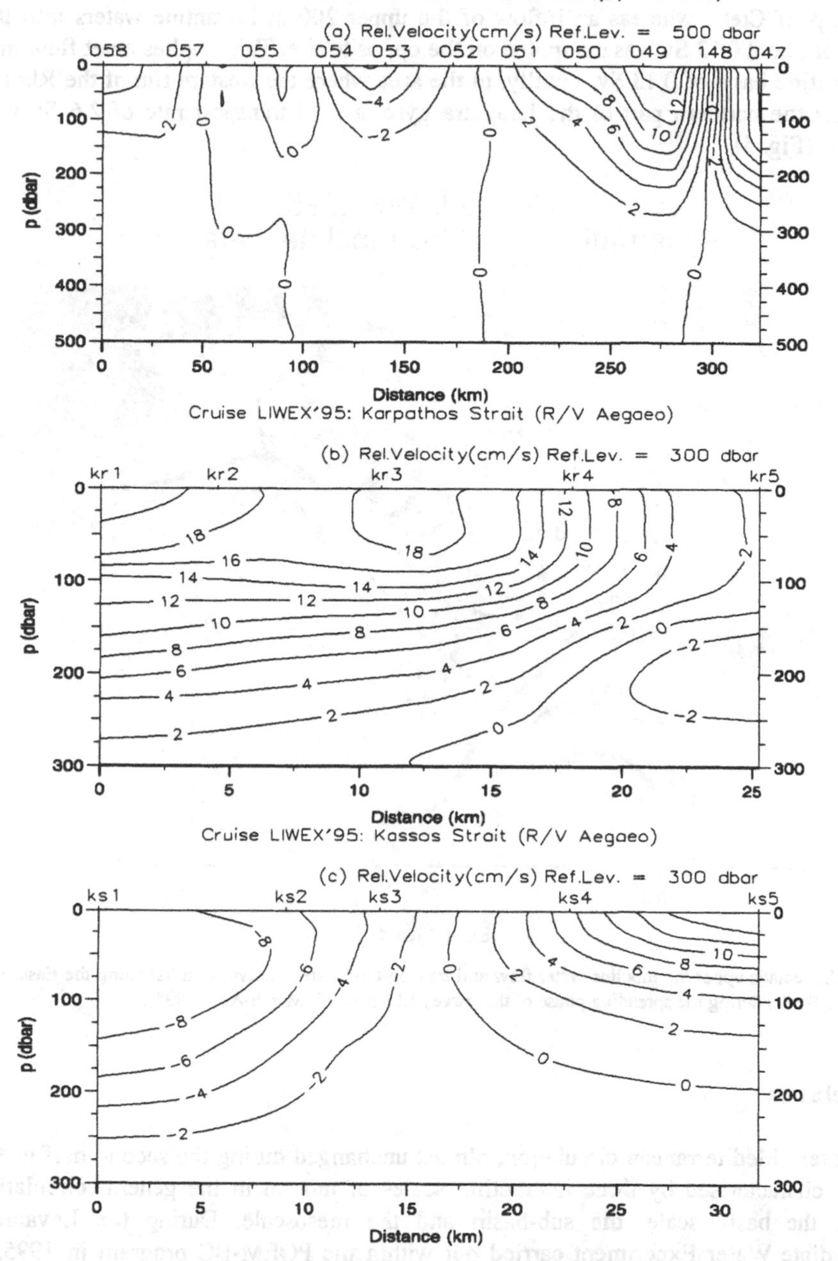


**Figure 4. Vertical distributions of the relative geostrophic velocity (in cm/s) along the sections (a) as in Fig. 3 (R/V Urania), (b) in the Karpathos Strait (R/V Aegaeo), and (c) in the Kassos Strait (R/V Aegaeo). The reference levels for the dynamical computation are indicated. See Fig. 5 for the position of the sections.**

manifest an outflow of the upper 300 m Aegean waters of about 0.26 Sv close to the eastern tip of Crete, whereas an inflow of the upper 200 m Levantine waters into the Aegean of about 0.13 Sv was computed on the opposite site. This implies a net flow into the Levantine basin of 0.13 Sv. Finally, in the area where the western rim of the Rhodes gyre joins the northern part of the Ierapetra gyre, a total transport rate of 2.6 Sv was estimated (Fig. 5).

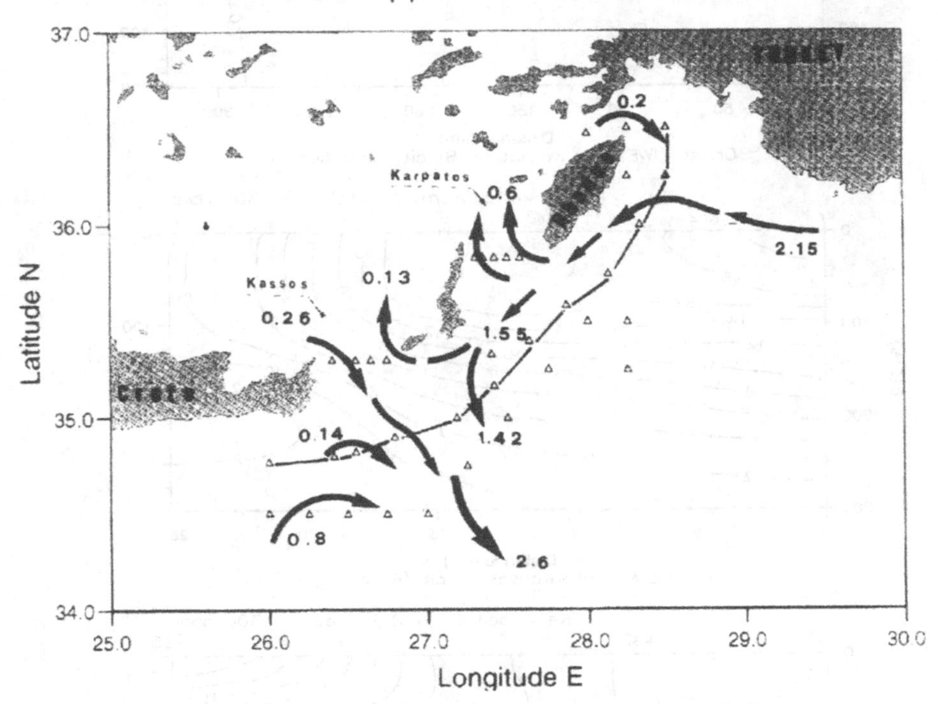


Figure 5. Schematic upper thermocline water flow and transport estimates (in Sv=$10^6$ $m^3/s$) along the Eastern Cretan Arc Straits during the spreading phase of the survey LIWEX '95 (March-April 1995).

## 4. Conclusion

The Eastern Mediterranean circulation, almost unchanged during the second half of the '80s, is characterised by three interacting scales of motion in the general circulation pattern, the basin scale, the sub-basin and the mesoscale. During the Levantine Intermediate Water Experiment carried out within the POEM-BC program in 1995, a detailed complex of the upper thermocline circulation pattern emerged in the Levantine basin that confirmed the significant changes that began in the early '90s [4]. Moreover, these changes concern also the flow regime through the straits of the Cretan Arc, associated with the climatic shift of the intermediate and deep thermohaline properties.

The permanent cyclonic Rhodes gyre is connected to the west Cyprus gyre forming a broad northern Levantine cyclonic circulation, extending from the eastern Cretan Arc to the west of Cyprus. The AMC meanders along the southern coasts of Turkey and transports westwards the upper 300m of the warm and saline Levantine waters at a rate of 2.1 Sv. It is the northern border of the large-scale North Levantine circulation. The contrasting system in the Southern Levantine is the broad anticyclonic flow region composed of three gyres, the Ierapetra, the Mersa-Matruh and the Shikmona. The MMJ, carrying MAW eastwards, seems to bifurcate at the entrance of the Levantine in the east Cretan Passage. Its first branch flows between the southern rim of the north Levantine cyclonic circulation and the northern peripheries of the Ierapetra becoming the Central Levantine Basin Current. The second branch encircles the Mersa-Matruh gyre. Two AMC branches enter the Aegean carrying Levantine waters, mainly through the Karpathos Strait at a rate of 0.6 Sv and through the Kassos Strait at a rate of 0.13 Sv. At the Rhodes Strait, where in the past was mostly detected the main AMC branch entering the Aegean [9], the Levantine water recirculates anticyclonically with an outflow from the Aegean of the order of 0.2 Sv. Finally, through the western part of the Kassos Strait an outflow of Aegean water of 0.26 Sv is estimated. Outside the Kassos Strait, we detected the Aegean waters as a tongue directed towards the Ierapetra anticyclone. Finally, a new regime in the deep waters was revealed. The warmer, saltier and denser waters of Aegean origin, that occupy the deepest parts of the basin, displaced the EMDW of Adriatic origin that is uplifted several hundreds meters. The latter are found shallow enough in the vicinity of the Cretan Straits to intrude into the Aegean, compensating the deep water outflow.

## ACKNOWLEDGMENTS

This work was carried out with the financial support from the Consiglio Nazionale delle Ricerche (CNR), Rome, contract no. 93.01306.02 to B. Manca. We acknowledge funding by governments, national agencies and Institutions of affiliation who made instruments and ships available to the POEM program. The R/V Meteor was funded by the Deutsche Forschungsgemeinschaft, the R/V Urania was funded by the CNR in Rome, and the R/V Aegaeo by the National Centre for Marine Research of Greece. We wish to thank the masters and crews of the ships for their skilful work during the surveys.

## 5. References

1. Robinson, A.R., Golnaraghi, M., Leslie, W.G., Artegiani, A., Hecht, A., Lazzone, E., Michelato, A., Sansone, E., Theocharis, A., and Unluata, U. (1991) Structure and variability of the Eastern Mediterranean general circulation, *Dynamics of Atmospheres and Oceans* **15**, 215-240.
2. POEM Group (1992) The general circulation of the Eastern Mediterranean, *Earth Science Reviews* **32**, 285-309.
3. Roether, W., Manca, B.B., Klein, B., Bregant, D., Georgopoulos, D., Beitzel, V, Kovacevic, V, and Luchetta, A. (1996) Recent changes in the Eastern Mediterranean deep water, *Science* **271**, 333-335.

4. Malanotte-Rizzoli, P., Manca B.B., Ribera d'Alcalà, M., Theocharis, A., Brenner, S., Budillon, G., and Ozsoy, E. (1998) The Eastern Mediterranean in the 80' and in the 90's: the transition in the intermediate and deep circulation, *Dyn. Atm and Oceans* (submitted).
5. Malanotte-Rizzoli, P., Robinson, A.R., Roether, W., Manca, B.B., Bergamasco, A., Brenner, S., Civitarese, G., Georgopoulos, D., Haley, P.J., Kioroglou,S., Kontoyannis, H., Kress, N., Latif, M.A., Leslie, W.G., Ozsoy, E., Ribera d'Alcalà, M., Salihoglu, I., Sansone, E., and Theocharis, A. (1996) Experiment in Eastern Mediterranean probes origin of deep water masses, *EOS* 77, p.305.
6. Ozsoy, E., Hecht, A., Unluata, U., Brenner, S., Sur, H.I., Bishop, J., Latif, M.A., Rozentraub, Z., and Oguz, T. (1993) A synthesis of the Levantine basin circulation and hydrography, 1985-1990, *Deep-Sea Research (Part II, Special Issue)* **40**, 1075-1120.
7. Klein, B., Roether, W., Manca, B.B., Bregant, D., Beitzel, V., Kovacevic, V., and Luchetta, A. (1998) The large deep water transient in the Eastern Mediterranean, *Deep-Sea Research* (accepted).
8. Theocharis, A., Balopoulos, E., Kioroglou, S., Kontoyannis, H., and Iona, A. (1988) A synthesis of the circulation and hydrography of the South Aegean Sea and Straits of the Cretan Arc (March 1994-January 1995), *Prog. In Oceanography* (accepted).
9. Theocharis, A., Georgopoulos, D., Lascaratos, A., and Nittis, K. (1993) Water masses and circulation in the central region of the Eastern Mediterranean: Eastern Ionian, South Aegean and Northwest Levantine, 1986-1987, *Deep-Sea Research (Part II, Special Issue)* **40**,1121-1142.

# FLOW PATTERNS OF THE MAIN WATER MASSES ACROSS TRANSVERSAL AREAS IN THE SOUTHERN ADRIATIC SEA: SEASONAL VARIABILITY

Beniamino MANCA and Alessandra GIORGETTI
*Osservatorio Geofisico Sperimentale, P.O. Box 2011, 34016 Trieste, Italy*

ABSTRACT. The results here presented regard four seasonal multidisciplinary surveys conducted in the Southern Adriatic basin, in the framework of the collaborative research program PRISMA (Programma di Ricerca e Sperimentazione del Mare Adriatico), with the aim of investigating the longitudinal fluxes across a series of significant transverse sections. The thermohaline properties of the main water masses and their seasonal and spatial variability over an entire annual cycle were analyzed. Five water masses take part in these processes; namely the Adriatic Surface Water (ASW), the Ionian Surface Water (ISW), the Levantine Intermediate Water (LIW), the Adriatic Deep Water (ADW), which overflows the Sill of Otranto to form the main component of the Eastern Mediterranean Deep Water, and finally the Northern Adriatic Dense Water (NADW). In order to follow the spatial extension of these water masses and their seasonal variability, the distribution of thermohaline properties along the transverse sections were considered. The major results concern the advection of the NADW, which has been observed to flow southward over the Pelagosa Sill and subsequently into the southernmost western shelf area, reaching the Otranto Strait. The horizontal thermohaline fields, objectively analyzed, show the main pathways of the above-mentioned water masses. Important signals in the seasonal variability of the baroclinic basin-wide circulation have been noted. Finally, a quantitative measure of the cross-sectional areas occupied by the intermediate and deep waters along the transverse sections located at the Gargano passage and Otranto Strait have been calculated and are discussed.

## 1. Introduction

In the framework of the national collaborative research program PRISMA (Programma di Ricerca e Sperimentazione del Mare Adriatico) aimed at studying longitudinal fluxes in the Adriatic, four seasonal multidisciplinary hydrological surveys were conducted from May 1995 to February 1996. The synoptic shipboard measurements, including high resolution CTD (Conductivity, Temperature and Depth) vertical profiles, as well as water sample collection for bio-chemical parameter determinations,



*P. Malanotte-Rizzoli and V.N. Eremeev (eds.),*
*The Eastern Mediterranean as a Laboratory Basin for the Assessment of Contrasting Ecosystems,* 495–506.

were conducted over four transverse sections in the Adriatic Sea, which separate the northernmost shelf area, the Middle-Adriatic depression, the deeper southern part of the basin, and finally the southern opening at the Otranto Strait by which the Adriatic communicates with the Ionian Sea. These recent observations made it possible to coherently define the water masses involved in the general circulation, in agreement with the regional oceanography, the water exchange dynamics under the topographic constraints, and their seasonal variability during an entire annual cycle. In fact, the morphology of the Adriatic Sea bottom, in combination with its location at middle-latitudes and the large amount of fresh water input in the northern shelf area, strongly influence the flow and the seasonal variability of the general circulation and dynamics. On the other hand, the Adriatic Sea is one of the regions where dense water formation processes take place. It has been identified as a prototype site of convection and dense water mass formation on the northern shelf area during the winter [7], and as a site of open-ocean deep convective movement in the southern basin [10]. The Adriatic Sea is a dilution basin and, at the same time, registers an annual heat loss of about 19-22 $W \cdot m^{-2}$ [2]; thus an inflow of Ionian water, warmer and saltier than the Adriatic, is documented to occur at the Otranto Strait.

In the past, the main aspects of the baroclinic circulation in the Southern Adriatic Sea have been described by various authors using very poor and sparse in situ data, by syntheses of previous knowledge [13, 4, 9, 1], or by a modeling approach [5, 6, 3], pointing out the importance of forcing mechanisms which drive the dynamics, and the coastal and basin-wide physical processes. Now, the recent observations have revealed the important seasonal variability.

The results here presented are concerned with the investigations conducted in the southern part of the Adriatic Sea, delimited to the north by the Pelagosa sill, about 200 m deep, and to the south by the sill at the Otranto Strait, about 800 m deep. The station network (*Fig. 1*) was designed with the aim of investigating the transport associated with the major water masses through cross-sections at the Gargano passage (transects G, H and I) and at the Otranto Strait (transects L, M and N). The hydrography in the large depression of the Southern Adriatic Pit, which extends down to 1200 m, and on the western shelf along the Italian shorelines, extending down to 200 m depth, were investigated as well. Throughout the paper, temperature is given as potential temperature (θ), the salinity is according to the practical salinity scale and the density is presented as potential density excess ($\sigma_\theta$), following the IAPSO recommendation [11].

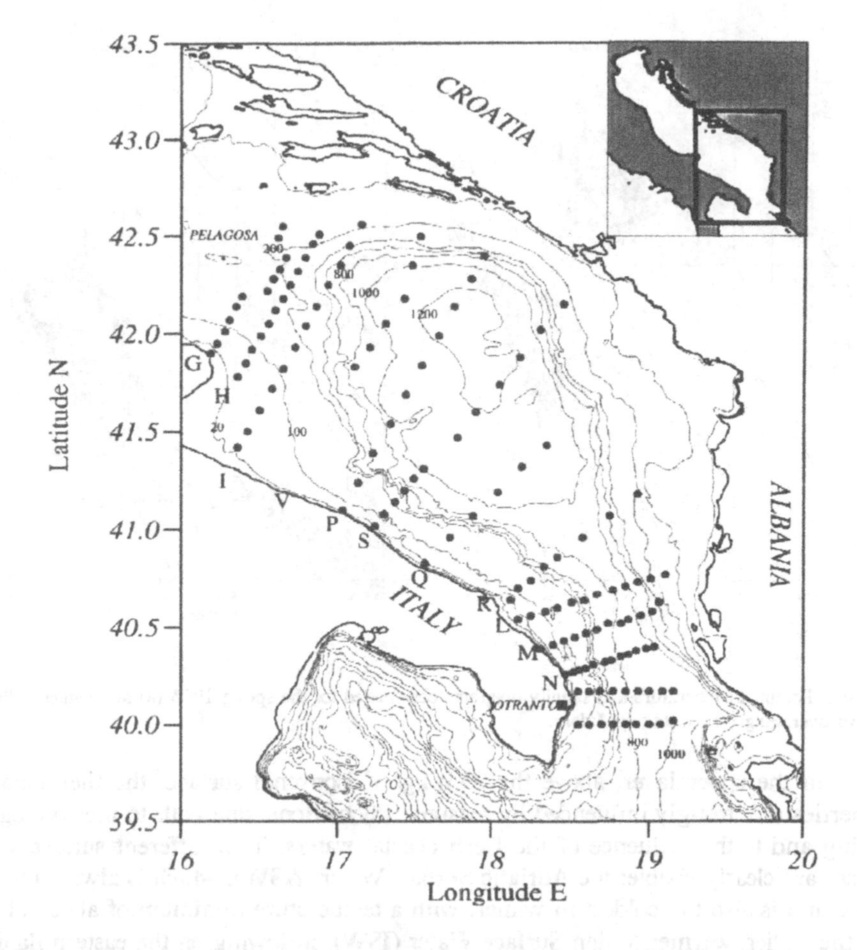


*Figure 1.* **Map showing the bathymetry of the Southern Adriatic sea and the combined network of hydrological stations occupied during the seasonal surveys. The letters on the western border of the basin denote the sections along which the stations were distributed. Depth contours are given in metres.**

## 2. Water mass properties and seasonal variability

The thermohaline properties of the main water masses in the Southern Adriatic Sea are analyzed by examining the Θ-S diagrams constructed for the data set collected during the spring 1995 and winter 1996 surveys (*Fig. 2*); that is to say the measurements conducted within the maximum time lag. They corresponds to the mixing and sinking phase (winter) and to the spreading phase (spring) of the deep water formation process. Five water masses affect the Southern Adriatic circulation, which is strongly influenced by the interaction between the northern part of the basin and the Ionian inflow through the Otranto Strait [12].

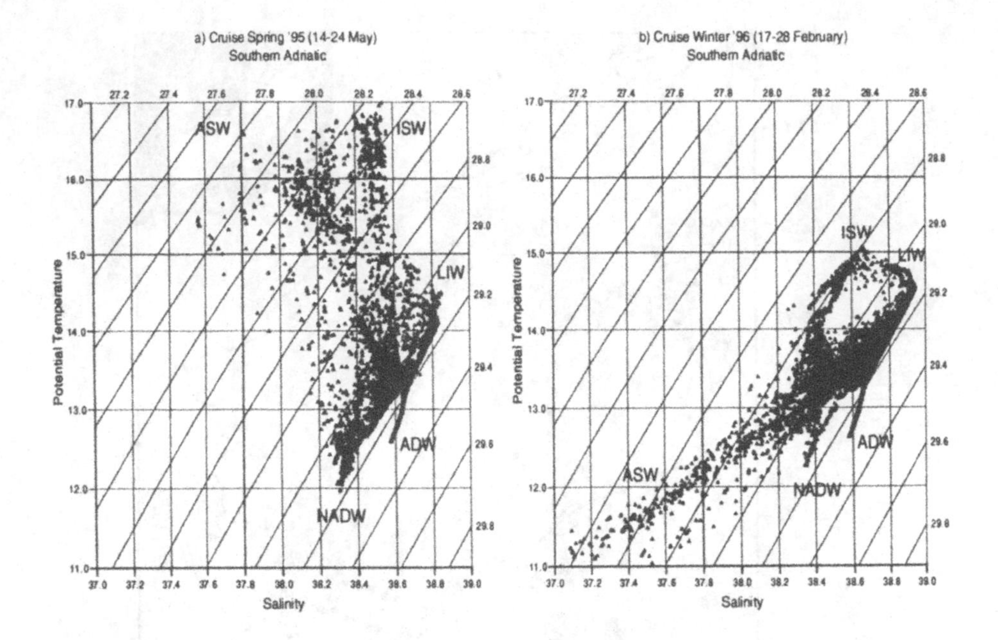


*Figure 2.* **Potential temperature and salinity diagrams constructed for the spring 1995 (a) and winter 1996 (b) surveys, averaging CTD data over 5 dbar.**

In the upper layer, above the 29.0 kg/m$^3$ isopycnal surface, the thermohaline properties are strongly influenced by seasonal fluctuations, due both to meteorological forcing and to the influence of the fresh coastal waters. Two different surface water masses are clearly visible: the Adriatic Surface Water (**ASW**), which is always the less saline and is also the coldest in winter, with a temperature minimum of about 11 °C, and the saltier, warmer Ionian Surface Water (**ISW**), inflowing on the eastern flank of the Otranto Strait.

The cluster of points in the range of potential density $\sigma_\theta$ = 29.0-29.18 kg/m$^3$, characterized by the highest salinity values, represents the Levantine Intermediate Water (**LIW**), which is partially subjected to seasonal variability. It's core has values of $\theta$ = 14.0 °C and S > 38.80. The data sets in May 1995, August 1995, November 1995 and February 1996 exhibit a continuing trend of increasing salinity in the LIW core, shifted toward higher temperature, which in August 1995 reached values as high as $\theta$ = 14.71 °C and S = 38.94.

The tail, within the same potential density range as the LIW core but much lower temperature and salinity values, detects the Northern Adriatic Dense Water (**NADW**). This water in the meridional sections has properties in the ranges $\theta$ = 12.3-12.5 °C and S = 38.3-38.4 in winter, and is abundant and cooler in spring.

The maximum density values, marking the tail in the Θ-S diagrams, represent the Adriatic Deep Water (**ADW**), which resides in the bottom layer of the Southern

Adriatic Pit, keeping the same hydrographic characteristics ($\theta < 13$ °C and S ≅ 38.60) throughout the year.

## 3. Vertical Sections Analysis

The vertical distributions of temperature and salinity along the transects show the spatial extension of the above mentioned water masses. The central sections at the Gargano passage and Otranto Strait for the spring '95 situation are shown in *Fig. 3*. They provide evidence of advection by the existing circulation.

At the Gargano transect (*Fig. 3a and b*) the thermal and salinity front (T < 13.3 °C; S < 38.45) delimits the NADW, which forms on the north shelf area during the winter. It reaches the meridional sections, prevalently during the following spring period, in the form of a bottom density driven current located against the western continental slope down to 200 m depth. The bottom layer is occupied by the LIW, which in modified form (S ≅ 38.65) reaches the Pelagosa sill and then, more diluted, flows towards the Middle Adriatic Sea. The surface layer down to 25 dbar is subjected to seasonal variability and strongly influenced by mesoscale activities, as evidenced by the temperature field. The salinity reveals a coastal surface boundary layer on the western side of relatively fresh southward flowing water separated from that in the open sea.

At the Otranto Strait (*Fig. 3c and d*) the ADW can be identified by the bottom temperature and salinity minimum, mostly against the deepest parts of the western continental slope, limited by the 13.2 °C isotherm and the 38.65 isohaline. The LIW is evident as a maximum in the salinity field lying between 100 and 400 m depth against the eastern part of the section. The LIW tongue with S ≥ 38.75 becomes more diluted (S ≅ 38.70) towards the middle of the Strait. The relatively fresh water on the western continental shelf marks the southward flow of the NADW. It also detaches from the shelf and flows in the intermediate layer (200-400 m), prevalently against the western shelf-break. Small packets of NADW filaments have consistently been found in the centre of the strait, indicating intense mesoscale activity induced by eddies traveling through the straits.

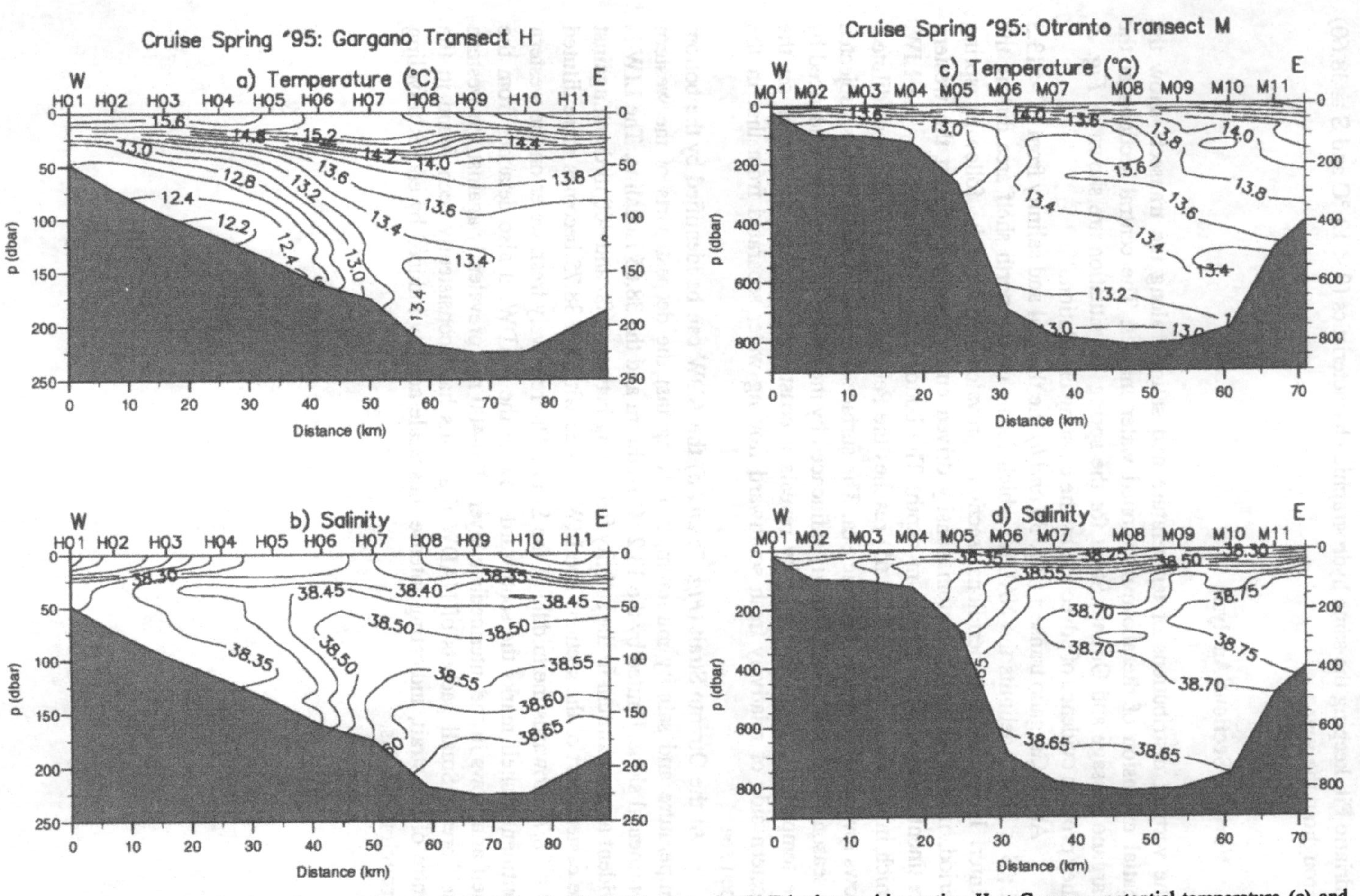


*Figure 3*. **Vertical distribution of potential temperature (a) and salinity (b) along the W-E hydrographic section H at Gargano; potential temperature (c) and salinity (d) along the W-E section M at Otranto during spring 1995.**

## 4. Horizontal Analysis

The distribution of salinity and density at surface for the spring 1995 survey are given in *Fig. 4*. The patterns show relatively fresh water, with salinities less than 38.0, which flows from the north along the western coast and leaves the Adriatic with a modified salinity of 38.15. It forms a strong zonal front with the saltier water of Ionian origin (S > 38.50 ) which intrudes into the Adriatic at the opening of the Otranto Strait. Fresh water from the Albanian rivers is also detected on the eastern side. The Ionian water mixes to become the salty water which is entrained cyclonically in the southern basin. Both inflowing and outflowing current patterns are modified by the transient mesoscale features. The density at the surface (*Fig. 4b*) gives an indication of the upper thermocline circulation, which, in combination with the salinity distribution, reflects essentially the upwelling of salty water from the intermediate layer. The central portion of the Southern Adriatic is occupied by the large cyclonic gyre, which is a permanent feature in that area, and the surface density shows the role of buoyancy loss to the atmosphere in the centre of the gyre.

Property maps on the isopycnal surface $\sigma_\theta = 29.10$ kg/m$^3$, the middle horizon chosen to characterize the spreading of the LIW and NADW, are shown in *Fig. 5*. The salinity distribution (*Fig. 5a*) indicates clearly the main pathways of the LIW. The LIW core with S > 38.85 enters through the Otranto Strait and protrudes more diluted toward the eastern portion of Southern Adriatic Sea, where it recirculates forming the northward branch of the cyclonic gyre. The LIW bifurcates upon reaching the northern shelf break at the Pelagosa sill. A branch is deflected southward by topographic constraints, while a second branch protrudes northwards. In fact, only a well homogenized modified LIW with S $\cong$ 38.65 extends over the Pelagosa sill, where an anticyclonic eddy seems to reinforce a coherent flow pattern transporting LIW over the northern shelf region. All these dynamic features are clearly visible in the isopycnal depth map of *Fig. 5b*. The return path in the west is established by the pressure gradient due to the relatively fresh NADW along the Italian coast. It flows along the western shelf and reduces in spatial extent moving southward. Upon reaching the meridional cross-section at the Otranto Strait, the NADW detaches from the coast and intrudes into the Northern Ionian at the shelf break, as is well documented in the vertical sections.

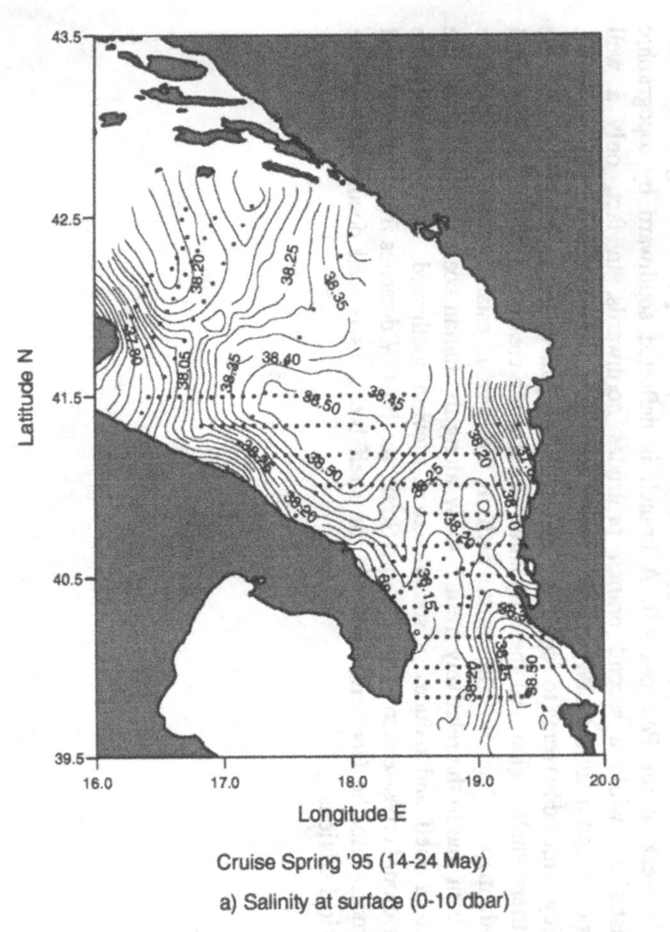


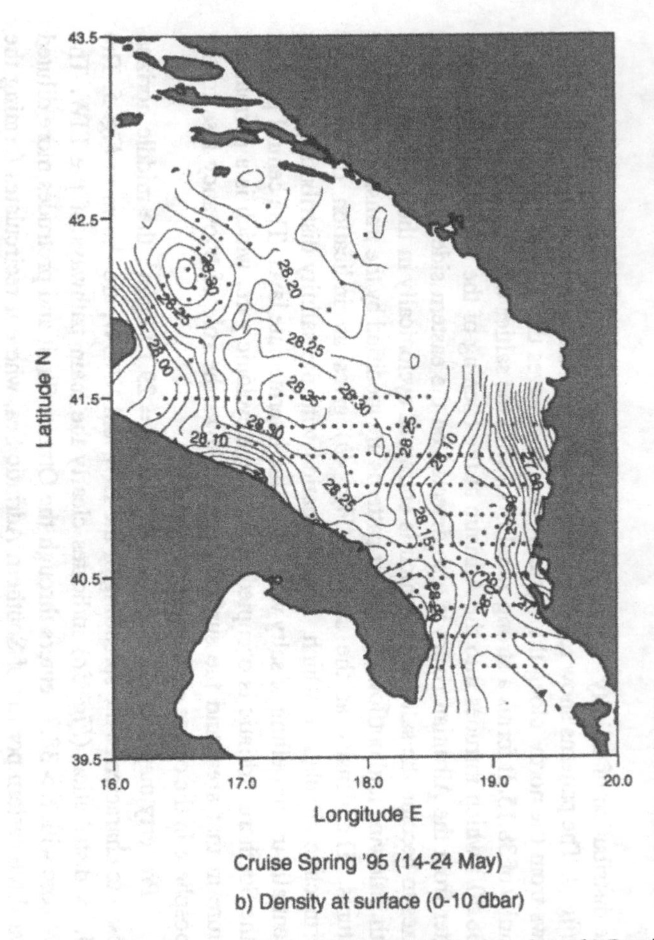


*Figure 4*. **Horizontal distributions of hydrographic properties in the Southern Adriatic for the spring 1995 survey: (a) Salinity at surface; (b) Density excess (kg/m$^{-3}$) at surface.**

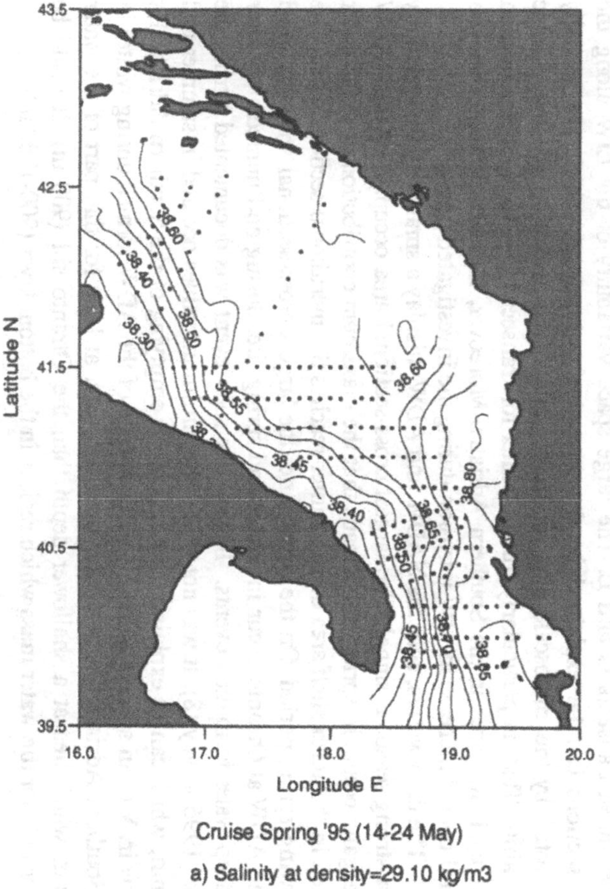


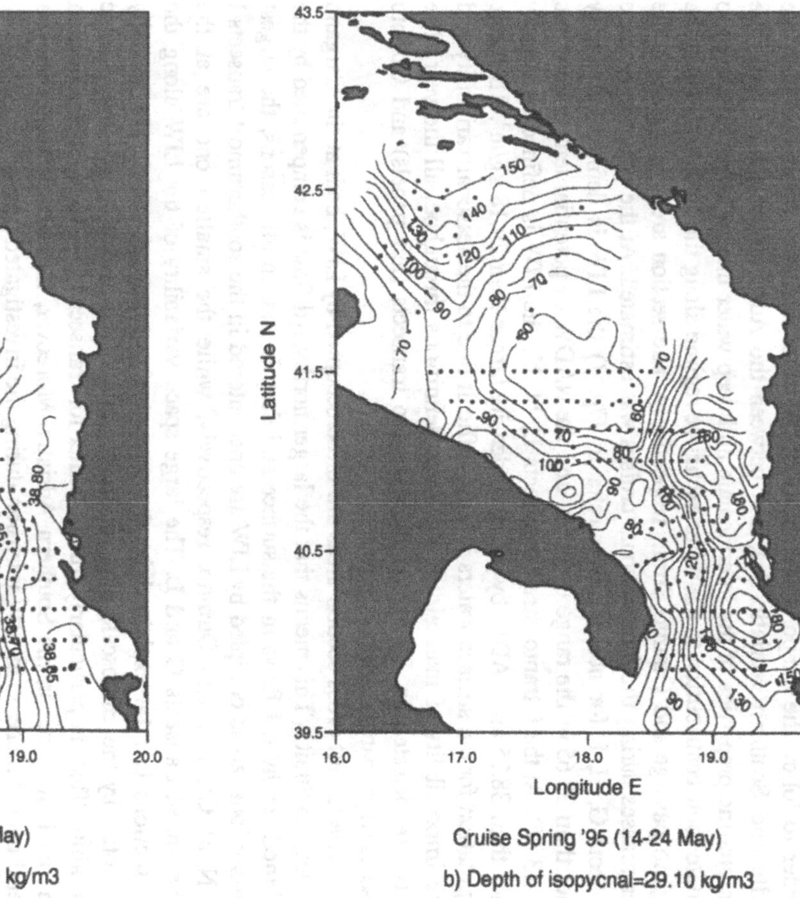


*Figure 5.* Salinity distributions on isopycnal surface 29.10 kg/m$^3$ (a) and Depth in m of the same isopycnal (b).

## 5. Discussion and Conclusions

In order to follow the temporal variability of the water exchange regime between the Middle and Southern Adriatic Sea, and between the Adriatic and Ionian Seas, the thermohaline properties of the intermediate and deep water masses were considered to evaluate their contribution to the water exchange regime along the cross-sections at the Gargano passage and Otranto Strait, respectively. The section areas occupied by the water masses during the four seasonal cruises was estimated. At the Gargano passage (transects G, H, I, for nomenclature refer to *Fig. 1*) the LIW is defined by salinity higher than 38.65 in the range 0-400 m, and the NADW by potential temperature less than 13 °C. At the Otranto Strait (transects L, M, N) the LIW is defined by salinity higher than 38.75 and ADW by potential density larger than 29.18 kg/m$^3$. The same computation for the surface waters is quite difficult due to the seasonal variability and to the impossibility of maintaining the same definitions throughout all the year. The results are depicted in *Fig. 6* for the Gargano transects (left panels) and Otranto transects (right panels), respectively.

The largest cross-section areas are those occupied by LIW, both at the Gargano and Otranto Straits. This means that the larger northward flow is compensated by the combined southward flows in the surface and deep layers. In all seasons, the largest cross-sectional areas occupied by LIW are encountered in the southernmost transects I and N at Gargano and Otranto, respectively; while the smallest one are at the northernmost transects G and L. The large space variability of the LIW along the cross-sections can be explained by the advective movements, which are partially interrupted by the mesoscale and sub-basin scale features, due to the topographic constraints. That is particularly true in winter for transect I, due to the intensification of the cyclonic gyres in the Southern Adriatic. Moreover, the average values show a general overall increase in LIW intrusion during the investigated period.

The deep waters, both the NADW and ADW, display a smaller space variability and a strong seasonal signal. The largest cross-sectional area occupied by the NADW at Gargano occurs in spring; this means that the maximum contribution of dense water formed in the north shelf area during winter reaches the meridional sections during the following spring period. On the other hand, the largest cross-sectional areas occupied by the ADW at Otranto occur in winter and spring, i.e. during and immediately after the deep water formation events. In fact, one such event was documented during the winter 1996 survey [8]. It was noted that the minimum values occured in summer and autumn, which can be explained in terms of the emptying of the Southern Adriatic Pit reservoir. We can speculate that the renewal time of the ADW formed during winter in the Southern Adriatic Pit is shorter than one year, at least for that part of the water column which lies at a shallower depth than the Otranto sill (900 m). It could be longer only for the water mass which resides in the bottom layer (900-1200 m).

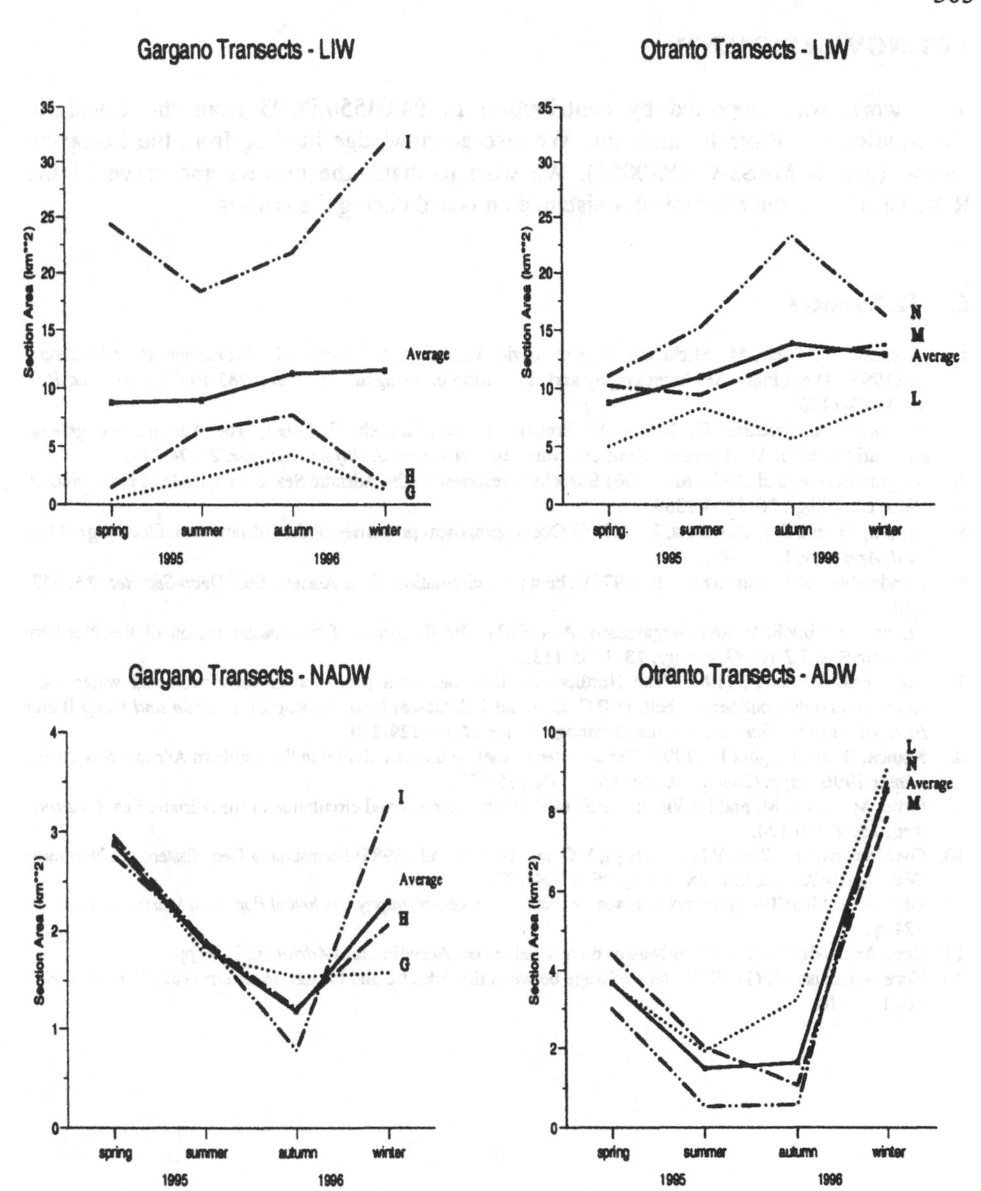


*Figure 6.* Section areas occupied by the intermediate and deep waters during the seasonal cruises at the Gargano transects G, H, I (for nomenclature refer to Fig. 1) and at the Otranto Strait transects L, M and N (refer to Fig. 1).

## ACKNOWLEDGMENTS

This work was supported by contribution n. 94.04550.PG03 from the Consiglio Nazionale delle Ricerche in Rome. We also acknowledge funding from the European Union (grants MAS3-CT96-0051). We wish to thank the masters and crews of the R.V. Urania for their technical assistance on board during the cruises.

## 6. References

1. Artegiani, A., Gacic, M., Michelato, A., Kovacevic, V., Russo, A., Paschini, E., Scarazzato, P. and Smircic, A. (1993) The Adriatic Sea hydrography and circulation in spring and autumn (1985-1987). *Deep-Sea Res.* **40**, 1143-1180.
2. Artegiani, A., Paschini, E., Russo, A., Bregant, D. and Raicich, F. (1997) The Adriatic Sea general circulation. Part I: Air-Sea interactions and water mass structure. *J. Phys. Oceanogr.* **27**, 74-114.
3. Bergamasco, A. and Gacic, M. (1996) Baroclinic response of the Adriatic Sea to an episode of bora wind. *J. Phys. Oceanogr.* **26**, 1354-1369.
4. Buljan, M. and Zore-Armanda, M. (1976) Oceanographical properties of the Adriatic Sea. *Oceanogr. Mar. Biol. Ann. Rev.* **14**, 11-98.
5. Hendershott, M.C. and Rizzoli, P. (1976) The winter circulation of the Adriatic Sea. *Deep-Sea Res.* **23**, 353-370.
6. Malanotte Rizzoli, P. and Bergamasco, A. (1983) The dynamics of the coastal region of the Northern Adriatic Sea. *J. Phys. Oceanogr.* **13**, 1105-1130.
7. Malanotte-Rizzoli, P. (1991) The Northern Adriatic Sea as a prototype of convection and water mass formation on the continental shelf, in P.C. Chu and J.C. Gascard (eds.), *Deep Convection and Deep Water Formation in the Oceans*, Elsevier Oceanogr. Series **57**, pp. 229-239.
8. Manca, B. and Bregant D. (1998) Dense water formation and circulation in the Southern Adriatic Sea during winter 1996. *Rapp. Comm. int. Mer Medit.* **35**, 176-177.
9. Orlic, M., Gacic, M. and La Violette, P.E. (1992) The currents and circulation of the Adriatic Sea. *Oceanol. Acta* **15**, 2, 109-124.
10. Ovchinnikov, I.M., Zats, V.I., Krvosheya, V.G. and Udodov, A.I. (1985) Formation of Deep Eastern Mediterranean Waters in the Adriatic Sea, *Oceanology* **25**, 6, 704-707.
11. UNESCO (1985) The international system of units (SI) in oceanography. *Technical Papers in Marine Science*, **45**, 124 pp.
12. Zore-Armanda, M. (1963) Les Masses d'eaux del la mer Adriatic. *Acta Adriat.* **X**, 3, 89 pp.
13. Zore-Armanda, M. (1969) Water exchange between the Adriatic and Eastern Mediterranean. *Deep-Sea Res.* **16**, 171-178.

# LITHOLOGY AND GEOCHEMISTRY OF THE SEDIMENTS ALONG THE EASTERN MEDITERRANEAN COAST

A. G. ROZANOV (1), K. M. SHIMKUS (2) and S. NOUREDDIN (3)
*(1) P. P. Shirshov Institute of Oceanology,Russian Academy of Sciences, 36 Nakhimovskny Avenue,Moscow,117851,Russia*
*(2) Southern Branch of P. P. Shirshov Institute of Oceanology,Russian Academy of Sciences,Gelendzhik-7,353470,Russia*
*(3) Higher Institute of Applied Sciences and Technology,Damascus, Syria*

## 1. Introduction

Area between the Cyprus and Syrian coast is a wide strait with a complicated relief of the floor and specific circulation of the waters ( Hall et al., 1994; Ovchinnikov and Fuoad, 1994 ) [ Figure 1 ]. Because of the dry climate geological material discharging from the watershed area of Cyprus, Syria and Turkey is week. From the other side the discharges of the Nile river penetrate here in spite of tremendously lardge distance ( Emelyanov, 1994 ). It is still uncertain in which regions of the area an influence of the Nile material is predominant and where it is weakened due to strengthening the role of local material. The latter is spiked by the products of weathering of the ophiolite massives wide spead both on the Cyprus and Syrian coasts. The results of the 27-th cruise of r/v "Vityaz" ( 1993 ) in this part of the Mediterranean Sea [ Figure 1] give additional possibility to clear up the situation.

Quaternary sedimentation in the studied area was under influence of glacial and interglacial oscillations of sedimentary level, sources and ways of transportation of sedimentary material. Climate changes followed regression and transgrassion periods are reflected in lithology and geochemestry of the sediments. These changes are also documented by the presented results.

## 2. Material and Methods

Grab ( about 0,25 $m^2$ ) and gravity core ( 62 mm ) were used to collect sediment samples on shelf, slope and depression depths. Station numbers are shown in Figure 1. Grain size analysis was carried out by the suspension balance method ( Prokoptzev, 1964 ). Macrocomponent composition ( $CaCO_3$, $C_{org}$ , Fe,



*P. Malanotte-Rizzoli and V.N. Eremeev (eds.),*
*The Eastern Mediterranean as a Laboratory Basin for the Assessment of Contrasting Ecosystems,* 507–521.

Mn, Ti ) was determined by the methods, accepted in the P. P. Shirshov Institute of Oceanology ( SIO ): $CaCO_3$, $C_{org}$ - dry, Fe, Mn, Ti - wet methods. Metal contents were determined by the flame atomic absorbtion spectroscopy. General mineral composition of the sediments and clay fraction ( < 2 mm ) was studied by X-ray diffraction method ( Zemmels et al., 1972 ), mineral composition of the coarse-silt fraction - by immersion liquid method.

Compositional-genetic and grain-size types of the sediments were classifided both according to the SIO system ( Bezrukov and Lisitzin, 1960 ) with little modifications accepted for Mediterranean sediments ( Shimkus, 1981; Emelyanov and Shimkus, 1986 ) and the Folk system ( Folk, 1954 ). Thus it is possible to compare Russian classification and the types of sediments presented in the new international maps of the Mediterranean sediments ( Emelyanov and Shimkus, 1995 ).

## 3. Types and Compositions of the Sediments

On the most part of Syria region of the Mediterranean Sea carbonate (30-50% $CaCO_3$) and carbonate-poor ( 10-30% $CaCO_3$ ) sediments predominate. Carbonate-rich ( 50-70% $CaCO_3$ ) and very carbonate-rich ( >70% ) sediments presented mostly by shelly detritus are spread in various areas of the shelf. The sediment stripe along the shelf edge was formed by washing out of the muds by bottom currents and probably marks the deposits of the ancient coastal zone. Several fields of sand and silt were revealed at different depths among fine-silt and silty-pelitic sediments within the shelf which width in studied area is not more than 10 km [ Figure 2 ].

In the coastal part of the shelf wellsorted carbonate sands and coarse silts are, in the lower part - organogenic carbonate-rich mild-sorted sands are spread [ Table 1 ].

There are carbonate pelitic and silty-pelitic muds on the bottom surface in the upper zone of the continental slope ( 250-600 m ). Single fields of fine-silt muds are mainly met in underwater valleys. Carbonate pelitic muds predominate deeper.

Surface sediments, touched by oxic processes at all depths, have a brown colour of various nuances. They are soft or even very soft, highly plastic and wet ( up to 68,3% $H_2O$ ).

## 4. Grain-Size Composition

Shallow-water mild-sorted sands contain substantial admixture of silty and pelitic material. Sands and coarse-silt fractious predominate in wellsorted coarse silts ( mudy sands ) along with a valuable admixture of pelitic material [ Table 1 ].

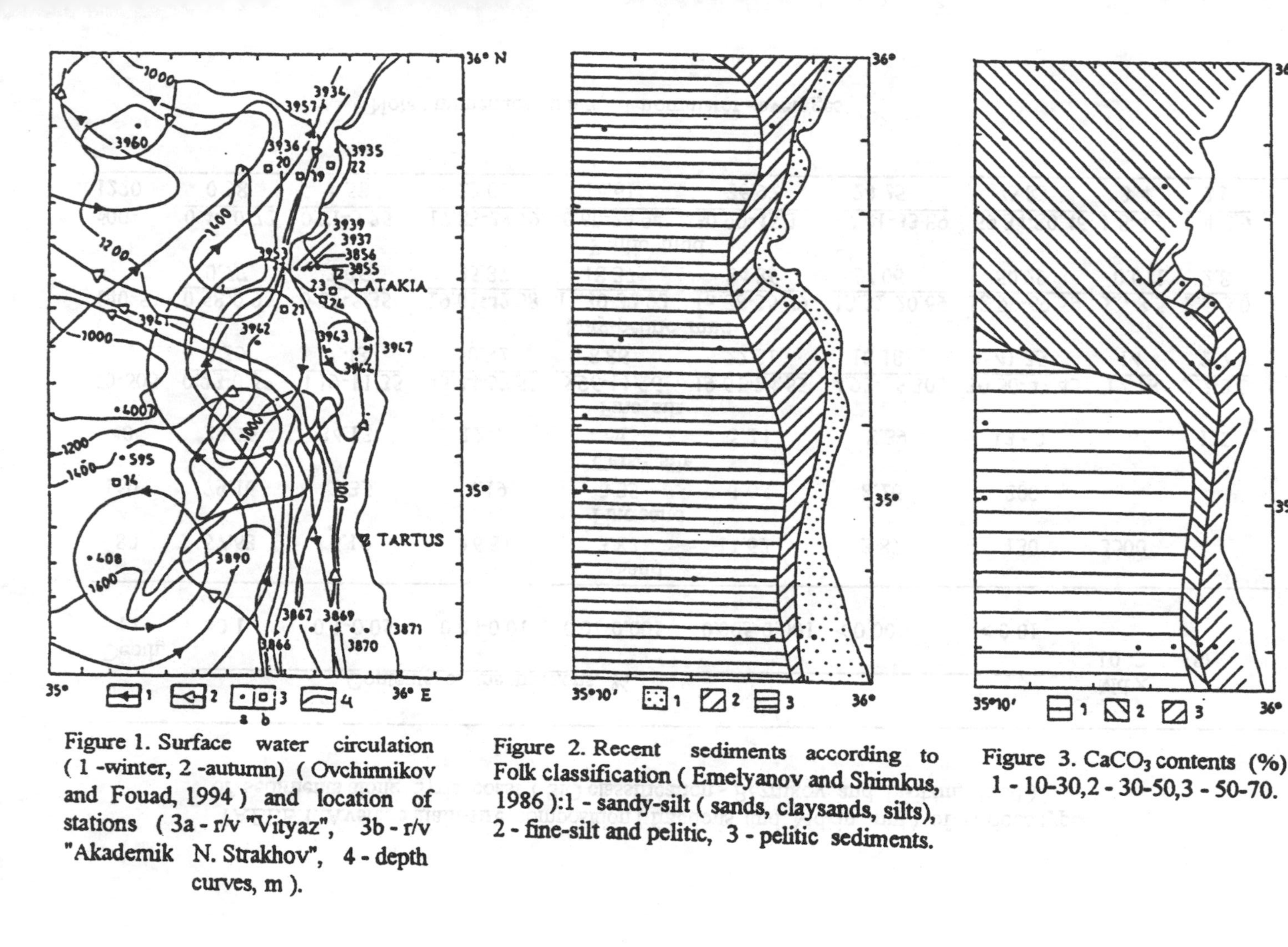


Figure 1. Surface water circulation ( 1 -winter, 2 -autumn) ( Ovchinnikov and Fouad, 1994 ) and location of stations ( 3a - r/v "Vityaz", 3b - r/v "Akademik N. Strakhov", 4 - depth curves, m ).

Figure 2. Recent sediments according to Folk classification ( Emelyanov and Shimkus, 1986 ):1 - sandy-silt ( sands, claysands, silts), 2 - fine-silt and pelitic, 3 - pelitic sediments.

Figure 3. $CaCO_3$ contents (%) 1 - 10-30,2 - 30-50,3 - 50-70.

TABLE 1. Average grain-size composition ( fractions and Md in mm ) of various types of sediments along Syria coast ( SIO classification - Bezrukov and Lisitzin, 1960 )

| Depth, m | Contents of the fractions, % | | | | | | | Md x 10 -3 | So |
|---|---|---|---|---|---|---|---|---|---|
| | >0.1 | 0.1-0.05 | 0.05-0.01 | 0.01-0.005 | 0.005-0.001 | < 0.001 | < 0.01 | | |
| | | | | Sand | | | | | |
| 80 | 57.63 | 6.19 | 16.81 | 3.87 | 11.62 | 3.87 | 130 | 3500 | |
| | | | | Fine sand | | | | | |
| 150 | 79.16 | 3.35 | 4.19 | 3.35 | 1.68 | 8.26 | 200 | | |
| | | | | Corse silt | | | | | |
| 40 | 39.75 | 34.13 | 12.49 | сл. | 5.24 | 8.39 | 13.63 | 85 | 1.7 |
| | | | | Fine silt | | | | | |
| 70-605 | 0.23-0.57<br>0.35 | 4.19-11.35<br>7.48 | 45.44-57.57<br>50.82 | 5.96-11.26<br>8.86 | 19.61-24.55<br>22.31 | 4.93-15.30<br>10.18 | 30.50-47.43<br>41.35 | 12-18<br>14 | 2.2-3.0<br>2.7 |
| | | | | Silty-pelitic mud | | | | | |
| 140-530 | 0.58-0.95<br>0.77 | 0.81-5.35<br>3.06 | 29.02-42.58<br>35.87 | 11.30-32.02<br>18.37 | 16.88-35.79<br>24.86 | 13.32-20.69<br>17.06 | 52.99-69.59<br>60.29 | 5.5-8.5<br>0.071 | 2.5-3.0<br>2.8 |
| | | | | Pelitic mud | | | | | |
| 500-1220 | 0.44-0.72<br>0.58 | 0.71-5.25<br>1.58 | 17.83-28.79<br>22.01 | 0.60-24.94<br>14.61 | 30.90-45.37<br>36.97 | 17.41-33.89<br>24.25 | 69.34-80.46<br>75.83 | 1.8-4.9<br>3.4 | 2.4-3.9<br>2.9 |

Note : numerator - limits, denominator - averages

Shallow-water fine-silty muds ( silts ) are well sorted, whereas deep-water fine-silty muds contain an increased admixture of pelitic material.

Silty-pelitic muds ( muds ) are characterized by double-summit histogram of the grain-size profile. The first ( highest ) pick relates to the fine-silt fraction, the second ( lower ) one corresponds to the mild-silt fraction.

In sutisfactory sorted silty-pelitic muds ( muds ) of the shallow-water region contentents of the 0,05-0,01 mm fraction are more high in compare with similar muds of the deep-water zone.

Grain-size profile of pelitic muds ( muds, clays ) is also characterized by double-summit histogram often with picks at the same fractions as in silty-pelitic muds. However mild-pelitic fractions usually predominate in them. In some sediments subcolloidal fractions amount to the close values [ Table 1 ].

In contrast to the continental slope on the bottom of relatively smoothed depressions pelitic muds ( clays ) are enriched in fractions < 0,01 mm ( 77.65-80.46% ) [ Table 1 ]. Maximum values were observed in the southern depressions which sedimentogenesis is significantly influenced by the Nile material.

## 5. Mineral Composition

X-ray diffraction studies [ Table 2 ] show that the recent sediment carbonate is presented predominantly by calcite, mainly low-magnesim calcite. Sobordinate role plays high-magnesium calcite that is entirely absent in coarse-silt and fine-silt muds of the shallow-water areas. In some places it is absent in the deep-sea areas too.

Biogenic origin of low-magnesium calcite was testified by microscope observations. It showed wide spread of foraminitera-cocolith material in muds. Silt material is presented mostly by foraminiferal shells. Carbonate of pelitic fraction consists mostly of coccolithes.

Quartz predominates in clastic siliceous materal, in some places plagioclase occures in valuable amounts : in one specimen of shallow-water fine-silty mud it predominatees over quartz ( Station 3947 ).

Potassium feld spar is scaresly met. The sum of clay minerals ranges from 41 to 79% [ Table 2 ]. Montomorillonate predominates, showing valuable oscilations, it is followed by illite, kaolinite and chlorite.

## 6. Chemical Composition

Shallow-water sediments are almost 50% calcarious [ Table 3 ]. In deep-sea muds $CaCO_3$ contain is lower ( somewhere to 30% ) [ Figure 2 ]. The contents of organic matter in shallow-water fine silty and silty-pelitic sediments is much higher than in deep-sea muds ( 0,88-0,90% as against 0,50-0,55% $C_{org}$ ; Table 3 ).

TABLE 2. Mineral composition of the sediments along Syria coast ( % dry weight, X - ray diffraction data )

| Station | Sea depth, m | Depth in core, cm | Ca | MgCa | Ar | D | Q | Pl | KFS | Cl | Oth |
|---|---|---|---|---|---|---|---|---|---|---|---|
| 3856 | 40 | 0-2 | 38 | - | - | 4 | 11 | 5 | - | 42 | - |
| 3866 | 1100 | 0-3 | 33 | - | 4 | 7 | 15 | - | - | 41 | - |
| 3867 | 500 | 0-2 | 25 | 8 | 1 | 11 | 9 | - | - | 44 | 2 |
| 3870 | 82 | 0-2 | 12 | 18 | - | 6 | 14 | - | - | 50 | - |
| 3941 | 1060 | 0-5 | 5 | 8 | - | 4 | 9 | - | - | 64 | - |
| 3942 | 1000 | 0-5 | 19 | 5 | 2 | 4 | 8 | 3 | 3 | 44 | 10 |
| 3943 | 500 | 0-2 | 22 | 7 | 2 | 8 | 8 | 3 | - | 47 | 3 |
| 3944 | 140 | 0-2 | 26 | 5 | 1 | 12 | 6 | 2 | 2 | 41 | 2 |
| 3947 | 70 | 0-1 | 26 | 7 | - | 7 | 7 | 12 | - | 51 | - |
| 3953 | 605 | 0-2 | 40 | - | - | 10 | 9 | + | - | 41 | - |
| 3960 | 1220 | 0-5 | 32 | 6 | - | 5 | 14 | + | - | 49 | - |
| 4007 | 1210 | 0-10 | 26 | 8 | 2 | 5 | 10 | + | - | 47 | 2 |
| 3869 | 150 | 0-20 | 10 | 6 | - | 5 | - | - | - | 79 | - |

Note : Ca - calcite, MgCa - magnesiun calcite, Ar - arogonite, D - dolomite. Q - quartz, Pl - plagioclase, KFS - potassium feldspar, Cl - clay minerals ( sum ), Oth - other minerals

Deep-sea muds contain several times more manganese in compare with shallow ones. Free carbonate material of shallow sediments south of Latakia ( Station 3941 and others ) is significantly enriched by Ni, Cr, Ti, V. Deep-sea sediments are enriched along with Mn by Cu too.

Terrigeneous free carbonate material of deep-sea pelitic muds north and north-west of Latakia ( Stations 20-21-22 ) is charecterized by maximum Ni contents and increased Ni/Cr ratios coming close to 3. This ratio in shallow sediments is substantialy higher.

Syria region sediments differ chemically from the Nile deposits by the lower contents of organic matter, Fe, Ti [ Table 3 ]. Contents of Ni is increasing towards the sediments of Syria watershed area.

## 7. Terrigeneous Sedimentogenesis

Syrian watershed area is characterized by dry climate. Rainfall amount is less than 700 mm/year. In summer hot winds blow from the Arabian and Syrian deserts bringing valuable amounts of eolian dust.

TABLE 3. Chemical composition of the sediments along Syria coast (% dry weight, Cu and following chemical elements - $10^{-4}$ % dry weight)

| Station | Sea depth, m | Depth in core, cm | $CaCO_3$ | $C_{org}$ | Fe | Mn | Ti | Al | Ba | Cu | Zn | Pb | Ni | Co | Cr | V | Mo |
|---|---|---|---|---|---|---|---|---|---|---|---|---|---|---|---|---|---|
| | | | | | | Fine silt | | | | | | | | | | | |
| 3947 | 70 | 0-1 | 51.10 | 0.90 | 4.54 | 0.034 | 0.70 | 4.18 | 0.16 | 26 | 70 | 36 | 83 | 35 | 133 | 85 | 3 |
| 3947 | 70 | 0-8 | 56.52 | 0.96 | 3.63 | 0.024 | 0.50 | 4.23 | 0.26 | 26 | 74 | 64 | 65 | 32 | 112 | 92 | |
| | | | | | | Silty-pelitic muds | | | | | | | | | | | |
| 3943 | 500 | 0-1 | 34.76 | 0.86 | 4.71 | 0.141 | 0.60 | 6.08 | 0.11 | 39 | 91 | 64 | 93 | 37 | 118 | 23 | - |
| 3944 | 140 | 0-3 | 37.05 | 0.75 | 4.91 | 0.171 | 0.59 | 6.09 | 0.30 | 41 | 79 | 60 | 90 | 30 | 112 | 77 | - |
| 3944 | 140 | 0-1 | 50.28 | 0.88 | 3.96 | 0.040 | 0.55 | 4.14 | 0.14 | 30 | 68 | 50 | 86 | 32 | 119 | 100 | <3 |
| | | | | | | Pelitic muds | | | | | | | | | | | |
| 3941 | 1060 | 0-5 | 36.21 | 0.50 | 4.60 | 0.174 | 0.45 | 5.55 | 0.11 | 54 | 76 | 50 | 100 | 32 | 114 | 54 | 4 |
| 3942 | 1000 | 0-1 | 33.71 | 0.55 | 5.45 | 0.207 | 0.48 | 5.44 | 0.10 | 60 | 87 | 64 | 93 | 30 | 100 | 100 | <3 |
| 4007 | 1210 | 0-10 | 29.79 | 0.55 | 5.33 | 0.192 | 0.55 | 5.92 | 0.16 | 56 | 84 | 50 | 90 | 32 | 97 | 162 | <3 |

Major part of river discharges are brought by the Oront river draining the eastern slope of the Ansaria Ridge and eroding the Neogene-Quaternary basalts and ophiolite deposits [ Figure 4 ] enriched by iron, manganese, copper, chromium and other metals ( Zitzmann, 1976 ). Few discharges are given by small rivers coming directly into the sea. Some of them dry up periodically. The northern part of the studied area is close to Turkey where the Jeihan and the Seihan rivers drain predominantly the Mesozoic-Cenozoic sedimentary complexes and some ophiolite massives. Anyway the most powerful sourse of terrigeneous material in the eastern part of the Mediterranean Sea is the Nile. The surface currents bring the most disperssed portions of the Nile alleuvium here.

Main features of space distribution of the terrigeneous material in the studied area are presented by the maps of distribution of heavy clastic minerals and metals. These are the marks of weathering production of the ophiolite complexes.

Heavy minerals, main part of which is presented by coarse-silt material of bottom sediments, are connected by origin with the nearby watershed. It is clear according to the maximum contents of omphibol, monoclinical and rhombic pyroxene, olivine, spinel, palagonite, epidote [ Figure 4 and 5 ]. Their root parts, crossing the shelf, are adjacent to the regions of the coast where products of ophiolite rock weathering are discharged. Hightened contents of these minerals form stripes, which cross the continental slope and farther follow to the adjacent depression. They evidence a centre directed transfer of the terrigenous material to the deep part of the sedimentary basin. That can be a conseqence of transfer by suspension fluxes along the underwater valleys. In the transfer the effect of surface water circulation is hardly seen. It means that terrigeneous sandy-silt material in the sediments of the studied area is formed predominantly by local material and the character of their distribution is mainly stipulated by near bottom processes.

It should be stressed that in some places the local material do play an important role in formation of clay fractions of the sediments. It is seen from square distribution both of clay minerals and some metals ( Ni, Zn, Cu, Co ), that are transported witthin clay materials and genetically connected with weathering products of ophiolite massives [ Figure 6 ].

Dry climate is favorable for eolian material transfer as well. To some extend the situation is clarified by quartz square distribution. It sharply differs from the distribution of plagioclase and demonstrates spotted character of maximum contents of quartz.

The largest field of increased quartz contents in corse-silt fractions is located in the southern deep-sea part of the studied area, the smaller one is adjacent to the northern part.

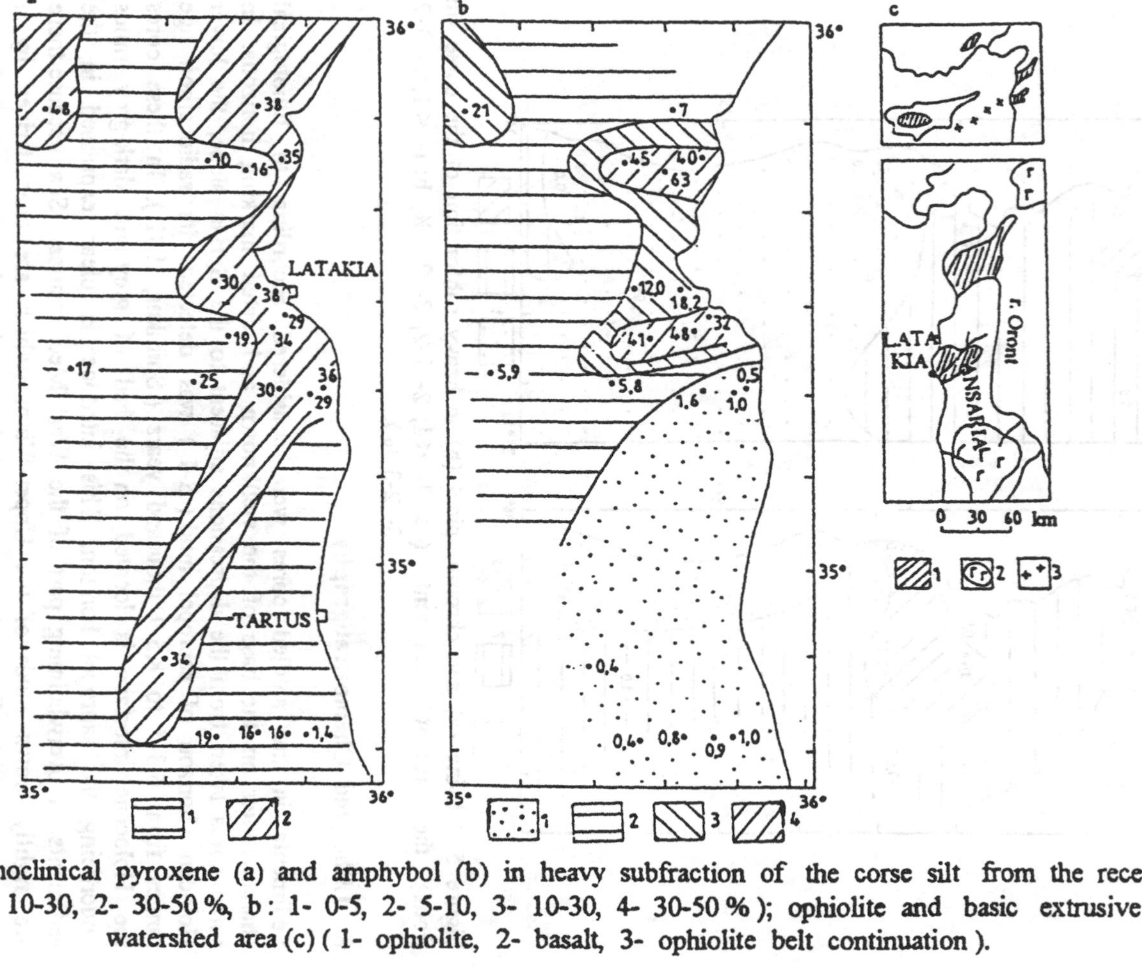


Figure 4. Monoclinical pyroxene (a) and amphybol (b) in heavy subfraction of the corse silt from the recent sediments ( a : 1- 10-30, 2- 30-50 %, b : 1- 0-5, 2- 5-10, 3- 10-30, 4- 30-50 % ); ophiolite and basic extrusive at the watershed area (c) ( 1- ophiolite, 2- basalt, 3- ophiolite belt continuation ).

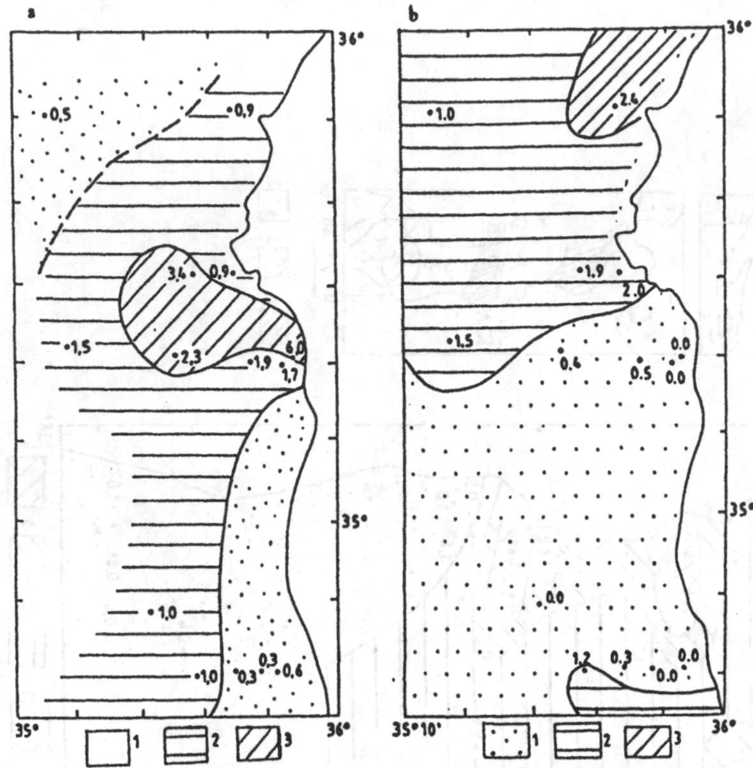


Figure 5. Olivine (a) and chrome spinel (b) in heavy subfraction of the corse silt from the recent sediments ( a : 1- < 1, 2- 1-2, 3- 2-6 %, b: 1- < 1, 2- 1-2, 3- 2-3 % ).

## 8. Litho- and Chemostratigraphy

Sediments in the studied cores were considered according to lithological characteristics on the base of the wellknown and dated marking horizons. In some cores taken from the depressions adjacent to the Syrian slope, the lower Holocene horizon of sapropel mud ( S-1 ) was detected. Its radiocarbon age ranges from 6,3-7 to 8,5-9 thousand years ( Shimkus, 1981 ). In these cores the Holocene interface is located on the bed of grey and dark-grey muds underlieing the sapropel horizon. This interface is clear expressed in the sediments of overwhelming part of the West Mediterranean Sea because there are brightly brown muds of the Upper-Würm (which are characterized by a very low content of organic matter and increased carbonate content) beneat the dark-grey muds. In Syria region the Upper-Würm deep-sea sediments mostly consist of the Nile material which is enriched by terrigenious organic matter and looks rather reduced. According to the specific changes of lithological composition most cores from the deep-sea areas exhibit stratigraphic sections which consist of the deposits of the Upper-Würm and Holocene. The former has not been fully opened. The cores taken from the

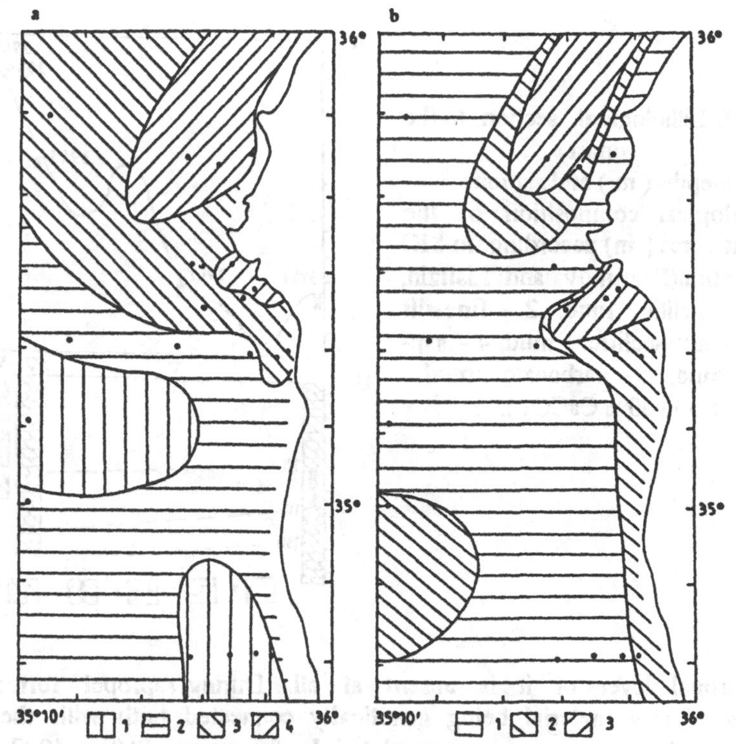


Figure 6. Nickel (a) and zinc (b) in the recent sediments (free carbonate matter) a : $n \times 10^{-4}$ % : 1- 100-150, 2- 150-200, 3- 200-300, 4-300-550 ; b : $n \times 10^{-4}$ % : 1- 80-100, 2- 100-150, 3- 150-200.

slope, do not contain the S-1 horison probably due to the presence of the Würm section only. The Holocene is absent here because of sliding processes. Stratigraphy of the shelf sediments is not quite clear. Probably there are the Holocene sediments only.

Formation of the sapropel mud took place in climatic conditions quite different from the recent ones. In that period everywhere in the nowaday arid zone of the Mediterranean sea, including the eastern coast, subtropical humid climate predominated ( Shimkus, 1981 ). As a consequence, both volumes of solid river discharges were enhanced and production of chemical weathering ( montmorillonite, iron and manganese hydroxides, other reactive forms of chemical elements ) was increased. Changes in Mediterranean climate and hydrodinamics in this period stipulated not an increase of terrigeneous organic fluxes only, but an increase of biological production as well. This all was accompanied by enhanced sedimentation and burial of organic matter in bottom sediments. The following reducing diagenesis was a course of formation of large amounts of pyrite which now dominates among heavy minerals in the sapropel mud.

The olivine was distructed more rapidly under chemical weathering in compare with other ophiolite minerals. Therefore its contents is deminished in

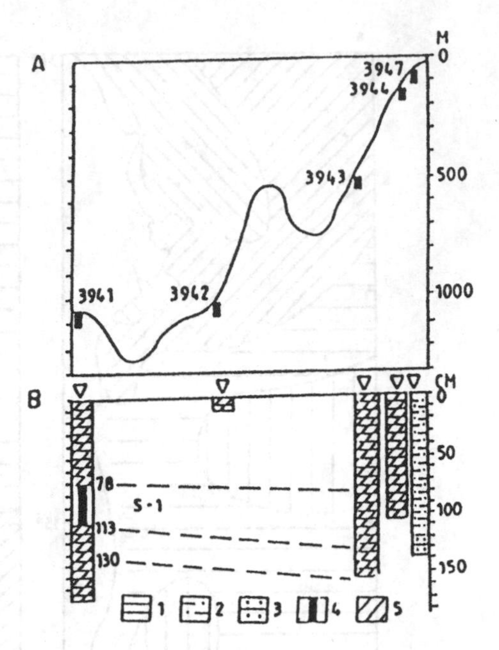


Figure 7. Lithological section to the coast
A. Depths ( m ) and stations
B. Lithological composition of the sediment cores (cm) according to SIO classification(Bezrukov and Lisitzin, 1960): 1 - pelitic mud, 2 - fine-silt mud, 3 - silty-pelitic mud, 4 - sapropel mud, 5 - carbonate mud ( 30 - 50% $CaCO_3$ ).

the sapropel layer or it is absent at all. During sapropel formation, large volumes of clay material, being genetically connected both with the Nile and Syria watershed rivers were accumulated. In the cores ( Station 4007, Figure 8 ) this layer is characterized by obvious predominance of montmorillonite ( smectite ) among clay minerals ( 51 - 79% ). Fluctuations in montmorillonite concentrations are explained by irregular fluxes of clay material delivered from the Syria watershed as a consequence of the climate changes during the sapropel mud formation.

Chemical data also evidence an importance of Syria rivers in the Late Quaternary sedimentogenesis in this region. For instance quantitative ratios Ni/Cr and Fe/Ti in the studied cores of the sediments are 1,5-2 times higher comparing with the data in the Nile alluvium of various grain-size composition. In the Nile alluvium these ratios are 0,32-0,64 and 4,13-7,04, but in the studied cores - 0,64-1,26 and 8,9-13,6 correspondingly. It is clear from distribution of the highest values of these ratios, that intensity of sedimentary material fluxes from Syria watershed increased in some periods of the Late Würm and Holocene. Such fluxes were most intensive during the sapropel formation.

Very sensitive indicator of the Nile material is vanadium which enriches the Nile alluvium. Increasing role of the Nile material in formation of the Late. Quaternary sediments in the Syria region is indicated by the higher contents of vanadium, which was mentioned in a single layer of the Upper Würm section and a number of the Holocene layers. Maximum vanadium concentrations in the core 4007 were met in the medium part of the sapropel

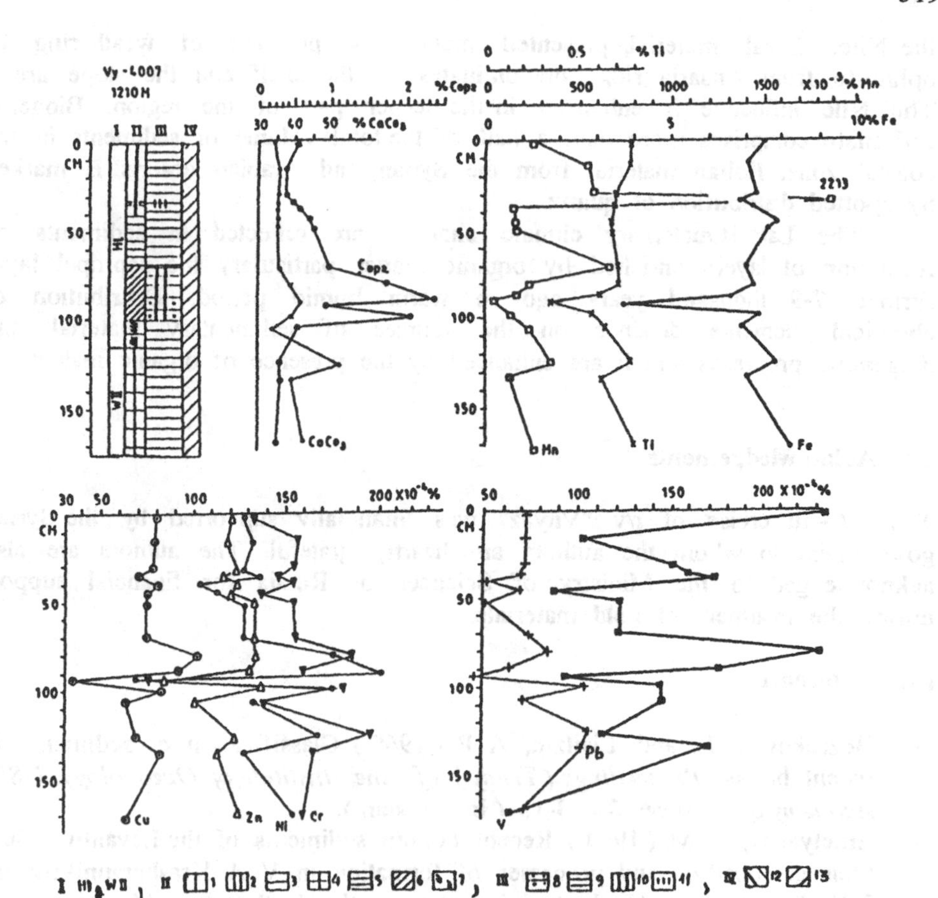


**Figure 8. Lithology and geochemistry of the sediments ( St. 4007 )**
**I Stratigraphy : Hl - Holocene, W II - Upper Würm**
**II Colour of the sediments : 1- brown, 2- dark brown, 3- grey, 4- brownish-grey, 5- dark grey, 6- dirty-dark green, 7- black spots**
**III Types of sediments : 8- silty-pelitic mud, 9- pelitic mud, 10- sapropel mud, 11- manganese mud ( > 0,2 % Mn )**
**IV $CaCO_3$ : 12- carbonate poor, 13- carbonate**
**Metal % are given on free carbonate matter.**

**mud whereas in the core 3941 - in its foot. Irregular vanadium distribution is explained by the changes in the circulation of surface waters in this strait region. As it was shown, contemporary currents in Syria region of the Mediterranean sea are very complicated and changeable.**

the Nile. Local material, presented mainly by products of weathering of ophiolite from Ansaria ridge, predominates on the shelf and the slope areas. The Nile influence is enhanced in the deeper part of the region. Biogenic carbonate comprises more than a half of the total volume of sediments in the coastal zone. Eolian material from the Syrian and Arabian deserts is marked by spotted distribution of quartz.

The Late Quaternary climate changes are reflected in sediments by formation of layers enriched by organic matter particulary the sapropel layer formed 7-9 thousand years ago in warm humid period. Distribution of chemical elements depends on the sources of sedimentary material and diagenetic processes which are enhanced by the presence of organic matter.

## 9. Acknowledgements

The 27 - th cruise of r/v "Vityaz" was financially supported by the Syrian government to whom the authors are heartly grateful. The authors are also acknowledged to the Ministry of Sciences of Russia for financial support during the treatment of field materials.

## 10. References


1. Bezrukov, P. L. and Lisitzin, A. P. ( 1960 ) Classification of sediments in recent basins, *Proceedings ( Trudy ) of the Institute of Oceanology, USSR Academy of Sciences* 32, 3-15 ( In Russian ).
2. Emelyanov, E. M. ( 1994 ) Recent bottom sediments of the Levantine Sea: their composition and processes of formation, in V. A. Krasheninnikov and J. K. Hall ( ed . ), *Geological Structure of the Northeastern Mediterranean Sea ( Cruise 5 of r/v "Akademik Strakhov" )*, Historical Productions - Hall, Jerusalem, pp. 141-158.
3. Emelyanov, E. M. and Shimkus, K. M. ( 1986 ) *Geochemistry and Sedimentology of the Mediterranean Sea*, D. Reider Publishing Company, Dordrecht.
4. Folk, R. L. ( 1954 ) The distination between grain size and mineral composition in sedimentary rock nomenclature, *J. Geology* 62, 344-359.
5. Hall, J. K., Udintsev, G. B. and Odinokov, Yu. Yu. ( 1994 ) The bottom relief of the Levantine Sea, in V. A. Krasheninnikov and J. K. Hall ( ed . ), *Geological Structure of the Northeastern Mediterranean Sea ( Cruise 5 of r/v "Akademik Strakhov" )*, Historical Productions - Hall, Jerusalem, pp. 5-32.
6. Moskalenko, L. V. and Ovchinnikov, I. M. ( 1991 ) Geostrophic water circu-

lation in The Mediterranean Sea, *Oceanologia* 31, 907-915 ( In Russian ).

7. Ovchinnikov, I. M. and Fouad, A. S. ( 1994 ) Study of winter circulation in the Eastern Mediterranean Sea along Syria coast, *Oceanologia* 34, 467-471 ( In Russian ).
8. Prokoptsew, N. G. ( 1964 ) On mechanical analysis of pelitic fractions in sea sediments ( suspension balance ), *Oceanologia* 4, 699-707 ( In Russian ).
9. Sevastjanov, V. F. ( 1968 ) Chemical element redistribution during redox processes in Mediterranean sediments, *Lithology and Mineral Resources*, №1, 3-15 ( In Russian ).
10. Shimkus, K. M. ( 1981 ) *Late Quaternary Sedimentation in the Mediterranean Sea*, Nauka Publishers, Moscow ( In Russian ).
11. Volkov, I. I., Rosanov, A. G. and Sokolov, V. S. ( 1975 ) Forms of manganese, iron and sulfur in sediments of Terranian Sea, in Ya. P. Molovitsky ( ed. ), *Hydrological and Geological Studies of the Mediterranean and Black Seas*, Institute of Oceanology, Moscow, pp. 257-284 ( In Russian ).
12. Zemmels, J., Cook, H. A. and Harthway, J. C. ( 1972 ) X-ray mineralogy studies, Leg 11, *Init. Rep. Deep - Sea Drilling Project* 11, 729.
13. Zitzmann, A. ( 1976 ) The iron ore deposits of Syria, in *Iron Ore Deposits of Europe etc.* 1, 299 - 301 ( Hannover ).